Algebra Review

Rules of Exponents

- **Multiplication of Exponents** To multiply a^m by a^n, *add* the exponents: $a^m a^n = a^{m+n}$.

- **Division of Exponents** To divide a^m by a^n, *subtract* the exponents: $\dfrac{a^m}{a^n} = a^{m-n}$.

- **Power of a Power** To find a power of a power, multiply the exponents: $(a^m)^n = a^{mn}$.

- **Product to a Power** To find $(ab)^n$, apply the exponent to *every* term inside the parentheses: $(ab)^n = a^n b^n$.

- **Quotient to a Power** To find $\left(\dfrac{a}{b}\right)^n$, apply the exponent to *both* the numerator and the denominator: $\left(\dfrac{a}{b}\right)^n = \dfrac{a^n}{b^n}$.

- **Zero Exponent** If a is any nonzero real number, then $a^0 = 1$.

- **Negative Exponent** If n is a natural number, and if $a \neq 0$, then $a^{-n} = \dfrac{1}{a^n}$.

- **Inversion Property** $\left(\dfrac{a}{b}\right)^{-n} = \left(\dfrac{b}{a}\right)^n$.

Radicals

- ***n*th root of *a*** If n is an even natural number and $a \geq 0$, or if n is an odd natural number, $\sqrt[n]{a} = a^{1/n}$.

- For all rational numbers m/n and all real numbers a for which $\sqrt[n]{a}$ exists, $a^{m/n} = (\sqrt[n]{a})^m = \sqrt[n]{a^m}$.

- For any real number a and any natural number n, $\sqrt[n]{a^n} = |a|$ if n is even and $\sqrt[n]{a^n} = a$ if n is odd.

- For all real numbers a and b, and for positive integers n for which all indicated roots exists, $\sqrt[n]{a} \cdot \sqrt[n]{b} = \sqrt[n]{ab}$ and $\dfrac{\sqrt[n]{a}}{\sqrt[n]{b}} = \sqrt[n]{\dfrac{a}{b}}$ $(b \neq 0)$.

Factoring

- **Difference of Squares** $x^2 - y^2 = (x + y)(x - y)$.

- **Perfect Squares** $x^2 + 2xy + y^2 = (x + y)^2$ and $x^2 - 2xy + y^2 = (x - y)^2$.

- **Difference of Cubes** $x^3 - y^3 = (x - y)(x^2 + xy + y^2)$.

- **Sum of Cubes** $x^3 + y^3 = (x + y)(x^2 - xy + y^2)$.

Quadratic Equation and Formulas

- **Square Root Property** If $b > 0$, then the solutions of $x^2 = b$ are $x = \sqrt{b}$ and $x = -\sqrt{b}$.

- **Quadratic Formula** The solutions of the quadratic equation $ax^2 + bx - c$, where $a \neq 0$,

  are given by $x = \dfrac{-b \pm \sqrt{b^2 - 4ac}}{2a}$.

- **Graph of a Quadratic Function** The graph of $f(x) = a(x - h)^2 + k$ is a parabola with vertex (h, k). It opens upward when $a > 0$ and downward when $a < 0$. The graph of

  $f(x) = ax^2 + bx + c$ is a parabola with vertex (h, k), where $h = \dfrac{-b}{2a}$ and $k = f(h)$.

Equations of Lines

- **Slope** The slope of the line m through the two points (x_1, y_1) and (x_2, y_2), where $x_1 \neq x_2$

  is $m = \dfrac{y_2 - y_1}{x_2 - x_1}$.

- The slope of every **horizontal** line is 0, and the slope of every **vertical** line is undefined.

- **Slope–Intercept Form** If a line has slope m and y-intercept b, then it is the graph of the equation $y = mx + b$.

- **Parallel Lines** Two nonvertical lines are parallel whenever they have the same slope.

- **Perpendicular Lines** Two nonvertical lines are perpendicular whenever the product of their slopes is -1 (in other words, they are negative reciprocals of each other).

- **Point–Slope Form** If a line has slope m and passes through the point (x_1, y_1), then $y - y_1 = m(x - x_1)$.

Properties of Logarithms

- **Definition of Logarithms to the Base a** $y = \log_a x$ means $a^y = x$.

- Let x and a be any positive real numbers, with $a \neq 1$, and r be any real number. Then $\log_a 1 = 0$; $\log_a a = 1$; $\log_a a^r = r$; $a^{\log_a x} = x$.

- **Product Property** $\log_a xy = \log_a x + \log_a y$

- **Quotient Property** $\log_a \dfrac{x}{y} = \log_a x - \log_a y$

- **Power Property** $\log_a x^r = r \log_a x$

Finite Mathematics
with Applications

IN THE MANAGEMENT, NATURAL, AND SOCIAL SCIENCES

Finite Mathematics with Applications

IN THE MANAGEMENT, NATURAL, AND SOCIAL SCIENCES

TWELFTH EDITION

Margaret L. Lial
American River College

Thomas W. Hungerford
Saint Louis University

John P. Holcomb, Jr.
Cleveland State University

Bernadette Mullins
Birmingham-Southern College

 Pearson

Director, Portfolio Management: Deirdre Lynch
Executive Editor: Jeff Weidenaar
Editorial Assistant: Jennifer Snyder
Content Producer: Kathleen A. Manley
Managing Producer: Karen Wernholm
Producer: Jean Choe
Manager, Courseware QA: Mary Durnwald
Manager, Content Development: Kristina Evans

Product Marketing Manager: Emily Ockay
Field Marketing Manager: Evan St. Cyr
Marketing Assistants: Shannon McCormack, Erin Rush
Senior Author Support/Technology Specialist: Joe Vetere
Manager, Rights and Permissions: Gina Cheselka
Text Design: Studio Montage

Production Coordination, Composition, and Illustrations: iEnergizer Aptara®, Ltd.
Cover Design: Studio Montage
Cover Image: Starman963/Shutterstock
About the Cover: The Metropol Parasol in Seville, Spain, features a honeycomb of wooden panels supported by concrete bases. It is the world's largest wooden structure.

Library of Congress Cataloging-in-Publication Data

Names: Lial, Margaret L., author. | Hungerford, Thomas W., author. | Holcomb, John P., Jr., author. | Mullins, Bernadette, author.
Title: Finite mathematics with applications : in the management, natural, and social sciences / Margaret L. Lial (American River College), Thomas W. Hungerford (Saint Louis University), John P. Holcomb, Jr. (Cleveland State University), Bernadette Mullins (Birmingham-Southern College).
Description: Twelfth edition. | Boston, MA : Pearson, [2018] | Includes indexes.
Identifiers: LCCN 2017051121| ISBN 9780134767611 | ISBN 0134767616
Subjects: LCSH: Mathematics—Textbooks. | Social sciences--Mathematics--Textbooks.
Classification: LCC QA37.3 .L544 2018 | DDC 567.9--dc23 LC record available at https://lccn.loc.gov/2017051121

2   18

ISBN 13:   978-0-13-476761-1
ISBN 10:   0-13-476761-6

We would like to dedicate this twelfth edition to Thomas W. Hungerford, who passed away from a sudden illness in 2014. Tom had a major impact on both our lives. For Bernadette, it was through his authorship of his classic graduate text, Algebra*; for John it was through working closely with Tom on the ninth and tenth editions of this text. Tom was exact and thorough, and he was a deep thinker of how students and teachers would read and use the material in his books. He is greatly missed.*

We would also like to thank Greg Webster and Stan Palla for their enthusiastic support of our work on this project. Without their understanding, we never would have been able to make enough time to complete the revision. We thank you from the bottoms of our hearts.

Contents

Preface

Finite Mathematics with Applications is an applications-centered text for students in business, management, and natural and social sciences. It covers the basics of college algebra, followed by topics in finite mathematics. The text can be used for a variety of different courses, and the only prerequisite is a basic course in algebra. Chapter 1 provides a thorough review of basic algebra for those students who need it.

It has been our primary goal to present sound mathematics in an understandable manner, proceeding from the familiar to new material, and from concrete examples to general rules and formulas. There is an ongoing focus on real-world problem solving, and almost every section includes relevant, contemporary applications.

New to This Edition

We have revised and added content, updated and added new applications, fine-tuned pedagogical devices, and evaluated and enhanced the exercise sets. In addition, both the functionality of MyLab™ Math and the resources within it have been greatly improved and expanded for this edition. These improvements were incorporated after careful consideration and much feedback from those who teach this course regularly. Following is a list of some of the more substantive revisions made to this edition.

- We updated or added real-world data for hundreds of examples and exercises. We have tried to make this text the most relevant and interesting of its kind, and the main way we do this is by immersing ourselves in the kinds of applications that we know from experience will motivate students. We realize that motivating reluctant learners is a major part of this course; the applications in this text are designed to give instructors a big advantage in facing this challenge.

- We analyzed aggregated student usage and performance data from MyLab Math for the previous edition of this text. The results of this analysis helped improve the quality and quantity of exercises that matter the most to instructors and students. Also, within exercise sets, we improved even-odd pairing of exercises and better gradated them by level.

- In Chapter 5, we changed the notation for financial formulas to match the notation used in the TVM Solver from the TI-84 calculator.

- We added weighted averages to Section 10.2 so that students can understand applications such as calculating their own final grade when different components of a course are weighted at different percentages.

- We moved the discussion about boxplots from Section 10.4 to Section 10.3 because boxplots are tools for visualizing variation, and variation is discussed in Section 10.3.

- Material in the previous edition on using the normal distribution to approximate the binomial distribution in Section 10.5 was consolidated and moved to Section 10.4. The normal approximation to the binomial is not as important a topic today as it once was because technology makes calculating exact binomial probabilities easy. However, the conceptual understanding of the ideas can be important for students to learn, so we condensed the material and put it in the section on the normal distribution.

- Labels for the applications in the previous edition were very general (e.g., Business, Life Sciences). In this edition, we made them more specific (e.g., Google Profits) to pique student interest and to allow students to find applications that relate to their specific major and areas of interest

The copy below is an example of the efforts put forth to update each chapter. **To see this type of detailed information for *all chapters* in the text, please see the "Features" portion of the Pearson online catalog for this text.**

New to Chapter 4

- In Section 4.1, updated application Examples 5 and 6 and Checkpoint 7 with data on wine consumption and assets of AIG. Replaced or updated ten of the exercises with current data on the assets of Prudential Financial, Inc.; Netflix costs; GDP for China and the U.S.; asset management; imports from Vietnam; and subprime mortgages.

- Replaced Examples 1 and 2 in Section 4.2 with current data on debt in the U.S. and on sales of single-family homes. Updated Example 4 with more recent data on infant mortality rates. Replaced Example 6 and Checkpoint 4 with a current example on the price of scrap steel. Replaced or updated eleven of the application exercises with data on wind power, oil production, office rent, personal consumption, Medicare expenditures, Chinese assets in banks, Internet access in China, seat-belt use, death rates, food assistance, and labor force participation.

- In Section 4.3, added a graph of several logarithmic functions of different bases to help students visualize logarithmic functions better. Replaced an application example with a current logarithmic model function on wind energy generated in the U.S. Updated or replaced eight of the application exercises with data on health insurance costs, dairy expenditures, credit union assets, border patrol budgets, opioid deaths, iPhone sales, and vehicle miles traveled.

- In Section 4.4, utilized color to indicate nonpossible solutions to logarithmic equations more clearly. Replaced Example 7 and Checkpoint 7 with a new example on new jobs added to the U.S. economy. Also added a new example and checkpoint (using data on the digital grocery market) to illustrate solving for x with logarithmic and exponential equations. Updated or replaced twelve of the application exercises with data on foreign earnings, nursing degrees, veterans' benefits, Snapchat users, wind energy, Japanese messaging, CVS Health earnings and revenue, the number of teachers in the U.S., Best Buy revenue, Twitter stock price, and outstanding loans in U.S. banks.

- In the Review Exercises, updated or replaced ten of the application exercises with current data on exports to Mexico, Royal Caribbean share price, the number of murders in Chicago, crude oil and coal futures, recent earthquakes, FedEx profits, Starbucks and Dunkin' Donuts App users, and bank capital.

- Updated Case Study examples and exercises with more recent data and graphs from gapminder.org.

New to MyLab Math

Many improvements have been made to the overall functionality of MyLab Math since the previous edition. Beyond that, however, we have also increased and improved the content specific to this text.

- Instructors now have **more exercises** than ever to choose from when assigning homework. Most new questions are application-oriented. There are approximately 3700 assignable exercises in MyLab Math for this text. New exercise types include:
  - Additional Conceptual Questions provide support for assessing concepts and vocabulary. Many of these questions are application-oriented.
  - Setup & Solve exercises require students to show how they set up a problem as well as the solution, which approximates more closely what is required of students on tests.

- The **videos are *all new***, and they feature veteran instructors Thomas Hartfield (University of North Georgia), Mike Rosenthal (Florida International University), and Kate Haynes (Delaware Technical Community College).

  ∘ Each section of the text now has an accompanying full lecture video. To make it easier for students to navigate to the content they need, each lecture video is segmented into shorter clips (labeled Introduction, Example, or Summary).

  ∘ Both the video lectures and video segments are assignable within MyLab Math. We have included a Guide to Video-Based Assignments within the Instructor Resources section of MyLab Math that allows you to assign exercises for each video.

  ∘ MathTalk and StatTalk videos highlight applications of the content of the course to business. The videos are supported by assignable exercises.

- A full suite of **Interactive Figures** has been added to support teaching and learning. The figures illustrate key concepts and allow manipulation. They have been designed to be used during lectures as well as by students working independently.

- An **Integrated Review** version of the MyLab Math course contains premade quizzes to assess the prerequisite skills needed for each chapter, plus personalized remediation for any gaps in skills that are identified.

- **Study Skills Modules** help students with the life skills that can make the difference between passing and failing.

- The **Graphing Calculator Manual** and **Excel Spreadsheet Manual**, both specific to this course, have been updated to support the TI-84 CE (color edition) and Excel 2016, respectively. Both manuals also contain additional topics to support the course.

- We heard from users that the Annotated Instructor Edition for the previous edition required too much flipping of pages to find answers, so MyLab Math now contains a downloadable **Instructor Answers document**—with *all answers in one place*. (This augments the downloadable Instructor Solutions Manual, which contains all solutions.)

Continued Pedagogical Support

- **Real-Data Examples and Explanations:** Real-data exercises have long been a popular and integral aspect of this text. A significant number of new real-data examples and exercises have also been introduced into the text. Applications are noted with a **green header** to indicate the subject of the problem so instructors or students can focus on applications that are in line with students' majors.

- **Balanced Approach:** Multiple representations of a topic (symbolic, numerical, graphical, verbal) are given when appropriate. However, we do not believe that all representations are useful for all topics, so effective alternatives are discussed only when they are likely to increase student understanding.

- **Strong Algebra Foundation:** The text begins with four thorough chapters of college algebra that can be used in a variety of ways based on the needs of the students and the goals of the course. Take advantage of the content in these chapters as needed so students will be more successful with later topics and future courses.

- **Help for Skill Gaps:** The Prerequisite Skills Test (for Chapters 1–4) at the front of the text can help students determine where remediation is needed. The text contains solutions to the test exercises to help students remediate any gaps in basic skills.

- **Checkpoint** exercises are marked with icons such as and provide an opportunity for students to stop, check their understanding of the specific concept at hand, and move forward with confidence. Answers to Checkpoint exercises are located at the end of the section to encourage students to work the problems before looking at the answers. (See pages 185 and 186.)

- **Caution** notes highlight common student difficulties or warn against frequently made mistakes. (See page 209.)
- **Exercises:** In addition to skill-based practice, conceptual, and application-based exercises, the text includes some specially marked exercises:
  - Writing Exercises ◣ (See page 188.)
  - Connection Exercises ⒸⒽ relate current topics to earlier sections (See page 213.)
  - Graphing Calculator Exercises ▦ (See page 205.)
  - Spreadsheet Exercises ▢ (See page 260.)
- **Example/Exercise Connection:** Selected exercises include a reference to related example(s) within the section (e.g., "See Examples 6 and 7") to facilitate what students do naturally when they use a book—that is, look for specific examples when they get stuck on a problem. Later exercises leave this information out and provide opportunities for mixed skill practice.
- **Graphing Calculators and Spreadsheets:** It is assumed that all students have a calculator that will handle exponential and logarithmic functions. Graphing calculator and spreadsheet references are highlighted in the text so that those who use the technology can easily incorporate it and those who do not can easily omit it. Examples and exercises that require a graphing calculator are marked with ▦ and those that require a spreadsheet are marked with ▢, making it obvious where technology is being included.
- **Technology Tips:** These tips are placed at appropriate points in the text to inform students of various features of their graphing calculator, spreadsheet, or other computer programs. Note that Technology Tips designed for TI-84 CE also apply to the TI-84 Plus, TI-83, and TI-Nspire.
- **End-of-chapter materials:** are designed to help students prepare for exams. These materials include a List of Key Terms and Symbols and Summary of Key Concepts, as well as a thorough set of Chapter Review Exercises.
- **Case Studies:** appear at the end of each chapter and offer contemporary, real-world applications of some of the mathematics presented in the chapter. Not only do these provide an opportunity for students to see the mathematics they are learning in action, but they also provide at least a partial answer to the question, "What is this stuff good for?" These have been expanded to include options for longer-term projects if the instructor should choose to use them.

Course Flexibility

The content of the text is divided into two parts:

1. College Algebra (Chapters 1–4)
2. Finite Mathematics (Chapters 5–10)

This coverage of the material offers flexibility, making the text appropriate for a variety of courses, including:

- **Finite Mathematics** (one semester or two quarters). Use as much of Chapters 1–4 as needed, and then go into Chapters 5–10 as time permits and local needs require.
- **College Algebra with Applications** (one semester or quarter). Use Chapters 1–8, with Chapters 7 and 8 being optional.

Pearson regularly produces custom versions of this text (and its accompanying MyLab Math course) to address the needs of specific course sequences. Custom versions can be produced for even smaller-enrollment courses due to advances in digital printing. Please contact your local Pearson representative for more details.

Chapter interdependence is as follows:

	Chapter	Prerequisite
1	Algebra and Equations	None
2	Graphs, Lines, and Inequalities	Chapter 1
3	Functions and Graphs	Chapters 1 and 2
4	Exponential and Logarithmic Functions	Chapter 3
5	Mathematics of Finance	Chapter 4
6	Systems of Linear Equations and Matrices	Chapters 1 and 2
7	Linear Programming	Chapters 3 and 6
8	Sets and Probability	None
9	Counting, Probability Distributions, and Further Topics in Probability	Chapter 8
10	Introduction to Statistics	Chapter 8

Acknowledgments

We wish to thank the following reviewers for their thoughtful feedback and helpful suggestions, without which we could not continue to improve this text.

John Altiere, Cleveland State University

Sviatoslav Archava, East Carolina University

Stephen Bast, Anne Arundel Community College

Phanuel Bediako, Delaware State University

Susan Bellini, Cleveland State University

Eric Erdmann, University of Minnesota, Duluth

Meghan Foster, American University

Lobna Mazzawi, Everett Community College

Thomas Milligan, University of Oklahoma

Margie Nowlin, Texas Christian University

David Stott, Sinclair Community College

Cong-Cong Xing, Nicholls State University

The following faculty members provided direction on the development of the MyLab Math course for this edition:

Vince Bander, Pierce College, Puyallup

Rachel Bates, Redlands Community College

Krista Blevins Cohlmia, Midland College

Larry Cook, Taft College

Stanislav Dubrovskiy, Sierra College

Jeffrey K. Dyess, Bishop State Community College

Shurron Farmer, University of the District of Columbia

Tonia Garrett, San Jacinto College

Abe Haje, Lone Star College, University Park

Ray Hendrickson, Bucks County Community College

MaryAlice Howe, Bucks County Community College

Edgar Jasso, North Seattle College

Dynechia Jones, Baton Rouge Community College

Tiffany Jones, Baylor University

Shelley Lenahan, University of Alabama, Huntsville

Steven Mark, Wilmington University

Maria Mathews, Baldwin Wallace University

Janet Noah, Brookhaven College

Margie Torres Nowlin, All Saints Episcopal School

Rose L. Pugh, Bellevue College

Peggy M. Slavik, Baldwin Wallace University

Kay Tekumalla, Prince George's Community College

R. Vafa, Temple University

Rick L. Wing, San Francisco State University

Cong-Cong Xing, Nicholls State University

Ju Zhou, Kutztown University

We gratefully acknowledge Birmingham-Southern College and Cleveland State University for their wholehearted support of this edition. Bernadette Mullins thanks two BSC students who checked new content: Amer Babi and Adam Alden Pratt.

We also thank our accuracy checkers, who did an excellent job of checking both text and exercise answers: Stephen Bast and Cong-Cong Xing. Thanks also to the supplements authors: Salvatore Sciandra, Chris True, and Stela Pudar-Hozo.

We want to thank the staff of Pearson Education for their assistance with and contributions to this book, particularly Jeff Weidenaar, Kathy Manley, Jenn Snyder, Emily Ockay, Kristina Evans, and Jean Choe. Finally, we wish to express our appreciation to Sherrill Redd of Aptara Corporation, who was a pleasure to work with.

John P. Holcomb, Jr.
Bernadette Mullins

MyLab™ Math Online Course

for *Finite Mathematics with Applications in the Management, Natural, and Social Sciences*, 12e

(access code required)

MyLab™ Math is available to accompany Pearson's market-leading text offerings. To give students a consistent tone, voice, and teaching method, each text's flavor and approach are tightly integrated throughout the accompanying MyLab Math course, making learning the material as seamless as possible.

PREPAREDNESS

One of the biggest challenges in applied math courses is making sure students are adequately prepared with the prerequisite skills needed to complete their course work successfully. MyLab Math supports students with just-in-time remediation and key-concept review.

NEW! Integrated Review Course   An Integrated Review version of the MyLab Math course contains premade, assignable quizzes to assess the prerequisite skills needed for each chapter, plus personalized remediation for any gaps in skills that are identified. Therefore, each student receives just the help that he or she needs—no more, no less.

Study Skills Modules   Study skills modules help students with the life skills that can make the difference between passing and failing.

DEVELOPING DEEPER UNDERSTANDING

MyLab Math provides content and tools that help students build a deeper understanding of course content than would otherwise be possible.

Exercises with Immediate Feedback

Homework and practice exercises for this text regenerate algorithmically to give students unlimited opportunity for practice and mastery. MyLab Math provides helpful feedback when students enter incorrect answers and includes the optional learning aids Help Me Solve This, View an Example, videos, and/or the eText.

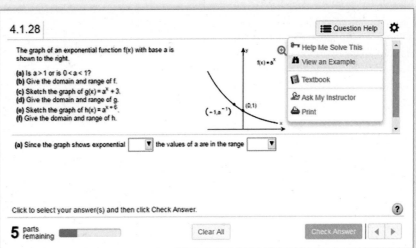

pearson.com/mylab/math

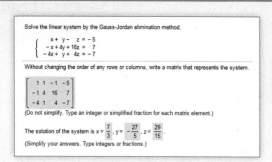

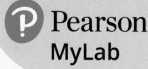

NEW! Setup & Solve Exercises

These exercises require students to show how they set up a problem as well as the solution, better mirroring what is required of students on tests.

NEW! Additional Conceptual Questions

Additional Conceptual Questions provide support for assessing concepts and vocabulary. Many of these questions are application-oriented. They are clearly labeled "Conceptual" in the Assignment Manager.

NEW! Interactive Figures

A full suite of Interactive Figures has been added to support teaching and learning. The figures illustrate key concepts and allow manipulation. They have been designed for use during lectures as well as by students working independently.

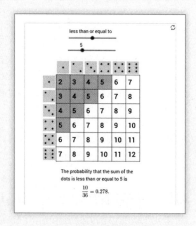

NEW! Instructional Videos

The instructional videos for this course are *all new*, featuring veteran instructors Thomas Hartfield (University of North Georgia) Mike Rosenthal (Florida International University), and Kate Haynes (Delaware Technical Community College).

- Each section of the text now has an accompanying full lecture video. To make it easier for students to navigate to the content they need, each lecture video is segmented into shorter clips (labeled Introduction, Example, or Summary).

- Both the video lectures and video segments are assignable within MyLab Math. We have included a Guide to Video-Based Assignments within the Instructor Resources section of MyLab Math that allows you to assign exercises for each video.

In addition, the MathTalk and StatTalk videos connect the content of the course to business and management applications. The videos are supported by assignable exercises.

pearson.com/mylab/math

Technology Manuals and Projects (downloadable)

- *Excel Spreadsheet Manual and Projects* by Stela Pudar-Hozo, Indiana University–Northwest
- *Graphing Calculator Manual and Projects* by Chris True, University of Nebraska

These manuals, both specific to this course, have been updated to support the TI-84 CE (color edition) and Excel 2016, respectively. Instructions are ordered by mathematical topic. The files can be downloaded from within MyLab Math.

Student's Solutions Manual (softcover and downloadable)
ISBN: 0-13-477635-6 | 978-0-13-477635-4
Written by Salvatore Sciandra from Niagara County Community College, the Student's Solutions Manual contains worked-out solutions to all the odd-numbered exercises and all Chapter Review and Case Studies. This manual is available in print and can be download from within MyLab Math.

A Complete eText
Students get unlimited access to the eText within any MyLab Math course using that edition of the textbook. The Pearson eText app allows existing subscribers to access their titles on an iPad or Android tablet for either online or offline viewing.

SUPPORTING INSTRUCTION

MyLab Math comes from an experienced partner with educational expertise and an eye on the future. It provides resources to help you assess and improve students' results at every turn, and unparalleled flexibility to create a course tailored to you and your students.

Learning Catalytics™
Now included in all MyLab Math courses, this student response tool uses students' smartphones, tablets, or laptops to engage them in more interactive tasks and thinking during lecture. Learning Catalytics fosters student engagement and peer-to-peer learning with real-time analytics. Access prebuilt exercises created specifically for this course.

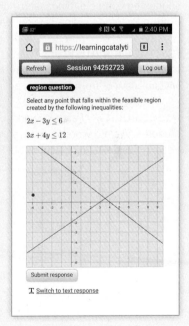

pearson.com/mylab/math

PowerPoint® Lecture Resources (downloadable)   Slides contain presentation resources such as figures, feature boxes, and key examples from the text. They can be downloaded from within MyLab Math or from Pearson's online catalog, **www.pearson.com**.

Comprehensive Gradebook   The gradebook includes enhanced reporting functionality such as item analysis and a reporting dashboard to allow you to manage your course efficiently. Student performance data is presented at the class, section, and program levels in an accessible, visual manner so you will have the information you need to keep your students on track.

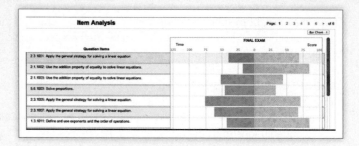

TestGen   TestGen® (**www.pearson.com/testgen**) enables instructors to build, edit, print, and administer tests using a computerized bank of questions developed to cover all the objectives of the text. TestGen is algorithmically based, allowing instructors to create multiple but equivalent versions of the same question or test with the click of a button. Instructors can also modify test bank questions or add new questions. The software and test bank are available for download from Pearson's online catalog, **www.pearson.com**. The questions are also assignable in MyLab Math.

Instructor's Solutions Manual (downloadable)   Written by Salvatore Sciandra from Niagara County Community College, the Instructor's Solutions Manual contains detailed solutions to all text exercises, suggested course outlines, and a chapter interdependence chart. It can be downloaded from within MyLab Math or from Pearson's online catalog, **www.pearson.com**.

Instructor's Answers (downloadable)   These handy chapter-by-chapter documents provide answers to all Student Edition exercises in one place for easy reference by instructors. They are downloadable from within MyLab Math or from Pearson's online catalog, **www.pearson.com**.

Accessibility   Pearson works continuously to ensure our products are as accessible as possible to all students. We are working toward achieving WCAG 2.0 Level AA and Section 508 standards, as expressed in the Pearson Guidelines for Accessible Educational Web Media, **www.pearson.com/mylab/math/accessibility**.

To the Student

The key to succeeding in this course is to remember that *mathematics is not a spectator sport*. You can't expect to learn mathematics without *doing* mathematics any more than you could learn to swim without getting wet. You must take an active role and use all the resources at your disposal: your instructor, your fellow students, this book, and the supplements that accompany this book. Following are some general tips on how to be successful in the course and some specific tips on how to get the most out of this text and supplementary resources.

Ask Questions! Remember the words of the great Hillel: "The bashful do not learn." There is no such thing as a "dumb question" (assuming, of course, that you have read the book and your class notes and attempted the homework). Your instructor will welcome questions that arise from a serious effort on your part. So get your money's worth: Ask questions!

Read the Book Interactively! There is more to a math textbook than just the exercise sets. Each section introduces topics carefully with many examples—both mathematical and contextual. Take note of the "Caution" and "Note" comments, and bookmark pages with key definitions or formulas. After reading the example, try the Checkpoint exercise that appears next to it in the margin to check your understanding of the concept. This will help you solidify your understanding or diagnose if you do not fully understand the concept. The answers to the Checkpoint exercises are right after the homework exercises in each section. Resist the temptation to flip to the answer until you've worked the problem completely!

Take Advantage of the Supplementary Material! Many resources are at your disposal within and outside the text. Take the time to interact with them and determine which resources suit your learning style best.

- If your instructor allows the use of graphing calculators and/or spreadsheets, work through the examples and exercises marked with ▦ or ▢. Some instructors may make this material part of the course, whereas others will not require you to use technology. If your instructor asks that you use technology, a Graphing Calculator Manual and an Excel Spreadsheet Manual are available in MyLab Math. In addition, there are ***Technology Tips*** throughout the text that describe the proper menu or keys to use for various procedures on a graphing calculator. Note that Technology Tips for the TI-84+ CE also apply to TI-83+, TI-Nspire, and usually TI-83.

- MyLab Math has a variety of types of resources to help you learn, including videos for every section of the text; Interactive Figures to help visualize difficult concepts; unlimited practice and assessment on newly learned or prerequisite skills; and access to the Student Solutions Manual, Graphing Calculator Manual, Excel Spreadsheet Manual, and a variety of helpful reference cards.

Do Your Homework! Whether the homework is paper and pencil or assigned online, you must practice what you have learned. This is your opportunity to practice those essential skills needed for passing this course and those skills needed for application in future courses or your career.

We wish you the best in your efforts throughout this course, in future courses, and beyond school.

John and Bernadette

Prerequisite Skills Test*

The following test is unlike your typical math test. Rather than testing your skills after you have worked on them, this test assesses skills that you should know from previous coursework and will use in this class. It is intended to diagnose any areas that you may need to remediate. Take advantage of the results of this test by checking your answers in Appendix B. The full solutions are included to remind you of the steps to answer the problem.

Find the most simplified answer for the given problems involving fractions:

1. $\dfrac{5}{2} - 6 =$

2. $\dfrac{1}{2} \div \dfrac{2}{5} =$

3. $\dfrac{1}{3} \div 3 =$

Simplify the given expression, keeping in mind the appropriate order of operations:

4. $7 + 2 - 3(2 \div 6) =$

5. $\dfrac{2 \times 3 + 12}{1 + 5} - 1 =$

Indicate whether each of the statements is true or false:

6. $\dfrac{4 + 3}{3} = 5$

7. $\dfrac{5}{7} + \dfrac{7}{5} = 1$

8. $\dfrac{3}{5} + 1 = \dfrac{6}{5}$

Translate each of the following written expressions into a corresponding mathematical statement. If possible, solve for the unknown value or values.

9. Alicia has n pairs of shoes. Manuel has two more pairs of shoes than Alicia. How many pairs of shoes does Manuel have?

10. David's age and Selina's age, when added together, equals 42. Selena is 6 years older than David. What are David's and Selina's ages?

Solve the following problem.

11. The price of a sweater, originally sold for $72, is reduced by 20%. What is the new sale price of the sweater?

Given the following rectangular coordinate system, graph and label the following points:

12. A: $(3, -2)$ B: $(4, 0)$ C: $(-2, -3)$

13. D: $(3, 5)$ E: $(-1, 4)$ F: $(-4, -5)$

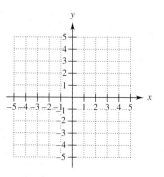

Round the following values as indicated:

14. (a) 4.27659 to the nearest tenth

    (b) 245.984 to the nearest unit (whole number)

15. (a) 16.38572 to the nearest hundredth

    (b) 1,763,304.42 to the nearest thousand

Write the number that corresponds with the given numerical statement:

16. (a) The Company's liabilities totaled 34 million dollars.

    (b) The total of investments was 2.2 thousand dollars.

17. (a) The population of a country is 17 hundred thousand.

    (b) The cost of the new airport could run as high as three and a quarter billion dollars.

Answer the following. If there is no solution, indicate that there is no solution and provide a reason.

18. $\dfrac{5}{0} =$

19. A car is traveling 60 miles per hour. Is the car traveling at 1 mile per minute?

20. Which number is greater, -9 or -900?

---

*Full Solutions to this test are provided in the Appendix B.

Algebra and Equations

1

CHAPTER OUTLINE

Mathematics is widely used in business, finance, and the biological, social, and physical sciences, from developing efficient production schedules for a factory to mapping the human genome. Mathematics also plays a role in determining interest on a loan from a bank, the growth of traffic on websites, and the study of falling objects.

Algebra and equations are the basic mathematical tools for handling many applications. Your success in this course will depend on your having the algebraic skills presented in this chapter.

1.1 The Real Numbers

Only real numbers will be used in this book.* The names of the most common types of real numbers are as follows.

> ### The Real Numbers
>
> **Natural (counting) numbers** $1, 2, 3, 4, \ldots$
> **Whole numbers** $0, 1, 2, 3, 4, \ldots$

---

*Not all numbers are real numbers. For example, $\sqrt{-1}$ is a number that is *not* a real number.

1

Integers	$\ldots, -3, -2, -1, 0, 1, 2, 3, \ldots$
Rational numbers	All numbers that can be written in the form p/q, where p and q are integers and $q \neq 0$
Irrational numbers	Real numbers that are not rational

As you can see, every natural number is a whole number, and every whole number is an integer. Furthermore, every integer is a rational number. For instance, the integer 7 can be written as the fraction $\frac{7}{1}$ and is therefore a rational number.

One example of an irrational number is π, the ratio of the circumference of a circle to its diameter. The number π can be approximated as $\pi \approx 3.14159$ ($\approx$ means "is approximately equal to"), but there is no rational number that is exactly equal to π.

EXAMPLE 1   What kind of number is each of the following?

(a) 6

Solution The number 6 is a natural number, a whole number, an integer, a rational number, and a real number.

(b) $\dfrac{3}{4}$

Solution This number is rational and real.

(c) 3π

Solution Because π is not a rational number, 3π is irrational and real.

✓**Checkpoint 1**

Name all the types of numbers that apply to the following.

(a) -2

(b) $-5/8$

(c) $\pi/5$

Answers to Checkpoint exercises are found at the end of the section.

All real numbers can be written in decimal form. A rational number, when written in decimal form, is either a terminating decimal, such as .5 or .128, or a repeating decimal, in which some block of digits eventually repeats forever, such as 1.3333 . . . or 4.7234234234 . . . .[†] Irrational numbers are decimals that neither terminate nor repeat.

When a calculator is used for computations, the answers it produces are often decimal *approximations* of the actual answers; they are accurate enough for most applications. To ensure that your final answer is as accurate as possible,

> *you should not round off any numbers during long calculator computations.*

It is usually OK to round off the final answer to a reasonable number of decimal places once the computation is finished.

The important basic properties of the real numbers are as follows.

Properties of the Real Numbers

For all real numbers, a, b, and c, the following properties hold true:

Commutative properties	$a + b = b + a$ and $ab = ba$.
Associative properties	$(a + b) + c = a + (b + c)$ and $(ab)c = a(bc)$.

---

*The use of Checkpoint exercises is explained in the "To the Student" section preceding this chapter.

[†]Some graphing calculators have a FRAC key that automatically converts some repeating decimals to fraction form.

Identity properties	There exists a unique real number 0, called the **additive identity,** such that

$$a + 0 = a \quad \text{and} \quad 0 + a = a.$$

There exists a unique real number 1, called the **multiplicative identity,** such that

$$a \cdot 1 = a \quad \text{and} \quad 1 \cdot a = a.$$

Inverse properties	For each real number a, there exists a unique real number $-a$, called the **additive inverse** of a, such that

$$a + (-a) = 0 \quad \text{and} \quad (-a) + a = 0.$$

If $a \neq 0$, there exists a unique real number $1/a$, called the **multiplicative inverse** of a, such that

$$a \cdot \frac{1}{a} = 1 \quad \text{and} \quad \frac{1}{a} \cdot a = 1.$$

Distributive property	$a(b + c) = ab + ac \quad \text{and} \quad (b + c)a = ba + ca.$

The next five examples illustrate the properties listed in the preceding box.

EXAMPLE 2   The commutative property says that the order in which you add or multiply two quantities doesn't matter.

(a) $(6 + x) + 9 = 9 + (6 + x) = 9 + (x + 6)$ **(b)** $5 \cdot (9 \cdot 8) = (9 \cdot 8) \cdot 5$

EXAMPLE 3   When the associative property is used, the order of the numbers does not change, but the placement of parentheses does.

(a) $4 + (9 + 8) = (4 + 9) + 8$ **(b)** $3(9x) = (3 \cdot 9)x$

✔**Checkpoint 2**

Name the property illustrated in each of the following examples.

(a) $(2 + 3) + 9 = (3 + 2) + 9$

(b) $(2 + 3) + 9 = 2 + (3 + 9)$

(c) $(2 + 3) + 9 = 9 + (2 + 3)$

(d) $(4 \cdot 6)p = (6 \cdot 4)p$

(e) $4(6p) = (4 \cdot 6)p$

EXAMPLE 4   By the identity properties,

(a) $-8 + 0 = -8$ **(b)** $(-9) \cdot 1 = -9.$

▦ **TECHNOLOGY TIP** To enter -8 on a calculator, use the negation key (labeled $(-)$ or $+/-$), *not* the subtraction key. On most one-line scientific calculators, key in $8 +/-$. On graphing calculators or two-line scientific calculators, key in either $(-) 8$ or $+/- 8$.

✔**Checkpoint 3**

Name the property illustrated in each of the following examples.

(a) $2 + 0 = 2$

(b) $-\frac{1}{4} \cdot (-4) = 1$

(c) $-\frac{1}{4} + \frac{1}{4} = 0$

(d) $1 \cdot \frac{2}{3} = \frac{2}{3}$

EXAMPLE 5   By the inverse properties, the statements in parts (a) through (d) are true.

(a) $9 + (-9) = 0$ **(b)** $-15 + 15 = 0$

(c) $-8 \cdot \left(\frac{1}{-8}\right) = 1$ **(d)** $\frac{1}{\sqrt{5}} \cdot \sqrt{5} = 1$

📄 **NOTE**   There is no real number x such that $0 \cdot x = 1$, so 0 has no multiplicative inverse.

| EXAMPLE 6 | By the distributive property,

(a) $9(6 + 4) = 9 \cdot 6 + 9 \cdot 4$

(b) $3(x + y) = 3x + 3y$

(c) $-8(m + 2) = (-8)(m) + (-8)(2) = -8m - 16$

(d) $(5 + x)y = 5y + xy.$ ✔4

✓**Checkpoint 4**

Use the distributive property to rewrite each of the following.

(a) $4(-2 + 5)$

(b) $2(a + b)$

(c) $-3(p + 1)$

(d) $(8 - k)m$

(e) $5x + 3x$

Order of Operations

Some complicated expressions may contain many sets of parentheses. To avoid ambiguity, the following procedure should be used.

Parentheses

Work separately above and below any fraction bar. Within each set of parentheses or square brackets, start with the innermost set and work outward.

| EXAMPLE 7 | Simplify: $[(3 + 2) - 7]5 + 2([6 \cdot 3] - 13)$.

Solution On each segment, work from the inside out:

$$[(3 + 2) - 7]5 + 2([6 \cdot 3] - 13)$$
$$= [5 - 7]5 + 2(18 - 13)$$
$$= [-2]5 + 2(5)$$
$$= -10 + 10 = 0.$$

Does the expression $2 + 4 \times 3$ mean

$$(2 + 4) \times 3 = 6 \times 3 = 18?$$

Or does it mean

$$2 + (4 \times 3) = 2 + 12 = 14?$$

To avoid this ambiguity, mathematicians have adopted the following rules (which are also followed by almost all scientific and graphing calculators).

Order of Operations

1. Find all powers and roots, working from left to right.

2. Do any multiplications or divisions in the order in which they occur, working from left to right.

3. Finally, do any additions or subtractions in the order in which they occur, working from left to right.

If sets of parentheses or square brackets are present, use the rules in the preceding box within each set, working from the innermost set outward.

According to these rules, multiplication is done *before* addition, so $2 + 4 \times 3 = 2 + 12 = 14$. Here are some additional examples.

EXAMPLE 8 Use the order of operations to evaluate each expression if $x = -2$, $y = 5$, and $z = -3$.

(a) $-4x^2 - 7y + 4z$

Solution Use parentheses when replacing letters with numbers:

$$-4x^2 - 7y + 4z = -4(-2)^2 - 7(5) + 4(-3)$$
$$= -4(4) - 7(5) + 4(-3) = -16 - 35 - 12 = -63.$$

(b) $\dfrac{2(x - 5)^2 + 4y}{z + 4} = \dfrac{2(-2 - 5)^2 + 4(5)}{-3 + 4}$

$$= \dfrac{2(-7)^2 + 20}{1}$$

$$= 2(49) + 20 = 118. \ \checkmark_5$$

✓ **Checkpoint 5**

Evaluate the following if $m = -5$ and $n = 8$.

(a) $-2mn - 2m^2$

(b) $\dfrac{4(n - 5)^2 - m}{m + n}$

EXAMPLE 9 Use a calculator to evaluate

$$\frac{-9(-3) + (-5)}{3(-4) - 5(2)}.$$

Solution Use extra parentheses (shown here in blue) around the numerator and denominator when you enter the number in your calculator, and be careful to distinguish the negation key from the subtraction key.

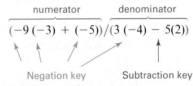

numerator   denominator

$(-9\,(-3) + (-5))/(3\,(-4) - 5(2))$

Negation key   Subtraction key

If you don't get -1 as the answer, then you are entering something incorrectly. $\checkmark_6$

✓ **Checkpoint 6**

Use a calculator to evaluate the following.

(a) $4^2 \div 8 + 3^2 \div 3$

(b) $[-7 + (-9)] \cdot (-4) - 8(3)$

(c) $\dfrac{-11 - (-12) - 4 \cdot 5}{4(-2) - (-6)(-5)}$

(d) $\dfrac{36 \div 4 \cdot 3 \div 9 + 1}{9 \div (-6) \cdot 8 - 4}$

Square Roots

There are two numbers whose square is 16, namely, 4 and -4. The positive one, 4, is called the **square root** of 16. Similarly, the square root of a nonnegative number d is defined to be the *nonnegative* number whose square is d; this number is denoted $\sqrt{d}$. For instance,

$$\sqrt{36} = 6 \text{ because } 6^2 = 36, \quad \sqrt{0} = 0 \text{ because } 0^2 = 0, \text{ and}$$
$$\sqrt{1.44} = 1.2 \text{ because } (1.2)^2 = 1.44.$$

No negative number has a square root that is a real number. For instance, there is no real number whose square is -4, so -4 has no square root.

Every nonnegative real number has a square root. Unless an integer is a perfect square (such as $64 = 8^2$), its square root is an irrational number. A calculator can be used to obtain a rational approximation of these square roots.

📓 **TECHNOLOGY TIP**

On one-line scientific calculators, $\sqrt{40}$ is entered as 40 $\sqrt{\ }$. On graphing calculators and two-line scientific calculators, key in $\sqrt{\ }$ 40 ENTER (or EXE).

EXAMPLE 10 Estimate each of the given quantities. Verify your estimates with a calculator.

(a) $\sqrt{40}$

Solution Since $6^2 = 36$ and $7^2 = 49$, $\sqrt{40}$ must be a number between 6 and 7. A typical calculator shows that $\sqrt{40} \approx 6.32455532$.

(b) $5\sqrt{7}$

Solution $\sqrt{7}$ is between 2 and 3 because $2^2 = 4$ and $3^2 = 9$, so $5\sqrt{7}$ must be a number between $5 \cdot 2 = 10$ and $5 \cdot 3 = 15$. A calculator shows that $5\sqrt{7} \approx 13.22875656.$ ✓7

❗ CAUTION If c and d are positive real numbers, then $\sqrt{c + d}$ is *not* equal to $\sqrt{c} + \sqrt{d}$. For example, $\sqrt{9 + 16} = \sqrt{25} = 5$, but $\sqrt{9} + \sqrt{16} = 3 + 4 = 7$.

The Number Line

The real numbers can be illustrated geometrically with a diagram called a **number line.** Each real number corresponds to exactly one point on the line and vice versa. A number line with several sample numbers located (or **graphed**) on it is shown in Figure 1.1. ✓8

Figure 1.1

When comparing the sizes of two real numbers, the following symbols are used.

Symbol	Read	Meaning
$a < b$	a is less than b.	a lies to the *left* of b on the number line.
$b > a$	b is greater than a.	b lies to the *right* of a on the number line.

Note that $a < b$ means the same thing as $b > a$. The inequality symbols are sometimes joined with the equals sign, as follows.

Symbol	Read	Meaning
$a \leq b$	a is less than or equal to b.	either $a < b$ or $a = b$
$b \geq a$	b is greater than or equal to a.	either $b > a$ or $b = a$

Only one part of an "either . . . or" statement needs to be true for the entire statement to be considered true. So the statement $3 \leq 7$ is true because $3 < 7$, and the statement $3 \leq 3$ is true because $3 = 3$.

EXAMPLE 11   Write *true* or *false* for each of the following.

(a) $8 < 12$

Solution This statement says that 8 is less than 12, which is true.

(b) $-6 > -3$

Solution The graph in Figure 1.2 shows that -6 is to the *left* of -3. Thus, $-6 < -3$, and the given statement is false.

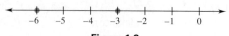

Figure 1.2

(c) $-2 \leq -2$

Solution Because $-2 = -2$, this statement is true. ✓9

✓ Checkpoint 7

Estimate each of the following.

(a) $\sqrt{73}$

(b) $\sqrt{22} + 3$

(c) Confirm your estimates in parts (a) and (b) with a calculator.

✓ Checkpoint 8

Draw a number line, and graph the numbers $-4, -1, 0, 1, 2.5,$ and $13/4$ on it.

▦ TECHNOLOGY TIP

If your graphing calculator has inequality symbols (usually located on the TEST menu), you can key in statements such as "$5 < 12$" or "$-2 \geq 3$." When you press ENTER, the calculator will display 1 if the statement is true and 0 if it is false.

✓ Checkpoint 9

Write *true* or *false* for the following.

(a) $-9 \leq -2$

(b) $8 > -3$

(c) $-14 \leq -20$

A number line can be used to draw the graph of a set of numbers, as shown in the next few examples.

EXAMPLE 12   Graph all real numbers x such that $1 < x < 5$.

Solution This graph includes all the real numbers between 1 and 5, not just the integers. Graph these numbers by drawing a heavy line from 1 to 5 on the number line, as in Figure 1.3. Parentheses at 1 and 5 show that neither of these points belongs to the graph. ✔10

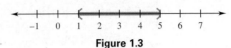

Figure 1.3

✓ **Checkpoint 10**

Graph all real numbers x such that

(a) $-5 < x < 1$

(b) $4 < x < 7$.

A set that consists of all the real numbers between two points, such as $1 < x < 5$ in Example 12, is called an **interval**. A special notation called **interval notation** is used to indicate an interval on the number line. For example, the interval including all numbers x such that $-2 < x < 3$ is written as $(-2, 3)$. The parentheses indicate that the numbers -2 and 3 are *not* included. If -2 and 3 are to be included in the interval, square brackets are used, as in $[-2, 3]$. The following chart shows several typical intervals, where $a < b$.

Intervals

Inequality	Interval Notation	Explanation
$a \leq x \leq b$	$[a, b]$	Both a and b are included.
$a \leq x < b$	$[a, b)$	a is included; b is not.
$a < x \leq b$	$(a, b]$	b is included; a is not.
$a < x < b$	(a, b)	Neither a nor b is included.

Interval notation is also used to describe sets such as the set of all numbers x such that $x \geq -2$. This interval is written $[-2, \infty)$. The set of all real numbers is written $(-\infty, \infty)$ in interval notation.

EXAMPLE 13   Graph the interval $[-2, \infty)$.

Solution Start at -2 and draw a heavy line to the right, as in Figure 1.4. Use a square bracket at -2 to show that -2 itself is part of the graph. The symbol ∞, read "infinity," *does not* represent a number. This notation simply indicates that *all* numbers greater than -2 are in the interval. Similarly, the notation $(-\infty, 2)$ indicates the set of all numbers x such that $x < 2$. ✔11

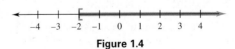

Figure 1.4

✓ **Checkpoint 11**

Graph all real numbers x in the given interval.

(a) $(-\infty, 4]$

(b) $[-2, 1]$

Absolute Value

The **absolute value** of a real number a is the distance from a to 0 on the number line and is written $|a|$. For example, Figure 1.5 shows that the distance from 9 to 0 on the number line is 9, so we have $|9| = 9$. The figure also shows that $|-9| = 9$, because the distance from -9 to 0 is also 9.

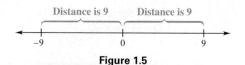

Figure 1.5

The facts that $|9| = 9$ and $|-9| = 9 = -(-9)$ suggest the following algebraic definition of absolute value.

Absolute Value

For any real number a,

$$|a| = a \qquad \text{if } a \geq 0$$
$$|a| = -a \qquad \text{if } a < 0.$$

The first part of the definition shows that $|0| = 0$ (because $0 \geq 0$). It also shows that the absolute value of any positive number a is the number itself, so $|a|$ is positive in such cases. The second part of the definition says that the absolute value of a negative number a is the *negative* of a. For instance, if $a = -5$, then $|-5| = -(-5) = 5$. So $|-5|$ is positive. The same thing works for any negative number—that is, its absolute value (the negative of a negative number) is positive. Thus, we can state the following:

For every nonzero real number a, the number $|a|$ is positive.

✓ **Checkpoint 12**

Find the following.

(a) $|-6|$

(b) $-|7|$

(c) $-|-2|$

(d) $|-3 - 4|$

(e) $|2 - 7|$

EXAMPLE 14   Evaluate $|8 - 9|$.

Solution   First, simplify the expression within the absolute-value bars:

$$|8 - 9| = |-1| = 1. \; ✓_{12}$$

1.1 Exercises

In Exercises 1 and 2, label the statement true or false. (See Example 1.)

1. Every integer is a rational number.

2. Every real number is an irrational number.

3. The decimal expansion of the irrational number π begins 3.141592653589793 . . . . Use your calculator to determine which of the following rational numbers is the best approximation for the irrational number π:

$$\frac{22}{7}, \quad \frac{355}{113}, \quad \frac{103{,}993}{33{,}102}, \quad \frac{2{,}508{,}429{,}787}{798{,}458{,}000}.$$

Your calculator may tell you that some of these numbers are equal to π, but that just indicates that the number agrees with π for as many decimal places as your calculator can handle (usually 10–14). No rational number is exactly equal to π.

Identify the properties that are illustrated in each of the following. (See Examples 2–6.)

4. $0 + (-7) = -7 + 0$

5. $6(t + 4) = 6t + 6 \cdot 4$

6. $3 + (-3) = (-3) + 3$

7. $-5 + 0 = -5$

8. $(-4) \cdot \left(\dfrac{-1}{4} \right) = 1$

9. $8 + (12 + 6) = (8 + 12) + 6$

10. $1 \cdot (-20) = -20$

11. How is the additive inverse property related to the additive identity property? the multiplicative inverse property to the multiplicative identity property?

12. Explain the distinction between the commutative and associative properties.

Evaluate each of the following if $p = -2$, $q = 3$, and $r = -5$. (See Examples 7–9.)

13. $-3(p + 5q)$

14. $2(q - r)$

15. $\dfrac{q + r}{q + p}$

16. $\dfrac{3q}{3p - 2r}$

Interest Rates *The nominal annual percentage rate (APR) reported by lenders has the formula APR = 12r, where r is the monthly interest rate. Find the APR when*

17. $r = 3.8$

18. $r = 0.8$

Find the monthly interest rate r when

19. $APR = 11$

20. $APR = 13.2$

Evaluate each expression, using the order of operations given in the text. (See Examples 7–9.)

21. $3 - 4 \cdot 5 + 5$

22. $8 - (-4)^2 - (-12)$

23. $(4 - 5) \cdot 6 + 6$

24. $\dfrac{2(3 - 7) + 4(8)}{4(-3) + (-3)(-2)}$

25. $8 - 4^2 - (-12)$

26. $-(3 - 5) - [2 - (3^2 - 13)]$

27. $\dfrac{2(-3) + 3/(-2) - 2/(-\sqrt{16})}{\sqrt{64} - 1}$

28. $\dfrac{6^2 - 3\sqrt{25}}{\sqrt{6^2 + 13}}$

Use a calculator to help you list the given numbers in order from smallest to largest. (See Example 10.)

29. $\dfrac{189}{37}, \quad \dfrac{4587}{691}, \quad \sqrt{47}, \quad 6.735, \quad \sqrt{27}, \quad \dfrac{2040}{523}$

30. $\dfrac{385}{117}, \quad \sqrt{10}, \quad \dfrac{187}{63}, \quad \pi, \quad \sqrt{\sqrt{85}}, \quad 2.9884$

Express each of the following statements in symbols, using $<, >, \leq,$ *or* $\geq$.

31. 12 is less than 18.5.

32. -2 is greater than -20.

33. x is greater than or equal to 5.7.

34. y is less than or equal to -5.

35. z is at most 7.5.

36. w is negative.

Fill in the blank with $<, =,$ *or* $>$ *so that the resulting statement is true.*

37. -6 _____ -2

38. 3.14 _____ π

39. $3/4$ _____ $.75$

40. $1/3$ _____ $.33$

Fill in the blank so as to produce two equivalent statements. For example, the arithmetic statement "a is negative" is equivalent to the geometric statement "the point a lies to the left of the point 0."

Arithmetic Statement	Geometric Statement
41. $a \geq b$	_____
42. _____	a lies c units to the right of b
43.	a lies between b and c, and to the right of c
44. a is positive	_____

Graph the given intervals on a number line. (See Examples 12 and 13.)

45. $(-8, -1)$

46. $[-1, 10]$

47. $(-2, 3]$

48. $[-2, 2)$

49. $(-2, \infty)$

50. $(-\infty, -2]$

Evaluate each of the following expressions (see Example 14).

51. $|-9| - |-12|$

52. $|8| - |-4|$

53. $-|-4| - |-1 - 14|$

54. $-|6| - |-12 - 4|$

In each of the following problems, fill in the blank with either $=, <,$ *or* $>$, *so that the resulting statement is true.*

55. $|5|$ _____ $|-5|$

56. $-|-4|$ _____ $|4|$

57. $|10 - 3|$ _____ $|3 - 10|$

58. $|6 - (-4)|$ _____ $|-4 - 6|$

59. $|-2 + 8|$ _____ $|2 - 8|$

60. $|3| \cdot |-5|$ _____ $|3(-5)|$

61. $|3 - 5|$ _____ $|3| - |5|$

62. $|-5 + 1|$ _____ $|-5| + |1|$

Write the expression without using absolute-value notation.

63. $|a - 7|$ if $a < 7$

64. $|b - c|$ if $b \geq c$

65. If a and b are any real numbers, is it always true that $|a + b| = |a| + |b|$? Explain your answer.

66. If a and b are any two real numbers, is it always true that $|a - b| = |b - a|$? Explain your answer.

67. For which real numbers b does $|2 - b| = |2 + b|$? Explain your answer.

68. **Health** Data from the National Health and Nutrition Examination Study estimates that 95% of adult heights (inches) are in the following ranges for females and males. (Data from: www.cdc.gov/nchs/nhanes.htm.)

Females	63.5 ± 8.4
Males	68.9 ± 9.3

Express the ranges as an absolute-value inequality in which x is the height of the person.

Consumer Price Index *The Consumer Price Index (CPI) tracks the cost of a typical sample of a consumer goods. The following table shows the percentage increase in the CPI for each year in a 10-year period.*

Year	2006	2007	2008	2009	2010
% Increase in CPI	3.2	4.1	.1	2.7	1.5

Year	2011	2012	2013	2014	2015
% Increase in CPI	3.0	1.7	1.5	1.6	.1

Let r denote the yearly percentage increase in the CPI. For each of the following inequalities, find the number of years during the given period that r satisfied the inequality. (Data from: U.S. Bureau of Labor Statistics.)

69. $r > 3.2$

70. $r < 3.2$

71. $r \leq 3.2$

72. $r \leq 1.5$

73. $r \geq 2.1$

74. $r \geq 1.3$

Stocks *The table presents the 2016 percentage change in the stock price for six well-known companies. (Data from: finance.yahoo.com.)*

Company	Percentage Change
American Express	+10.6%
Coca-Cola	−1.0%
ExxonMobil	+14.9%
Hewlett Packard	+63.1%
Ford	−5.7%
Red Robin Gourmet Burgers	−8.0%

Suppose that we wish to determine the difference in percentage change between two of the companies in the table, and suppose that we are interested only in the magnitude, or absolute value, of this difference. Then we subtract the two entries and find the absolute value. For example, the difference in percentage change of stock price for American Express and Ford is $|10.6\% - (-5.7\%)| = 16.3\%$.

Find the absolute value of the difference between the two indicated changes in stock price.

75. American Express and ExxonMobil

76. Hewlett Packard and Red Robin

77. Coca-Cola and Hewlett Packard

78. American Express and Ford

79. Ford and Red Robin

80. Coca-Cola and Ford

Disposable Income *The following graph shows the real per capita amount of disposable income (in thousands of dollars) in the United States. (Data from: Federal Reserve Bank of St. Louis.)*

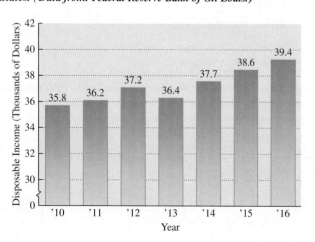

For each of the following, determine the years for which the expression is true, where x is the per capita disposable income.

81. $|x - 37,000| > 1000$

82. $|x - 32,000| > 3500$

83. $|x - 36,500| \leq 2200$

84. $|x - 38,500| \leq 1000$

✓**Checkpoint Answers**

1. (a) Integer, rational, real

   (b) Rational, real

   (c) Irrational, real

2. (a) Commutative property

   (b) Associative property

   (c) Commutative property

   (d) Commutative property

   (e) Associative property

3. (a) Additive identity property

   (b) Multiplicative inverse property

   (c) Additive inverse property

   (d) Multiplicative identity property

4. (a) $4(-2) + 4(5) = 12$

   (b) $2a + 2b$

   (c) $-3p - 3$

   (d) $8m - km$

   (e) $(5 + 3)x = 8x$

5. (a) 30   (b) $\dfrac{41}{3}$

6. (a) 5   (b) 40

   (c) $\dfrac{19}{38} = \dfrac{1}{2} = .5$ (d) $-\dfrac{1}{4} = -.25$

7. (a) Between 8 and 9

   (b) Between 7 and 8

   (c) 8.5440; 7.6904

8.

9. (a) True   (b) True   (c) False

10. (a)   (b)

11. (a)   (b)

12. (a) 6   (b) −7

   (c) −2   (d) 7

   (e) 5

1.2 Polynomials

Polynomials are the fundamental tools of algebra and will play a central role in this course. In order to do polynomial arithmetic, you must first understand exponents. So we begin with them. You are familiar with the usual notation for squares and cubes, such as:

$$5^2 = 5 \cdot 5 \quad \text{and} \quad 6^3 = 6 \cdot 6 \cdot 6.$$

We now extend this convenient notation to other cases.

> If n is a natural number and a is any real number, then
>
> $$a^n \qquad \text{denotes the product} \qquad a \cdot a \cdot a \cdots a \, (n \text{ factors}).$$
>
> The number a is the **base,** and the number n is the **exponent.**

EXAMPLE 1 4^6, which is read "four to the sixth," or "four to the sixth power," is the number

$$4 \cdot 4 \cdot 4 \cdot 4 \cdot 4 \cdot 4 = 4096.$$

Similarly, $(-5)^3 = (-5)(-5)(-5) = -125$, and

$$\left(\frac{3}{2}\right)^4 = \frac{3}{2} \cdot \frac{3}{2} \cdot \frac{3}{2} \cdot \frac{3}{2} = \frac{81}{16}.$$

EXAMPLE 2 Use a calculator to approximate the given expressions.

(a) $(1.2)^8$

Solution Key in 1.2 and then use the x^y key (labeled $\wedge$ on some calculators); finally, key in the exponent 8. The calculator displays the (exact) answer 4.29981696.

(b) $\left(\frac{12}{7}\right)^{23}$

Solution Don't compute 12/7 separately. Use parentheses and key in (12/7), followed by the x^y key and the exponent 23 to obtain the approximate answer 242,054.822. ✓₁

⚠ **CAUTION** A common error in using exponents occurs with expressions such as $4 \cdot 3^2$. The exponent of 2 applies only to the base 3, so that

$$4 \cdot 3^2 = 4 \cdot 3 \cdot 3 = 36.$$

On the other hand,

$$(4 \cdot 3)^2 = (4 \cdot 3)(4 \cdot 3) = 12 \cdot 12 = 144,$$

so

$$4 \cdot 3^2 \neq (4 \cdot 3)^2.$$

Be careful to distinguish between expressions like -2^4 and $(-2)^4$:

$$-2^4 = -(2^4) = -(2 \cdot 2 \cdot 2 \cdot 2) = -16$$
$$(-2)^4 = (-2)(-2)(-2)(-2) = 16,$$

so

$$-2^4 \neq (-2)^4. \quad ✓_2$$

✓ **Checkpoint 1**

Evaluate the following.

(a) 6^3

(b) 5^{12}

(c) 1^9

(d) $\left(\dfrac{7}{5}\right)^8$

✓ **Checkpoint 2**

Evaluate the following.

(a) $3 \cdot 6^2$

(b) $5 \cdot 4^3$

(c) -3^6

(d) $(-3)^6$

(e) $-2 \cdot (-3)^5$

By the definition of an exponent,

$$3^4 \cdot 3^2 = (3 \cdot 3 \cdot 3 \cdot 3)(3 \cdot 3) = 3^6.$$

Since $6 = 4 + 2$, we can write the preceding equation as $3^4 \cdot 3^2 = 3^{4+2}$. This result suggests the following fact, which applies to any real number a and natural numbers m and n.

Multiplication with Exponents

To multiply a^m by a^n, *add* the exponents:

$$a^m \cdot a^n = a^{m+n}.$$

EXAMPLE 3   Verify each of the following simplifications.

(a) $7^4 \cdot 7^6 = 7^{4+6} = 7^{10}$

(b) $(-2)^3 \cdot (-2)^5 = (-2)^{3+5} = (-2)^8$

(c) $(3k)^2 \cdot (3k)^3 = (3k)^5$

(d) $(m + n)^2 \cdot (m + n)^5 = (m + n)^7$ ✔₃

✔ **Checkpoint 3**

Simplify the following.

(a) $5^3 \cdot 5^6$

(b) $(-3)^4 \cdot (-3)^{10}$

(c) $(5p)^2 \cdot (5p)^8$

The multiplication property of exponents has a convenient consequence. By definition,

$$(5^2)^3 = 5^2 \cdot 5^2 \cdot 5^2 = 5^{2+2+2} = 5^6.$$

Note that $2 + 2 + 2$ is $3 \cdot 2 = 6$. This is an example of a more general fact about any real number a and natural numbers m and n.

Power of a Power

To find a power of a power, $(a^m)^n$, *multiply* the exponents:

$$(a^m)^n = a^{mn}.$$

EXAMPLE 4   Verify the following computations.

(a) $(x^3)^4 = x^{3 \cdot 4} = x^{12}$.

(b) $[(-3)^5]^3 = (-3)^{5 \cdot 3} = (-3)^{15}$.

(c) $[(6z)^4]^4 = (6z)^{4 \cdot 4} = (6z)^{16}$. ✔₄

✔ **Checkpoint 4**

Compute the following.

(a) $(6^3)^7$

(b) $[(4k)^5]^6$

It will be convenient to give a zero exponent a meaning. If the multiplication property of exponents is to remain valid, we should have, for example, $3^5 \cdot 3^0 = 3^{5+0} = 3^5$. But this will be true only when $3^0 = 1$. So we make the following definition.

✔ **Checkpoint 5**

Evaluate the following.

(a) 17^0

(b) 30^0

(c) $(-10)^0$

(d) $-(12)^0$

Zero Exponent

If a is any nonzero real number, then

$$a^0 = 1.$$

For example, $6^0 = 1$ and $(-9)^0 = 1$. Note that 0^0 is *not* defined. ₅

Polynomials

A **polynomial** is an algebraic expression such as

$$5x^4 + 2x^3 + 6x, \qquad 8m^3 + 9m^2 + \frac{3}{2}m + 3, \qquad -10p, \qquad \text{or} \qquad 8.$$

The letter used is called a **variable,** and a polynomial is a sum of **terms** of the form

$$(\text{constant}) \times (\text{nonnegative integer power of the variable}).$$

We assume that $x^0 = 1$, $m^0 = 1$, etc., so terms such as 3 or 8 may be thought of as $3x^0$ and $8x^0$, respectively. The constants that appear in each term of a polynomial are called the **coefficients** of the polynomial. The coefficient of x^0 is called the **constant term.**

> **EXAMPLE 5**   Identify the coefficients and the constant term of the given polynomials.
>
> **(a)** $5x^2 - x + 12$
>
> **Solution** The coefficients are 5, -1, and 12, and the constant term is 12.
>
> **(b)** $7x^3 + 2x - 4$
>
> **Solution** The coefficients are 7, 0, 2, and -4, because the polynomial can be written $7x^3 + 0x^2 + 2x - 4$. The constant term is -4.

A polynomial that consists only of a constant term, such as 15, is called a **constant polynomial.** The **zero polynomial** is the constant polynomial 0. The **degree** of a polynomial is the *exponent* of the highest power of x that appears with a *nonzero* coefficient, and the nonzero coefficient of this highest power of x is the **leading coefficient** of the polynomial. For example,

Polynomial	Degree	Leading Coefficient	Constant Term
$6x^7 + 4x^3 + 5x^2 - 7x + 10$	7	6	10
$-x^4 + 2x^3 + \frac{1}{2}$	4	-1	$\frac{1}{2}$
x^3	3	1	0
12	0	12	12

The degree of the zero polynomial is *not defined*, since no exponent of x occurs with a nonzero coefficient. First-degree polynomials are often called **linear polynomials.** Second- and third-degree polynomials are called **quadratics** and **cubics**, respectively. ☑ 6

✓ Checkpoint 6

Find the degree of each polynomial.

(a) $x^4 - x^2 + x + 5$

(b) $7x^5 + 6x^3 - 3x^8 + 2$

(c) 17

(d) 0

Addition and Subtraction

Two terms having the same variable with the same exponent are called **like terms;** other terms are called **unlike terms.** Polynomials can be added or subtracted by using the distributive property to combine like terms. Only like terms can be combined. For example,

$$12y^4 + 6y^4 = (12 + 6)y^4 = 18y^4$$

and

$$-2m^2 + 8m^2 = (-2 + 8)m^2 = 6m^2.$$

The polynomial $8y^4 + 2y^5$ has unlike terms, so it cannot be further simplified.

In more complicated cases of addition, you may have to eliminate parentheses, use the commutative and associative laws to regroup like terms, and then combine them.

EXAMPLE 6   Add the following polynomials.

(a) $(8x^3 - 4x^2 + 6x) + (3x^3 + 5x^2 - 9x + 8)$

Solution $8x^3 - 4x^2 + 6x + 3x^3 + 5x^2 - 9x + 8$ Eliminate parentheses.

$\qquad = \underbrace{(8x^3 + 3x^3)} + \underbrace{(-4x^2 + 5x^2)} + \underbrace{(6x - 9x)} + 8$ Group like terms.

$\qquad = 11x^3 + x^2 - 3x + 8$ Combine like terms.

(b) $(-4x^4 + 6x^3 - 9x^2 - 12) + (-3x^3 + 8x^2 - 11x + 7)$

Solution

$\qquad -4x^4 + 6x^3 - 9x^2 - 12 - 3x^3 + 8x^2 - 11x + 7$ Eliminate parentheses.

$\qquad = -4x^4 + \underbrace{(6x^3 - 3x^3)} + \underbrace{(-9x^2 + 8x^2)} - 11x + \underbrace{(-12 + 7)}$ Group like terms.

$\qquad = -4x^4 + 3x^3 - x^2 - 11x - 5$ Combine like terms.

Care must be used when parentheses are preceded by a minus sign. For example, we know that

$$-(4 + 3) = -(7) = -7 = -4 - 3.$$

If you simply delete the parentheses in $-(4 + 3)$, you obtain $-4 + 3 = -1$, which is the wrong answer. This fact and the preceding examples illustrate the following rules.

Rules for Eliminating Parentheses

Parentheses preceded by a plus sign (or no sign) may be deleted.

Parentheses preceded by a minus sign may be deleted *provided that* the sign of every term within the parentheses is changed.

EXAMPLE 7   Subtract: $(2x^2 - 11x + 8) - (7x^2 - 6x + 2)$.

Solution $2x^2 - 11x + 8 - 7x^2 + 6x - 2$ Eliminate parentheses.

$\qquad = (2x^2 - 7x^2) + (-11x + 6x) + (8 - 2)$ Group like terms.

$\qquad = -5x^2 - 5x + 6$ Combine like terms. ✓7

Multiplication

The distributive property is also used to multiply polynomials. For example, the product of $8x$ and $6x - 4$ is found as follows:

$$8x(6x - 4) = 8x(6x) - 8x(4)$$ Distributive property

$$= 48x^2 - 32x.$$ $x \cdot x = x^2$

EXAMPLE 8   Use the distributive property to find each product.

(a) $2p^3(3p^2 - 2p + 5) = 2p^3(3p^2) + 2p^3(-2p) + 2p^3(5)$

$\qquad\qquad\qquad\qquad = 6p^5 - 4p^4 + 10p^3$

(b) $(3k - 2)(k^2 + 5k - 4) = 3k(k^2 + 5k - 4) - 2(k^2 + 5k - 4)$

$\qquad\qquad\qquad\qquad\quad = 3k^3 + 15k^2 - 12k - 2k^2 - 10k + 8$

$\qquad\qquad\qquad\qquad\quad = 3k^3 + 13k^2 - 22k + 8$ ✓8

✓**Checkpoint 7**

Add or subtract as indicated.

(a) $(-2x^2 + 7x + 9)$
$\quad + (3x^2 + 2x - 7)$

(b) $(4x + 6) - (13x - 9)$

(c) $(9x^3 - 8x^2 + 2x)$
$\quad - (9x^3 - 2x^2 - 10)$

✓**Checkpoint 8**

Find the following products.

(a) $-6r(2r - 5)$

(b) $(8m + 3) \cdot (m^4 - 2m^2 + 6m)$

EXAMPLE 9   The product $(2x - 5)(3x + 4)$ can be found by using the distributive property twice:

$$(2x - 5)(3x + 4) = 2x(3x + 4) - 5(3x + 4)$$
$$= \underline{2x \cdot 3x} + \underline{2x \cdot 4} + \underline{(-5) \cdot 3x} + \underline{(-5) \cdot 4}$$
$$= 6x^2 \quad + \quad 8x \quad - \quad 15x \quad - \quad 20$$
$$= 6x^2 \quad - \quad\quad\quad 7x \quad\quad\quad - \quad 20.$$

Observe the pattern in the second line of Example 9 and its relationship to the terms being multiplied.

$$(2x - 5)(3x + 4) = 2x \cdot 3x + 2x \cdot 4 + (-5) \cdot 3x + (-5) \cdot 4$$

First terms

$(2x - 5)(3x + 4)$ Outside terms

$(2x - 5)(3x + 4)$ Inside terms

$(2x - 5)(3x + 4)$ Last terms

This pattern is easy to remember by using the acronym **FOIL** (**F**irst, **O**utside, **I**nside, **L**ast). The FOIL method makes it easy to find products such as this one mentally, without the necessity of writing out the intermediate steps.

EXAMPLE 10   Use FOIL to find the product of the given polynomials.

(a) $(3x + 2)(x + 5) = 3x^2 + 15x + 2x + 10 = 3x^2 + 17x + 10$

First   Outside   Inside   Last

(b) $(x + 3)^2 = (x + 3)(x + 3) = x^2 + 3x + 3x + 9 = x^2 + 6x + 9$

(c) $(2x + 1)(2x - 1) = 4x^2 - 2x + 2x - 1 = 4x^2 - 1$ **9**

Checkpoint 9

Use FOIL to find these products.

(a) $(5k - 1)(2k + 3)$

(b) $(7z - 3)(2z + 5)$

Applications

In business, the *revenue* from the sales of an item is given by

Revenue = (price per item) × (number of items sold).

The *cost* to manufacture and sell these items is given by

Cost = Fixed Costs + Variable Costs,

where the fixed costs include such things as buildings and machinery (which do not depend on how many items are made) and variable costs include such things as labor and materials (which vary, depending on how many items are made). Then

Profit = Revenue − Cost.

EXAMPLE 11   **Profit** A manufacturer of scientific calculators sells calculators for $12 each (wholesale) and can produce a maximum of 150,000. The variable cost of producing x thousand calculators is $6995x - 7.2x^2$ dollars, and the fixed costs for the manufacturing operation are $230,000. If x thousand calculators are manufactured and sold, find expressions for the revenue, cost, and profit.

Solution If x thousand calculators are sold at $12 each, then

Revenue = (price per item) × (number of items sold)
$$R = 12 \times 1000x = 12{,}000x,$$

✓ **Checkpoint 10**

Suppose revenue is given by $7x^2 - 3x$, fixed costs are $500, and variable costs are given by $3x^2 + 5x - 25$. Write an expression for

(a) Cost

(b) Profit

where $x \le 150$ (because only 150,000 calculators can be made). The variable cost of making x thousand calculators is $6995x - 7.2x^2$, so that

$$\text{Cost} = \text{Fixed Costs} + \text{Variable Costs},$$
$$C = 230{,}000 + (6995x - 7.2x^2) \quad (x \le 150).$$

Therefore, the profit is given by

$$P = R - C = 12{,}000x - (230{,}000 + 6995x - 7.2x^2)$$
$$= 12{,}000x - 230{,}000 - 6995x + 7.2x^2$$
$$P = 7.2x^2 + 5005x - 230{,}000 \quad (x \le 150). \quad ✓_{10}$$

1.2 Exercises

Use a calculator to approximate these numbers. (See Examples 1 and 2.)

1. 11.2^6

2. $(-6.54)^{11}$

3. $(-18/7)^6$

4. $(5/9)^7$

5. Explain how the value of -3^2 differs from $(-3)^2$. Do -3^3 and $(-3)^3$ differ in the same way? Why or why not?

6. Describe the steps used to multiply 4^3 and 4^5. Is the product of 4^3 and 3^4 found in the same way? Explain.

Simplify each of the given expressions. Leave your answers in exponential notation. (See Examples 3 and 4.)

7. $4^2 \cdot 4^3$

8. $(-4)^4 \cdot (-4)^6$

9. $(-6)^2 \cdot (-6)^5$

10. $(2z)^5 \cdot (2z)^6$

11. $[(5u)^4]^7$

12. $(6y)^3 \cdot [(6y)^5]^4$

List the degree of the given polynomial, its coefficients, and its constant term. (See Example 5.)

13. $6.2x^4 - 5x^3 + 4x^2 - 3x + 3.7$

14. $6x^7 + 4x^6 - x^3 + x$

State the degree of the given polynomial.

15. $1 + x + 2x^2 + 3x^3$

16. $5x^4 - 4x^5 - 6x^3 + 7x^4 - 2x + 8$

Add or subtract as indicated. (See Examples 6 and 7.)

17. $(3x^3 + 2x^2 - 5x) + (-4x^3 - x^2 - 8x)$

18. $(-2p^3 - 5p + 7) + (-4p^2 + 8p + 2)$

19. $(-4y^2 - 3y + 8) - (2y^2 - 6y + 2)$

20. $(7b^2 + 2b - 5) - (3b^2 + 2b - 6)$

21. $(2x^3 + 2x^2 + 4x - 3) - (2x^3 + 8x^2 + 1)$

22. $(3y^3 + 9y^2 - 11y + 8) - (-4y^2 + 10y - 6)$

Find each of the given products. (See Examples 8–10.)

23. $-9m(2m^2 + 6m - 1)$

24. $2a(4a^2 - 6a + 8)$

25. $(3z + 5)(4z^2 - 2z + 1)$

26. $(2k + 3)(4k^3 - 3k^2 + k)$

27. $(6k - 1)(2k + 3)$

28. $(8r + 3)(r - 1)$

29. $(3y + 5)(2y + 1)$

30. $(5r - 3s)(5r - 4s)$

31. $(9k + q)(2k - q)$

32. $(.012x - .17)(.3x + .54)$

33. $(6.2m - 3.4)(.7m + 1.3)$

34. $2p - 3[4p - (8p + 1)]$

35. $5k - [k + (-3 + 5k)]$

36. $(3x - 1)(x + 2) - (2x + 5)^2$

Profit   *Find expressions for the revenue, cost, and profit from selling x thousand items. (See Example 11.)*

Item Price	Fixed Costs	Variable Costs
37. $5.00	$200,000	$1800x$
38. $8.50	$225,000	$4200x$

39. **Profit** Beauty Works sells its cologne wholesale for $9.75 per bottle. The variable costs of producing x thousand bottles is $-3x^2 + 3480x - 325$ dollars, and the fixed costs of manufacturing are $260,000. Find expressions for the revenue, cost, and profit from selling x thousand items.

40. **Profit** A self-help guru sells her book *Be Happy in 45 Easy Steps* for $23.50 per copy. Her fixed costs are $145,000 and the estimate of the variable cost of printing, binding, and distributing x thousand books is given by $-4.2x^2 + 3220x - 425$ dollars. Find expressions for the revenue, cost, and profit from selling x thousand copies of the book.

Work these problems.

Starbucks Earnings   *The accompanying bar graph shows the net earnings (in millions of dollars) for the Starbucks Corporation. The polynomial*

$$4.79x^3 - 122.5x^2 + 1104x - 2863$$

gives a good approximation of Starbuck's net earnings in year x, where x = 7 corresponds to 2007, and so on (7 ≤ x ≤ 16). *For each of the given years,*

(a) *use the bar graph to determine the net earnings;*

(b) *use the polynomial to determine the net earnings.*

(Data from: www.morningstar.com.)

41. 2007   **42.** 2015

43. 2012   **44.** 2013

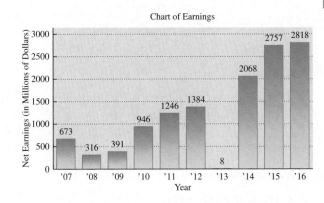

Assuming that the polynomial approximation in Exercises 41–44 remains accurate in later years, use it to estimate Starbucks's net earnings in each of the following years.

45. 2017   **46.** 2018   **47.** 2019

48. Do the estimates in Exercises 45–47 seem plausible? Explain.

Starbucks Costs *The costs (in millions of dollars) for the Starbucks Corporation can be approximated by the polynomial* $9.5x^3 - 401.6x^2 + 6122x - 25{,}598$, *where x = 7 corresponds to the year 2007. Determine whether each of the given statements is true or false. (Data from: www.morningstar.com.)*

49. The costs were higher than $5000 million in 2010.

50. The costs were higher than $7500 million in 2015.

51. The costs were higher in 2012 than in 2015.

52. The costs were lower in 2011 than in 2016.

PepsiCo Profits *The profits (in millions of dollars) for PepsiCo Inc. can be approximated by the polynomial* $-72.85x^3 + 2082x^2 -$

$16{,}532x + 59{,}357$, *where x = 7 corresponds to the year 2007. (Data from: www.morningstar.com.) Find the approximate profits for the following years.*

53. 2007   **54.** 2010   **55.** 2012   **56.** 2015

Determine whether each of the given statements is true or false.

57. Were profits higher in 2013 or 2009?

58. Were profits higher in 2012 or 2015?

Business *Use the table feature of a graphing calculator or use a spreadsheet to make a table of values for the profit expression in Example 11, with x = 0, 5, 10, . . . , 150. Use the table to answer the following questions.*

59. What is the profit or loss (negative profit) when 25,000 calculators are sold? when 60,000 are sold? Explain these answers.

60. Approximately how many calculators must be sold in order for the company to make a profit?

61. What is the profit from selling 100,000 calculators?

62. What is the profit from selling 150,000 calculators?

✓**Checkpoint Answers**

1. (a) 216   (b) 244,140,625
    (c) 1   (d) 14.75789056

2. (a) 108   (b) 320   (c) −729
    (d) 729   (e) 486

3. (a) 5^9 (b) $(-3)^{14}$ (c) $(5p)^{10}$

4. (a) 6^{21} (b) $(4k)^{30}$

5. (a) 1   (b) 1   (c) 1   (d) −1

6. (a) 4   (b) 8   (c) 0   (d) Not defined

7. (a) $x^2 + 9x + 2$
    (b) $-9x + 15$
    (c) $-6x^2 + 2x + 10$

8. (a) $-12r^2 + 30r$
    (b) $8m^5 + 3m^4 - 16m^3 + 42m^2 + 18m$

9. (a) $10k^2 + 13k - 3$ (b) $14z^2 + 29z - 15$

10. (a) $C = 3x^2 + 5x + 475$ (b) $P = 4x^2 - 8x - 475$

1.3 Factoring

The number 18 can be written as a product in several ways: $9 \cdot 2$, $(-3)(-6)$, $1 \cdot 18$, etc. The numbers in each product (9, 2, −3, etc.) are called **factors,** and the process of writing 18 as a product of factors is called **factoring.** Thus, factoring is the reverse of multiplication.

Factoring of polynomials is a means of simplifying many expressions and of solving certain types of equations. As is the usual custom, factoring of polynomials in this text will be restricted to finding factors with *integer* coefficients (otherwise there may be an infinite number of possible factors).

Greatest Common Factor

The algebraic expression $15m + 45$ is made up of two terms: $15m$ and 45. Each of these terms has 15 as a factor. In fact, $15m = 15 \cdot m$ and $45 = 15 \cdot 3$. By the distributive property,

$$15m + 45 = 15 \cdot m + 15 \cdot 3 = 15(m + 3).$$

Both 15 and $m + 3$ are factors of $15m + 45$. Since 15 is a factor of all terms of $15m + 45$ and is the largest such number, it is called the **greatest common factor** for the polynomial $15m + 45$. The process of writing $15m + 45$ as $15(m + 3)$ is called **factoring out** the greatest common factor.

EXAMPLE 1   Factor out the greatest common factor.

(a) $12p - 18q$

Solution Both $12p$ and $18q$ are divisible by 6, and

$$12p - 18q = 6 \cdot 2p - 6 \cdot 3q$$
$$= 6(2p - 3q).$$

(b) $8x^3 - 9x^2 + 15x$

Solution Each of these terms is divisible by x:

$$8x^3 - 9x^2 + 15x = (8x^2) \cdot x - (9x) \cdot x + 15 \cdot x$$
$$= x(8x^2 - 9x + 15).$$

(c) $5(4x - 3)^3 + 2(4x - 3)^2$

Solution The quantity $(4x - 3)^2$ is a common factor. Factoring it out gives

$$5(4x - 3)^3 + 2(4x - 3)^2 = (4x - 3)^2[5(4x - 3) + 2]$$
$$= (4x - 3)^2(20x - 15 + 2)$$
$$= (4x - 3)^2(20x - 13).$$ ✓1

✓**Checkpoint 1**

Factor out the greatest common factor.

(a) $12r + 9k$

(b) $75m^2 + 100n^2$

(c) $6m^4 - 9m^3 + 12m^2$

(d) $3(2k + 1)^3 + 4(2k + 1)^4$

Factoring Quadratics

If we multiply two first-degree polynomials, the result is a quadratic. For instance, using FOIL, we see that $(x + 1)(x - 2) = x^2 - x - 2$. Since factoring is the reverse of multiplication, factoring quadratics requires using FOIL backward.

EXAMPLE 2   Factor $x^2 + 9x + 18$.

Solution We must find integers b and d such that

$$x^2 + 9x + 18 = (x + b)(x + d)$$
$$= x^2 + dx + bx + bd$$
$$x^2 + 9x + 18 = x^2 + (b + d)x + bd.$$

Since the constant coefficients on each side of the equation must be equal, we must have $bd = 18$; that is, b and d are factors of 18. Similarly, the coefficients of x must be the same, so that $b + d = 9$. The possibilities are summarized in this table:

Factors b, d of 18	Sum $b + d$
$18 \cdot 1$	$18 + 1 = 19$
$9 \cdot 2$	$9 + 2 = 11$
$6 \cdot 3$	$6 + 3 = 9$

✓ **Checkpoint 2**

Factor the following.

(a) $r^2 + 7r + 10$

(b) $x^2 + 4x + 3$

(c) $y^2 + 6y + 8$

There is no need to list negative factors, such as $(-3)(-6)$, because their sum is negative. The table suggests that 6 and 3 will work. Verify that

$$(x + 6)(x + 3) = x^2 + 9x + 18. \; ✓_2$$

EXAMPLE 3   Factor $x^2 + 3x - 10$.

Solution As in Example 2, we must find factors b and d whose product is -10 (the constant term) and whose sum is 3 (the coefficient of x). The following table shows the possibilities.

Factors b, d of -10	Sum $b + d$
$1(-10)$	$1 + (-10) = -9$
$(-1)10$	$-1 + 10 = 9$
$2(-5)$	$2 + (-5) = -3$
$(-2)5$	$-2 + 5 = 3$

The only factors with product -10 and sum 3 are -2 and 5. So the correct factorization is

$$x^2 + 3x - 10 = (x - 2)(x + 5),$$

as you can readily verify.

It is usually not necessary to construct tables as was done in Examples 2 and 3—you can just mentally check the various possibilities. The approach used in Examples 2 and 3 (with minor modifications) also works for factoring quadratic polynomials whose leading coefficient is not 1.

EXAMPLE 4   Factor $4y^2 - 11y + 6$.

Solution We must find integers a, b, c, and d such that

$$4y^2 - 11y + 6 = (ay + b)(cy + d)$$
$$= acy^2 + ady + bcy + bd$$
$$4y^2 - 11y + 6 = acy^2 + (ad + bc)y + bd.$$

Since the coefficients of y^2 must be the same on both sides, we see that $ac = 4$. Similarly, the constant terms show that $bd = 6$. The positive factors of 4 are 4 and 1 or 2 and 2. Since the middle term is negative, we consider only negative factors of 6. The possibilities are -2 and -3 or -1 and -6. Now we try various arrangements of these factors until we find one that gives the correct coefficient of y:

$$(2y - 1)(2y - 6) = 4y^2 - 14y + 6 \quad \text{Incorrect}$$
$$(2y - 2)(2y - 3) = 4y^2 - 10y + 6 \quad \text{Incorrect}$$
$$(y - 2)(4y - 3) = 4y^2 - 11y + 6. \quad \text{Correct}$$

The last trial gives the correct factorization. ✓$_3$

✓ **Checkpoint 3**

Factor the following.

(a) $x^2 - 4x + 3$

(b) $2y^2 - 5y + 2$

(c) $6z^2 - 13z + 6$

EXAMPLE 5   Factor $6p^2 - 7pq - 5q^2$.

Solution Again, we try various possibilities. The positive factors of 6 could be 2 and 3 or 1 and 6. As factors of -5, we have only -1 and 5 or -5 and 1. Try different combinations of these factors until the correct one is found:

$$(2p - 5q)(3p + q) = 6p^2 - 13pq - 5q^2 \quad \text{Incorrect}$$
$$(3p - 5q)(2p + q) = 6p^2 - 7pq - 5q^2. \quad \text{Correct}$$

So $6p^2 - 7pq - 5q^2$ factors as $(3p - 5q)(2p + q)$. ✓$_4$

✓ **Checkpoint 4**

Factor the following.

(a) $r^2 - 5r - 14$

(b) $3m^2 + 5m - 2$

(c) $6p^2 + 13pq - 5q^2$

NOTE In Examples 2–4, we chose positive factors of the positive first term. Of course, we could have used two negative factors, but the work is easier if positive factors are used.

EXAMPLE 6   Factor $x^2 + x + 3$.

Solution There are only two ways to factor 3, namely, $3 = 1 \cdot 3$ and $3 = (-1)(-3)$. They lead to these products:

$$(x + 1)(x + 3) = x^2 + 4x + 3 \qquad \text{Incorrect}$$
$$(x - 1)(x - 3) = x^2 - 4x + 3. \qquad \text{Incorrect}$$

Therefore, this polynomial cannot be factored.

Factoring Patterns

In some cases, you can factor a polynomial with a minimum amount of guesswork by recognizing common patterns. The easiest pattern to recognize is the *difference of squares*.

Difference of Squares

$$x^2 - y^2 = (x + y)(x - y).$$

To verify the accuracy of the preceding equation, multiply out the right side.

EXAMPLE 7   Factor each of the following.

(a) $4m^2 - 9$

Solution Notice that $4m^2 - 9$ is the difference of two squares, since $4m^2 = (2m)^2$ and $9 = 3^2$. Use the pattern for the difference of two squares, letting $2m$ replace x and 3 replace y. Then the pattern $x^2 - y^2 = (x + y)(x - y)$ becomes

$$4m^2 - 9 = (2m)^2 - 3^2$$
$$= (2m + 3)(2m - 3).$$

(b) $128p^2 - 98q^2$

Solution First factor out the common factor of 2:

$$128p^2 - 98q^2 = 2(64p^2 - 49q^2)$$
$$= 2[(8p)^2 - (7q)^2]$$
$$= 2(8p + 7q)(8p - 7q).$$

(c) $x^2 + 36$

Solution The sum of two perfect squares cannot be factored with real numbers. To understand why this is true, let us say for the sake of argument that $x^2 + 36$ can be factored with real numbers. Then there would exist two real numbers a and b where we could write

$$x^2 + 36 = (x + a)(x + b).$$

The above expression is an equality. This means that for any real number that we substitute for x, the expressions on each side of the equals sign must evaluate to the same value. However, if we let $x = -a$, then we obtain:

$$(-a)^2 + 36 = (-a + a)(-a + b) \qquad \text{Substituting } x = -a.$$
$$(-a)^2 + 36 = 0(-a + b) \qquad \text{Additive inverse of } a$$
$$(-a)^2 + 36 = 0. \qquad \text{Multiplication by 0}$$

The above is problematic. The left-hand side of the equation has to be greater than 0 because $(-a)^2$ has to be 0 or positive, and then, when added to 36, it will remain greater than 0. As we see above, the right-hand side has to be 0. This is a contradiction. Thus, our assumption that $x^2 + 36$ can be factored with real numbers must be incorrect.

When we see an expression of the form $x^2 + a^2$, we call this a *sum of two squares*. Such expressions cannot be factored with real numbers.

(d) $(x - 2)^2 - 49$

Solution Since $49 = 7^2$, this is a difference of two squares. So it factors as follows:

$$(x - 2)^2 - 49 = (x - 2)^2 - 7^2$$
$$= [(x - 2) + 7][(x - 2) - 7]$$
$$= (x + 5)(x - 9). \quad \checkmark_{5}$$

✓ **Checkpoint 5**

Factor the following.

(a) $9p^2 - 49$

(b) $y^2 + 100$

(c) $(x + 3)^2 - 64$

Another common pattern is the *perfect square*. Verify each of the following factorizations by multiplying out the right side.

Perfect Squares

$$x^2 + 2xy + y^2 = (x + y)^2$$
$$x^2 - 2xy + y^2 = (x - y)^2$$

Whenever you have a quadratic whose first and last terms are squares, it *may* factor as a perfect square. The key is to look at the middle term. To have a perfect square whose first and last terms are x^2 and y^2, the middle term must be $\pm 2xy$. To avoid errors, always check this.

EXAMPLE 8   Factor each polynomial, if possible.

(a) $16p^2 - 40pq + 25q^2$

Solution The first and last terms are squares, namely, $16p^2 = (4p)^2$ and $25q^2 = (5q)^2$. So the second perfect-square pattern, with $x = 4p$ and $y = 5q$, might work. To have a perfect square, the middle term $-40pq$ must equal $-2(4p)(5q)$, which it does. So the polynomial factors as

$$16p^2 - 40pq + 25q^2 = (4p - 5q)(4p - 5q),$$

as you can easily verify.

(b) $9u^2 + 5u + 1$

Solution Again, the first and last terms are squares: $9u^2 = (3u)^2$ and $1 = 1^2$. The middle term is positive, so the first perfect-square pattern might work, with $x = 3u$ and $y = 1$. To have a perfect square, however, the middle term would have to be $2(3u) \cdot 1 = 6u$, which is *not* the middle term of the given polynomial. So it is not a perfect square—in fact, it cannot be factored.

✓ **Checkpoint 6**

Factor.

(a) $4m^2 + 4m + 1$

(b) $25z^2 - 80zt + 64t^2$

(c) $9x^2 + 15x + 25$

(c) $169x^2 + 104xy^2 + 16y^4$

Solution This polynomial may be factored as $(13x + 4y^2)^2$, since $169x^2 = (13x)^2$, $16y^4 = (4y^2)^2$, and $2(13x)(4y^2) = 104xy^2$. $\quad \checkmark_{6}$

EXAMPLE 9   Factor each of the given polynomials.

(a) $12x^2 - 26x - 10$

Solution Look first for a greatest common factor. Here, the greatest common factor is 2: $12x^2 - 26x - 10 = 2(6x^2 - 13x - 5)$. Now try to factor $6x^2 - 13x - 5$. Possible factors of 6 are 3 and 2 or 6 and 1. The only factors of -5 are -5 and 1 or 5 and -1. Try various combinations. You should find that the quadratic factors as $(3x + 1)(2x - 5)$. Thus,

$$12x^2 - 26x - 10 = 2(3x + 1)(2x - 5).$$

(b) $4z^2 + 12z + 9 - w^2$

Solution There is no common factor here, but notice that the first three terms can be factored as a perfect square:

$$4z^2 + 12z + 9 - w^2 = (2z + 3)^2 - w^2.$$

Written in this form, the expression is the difference of squares, which can be factored as follows:

$$(2z + 3)^2 - w^2 = [(2z + 3) + w][(2z + 3) - w]$$
$$= (2z + 3 + w)(2z + 3 - w).$$

(c) $16a^2 - 100 - 48ac + 36c^2$

Solution Factor out the greatest common factor of 4 first:

$$16a^2 - 100 - 48ac + 36c^2 = 4[4a^2 - 25 - 12ac + 9c^2]$$
$$= 4[(4a^2 - 12ac + 9c^2) - 25] \quad \text{Rearrange terms and group.}$$
$$= 4[(2a - 3c)^2 - 25] \quad \text{Factor.}$$
$$= 4(2a - 3c + 5)(2a - 3c - 5). \quad \text{Factor the difference of squares.}$$

✓**Checkpoint 7**

Factor the following.
(a) $6x^2 - 27x - 15$
(b) $9r^2 + 12r + 4 - t^2$
(c) $18 - 8xy - 2y^2 - 8x^2$

🛑 **CAUTION** Remember always to look first for a greatest common factor.

Higher Degree Polynomials

Polynomials of degree greater than 2 are often difficult to factor. However, factoring is relatively easy in two cases: *the difference and the sum of cubes*. By multiplying out the right side, you can readily verify each of the following factorizations.

Difference and Sum of Cubes

$$x^3 - y^3 = (x - y)(x^2 + xy + y^2)$$
$$x^3 + y^3 = (x + y)(x^2 - xy + y^2)$$

EXAMPLE 10   Factor each of the following polynomials.

(a) $k^3 - 8$

Solution Since $8 = 2^3$, use the pattern for the difference of two cubes to obtain

$$k^3 - 8 = k^3 - 2^3 = (k - 2)(k^2 + 2k + 4).$$

(b) $m^3 + 125$

Solution $m^3 + 125 = m^3 + 5^3 = (m + 5)(m^2 - 5m + 25)$

(c) $8k^3 - 27z^3$

Solution $8k^3 - 27z^3 = (2k)^3 - (3z)^3 = (2k - 3z)(4k^2 + 6kz + 9z^2)$ ✓8

✓Checkpoint 8

Factor the following.
(a) $a^3 + 1000$
(b) $z^3 - 64$
(c) $1000m^3 - 27z^3$

Substitution and appropriate factoring patterns can sometimes be used to factor higher degree expressions.

EXAMPLE 11   Factor the following polynomials.

(a) $x^8 + 4x^4 + 3$

Solution The idea is to make a substitution that reduces the polynomial to a quadratic or cubic that we can deal with. Note that $x^8 = (x^4)^2$. Let $u = x^4$. Then

$$
\begin{aligned}
x^8 + 4x^4 + 3 &= (x^4)^2 + 4x^4 + 3 && \text{Power of a power} \\
&= u^2 + 4u + 3 && \text{Substitute } x^4 = u. \\
&= (u + 3)(u + 1) && \text{Factor.} \\
&= (x^4 + 3)(x^4 + 1). && \text{Substitute } u = x^4.
\end{aligned}
$$

(b) $x^4 - y^4$

Solution Note that $x^4 = (x^2)^2$, and similarly for the y term. Let $u = x^2$ and $v = y^2$. Then

$$
\begin{aligned}
x^4 - y^4 &= (x^2)^2 - (y^2)^2 && \text{Power of a power} \\
&= u^2 - v^2 && \text{Substitute } x^2 = u \text{ and } y^2 = v. \\
&= (u + v)(u - v) && \text{Difference of squares} \\
&= (x^2 + y^2)(x^2 - y^2) && \text{Substitute } u = x^2 \text{ and } v = y^2. \\
&= (x^2 + y^2)(x + y)(x - y). && \text{Difference of squares} ✓9
\end{aligned}
$$

✓Checkpoint 9

Factor each of the following.
(a) $2x^4 + 5x^2 + 2$
(b) $3x^4 - x^2 - 2$

Once you understand Example 11, you can often factor without making explicit substitutions.

EXAMPLE 12   Factor $256k^4 - 625m^4$.

Solution Use the difference of squares twice, as follows:

$$
\begin{aligned}
256k^4 - 625m^4 &= (16k^2)^2 - (25m^2)^2 \\
&= (16k^2 + 25m^2)(16k^2 - 25m^2) \\
&= (16k^2 + 25m^2)(4k + 5m)(4k - 5m). ✓10
\end{aligned}
$$

✓Checkpoint 10

Factor $81x^4 - 16y^4$.

1.3 Exercises

Factor out the greatest common factor in each of the given polynomials. (See Example 1.)

1. $12x^2 - 24x$

2. $5y - 65xy$

3. $r^3 - 5r^2 + r$

4. $t^3 + 3t^2 + 8t$

5. $6z^3 - 12z^2 + 18z$

6. $5x^3 + 55x^2 + 10x$

7. $3(2y - 1)^2 + 7(2y - 1)^3$

8. $(3x + 7)^5 - 4(3x + 7)^3$

9. $3(x + 5)^4 + (x + 5)^6$

10. $3(x + 6)^2 + 6(x + 6)^4$

Factor the polynomial. (See Examples 2 and 3.)

11. $x^2 + 5x + 4$

12. $u^2 + 7u + 6$

13. $x^2 + 7x + 12$

14. $y^2 + 8y + 12$

15. $x^2 + x - 6$

16. $x^2 + 4x - 5$

17. $x^2 + 2x - 3$

18. $y^2 + y - 12$

19. $x^2 - 3x - 4$

20. $u^2 - 2u - 8$

21. $z^2 - 9z + 14$ **22.** $w^2 - 6w - 16$

23. $z^2 + 10z + 24$ **24.** $r^2 + 16r + 60$

Factor the polynomial. (See Examples 4–6.)

25. $2x^2 - 9x + 4$ **26.** $3w^2 - 8w + 4$

27. $15p^2 - 23p + 4$ **28.** $8x^2 - 14x + 3$

29. $4z^2 - 16z + 15$ **30.** $12y^2 - 29y + 15$

31. $6x^2 - 5x - 4$ **32.** $12z^2 + z - 1$

33. $10y^2 + 21y - 10$ **34.** $15u^2 + 4u - 4$

35. $6x^2 + 5x - 4$ **36.** $12y^2 + 7y - 10$

Factor each polynomial completely. Factor out the greatest common factor as necessary. (See Examples 2–9.)

37. $3a^2 + 2a - 5$ **38.** $6a^2 - 48a - 120$

39. $x^2 - 81$ **40.** $x^2 + 17xy + 72y^2$

41. $9p^2 - 12p + 4$ **42.** $3r^2 - r - 2$

43. $r^2 + 3rt - 10t^2$ **44.** $2a^2 + ab - 6b^2$

45. $m^2 - 8mn + 16n^2$ **46.** $8k^2 - 16k - 10$

47. $4u^2 + 12u + 9$ **48.** $9p^2 - 16$

49. $25p^2 - 10p + 4$ **50.** $10x^2 - 17x + 3$

51. $4r^2 - 9v^2$ **52.** $x^2 + 3xy - 28y^2$

53. $x^2 + 4xy + 4y^2$ **54.** $16u^2 + 12u - 18$

55. $3a^2 - 13a - 30$ **56.** $3k^2 + 2k - 8$

57. $21m^2 + 13mn + 2n^2$ **58.** $81y^2 - 100$

59. $y^2 - 4yz - 21z^2$ **60.** $49a^2 + 9$

61. $121x^2 - 64$ **62.** $4z^2 + 56zy + 196y^2$

Factor each of these polynomials. (See Example 10.)

63. $a^3 - 64$ **64.** $b^3 + 216$

65. $8r^3 - 27s^3$ **66.** $1000p^3 + 27q^3$

67. $64m^3 + 125$ **68.** $216y^3 - 343$

69. $1000y^3 - z^3$ **70.** $125p^3 + 8q^3$

Factor each of these polynomials. (See Examples 11 and 12.)

71. $x^4 + 5x^2 + 6$ **72.** $y^4 + 7y^2 + 10$

73. $b^4 - b^2$ **74.** $z^4 - 3z^2 - 4$

75. $x^4 - x^2 - 12$ **76.** $4x^4 + 27x^2 - 81$

77. $16a^4 - 81b^4$ **78.** $x^6 - y^6$

79. $x^8 + 8x^2$ **80.** $x^9 - 64x^3$

81. When asked to factor $6x^4 - 3x^2 - 3$ completely, a student gave the following result:

$$6x^4 - 3x^2 - 3 = (2x^2 + 1)(3x^2 - 3).$$

Is this answer correct? Explain why.

82. When can the sum of two squares be factored? Give examples.

83. Explain why $(x + 2)^3$ is not the correct factorization of $x^3 + 8$, and give the correct factorization.

84. Describe how factoring and multiplication are related. Give examples.

✓ Checkpoint Answers

1. **(a)** $3(4r + 3k)$ **(b)** $25(3m^2 + 4n^2)$
   (c) $3m^2(2m^2 - 3m + 4)$ **(d)** $(2k + 1)^3(7 + 8k)$

2. **(a)** $(r + 2)(r + 5)$ **(b)** $(x + 3)(x + 1)$
   (c) $(y + 2)(y + 4)$

3. **(a)** $(x - 3)(x - 1)$ **(b)** $(2y - 1)(y - 2)$
   (c) $(3z - 2)(2z - 3)$

4. **(a)** $(r - 7)(r + 2)$ **(b)** $(3m - 1)(m + 2)$
   (c) $(2p + 5q)(3p - q)$

5. **(a)** $(3p + 7)(3p - 7)$ **(b)** Cannot be factored
   (c) $(x + 11)(x - 5)$

6. **(a)** $(2m + 1)^2$ **(b)** $(5z - 8t)^2$
   (c) Does not factor

7. **(a)** $3(2x + 1)(x - 5)$
   (b) $(3r + 2 + t)(3r + 2 - t)$
   (c) $2(3 - 2x - y)(3 + 2x + y)$

8. **(a)** $(a + 10)(a^2 - 10a + 100)$
   (b) $(z - 4)(z^2 + 4z + 16)$
   (c) $(10m - 3z)(100m^2 + 30mz + 9z^2)$

9. **(a)** $(2x^2 + 1)(x^2 + 2)$
   (b) $(3x^2 + 2)(x + 1)(x - 1)$

10. $(9x^2 + 4y^2)(3x + 2y)(3x - 2y)$

1.4 Rational Expressions

A **rational expression** is an expression that can be written as the quotient of two polynomials, such as

$$\frac{8}{x - 1}, \qquad \frac{3x^2 + 4x}{5x - 6}, \qquad \text{and} \qquad \frac{2y + 1}{y^4 + 8}.$$

It is sometimes important to know the values of the variable that make the denominator 0 (in which case the quotient is not defined). For example, 1 cannot be used as a replacement for x in the first expression above and 6/5 cannot be used in the second one, since these values make the respective denominators equal 0. *Throughout this section, we assume that all denominators are nonzero,* which means that some replacement values for the variables may have to be excluded. ✔₁

✓ Checkpoint 1

What value of the variable makes each denominator equal 0?

(a) $\dfrac{5}{x - 3}$

(b) $\dfrac{2x - 3}{4x - 1}$

(c) $\dfrac{x + 2}{x}$

(d) Why do we need to determine these values?

Simplifying Rational Expressions

A key tool for simplification is the following fact.

> ## Cancellation Property
>
> For all expressions P, Q, and S, with $Q \neq 0$ and $S \neq 0$,
>
> $$\frac{PS}{QS} = \frac{P}{Q}.$$

EXAMPLE 1 Write each of the following rational expressions in lowest terms (so that the numerator and denominator have no common factor with integer coefficients except 1 or -1).

(a) $\dfrac{12m}{-18}$

Solution Both $12m$ and -18 are divisible by 6. By the cancellation property,

$$\frac{12m}{-18} = \frac{2m \cdot 6}{-3 \cdot 6}$$

$$= \frac{2m}{-3}$$

$$= -\frac{2m}{3}.$$

(b) $\dfrac{8x + 16}{4}$

Solution Factor the numerator and cancel:

$$\frac{8x + 16}{4} = \frac{8(x + 2)}{4} = \frac{4 \cdot 2(x + 2)}{4} = \frac{2(x + 2)}{1} = 2(x + 2).$$

The answer could also be written as $2x + 4$ if desired.

(c) $\dfrac{k^2 + 7k + 12}{k^2 + 2k - 3}$

Solution Factor the numerator and denominator and cancel:

$$\frac{k^2 + 7k + 12}{k^2 + 2k - 3} = \frac{(k + 4)(k + 3)}{(k - 1)(k + 3)} = \frac{k + 4}{k - 1}.$$ ✔₂

✓ Checkpoint 2

Write each of the following in lowest terms.

(a) $\dfrac{12k + 36}{18}$

(b) $\dfrac{15m + 30m^2}{5m}$

(c) $\dfrac{2p^2 + 3p + 1}{p^2 + 3p + 2}$

Multiplication and Division

The rules for multiplying and dividing rational expressions are the same fraction rules you learned in arithmetic.

Multiplication and Division Rules

For all expressions P, Q, R, and S, with $Q \neq 0$ and $S \neq 0$,

$$\frac{P}{Q} \cdot \frac{R}{S} = \frac{PR}{QS}$$

and

$$\frac{P}{Q} \div \frac{R}{S} = \frac{P}{Q} \cdot \frac{S}{R} \qquad (R \neq 0).$$

EXAMPLE 2

(a) Multiply $\dfrac{2}{3} \cdot \dfrac{y}{5}$.

Solution Use the multiplication rule. Multiply the numerators and then the denominators:

$$\frac{2}{3} \cdot \frac{y}{5} = \frac{2 \cdot y}{3 \cdot 5} = \frac{2y}{15}.$$

The result, $2y/15$, is in lowest terms.

(b) Multiply $\dfrac{3y + 9}{6} \cdot \dfrac{18}{5y + 15}$.

Solution Factor where possible:

$$\frac{3y + 9}{6} \cdot \frac{18}{5y + 15} = \frac{3(y + 3)}{6} \cdot \frac{18}{5(y + 3)}$$

$$= \frac{3 \cdot 18(y + 3)}{6 \cdot 5(y + 3)} \qquad \text{Multiply numerators and denominators.}$$

$$= \frac{3 \cdot 6 \cdot 3(y + 3)}{6 \cdot 5(y + 3)} \qquad 18 = 6 \cdot 3$$

$$= \frac{3 \cdot 3}{5} \qquad \text{Write in lowest terms.}$$

$$= \frac{9}{5}.$$

(c) Multiply $\dfrac{m^2 + 5m + 6}{m + 3} \cdot \dfrac{m^2 + m - 6}{m^2 + 3m + 2}$.

Solution Factor numerators and denominators:

$$\frac{(m + 2)(m + 3)}{m + 3} \cdot \frac{(m - 2)(m + 3)}{(m + 2)(m + 1)} \qquad \text{Factor.}$$

$$= \frac{(m + 2)(m + 3)(m - 2)(m + 3)}{(m + 3)(m + 2)(m + 1)} \qquad \text{Multiply.}$$

$$= \frac{(m - 2)(m + 3)}{m + 1} \qquad \text{Lowest terms}$$

$$= \frac{m^2 + m - 6}{m + 1}. \qquad \boxed{\checkmark_3}$$

✓ **Checkpoint 3**

Multiply.

(a) $\dfrac{3r^2}{5} \cdot \dfrac{20}{9r}$

(b) $\dfrac{y - 4}{y^2 - 2y - 8} \cdot \dfrac{y^2 - 4}{3y}$

EXAMPLE 3

(a) Divide $\dfrac{8x}{5} \div \dfrac{11x^2}{20}$.

Solution Invert the second expression and multiply (division rule):

$$\frac{8x}{5} \div \frac{11x^2}{20} = \frac{8x}{5} \cdot \frac{20}{11x^2} \qquad \text{Invert and multiply.}$$

$$= \frac{8x \cdot 20}{5 \cdot 11x^2} \qquad \text{Multiply.}$$

$$= \frac{32}{11x}. \qquad \text{Lowest terms}$$

(b) Divide $\dfrac{9p - 36}{12} \div \dfrac{5(p - 4)}{18}$.

Solution We have

$$\frac{9p - 36}{12} \cdot \frac{18}{5(p - 4)} \qquad \text{Invert and multiply.}$$

$$= \frac{9(p - 4)}{12} \cdot \frac{18}{5(p - 4)} \qquad \text{Factor.}$$

$$= \frac{27}{10}. \qquad \text{Cancel, multiply, and write in lowest terms.} \quad \text{✓}\ 4$$

✓ **Checkpoint 4**

Divide.

(a) $\dfrac{5m}{16} \div \dfrac{m^2}{10}$

(b) $\dfrac{2y - 8}{6} \div \dfrac{5y - 20}{3}$

(c) $\dfrac{m^2 - 2m - 3}{m(m + 1)} \div \dfrac{m + 4}{5m}$

Addition and Subtraction

As you know, when two numerical fractions have the same denominator, they can be added or subtracted. The same rules apply to rational expressions.

Addition and Subtraction Rules

For all expressions P, Q, R, with $Q \neq 0$,

$$\frac{P}{Q} + \frac{R}{Q} = \frac{P + R}{Q}$$

and

$$\frac{P}{Q} - \frac{R}{Q} = \frac{P - R}{Q}.$$

EXAMPLE 4   Add or subtract as indicated.

(a) $\dfrac{4}{5k} + \dfrac{11}{5k}$

Solution Since the denominators are the same, we add the numerators:

$$\frac{4}{5k} + \frac{11}{5k} = \frac{4 + 11}{5k} = \frac{15}{5k} \qquad \text{Addition rule}$$

$$= \frac{3}{k}. \qquad \text{Lowest terms}$$

(b) $\dfrac{2x^2 + 3x + 1}{x^5 + 1} - \dfrac{x^2 - 7x}{x^5 + 1}$

Solution The denominators are the same, so we subtract numerators, paying careful attention to parentheses:

$$\dfrac{2x^2 + 3x + 1}{x^5 + 1} - \dfrac{x^2 - 7x}{x^5 + 1} = \dfrac{(2x^2 + 3x + 1) - (x^2 - 7x)}{x^5 + 1} \qquad \text{Subtraction rule}$$

$$= \dfrac{2x^2 + 3x + 1 - x^2 - (-7x)}{x^5 + 1} \qquad \text{Subtract numerators.}$$

$$= \dfrac{2x^2 + 3x + 1 - x^2 + 7x}{x^5 + 1}$$

$$= \dfrac{x^2 + 10x + 1}{x^5 + 1}. \qquad \text{Simplify the numerator.}$$

When fractions do not have the same denominator, you must first find a common denominator before you can add or subtract. A common denominator is a denominator that has each fraction's denominator as a factor.

EXAMPLE 5 Add or subtract as indicated.

(a) $\dfrac{7}{p^2} + \dfrac{9}{2p} + \dfrac{1}{3p^2}$

Solution These three denominators are different, so we must find a common denominator that has each of p^2, $2p$, and $3p^2$ as factors. Observe that $6p^2$ satisfies these requirements. Use the cancellation property to rewrite each fraction as one that has $6p^2$ as its denominator and then add them:

$$\dfrac{7}{p^2} + \dfrac{9}{2p} + \dfrac{1}{3p^2} = \dfrac{6 \cdot 7}{6 \cdot p^2} + \dfrac{3p \cdot 9}{3p \cdot 2p} + \dfrac{2 \cdot 1}{2 \cdot 3p^2} \qquad \text{Cancellation property}$$

$$= \dfrac{42}{6p^2} + \dfrac{27p}{6p^2} + \dfrac{2}{6p^2}$$

$$= \dfrac{42 + 27p + 2}{6p^2} \qquad \text{Addition rule}$$

$$= \dfrac{27p + 44}{6p^2}. \qquad \text{Simplify.}$$

(b) $\dfrac{k^2}{k^2 - 1} - \dfrac{2k^2 - k - 3}{k^2 + 3k + 2}$

Solution Factor the denominators to find a common denominator:

$$\dfrac{k^2}{k^2 - 1} - \dfrac{2k^2 - k - 3}{k^2 + 3k + 2} = \dfrac{k^2}{(k + 1)(k - 1)} - \dfrac{2k^2 - k - 3}{(k + 1)(k + 2)}.$$

A common denominator here is $(k + 1)(k - 1)(k + 2)$, because each of the preceding denominators is a factor of this common denominator. Write each fraction with the common denominator:

$$\frac{k^2}{(k + 1)(k - 1)} - \frac{2k^2 - k - 3}{(k + 1)(k + 2)}$$

$$= \frac{k^2(k + 2)}{(k + 1)(k - 1)(k + 2)} - \frac{(2k^2 - k - 3)(k - 1)}{(k + 1)(k - 1)(k + 2)}$$

$$= \frac{k^3 + 2k^2 - (2k^2 - k - 3)(k - 1)}{(k + 1)(k - 1)(k + 2)} \qquad \text{Subtract fractions.}$$

$$= \frac{k^3 + 2k^2 - (2k^3 - 3k^2 - 2k + 3)}{(k + 1)(k - 1)(k + 2)} \qquad \text{Multiply } (2k^2\text{-}k\text{-}3)(k\text{-}1).$$

$$= \frac{k^3 + 2k^2 - 2k^3 + 3k^2 + 2k - 3}{(k + 1)(k - 1)(k + 2)} \qquad \text{Polynomial subtraction}$$

$$= \frac{-k^3 + 5k^2 + 2k - 3}{(k + 1)(k - 1)(k + 2)}. \qquad \text{Combine terms. } \checkmark_5$$

✓ Checkpoint 5

Add or subtract.

(a) $\dfrac{3}{4r} + \dfrac{8}{3r}$

(b) $\dfrac{1}{m - 2} - \dfrac{3}{2(m - 2)}$

(c) $\dfrac{p + 1}{p^2 - p} - \dfrac{p^2 - 1}{p^2 + p - 2}$

Complex Fractions

Any quotient of rational expressions is called a **complex fraction.** Complex fractions are simplified as demonstrated in Example 6.

EXAMPLE 6   Simplify the complex fraction

$$\frac{6 - \dfrac{5}{k}}{1 + \dfrac{5}{k}}.$$

Solution Multiply both numerator and denominator by the common denominator k:

$$\frac{6 - \dfrac{5}{k}}{1 + \dfrac{5}{k}} = \frac{k\left(6 - \dfrac{5}{k}\right)}{k\left(1 + \dfrac{5}{k}\right)} \qquad \text{Multiply by } \dfrac{k}{k}.$$

$$= \frac{6k - k\left(\dfrac{5}{k}\right)}{k + k\left(\dfrac{5}{k}\right)} \qquad \text{Distributive property}$$

$$= \frac{6k - 5}{k + 5}. \qquad \text{Simplify.}$$

1.4 Exercises

Write each of the given expressions in lowest terms. Factor as necessary. (See Example 1.)

1. $\dfrac{8x^2}{56x}$

2. $\dfrac{27m}{81m^3}$

3. $\dfrac{25p^2}{35p^3}$

4. $\dfrac{18y^4}{24y^2}$

5. $\dfrac{5m + 15}{4m + 12}$

6. $\dfrac{10z + 5}{20z + 10}$

7. $\dfrac{4(w - 3)}{(w - 3)(w + 6)}$

8. $\dfrac{-6(x + 2)}{(x + 4)(x + 2)}$

9. $\dfrac{3y^2 - 12y}{9y^3}$

10. $\dfrac{15k^2 + 45k}{9k^2}$

11. $\dfrac{m^2 - 4m + 4}{m^2 + m - 6}$

12. $\dfrac{r^2 - r - 6}{r^2 + r - 12}$

13. $\dfrac{x^2 + 2x - 3}{x^2 - 1}$

14. $\dfrac{z^2 + 4z + 4}{z^2 - 4}$

Multiply or divide as indicated in each of the exercises. Write all answers in lowest terms. (See Examples 2 and 3.)

15. $\dfrac{3a^2}{64} \cdot \dfrac{8}{2a^3}$

16. $\dfrac{2u^2}{8u^4} \cdot \dfrac{10u^3}{9u}$

17. $\dfrac{7x}{11} \div \dfrac{14x^3}{66y}$

18. $\dfrac{6x^2y}{2x} \div \dfrac{21xy}{y}$

19. $\dfrac{2a + b}{3c} \cdot \dfrac{15}{4(2a + b)}$

20. $\dfrac{4(x + 2)}{w} \cdot \dfrac{3w^2}{8(x + 2)}$

21. $\dfrac{15p - 3}{6} \div \dfrac{10p - 2}{3}$

22. $\dfrac{2k + 8}{6} \div \dfrac{3k + 12}{3}$

23. $\dfrac{9y - 18}{6y + 12} \cdot \dfrac{3y + 6}{15y - 30}$

24. $\dfrac{12r + 24}{36r - 36} \div \dfrac{6r + 12}{8r - 8}$

25. $\dfrac{4a + 12}{2a - 10} \div \dfrac{a^2 - 9}{a^2 - a - 20}$

26. $\dfrac{6r - 18}{9r^2 + 6r - 24} \cdot \dfrac{12r - 16}{4r - 12}$

27. $\dfrac{k^2 - k - 6}{k^2 + k - 12} \cdot \dfrac{k^2 + 3k - 4}{k^2 + 2k - 3}$

28. $\dfrac{n^2 - n - 6}{n^2 - 2n - 8} \div \dfrac{n^2 - 9}{n^2 + 7n + 12}$

29. In your own words, explain how to find the least common denominator of two fractions.

30. Describe the steps required to add three rational expressions. You may use an example to illustrate.

Add or subtract as indicated in each of the following. Write all answers in lowest terms. (See Example 4.)

31. $\dfrac{2}{7z} - \dfrac{1}{5z}$

32. $\dfrac{4}{3z} - \dfrac{5}{4z}$

33. $\dfrac{r + 2}{3} - \dfrac{r - 2}{3}$

34. $\dfrac{3y - 1}{8} - \dfrac{3y + 1}{8}$

35. $\dfrac{4}{x} + \dfrac{1}{5}$

36. $\dfrac{6}{r} - \dfrac{3}{4}$

37. $\dfrac{1}{m - 1} + \dfrac{2}{m}$

38. $\dfrac{8}{y + 2} - \dfrac{3}{y}$

39. $\dfrac{7}{b + 2} + \dfrac{2}{5(b + 2)}$

40. $\dfrac{4}{3(k + 1)} + \dfrac{3}{k + 1}$

41. $\dfrac{2}{5(k - 2)} + \dfrac{5}{4(k - 2)}$

42. $\dfrac{11}{3(p + 4)} - \dfrac{5}{6(p + 4)}$

43. $\dfrac{2}{x^2 - 4x + 3} + \dfrac{5}{x^2 - x - 6}$

44. $\dfrac{3}{m^2 - 3m - 10} + \dfrac{7}{m^2 - m - 20}$

45. $\dfrac{2y}{y^2 + 7y + 12} - \dfrac{y}{y^2 + 5y + 6}$

46. $\dfrac{-r}{r^2 - 10r + 16} - \dfrac{3r}{r^2 + 2r - 8}$

In each of the exercises in the next set, simplify the complex fraction. (See Example 6.)

47. $\dfrac{1 + \dfrac{1}{x}}{1 - \dfrac{1}{x}}$

48. $\dfrac{2 - \dfrac{2}{y}}{2 + \dfrac{2}{y}}$

49. $\dfrac{\dfrac{1}{x + h} - \dfrac{1}{x}}{h}$

50. $\dfrac{\dfrac{1}{(x + h)^2} - \dfrac{1}{x^2}}{h}$

Work these problems.

Natural Science *Each figure in the following exercises is a dartboard. The probability that a dart which hits the board lands in the shaded area is the fraction*

$$\frac{\text{area of the shaded region}}{\text{area of the dartboard}}.$$

Note: The area of a circle is πr^2, the area of a square is bh, and the area of a triangle is $\dfrac{1}{2}bh$.

(a) *Express the probability as a rational expression in x.*

(b) *Then reduce the expression to lowest terms.*

51.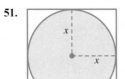

52.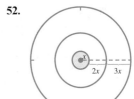

53.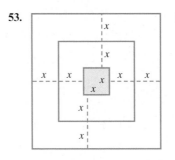

54.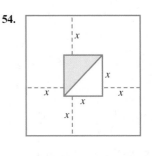

Costs *In Example 11 of Section 1.2, we saw that the cost C of producing x thousand calculators is given by*

$$C = -7.2x^2 + 6995x + 230{,}000 \quad (x \le 150).$$

55. The average cost per calculator is the total cost C divided by the number of calculators produced. Write a rational expression that gives the average cost per calculator when x thousand are produced.

56. Find the average cost per calculator for each of these production levels: 20,000, 50,000, and 125,000.

Superbowl Advertising Costs *The cost (in millions of dollars) for a 30-second ad during the TV broadcast of the Superbowl can be approximated by*

$$\frac{.505x^2 - 4.587x + 27.6}{x + 1}$$

where $x = 6$ *corresponds to the year 2006. (Data from: forbes.com and statistica.com.)*

57. How much did an ad cost in 2013?

58. How much did an ad cost in 2016?

59. If this trend continues, will the cost of an ad reach $7 million by 2020?

60. If this trend continues, will the cost of an ad reach $8 million by 2022?

Health Care Costs *The average company cost per hour of an employee's health insurance in year x is approximated by*

$$\frac{.049x^2 + 2.40x + .83}{x + 2},$$

where $x = 4$ *corresponds to the year 2004. (Data from: U.S. Bureau of Labor Statistics.)*

61. What is the hourly health insurance cost in 2011?

62. What is the hourly health insurance cost in 2017?

63. Assuming that this model remains accurate and that an employee works 2100 hours per year, what is the annual company cost of her health care insurance in 2020?

64. Will annual costs reach $10,000 by 2023?

✓**Checkpoint Answers**

1. (a) 3   (b) 1/4   (c) 0
   (d) Because division by 0 is undefined

2. (a) $\frac{2(k + 3)}{3}$ or $\frac{2k + 6}{3}$
   (b) $3(1 + 2m)$ or $3 + 6m$ (c) $\frac{2p + 1}{p + 2}$

3. (a) $\frac{4r}{3}$ (b) $\frac{y - 2}{3y}$

4. (a) $\frac{25}{8m}$ (b) $\frac{1}{5}$ (c) $\frac{5(m - 3)}{m + 4}$

5. (a) $\frac{41}{12r}$ (b) $\frac{-1}{2(m - 2)}$ (c) $\frac{-p^3 + p^2 + 4p + 2}{p(p - 1)(p + 2)}$

1.5   Exponents and Radicals

Exponents were introduced in Section 1.2. In this section, the definition of exponents will be extended to include negative exponents and rational-number exponents such as 1/2 and 7/3.

Integer Exponents

Positive-integer and zero exponents were defined in Section 1.2, where we noted that

$$a^m \cdot a^n = a^{m+n}$$

for nonnegative integers m and n. Now we develop an analogous property for quotients. By definition,

$$\frac{6^5}{6^2} = \frac{6 \cdot 6 \cdot 6 \cdot 6 \cdot 6}{6 \cdot 6} = 6 \cdot 6 \cdot 6 = 6^3.$$

Because there are 5 factors of 6 in the numerator and 2 factors of 6 in the denominator, the quotient has $5 - 2 = 3$ factors of 6. In general, we can make the following statement, which applies to any real number a and nonnegative integers m and n with $m > n$.

Division with Exponents

To divide a^m by a^n, *subtract* the exponents (assuming $a \neq 0$):

$$\frac{a^m}{a^n} = a^{m-n}.$$

EXAMPLE 1 Compute each of the following.

(a) $\dfrac{5^7}{5^4} = 5^{7-4} = 5^3$.

(b) $\dfrac{(-8)^{10}}{(-8)^5} = (-8)^{10-5} = (-8)^5$.

(c) $\dfrac{(3c)^9}{(3c)^3} = (3c)^{9-3} = (3c)^6$. ✓1

✓ **Checkpoint 1**

Evaluate each of the following.

(a) $\dfrac{2^{14}}{2^5}$

(b) $\dfrac{(-5)^9}{(-5)^5}$

(c) $\dfrac{(xy)^{17}}{(xy)^{12}}$

When an exponent applies to the product of two numbers, such as $(7 \cdot 19)^3$, use the definitions carefully. For instance,

$$(7 \cdot 19)^3 = (7 \cdot 19)(7 \cdot 19)(7 \cdot 19) = 7 \cdot 7 \cdot 7 \cdot 19 \cdot 19 \cdot 19 = 7^3 \cdot 19^3.$$

In other words, $(7 \cdot 19)^3 = 7^3 \cdot 19^3$. This is an example of the following fact, which applies to any real numbers a and b and any nonnegative-integer exponent n.

Product to a Power

To find $(ab)^n$, apply the exponent to *every* factor inside the parentheses:

$$(ab)^n = a^n b^n.$$

⊘ **CAUTION** A common mistake is to write an expression such as $(2x)^5$ as $2x^5$, rather than the correct answer $(2x)^5 = 2^5 x^5 = 32x^5$.

Analogous conclusions are valid for quotients (where a and b are any real numbers with $b \neq 0$ and n is a nonnegative-integer exponent).

Quotient to a Power

To find $\left(\dfrac{a}{b}\right)^n$, (assuming $b \neq 0$), apply the exponent to both numerator and denominator:

$$\left(\frac{a}{b}\right)^n = \frac{a^n}{b^n}.$$

EXAMPLE 2 Compute each of the following.

(a) $(5y)^3 = 5^3 y^3 = 125y^3$ Product to a power

(b) $(c^2 d^3)^4 = (c^2)^4 (d^3)^4$ Product to a power

$\qquad = c^8 d^{12}$ Power of a power

(c) $\left(\dfrac{x}{2}\right)^6 = \dfrac{x^6}{2^6} = \dfrac{x^6}{64}$ Quotient to a power

(d) $\left(\dfrac{a^4}{b^3}\right)^3 = \dfrac{(a^4)^3}{(b^3)^3}$ Quotient to a power

$\qquad = \dfrac{a^{12}}{b^9}$ Power of a power

(e) $\left(\dfrac{(rs)^3}{r^4}\right)^2$

Solution Use several of the preceding properties in succession:

$$\left(\frac{(rs)^3}{r^4}\right)^2 = \left(\frac{r^3 s^3}{r^4}\right)^2 \qquad \text{Product to a power in numerator}$$

$$= \left(\frac{s^3}{r}\right)^2 \qquad \text{Cancel.}$$

$$= \frac{(s^3)^2}{r^2} \qquad \text{Quotient to a power}$$

$$= \frac{s^6}{r^2}. \qquad \text{Power of a power in numerator}$$

As is often the case, there is another way to reach the last expression. You should be able to supply the reasons for each of the following steps:

$$\left(\frac{(rs)^3}{r^4}\right)^2 = \frac{[(rs)^3]^2}{(r^4)^2} = \frac{(rs)^6}{r^8} = \frac{r^6 s^6}{r^8} = \frac{s^6}{r^2}. \quad \boxed{\checkmark}_2$$

✓ **Checkpoint 2**

Compute each of the following.

(a) $(3x)^4$

(b) $(r^2 s^5)^6$

(c) $\left(\dfrac{2}{z}\right)^5$

(d) $\left(\dfrac{3a^5}{(ab)^3}\right)^2$

Negative Exponents

The next step is to define negative-integer exponents. If they are to be defined in such a way that the quotient rule for exponents remains valid, then we must have, for example,

$$\frac{3^2}{3^4} = 3^{2-4} = 3^{-2}.$$

However,

$$\frac{3^2}{3^4} = \frac{3 \cdot 3}{3 \cdot 3 \cdot 3 \cdot 3} = \frac{1}{3^2},$$

which suggests that 3^{-2} should be defined to be $1/3^2$. Thus, we have the following definition of a negative exponent.

> ## Negative Exponent
>
> If n is a natural number, and if $a \neq 0$, then
>
> $$a^{-n} = \frac{1}{a^n}.$$

✓ **Checkpoint 3**

Evaluate the following.

(a) 6^{-2}

(b) -6^{-3}

(c) 3^{-4}

(d) $\left(\dfrac{5}{8}\right)^{-1}$

EXAMPLE 3 Evaluate the following.

(a) $3^{-2} = \dfrac{1}{3^2} = \dfrac{1}{9}.$

(b) $5^{-4} = \dfrac{1}{5^4} = \dfrac{1}{625}.$

(c) $x^{-1} = \dfrac{1}{x^1} = \dfrac{1}{x}.$

(d) $-4^{-2} = -\dfrac{1}{4^2} = -\dfrac{1}{16}.$

(e) $\left(\dfrac{3}{4}\right)^{-1} = \dfrac{1}{\left(\dfrac{3}{4}\right)^1} = \dfrac{1}{\dfrac{3}{4}} = \dfrac{4}{3}.$ $\boxed{\checkmark}_3$

There is a useful property that makes it easy to raise a fraction to a negative exponent. Consider, for example,

$$\left(\frac{2}{3}\right)^{-4} = \frac{1}{\left(\frac{2}{3}\right)^4} = \frac{1}{\left(\frac{2^4}{3^4}\right)} = 1 \cdot \frac{3^4}{2^4} = \left(\frac{3}{2}\right)^4.$$

This example is easily generalized to the following property (in which a/b is a nonzero fraction and n a positive integer).

Inversion Property

$$\left(\frac{a}{b}\right)^{-n} = \left(\frac{b}{a}\right)^{n}.$$

EXAMPLE 4   Use the inversion property to compute each of the following.

(a) $\left(\frac{2}{5}\right)^{-3} = \left(\frac{5}{2}\right)^{3} = \frac{5^3}{2^3} = \frac{125}{8}.$

(b) $\left(\frac{3}{x}\right)^{-5} = \left(\frac{x}{3}\right)^{5} = \frac{x^5}{3^5} = \frac{x^5}{243}.$ ✓ 4

✓ **Checkpoint 4**

Compute each of the following.

(a) $\left(\frac{5}{8}\right)^{-1}$

(b) $\left(\frac{1}{2}\right)^{-5}$

(c) $\left(\frac{a^2}{b}\right)^{-3}$

When keying in negative exponents on a calculator, be sure to use the negation key (labeled $(-)$ or $+/-$), not the subtraction key. Calculators normally display answers as decimals, as shown in Figure 1.6. Some graphing calculators have a FRAC key that converts these decimals to fractions, as shown in Figure 1.7.

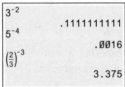

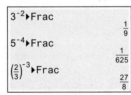

Figure 1.6   **Figure 1.7**

📇 TECHNOLOGY TIP

The FRAC key is in the MATH menu of TI graphing calculators. Fractions can be displayed on some graphing calculators by changing the number display format (in the MODES menu) to "fraction" or "exact."

Roots and Rational Exponents

There are two numbers whose square is 16: 4 and -4. As we saw in Section 1.1, the nonnegative one, 4, is called the *square root* (or second root) of 16. Similarly, there are two numbers whose fourth power is 16: 2 and -2. We call 2 the **fourth root** of 16. This suggests the following generalization.

If a is a nonnegative real number and n is an even natural number, the **nth root of a** is the nonnegative real number whose nth power is a.

All nonnegative numbers have nth roots for every natural number n, but *no negative number has a real, even nth root*. For example, there is no real number whose square is -16, so -16 has no square root.

We say that the **cube root** (or third root) of 8 is 2 because $2^3 = 8$. Similarly, since $(-2)^3 = -8$, we say that -2 is the cube root of -8. Again, we can make the following generalization.

If a is a real number and n is an odd natural number, the **nth root of a** is the real number whose nth power is a.

Every real number has an nth root for every *odd* natural number n.

We can now define rational exponents. If they are to have the same properties as integer exponents, we want $a^{1/2}$ to be a number such that

$$(a^{1/2})^2 = a^{1/2} \cdot a^{1/2} = a^{1/2 + 1/2} = a^1 = a.$$

Thus, $a^{1/2}$ should be a number whose square is a, and it is reasonable to *define* $a^{1/2}$ to be the square root of a (if it exists). Similarly, $a^{1/3}$ is defined to be the cube root of a, and we have the following definition.

If a is a real number and n is a natural number, then

$a^{1/n}$ is defined to be the nth root of a (if it exists).

EXAMPLE 5   Examine the reasoning used to evaluate the following roots.

(a) $36^{1/2} = 6$ because $6^2 = 36$.

(b) $100^{1/2} = 10$ because $10^2 = 100$.

(c) $-(225^{1/2}) = -15$ because $15^2 = 225$.

(d) $625^{1/4} = 5$ because $5^4 = 625$.

(e) $(-1296)^{1/4}$ is not a real number.

(f) $-1296^{1/4} = -6$ because $6^4 = 1296$.

(g) $(-27)^{1/3} = -3$ because $(-3)^3 = -27$.

(h) $-32^{1/5} = -2$ because $2^5 = 32$. ✔ 5

✓ **Checkpoint 5**

Evaluate the following.

(a) $16^{1/2}$

(b) $16^{1/4}$

(c) $-256^{1/2}$

(d) $(-256)^{1/2}$

(e) $-8^{1/3}$

(f) $243^{1/5}$

A calculator can be used to evaluate expressions with fractional exponents. Whenever it is easy to do so, enter the fractional exponents in their equivalent decimal form. For instance, to find $625^{1/4}$, enter $625^{.25}$ into the calculator. When the decimal equivalent of a fraction is an infinitely repeating decimal, however, it is best to enter the fractional exponent directly. If you use a shortened decimal approximation (such as .333 for $1/3$), you will not get the correct answers. Compare the incorrect answers in Figure 1.9 with the correct ones in Figure 1.8.

$27^{\frac{1}{3}}$	
	3

Figure 1.8

$27^{.333}$	
	2.996705973

Figure 1.9

For other rational exponents, the symbol $a^{m/n}$ should be defined so that the properties for exponents still hold. For example, by the product property, we want

$$(a^{1/3})^2 = a^{1/3} \cdot a^{1/3} = a^{1/3 + 1/3} = a^{2/3}.$$

This result suggests the following definition.

> For all integers m and all positive integers n, and for all real numbers a for which $a^{1/n}$ is a real number,
>
> $$a^{m/n} = (a^{1/n})^m.$$

EXAMPLE 6   Verify each of the following calculations.

(a) $27^{2/3} = (27^{1/3})^2 = 3^2 = 9.$

(b) $32^{2/5} = (32^{1/5})^2 = 2^2 = 4.$

(c) $64^{4/3} = (64^{1/3})^4 = 4^4 = 256.$

(d) $25^{3/2} = (25^{1/2})^3 = 5^3 = 125.$

✓**Checkpoint 6**

Evaluate the following.

(a) $16^{3/4}$

(b) $25^{5/2}$

(c) $32^{7/5}$

(d) $100^{3/2}$

⚠ CAUTION   When the base is negative, as in $(-8)^{2/3}$, some calculators produce an error message. On such calculators, you should first compute $(-8)^{1/3}$ and then square the result; that is, compute $[(-8)^{1/3}]^2$.

Since every terminating decimal is a rational number, decimal exponents now have a meaning. For instance, $5.24 = \frac{524}{100}$, so $3^{5.24} = 3^{524/100}$, which is easily approximated by a calculator (Figure 1.10).

Rational exponents were defined so that one of the familiar properties of exponents remains valid. In fact, it can be proved that *all* of the rules developed earlier for integer exponents are valid for rational exponents. The following box summarizes these rules, which are illustrated in Examples 7–9.

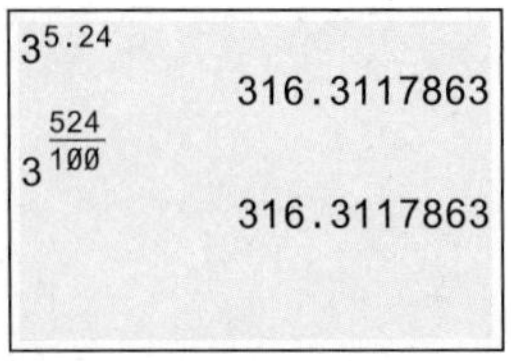

```
3^5.24
              316.3117863
 524
3 100
              316.3117863
```

Figure 1.10

Properties of Exponents

For any rational numbers m and n, and for any real numbers a and b for which the following exist,

(a) $a^m \cdot a^n = a^{m+n}$ Product property

(b) $\dfrac{a^m}{a^n} = a^{m-n}$ Quotient property

(c) $(a^m)^n = a^{mn}$ Power of a power

(d) $(ab)^m = a^m \cdot b^m$ Product to a power

(e) $\left(\dfrac{a}{b}\right)^m = \dfrac{a^m}{b^m}$ Quotient to a power

(f) $a^0 = 1$ Zero exponent

(g) $a^{-n} = \dfrac{1}{a^n}$ Negative exponent

(h) $\left(\dfrac{a}{b}\right)^{-n} = \left(\dfrac{b}{a}\right)^n.$ Inversion property

The power-of-a-power property provides another way to compute $a^{m/n}$ (when it exists):

$$a^{m/n} = a^{m(1/n)} = (a^m)^{1/n}. \tag{1}$$

For example, we can now find $4^{3/2}$ in two ways:

$$4^{3/2} = (4^{1/2})^3 = 2^3 = 8 \quad \text{or} \quad 4^{3/2} = (4^3)^{1/2} = 64^{1/2} = 8.$$

Definition of $a^{m/n}$ Statement (1)

EXAMPLE 7   Simplify each of the following expressions.

(a) $7^{-4} \cdot 7^6 = 7^{-4+6} = 7^2 = 49.$ Product property

(b) $5x^{2/3} \cdot 2x^{1/4} = 10x^{2/3}x^{1/4}$

$$= 10x^{2/3+1/4} \quad \text{Product property}$$

$$= 10x^{11/12}. \qquad \frac{2}{3} + \frac{1}{4} = \frac{8}{12} + \frac{3}{12} = \frac{11}{12}$$

(c) $\dfrac{9^{14}}{9^{-6}} = 9^{14-(-6)} = 9^{20}.$ Quotient property

(d) $\dfrac{c^5}{2c^{4/3}} = \dfrac{1}{2} \cdot \dfrac{c^5}{c^{4/3}}$

$$= \frac{1}{2}c^{5-4/3} \qquad \text{Quotient property}$$

$$= \frac{1}{2}c^{11/3} = \frac{c^{11/3}}{2}. \qquad 5 - \frac{4}{3} = \frac{15}{3} - \frac{4}{3} = \frac{11}{3}$$

(e) $\dfrac{27^{1/3} \cdot 27^{5/3}}{27^3} = \dfrac{27^{1/3+5/3}}{27^3}$ Product property

$$= \frac{27^2}{27^3} = 27^{2-3} \qquad \text{Quotient property}$$

$$= 27^{-1} = \frac{1}{27}. \qquad \text{Definition of negative exponent} \quad \boxed{\checkmark\,7}$$

✓**Checkpoint 7**

Simplify each of the following.

(a) $9^7 \cdot 9^{-5}$

(b) $3x^{1/4} \cdot 5x^{5/4}$

(c) $\dfrac{8^7}{8^{-3}}$

(d) $\dfrac{5^{2/3} \cdot 5^{-4/3}}{5^2}$

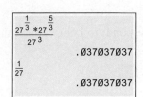

Figure 1.11

You can use a calculator to check numerical computations, such as those in Example 7, by computing the left and right sides separately and confirming that the answers are the same in each case. Figure 1.11 shows this technique for part (e) of Example 7.

EXAMPLE 8   Perform the indicated operations.

(a) $(2^{-3})^{-4/7} = 2^{(-3)(-4/7)} = 2^{12/7}.$ Power of a power

(b) $\left(\dfrac{3m^{5/6}}{y^{3/4}}\right)^2 = \dfrac{(3m^{5/6})^2}{(y^{3/4})^2}$ Quotient to a power

$$= \frac{3^2(m^{5/6})^2}{(y^{3/4})^2} \qquad \text{Product to a power}$$

$$= \frac{9m^{(5/6)2}}{y^{(3/4)2}} \qquad \text{Power of a power}$$

$$= \frac{9m^{5/3}}{y^{3/2}}. \qquad \frac{5}{6} \cdot 2 = \frac{10}{6} = \frac{5}{3} \text{ and } \frac{3}{4} \cdot 2 = \frac{3}{2}$$

(c) $m^{2/3}(m^{7/3} + 2m^{1/3}) = m^{2/3}m^{7/3} + m^{2/3}2m^{1/3}$ Distributive property

$$= m^{2/3+7/3} + 2m^{2/3+1/3} = m^3 + 2m. \qquad \text{Product rule} \quad \boxed{\checkmark\,8}$$

✓**Checkpoint 8**

Simplify each of the following.

(a) $(7^{-4})^{-2} \cdot (7^4)^{-2}$

(b) $\dfrac{c^4 c^{-1/2}}{c^{3/2}d^{1/2}}$

(c) $a^{5/8}(2a^{3/8} + a^{-1/8})$

EXAMPLE 9 Simplify each expression in parts (a)–(c). Give answers with only positive exponents.

(a) $\dfrac{(m^3)^{-2}}{m^4} = \dfrac{m^{-6}}{m^4} = m^{-6-4} = m^{-10} = \dfrac{1}{m^{10}}$.

(b) $6y^{2/3} \cdot 2y^{-1/2} = 12y^{2/3-1/2} = 12y^{1/6}$.

(c) $\dfrac{x^{1/2}(x-2)^{-3}}{5(x-2)} = \dfrac{x^{1/2}}{5} \cdot \dfrac{(x-2)^{-3}}{x-2} = \dfrac{x^{1/2}}{5} \cdot (x-2)^{-3-1}$

$\qquad = \dfrac{x^{1/2}}{5} \cdot \dfrac{1}{(x-2)^4} = \dfrac{x^{1/2}}{5(x-2)^4}$.

(d) Write $a^{-1} + b^{-1}$ as a single quotient.

Solution Be careful here. $a^{-1} + b^{-1}$ does *not* equal $(a+b)^{-1}$; the exponent properties deal only with products and quotients, not with sums. However, using the definition of negative exponents and addition of fractions, we have

$$a^{-1} + b^{-1} = \frac{1}{a} + \frac{1}{b} = \frac{b+a}{ab}. \quad \boxed{9}$$

We can use some expressions of the form ax^b where both a and b are constants and b is also an exponent.

EXAMPLE 10 **College Students** The total number of students (in millions) attending institutes of higher education can be approximated by the expression

$$6.1x^{.337} \quad (x \geq 10),$$

where $x = 10$ corresponds to the year 1990. Find the approximate number of students enrolled in higher education in 2017. (Data from: U.S. National Center for Education Statistics.)

Solution Since 2017 is 27 years after 1990 and $x = 10$ corresponds to 1990, we have that $x = 27 + 10 = 37$ corresponds to 2017. We then obtain

$$6.1(37)^{.337} \approx 20.6 \text{ million students.} \quad \boxed{10}$$

Radicals

Earlier, we denoted the nth root of a as $a^{1/n}$. An alternative notation for nth roots uses the radical symbol $\sqrt[n]{\ }$.

> If n is an even natural number and $a \geq 0$, or if n is an odd natural number,
> $$\sqrt[n]{a} = a^{1/n}.$$

In the radical expression $\sqrt[n]{a}$, a is called the *radicand* and n is called the *index*. When $n = 2$, the familiar square-root symbol $\sqrt{a}$ is used instead of $\sqrt[2]{a}$.

EXAMPLE 11 Simplify the following radicals.

(a) $\sqrt[4]{16} = 16^{1/4} = 2$. **(b)** $\sqrt[5]{-32} = -2$.

(c) $\sqrt[3]{1000} = 10$. **(d)** $\sqrt[6]{\dfrac{64}{729}} = \left(\dfrac{64}{729}\right)^{1/6} = \dfrac{64^{1/6}}{729^{1/6}} = \dfrac{2}{3}$. $\boxed{11}$

✓ **Checkpoint 9**

Simplify the given expressions. Give answers with only positive exponents.

(a) $(3x^{2/3})(2x^{-1})(y^{-1/3})^2$

(b) $\dfrac{(t^{-1})^2}{t^{-5}}$

(c) $\left(\dfrac{2k^{1/3}}{p^{5/4}}\right)^2 \cdot \left(\dfrac{4k^{-2}}{p^5}\right)^{3/2}$

(d) $x^{-1} - y^{-2}$

✓ **Checkpoint 10**

Assuming the model from Example 10 remains accurate, find the number of students for 2020.

✓ **Checkpoint 11**

Simplify.

(a) $\sqrt[3]{27}$

(b) $\sqrt[4]{625}$

(c) $\sqrt[6]{64}$

(d) $\sqrt[3]{\dfrac{64}{125}}$

Recall that $a^{m/n} = (a^{1/n})^m$ by definition and $a^{m/n} = (a^m)^{1/n}$ by statement (**1**) on page 36 (provided that all terms are defined). We translate these facts into radical notation as follows.

> For all rational numbers m/n and all real numbers a for which $\sqrt[n]{a}$ exists,
> $$a^{m/n} = (\sqrt[n]{a})^m \quad \text{or} \quad a^{m/n} = \sqrt[n]{a^m}.$$

Notice that $\sqrt[n]{x^n}$ cannot be written simply as x when n is even. For example, if $x = -5$, then
$$\sqrt{x^2} = \sqrt{(-5)^2} = \sqrt{25} = 5 \ne x.$$
However, $|-5| = 5$, so that $\sqrt{x^2} = |x|$ when x is -5. This relationship is true in general.

> For any real number a and any natural number n,
> $$\sqrt[n]{a^n} = |a| \quad \text{if } n \text{ is even}$$
> and
> $$\sqrt[n]{a^n} = a \quad \text{if } n \text{ is odd}.$$

To avoid the difficulty that $\sqrt[n]{a^n}$ is not necessarily equal to a, we shall assume that all variables in radicands represent only nonnegative numbers, as they usually do in applications. The properties of exponents can be written with radicals as follows.

> For all real numbers a and b, and for natural numbers n for which all indicated roots exist,
> (a) $\sqrt[n]{a} \cdot \sqrt[n]{b} = \sqrt[n]{ab}$ and
> (b) $\dfrac{\sqrt[n]{a}}{\sqrt[n]{b}} = \sqrt[n]{\dfrac{a}{b}} \quad (b \ne 0).$

EXAMPLE 12   Simplify the following expressions.

(a) $\sqrt{6} \cdot \sqrt{54} = \sqrt{6 \cdot 54} = \sqrt{324} = 18.$

Alternatively, simplify $\sqrt{54}$ first:
$$\sqrt{6} \cdot \sqrt{54} = \sqrt{6} \cdot \sqrt{9 \cdot 6}$$
$$= \sqrt{6} \cdot 3\sqrt{6} = 3 \cdot 6 = 18.$$

(b) $\sqrt{\dfrac{7}{64}} = \dfrac{\sqrt{7}}{\sqrt{64}} = \dfrac{\sqrt{7}}{8}.$

(c) $\sqrt{75} - \sqrt{12}.$

Solution   Note that $12 = 4 \cdot 3$ and that 4 is a perfect square. Similarly, $75 = 25 \cdot 3$ and 25 is a perfect square. Consequently,

$$\sqrt{75} - \sqrt{12} = \sqrt{25 \cdot 3} - \sqrt{4 \cdot 3} \qquad \text{Factor.}$$
$$= \sqrt{25}\sqrt{3} - \sqrt{4}\sqrt{3} \qquad \text{Property (a)}$$
$$= 5\sqrt{3} - 2\sqrt{3} = 3\sqrt{3}. \qquad \text{Simplify.} \; \checkmark_{12}$$

✓ **Checkpoint 12**

Simplify.

(a) $\sqrt{3} \cdot \sqrt{27}$

(b) $\sqrt{\dfrac{3}{49}}$

(c) $\sqrt{50} + \sqrt{72}$

⚠ **CAUTION** When a and b are nonzero real numbers,

$$\sqrt[n]{a + b} \text{ is NOT equal to } \sqrt[n]{a} + \sqrt[n]{b}.$$

For example,

$$\sqrt{9 + 16} = \sqrt{25} = 5, \text{ but } \sqrt{9} + \sqrt{16} = 3 + 4 = 7,$$

so $\sqrt{9 + 16} \neq \sqrt{9} + \sqrt{16}$.

Multiplying radical expressions is much like multiplying polynomials.

EXAMPLE 13 Perform the following multiplications.

(a) $(\sqrt{2} + 3)(\sqrt{8} - 5) = \sqrt{2}(\sqrt{8}) - \sqrt{2}(5) + 3\sqrt{8} - 3(5)$ FOIL

$$= \sqrt{16} - 5\sqrt{2} + 3(2\sqrt{2}) - 15$$

$$= 4 - 5\sqrt{2} + 6\sqrt{2} - 15$$

$$= -11 + \sqrt{2}.$$

(b) $(\sqrt{7} - \sqrt{10})(\sqrt{7} + \sqrt{10}) = (\sqrt{7})^2 - (\sqrt{10})^2$

$$= 7 - 10 = -3. \; ✓_{13}$$

✔ **Checkpoint 13**

Multiply.

(a) $(\sqrt{5} - \sqrt{2})(3 + \sqrt{2})$

(b) $(\sqrt{3} + \sqrt{7})(\sqrt{3} - \sqrt{7})$

Rationalizing Denominators and Numerators

Before the invention of calculators, it was customary to **rationalize the denominators** of fractions (that is, write equivalent fractions with no radicals in the denominator), because this made many computations easier. Although there is no longer a computational reason to do so, rationalization of denominators (and sometimes numerators) is still used today to simplify expressions and to derive useful formulas.

EXAMPLE 14 Rationalize each denominator.

(a) $\dfrac{4}{\sqrt{3}}$

Solution The key is to multiply by 1, with 1 written as a radical fraction:

$$\frac{4}{\sqrt{3}} = \frac{4}{\sqrt{3}} \cdot 1 = \frac{4}{\sqrt{3}} \cdot \frac{\sqrt{3}}{\sqrt{3}} = \frac{4\sqrt{3}}{3}.$$

(b) $\dfrac{1}{3 - \sqrt{2}}$

Solution The same technique works here, using $1 = \dfrac{3 + \sqrt{2}}{3 + \sqrt{2}}$:

$$\frac{1}{3 - \sqrt{2}} = \frac{1}{3 - \sqrt{2}} \cdot 1 = \frac{1}{3 - \sqrt{2}} \cdot \frac{3 + \sqrt{2}}{3 + \sqrt{2}} = \frac{3 + \sqrt{2}}{(3 - \sqrt{2})(3 + \sqrt{2})}$$

$$= \frac{3 + \sqrt{2}}{9 - 2} = \frac{3 + \sqrt{2}}{7}. \; ✓_{14}$$

✔ **Checkpoint 14**

Rationalize the denominator.

(a) $\dfrac{2}{\sqrt{5}}$

(b) $\dfrac{1}{2 + \sqrt{3}}$

EXAMPLE 15 Rationalize the numerator of

$$\frac{2 + \sqrt{5}}{1 + \sqrt{3}}.$$

Solution As in Example 14(b), we must write 1 as a suitable fraction. Since we want to rationalize the numerator here, we multiply by the fraction $1 = \dfrac{2 - \sqrt{5}}{2 - \sqrt{5}}$:

$$\frac{2 + \sqrt{5}}{1 + \sqrt{3}} = \frac{2 + \sqrt{5}}{1 + \sqrt{3}} \cdot \frac{2 - \sqrt{5}}{2 - \sqrt{5}} = \frac{4 - 5}{2 - \sqrt{5} + 2\sqrt{3} - \sqrt{3}\sqrt{5}}$$

$$= \frac{-1}{2 - \sqrt{5} + 2\sqrt{3} - \sqrt{15}}.$$

1.5 Exercises

Perform the indicated operations and simplify your answer. (See Examples 1 and 2.)

1. $\dfrac{7^5}{7^3}$

2. $\dfrac{(-6)^{14}}{(-6)^6}$

3. $(4c)^2$

4. $(-2x)^4$

5. $\left(\dfrac{2}{x}\right)^5$

6. $\left(\dfrac{5}{xy}\right)^3$

7. $(3u^2)^3(2u^3)^2$

8. $\dfrac{(5v^2)^3}{(2v)^4}$

Perform the indicated operations and simplify your answer, which should not have any negative exponents. (See Examples 3 and 4.)

9. 7^{-1}

10. 10^{-3}

11. -6^{-5}

12. $(-x)^{-4}$

13. $(-y)^{-3}$

14. $\left(\dfrac{1}{6}\right)^{-2}$

15. $\left(\dfrac{4}{3}\right)^{-2}$

16. $\left(\dfrac{x}{y^2}\right)^{-2}$

17. $\left(\dfrac{a}{b^3}\right)^{-1}$

18. Explain why $-2^{-4} = -1/16$, but $(-2)^{-4} = 1/16$.

Evaluate each expression. Write all answers without exponents. Round decimal answers to two places. (See Examples 5 and 6.)

19. $49^{1/2}$

20. $8^{1/3}$

21. $(5.71)^{1/4}$

22. $12^{5/2}$

23. $-64^{2/3}$

24. $-64^{3/2}$

25. $(8/27)^{-4/3}$

26. $(27/64)^{-1/3}$

Simplify each expression. Write all answers using only positive exponents. (See Example 7.)

27. $\dfrac{5^{-3}}{4^{-2}}$

28. $\dfrac{7^{-4}}{7^{-3}}$

29. $4^{-3} \cdot 4^6$

30. $9^{-9} \cdot 9^{10}$

31. $\dfrac{4^{10} \cdot 4^{-6}}{4^{-4}}$

32. $\dfrac{5^{-4} \cdot 5^6}{5^{-1}}$

Simplify each expression. Assume all variables represent positive real numbers. Write answers with only positive exponents. (See Examples 8 and 9.)

33. $\dfrac{z^6 \cdot z^2}{z^5}$

34. $\dfrac{k^6 \cdot k^9}{k^{12}}$

35. $\dfrac{3^{-1}(p^{-2})^3}{3p^{-7}}$

36. $\dfrac{(5x^3)^{-2}}{x^4}$

37. $(q^{-5}r^3)^{-1}$

38. $(2y^2z^{-2})^{-3}$

39. $(2p^{-1})^3 \cdot (5p^2)^{-2}$

40. $(4^{-1}x^3)^{-2} \cdot (3x^{-3})^4$

41. $(2p)^{1/2} \cdot (2p^3)^{1/3}$

42. $(5k^2)^{3/2} \cdot (5k^{1/3})^{3/4}$

43. $p^{2/3}(2p^{1/3} + 5p)$

44. $3x^{3/2}(2x^{-3/2} + x^{3/2})$

45. $\dfrac{(x^2)^{1/3}(y^2)^{2/3}}{3x^{2/3}y^2}$

46. $\dfrac{(c^{1/2})^3(d^3)^{1/2}}{(c^3)^{1/4}(d^{1/4})^3}$

47. $\dfrac{(7a)^2(5b)^{3/2}}{(5a)^{3/2}(7b)^4}$

48. $\dfrac{(4x)^{1/2}\sqrt{xy}}{x^{3/2}y^2}$

49. $x^{1/2}(x^{2/3} - x^{4/3})$

50. $x^{1/2}(3x^{3/2} + 2x^{-1/2})$

51. $(x^{1/2} + y^{1/2})(x^{1/2} - y^{1/2})$

52. $(x^{1/3} + y^{1/2})(2x^{1/3} - y^{3/2})$

Match the rational-exponent expression in Column I with the equivalent radical expression in Column II. Assume that x is not zero.

I	II
53. $(-3x)^{1/3}$	**(a)** $\dfrac{3}{\sqrt[3]{x}}$
54. $-3x^{1/3}$	**(b)** $-3\sqrt[3]{x}$
55. $(-3x)^{-1/3}$	**(c)** $\dfrac{1}{\sqrt[3]{3x}}$
56. $-3x^{-1/3}$	**(d)** $\dfrac{-3}{\sqrt[3]{x}}$
57. $(3x)^{1/3}$	**(e)** $3\sqrt[3]{x}$
58. $3x^{-1/3}$	**(f)** $\sqrt[3]{-3x}$
59. $(3x)^{-1/3}$	**(g)** $\sqrt[3]{3x}$
60. $3x^{1/3}$	**(h)** $\dfrac{1}{\sqrt[3]{-3x}}$

Simplify each of the given radical expressions. (See Examples 11–13.)

61. $\sqrt[3]{125}$

62. $\sqrt[6]{64}$

63. $\sqrt[4]{625}$

64. $\sqrt[7]{-128}$

65. $\sqrt{63}\sqrt{7}$

66. $\sqrt[3]{81} \cdot \sqrt[3]{9}$

67. $\sqrt{81 - 4}$

68. $\sqrt{49 - 16}$

69. $\sqrt{5}\sqrt{15}$

70. $\sqrt{8}\sqrt{96}$

71. $\sqrt{50} - \sqrt{72}$

72. $\sqrt{75} + \sqrt{192}$

73. $5\sqrt{20} - \sqrt{45} + 2\sqrt{80}$

74. $(\sqrt{3} + 2)(\sqrt{3} - 2)$

75. $(\sqrt{5} + \sqrt{2})(\sqrt{5} - \sqrt{2})$

76. What is wrong with the statement $\sqrt[3]{4} \cdot \sqrt[3]{4} = 4$?

Rationalize the denominator of each of the given expressions. (See Example 14.)

77. $\dfrac{3}{1 - \sqrt{2}}$

78. $\dfrac{2}{1 + \sqrt{5}}$

79. $\dfrac{9 - \sqrt{3}}{3 - \sqrt{3}}$

80. $\dfrac{\sqrt{3} - 1}{\sqrt{3} - 2}$

Rationalize the numerator of each of the given expressions. (See Example 15.)

81. $\dfrac{3 - \sqrt{2}}{3 + \sqrt{2}}$

82. $\dfrac{1 + \sqrt{7}}{2 - \sqrt{3}}$

The following exercises are applications of exponentiation and radicals.

83. **Lot Size** The theory of economic lot size shows that, under certain conditions, the number of units to order to minimize total cost is

$$x = \sqrt{\dfrac{kM}{f}},$$

where k is the cost to store one unit for one year, f is the (constant) setup cost to manufacture the product, and M is the total number of units produced annually. Find x for the following values of f, k, and M.

(a) $k = \$1, f = \$500, M = 100{,}000$

(b) $k = \$3, f = \$7, M = 16{,}700$

(c) $k = \$1, f = \$5, M = 16{,}800$

84. **Health** The threshold weight T for a person is the weight above which the risk of death increases greatly. One researcher found that the threshold weight in pounds for men aged 40–49 is related to height in inches by the equation $h = 12.3T^{1/3}$. What height corresponds to a threshold of 216 pounds for a man in this age group?

Box Office *The annual domestic revenue (in billions of dollars) generated by the sale of movie tickets can be approximated by the expression*

$$6.67x^{.188} \quad (x \geq 5),$$

where $x = 5$ corresponds to 2005. Assuming the model remains accurate, approximate the revenue in the following years. (Data from: www.the-numbers.com.)

85. 2015

86. 2018

87. 2020

88. 2023

Facebook Revenue *The quarterly advertising revenue (in billions of dollars) earned by Facebook can be approximated by the expression*

$$1.08x^{.527} \quad (x \geq 1),$$

where $x = 1$ corresponds to the first quarter of the year 2013. Assuming the model continues to be accurate, find the approximate quarterly revenue for the following quarters. (Data from: The Wall Street Journal.)

89. Q1 2016 ($x = 13$)

90. Q2 2017 ($x = 18$)

91. Q4 2018

92. Q3 2019

Part-Time College Students *The number of students attending college or university (in millions) on a part-time basis can be approximated by the expression*

$$2.72x^{.238}, \quad (x \geq 10),$$

where $x = 10$ corresponds to the year 1990. Assuming the model remains accurate, find the number of part-time students for the following years. (Data from: U.S. National Center for Education Statistics.)

93. 2009

94. 2014

95. 2018

96. 2023

✓ Checkpoint Answers

1. (a) 2^9 (b) $(-5)^4$ (c) $(xy)^5$

2. (a) $81x^4$ (b) $r^{12}s^{30}$

(c) $\dfrac{32}{z^5}$ (d) $\dfrac{9a^4}{b^6}$

3. (a) $1/36$ (b) $-1/216$

(c) $-1/81$ (d) $8/5$

4. (a) $8/5$ (b) 32 (c) b^3/a^6

5. (a) 4 (b) 2

(c) -16 (d) Not a real number

(e) -2 (f) 3

6. (a) 8 (b) 3125

(c) 128 (d) 1000

7. (a) 81 (b) $15x^{3/2}$

(c) 8^{10} (d) $5^{-8/3}$ or $1/5^{8/3}$

8. (a) 1 (b) $c^2/d^{1/2}$ (c) $2a + a^{1/2}$

9. (a) $\dfrac{6}{x^{1/3}y^{2/3}}$ (b) t^3

(c) $32/(p^{10}k^{7/3})$ (d) $\dfrac{y^2 - x}{xy^2}$

10. About 21.1 million

11. (a) 3 (b) 5

(c) 2 (d) $4/5$

12. (a) 9 (b) $\dfrac{\sqrt{3}}{7}$ (c) $11\sqrt{2}$

13. (a) $3\sqrt{5} + \sqrt{10} - 3\sqrt{2} - 2$ (b) -4

14. (a) $\dfrac{2\sqrt{5}}{5}$ (b) $2 - \sqrt{3}$

1.6 First-Degree Equations

An **equation** is a statement that two mathematical expressions are equal; for example,

$$5x - 3 = 13, \qquad 8y = 4, \qquad \text{and} \qquad -3p + 5 = 4p - 8$$

are equations.

The letter in each equation is called the variable. This section concentrates on **first-degree equations,** which are equations that involve only constants and the first power of the variable. All of the equations displayed above are first-degree equations, but neither of the following equations is of first degree:

$$2x^2 = 5x + 6 \qquad \text{(the variable has an exponent greater than 1);}$$
$$\sqrt{x + 2} = 4 \qquad \text{(the variable is under the radical).}$$

A **solution** of an equation is a number that can be substituted for the variable in the equation to produce a true statement. For example, substituting the number 9 for x in the equation $2x + 1 = 19$ gives

$$2x + 1 = 19$$
$$2(9) + 1 \overset{?}{=} 19 \qquad \text{Let } x = 9.$$
$$18 + 1 = 19. \qquad \text{True}$$

✓ **Checkpoint 1**

Is -4 a solution of the equations in parts (a) and (b)?

(a) $3x + 5 = -7$

(b) $2x - 3 = 5$

(c) Is there more than one solution of the equation in part (a)?

This true statement indicates that 9 is a solution of $2x + 1 = 19$. ✓₁

The following properties are used to solve equations.

Properties of Equality

1. The same number may be added to or subtracted from both sides of an equation:

If $a = b$, then $a + c = b + c$ and $a - c = b - c$.

2. Both sides of an equation may be multiplied or divided by the same nonzero number:

If $a = b$ and $c \neq 0$, then $ac = bc$ and $\dfrac{a}{c} = \dfrac{b}{c}$.

EXAMPLE 1 Solve the equation $5x - 3 = 12$.

Solution Using the first property of equality, add 3 to both sides. This isolates the term containing the variable on one side of the equation:

$$5x - 3 = 12$$
$$5x - 3 + 3 = 12 + 3 \qquad \text{Add 3 to both sides.}$$
$$5x = 15.$$

Now arrange for the coefficient of x to be 1 by using the second property of equality:

$$5x = 15$$
$$\frac{5x}{5} = \frac{15}{5} \qquad \text{Divide both sides by 5.}$$
$$x = 3.$$

✓ **Checkpoint 2**

Solve the following.

(a) $3p - 5 = 19$

(b) $4y + 3 = -5$

(c) $-2k + 6 = 2$

The solution of the original equation, $5x - 3 = 12$, is 3. Check the solution by substituting 3 for x in the original equation. ✓₂

EXAMPLE 2   Solve $2k + 3(k - 4) = 2(k - 3)$.

Solution First, simplify the equation by using the distributive property on the left-side term $3(k - 4)$ and right-side term $2(k - 3)$:

$$2k + 3(k - 4) = 2(k - 3)$$
$$2k + 3k - 12 = 2(k - 3) \qquad \text{Distributive property on left side}$$
$$2k + 3k - 12 = 2k - 6 \qquad \text{Distributive property on right side}$$
$$5k - 12 = 2k - 6. \qquad \text{Collect like terms on left side.}$$

One way to proceed is to add $-2k$ to both sides:

$$5k - 12 + (-2k) = 2k - 6 + (-2k) \qquad \text{Add } -2k \text{ to both sides.}$$
$$3k - 12 = -6$$
$$3k - 12 + 12 = -6 + 12 \qquad \text{Add 12 to both sides.}$$
$$3k = 6$$
$$\frac{1}{3}(3k) = \frac{1}{3}(6) \qquad \text{Multiply both sides by } \frac{1}{3}.$$
$$k = 2.$$

The solution is 2. Check this result by substituting 2 for k in the original equation. ✔️₃

✔️ **Checkpoint 3**

Solve the following.

(a) $3(m - 6) + 2(m + 4)$
    $= 4m - 2$

(b) $-2(y + 3) + 4y$
    $= 3(y + 1) - 6$

EXAMPLE 3   **IRAs** The total amount of money y (in trillions of dollars) invested in individual retirement accounts (IRAs) in the United States in year x can be approximated by the equation

$$5(x - 2000) = 15y - 30.$$

Assuming this equation remains valid, use a calculator to determine the year when the total assets are \$9 trillion. (Data from: ProQuest Statistical Abstract of the United States: 2016.)

Solution Let $y = 9$ in the equation and solve for x. To avoid any rounding errors in the intermediate steps, it is often a good idea to do all the algebra first, before using the calculator:

$$5(x - 2000) = 15y - 30$$
$$5(x - 2000) = 15(9) - 30 \qquad \text{Substitute } y = 9.$$
$$5x - 5(2000) = 15(9) - 30 \qquad \text{Distributive property}$$
$$5x = 15(9) - 30 + 5(2000) \qquad \text{Add } 5(2000) \text{ to both sides.}$$
$$x = \frac{15(9) - 30 + 5(2000)}{5}. \qquad \text{Divide both sides by 5.}$$

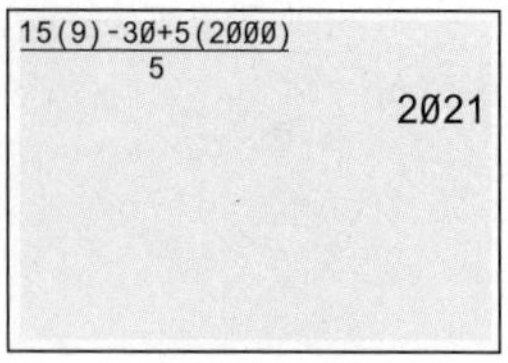

```
15(9)-30+5(2000)
          5
                        2021
```

Figure 1.12

Now use a calculator to determine that $x = 2021$, as shown in Figure 1.12. So in 2021, the total invested in IRAs is \$9 trillion. ✔️₄

✔️ **Checkpoint 4**

In Example 3, what year were the total IRA assets \$6 trillion?

The next three examples show how to simplify the solution of first-degree equations involving fractions. We solve these equations by multiplying both sides of the equation by a *common denominator.* This step will eliminate the fractions.

EXAMPLE 4   Solve

$$\frac{r}{10} - \frac{2}{15} = \frac{3r}{20} - \frac{1}{5}.$$

Solution Here, the denominators are 10, 15, 20, and 5. Each of these numbers is a factor of 60; therefore, 60 is a common denominator. Multiply both sides of the equation by 60:

$$60\left(\frac{r}{10} - \frac{2}{15}\right) = 60\left(\frac{3r}{20} - \frac{1}{5}\right)$$

$$60\left(\frac{r}{10}\right) - 60\left(\frac{2}{15}\right) = 60\left(\frac{3r}{20}\right) - 60\left(\frac{1}{5}\right) \qquad \text{Distributive property}$$

$$6r - 8 = 9r - 12$$

$$6r - 8 + (-6r) + 12 = 9r - 12 + (-6r) + 12 \qquad \text{Add } -6r \text{ and 12 to both sides.}$$

$$4 = 3r$$

$$r = \frac{4}{3}. \qquad \text{Multiply both sides by } 1/3.$$

Check this solution in the original equation.

✓ Checkpoint 5

Solve the following.

(a) $\dfrac{x}{2} - \dfrac{x}{4} = 6$

(b) $\dfrac{2x}{3} + \dfrac{1}{2} = \dfrac{x}{4} - \dfrac{9}{2}$

⚠ CAUTION Multiplying *both* sides of an *equation* by a number to eliminate fractions is valid. But multiplying a single fraction by a number to simplify it is not valid. For instance, multiplying $\dfrac{3x}{8}$ by 8 *changes* it to $3x$, which is *not equal to* $\dfrac{3x}{8}$.

The second property of equality (page 43) applies only to *nonzero* quantities. Multiplying or dividing both sides of an equation by a quantity involving the variable (which might be zero for some values) may lead to an **extraneous solution**—that is, a number that does not satisfy the original equation. To avoid errors in such situations, always *check your solutions in the original equation.*

EXAMPLE 5   Solve

$$\frac{4}{3(k + 2)} - \frac{k}{3(k + 2)} = \frac{5}{3}.$$

Solution Multiply both sides of the equation by the common denominator $3(k + 2)$. Here, $k \neq -2$, since $k = -2$ would give a 0 denominator, making the fraction undefined. So, we have

$$3(k + 2) \cdot \frac{4}{3(k + 2)} - 3(k + 2) \cdot \frac{k}{3(k + 2)} = 3(k + 2) \cdot \frac{5}{3}.$$

Simplify each side and solve for k:

$$4 - k = 5(k + 2)$$

$$4 - k = 5k + 10 \qquad \text{Distributive property}$$

$$4 - k + k = 5k + 10 + k \qquad \text{Add } k \text{ to both sides.}$$

$$4 = 6k + 10$$

$$4 + (-10) = 6k + 10 + (-10) \qquad \text{Add } -10 \text{ to both sides.}$$

$$-6 = 6k$$

$$-1 = k. \qquad \text{Multiply both sides by } \frac{1}{6}.$$

The solution is -1. Substitute -1 for k in the original equation as a check:

$$\frac{4}{3(-1 + 2)} - \frac{-1}{3(-1 + 2)} \overset{?}{=} \frac{5}{3}$$

$$\frac{4}{3} - \frac{-1}{3} \overset{?}{=} \frac{5}{3}$$

$$\frac{5}{3} = \frac{5}{3}.$$

✓ Checkpoint 6

Solve the equation

$$\frac{5p + 1}{3(p + 1)} = \frac{3p - 3}{3(p + 1)} + \frac{9p - 3}{3(p + 1)}.$$

The check shows that -1 is the solution. ✓ 6

EXAMPLE 6 Solve

$$\frac{3x - 4}{x - 2} = \frac{x}{x - 2}.$$

Solution Multiplying both sides by $x - 2$ produces

$$3x - 4 = x$$
$$2x - 4 = 0 \qquad \text{Subtract } x \text{ from both sides.}$$
$$2x = 4 \qquad \text{Add 4 to both sides.}$$
$$x = 2. \qquad \text{Divide both sides by 2.}$$

Substituting 2 for x in the original equation produces fractions with 0 denominators. Since division by 0 is not defined, $x = 2$ is an extraneous solution. So the original equation has no solution. ✓ 7

Sometimes an equation with several variables must be solved for one of the variables. This process is called **solving for a specified variable.**

EXAMPLE 7 Solve for x: $3(ax - 5a) + 4b = 4x - 2$.

Solution Use the distributive property to get

$$3ax - 15a + 4b = 4x - 2.$$

Treat x as the variable and the other letters as constants. Get all terms with x on one side of the equation and all terms without x on the other side:

$$3ax - 4x = 15a - 4b - 2 \qquad \text{Isolate terms with } x \text{ on the left.}$$
$$(3a - 4)x = 15a - 4b - 2 \qquad \text{Distributive property}$$
$$x = \frac{15a - 4b - 2}{3a - 4} \qquad \text{Multiply both sides by } \frac{1}{3a - 4}.$$

The final equation is solved for x, as required. ✓ 8

Absolute-Value Equations

Recall from Section 1.1 that the absolute value of a number a is either a or $-a$, whichever one is nonnegative. For instance, $|4| = 4$ and $|-7| = -(-7) = 7$.

EXAMPLE 8 Solve $|x| = 3$.

Solution Since $|x|$ is either x or $-x$, the equation says that

$$x = 3 \qquad \text{or} \qquad -x = 3$$
$$x = -3.$$

The solutions of $|x| = 3$ are 3 and -3.

EXAMPLE 9 Solve $|p - 4| = 2$.

Solution Since $|p - 4|$ is either $p - 4$ or $-(p - 4)$, we have

$$p - 4 = 2 \qquad \text{or} \qquad -(p - 4) = 2$$
$$p = 6 \qquad\qquad -p + 4 = 2$$
$$-p = -2$$
$$p = 2,$$

so that 6 and 2 are possible solutions. Checking them in the original equation shows that both are solutions. ✓ 9

✓ **Checkpoint 7**

Solve each equation.

(a) $\dfrac{3p}{p + 1} = 1 - \dfrac{3}{p + 1}$

(b) $\dfrac{8y}{y - 4} = \dfrac{32}{y - 4} - 3$

✓ **Checkpoint 8**

Solve for x.

(a) $2x - 7y = 3xk$

(b) $8(4 - x) + 6p = -5k - 11yx$

✓ **Checkpoint 9**

Solve each equation.

(a) $|y| = 9$

(b) $|r + 3| = 1$

(c) $|2k - 3| = 7$

EXAMPLE 10   Solve $|4m - 3| = |m + 6|$.

Solution To satisfy the equation, the quantities in absolute-value bars must either be equal or be negatives of one another. That is,

$$4m - 3 = m + 6 \qquad \text{or} \qquad 4m - 3 = -(m + 6)$$
$$3m = 9 \qquad\qquad\qquad 4m - 3 = -m - 6$$
$$m = 3 \qquad\qquad\qquad 5m = -3$$
$$m = -\frac{3}{5}.$$

Check that the solutions for the original equation are 3 and $-3/5$. ✓₁₀

✓ **Checkpoint 10**

Solve each equation.

(a) $|r + 6| = |2r + 1|$

(b) $|5k - 7| = |10k - 2|$

Applications

One of the main reasons for learning mathematics is to be able to use it to solve practical problems. There are no hard-and-fast rules for dealing with real-world applications, except perhaps to use common sense. However, you will find it much easier to deal with such problems if you do not try to do everything at once. After reading the problem carefully, attack it in stages, as suggested in the following guidelines.

Solving Applied Problems

Step 1   Read the problem carefully, focusing on the facts you are given and the unknown values you are asked to find. Look up any words you do not understand. You may have to read the problem more than once, until you understand exactly what you are being asked to do.

Step 2   Identify the unknown. (If there is more than one, choose one of them, and see Step 3 for what to do with the others.) Name the unknown with some variable that you *write down*. Many students try to skip this step. They are eager to get on with the writing of the equation. But this is an important step. If you do not know what the variable represents, how can you write a meaningful equation or interpret a result?

Step 3   Decide on a variable expression to represent any other unknowns in the problem. For example, if x represents the width of a rectangle, and you know that the length is one more than twice the width, then *write down* the fact that the length is $1 + 2x$.

Step 4   Draw a sketch or make a chart, if appropriate, showing the information given in the problem.

Step 5   Using the results of Steps 1–4, write an equation that expresses a condition that must be satisfied.

Step 6   Solve the equation.

Step 7   Check the solution in the words of the *original problem*, not just in the equation you have written.

EXAMPLE 11   **Finance** A financial manager has $15,000 to invest for her company. She plans to invest part of the money in tax-free bonds at 5% interest and the remainder in taxable bonds at 8%. She wants to earn $1020 per year in interest from the investments. Find the amount she should invest at each rate.

Solution

Step 1   We are asked to find how much of the $15,000 should be invested at 5% and how much at 8%, in order to earn the required interest.

Step 2   Let x represent the amount to be invested at 5%.

Step 3   After x dollars are invested, the remaining amount is $15,000 - x$ dollars, which is to be invested at 8%.

Step 4   To find the amount of interest that is earned for one year, we multiply the rate (converted to a decimal) by the amount invested. For instance, when we invest x dollars at 5%, we convert 5% to .05 and then multiply by x. This means that we calculate $.05x$. If we invest the remaining dollars at 8%, we have $.08(15,000 - x)$. The given information is summarized in the following chart.

Investment	Amount Invested	Interest Rate	Interest Earned in One Year
Tax-free bonds	x	5% = .05	.05x
Taxable bonds	15,000 − x	8% = .08	.08(15,000 − x)
Totals	15,000		1020

Step 5   Because the total interest is to be $1020, the last column of the table shows that

$$.05x + .08(15{,}000 - x) = 1020.$$

Step 6   Solve the preceding equation as follows:

$$.05x + .08(15{,}000 - x) = 1020$$
$$.05x + .08(15{,}000) - .08x = 1020$$
$$.05x + 1200 - .08x = 1020$$
$$-.03x = -180$$
$$x = 6000.$$

The manager should invest $6000 at 5% and $15,000 - $6000 = $9000 at 8%.

Step 7   Check these results in the original problem. If $6000 is invested at 5%, the interest is $.05(6000) = \$300$. If $9000 is invested at 8%, the interest is $.08(9000) = \$720$. So the total interest is $300 + $720 = $1020 as required ✓11

✓ **Checkpoint 11**

An investor owns two pieces of property. One, worth twice as much as the other, returns 6% in annual interest, while the other returns 4%. Find the value of each piece of property if the total annual interest earned is $8000.

EXAMPLE 12   **Music Streaming** According to billboard.com, Apple Music acquired 10 million subscribers in the beginning of 2016. This number took only 6 months to reach, whereas Spotify took 6 years (72 months) to hit that same milestone. Apple Music's average rate of acquiring subscribers per month was 1.53 million more than Spotify's average rate per month. Find the average rate of subscriber acquisition per month for each company.

Solution

Step 1   We must find the subscriber acquisition rate for Apple Music and Spotify.

Step 2   Let x represent the rate of subscriber acquisition rate for Spotify.

Step 3   The acquisition rate for Apple Music is 1.53 million more than Spotify, so its rate is $x + 1.53$.

Step 4   In general, the number of subscribers can be found by the formula:

Number of subscribers = number of months × rate of subscribers per month

For Spotify, the number of subscribers is $72x$ and for Apple Music, the number of subscribers is $6(x + 1.53)$.

We can collect these facts to make a chart that organizes the information given in the problem.

	Months	Rate	Number of Subscribers
Spotify	72	x	$72x$
Apple Music	6	$x + 1.53$	$6(x + 1.53)$

Step 5   Because both companies obtained the *same number of subscribers*, the equation is

$$72x = 6(x + 1.53).$$

Step 6   If we distribute the 6 through the quantity on the right side of the equation, we obtain

$$72x = 6x + 9.18$$
$$66x = 9.18 \qquad \text{Add } -6x \text{ to each side.}$$
$$x \approx .139. \qquad \text{Divide each side by 66.}$$

Step 7   Since x represents Spotify's rate, Spotify acquired, on average .139 million (139,000) users per month. Apple Music's rate is $x + 1.53$, or $.139 + 1.53 = 1.669$ million users (1,669,000) per month. ☑12

✓**Checkpoint 12**

In Example 12, suppose Apple Music had taken 9 months and Spotify had taken 75 months to reach the same number of subscribers. Assuming that Apple Music had acquired subscribers at a rate of 1.35 million per month more than Spotify, find the average rate at which Spotify and Apple Music had acquired subscribers.

EXAMPLE 13   **Business** An oil company needs to fill orders for 89-octane gas, but has only 87- and 93-octane gas on hand. The octane rating is the percentage of isooctane in the standard fuel. How much of each type should be mixed together to produce 100,000 gallons of 89-octane gas?

Solution

Step 1   We must find how much 87-octane gas and how much 93-octane gas are needed for the 100,000 gallon mixture.

Step 2   Let x be the amount of 87-octane gas.

Step 3   Then $100,000 - x$ is the amount of 93-octane gas.

Step 4   We can summarize the relevant information in a chart.

Type of Gas	Quantity	% Isooctane	Amount of Isooctane
87-octane	x	87%	$.87x$
93-octane	$100,000 - x$	93%	$.93(100,000 - x)$
Mixture	100,000	89%	$.89(100,000)$

Step 5   The amount of isooctane satisfies this equation:

$$.87x + .93(100,000 - x) = .89(100,000).$$

Step 6   Solving this equation yields

$$.87x + 93,000 - .93x = .89(100,000) \quad \text{Distribute the left side.}$$
$$.87x + 93,000 - .93x = 89,000 \qquad \text{Multiply on the right side.}$$
$$-.06x = -4000 \qquad \text{Combine terms and add } -93,000 \text{ to each side.}$$
$$x = \frac{-4000}{-.06} \approx 66,667.$$

Step 7   So the distributor should mix 66,667 gallons of 87-octane gas with $100,000 - 66,667 = 33,333$ gallons of 93-octane gas. Then the amount of isooctane in the mixture is

$$.87(66,667) + .93(33,333)$$
$$\approx 58,000 + 31,000$$
$$= 89,000.$$

✓**Checkpoint 13**

How much 89-octane gas and how much 94-octane gas are needed to produce 1500 gallons of 91-octane gas?

Hence, the octane rating of the mixture is $\dfrac{89,000}{100,000} = .89$ as required. ☑13

1.6 Exercises

Solve each equation. (See Examples 1–6.)

1. $3x + 8 = 20$

2. $4 - 5y = 19$

3. $.6k - .3 = .5k + .4$

4. $2.5 + 5.04m = 8.5 - .06m$

5. $2a - 1 = 4(a + 1) + 7a + 5$

6. $3(k - 2) - 6 = 4k - (3k - 1)$

7. $2[x - (3 + 2x) + 9] = 3x - 8$

8. $-2[4(k + 2) - 3(k + 1)] = 14 + 2k$

9. $\dfrac{3x}{5} - \dfrac{4}{5}(x + 1) = 2 - \dfrac{3}{10}(3x - 4)$

10. $\dfrac{4}{3}(x - 2) - \dfrac{1}{2} = 2\left(\dfrac{3}{4}x - 1\right)$

11. $\dfrac{5y}{6} - 8 = 5 - \dfrac{2y}{3}$

12. $\dfrac{x}{2} - 3 = \dfrac{3x}{5} + 1$

13. $\dfrac{m}{2} - \dfrac{1}{m} = \dfrac{6m + 5}{12}$

14. $-\dfrac{3k}{2} + \dfrac{9k - 5}{6} = \dfrac{11k + 8}{k}$

15. $\dfrac{4}{x - 3} - \dfrac{8}{2x + 5} + \dfrac{3}{x - 3} = 0$

16. $\dfrac{5}{2p + 3} - \dfrac{3}{p - 2} = \dfrac{4}{2p + 3}$

17. $\dfrac{3}{2m + 4} = \dfrac{1}{m + 2} - 2$

18. $\dfrac{8}{3k - 9} - \dfrac{5}{k - 3} = 4$

Use a calculator to solve each equation. Round your answer to the nearest hundredth. (See Example 3.)

19. $9.06x + 3.59(8x - 5) = 12.07x + .5612$

20. $-5.74(3.1 - 2.7p) = 1.09p + 5.2588$

21. $\dfrac{2.63r - 8.99}{1.25} - \dfrac{3.90r - 1.77}{2.45} = r$

22. $\dfrac{8.19m + 2.55}{4.34} - \dfrac{8.17m - 9.94}{1.04} = 4m$

Solve each equation for x. (See Example 7.)

23. $4(a + x) = b - a + 2x$

24. $(3a - b) - bx = a(x - 2)$ $(a \neq -b)$

25. $5(b - x) = 2b + ax$ $(a \neq -5)$

26. $bx - 2b = 2a - ax$

Solve each equation for the specified variable. Assume that all denominators are nonzero. (See Example 7.)

27. $PV = k$ for V

28. $i = prt$ for p

29. $V = V_0 + gt$ for g

30. $S = S_0 + gt^2 + k$ for g

31. $A = \dfrac{1}{2}(B + b)h$ for B

32. $C = \dfrac{5}{9}(F - 32)$ for F

Solve each equation. (See Examples 8–10.)

33. $|2h - 1| = 5$

34. $|4m - 3| = 12$

35. $|6 + 2p| = 10$

36. $|-5x + 7| = 15$

37. $\left|\dfrac{5}{r - 3}\right| = 10$

38. $\left|\dfrac{3}{2h - 1}\right| = 4$

Solve the following applied problems.

Natural Science *The equation that relates Fahrenheit temperature F to Celsius temperature C is*

$$C = \dfrac{5}{9}(F - 32).$$

Find the Fahrenheit temperature corresponding to these Celsius temperatures.

39. -5 **40.** -15

41. 22   **42.** 36

Federal Debt *The gross federal debt y (in trillions of dollars) in year x can be approximated by*

$$y = 1.15x + 1.62,$$

where x is the number of years after 2000. (Data from: www.treasury-direct.gov.) Find the federal debt in the following years.

43. 2010   **44.** 2015

Assuming the trend continues, in what year will the federal debt be the amount given in each instance below?

45. 22.32   **46.** 24.62

47. 25.77   **48.** 30.37

Health Economics *The total health care expenditures E in the United States (in trillions of dollars) can be approximated by*

$$E = .108x + 1.517,$$

where x is the number of years after 2000. Assuming the trend continues, determine the year in which health care expenditures are at the given level. (Data from: U.S. Centers for Medicare and Medicaid Services.)

49. $2.705 trillion

50. $3.029 trillion

51. $3.461 trillion

52. $3.893 trillion

Social Security *The amount y (in billions of dollars) of income taken in by the U.S. Social Security Administration for old-age and survivor's insurance in year x can be approximated by the equation*

$$114.8(x - 2010) = 5y - 3390.5.$$

Assuming the trend continues, find the year in which the given income amounts occurred. (Data from: www.ssa.gov.)

53. 746.98

54. 815.86

55. 907.7

56. 1022.5

Finance Charges *When a loan is paid off early, a portion of the finance charge must be returned to the borrower. By one method of calculating the finance charge (called the rule of 78), the amount of unearned interest (finance charge to be returned) is given by*

$$u = f \cdot \frac{n(n + 1)}{q(q + 1)},$$

where u represents unearned interest, f is the original finance charge, n is the number of payments remaining when the loan is paid off, and q is the original number of payments. Find the amount of the unearned interest in each of the given cases.

57. Original finance charge = $800, loan scheduled to run 36 months, paid off with 18 payments remaining

58. Original finance charge = $1400, loan scheduled to run 48 months, paid off with 12 payments remaining

Investing *Solve the following investment problems. (See Example 11.)*

59. Joe Gonzalez received $52,000 profit from the sale of some land. He invested part at 5% interest and the rest at 4% interest. He earned a total of $2290 interest per year. How much did he invest at 5%?

60. Weijen Luan invests $20,000 received from an insurance settlement in two ways: some at 6% and some at 4%. Altogether, she makes $1040 per year in interest. How much is invested at 4%?

61. Maria Martinelli bought two plots of land for a total of $120,000. On the first plot, she made a profit of 15%. On the second, she lost 10%. Her total profit was $5500. How much did she pay for each piece of land?

62. Suppose $20,000 is invested at 5%. How much additional money must be invested at 4% to produce a yield of 4.8% on the entire amount invested?

Solve the given applied problems. (See Example 12.)

Social Media *According to data from businessinsider.com, two social media companies, Twitter and Instagram, had the same number of active users in the fourth quarter of 2014. Twitter took 35 quarters to acquire these users, while Instagram took 21 quarters. Instagram's rate of active user growth was 5.7 million more per quarter, on average, than Twitter's average growth per quarter.*

63. Find the average rate of growth of Twitter.

64. Find the average rate of growth of Instagram.

65. Approximately how many active users did Twitter acquire in 35 quarters?

66. Approximately how many active users did Instagram acquire in 21 quarters?

Natural Science *Using the same assumptions about octane ratings as in Example 13, solve the following problems.*

67. How many liters of 94-octane gasoline should be mixed with 200 liters of 99-octane gasoline to get a mixture that is 97-octane gasoline?

68. A service station has 92-octane and 98-octane gasoline. How many liters of each gasoline should be mixed to provide 12 liters of 96-octane gasoline for a chemistry experiment?

✓ Checkpoint Answers

1. (a) Yes   (b) No   (c) No

2. (a) 8   (b) -2 (c) 2

3. (a) 8   (b) -3

4. 2012

5. (a) 24   (b) -12

6. 1

7. Neither equation has a solution.

8. (a) $x = \dfrac{7y}{2 - 3k}$ (b) $x = \dfrac{5k + 32 + 6p}{8 - 11y}$

9. (a) $9, -9$ (b) $-2, -4$ (c) $5, -2$

10. (a) $5, -7/3$ (b) $-1, 3/5$

11. 6% return: $100,000; 4% return: $50,000

12. (a) Spotify: .184 million per month;
       (b) Apple Music: 1.534 million per month

13. 900 gallons of 89-octane gas; 600 gallons of 94-octane gas

1.7   Quadratic Equations

An equation that can be written in the form

$$ax^2 + bx + c = 0,$$

where a, b, and c are real numbers with $a \neq 0$, is called a **quadratic equation.** For example, each of

$$2x^2 + 3x + 4 = 0, \qquad x^2 = 6x - 9, \qquad 3x^2 + x = 6, \qquad \text{and} \qquad x^2 = 5$$

is a quadratic equation. A solution of an equation that is a real number is said to be a **real solution** of the equation.

One method of solving quadratic equations is based on the following property of real numbers.

Zero-Factor Property

If a and b are real numbers, with $ab = 0$, then $a = 0$ or $b = 0$ or both.

EXAMPLE 1   Solve the equation $(x - 4)(3x + 7) = 0$.

Solution By the zero-factor property, the product $(x - 4)(3x + 7)$ can equal 0 only if at least one of the factors equals 0. That is, the product equals zero only if $x - 4 = 0$ or $3x + 7 = 0$. Solving each of these equations separately will give the solutions of the original equation:

$$x - 4 = 0 \qquad \text{or} \qquad 3x + 7 = 0$$
$$x = 4 \qquad \text{or} \qquad 3x = -7$$
$$x = -\frac{7}{3}.$$

The solutions of the equation $(x - 4)(3x + 7) = 0$ are 4 and $-7/3$. Check these solutions by substituting them into the original equation. ✔1

✓ **Checkpoint 1**

Solve the following equations.

(a) $(y - 6)(y + 2) = 0$

(b) $(5k - 3)(k + 5) = 0$

(c) $(2r - 9)(3r + 5) \cdot (r + 3) = 0$

EXAMPLE 2   Solve $6r^2 + 7r = 3$.

Solution Rewrite the equation as

$$6r^2 + 7r - 3 = 0.$$

Now factor $6r^2 + 7r - 3$ to get

$$(3r - 1)(2r + 3) = 0.$$

By the zero-factor property, the product $(3r - 1)(2r + 3)$ can equal 0 only if

$$3r - 1 = 0 \qquad \text{or} \qquad 2r + 3 = 0.$$

Solving each of these equations separately gives the solutions of the original equation:

$$3r = 1 \quad \text{or} \quad 2r = -3$$
$$r = \frac{1}{3} \qquad\qquad r = -\frac{3}{2}.$$

Verify that both $1/3$ and $-3/2$ are solutions by substituting them into the original equation. ✔2

An equation such as $x^2 = 5$ has two solutions: $\sqrt{5}$ and $-\sqrt{5}$. This fact is true in general.

Square-Root Property

If $b > 0$, then the solutions of $x^2 = b$ are $\sqrt{b}$ and $-\sqrt{b}$.

The two solutions are sometimes abbreviated $\pm\sqrt{b}$.

> ✔ **Checkpoint 2**
>
> Solve each equation by factoring.
>
> (a) $y^2 + 3y = 10$
>
> (b) $2r^2 + 9r = 5$
>
> (c) $4k^2 = 9k$

EXAMPLE 3   Solve each equation.

(a) $m^2 = 17$

Solution By the square-root property, the solutions are $\sqrt{17}$ and $-\sqrt{17}$, abbreviated $\pm\sqrt{17}$.

(b) $(y - 4)^2 = 11$

Solution Using a generalization of the square-root property, we work as follows.

$$(y - 4)^2 = 11$$
$$y - 4 = \sqrt{11} \quad \text{or} \quad y - 4 = -\sqrt{11}$$
$$y = 4 + \sqrt{11} \qquad\qquad y = 4 - \sqrt{11}.$$

Abbreviate the solutions as $4 \pm \sqrt{11}$. ✔3

> ✔ **Checkpoint 3**
>
> Solve each equation by using the square-root property.
>
> (a) $p^2 = 21$
>
> (b) $(m + 7)^2 = 15$
>
> (c) $(2k - 3)^2 = 5$

When a quadratic equation cannot be easily factored, it can be solved by using the following formula, which you should memorize.*

Quadratic Formula

The solutions of the quadratic equation $ax^2 + bx + c = 0$, where $a \neq 0$, are given by

$$x = \frac{-b \pm \sqrt{b^2 - 4ac}}{2a}.$$

🛑 **CAUTION** When using the quadratic formula, remember that the equation must be in the form $ax^2 + bx + c = 0$. Also, notice that the fraction bar in the quadratic formula extends under *both* terms in the numerator. Be sure to add $-b$ to $\pm\sqrt{b^2 - 4ac}$ *before* dividing by $2a$.

---

*A proof of the quadratic formula can be found in many College Algebra books.

EXAMPLE 4   Solve $x^2 + 1 = 4x$.

Solution First add $-4x$ to both sides to get 0 alone on the right side:

$$x^2 - 4x + 1 = 0.$$

Now identify the values of a, b, and c. Here, $a = 1$, $b = -4$, and $c = 1$. Substitute these numbers into the quadratic formula to obtain

$$x = \frac{-(-4) \pm \sqrt{(-4)^2 - 4(1)(1)}}{2(1)}$$

$$= \frac{4 \pm \sqrt{16 - 4}}{2}$$

$$= \frac{4 \pm \sqrt{12}}{2}$$

$$= \frac{4 \pm 2\sqrt{3}}{2} \qquad \sqrt{12} = \sqrt{4 \cdot 3} = \sqrt{4} \cdot \sqrt{3} = 2\sqrt{3}$$

$$= \frac{2(2 \pm \sqrt{3})}{2} \qquad \text{Factor } 4 \pm 2\sqrt{3}.$$

$$x = 2 \pm \sqrt{3}. \qquad \text{Cancel 2.}$$

The $\pm$ sign represents the two solutions of the equation. First use $+$ and then use $-$ to find each of the solutions: $2 + \sqrt{3}$ and $2 - \sqrt{3}$. ✓ 4

Example 4 shows that the quadratic formula produces exact solutions. In many real-world applications, however, you must use a calculator to find decimal approximations of the solutions. The approximate solutions in Example 4 are

$$x = 2 + \sqrt{3} \approx 3.732050808 \qquad \text{and} \qquad x = 2 - \sqrt{3} \approx .2679491924,$$

as shown in Figure 1.13.

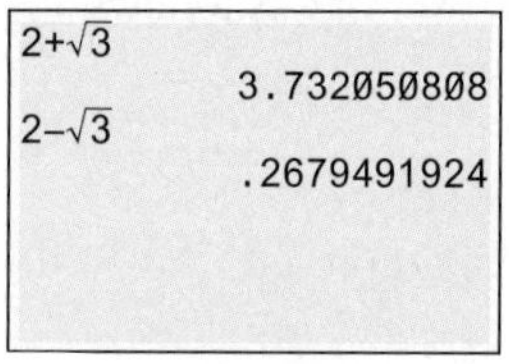

EXAMPLE 5   **Ford Revenue** Many companies in recent years saw their revenue follow a parabolic pattern because of the recession that began in 2008. One such example is Ford Motor Corp. The revenue R for Ford (in millions of dollars) can be approximated by the equation

$$R = 1.451x^2 - 31.84x + 306, \quad (6 \leq x \leq 15)$$

where $x = 6$ corresponds to the year 2006. Use the quadratic formula and a calculator to find the first full year in which revenue recovered back to at least \$145 million. (Data from: www.morningstar.com.)

Solution To find the year x, solve the equation above when $R = 145$:

$$1.451x^2 - 31.84x + 306 = R$$

$$1.451x^2 - 31.84x + 306 = 145 \qquad \text{Substitute } R = 145.$$

$$1.451x^2 - 31.84x + 161 = 0. \qquad \text{Subtract 145 from both sides.}$$

If you are using a scientific calculator, first compute the radical part to apply the quadratic formula:

$$\sqrt{b^2 - 4ac} = \sqrt{(-31.84)^2 - 4(1.451)(161)} = \sqrt{79.3416}.$$

Then store $\sqrt{79.3416}$ (which we denote by M) in the calculator memory. (Check your instruction manual for how to store and recall numbers.) By the quadratic formula, the exact solutions to the equation are

$$x = \frac{-b \pm \sqrt{b^2 - 4ac}}{2a} = \frac{-b \pm M}{2a} = \frac{-(-31.84) \pm M}{2(1.451)} = \frac{31.84 \pm M}{2.902}.$$

✓ **Checkpoint 4**

Use the quadratic formula to solve each equation.

(a) $x^2 - 2x = 2$

(b) $u^2 - 6u + 4 = 0$

```
2+√3
            3.732050808
2−√3
            .2679491924
```

Figure 1.13

Figure 1.14

✓ **Checkpoint 5**

Use the equation of Example 5 to find the year in which revenue was first $137 million.

▦ **TECHNOLOGY TIP**

You can approximate the solutions of quadratic equations on a graphing calculator by using a built-in quadratic equation solver if your calculator has one. Then you need only enter the values of the coefficients a, b, and c to obtain the approximate solutions.

✓ **Checkpoint 6**

Solve each equation.

(a) $9k^2 - 6k + 1 = 0$

(b) $4m^2 + 28m + 49 = 0$

(c) $2x^2 - 5x + 5 = 0$

✓ **Checkpoint 7**

Use the discriminant to determine the number of real solutions of each equation.

(a) $x^2 + 8x + 3 = 0$

(b) $2x^2 + x + 3 = 0$

(c) $x^2 - 194x + 9409 = 0$

The approximate solutions are

$$\frac{31.84 + M}{2.902} \approx 14.04 \quad \text{and} \quad \frac{31.84 - M}{2.902} \approx 7.90.$$

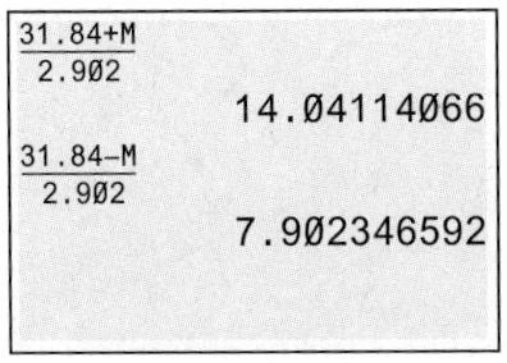

If you are using a graphing calculator, you can find the same results, as shown in Figure 1.14. We were told that revenue had recovered, so we use the solution later in time. The question also wants to know the "first full year" revenue recovered, and so we do not follow regular rounding rules. In this case round up to $x = 15$, which corresponds to 2015. ✓₅

EXAMPLE 6   Solve $9x^2 - 30x + 25 = 0$.

Solution Applying the quadratic formula with $a = 9$, $b = -30$, and $c = 25$, we have

$$x = \frac{-(-30) \pm \sqrt{(-30)^2 - 4(9)(25)}}{2(9)}$$

$$= \frac{30 \pm \sqrt{900 - 900}}{18} = \frac{30 \pm 0}{18} = \frac{30}{18} = \frac{5}{3}.$$

Therefore, the given equation has only one real solution. The fact that the solution is a rational number indicates that this equation could have been solved by factoring.

EXAMPLE 7   Solve $x^2 - 6x + 10 = 0$.

Solution Apply the quadratic formula with $a = 1$, $b = -6$, and $c = 10$:

$$x = \frac{-(-6) \pm \sqrt{(-6)^2 - 4(1)(10)}}{2(1)}$$

$$= \frac{6 \pm \sqrt{36 - 40}}{2}$$

$$= \frac{6 \pm \sqrt{-4}}{2}.$$

Since no negative number has a square root in the real-number system, $\sqrt{-4}$ is not a real number. Hence, the equation has no real solutions. ✓₆

Examples 4–7 show that the number of solutions of the quadratic equation $ax^2 + bx + c = 0$ is determined by $b^2 - 4ac$, the quantity under the radical, which is called the **discriminant** of the equation. ✓₇

The Discriminant

The equation $ax^2 + bx + c = 0$ has either two, or one, or no real solutions.

If $b^2 - 4ac > 0$, there are two real solutions. (*Examples 4 and 5*)

If $b^2 - 4ac = 0$, there is one real solution. (*Example 6*)

If $b^2 - 4ac < 0$, there are no real solutions. (*Example 7*)

Applications

Quadratic equations arise in a variety of settings, as illustrated in the next set of examples. Example 8 depends on the following useful fact from geometry.

The Pythagorean Theorem

In a right triangle with legs of lengths a and b and hypotenuse of length c,

$$a^2 + b^2 = c^2.$$

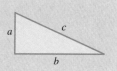

EXAMPLE 8   **Screen Size** The size of a flat screen television is the diagonal measurement of its screen. The height of the screen is approximately 63% of its width. Kathy claims that John's 39-inch flat screen has only half of the viewing area that her 46-inch flat screen television has. Is Kathy right?

Solution First, find the area of Kathy's screen. Let x be its width. Then its height is 63% of x (that is, $.63x$, as shown in Figure 1.15). By the Pythagorean theorem,

$$x^2 + (.63x)^2 = 46^2$$
$$x^2 + .3969x^2 = 2116 \qquad \text{Expand } (.63x)^2 \text{ and } 46^2.$$
$$(1 + .3969)x^2 = 2116 \qquad \text{Distributive property}$$
$$1.3969x^2 = 2116$$
$$x^2 = 1514.78 \qquad \text{Divide both sides by 1.3969.}$$
$$x = \pm\sqrt{1514.78} \qquad \text{Square-root property}$$
$$x \approx \pm38.9.$$

We can ignore the negative solution, since x is a width. Thus, the width is 38.9 inches and the height is $.63x = .63(38.9) \approx 24.5$ inches, so the area is

$$\text{Area} = \text{width} \times \text{height} = 38.9 \times 24.5 = 953.1 \text{ square inches.}$$

Next we find the area of John's television in a similar manner. Let y be the width of John's television. Then its height is 63% of y. By the Pythagorean theorem,

$$y^2 + (.63y)^2 = 39^2$$
$$y^2 + .3969y^2 = 1521 \qquad \text{Expand } (.63y)^2 \text{ and } 39^2.$$
$$(1 + .3969)y^2 = 1521 \qquad \text{Distributive property}$$
$$1.3969y^2 = 1521$$
$$y^2 = 1088.84 \qquad \text{Divide both sides by 1.3969.}$$
$$y = \pm\sqrt{1088.84} \qquad \text{Square-root property}$$
$$y \approx \pm33.0.$$

Again, we can ignore the negative solution, since y is a width. Thus, the width is 33.0 inches and the height is $.63y = .63(33.0) \approx 20.8$ inches, so the area is

$$\text{Area} = \text{width} \times \text{height} = 33.0 \times 20.8 = 686.4 \text{ square inches.}$$

Since half of 953.1 square inches (the area of Kathy's television) is 476.55 square inches, Kathy is wrong because John's viewing area is 686.4 square inches. ✓8

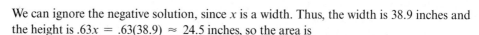

Figure 1.15

✓**Checkpoint 8**

If John's television was only a 26-inch flat screen television, would Kathy have been right?

EXAMPLE 9   **Landscaping** A landscape architect wants to make an exposed gravel border of uniform width around a small shed behind a company plant. The shed is 10 feet by 6 feet. He has enough gravel to cover 36 square feet. How wide should the border be?

Solution A sketch of the shed with border is given in Figure 1.16. Let x represent the width of the border. Then the width of the large rectangle is $6 + 2x$, and its length is $10 + 2x$.

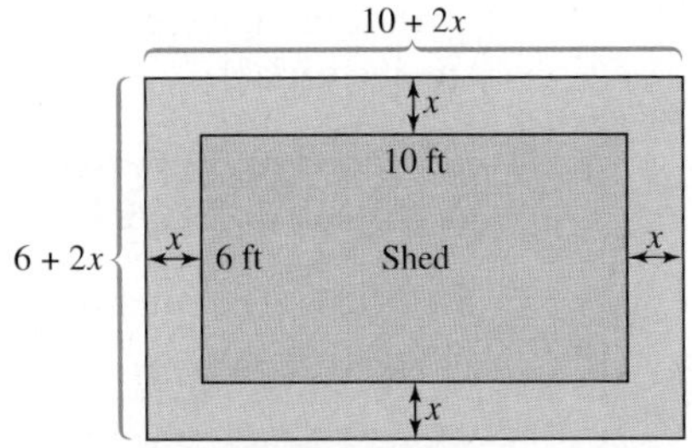

Figure 1.16

We must write an equation relating the given areas and dimensions. The area of the large rectangle is $(6 + 2x)(10 + 2x)$. The area occupied by the shed is $6 \cdot 10 = 60$. The area of the border is found by subtracting the area of the shed from the area of the large rectangle. This difference should be 36 square feet, giving the equation

$$(6 + 2x)(10 + 2x) - 60 = 36.$$

Solve this equation with the following sequence of steps:

$60 + 32x + 4x^2 - 60 = 36$	Multiply out left side.
$4x^2 + 32x - 36 = 0$	Simplify.
$x^2 + 8x - 9 = 0$	Divide both sides by 4.
$(x + 9)(x - 1) = 0$	Factor.
$x + 9 = 0 \quad$ or $\quad x - 1 = 0$	Zero-factor property.
$x = -9 \quad$ or $\quad x = 1.$	

The number -9 cannot be the width of the border, so the solution is to make the border 1 foot wide. ✓ 9

✓ **Checkpoint 9**

The length of a picture is 2 inches more than the width. It is mounted on a mat that extends 2 inches beyond the picture on all sides. What are the dimensions of the picture if the area of the mat is 99 square inches?

EXAMPLE 10 **Physical Science** If an object is thrown upward, dropped, or thrown downward and travels in a straight line subject only to gravity (with wind resistance ignored), the height h of the object above the ground (in feet) after t seconds is given by

$$h = -16t^2 + v_0 t + h_0,$$

where h_0 is the height of the object when $t = 0$ and v_0 is the initial velocity at time $t = 0$. The value of v_0 is taken to be positive if the object moves upward and negative if it moves downward. Suppose that a golf ball is thrown downward from the top of a 625-foot-high building with an initial velocity of 65 feet per second. How long does it take to reach the ground?

Solution In this case, $h_0 = 625$ (the height of the building) and $v_0 = -65$ (negative because the ball is thrown downward). The object is on the ground when $h = 0$, so we must solve the equation

$$h = -16t^2 + v_0 t + h_0$$
$$0 = -16t^2 - 65t + 625. \qquad \text{Let } h = 0, v_0 = -65, \text{ and } h_0 = 625.$$

Using the quadratic formula and a calculator, we see that

$$t = \frac{-(-65) \pm \sqrt{(-65)^2 - 4(-16)(625)}}{2(-16)} = \frac{65 \pm \sqrt{44{,}225}}{-32} \approx \begin{cases} -8.60 \\ \text{or} \\ 4.54 \end{cases}.$$

Only the positive answer makes sense in this case. So it takes about 4.54 seconds for the ball to reach the ground.

In some applications, it may be necessary to solve an equation in several variables for a specific variable.

EXAMPLE 11 Solve $v = mx^2 + x$ for x. (Assume that m and v are positive.)

Solution The equation is quadratic in x because of the x^2 term. Before we use the quadratic formula, we write the equation in standard form:

$$v = mx^2 + x$$
$$0 = mx^2 + x - v.$$

Let $a = m$, $b = 1$, and $c = -v$. Then the quadratic formula gives

$$x = \frac{-1 \pm \sqrt{1^2 - 4(m)(-v)}}{2m}$$

$$x = \frac{-1 \pm \sqrt{1 + 4mv}}{2m}.$$ ✓10

✓**Checkpoint 10**

Solve each of the given equations for the indicated variable. Assume that all variables are positive.

(a) $k = mp^2 - bp$ for p

(b) $r = \dfrac{APk^2}{3}$ for k

1.7 Exercises

Use factoring to solve each equation. (See Examples 1 and 2.)

1. $(x + 4)(x - 14) = 0$
2. $(p - 16)(p - 5) = 0$
3. $x(x + 6) = 0$
4. $x^2 - 2x = 0$
5. $2z^2 = 4z$
6. $x^2 - 64 = 0$
7. $y^2 + 15y + 56 = 0$
8. $k^2 - 4k - 5 = 0$
9. $2x^2 = 7x - 3$
10. $2 = 15z^2 + z$
11. $6r^2 + r = 1$
12. $3y^2 = 16y - 5$
13. $2m^2 + 20 = 13m$
14. $6a^2 + 17a + 12 = 0$
15. $m(m + 7) = -10$
16. $z(2z + 7) = 4$
17. $9x^2 - 16 = 0$
18. $36y^2 - 49 = 0$
19. $16x^2 - 16x = 0$
20. $12y^2 - 48y = 0$

Solve each equation by using the square-root property. (See Example 3.)

21. $(r - 2)^2 = 7$
22. $(b + 4)^2 = 27$
23. $(4x - 1)^2 = 20$
24. $(3t + 5)^2 = 11$

Use the quadratic formula to solve each equation. If the solutions involve square roots, give both the exact and approximate solutions. (See Examples 4–7.)

25. $2x^2 + 7x + 1 = 0$
26. $3x^2 - x - 7 = 0$
27. $4k^2 + 2k = 1$
28. $r^2 = 3r + 5$
29. $5y^2 + 5y = 2$
30. $2z^2 + 3 = 8z$
31. $6x^2 + 6x + 4 = 0$
32. $3a^2 - 2a + 2 = 0$
33. $2r^2 + 3r - 5 = 0$
34. $8x^2 = 8x - 3$
35. $2x^2 - 7x + 30 = 0$
36. $3k^2 + k = 6$
37. $1 + \dfrac{7}{2a} = \dfrac{15}{2a^2}$
38. $5 - \dfrac{4}{k} - \dfrac{1}{k^2} = 0$

Use the discriminant to determine the number of real solutions of each equation. You need not solve the equations.

39. $25t^2 + 49 = 70t$
40. $9z^2 - 12z = 1$
41. $13x^2 + 24x - 5 = 0$
42. $20x^2 + 19x + 5 = 0$

Use a calculator and the quadratic formula to find approximate solutions of each equation. (See Example 5.)

43. $4.42x^2 - 10.14x + 3.79 = 0$
44. $3x^2 - 82.74x + 570.4923 = 0$
45. $7.63x^2 + 2.79x = 5.32$
46. $8.06x^2 + 25.8726x = 25.047256$

Solve the following problems. (See Examples 3, 4, and 5.)

Trucking Pay *According to data from the National Transportation Institute, the average annual pay for truckers P (in thousands of dollars) can be approximated by*

$$P = .396x^2 - 7.09x + 74.9 \qquad (5 \le x \le 15),$$

where x is the number of years since 2000. Find the pay for the following years.

47. 2010
48. 2015

Assuming the trend continues, find the following.

49. The first full year when annual pay was below $48 thousand.
50. The first full year when annual pay recovered to be above $48 thousand.
51. The first full year when annual pay recovered to be above $55 thousand.
52. The first full year when annual pay recovered to be above $60 thousand.

Oil Prices *According to data from the U.S. Energy Information Administration, the price O in real dollars for an imported barrel of crude oil can be approximated by*

$$O = .867x^2 - 9.31x + 63.32 \qquad (1 \le x \le 8),$$

where $x = 1$ corresponds to the first quarter of 2015, and so forth. Find the price of imported crude oil for the following quarters.

53. Third quarter of 2015

54. Second quarter of 2016

55. Determine the most recent full quarter where the price was at least $45.

56. Determine the most recent full quarter where the price was at least $55.

Solve the following problems. (See Examples 8 and 9).

57. **Physical Science** A 13-foot-long ladder leans on a wall, as shown in the accompanying figure. The bottom of the ladder is 5 feet from the wall. If the bottom is pulled out 2 feet farther from the wall, how far does the top of the ladder move down the wall? [*Hint:* Draw pictures of the right triangle formed by the ladder, the ground, and the wall before and after the ladder is moved. In each case, use the Pythagorean theorem to find the distance from the top of the ladder to the ground.]

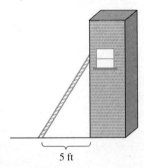

5 ft

58. **Physical Science** A 15-foot-long pole leans against a wall. The bottom is 9 feet from the wall. How much farther should the bottom be pulled away from the wall so that the top moves the same amount down the wall?

59. **Physical Science** Two trains leave the same city at the same time, one going north and the other east. The eastbound train travels 20 mph faster than the northbound one. After 5 hours, the trains are 300 miles apart. Determine the speed of each train, using the following steps.

    (a) Let x denote the speed of the northbound train. Express the speed of the eastbound train in terms of x.

    (b) Write expressions that give the distance traveled by each train after 5 hours.

    (c) Use part (b) and the fact that the trains are 300 miles apart after 5 hours to write an equation. (A diagram of the situation may help.)

    (d) Solve the equation and determine the speeds of the trains.

60. **Physical Science** Tyrone and Miguel have received walkie-talkies for Christmas. If they leave from the same point at the same time, Tyrone walking north at 2.5 mph and Miguel walking east at 3 mph, how long will they be able to talk to each other if the range of the walkie-talkies is 4 miles? Round your answer to the nearest minute.

61. **Landscaping** An ecology center wants to set up an experimental garden. It has 300 meters of fencing to enclose a rectangular area of 5000 square meters. Find the length and width of the rectangle as follows.

    (a) Let $x =$ the length and write an expression for the width.

    (b) Write an equation relating the length, width, and area, using the result of part (a).

    (c) Solve the problem.

62. **Landscaping** A landscape architect has included a rectangular flower bed measuring 9 feet by 5 feet in her plans for a new building. She wants to use two colors of flowers in the bed, one in the center and the other for a border of the same width on all four sides. If she can get just enough plants to cover 24 square feet for the border, how wide can the border be?

63. **Physical Science** Joan wants to buy a rug for a room that is 12 feet by 15 feet. She wants to leave a uniform strip of floor around the rug. She can afford 108 square feet of carpeting. What dimensions should the rug have?

64. **Indy 500** In 2016, Alexander Rossi won the 500-mile Indianapolis 500 race. His speed (rate) was, on average, 92 miles per hour faster than that of the 1911 winner, Ray Harroun. Rossi completed the race in 3.72 hours less time than Harroun. Find Harroun's and Rossi's rates to the nearest tenth.

Physical Science *Use the height formula in Example 10 to work the given problems. Note that an object that is dropped (rather than thrown downward) has initial velocity $v_0 = 0$.*

65. How long does it take a baseball to reach the ground if it is dropped from the top of a 625-foot-high building? Compare the answer with that in Example 10.

66. After the baseball in Exercise 65 is dropped, how long does it take for the ball to fall 196 feet? (*Hint:* How high is the ball at that time?)

67. You are standing on a cliff that is 200 feet high. How long will it take a rock to reach the ground if

    (a) you drop it?

    (b) you throw it downward at an initial velocity of 40 feet per second?

    (c) How far does the rock fall in 2 seconds if you throw it downward with an initial velocity of 40 feet per second?

68. A rocket is fired straight up from ground level with an initial velocity of 800 feet per second.

    (a) How long does it take the rocket to rise 3200 feet?

    (b) When will the rocket hit the ground?

69. A ball is thrown upward from ground level with an initial velocity of 64 ft per second. In how many seconds will the ball be at the given height?

    (a) 64 ft   (b) 39 ft

    (c) Why are two answers possible in part (b)?

70. **Physical Science** A ball is thrown upward from ground level with an initial velocity of 100 feet per second. In how many seconds will the ball first reach the given height?

    (a) 50 ft   (b) 35 ft

Solve each of the given equations for the indicated variable. Assume that all denominators are nonzero and that all variables represent positive real numbers. (See Example 11.)

71. $S = \dfrac{1}{2}gt^2$ for t

72. $a = \pi r^2$ for r

73. $L = \dfrac{d^4 k}{h^2}$ for h

74. $F = \dfrac{kMv^2}{r}$ for v

75. $P = \dfrac{E^2 R}{(r + R)^2}$ for R

76. $S = 2\pi rh + 2\pi r^2$ for r

3. **(a)** $\pm\sqrt{21}$ **(b)** $-7 \pm \sqrt{15}$ **(c)** $(3 \pm \sqrt{5})/2$

4. **(a)** $x = 1 + \sqrt{3}$ or $1 - \sqrt{3}$
   (b) $u = 3 + \sqrt{5}$ or $3 - \sqrt{5}$

5. Just at the end of 2008

6. **(a)** $1/3$ **(b)** $-7/2$ **(c)** No real solutions

7. **(a)** 2   **(b)** 0   **(c)** 1

8. Yes, area $= 304.92$ sq. in.

9. 5 inches by 7 inches

10. **(a)** $p = \dfrac{b \pm \sqrt{b^2 + 4mk}}{2m}$

   (b) $k = \pm\sqrt{\dfrac{3r}{AP}}$ or $\dfrac{\pm\sqrt{3rAP}}{AP}$

✓ Checkpoint Answers

1. **(a)** $6, -2$ **(b)** $3/5, -5$ **(c)** $9/2, -5/3, -3$

2. **(a)** $2, -5$ **(b)** $1/2, -5$ **(c)** $9/4, 0$

CHAPTER 1 Summary and Review

Key Terms and Symbols

1.1 $\approx$ is approximately equal to
π pi
$|a|$ absolute value of a
real number
natural (counting) number
whole number
integer
rational number
irrational number
properties of real numbers
order of operations
square roots
number line
interval
interval notation
absolute value

1.2 a^n a to the power n
exponent or power

multiplication with exponents
power of a power rule
zero exponent
base
polynomial
variable
coefficient
term
constant term
degree of a polynomial
zero polynomial
leading coefficient
quadratics
cubics
like terms
FOIL
revenue
fixed cost

variable cost
profit

1.3 factor
factoring
greatest common factor
difference of squares
perfect squares
sum and difference of cubes

1.4 rational expression
cancellation property
operations with rational expressions
complex fraction

1.5 $a^{1/n}$ nth root of a
$\sqrt{a}$ square root of a
$\sqrt[n]{a}$ nth root of a
properties of exponents
radical
radicand

index
rationalizing the denominator
rationalizing the numerator

1.6 first-degree equation
solution of an equation
properties of equality
extraneous solution
solving for a specified variable
absolute-value equations
solving applied problems

1.7 quadratic equation
real solution
zero-factor property
square-root property
quadratic formula
discriminant
Pythagorean theorem

Chapter 1 Key Concepts

Factoring

$$x^2 + 2xy + y^2 = (x + y)^2$$
$$x^2 - 2xy + y^2 = (x - y)^2$$
$$x^2 - y^2 = (x + y)(x - y)$$

$$x^3 - y^3 = (x - y)(x^2 + xy + y^2)$$
$$x^3 + y^3 = (x + y)(x^2 - xy + y^2)$$

Properties of Radicals

Let a and b be real numbers, n be a positive integer, and m be any integer for which the given relationships exist. Then

$$a^{m/n} = \sqrt[n]{a^m} = (\sqrt[n]{a})^m;$$ $$\sqrt[n]{a^n} = |a| \text{ if } n \text{ is even};$$ $$\sqrt[n]{a^n} = a \text{ if } n \text{ is odd};$$

$$\sqrt[n]{a} \cdot \sqrt[n]{b} = \sqrt[n]{ab};$$ $$\frac{\sqrt[n]{a}}{\sqrt[n]{b}} = \sqrt[n]{\frac{a}{b}} \quad (b \neq 0).$$

Properties of Exponents

Let a, b, r, and s be any real numbers for which the following exist. Then

$$a^{-r} = \frac{1}{a^r} \qquad\qquad a^0 = 1 \qquad\qquad \left(\frac{a}{b}\right)^r = \frac{a^r}{b^r}$$

$$a^r \cdot a^s = a^{r+s} \qquad\qquad (a^r)^s = a^{rs} \qquad\qquad a^{1/r} = \sqrt[r]{a}$$

$$\frac{a^r}{a^s} = a^{r-s} \qquad\qquad (ab)^r = a^r b^r \qquad\qquad \left(\frac{a}{b}\right)^{-r} = \left(\frac{b}{a}\right)^r$$

Absolute Value

Assume that a and b are real numbers with $b > 0$.

The solutions of $|a| = b$ or $|a| = |b|$ are $a = b$ or $a = -b$.

Quadratic Equations

Facts needed to solve quadratic equations (in which a, b, and c are real numbers):

Factoring If $ab = 0$, then $a = 0$ or $b = 0$ or both.

Square-Root Property If $b > 0$, then the solutions of $x^2 = b$ are $\sqrt{b}$ and $-\sqrt{b}$.

Quadratic Formula The solutions of $ax^2 + bx + c = 0$ (with $a \neq 0$) are

$$x = \frac{-b \pm \sqrt{b^2 - 4ac}}{2a}.$$

Discriminant There are two real solutions of $ax^2 + bx + c = 0$ if $b^2 - 4ac > 0$, one real solution if $b^2 - 4ac = 0$, and no real solutions if $b^2 - 4ac < 0$.

Chapter 1 Review Exercises

Name the numbers from the list $-12, -6, -9/10, -\sqrt{7}, -\sqrt{4},$
$0, 1/8, \pi/4, 6,$ *and* $\sqrt{11}$ *that are*

1. whole numbers
2. integers
3. rational numbers
4. irrational numbers

Identify the properties of real numbers that are illustrated in each of the following expressions.

5. $9[(-3)4] = 9[4(-3)]$

6. $7(4 + 5) = (4 + 5)7$

7. $6(x + y - 3) = 6x + 6y + 6(-3)$

8. $11 + (5 + 3) = (11 + 5) + 3$

Express each statement in symbols.

9. x is at least 9.
10. x is negative.

Write the following numbers in numerical order from smallest to largest.

11. $|6 - 4|, -|-2|, |8 + 1|, -|3 - (-2)|$

12. $\sqrt{7}, -\sqrt{8}, -|\sqrt{16}|, |-\sqrt{12}|$

Write the following without absolute-value bars.

13. $7 - |-8|$
14. $|-3| - |-9 + 6|$

Graph each of the following on a number line.

15. $x \geq -3$
16. $-4 < x \leq 6$

Use the order of operations to simplify each of the following.

17. $\dfrac{-9 + (-6)(-3) \div 9}{6 - (-3)}$
18. $\dfrac{20 \div 4 \cdot 2 \div 5 - 1}{-9 - (-3) - 12 \div 3}$

Perform each of the indicated operations.

19. $(3x^4 - x^2 + 5x) - (-x^4 + 3x^2 - 6x)$

20. $(-8y^3 + 8y^2 - 5y) - (2y^3 + 4y^2 - 10)$

21. $(5k - 2h)(5k + 2h)$
22. $(2r - 5y)(2r + 5y)$

23. $(3x + 4y)^2$
24. $(2a - 5b)^2$

Factor each of the following as completely as possible.

25. $2kh^2 - 4kh + 5k$
26. $2m^2n + 6mn^2 + 16n^2$

27. $5a^4 + 12a^3 + 4a^2$
28. $24x^3 + 4x^2 - 4x$

29. $144p^2 - 169q^2$
30. $81z^2 - 25x^2$

31. $27y^3 - 1$
32. $125a^3 + 216$

Perform each operation.

33. $\dfrac{3x}{5} \cdot \dfrac{45x}{12}$
34. $\dfrac{5k^2}{24} - \dfrac{70k}{36}$

35. $\dfrac{c^2 - 3c + 2}{2c(c - 1)} \div \dfrac{c - 2}{8c}$

36. $\dfrac{p^3 - 2p^2 - 8p}{3p(p^2 - 16)} \div \dfrac{p^2 + 4p + 4}{9p^2}$

37. $\dfrac{2m^2 - 4m + 2}{m^2 - 1} \div \dfrac{6m + 18}{m^2 + 2m - 3}$

38. $\dfrac{x^2 + 6x + 5}{4(x^2 + 1)} \cdot \dfrac{2x(x + 1)}{x^2 - 25}$

Simplify each of the given expressions. Write all answers without negative exponents. Assume that all variables represent positive real numbers.

39. 5^{-3}
40. 10^{-2}

41. -8^0
42. $\left(-\dfrac{5}{6}\right)^{-2}$

43. $4^6 \cdot 4^{-3}$
44. $7^{-5} \cdot 7^{-2}$

45. $\dfrac{8^{-5}}{8^{-4}}$
46. $\dfrac{6^{-3}}{6^4}$

47. $5^{-1} + 2^{-1}$

48. $5^{-2} + 5^{-1}$

49. $\dfrac{5^{1/3} \cdot 5^{1/2}}{5^{3/2}}$

50. $\dfrac{2^{3/4} \cdot 2^{-1/2}}{2^{1/4}}$

51. $(3a^2)^{1/2} \cdot (3^2a)^{3/2}$

52. $(4p)^{2/3} \cdot (2p^3)^{3/2}$

Simplify each of the following expressions.

53. $\sqrt[3]{27}$

54. $\sqrt[6]{-64}$

55. $\sqrt[3]{54p^3q^5}$

56. $\sqrt[4]{64a^5b^3}$

57. $3\sqrt{3} - 12\sqrt{12}$

58. $8\sqrt{7} + 2\sqrt{63}$

Rationalize each denominator.

59. $\dfrac{\sqrt{3}}{1 + \sqrt{2}}$

60. $\dfrac{4 + \sqrt{2}}{4 - \sqrt{5}}$

Solve each equation.

61. $3x - 4(x - 2) = 2x + 9$

62. $4y + 9 = -3(1 - 2y) + 5$

63. $\dfrac{2m}{m - 3} = \dfrac{6}{m - 3} + 4$

64. $\dfrac{15}{k + 5} = 4 - \dfrac{3k}{k + 5}$

Solve for x.

65. $8ax - 3 = 2x$

66. $b^2x - 2x = 4b^2$

Solve each equation.

67. $\left|\dfrac{2 - y}{5}\right| = 8$

68. $|4k + 1| = |6k - 3|$

Find all real solutions of each equation.

69. $(b + 7)^2 = 5$

70. $(2p + 1)^2 = 7$

71. $2p^2 + 3p = 2$

72. $2y^2 = 15 + y$

73. $2q^2 - 11q = 21$

74. $3x^2 + 2x = 16$

75. $6k^4 + k^2 = 1$

76. $21p^4 = 2 + p^2$

Solve each equation for the specified variable.

77. $p = \dfrac{E^2R}{(r + R)^2}$ for r

78. $p = \dfrac{E^2R}{(r + R)^2}$ for E

79. $K = s(s - a)$ for s

80. $kz^2 - hz - t = 0$ for z

Work these problems.

81. Stock Prices The percent change in stock price over the course of 2016 for JCPenny was +27%. For Sears, the percent change in stock price was −55%. Find the difference in the percent change for these two companies in 2016. (Data from: www.morningstar.com.)

82. Stock Prices The two worst performers from the Dow Jones Index in 2016 were Nike and Coca-Cola. Nike's stock price decreased 22% and Coca-Cola's decreased 2.5%. Find the difference in the percent change for these two companies in 2016. (Data from: www.morningstar.com.)

83. Computer Prices. A new laptop computer is on sale for 20% off. The sale price is $1229. What was the original price?

84. Stock Prices. The stock price for eBay increased 27.3% in the last six months of 2016. If the price was $29.83 a share at the end of the year, what was the price six months earlier?

Government Spending *The amount of outlays O of the U.S. government per year (in trillions of dollars) can be approximated by the equation*

$$O = .11x + 2.2,$$

where x = 6 corresponds to the year 2006. Find the year in which the outlays were the following. (Data from: U.S. Office of Management and Budget.)

85. $3.63 trillion

86. $3.96 trillion

Oil Output *The crude oil output (in millions of barrels per day) from Russia can be approximated by the equation*

$$O = .2x + 8,$$

where x = 10 corresponds to the year 2010. In what year was output at the following levels? (Data from: The Wall Street Journal.)

87. 10.8 million barrels per day

88. 11.4 million barrels per day

Vehicle Sales *The number (in millions) of new vehicles V sold in the United States can be approximated by the equation*

$$V = .04x^2 - .47x + 18.38,$$

where x = 1 corresponds to the month of January in 2016. (Data from tradingeconomics.com.)

89. Find the number of vehicles sold in July 2016.

90. Find the number of vehicles sold in October 2016.

91. Find the first full month of the year in which sales were below 17.5 million.

92. Find the first full month where sales recovered to 17.5 million.

Aircraft Departures *The number (in millions) of flight departures F within the United States per year can be approximated by the equation*

$$F = \dfrac{.061x^2 + 6.09x + 38.8}{x + 1},$$

where x = 6 corresponds to the year 2006. (Data from: U.S. Department of Transportation.)

93. How many flights occurred in 2013?

94. If trends continue, how many flights will occur in 2020?

95. In what year did the number of flights first fall below 9.2 million?

96. If trends continue, in what year will the number of flights first fall below 8.9 million?

Research Spending *The total amount (in billions of dollars) spent by the U.S. government on basic research R can be approximated by*

$$R = 16.2x^{.26},$$

where x = 10 corresponds to the year 2010. (Data from: nsf.gov/statistics.) Find the spending in the following years.

97. 2012

98. 2017

99. **Finance** Kinisha borrowed $2000 from a credit union at 12% annual interest and borrowed $500 from her aunt at 7% annual interest. What single rate of interest on $2500 results in the same total amount of interest?

100. **Business** A butcher makes a blend of stew meat from beef that sells for $2.80 a pound and pork that sells for $3.25 a pound. How many pounds of each meat should be used to get 30 pounds of a mix that can be sold for $3.10 a pound?

Federal Debt *The ratio of the U.S. federal debt D as a percentage of gross domestic product (GDP) can be approximated by the equation*

$$D = .047x^2 - 7.69x + 341.6,$$

where $x = 50$ *corresponds to the year 1950. (Data from: U.S. Office of Management and Budget.)*

101. What is the ratio of federal debt to GDP in 2012?

102. What is the most recent year in which the ratio of federal debt to GDP was approximately 55%?

103. **Landscaping** A concrete mixer needs to know how wide to make a walk around a rectangular fountain that is 10 by 15 feet. She has enough concrete to cover 200 square feet. To the nearest tenth of a foot, how wide should the walk be in order to use up all the concrete?

104. **Landscaping** A new homeowner needs to put up a fence to make his yard secure. The house itself forms the south boundary and the garage forms the west boundary of the property, so he needs to fence only 2 sides. The area of the yard is 4000 square feet. He has enough material for 160 feet of fence. Find the length and width of the yard (to the nearest foot).

105. **Physical Science** A projectile is launched from the ground vertically upward at a rate of 150 feet per second. How many seconds will it take for the projectile to be 200 feet from the ground on its downward trip?

106. **Physical Science** Suppose a tennis ball is thrown downward from the top of a 700-foot-high building with an initial velocity of 55 feet per second. How long does it take to reach the ground?

Case Study 1 Energy Efficiency and Long-Term Cost Savings

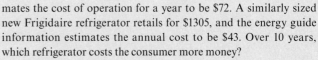

In a series of pioneering studies, researchers conducted famous experiments in which children were tested regarding their ability for delayed gratification.[*] In these experiments, a child would be offered a marshmallow, cookie, or pretzel and told that if they did not eat the treat, they could have two of the treats when the researcher returned (after about 15 minutes). Of course, some children would eat the treat immediately, and others would wait and receive two of the treats.

Similar studies occur in studying consumer behavior among adults. Often when making purchases, consumers will purchase an appliance that initially costs less, but in the long run costs more money when one accounts for the operating expense. Imagine trying to decide whether to buy a new furnace for $4500 that is 80% efficient versus a new furnace that is 90% efficient that costs $5700. Each furnace is expected to last at least 20 years. The furnace with the 90% efficiency is estimated to cost $1800 a year to operate, while the 80% efficiency furnace is estimated to cost $2100 a year to operate—a difference of $300 a year. Thus, the original price differential of $1200 would be made up in four years. Over the course of 16 additional years, the 90% efficiency furnace would end up saving the buyer $4800.

For other appliances, the difference in total expenditures may not be as dramatic, but can still be substantial. The energy guide tags that often accompany new appliances help make it easier for consumers "to do the math" by communicating the yearly operating costs of an appliance. For example, a Kenmore new side-by-side 25.4 cubic foot refrigerator retails for $1128. The energy guide estimates the cost of operation for a year to be $72. A similarly sized new Frigidaire refrigerator retails for $1305, and the energy guide information estimates the annual cost to be $43. Over 10 years, which refrigerator costs the consumer more money?

We can write mathematical expressions to answer this question. With an initial cost of $1128, and an annual operating cost of $72, for x years of operation the cost C for the Kenmore refrigerator is

$$C = 1128 + 72x$$

For 10 years, the Kenmore would cost $C = 1128 + 72(10) = 1848. For the Frigidaire model, the cost is

$$C = 1305 + 43x.$$

For 10 years, the Frigidaire would cost $C = 1305 + 43(10) = 1735.

Thus, the Frigidaire refrigerator costs $113 less in total costs over 10 years.

Behavior when a consumer seeks the up-front savings sounds very similar to the delayed gratification studies among children. These kinds of behavior are of interest to both psychologists and economists. The implications are also very important in the marketing of energy-efficient appliances that often do have a higher initial cost, but save consumers money over the lifetime of the product.

One tool to help consumers make the best choice in the long run is simply to take the time to "just do the math," as the common expression goes. The techniques of this chapter can be applied to do just that!

[*]Mischel, Walter, Ebbe B. Ebbesen, and Antonette Raskoff Zeiss. (1972). "Cognitive and attentional mechanisms in delay of gratification." *Journal of Personality and Social Psychology*, pp. 204–218.

Exercises

1. On the homedepot.com website, a General Electric 40-gallon electric hot water tank retails for $218 and the estimated annual cost of operation is $508. Write an expression for the cost to buy and run the hot water tank for *x* years.

2. On the homedepot.com website, a General Electric 40-gallon gas hot water tank retails for $328 and the estimated annual cost of operation is $309. Write an expression for the cost to buy and run the hot water tank for *x* years.

3. Over ten years, does the electric or gas hot water tank cost more? By how much?

4. In how many years will the total for the two hot water tanks be equal?

5. On the Lowes website, a Maytag 25-cubic-foot refrigerator was advertised for $1529.10 with an estimated cost per month in electricity of $50 a year. Write an expression for the cost to buy and run the refrigerator for *x* years.

6. On the HomeDepot website, an LG Electronics 25-cubic-foot refrigerator was advertised for $1618.20 with an estimated cost per month in electricity of $44 a year. Write an expression for the cost to buy and run the refrigerator for *x* years.

7. Over ten years, which refrigerator (the Maytag or the LG) costs the most? By how much?

8. In how many years will the total for the two refrigerators be equal?

Extended Project

A high-end real estate developer needs to install 50 new washing machines in a building she leases to tenants (with utilities included in the monthly rent). She knows that energy-efficient front loaders cost less per year to run, but they can be much more expensive than traditional top loaders. Investigate by visiting an appliance store or using the Internet, the prices and energy efficiency of two comparable front-loading and top-loading washing machines.

1. Estimate how many years it will take for the total expenditure (including purchase and cost of use) for the 50 front-loading machines to equal the total expenditure for the 50 top-loading machines.

2. The energy guide ratings use data that is often several years old. If the owner believes that current energy costs are 20% higher than what is indicated with the energy guide, redo the calculations for (1) above and determine if it will take less time for the two total expenditures to be equal.

Graphs, Lines, and Inequalities

2

CHAPTER

CHAPTER OUTLINE

2.1 Graphs
2.2 Equations of Lines
2.3 Linear Models
2.4 Linear Inequalities
2.5 Polynomial and Rational Inequalities

CASE STUDY 2
Using Extrapolation for Prediction

Data from current and past events is often a useful tool in business and in the social and health sciences. Gathering data is the first step in developing mathematical models that can be used to analyze a situation and predict future performance. In this chapter, for example, we explore linear models in transportation, stock price, and higher education enrollment.

Graphical representations of data are commonly used in business and in the health and social sciences. Lines, equations, and inequalities play an important role in developing mathematical models from such data. This chapter presents both algebraic and graphical methods for dealing with these topics.

2.1 Graphs

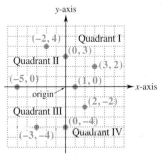

Figure 2.1

Just as the number line associates the points on a line with real numbers, a similar construction in two dimensions associates points in a plane with *ordered pairs* of real numbers. A **Cartesian coordinate system,** as shown in Figure 2.1, consists of a horizontal number line (usually called the **x-axis**) and a vertical number line (usually called the **y-axis**). The point where the number lines meet is called the **origin.** Each point in a Cartesian coordinate system is labeled with an **ordered pair** of real numbers, such as $(-2, 4)$ or $(3, 2)$. Several points and their corresponding ordered pairs are shown in Figure 2.1.

For the point labeled $(-2, 4)$, for example, -2 is the **x-coordinate** and 4 is the **y-coordinate.** You can think of these coordinates as directions telling you how to move to this point from the origin: You go 2 horizontal units to the left (x-coordinate) and 4 vertical units upward (*y*-coordinate). From now on, instead of referring to "the point labeled by the ordered pair $(-2, 4)$," we will say "the point $(-2, 4)$." ✓₁

EXAMPLE 1 **Unemployment Rate** The percentage of unemployed workers in the U.S. labor force in January of recent years is shown in the following table. (Data from: U.S. Department of Labor.)

Year	2012	2013	2014	2015	2016	2017
Rate	8.3	8.0	6.6	5.7	4.9	4.8

Graph the data points from the table with $x = 12$ corresponding to the year 2012.

Solution To make our graph, we need to determine the appropriate "window" that will display our data. The years go from 2012 to 2017, and using the correspondence, we will want our *x*-axis to go from 12 to 17. We see that our *y*-axis will need to go as high as 8.3. If we use a maximum value of 9, we can then plot the ordered pairs:

$$(12, 8.3), (13, 8.0), (14, 6.6), (15, 5.7), (16, 4.9), (17, 4.8),$$

as we see in Figure 2.2.

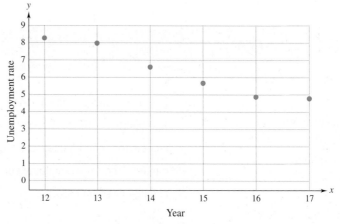

Figure 2.2

The *x*-axis and the *y*-axis divide the plane into four parts, or **quadrants,** which are numbered as shown in Figure 2.1. The points on the coordinate axes belong to no quadrant.

Equations and Graphs

A **solution of an equation** in two variables, such as

$$y = -2x + 3$$

or

$$y = x^2 + 7x - 2,$$

is an ordered pair of numbers such that the substitution of the first number for *x* and the second number for *y* produces a true statement.

EXAMPLE 2 Which of the following are solutions of $y = -2x + 3$?

(a) $(2, -1)$

Solution This is a solution of $y = -2x + 3$ because "$-1 = -2 \cdot 2 + 3$" is a true statement.

(b) $(4, 7)$

Solution Since $-2 \cdot 4 + 3 = -5$, and not 7, the ordered pair $(4, 7)$ is not a solution of $y = -2x + 3$. ✓₂

Equations in two variables, such as $y = -2x + 3$, typically have an infinite number of solutions. To find one, choose a number for x and then compute the value of y that produces a solution. For instance, if $x = 5$, then $y = -2 \cdot 5 + 3 = -7$, so that the pair $(5, -7)$ is a solution of $y = -2x + 3$. Similarly, if $x = 0$, then $y = -2 \cdot 0 + 3 = 3$, so that $(0, 3)$ is also a solution.

The **graph** of an equation in two variables is the set of points in the plane whose coordinates (ordered pairs) are solutions of the equation. Thus, the graph of an equation is a picture of its solutions. Since a typical equation in two variables has infinitely many solutions, its graph has infinitely many points.

> **EXAMPLE 3** Sketch the graph of $y = -2x + 5$.
>
> **Solution** Since we cannot plot infinitely many points, we construct a table of y-values for a reasonable number of x-values, plot the corresponding points, and make an "educated guess" about the rest. The table of values and points in Figure 2.3a suggests that the graph is a straight line, as shown in Figure 2.3b.

✓ **Checkpoint 3**

Graph $x = 5y$.

x	$-2x + 5$
-1	7
0	5
2	1
4	-3
5	-5

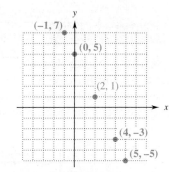

Figure 2.3a

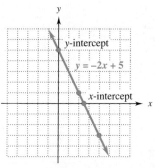

Figure 2.3b

An **x-intercept** of a graph is the x-coordinate of a point where the graph intersects the x-axis. The y-coordinate of this point is 0, since it is on the x-axis. Consequently, to find the x-intercepts of the graph of an equation, set $y = 0$ and solve for x. For instance, in Example 3, the x-intercept of the graph of $y = -2x + 5$ (see Figure 2.3b) is found by setting $y = 0$ and solving for x:

$$0 = -2x + 5$$
$$2x = 5$$
$$x = \frac{5}{2}.$$

Similarly, a **y-intercept** of a graph is the y-coordinate of a point where the graph intersects the y-axis. The x-coordinate of this point is 0. The y-intercepts are found by setting $x = 0$ and solving for y. For example, the graph of $y = -2x + 5$ in Figure 2.3b has y-intercept 5.

✓ **Checkpoint 4**

Find the x- and y-intercepts of the graphs of these equations.

(a) $3x + 4y = 12$

(b) $5x - 2y = 8$

> **EXAMPLE 4** Find the x- and y-intercepts of the graph of $y = x^2 - 2x - 8$, and sketch the graph.
>
> **Solution** To find the y-intercept, set $x = 0$ and solve for y:
>
> $$y = x^2 - 2x - 8 = 0^2 - 2 \cdot 0 - 8 = -8.$$
>
> The y-intercept is -8. To find the x-intercept, set $y = 0$ and solve for x.
>
> $$x^2 - 2x - 8 = y$$
> $$x^2 - 2x - 8 = 0 \qquad \text{Set } y = 0.$$
> $$(x + 2)(x - 4) = 0 \qquad \text{Factor,}$$
> $$x + 2 = 0 \quad \text{or} \quad x - 4 = 0 \qquad \text{Zero-factor property}$$
> $$x = -2 \quad \text{or} \qquad x = 4$$

The x-intercepts are -2 and 4. Now make a table, using both positive and negative values for x, and plot the corresponding points, as in Figure 2.4. These points suggest that the entire graph looks like Figure 2.5.

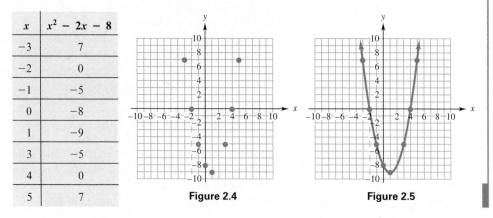

x	$x^2 - 2x - 8$
-3	7
-2	0
-1	-5
0	-8
1	-9
3	-5
4	0
5	7

Figure 2.4   **Figure 2.5**

EXAMPLE 5   Sketch the graph of $y = \sqrt{x + 2}$.

Solution Notice that $\sqrt{x + 2}$ is a real number only when $x + 2 \geq 0$—that is, when $x \geq -2$. Furthermore, $y = \sqrt{x + 2}$ is always nonnegative. Hence, all points on the graph lie on or above the x-axis and on or to the right of $x = -2$. Computing some typical values, we obtain the graph in Figure 2.6.

x	$\sqrt{x + 2}$
-2	0
0	$\sqrt{2} \approx 1.414$
2	2
5	$\sqrt{7} \approx 2.646$
7	3
9	$\sqrt{11} \approx 3.317$

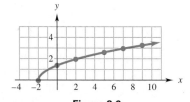

Figure 2.6

Example 3 shows that the solution of the equation $-2x + 5 = 0$ is the x-intercept of the graph of $y = -2x + 5$. Example 4 shows that the solutions of the equation $x^2 - 2x - 8 = 0$ are the x-intercepts of the graph $y = x^2 - 2x - 8$. Similar facts hold in the general case.

Intercepts and Equations

The real solutions of a one-variable equation of the form

$$\text{expression in } x = 0$$

are the x-intercepts of the graph of

$$y = \text{same expression in } x.$$

Graph Reading

Information is often given in graphical form, so you must be able to read and interpret graphs—that is, translate graphical information into statements in English.

EXAMPLE 6 **Finance** Newspapers and websites summarize activity of the S&P 500 Index in graphical form. The results for the 21 trading days for the month of December, 2016 are displayed in Figure 2.7. The first coordinate of each point on the graph is the trading day in December, and the second coordinate represents the closing price of the S&P 500 on that day. (Data from: www.morningstar.com.)

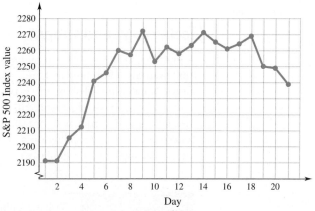

Figure 2.7

(a) What was the value of the S&P 500 Index on day 4 and day 15?

Solution If we examine the point when $x = 4$, we see that the y-value is a bit higher than the tick mark value of 2210. So we approximate the value for y as 2212, and thus the value of the S&P 500 for trading day 4 of December 2016 was about 2212. Similarly, when $x = 15$, we see the y-value is roughly halfway between the tick mark values of 2260 and 2270. Thus, we approximate the value of the S&P 500 index to be 2265 on trading day 15.

(b) On which trading days was the value of the index above 2240?

Solution Look for points whose second coordinates are greater than 2240, that is, points that lie above the grid line for 2240 in Figure 2.7. The first coordinates of these points are the days when the index value was above 2240. We see these days occurred on days 5 through 20. ◢5

The next example deals with the basic business relationship that was introduced in Section 1.2:

$$\text{Profit} = \text{Revenue} - \text{Cost}.$$

✓ **Checkpoint 5**

From Figure 2.7 determine when the S&P 500 had its highest point and its lowest point. What where the index values on those days?

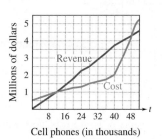

Figure 2.8

EXAMPLE 7 **Revenue, Cost, and Profit** Monthly revenue and costs for the Webster Cell Phone Company are determined by the number t of phones produced and sold, as shown in Figure 2.8.

(a) How many phones should be produced each month if the company is to make a profit (assuming that all phones produced are sold)?

Solution Profit is revenue minus cost, so the company makes a profit whenever revenue is greater than cost—that is, when the revenue graph is above the cost graph. Figure 2.8 shows that this occurs between $t = 12$ and $t = 48$—that is, when 12,000 to 48,000 phones are produced. If the company makes fewer than 12,000 phones, it will lose money (because costs will be greater than revenue.) It also loses money by making more than 48,000 phones. (One reason might be that high production levels require large amounts of overtime pay, which drives costs up too much.)

(b) Is it more profitable to make 40,000 or 44,000 phones?

Solution On the revenue graph, the point with first coordinate 40 has second coordinate of approximately 3.7, meaning that the revenue from 40,000 phones is about

3.7 million dollars. The point with first coordinate 40 on the cost graph is (40, 2), meaning that the cost of producing 40,000 phones is 2 million dollars. Therefore, the profit on 40,000 phones is about $3.7 - 2 = 1.7$ million dollars. For 44,000 phones, we have the approximate points (44, 4) on the revenue graph and (44, 3) on the cost graph. So the profit on 44,000 phones is $4 - 3 = 1$ million dollars. Consequently, it is more profitable to make 40,000 phones. ✓ 6

✓**Checkpoint 6**

In Example 7, find the profit from making

(a) 32,000 phones;

(b) 4000 phones.

Technology and Graphs

A graphing calculator or computer graphing program follows essentially the same procedure used when graphing by hand: The calculator selects a large number of x-values (95 or more), equally spaced along the x-axis, and plots the corresponding points, simultaneously connecting them with line segments. Calculator-generated graphs are generally quite accurate, although they may not appear as smooth as hand-drawn ones. The next example illustrates the basics of graphing on a graphing calculator. (Computer graphing software and cell phone apps operate similarly.)

EXAMPLE 8 Use a graphing calculator to sketch the graph of the equation

$$2x^3 - 2y - 10x + 2 = 0.$$

Solution *First, set the* **viewing window**—the portion of the coordinate plane that will appear on the screen. Press the WINDOW key (labeled RANGE or PLOT-SETUP on some calculators) and enter the appropriate numbers, as in Figure 2.9 (which shows the screen from a TI-84+; other calculators are similar). Then the calculator will display the portion of the plane inside the dashed lines shown in Figure 2.10—that is, the points (x, y) with $-9 \le x \le 9$ and $-6 \le y \le 6$.

In Figure 2.9, we have set Xscl = 2 and Yscl = 1, which means the **tick marks** on the x-axis are two units apart and the tick marks on the y-axis are one unit apart (as shown in Figure 2.10).

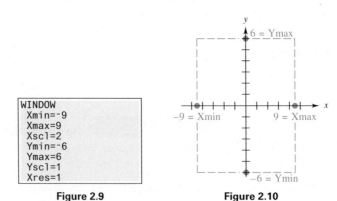

| Figure 2.9 | Figure 2.10 |

Second, enter the equation to be graphed in the equation memory. To do this, you must first solve the equation for y (because a calculator accepts only equations of the form $y =$ expression in x):

$$2y = 2x^3 - 10x + 2$$
$$y = x^3 - 5x + 1.$$

Now press the Y = key (labeled SYMB on some calculators) and enter the equation, using the "variable key" for x. (This key is labeled X, T, θ, n or X, θ, T or x-VAR, depending on the calculator.) Figure 2.11 shows the equation entered on a TI-84+; other calculators are similar. Now press GRAPH (or PLOT or DRW on some calculators), and obtain Figure 2.12.

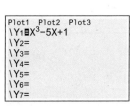

Figure 2.11

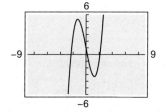

Figure 2.12

Finally, if necessary, change the viewing window to obtain a more readable graph. It is difficult to see the y-intercept in Figure 2.12, so press WINDOW and change the viewing window (Figure 2.13); then press GRAPH to obtain Figure 2.14, in which the y-intercept at $y = 1$ is clearly shown. (It isn't necessary to reenter the equation.) ✓7

✓**Checkpoint 7**

Use a graphing calculator to graph $y = 18x - 3x^3$ in the following viewing windows:

(a) $-10 \le x \le 10$ and $-10 \le y \le 10$ with $Xscl = 1, Yscl = 1$;

(b) $-5 \le x \le 5$ and $-20 \le y \le 20$ with $Xscl = 1, Yscl = 5$.

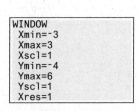

Figure 2.13

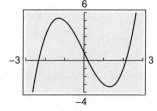

Figure 2.14

📇 Technology Tools

In addition to graphing equations, graphing calculators (and graphing software) provide convenient tools for solving equations and reading graphs. For example, when you have graphed an equation, you can readily determine the points the calculator plotted. Press **trace** (a cursor will appear on the graph), and use the left and right arrow keys to move the cursor along the graph. The coordinates of the point the cursor is on appear at the bottom of the screen.

Recall that the solutions of an equation, such as $x^3 - 5x + 1 = 0$, are the x-intercepts of the graph of $y = x^3 - 5x + 1$. (See the box on page 68.) A **graphical root finder** enables you to find these x-intercepts and thus to solve the equation.

> **EXAMPLE 9**   Use a graphical root finder to solve $x^3 - 5x + 1 = 0$.

Solution First, graph $y = x^3 - 5x + 1$. The x-intercepts of this graph are the solutions of the equation. To find these intercepts, look for "root" or "zero" in the appropriate menu.* Check your instruction manual for the proper syntax. A typical root finder (see Figure 2.15) shows that two of the solutions (x-intercepts) are $x \approx .2016$ and $x \approx 2.1284$. For the third solution, see Checkpoint 8. ✓8

✓**Checkpoint 8**

Use a graphical root finder to approximate the third solution of the equation in Example 8.

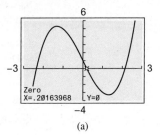

(a)

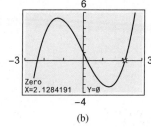

(b)

Figure 2.15

Many graphs have peaks and valleys (for instance, the graphs in Figure 2.15). A **maximum/minimum finder** provides accurate approximations of the locations of the "tops" of the peaks and the "bottoms" of the valleys.

---

*CALC on TI-84+, GRAPH MATH on TI-86.

2.1 Exercises

State the quadrant in which each point lies.

1. $(1, -2), (-2, 1), (3, 4), (-5, -6)$

2. $(\pi, 2), (3, -\sqrt{2}), (4, 0), (-\sqrt{3}, \sqrt{3})$

Determine whether the given ordered pair is a solution of the given equation. (See Example 2.)

3. $(1, -3); 3x - y - 6 = 0$

4. $(2, -1); x^2 + y^2 - 6x + 8y = -15$

5. $(3, 4); (x - 2)^2 + (y + 2)^2 = 6$

6. $(1, -1); \dfrac{x^2}{2} + \dfrac{y^2}{3} = -4$

Sketch the graph of each of these equations. (See Example 3.)

7. $4y + 3x = 12$ **8.** $2x + 7y = 14$

9. $8x + 3y = 12$ **10.** $9y - 4x = 12$

11. $x = 2y + 3$ **12.** $x - 3y = 0$

List the x-intercepts and y-intercepts of each graph.

13.

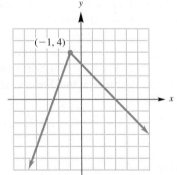

14.

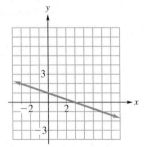

15.

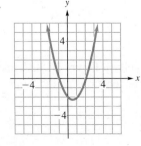

16.

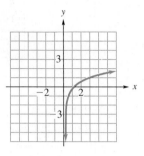

Find the x-intercepts and y-intercepts of the graph of each equation. You need not sketch the graph. (See Example 4.)

17. $3x + 4y = 12$ **18.** $x - 2y = 5$

19. $2x - 3y = 24$ **20.** $3x + y = 4$

21. $y = x^2 - 9$ **22.** $y = x^2 + 4$

23. $y = x^2 + x - 20$ **24.** $y = 5x^2 + 6x + 1$

25. $y = 2x^2 - 5x + 7$ **26.** $y = 3x^2 + 4x - 4$

Sketch the graph of the equation. (See Examples 3–5.)

27. $y = x^2$ **28.** $y = x^2 + 2$

29. $y = x^2 - 3$ **30.** $y = 2x^2$

31. $y = x^2 - 6x + 5$ **32.** $y = x^2 + 2x - 3$

33. $y = x^3$ **34.** $y = x^3 - 3$

35. $y = \sqrt{x + 4}$ **36.** $y = \sqrt{x - 2}$

Sketch the graph of the information in the following tables. (See Example 1.)

37. Military Construction The following table gives the amount of annual spending in the U.S. defense budget (in billions of dollars) for military construction. Let $x = 12$ correspond to the year 2012. (Data from: U.S. Office of Management and Budget.)

Year	2012	2013	2014	2015	2016
Spending	11.4	8.1	8.4	5.7	7.0

38. Military Procurement The following table gives the amount of annual spending in the U.S. defense budget (in billions of dollars) for military procurement. Let $x = 12$ correspond to the year 2012. (Data from: U.S. Office of Management and Budget.)

Year	2012	2013	2014	2015	2016
Spending	118	98	100	103	115

39. Coal Mine Fatalities The number of fatalities in coal mines in the United States for recent years is given in the following table. Let $x = 10$ correspond to the year 2010. (Data from: U.S. Mine Safety and Health Administration.)

Year	2010	2011	2012	2013	2014	2015	2016
Number	48	20	20	20	16	12	9

40. All Mining Fatalities The number of fatalities in all mining activities in the United States for recent years is given in the table. Let $x = 10$ correspond to the year 2010. (Data from: U.S. Mine Safety and Health Administration.)

Year	2010	2011	2012	2013	2014	2015	2016
Number	71	36	36	42	46	29	26

ATM Fees *An article in The Wall Street Journal on October 4, 2015, spoke of rising automated teller (ATM) fees. The graph below shows the average costs of noncustomer, out-of-network, and total fees from 1998 to 2016. (Data from: The Wall Street Journal and Bankrate.com.) (See Examples 6 and 7.)*

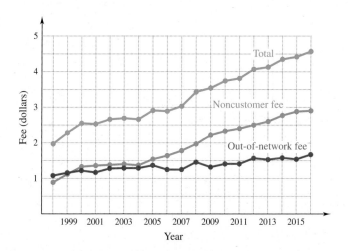

41. In what year were noncustomer and out-of-network fees almost the same amount?

42. In what year did the average out-of-network fee exceed $1.50 for the first time?

43. Which year was the last year that the average noncustomer fee was below $2.50?

44. Which year marked the first time that the average total fees rose above $3.00?

Business *Use the revenue and cost graphs for the Webster Cell Phone Company in Example 7 to do Exercises 45–48.*

45. Find the approximate cost of manufacturing the given number of phones.

(a) 20,000   (b) 36,000   (c) 48,000

46. Find the approximate revenue from selling the given number of phones.

(a) 12,000   (b) 24,000   (c) 36,000

47. Find the approximate profit from manufacturing the given number of phones.

(a) 20,000   (b) 28,000   (c) 36,000

48. The company must replace its aging machinery with better, but much more expensive, machines. In addition, raw material prices increase, so that monthly costs go up by $250,000.

Owing to competitive pressure, phone prices cannot be increased, so revenue remains the same. Under these new circumstances, find the approximate profit from manufacturing the given number of phones.

(a) 20,000   (b) 36,000   (c) 40,000

Business *The graph below gives the annual per-person retail availability of beef, chicken, and pork (in pounds) from 2007 to 2014. Use the graph to answer the following questions. (Data from: U.S. Department of Agriculture.)*

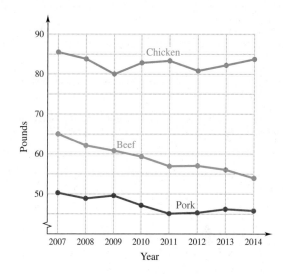

49. What is the approximate annual per-person retail availability of beef, chicken, and pork in 2012?

50. In what year was annual availability of beef the highest?

51. In what year was annual availability of chicken the lowest?

52. How much did the annual availability of beef decrease from 2007 to 2014?

Netflix and Apple Stock Prices *The graph below shows the opening share prices (in dollars) for Netflix, Inc., and Apple Inc. for the 21 trading days in December 2016. Use this graph to answer Exercises 53–62.*

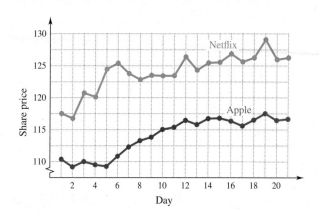

53. What is the approximate share price of Apple stock on day 10?

54. What is the approximate share price of Netflix stock on day 8?

55. On which days is the share price of Netflix stock above $125?

56. On which days is the share price of Apple stock below $115?

57. On what day was the Apple share price at its highest, and what was the value?

58. On what day was the Netflix share price at its highest, and what was the value?

59. On what day was the Netflix share price at its lowest, and what was the value?

60. On what day was the Apple share price at its lowest, and what was the value?

61. Over the course of the month, what was the gain in share price for Netflix?

62. Over the course of the month, what was the gain in share price for Apple?

Use a graphing calculator to find the graph of the equation. (See Example 8.)

63. $y = x^2 + x + 1$

64. $y = 2 - x - x^2$

65. $y = (x - 3)^3$

66. $y = x^3 + 2x^2 + 2$

67. $y = x^3 - 3x^2 + x - 1$

68. $y = x^4 - 5x^2 - 2$

Use a graphing calculator for Exercises 69–70.

69. Graph $y = x^4 - 2x^3 + 2x$ in a window with $-3 \leq x \leq 3$. Is the "flat" part of the graph near $x = 1$ really a horizontal line segment? (*Hint*: Use the trace feature to move along the "flat" part and watch the y-coordinates. Do they remain the same [as they should on a horizontal segment]?)

70. **(a)** Graph $y = x^4 - 2x^3 + 2x$ in the **standard window** (the one with $-10 \leq x \leq 10$ and $-10 \leq y \leq 10$). Use the trace feature to approximate the coordinates of the lowest point on the graph.

  (b) Use a minimum finder to obtain an accurate approximation of the lowest point. How does this compare with your answer in part (a)?

Use a graphing calculator to approximate all real solutions of the equation. (See Example 9.)

71. $x^3 - 3x^2 + 5 = 0$

72. $x^3 + x - 1 = 0$

73. $2x^3 - 4x^2 + x - 3 = 0$

74. $6x^3 - 5x^2 + 3x - 2 = 0$

75. $x^5 - 6x + 6 = 0$

76. $x^3 - 3x^2 + x - 1 = 0$

✓ **Checkpoint Answers**

1.

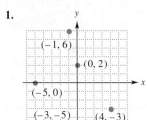

2. (a) and (c)

3.

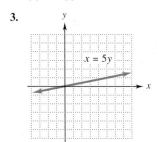

4. **(a)** x-intercept 4, y-intercept 3

  (b) x-intercept $8/5$, y-intercept -4

5. The highest was 2272 on Day 9
   The lowest was 2191 on Days 1 and 2

6. **(a)** About $1,200,000 (rounded)

  (b) About $-$500,000 (that is, a loss of $500,000)

7. **(a)**

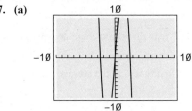

  (b)

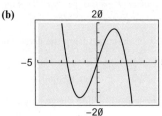

8. $x \approx -2.330059$

2.2 Equations of Lines

Straight lines, which are the simplest graphs, play an important role in a wide variety of applications. They are considered here from both a geometric and an algebraic point of view.

The key geometric feature of a nonvertical straight line is how steeply it rises or falls as you move from left to right. The "steepness" of a line can be represented numerically by a number called the *slope* of the line.

To see how the slope is defined, start with Figure 2.16, which shows a line passing through the two different points $(x_1, y_1) = (-5, 6)$ and $(x_2, y_2) = (4, -7)$. The difference in the two x-values,

$$x_2 - x_1 = 4 - (-5) = 9,$$

is called the **change in x.** Similarly, the **change in y** is the difference in the two y-values:

$$y_2 - y_1 = -7 - 6 = -13.$$

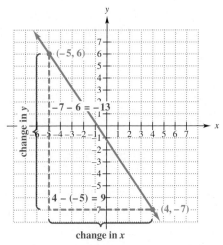

Figure 2.16

The **slope** of the line through the two points (x_1, y_1) and (x_2, y_2), where $x_1 \neq x_2$, is defined as the quotient of the change in y and the change in x:

$$\textbf{slope} = \frac{\textbf{change in } y}{\textbf{change in } x} = \frac{y_2 - y_1}{x_2 - x_1}.$$

The slope of the line in Figure 2.16 is

$$\text{slope} = \frac{-7 - 6}{4 - (-5)} = -\frac{13}{9}.$$

Using similar triangles from geometry, we can show that the slope is independent of the choice of points on the line. That is, the same value of the slope will be obtained for *any* choice of two different points on the line.

EXAMPLE 1   Find the slope of the line through the points $(-6, 8)$ and $(5, 4)$.

Solution   Let $(x_1, y_1) = (-6, 8)$ and $(x_2, y_2) = (5, 4)$. Use the definition of slope as follows:

$$\text{slope} = \frac{y_2 - y_1}{x_2 - x_1} = \frac{4 - 8}{5 - (-6)} = \frac{-4}{11} = -\frac{4}{11}.$$

The slope can also by found by letting $(x_1, y_1) = (5, 4)$ and $(x_2, y_2) = (-6, 8)$. In that case,

$$\text{slope} = \frac{y_2 - y_1}{x_2 - x_1} = \frac{8 - 4}{-6 - 5} = \frac{4}{-11} = -\frac{4}{11},$$

which is the same answer.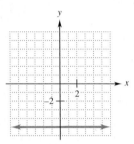

✓ Checkpoint 1

Find the slope of the line through the following pairs of points.

(a) $(5, 9), (-5, -3)$

(b) $(-4, 2), (-2, -7)$

⚠ CAUTION When finding the slope of a line, be careful to subtract the *x*-values and the *y*-values in the same order. For example, with the points (4, 3) and (2, 9), if you use 9 − 3 for the numerator, you must use 2 − 4 (*not* 4 − 2) for the denominator.

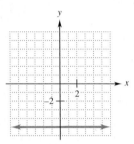

Figure 2.17

> **EXAMPLE 2**   Find the slope of the horizontal line in Figure 2.17.
>
> **Solution** Every point on the line has the same *y*-coordinate, −5. Choose any two of them to compute the slope, say, $(x_1, y_1) = (-3, -5)$ and $(x_2, y_2) = (2, -5)$:
>
> $$\text{slope} = \frac{-5 - (-5)}{2 - (-3)}$$
>
> $$= \frac{0}{5}$$
>
> $$= 0.$$

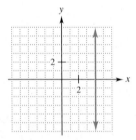

Figure 2.18

> **EXAMPLE 3**   What is the slope of the vertical line in Figure 2.18?
>
> **Solution** Every point on the line has the same *x*-coordinate, 4. If we attempt to compute the slope with two of these points, say, $(x_1, y_1) = (4, -2)$ and $(x_2, y_2) = (4, 1)$, we obtain
>
> $$\text{slope} = \frac{1 - (-2)}{4 - 4}$$
>
> $$= \frac{3}{0}.$$

Division by 0 is not defined, so the slope of this line is undefined.

The arguments used in Examples 2 and 3 work in the general case and lead to the following conclusion.

> The slope of every horizontal line is 0.
>
> The slope of every vertical line is undefined.

Slope–Intercept Form

The slope can be used to develop an algebraic description of nonvertical straight lines. Assume that a line with slope *m* has *y*-intercept *b*, so that it goes through the point (0, *b*). (See Figure 2.19.) Let (*x*, *y*) be any point on the line other than (0, *b*). Using the definition of slope with the points (0, *b*) and (*x*, *y*) gives

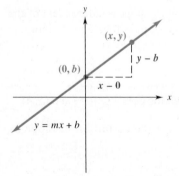

Figure 2.19

$$m = \frac{y - b}{x - 0}$$

$$m = \frac{y - b}{x}$$

$$mx = y - b \qquad \text{Multiply both sides by } x.$$

$$y = mx + b. \qquad \text{Add } b \text{ to both sides. Reverse the equation.}$$

In other words, the coordinates of any point on the line satisfy the equation $y = mx + b$.

Slope–Intercept Form

If a line has slope m and y-intercept b, then it is the graph of the equation

$$y = mx + b.$$

This equation is called the **slope–intercept form** of the equation of the line.

EXAMPLE 4   Find an equation for the line with y-intercept $7/2$ and slope $-5/2$.

Solution Use the slope–intercept form with $b = 7/2$ and $m = -5/2$:

$$y = mx + b$$
$$y = -\frac{5}{2}x + \frac{7}{2}. \quad ✓_2$$

Checkpoint 2

Find an equation for the line with

(a) y-intercept -3 and slope $2/3$;

(b) y-intercept $1/4$ and slope $-3/2$.

EXAMPLE 5   Find the equation of the horizontal line with y-intercept 3.

Solution The slope of the line is 0 (why?) and its y-intercept is 3, so its equation is

$$y = mx + b$$
$$y = 0x + 3$$
$$y = 3.$$

The argument in Example 5 also works in the general case.

If k is a constant, then the graph of the equation $y = k$ is the horizontal line with y-intercept k.

EXAMPLE 6   Find the slope and y-intercept for each of the following lines.

(a) $5x - 3y = 1$

Solution Solve for y:

$$5x - 3y = 1$$
$$-3y = -5x + 1 \qquad \text{Subtract } 5x \text{ from both sides.}$$
$$y = \frac{5}{3}x - \frac{1}{3}. \qquad \text{Divide both sides by } -3.$$

This equation is in the form $y = mx + b$, with $m = 5/3$ and $b = -1/3$. So the slope is $5/3$ and the y-intercept is $-1/3$.

(b) $-9x + 6y = 2$

Solution Solve for y:

$$-9x + 6y = 2$$
$$6y = 9x + 2 \qquad \text{Add } 9x \text{ to both sides.}$$
$$y = \frac{3}{2}x + \frac{1}{3}. \qquad \text{Divide both sides by } 6.$$

Checkpoint 3

Find the slope and y-intercept of

(a) $x + 4y = 6$;

(b) $3x - 2y = 1$.

The slope is $3/2$ (the coefficient of x), and the y-intercept is $1/3$. ✓_3

The slope–intercept form can be used to show how the slope measures the steepness of a line. Consider the straight lines *A*, *B*, *C*, and *D* given by the following equations, where each has *y*-intercept 0 and slope as indicated:

A: y = .5x;	*B: y = x;*	*C: y = 3x;*	*D: y = 7x.*
Slope .5	Slope 1	Slope 3	Slope 7

For these lines, Figure 2.20a shows that the bigger the slope, the more steeply the line rises from left to right.

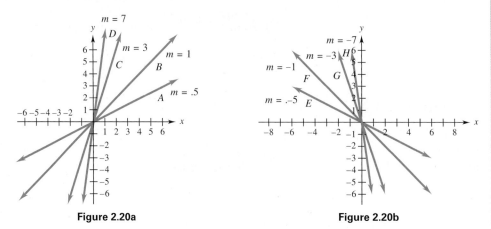

Figure 2.20a   **Figure 2.20b**

If we consider straight lines, *E, F, G,* and *H* given by the following equations

E: y = −.5x;	*F: y = −x;*	*G: y = −3x;*	*H: y = −7x.*
Slope −.5	Slope −1	Slope −3	Slope −7

For these lines, Figure 2.20b shows the more negative the slope, the more steeply the line falls from left to right. ✔️4

The preceding discussion and Checkpoint 4 may be summarized as follows.

Direction of Line (moving from left to right)	Slope
Upward	**Positive** (larger for steeper lines)
Horizontal	**0**
Downward	**Negative** (larger in absolute value for steeper lines)
Vertical	**Undefined**

✔️**Checkpoint 4**

(a) List the slopes of the following lines:

$E: y = -.3x;$ $F: y = -x;$
$G: y = -2x;$ $H: y = -5x.$

(b) Graph all four lines on the same set of axes.

(c) How are the slopes of the lines related to their steepness?

EXAMPLE 7   Sketch the graph of $x + 2y = 5$, and label the intercepts.

Solution Find the *x*-intercept by setting $y = 0$ and solving for *x*:

$$x + 2 \cdot 0 = 5$$
$$x = 5.$$

The *x*-intercept is 5, and (5, 0) is on the graph. The *y*-intercept is found similarly, by setting $x = 0$ and solving for *y*:

$$0 + 2y = 5$$
$$y = 5/2.$$

The *y*-intercept is 5/2, and (0, 5/2) is on the graph. The points (5, 0) and (0, 5/2) can be used to sketch the graph, as shown on the following page (Figure 2.21). ✔️5

✔️**Checkpoint 5**

Graph the given lines and label the intercepts.

(a) $3x + 4y = 12$

(b) $5x - 2y = 8$

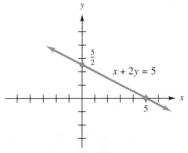

Figure 2.21

⊞ **TECHNOLOGY TIP** To graph a linear equation on a graphing calculator, you must first put the equation in slope–intercept form $y = mx + b$ so that it can be entered in the equation memory (called the Y = list on some calculators). Vertical lines cannot be graphed on most calculators.

Slopes of Parallel and Perpendicular Lines

We shall assume the following facts without proof. The first one is a consequence of the fact that the slope measures steepness and that parallel lines have the same steepness.

> Two nonvertical lines are **parallel** whenever they have the same slope.
>
> Two nonvertical lines are **perpendicular** whenever the product of their slopes is −1.
>
> *Note:* An equivalent definition for **perpendicular lines** is nonvertical lines whose slopes are negative reciprocals of one another.

EXAMPLE 8 Determine whether each of the given pairs of lines are *parallel*, *perpendicular*, or *neither*.

(a) $2x + 3y = 5$ and $4x + 5 = -6y$.

Solution Put each equation in slope–intercept form by solving for y:

$$3y = -2x + 5 \qquad -6y = 4x + 5$$
$$y = -\frac{2}{3}x + \frac{5}{3} \qquad y = -\frac{2}{3}x - \frac{5}{6}.$$

In each case, the slope (the coefficient of x) is $-2/3$, so the lines are parallel. See Figure 2.22(a).

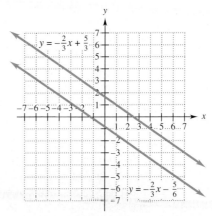

Figure 2.22(a)

(b) $3x = y + 7$ and $x + 3y = 4$.

Solution Put each equation in slope–intercept form to determine the slope of the associated line:

$$3x = y + 7 \qquad\qquad 3y = -x + 4$$

$$y = 3x - 7 \qquad\qquad y = -\frac{1}{3}x + \frac{4}{3}$$

$$\text{slope 3} \qquad\qquad \text{slope } -1/3.$$

The slopes of 3 and $-1/3$ are negative reciprocals of one another. For this reason, and because $3(-1/3) = -1$, these lines are perpendicular. See Figure 2.22(b).

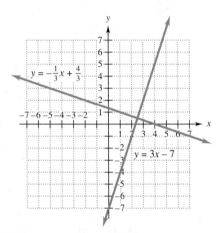

Figure 2.22(b)

(c) $x + y = 4$ and $x - 2y = 3$.

Solution Verify that the slope of the first line is -1 and the slope of the second is $1/2$. The slopes are not equal and their product is not -1, so the lines are neither parallel nor perpendicular. See Figure 2.22(c).

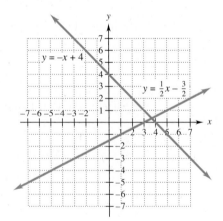

Figure 2.22(c)

✓ **Checkpoint 6**

Tell whether the lines in each of the following pairs are *parallel*, *perpendicular*, or *neither*.

(a) $x - 2y = 6$ and $2x + y = 5$

(b) $3x + 4y = 8$ and $x + 3y = 2$

(c) $2x - y = 7$ and $2y = 4x - 5$

⊞ TECHNOLOGY TIP Perpendicular lines may not appear perpendicular on a graphing calculator unless you use a *square window*—a window in which a one-unit segment on the *y*-axis is the same length as a one-unit segment on the *x*-axis. To obtain such a window on most calculators, use a viewing window in which the *y*-axis is about two-thirds as long as the *x*-axis. The SQUARE (or ZSQUARE) key in the ZOOM menu will change the current window to a square window by automatically adjusting the length of one of the axes.

Point–Slope Form

The slope–intercept form of the equation of a line is usually the most convenient for graphing and for understanding how slopes and lines are related. However, it is not always the best way to *find* the equation of a line. In many situations (particularly in calculus), the slope and a point on the line are known and you must find the equation of the line. In such cases, the best method is to use the *point–slope form*, which we now explain.

Suppose that a line has slope m and that (x_1, y_1) is a point on the line. Let (x, y) represent any other point on the line. Since m is the slope, then, by the definition of slope,

$$\frac{y - y_1}{x - x_1} = m.$$

Multiplying both sides by $x - x_1$ yields

$$y - y_1 = m(x - x_1).$$

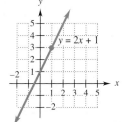

Figure 2.23

Point–Slope Form

If a line has slope m and passes through the point (x_1, y_1), then

$$y - y_1 = m(x - x_1)$$

is the **point–slope form** of the equation of the line.

EXAMPLE 9   Find the equation of the line satisfying the given conditions.

(a) Slope 2; the point $(1, 3)$ is on the line.

Solution Use the point–slope form with $m = 2$ and $(x_1, y_1) = (1, 3)$. Substitute $x_1 = 1$, $y_1 = 3$, and $m = 2$ into the point–slope form of the equation.

$$y - y_1 = m(x - x_1)$$
$$y - 3 = 2(x - 1). \qquad \text{Point–slope form}$$

For some purposes, this form of the equation is fine; in other cases, you may want to rewrite it in the slope–intercept form.

Using algebra, we obtain the slope–intercept form of this equation:

$$y - 3 = 2(x - 1)$$
$$y - 3 = 2x - 2 \qquad \text{Distributive property}$$
$$y = 2x + 1. \qquad \text{Add 3 to each side to obtain slope-intercept form.}$$

See Figure 2.23 for the graph.

(b) Slope -3; the point $(-4, 1)$ is on the line.

Solution Use the point–slope form with $m = -3$ and $(x_1, y_1) = (-4, 1)$:

$$y - y_1 = m(x - x_1)$$
$$y - 1 = -3[x - (-4)]. \qquad \text{Point–slope form}$$

Using algebra, we obtain the slope–intercept form of this equation:

$$y - 1 = -3(x + 4)$$
$$y - 1 = -3x - 12 \qquad \text{Distributive property}$$
$$y = -3x - 11. \qquad \text{Slope–intercept form}$$

See Figure 2.24.

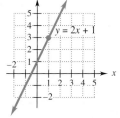

Figure 2.24

✓Checkpoint 7

Find both the point–slope and the slope–intercept form of the equation of the line having the given slope and passing through the given point.

(a) $m = -3/5, (5, -2)$

(b) $m = 1/3, (6, 8)$

The point–slope form can also be used to find an equation of a line, given two different points on the line. The procedure is shown in the next example.

EXAMPLE 10 Find an equation of the line through $(5, 4)$ and $(-10, -2)$.

Solution Begin by using the definition of the slope to find the slope of the line that passes through the two points:

$$\text{slope} = m = \frac{-2 - 4}{-10 - 5} = \frac{-6}{-15} = \frac{2}{5}.$$

Use $m = 2/5$ and either of the given points in the point–slope form. If $(x_1, y_1) = (5, 4)$, then

$$y - y_1 = m(x - x_1)$$

$$y - 4 = \frac{2}{5}(x - 5) \qquad \text{Let } y_1 = 4, m = \frac{2}{5}, \text{ and } x_1 = 5.$$

$$y - 4 = \frac{2}{5}x - \frac{10}{5} \qquad \text{Distributive property}$$

$$y = \frac{2}{5}x + 2. \qquad \text{Add 4 to both sides and simplify.}$$

✓ **Checkpoint 8**

Find an equation of the line through

(a) $(2, 3)$ and $(-4, 6)$;

(b) $(-8, 2)$ and $(3, -6)$.

See Figure 2.25. Check that the results are the same when $(x_1, y_1) = (-10, -2)$. ✓ 8

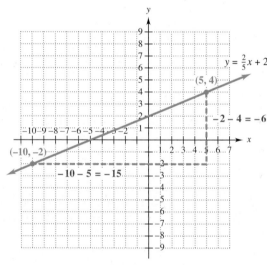

Figure 2.25

Vertical Lines

The equation forms we just developed do not apply to vertical lines, because the slope is not defined for such lines. However, vertical lines can easily be described as graphs of equations.

EXAMPLE 11 Find the equation whose graph is the vertical line in Figure 2.26.

Solution Every point on the line has x-coordinate -1 and hence has the form $(-1, y)$. Thus, every point is a solution of the equation $x + 0y = -1$, which is usually written simply as $x = -1$. Note that -1 is the x-intercept of the line.

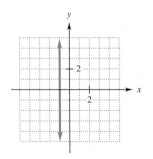

Figure 2.26

The argument in Example 11 also works in the general case.

If k is a constant, then the graph of the equation $x = k$ is the vertical line with x-intercept k.

Linear Equations

An equation in two variables whose graph is a straight line is called a **linear equation.** Linear equations have a variety of forms, as summarized in the following table.

Equation	Description
$x = k$	**Vertical line,** x-intercept k, no y-intercept (unless $x = 0$), undefined slope
$y = k$	**Horizontal line,** y-intercept k, no x-intercept (unless $y = 0$), slope 0
$y = mx + b$	**Slope–intercept form,** slope m, y-intercept b
$y - y_1 = m(x - x_1)$	**Point–slope form,** slope m, the line passes through (x_1, y_1)
$ax + by = c$	**General form.** If $a \neq 0$ and $b \neq 0$, the line has x-intercept c/a, y-intercept c/b, and slope $-a/b$.

Note that every linear equation can be written in general form. For example, $y = 4x - 5$ can be written in general form as $4x - y = 5$, and $x = 6$ can be written in general form as $x + 0y = 6$.

Applications

Many relationships are linear or almost linear, so that they can be approximated by linear equations.

EXAMPLE 12   **Emerald Prices** The price of emeralds has surged in recent years. The average price per carat for high-quality rough emeralds can be approximated by the linear equation

$$y = 10.2x - 84.4,$$

where $x = 9$ corresponds to the year 2009. The graph appears in Figure 2.27. (Data from: *The Wall Street Journal.*)

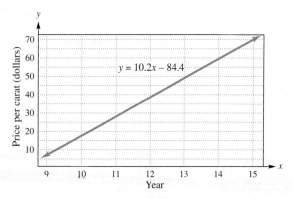

Figure 2.27

(a) What was the approximate price per carat for emeralds in 2014?

Solution Substitute $x = 14$ in the equation and compute y:

$$y = 10.2x - 84.4$$
$$y = 10.2(14) - 84.4 = 58.4.$$

The approximate price per carat was $58.40.

(b) What was the first full year that the price exceeded $50?

Solution Substitute $y = 50$ in the equation and solve for x:

$$50 = 10.2x - 84.4$$
$$50 + 84.4 = 10.2x \qquad \text{Add 84.4 to each side.}$$
$$x = \frac{50 + 84.4}{10.2} \qquad \text{Divide each side by 10.2 and reverse the equation.}$$
$$x \approx 13.2.$$

This implies that the first full year with a price $50 or higher was 2014. Note that we *round up* the year to 2014 to find the first time when the price was $50 for the *full year* (if we substitute $x = 13$ into the equation, we get $y = 10.2(13) - 84.4 = 48.2$, so the price was not $50 for the full year in 2013).

EXAMPLE 13   **Video Streaming** According to data from *The Wall Street Journal*, the online video audience in China was approximately 200 million people in the year 2008. In 2015, the audience had grown to 461.2 million.

(a) Find a linear equation for these data.

Solution Measure time along the x-axis and viewership along the y-axis. Then the x-coordinate of each point is a year and the y-coordinate is the viewership in that year. For convenience, let $x = 8$ correspond to 2008. The given data points are $(8, 200)$ and $(15, 461.2)$. The slope of the line joining these two points is

$$\frac{461.2 - 200}{15 - 8} = \frac{261.2}{7} \approx 37.3.$$

We can now use the point–slope form of a line with the point $(8, 200)$ to obtain the equation of the line:

$$y - 200 = 37.3(x - 8) \qquad \text{Point–slope form}$$
$$y - 200 = 37.3x - 298.4 \qquad \text{Distributive property}$$
$$y = 37.3x - 98.4. \qquad \text{Add 200 to each side.}$$

Figure 2.28 shows the derived equation.

(b) Use this equation to estimate the viewership in China in 2014.

Solution Because 2014 corresponds to $x = 14$, let $x = 14$ in the equation part of (a). Then

$$y = 37.3(14) - 98.4 = 423.8.$$

The viewership was approximately 423.8 million.

(c) Assuming the equation remains valid beyond 2015, estimate the viewership in 2019.

Solution The year 2019 corresponds to $x = 19$, so the viewership is

$$y = 37.3(19) - 98.4 = 610.3.$$

The viewership is estimated to be 610.3 million. ☑ 9

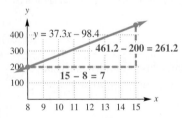

Figure 2.28

✓ **Checkpoint 9**

The online viewership in India was estimated to be 61.5 million in 2014. In 2016, the online viewership was estimated to be 81.7 million. (Data from: Statistica.com and cxotoday.com.)

(a) Let $x = 14$ correspond to the year 2014, and find a linear equation for the given data.

(b) Assuming that your equation remains accurate, estimate the online viewership in India in 2020.

2.2 Exercises

Find the slope of the given line, if it is defined. (See Examples 1–3.)

1. The line through (2, 5) and (0, 8)

2. The line through (9, 0) and (12, 12)

3. The line through (−4, 14) and (3, 0)

4. The line through (−5, −2) and (−4, 11)

5. The line through the origin and (−4, 10)

6. The line through the origin and (8, −2)

7. The line through (−1, 4) and (−1, 6)

8. The line through (−3, 5) and (2, 5)

Find an equation of the line with the given y-intercept and slope m.
(See Examples 4 and 5.)

9. 5, $m = 4$

10. −3, $m = −7$

11. 1.5, $m = −2.3$

12. −4.5, $m = 2.5$

13. 4, $m = −3/4$

14. −3, $m = 4/3$

Find the slope m and the y-intercept b of the line whose equation is given. (See Example 6.)

15. $2x − y = 9$

16. $x + 2y = 7$

17. $6x = 2y + 4$

18. $4x + 3y = 24$

19. $6x − 9y = 16$

20. $4x + 2y = 0$

21. $2x − 3y = 0$

22. $y = 7$

23. $x = y − 5$

24. On one graph, sketch six straight lines that meet at a single point and satisfy this condition: one line has slope 0, two lines have positive slope, two lines have negative slope, and one line has undefined slope.

25. For which of the line segments in the figure is the slope

    (a) largest?

    (b) smallest?

    (c) largest in absolute value?

    (d) closest to 0?

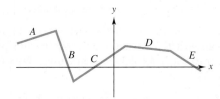

26. Match each equation with the line that most closely resembles its graph. (*Hint:* Consider the signs of m and b in the slope–intercept form.)

    (a) $y = 3x + 2$

    (b) $y = −3x + 2$

    (c) $y = 3x − 2$

    (d) $y = −3x − 2$

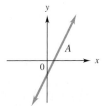
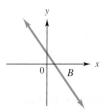
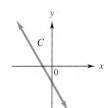
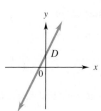

Sketch the graph of the given equation and label its intercepts. (See Example 7.)

27. $2x − y = −2$

28. $2y + x = 4$

29. $2x + 3y = 4$

30. $−5x + 4y = 3$

31. $4x − 5y = 2$

32. $3x + 2y = 8$

Determine whether each pair of lines is parallel, perpendicular, or neither. (See Example 8.)

33. $4x − 3y = 6$ and $3x + 4y = 8$

34. $2x − 5y = 7$ and $15y − 5 = 6x$

35. $3x + 2y = 8$ and $6y = 5 − 9x$

36. $x − 3y = 4$ and $y = 1 − 3x$

37. $4x = 2y + 3$ and $2y = 2x + 3$

38. $2x − y = 6$ and $x − 2y = 4$

39. (a) Find the slope of each side of the triangle with vertices (9, 6), (−1, 2), and (1, −3).

    (b) Is this triangle a right triangle? (*Hint:* Are two sides perpendicular?)

40. (a) Find the slope of each side of the quadrilateral with vertices (−5, −2), (−3, 1), (3, 0), and (1, −3).

    (b) Is this quadrilateral a parallelogram? (*Hint:* Are opposite sides parallel?)

Find an equation of the line with slope m that passes through the given point. Put the answer in slope–intercept form. (See Example 9.)

41. $(−3, 2), m = −2/3$

42. $(−5, −2), m = 4/5$

43. $(2, 3), m = 3$

44. $(3, −4), m = −1/4$

45. $(10, 1), m = 0$

46. $(−3, −9), m = 0$

47. $(−2, 12)$, undefined slope

48. $(1, 1)$, undefined slope

Find an equation of the line that passes through the given points. (See Example 10.)

49. $(−1, 1)$ and $(2, 7)$

50. $(2, 5)$ and $(0, 6)$

51. $(1, 2)$ and $(3, 9)$

52. $(−1, −2)$ and $(2, −1)$

Find an equation of the line satisfying the given conditions.

53. Through the origin with slope 5

54. Through the origin and horizontal

55. Through (6, 8) and vertical

56. Through (7, 9) and parallel to $y = 6$

57. Through (3, 4) and parallel to $4x - 2y = 5$

58. Through (6, 8) and perpendicular to $y = 2x - 3$

59. x-intercept 6; y-intercept -6

60. Through $(-5, 2)$ and parallel to the line through (1, 2) and (4, 3)

61. Through $(-1, 3)$ and perpendicular to the line through (0, 1) and (2, 3)

62. y-intercept 3 and perpendicular to $2x - y + 6 = 0$

Depreciation *The lost value of equipment over a period of time is called depreciation. The simplest method for calculating depreciation is straight-line depreciation. The annual straight-line depreciation D of an item that cost x dollars with a useful life of n years is $D = (1/n)x$. Find the depreciation for items with the given characteristics.*

63. Cost: $15,965; life 12 yr

64. Cost: $41,762; life 15 yr

65. Cost: $201,457; life 30 yr

66. Ral Corp. has an incentive compensation plan under which a branch manager receives 10% of the branch's income after deduction of the bonus, but before deduction of income tax. The income of a particular branch before the bonus and income tax was $165,000. The tax rate was 30%. What was the amount of the bonus?

Light Beer Sales *Data from The Wall Street Journal indicate that the percent change in beer shipments since the year 2007 for Miller Lite beer can be approximated by*

$$y = -3.9x + 27.3,$$

where $x = 7$ corresponds to the year 2007. Find the percent change since 2007 that occurred by the following years.

67. 2010   68. 2014

Assume the model remains accurate, and find the first full year in which the percent change reaches the following values.

69. -32% 70. -40%

Global Malaria Cases *Data from The Wall Street Journal indicate the number of global malaria cases has risen sharply since the year 2000. The equation*

$$y = 5.6x + 52$$

approximates the number of global malaria cases y (in millions), where $x = 0$ corresponds to the year 2000. Find the number of global malaria cases in the following years.

71. 2007   72. 2015

Assume the model remains accurate, and find the first full year in which the number of global malaria cases per year achieves the following values.

73. 150 million   **74.** 175 million

Olympic 5000-Meters *The accompanying graph shows the winning time (in minutes) at the Olympic Games from 1952 to 2016 for the men's 5000-meter run, together with a linear approximation of these data. (Data from: The World Almanac and Book of Facts: 2017.)*

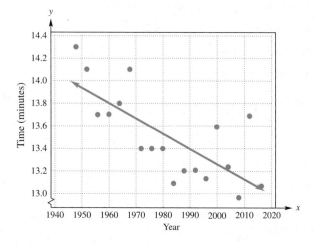

The equation for the linear approximation is $y = -.0135x + 40.32$, where x is the actual year.

75. What does the linear approximation predict for the winning time for the 2016 Olympics for this event? The actual time was 13.06 minutes. How close was the mathematical model's estimate?

76. What does the slope of the line represent? Why is it negative?

77. **Luxury Handbags** Michael Kors now dominates the handbag market in North America. In 2011, the company had 5% of the market; in 2015, it had 23%. (Data from: *The Wall Street Journal.*)

    (a) Write a linear equation expressing the percentage of the luxury handbag market y that Michael Kors had x years after 2000.

    (b) What was the percentage of the market that Michael Kors had in 2013?

    (c) If the trend continues, what will be the first full year in which Michael Kors had 30% of the market?

78. **Cocoa Production** In 2005, global cocoa bean production was approximately 3.3 million metric tons. In 2015, production was approximately 4.1 million metric tons. (Data from: *The Wall Street Journal.*)

    (a) Write a linear equation expressing cocoa bean production y in terms of the number of years x after 2000.

    (b) What was the cocoa bean production in 2012?

    (c) If the trend continues, what will be the first full year in which cocoa bean production exceeds 5 million metric tons?

79. **Employee Health Costs** In the year 2010, the average annual cost to employees for employer family health coverage was approximately $3500. In 2015, that amount was approximately $5000. (Data from: *The Wall Street Journal*.)

    (a) Write a linear equation expressing the average annual cost to employees *y* in terms of the number of years *x* after 2000.

    (b) What was the average annual cost in 2014?

    (c) If the trend continues, what will be the first full year in which costs exceed $6000?

80. **Employer Health Costs** In the year 2010, the average annual cost to employers for employer family health coverage was approximately $10,000. In 2015, that amount was approximately $17,000. (Data from: *The Wall Street Journal*.)

    (a) Write a linear equation expressing the average annual cost to employers *y* in terms of the number of years *x* after 2000.

    (b) What was the average annual cost in 2012?

    (c) If the trend continues, what will be the first full year in which costs exceed $25,000?

✓ **Checkpoint Answers**

1. **(a)** $6/5$ **(b)** $-9/2$

2. **(a)** $y = \dfrac{2}{3}x - 3$ **(b)** $y = -\dfrac{3}{2}x + \dfrac{1}{4}$

3. **(a)** Slope $-1/4$; *y*-intercept $3/2$

    (b) Slope $3/2$; *y*-intercept $-1/2$

4. **(a)** Slope of $E = -.3$; slope of $F = -1$; slope of $G = -2$;
    slope of $H = -5$.

(b)

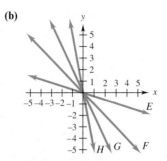

(c) The larger the slope in absolute value, the more steeply the line falls from left to right.

5. **(a)**

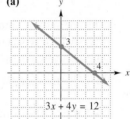

$3x + 4y = 12$

(b)

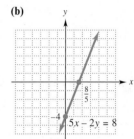

$5x - 2y = 8$

6. **(a)** Perpendicular **(b)** Neither   **(c)** Parallel

7. **(a)** $y + 2 = -\dfrac{3}{5}(x - 5); y = -\dfrac{3}{5}x + 1$.

    (b) $y - 8 = \dfrac{1}{3}(x - 6); y = \dfrac{1}{3}x + 6$.

8. **(a)** $y = -\dfrac{1}{2}x + 4$ **(b)** $y = -\dfrac{8}{11}x - \dfrac{42}{11}$

9. **(a)** $y = 10.1x - 79.9$ **(b)** 122.1 million

2.3 Linear Models

In business and science, it is often necessary to make judgments on the basis of data from the past. For instance, a stock analyst might use a company's profits in previous years to estimate the next year's profits. Or a life insurance company might look at life expectancies of people born in various years to predict how much money it should expect to pay out in the next year.

In such situations, the available data is used to construct a mathematical model, such as an equation or a graph, which is used to approximate the likely outcome in cases where complete data is not available. In this section, we consider applications in which the data can be modeled by a linear equation.

The simplest way to construct a linear model is to use the line determined by two of the data points, as illustrated in the following example.

EXAMPLE 1 **Full-Time Faculty Members** The number of full-time faculty members at four-year colleges and universities (in thousands) in selected years is shown in the following table. (Data from: U.S. National Center for Education Statistics.)

Year	2005	2007	2009	2011	2013
Number of Faculty Members	917	990	1038	1115	1152

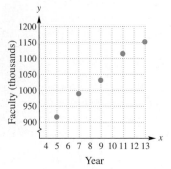

Figure 2.29

(a) Let $x = 5$ correspond to the year 2005. Plot the points (x, y) where x is the year and y is the number of full-time faculty members.

Solution The data points are (5, 917), (7, 990), (9, 1038), (11, 1115), and (13, 1152), as shown in Figure 2.29.

(b) Use the data points (9, 1038) and (11, 1115) to find a line that models these data.

Solution The slope of the line through (9, 1038) and (11, 1115) is $\dfrac{1115 - 1038}{11 - 9} = \dfrac{77}{2} = 38.5$.

Using the point (9, 1038) and the slope of 38.5, we find the equation of this line is

$$y - 1038 = 38.5(x - 9) \qquad \text{Point–slope form for the equation of a line}$$
$$y - 1038 = 38.5x - 346.5 \qquad \text{Distributive property}$$
$$y = 38.5x + 691.5. \qquad \text{Slope–intercept form}$$

The line and the data points are shown in Figure 2.30. Although the line fits the two points that we used to calculate the slope perfectly, it seems to underestimate the first two points and overestimate the last point.

(c) Use the data points (5, 917) and (7, 990) to find a line that models these data.

Solution The slope of the line through (5, 917) and (7, 990) is $\dfrac{990 - 917}{7 - 5} = \dfrac{73}{2} = 36.5$.

Using the point (5, 917) and the slope of 36.5, we find the equation of this line is

$$y - 917 = 36.5(x - 5) \qquad \text{Point–slope form for the equation of a line}$$
$$y - 917 = 36.5x - 182.5 \qquad \text{Distributive property}$$
$$y = 36.5x + 734.5. \qquad \text{Slope–intercept form}$$

The line and the data points are shown in Figure 2.31. Although the line fits the first two points perfectly, it seems to overestimate all three latter points. ✓₁

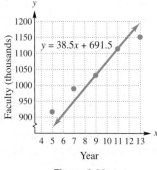

Figure 2.30

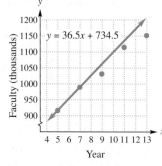

Figure 2.31

✓ **Checkpoint 1**

Use the points (5, 917) and (9, 1038) to find another model for the data in Example 1.

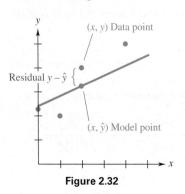

Figure 2.32

Opinions may vary as to which of the lines in Example 1 best fits the data. To make a decision, we might try to measure the amount of error in each model. One way to do this is to compute the difference between the number of faculty y and the amount $\hat{y}$ given by the model. If the data point is (x, y) and the corresponding point on the line is $(x, \hat{y})$, then the difference $y - \hat{y}$ measures the error in the model for that particular value of x. The number $y - \hat{y}$ is called a **residual**. As shown in Figure 2.32, the residual $y - \hat{y}$ is the vertical distance from the data point to the line (positive when the data point is above the line, negative when it is below the line, and 0 when it is on the line).

One way to determine how well a line fits the data points is to compute the sum of its residuals—that is, the sum of the individual errors. Unfortunately, however, the sum of the residuals of two different lines might be equal, thwarting our effort to decide which is the better fit. Furthermore, the residuals may sum to 0, which doesn't mean that there is no error, but only that the positive and negative errors (which might be quite large) add up to 0. (See Exercise 11 at the end of this section for an example.)

To avoid this difficulty, mathematicians use the sum of the *squares* of the residuals to measure how well a line fits the data points. When the sum of the squares is used, a smaller sum means a smaller overall error and hence a better fit. The error is 0 only when all the data points lie on the line (a perfect fit).

EXAMPLE 2 **Full-Time Faculty Members** Two linear models for the number of full-time faculty members at four-year colleges and universities were constructed in Example 1:

$$y = 38.5x + 691.5 \quad \text{and} \quad y = 36.5x + 734.5.$$

For each model, determine the five residuals, square each residual, and sum the squares of the residuals.

Solution The information for each model is summarized in the following tables.

$$y = 38.5x + 691.5$$

Data Point (x, y)	Model Point $(x, \hat{y})$	Residual $y - \hat{y}$	Squared Residual $(y - \hat{y})^2$
(5, 917)	(5, 884)	33	1089
(7, 990)	(7, 961)	29	841
(9, 1038)	(9, 1038)	0	0
(11, 1115)	(11, 1115)	0	0
(13, 1152)	(13, 1192)	−40	1600
			Sum = 3530

$$y = 36.5x + 734.5$$

Data Point (x, y)	Model Point $(x, \hat{y})$	Residual $y - \hat{y}$	Squared Residual $(y - \hat{y})^2$
(5, 917)	(5, 917)	0	0
(7, 990)	(7, 990)	0	0
(9, 1038)	(9, 1063)	−25	625
(11, 1115)	(11, 1136)	−21	441
(13, 1152)	(13, 1209)	−57	3249
			Sum = 4315

According to this measure of the error, the line $y = 38.5x + 691.5$ is a better fit for these data because the sum of the squares of its residuals is smaller than the sum of the squares of the residuals for $y = 36.5x + 734.5$. ✓2

✓**Checkpoint 2**

Another model for the data in Examples 1 and 2 is $y = 37x + 702$. Use this line to find

(a) the residuals and

(b) the sum of the squares of the residuals.

(c) Does this line fit the data better than the two lines in Example 2?

Linear Regression (Optional)*

Mathematical techniques from multivariable calculus can be used to prove the following result.

> For any set of data points, there is one, and only one, line for which the sum of the squares of the residuals is as small as possible.

---

*Examples 3–6 require either a graphing calculator or a spreadsheet program.

This *line of best fit* is called the **least-squares regression line,** and the computational process for finding its equation is called **linear regression.** Linear-regression formulas are quite complicated and require a large amount of computation. Fortunately, most graphing calculators and spreadsheet programs can do linear regression quickly and easily.

📄 **NOTE** The process outlined here works for most TI graphing calculators. Other graphing calculators and spreadsheet programs operate similarly, but check your instruction manual.

EXAMPLE 3 **Full-Time Faculty Members** Recall the number of full-time faculty members at four-year colleges and universities (in thousands) in selected years was as follows.

Year	2005	2007	2009	2011	2013
Number of Faculty Members	917	990	1038	1115	1152

Use a graphing calculator or spreadsheet to do the following:

(a) Plot the data points with $x = 5$ corresponding to the year 2005.

Solution The data points are $(5, 917)$, $(7, 990)$, $(9, 1038)$, $(11, 1115)$, and $(13, 1152)$. Press STAT EDIT to bring up the statistics editor. Enter the x-coordinates as list L_1 and the corresponding y-coordinates as L_2, as shown in Figure 2.33. To plot the data points, go to the STAT PLOT menu, choose a plot (here it is Plot 1), choose ON, and enter the lists L_1 and L_2 as shown in Figure 2.34. Then set the viewing window as usual and press GRAPH to produce the plot in Figure 2.35.

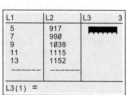

Figure 2.33

Figure 2.34

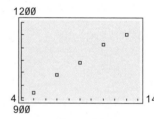

Figure 2.35

(b) Find the least squares regression line for these data.

Solution Go to the STAT CALC menu and choose LINREG, which returns you to the home screen. As shown in Figure 2.36, enter the list names and the place where the equation of the regression line should be stored (here, Y_1 is chosen; it is on the VARS Y-VARS FUNCTION menu); then press ENTER. Figure 2.37 shows that the equation of the regression line is

$$y = 29.75x + 774.65$$

✓ **Checkpoint 3**

Use the least-squares regression line $y = 29.75x + 774.65$ and the data points of Example 3 to find

(a) the residuals and

(b) the sum of the squares of the residuals.

(c) How does this line compare with those in Example 1 and Checkpoint 2?

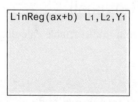

Figure 2.36

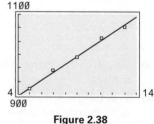

Figure 2.37

Figure 2.38

(c) Graph the data points and the regression line on the same screen.

Solution Press GRAPH to see the line plotted with the data points (Figure 2.38).

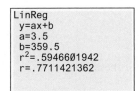

LinReg
y=ax+b
a=3.5
b=359.5
r^2=.5946601942
r=.7711421362

Figure 2.39

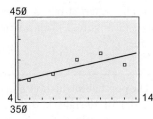

Figure 2.40

✓**Checkpoint 4**

Using only the data from 2005, 2007, and 2009 in Example 4, find the equation of the least-squares regression line. Round the coefficients to two decimal places.

EXAMPLE 4 **Full-Time Faculty Members** The following table gives the number (in thousands) of full-time faculty at two-year colleges. (Data from: U.S. National Center for Education Statistics.)

Year	2005	2007	2009	2011	2013
Number of Faculty	373	380	400	408	394

(a) Let $x = 5$ correspond to the year 2005. Use a graphing calculator or spreadsheet program to find the least-squares regression line that models the data in the table.

Solution The data points are (5, 373), (7, 380), (9, 400), (11, 408), and (13, 394). Enter the x-coordinates as list L_1 and the corresponding y-coordinates as list L_2 in a graphing calculator and then find the regression line as in Figure 2.39.

(b) Plot the data points and the regression line on the same screen.

Solution See Figure 2.40, which shows that the line is a reasonable model for the data.

(c) Assuming the trend continues, estimate the number of faculty in the year 2020.

Solution The year 2020 corresponds to $x = 20$. Substitute $x = 20$ into the regression-line equation:

$$y = 3.5(20) + 359.5 = 429.5.$$

This model estimates that the number of full-time faculty at two-year colleges will be approximately 429,500 in the year 2020. ✓ 4

Correlation

Although the "best fit" line can always be found by linear regression, it may not be a good model. For instance, if the data points are widely scattered, no straight line will model the data accurately. We calculate a value called the **correlation coefficient** to assess the fit of the regression line to the scatterplot. It measures how closely the data points fit the regression line and thus indicates how good the regression line is for predictive purposes.

The correlation coefficient r is always between -1 and 1. When $r = \pm 1$, the data points all lie on the regression line (a perfect fit). When the absolute value of r is close to 1, the line fits the data quite well, and when r is close to 0, the line is a poor fit for the data (but some other curve might be a good fit). Figure 2.41 shows how the value of r varies, depending on the pattern of the data points.

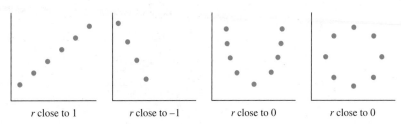

r close to 1   r close to −1   r close to 0   r close to 0

Figure 2.41

EXAMPLE 5 **Unemployment Rate** The percentage of unemployed workers in the U.S. labor force in January of recent years is shown in the table. (Data from: U.S. Department of Labor.)

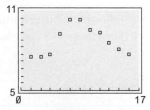

Figure 2.42

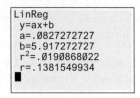

Figure 2.43

✓ **Checkpoint 5**

Using only the data from 2010 and later in Example 5, find

(a) the equation of the least-squares regression line and

(b) the correlation coefficient.

(c) How well does this line fit the data?

Year	Rate	Year	Rate
2006	4.7	2012	8.3
2007	4.6	2013	8.0
2008	5.0	2014	6.6
2009	7.8	2015	5.7
2010	9.8	2016	4.9
2011	9.7		

Determine whether or not a linear equation is a good model for these data.

Solution Let $x = 6$ correspond to the year 2006, and plot the data points $(6, 4.7)$, and so on, either by hand or with a graphing calculator, as in Figure 2.42. They do not form a linear pattern (because unemployment rates tend to rise and fall). Alternatively, you can compute the regression equation for these data, as in Figure 2.43. Figure 2.43 shows the correlation coefficient is $r \approx .138$, which is a relatively low value. ✓ 5

EXAMPLE 6   **Higher Education Enrollment** The population (in thousands) and enrollment in all higher education institutions (in thousands) for 10 states in the year 2013 is given in the table below. (Data from: U.S. Center for Education Statistics and the U.S. Census Bureau.)

State	Population	Enrollment	State	Population	Enrollment
Alabama	4833	306	Missouri	6044	438
Colorado	5268	359	New Jersey	8899	437
Georgia	9992	533	Ohio	11,571	697
Iowa	3090	340	Oklahoma	8260	584
Maryland	5929	364	South Carolina	4775	258

(a) Use a graphing calculator or spreadsheet program to find a linear model that uses state population (x) to predict the enrollment in higher education (y).

Solution The least squares regression line (with coefficients rounded) is

$$y = .045x + 121.7,$$

as shown in Figure 2.44. The correlation coefficient is $r \approx .886$, which is quite high, and Figure 2.45 shows a linear trend. Thus, the line fits these data well.

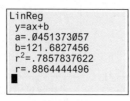

Figure 2.44

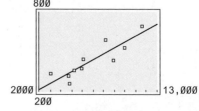

Figure 2.45

(b) Michigan had a population of 9,896 thousand in 2013. Predict its enrollment in higher education.

Solution Let $x = 9896$ in the regression equation:

$$y = .045(9896) + 121.7 = 567.02.$$

Therefore, the estimated higher education enrollment for Michigan would be 567,020 students.

(c) According to this model, what is the state population that would predict a higher education enrollment of 800 thousand students?

Solution Let $y = 800$ and solve the regression equation for x:

$$800 = .045x + 121.7 \qquad \text{Let } y = 800.$$
$$678.3 = .045x \qquad \text{Subtract 121.7 from both sides.}$$
$$x \approx 15073.333. \qquad \text{Divide both sides by .045.}$$

Thus, the model predicts that a state with a population of 15,073,333 has an enrollment of 800,000 in higher education.

NOTE The value r^2 seen on the TI calculator output is known as the coefficient of determination. It is simply the value of the correlation coefficient squared and $0 \le r^2 \le 1$. One interpretation of its value is that it describes the proportion of variation in the y-variable that is explained by variation in the x-variable.

2.3 Exercises

1. **Temperature** The following table shows equivalent Fahrenheit and Celsius temperatures:

Degrees Fahrenheit	32	68	104	140	176	212
Degrees Celsius	0	20	40	60	80	100

(a) Choose any two data points and use them to construct a linear equation that models the data, with x being Fahrenheit and y being Celsius.

(b) Use the model in part (a) to find the Celsius temperature corresponding to

50° Fahrenheit and 75° Fahrenheit.

Temperature *Use the linear equation derived in Exercise 1 to work the following problems.*

2. Convert each temperature.

(a) 58°F to Celsius   (b) 50°C to Fahrenheit

(c) −10°C to Fahrenheit   (d) −20°F to Celsius

3. According to the *World Almanac and Book of Facts*, 2008, Venus is the hottest planet, with a surface temperature of 867° Fahrenheit. What is this temperature in degrees Celsius?

4. Find the temperature at which Celsius and Fahrenheit temperatures are numerically equal.

In each of the next set of problems, assume that the data can be modeled by a straight line and that the trend continues indefinitely. Use two data points to find such a line and then estimate the requested quantities. (See Example 1.)

5. **CPI** The Consumer Price Index (CPI), which measures the cost of a typical package of consumer goods, was 224.9 in the year 2011 and 240.0 in 2016. Let $x = 11$ correspond to the year 2011 and estimate the CPI in 2012 and 2015. (Data from: U.S. Bureau of Labor Statistics.)

6. **Tax Returns** The approximate number (in millions) of individual electronically filed tax returns filed with the Internal Revenue Service (IRS) in the year 2005 was 68.5, and it was 123.7 in 2016. Let $x = 5$ correspond to the year 2005, and estimate the number of electronically filed returns in 2010 and 2014. *Hint:* Round slope to two decimal places. (Data from: U.S. Internal Revenue Service.)

7. **Educational Services** The number (in millions) of employees working in educational services was 12.3 in 2005 and 13.3 in 2014. Let $x = 5$ correspond to the year 2005, and estimate the number of employees in 2012. *Hint:* Round slope to two decimal places. (Data from: U.S. Census Bureau.)

8. **Leisure and Hospitality** The number (in millions) of employees working in the leisure and hospitality industries was 12.1 in 2005 and 13.5 in 2014. Let $x = 5$ correspond to the year 2005, and estimate the number of employees in 2013. *Hint:* Round slope to two decimal places. (Data from: U.S. Census Bureau.)

9. **Baseball** Suppose a baseball is thrown at 85 miles per hour. The ball will travel 320 feet when hit by a bat swung at 50 miles per hour and will travel 440 feet when hit by a bat swung at 80 miles per hour. Let y be the number of feet traveled by the ball when hit by a bat swung at x miles per hour. (*Note*: The preceding data are valid for $50 \le x \le 90$, where the bat is 35 inches long, weighs 32 ounces, and strikes a waist-high pitch so that the place of the swing lies at 10° from the diagonal). [Data from: Robert K. Adair, *The Physics of Baseball* (HarperCollins, 1990)]. How much farther will a ball travel for each mile-per-hour increase in the speed of the bat?

10. **Skiing** Ski resorts require large amounts of water in order to make snow. Snowmass Ski Area in Colorado plans to pump at least 1120 gallons of water per minute for at least 12 hours a day from Snowmass Creek between mid-October and late December. Environmentalists are concerned about the effects on the ecosystem. Find the minimum amount of water pumped in 30 days.

[*Hint*: Let y be the total number of gallons pumped x days after pumping begins. Note that $(0, 0)$ is on the graph of the equation.] (Data from: York Snow, Inc.)

In each of the next two problems, two linear models are given for the data. For each model,

(a) *find the residuals and their sum;*

(b) *find the sum of the squares of the residuals;*

(c) *decide which model is the better fit. (See Example 2.)*

11. Cisco Sales The following table shows the estimated sales (in billions of dollars) in China for Cisco Systems, Inc. for several years.

Year	2006	2008	2010	2012	2014
Sales	.7	1.22	1.53	2.02	1.81

Let $x = 6$ correspond to the year 2006. Two equations that model these data are $y = .07x + .83$ and $y = .2x - .38$ (Data from: *The Wall Street Journal*.)

12. International Trade The following table shows South Korea's petrochemical exports to China (in billions of dollars) for several years.

Year	2010	2011	2012	2013	2014
Petrochemical exports	16	21	21	23	21.4

Let $x = 10$ correspond to the year 2010. Two equations that model these data are $y = 1.35x + 2.5$ and $y = 2x - 3$ (Data from: *The Wall Street Journal*.)

In each of the following problems, determine whether a straight line is a good model for the data. Do this visually by plotting the data points and by finding the correlation coefficient for the least-squares regression line. (See Examples 5 and 6.)

13. Credit Unions The following table shows the number of operating federal credit unions in the United States for several years.

Year	2011	2012	2013	2014	2015
Number of federal credit unions	4444	4268	4100	3924	3762

Let $x = 11$ correspond to the year 2011. (Data from: Credit Union National Association.)

14. Gas Prices The following tables shows the average price (in dollars) for a gallon of regular unleaded gasoline on the first weekday of selected months in the year 2015. Let $x = 1$ correspond to the month of January, $x = 3$ correspond to the month of March, and so on. (Data from: U.S. Energy Information Administration.)

Month	January	March	May	July	September	November
Price per gallon	$2.14	$2.36	$2.52	$2.70	$2.34	$2.17

In Exercises 15–18 find the required linear model using leastsquares regression. (See Examples 3–6.)

15. Credit Unions Use the data on the number of federal credit unions in Exercise 13.

(a) Find a linear model for these data with $x = 11$ corresponding to the year 2011.

(b) Assuming the trend continues, estimate the number of federal credit unions in the year 2018.

16. Student Loans The accompanying table shows the approximate amount (in billions of dollars) of student loans in repayment in various quarters of 2015 and 2016. Let $x = 1$ correspond to the first quarter of 2015. (Data from: *The Wall Street Journal*.)

Quarter	1	2	3	4	5	6	7
Loans in repayment	510	525	570	580	610	630	650

(a) Find a linear model for these data.

(b) Assuming the trend continues, estimate the amount of loans in repayment in quarter 8.

17. Google Ad Revenue The accompanying table shows the approximate amount (in billions of dollars) of Google advertising revenue in various quarters of 2015 and 2016. Let $x = 1$ correspond to the first quarter of 2015. (Data from: *The Wall Street Journal*.)

Quarter	1	2	3	4	5	6	7	8
Ad revenue	15.1	16.2	16.9	18.0	17.5	18.0	19.2	22.4

(a) Find a linear model or these data.

(b) Assuming the trend continues, estimate the amount of ad revenue in the third quarter of 2017.

18. Blood Pressure Researchers wish to determine if a relationship exists between age and systolic blood pressure (SBP). The accompanying table gives the age and systolic blood pressure for 10 adults.

Age (x)	35	37	42	46	53	57	61	65	69	74
SBP (y)	102	127	120	131	126	137	140	130	148	147

(a) Find a linear model for these data using age (x) to predict systolic blood pressure (y).

(b) Use the results from part (a) to predict the systolic blood pressure for a person who is 42-years-old, 53-years-old, and 69-years-old. How well do the actual data agree with the predicted values?

(c) Use the results from part (a) to predict the systolic blood pressure for a person who is 45-years-old and 70-years-old.

Work these problems.

19. Street and Highway Construction The total amount of construction spending on highways and streets (in billions of dollars) is given in the accompanying table for the first six months of 2016.

Month	January	February	March	April	May	June
Spending	97	94	94	90	89	88

(a) Find the least squares regression line that models these data, with $x = 1$ corresponding to January 2016.

(b) Assuming the trend continues, estimate the amount of spending in December 2016.

20. **Cable Subscribers** The number of basic cable television subscribers (in millions) is shown in the following table for various years. (Data from: ProQuest Statistical Abstract of the United States 2016.)

Year	2006	2008	2010	2012	2014
Subscribers	65.8	64.8	61.5	57.8	54.4

(a) Find the least-squares regression line that models these data with $x = 6$ corresponding to the year 2006.

(b) Find the number of subscribers for the year 2013.

(c) If the trend continues indefinitely, determine the first full year when there will be at least 50 million subscribers.

(d) Find the correlation coefficient.

21. **Software Publishers** The revenue per quarter for the software publishing industry (in billions of dollars) for the 8 quarters of 2015 and 2016 is displayed in the accompanying table. Let $x = 1$ correspond to the first quarter of 2015. (Data from: U.S. Census Bureau.)

Quarter	1	2	3	4	5	6	7	8
Revenue	47.4	48.9	49.8	49.3	49.8	49.8	49.6	50.4

(a) Find the least-squares regression line that models these data.

(b) If the trend continues indefinitely, find the revenue for the second quarter of 2017.

(c) If the trend continues indefinitely, what is the first full quarter when revenue exceeds $52 billion?

(d) Find the correlation coefficient.

22. **Life Expectancy** The following table shows men's and women's life expectancy at birth (in years) for selected birth years in the United States: (Data from: U.S. Center for National Health Statistics.)

Birth Year	Life Expectancy	
	Men	Women
1970	67.1	74.7
1975	68.8	76.6
1980	70.0	77.4
1985	71.1	78.2
1990	71.8	78.8
1995	72.5	78.9
2000	74.1	79.3
2005	75.0	80.1
2010	76.2	81.0
2015	77.1	81.7

(a) Find the least-squares regression line for the men's data, with $x = 70$ corresponding to 1970.

(b) Find the least-squares regression line for the women's data, with $x = 70$ corresponding to 1970.

(c) Suppose life expectancy continues to increase as predicted by the equations in parts (a) and (b). Will men's life expectancy ever be higher than women's? If so, in what birth year will this occur?

✓**Checkpoint Answers**

1. $y = 30.25x + 765.75$

2. (a) 30, 29, 3, 6, −31

   (b) 2747

   (c) Yes, because the sum of the squares of the residuals is lower.

3. (a) −6.4, 7.1, −4.4, 13.1, −9.4

   (b) 370.7

   (c) It fits the data best because the sum of the squares of its residuals is smallest.

4. $y = 6.75x + 337.08$

5. (a) $y = .87x + 18.9$

   (b) $r \approx -.988$

   (c) It fits reasonably well because $|r|$ is very close to 1 and the pattern is linear.

2.4 Linear Inequalities

An **inequality** is a statement that one mathematical expression is greater than (or less than) another. Inequalities are very important in applications. For example, a company wants revenue to be *greater than* costs and must use *no more than* the total amount of capital or labor available.

Inequalities may be solved by algebraic or geometric methods. In this section, we shall concentrate on algebraic methods for solving **linear inequalities**, such as

$$4 - 3x \le 7 + 2x \quad \text{and} \quad -2 < 5 + 3m < 20,$$

and absolute-value inequalities, such as $|x - 2| < 5$. The following properties are the basic algebraic tools for working with inequalities.

> ## Properties of Inequality
>
> For real numbers a, b, and c,
>
> **(a)** if $a < b$, then $a + c < b + c$;
>
> **(b)** if $a < b$, and if $c > 0$, then $ac < bc$;
>
> **(c)** if $a < b$, and if $c < 0$, then $ac > bc$.

Throughout this section, definitions are given only for $<$, but they are equally valid for $>$, $\leq$, and $\geq$.

⚠ **CAUTION** Pay careful attention to part (c) in the previous box; if both sides of an inequality are multiplied by a negative number, the direction of the inequality symbol must be reversed. For example, starting with the true statement $-3 < 5$ and multiplying both sides by the positive number 2 gives

$$-3 \cdot 2 < 5 \cdot 2,$$

or

$$-6 < 10,$$

still a true statement. However, starting with $-3 < 5$ and multiplying both sides by the negative number -2 gives a true result only if the direction of the inequality symbol is reversed:

$$-3(-2) > 5(-2)$$
$$6 > -10. \; ✓_1$$

✓ **Checkpoint 1**

(a) First multiply both sides of $-6 < -1$ by 4, and then multiply both sides of $-6 < -1$ by -7.

(b) First multiply both sides of $9 \geq -4$ by 2, and then multiply both sides of $9 \geq -4$ by -5.

(c) First add 4 to both sides of $-3 < -1$, and then add -6 to both sides of $-3 < -1$.

EXAMPLE 1   Solve $3x + 5 > 11$. Graph the solution.

Solution First add -5 to both sides:

$$3x + 5 + (-5) > 11 + (-5)$$
$$3x > 6.$$

Now multiply both sides by $1/3$:

$$\frac{1}{3}(3x) > \frac{1}{3}(6)$$
$$x > 2.$$

(Why was the direction of the inequality symbol not changed?) In interval notation (introduced in Section 1.1), the solution is the interval $(2, \infty)$, which is graphed on the number line in Figure 2.46. The parenthesis at 2 shows that 2 is not included in the solution.

Figure 2.46

✓ **Checkpoint 2**

Solve these inequalities. Graph each solution.

(a) $5z - 11 < 14$

(b) $-3k \leq -12$

(c) $-8y \geq 32$

As a partial check, note that 0, which is not part of the solution, makes the inequality false, while 3, which is part of the solution, makes it true:

$$\overset{?}{3(0) + 5 > 11} \qquad\qquad \overset{?}{3(3) + 5 > 11}$$
$$5 > 11 \quad \text{False} \qquad\qquad 14 > 11. \quad \text{True} \; ✓_2$$

EXAMPLE 2   Solve $4 - 3x \leq 7 + 2x$.

Solution   Add -4 to both sides:

$$4 - 3x + (-4) \leq 7 + 2x + (-4)$$
$$-3x \leq 3 + 2x.$$

Add $-2x$ to both sides (remember that *adding* to both sides never changes the direction of the inequality symbol):

$$-3x + (-2x) \leq 3 + 2x + (-2x)$$
$$-5x \leq 3.$$

Multiply both sides by $-1/5$. Since $-1/5$ is negative, change the direction of the inequality symbol:

$$-\frac{1}{5}(-5x) \geq -\frac{1}{5}(3)$$

$$x \geq -\frac{3}{5}.$$

Figure 2.47 shows a graph of the solution, $[-3/5, \infty)$. The bracket in Figure 2.47 shows that $-3/5$ is included in the solution. ✓ 3

$-\frac{3}{5}$

Figure 2.47

✓ **Checkpoint 3**

Solve these inequalities. Graph each solution.

(a) $8 - 6t \geq 2t + 24$

(b) $-4r + 3(r + 1) < 2r$

EXAMPLE 3   Solve $-2 < 5 + 3m < 20$. Graph the solution.

Solution   The inequality $-2 < 5 + 3m < 20$ says that $5 + 3m$ is between -2 and 20. We can solve this inequality with an extension of the properties given at the beginning of this section. Work as follows, first adding -5 to each part:

$$-2 + (-5) < 5 + 3m + (-5) < 20 + (-5)$$
$$-7 < 3m < 15.$$

Now multiply each part by $1/3$:

$$-\frac{7}{3} < m < 5.$$

A graph of the solution, $(-7/3, 5)$, is given in Figure 2.48. ✓ 4

$-\frac{7}{3}$ 5

Figure 2.48

✓ **Checkpoint 4**

Solve each of the given inequalities. Graph each solution.

(a) $9 < k + 5 < 13$

(b) $-6 \leq 2z + 4 \leq 12$

EXAMPLE 4   **Temperature Conversion** The formula for converting from Celsius to Fahrenheit temperature is

$$F = \frac{9}{5}C + 32.$$

What Celsius temperature range corresponds to the range from $32°F$ to $77°F$?

Solution   The Fahrenheit temperature range is $32 < F < 77$. Since $F = (9/5)C + 32$, we have

$$32 < \frac{9}{5}C + 32 < 77.$$

Solve the inequality for C:

$$32 < \frac{9}{5}C + 32 < 77$$

$$0 < \frac{9}{5}C < 45 \qquad \text{Subtract 32 from each part.}$$

$$\frac{5}{9} \cdot 0 < \frac{5}{9} \cdot \frac{9}{5}C < \frac{5}{9} \cdot 45 \qquad \text{Multiply each part by } \frac{5}{9}.$$

$$0 < C < 25.$$

✓ **Checkpoint 5**

In Example 4, what Celsius temperatures correspond to the range from 5°F to 95°F?

The corresponding Celsius temperature range is 0°C to 25°C. ✓₅

A product will break even or produce a profit only if the revenue R from selling the product at least equals the cost C of producing it—that is, if $R \geq C$.

EXAMPLE 5   **Revenue and Cost** A company analyst has determined that the cost to produce and sell x units of a certain product is $C = 20x + 1000$. The revenue for that product is $R = 70x$. Find the values of x for which the company will break even or make a profit on the product.

Solution Solve the inequality $R \geq C$:

$$R \geq C$$
$$70x \geq 20x + 1000 \qquad \text{Let } R = 70x \text{ and } C = 20x + 1000.$$
$$50x \geq 1000 \qquad \text{Subtract } 20x \text{ from both sides.}$$
$$x \geq 20. \qquad \text{Divide both sides by 50.}$$

The company must produce and sell 20 items to break even and more than 20 to make a profit.

EXAMPLE 6   **Break-Even Point** A pretzel manufacturer can sell a 6-ounce bag of pretzels to a wholesaler for $.35 a bag. The variable cost of producing each bag is $.25 per bag, and the fixed cost for the manufacturing operation is $110,000.* How many bags of pretzels need to be sold in order to break even or earn a profit?

Solution Let x be the number of bags produced. Then the revenue equation is

$$R = .35x,$$

and the cost is given by

$$\text{Cost} = \text{Fixed Costs} + \text{Variable Costs}$$
$$C = 110,000 + .25x.$$

We now solve the inequality $R \geq C$:

$$R \geq C$$
$$.35x \geq 110,000 + .25x$$
$$.1x \geq 110,000$$
$$x \geq 1,110,000.$$

The manufacturer must produce and sell 1,110,000 bags of pretzels to break even and more than that to make a profit.

_____

*Variable costs, fixed costs, and revenue were discussed on p. 15.

Absolute-Value Inequalities

You may wish to review the definition of absolute value in Section 1.1 before reading the following examples, which show how to solve inequalities involving absolute values.

EXAMPLE 7 Solve each inequality.

(a) $|x| < 5$

Solution Because absolute value gives the distance from a number to 0, the inequality $|x| < 5$ is true for all real numbers whose distance from 0 is less than 5. This includes all numbers between -5 and 5, or numbers in the interval $(-5, 5)$. A graph of the solution is shown in Figure 2.49.

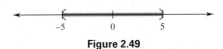

Figure 2.49

(b) $|x| > 5$

Solution The solution of $|x| > 5$ is given by all those numbers whose distance from 0 is *greater* than 5. This includes the numbers satisfying $x < -5$ or $x > 5$. A graph of the solution, all numbers in

$$(-\infty, -5) \quad \text{or} \quad (5, \infty),$$

is shown in Figure 2.50. ✓ 6

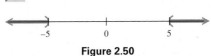

Figure 2.50

The preceding examples suggest the following generalizations.

> Assume that a and b are real numbers and that b is positive.
>
> **1.** Solve $|a| < b$ by solving $-b < a < b$.
> **2.** Solve $|a| > b$ by solving $a < -b$ or $a > b$.

EXAMPLE 8 Solve $|x - 2| < 5$.

Solution Replace a with $x - 2$ and b with 5 in property (1) in the box above. Now solve $|x - 2| < 5$ by solving the inequality

$$-5 < x - 2 < 5.$$

Add 2 to each part, getting the solution

$$-3 < x < 7,$$

which is graphed in Figure 2.51. ✓ 7

EXAMPLE 9 Solve $|2 - 7m| - 1 > 4$.

Solution First add 1 to both sides:

$$|2 - 7m| > 5$$

Now use property (2) from the preceding box to solve $|2 - 7m| > 5$ by solving the inequality

$$2 - 7m < -5 \quad \text{or} \quad 2 - 7m > 5.$$

✓ **Checkpoint 6**

Solve each inequality. Graph each solution.

(a) $|x| \le 1$
(b) $|y| \ge 3$

Figure 2.51

✓ **Checkpoint 7**

Solve each inequality. Graph each solution.

(a) $|p + 3| < 4$
(b) $|2k - 1| \le 7$

Solve each part separately:

$$-7m < -7 \qquad \text{or} \qquad -7m > 3$$

$$m > 1 \qquad \text{or} \qquad m < -\frac{3}{7}.$$

The solution, all numbers in $\left(-\infty, -\frac{3}{7}\right)$ or $(1, \infty)$, is graphed in Figure 2.52. ✓ 8

Figure 2.52

✓ **Checkpoint 8**

Solve each inequality. Graph each solution.

(a) $|y - 2| > 5$

(b) $|3k - 1| \geq 2$

(c) $|2 + 5r| - 4 \geq 1$

✓ **Checkpoint 9**

Solve each inequality.

(a) $|5m - 3| > -10$

(b) $|6 + 5a| < -9$

(c) $|8 + 2r| > 0$

EXAMPLE 10 Solve $|3 - 7x| \geq -8$.

Solution The absolute value of a number is always nonnegative. Therefore, $|3 - 7x| \geq -8$ is always true, so the solution is the set of all real numbers. Note that the inequality $|3 - 7x| \leq -8$ has no solution, because the absolute value of a quantity can never be less than a negative number. ✓ 9

2.4 Exercises

1. Explain how to determine whether a parenthesis or a bracket is used when graphing the solution of a linear inequality.

2. The three-part inequality $p < x < q$ means "p is less than x and x is less than q." Which one of the given inequalities is not satisfied by any real number x? Explain why.

   (a) $-3 < x < 5$ (b) $0 < x < 4$

   (c) $-7 < x < -10$ (d) $-3 < x < -2$

Solve each inequality and graph each solution. (See Examples 1–3.)

3. $-8k \leq 32$ 4. $-4a \leq 36$

5. $-2b > 0$ 6. $6 - 6z < 0$

7. $3x + 4 \leq 14$ 8. $2y - 7 < 9$

9. $-5 - p \geq 3$ 10. $5 - 3r \leq -4$

11. $7m - 5 < 2m + 10$ 12. $6x - 2 > 4x - 10$

13. $m - (4 + 2m) + 3 < 2m + 2$

14. $2p - (3 - p) \leq -7p - 2$

15. $-2(3y - 8) \geq 5(4y - 2)$

16. $5r - (r + 2) \geq 3(r - 1) + 6$

17. $3p - 1 < 6p + 2(p - 1)$

18. $x + 5(x + 1) > 4(2 - x) + x$

19. $-7 < y - 2 < 5$ 20. $-3 < m + 6 < 2$

21. $8 \leq 3r + 1 \leq 16$ 22. $-6 < 2p - 3 \leq 5$

23. $-4 \leq \dfrac{2k - 1}{3} \leq 2$ 24. $-1 \leq \dfrac{5y + 2}{3} \leq 4$

25. $\dfrac{3}{5}(2p + 3) \geq \dfrac{1}{10}(5p + 1)$

26. $\dfrac{8}{3}(z - 4) \leq \dfrac{2}{9}(3z + 2)$

In the following exercises, write a linear inequality that describes the given graph.

27.
```
←++++++++++[+++++→
 -6 -4 -2  0  2  4  6
```

28.
```
←+++++→++++++++++→
 -6 -4 -2  0  2  4  6
```

29.
```
←+++(++++++++++]++→
 -6 -4 -2  0  2  4  6
```

30.
```
←++[++++++++++]++→
 -6 -4 -2  0  2  4  6
```

Break-Even Point *In Exercises 31–36, find all values of x for which the given products will at least break even. (See Examples 5 and 6.)*

31. $C = 50x + 6000; R = 65x.$

32. $C = 100x + 6000; R = 500x.$

33. $C = 85x + 1000; R = 105x$

34. $C = 70x + 500; R = 60x$

35. $C = 1000x + 5000; R = 900x$

36. $C = 25{,}000x + 21{,}700{,}000; R = 102{,}500x$

Solve each inequality. Graph each solution. (See Examples 7–10.)

37. $|p| > 7$ 38. $|m| < 2$

39. $|r| \leq 5$ 40. $|a| < -2$

41. $|b| > -5$

42. $|2x + 5| < 1$

43. $\left| x - \dfrac{1}{2} \right| < 2$

44. $|3z + 1| \geq 4$

45. $|8b + 5| \geq 7$

46. $\left| 5x + \dfrac{1}{2} \right| - 2 < 5$

Work these problems.

Temperature *The given inequality describes the monthly average high daily temperature T in degrees Fahrenheit in the given location. (Data from: Weatherbase.com.) What range of temperatures corresponds to the inequality?*

47. $|T - 83| \leq 7$; Miami, Florida

48. $|T - 63| \leq 27$; Boise, Idaho

49. $|T - 61| \leq 21$; Flagstaff, Arizona

50. $|T - 43| \leq 22$; Anchorage, Alaska

51. Engagement at Work According to the polling organization Gallup.com, 35.5% of American workers are meaningfully engaged at work. Taking into account the margin of error, we can write this result as $|P - 35.5| \leq 3$, where P is the percentage of American workers engaged at work. Find the range of estimates for P.

52. Social Science When administering a standard intelligence quotient (IQ) test, we expect about one-third of the scores to be more than 12 units above 100 or more than 12 units below 100. Describe this situation by writing an absolute value inequality.

53. Social Science A Gallup poll found that among Americans aged 18–34 years who consume alcohol, from 35% to 43% prefer beer. Let B represent the percentage of American alcohol consumers who prefer beer. Write the preceding information as an inequality. (Data from: Gallup.com.)

54. Social Science A Gallup poll in January 2017 found that from 13.1% to 15.1% of Americans considered themselves underemployed (working part-time or not working at all, but wanting to work full-time). Let U represent the percentage of underemployed workers. Write the preceding information as an inequality. (Data from: Gallup.com.)

55. Income Tax The following table shows the 2017 federal income tax for a single person. (Data from: Internal Revenue Service.)

If Taxable Income Is Over	But Not Over	Tax Rate Is
$0	$9325	10%
$9325	$37,950	15%
$37,950	$91,900	25%
$91,900	$191,650	28%
$191,650	$416,700	33%
$416,700	$418,400	35%
$418,400	No Limit	39.6%

Let x denote the taxable income. Write each of the seven income ranges in the table as an inequality.

56. Income Tax The following table shows the 2016 New York state income tax for a single person. (Data from: www.tax-brackets.org.)

If Taxable Income Is Over	But Not Over	Tax Rate Is
$0	$8450	4.00%
$8450	$11,650	4.50%
$11,650	$13,850	5.25%
$13,850	$21,300	5.90%
$21,300	$80,150	6.45%
$80,150	$214,000	6.65%
$214,000	$1,070,350	6.85%
$1,070,350	No Limit	8.82%

Let x denote the taxable income. Write each of the eight income ranges in the table as an inequality.

✓**Checkpoint Answers**

1. (a) $-24 < -4$; $42 > 7$

(b) $18 \geq -8$; $-45 \leq 20$

(c) $1 < 3$; $-9 < -7$

2. (a) $z < 5$

(b) $k \geq 4$

(c) $y \leq -4$

3. (a) $t \leq -2$

(b) $r > 1$

4. (a) $4 < k < 8$

(b) $-5 \leq z \leq 4$

5. $-15°C$ to $35°C$

6. (a) $[-1, 1]$

(b) All numbers in $(-\infty, -3]$ or $[3, \infty)$

7. (a) $(-7, 1)$

(b) $[-3, 4]$

8. (a) All numbers in $(-\infty, -3)$ or $(7, \infty)$

(b) All numbers in $\left(-\infty, -\dfrac{1}{3}\right]$ or $[1, \infty)$

(c) All numbers in $\left(-\infty, -\dfrac{7}{5}\right]$ or $\left[\dfrac{3}{5}, \infty\right)$

9. (a) All real numbers

(b) No solution

(c) All real numbers except -4

2.5 Polynomial and Rational Inequalities

This section deals with the solution of polynomial and rational inequalities, such as

$$r^2 + 3r - 4 \geq 0, \quad x^3 - x \leq 0, \quad \text{and} \quad \frac{2x - 1}{3x + 4} < 5.$$

We shall concentrate on algebraic solution methods, but to understand why these methods work, we must first look at such inequalities from a graphical point of view.

> **EXAMPLE 1**   Use the graph of $y = x^3 - 5x^2 + 2x + 8$ in Figure 2.53 to solve each of the given inequalities.
>
> **(a)** $x^3 - 5x^2 + 2x + 8 > 0$
>
> **Solution** Each point on the graph has coordinates of the form $(x, x^3 - 5x^2 + 2x + 8)$. The number x is a solution of the inequality exactly when the second coordinate of this point is positive—that is, when the point lies *above* the x-axis. So to solve the inequality, we need only find the first coordinates of points on the graph that are above the x-axis. This information can be read from Figure 2.53. The graph is above the x-axis when $-1 < x < 2$ and when $x > 4$. Therefore, the solutions of the inequality are all numbers x in the interval $(-1, 2)$ or the interval $(4, \infty)$.
>
> **(b)** $x^3 - 5x^2 + 2x + 8 < 0$
>
> **Solution** The number x is a solution of the inequality exactly when the second coordinate of the point $(x, x^3 - 5x^2 + 2x + 8)$ on the graph is negative—that is, when the point lies *below* the x-axis. Figure 2.53 shows that the graph is below the x-axis when $x < -1$ and when $2 < x < 4$. Hence, the solutions are all numbers x in the interval $(-\infty, -1)$ or the interval $(2, 4)$.

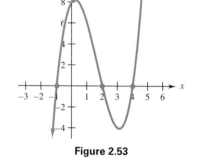

Figure 2.53

The solution process in Example 1 depends only on knowing the graph and its x-intercepts (that is, the points where the graph intersects the x-axis). This information can often be obtained algebraically, without doing any graphing.

Steps for Solving an Inequality Involving a Polynomial

1. Rewrite the inequality so that all the terms are on the left side and 0 is on the right side.
2. Find the x-intercepts by setting $y = 0$ and solving for x.
3. Divide the x-axis (number line) into regions using the solutions found in Step 2.
4. Test a point in each region by choosing a value for x and substituting it into the equation for y.
5. Determine which regions satisfy the original inequality and graph the solution.

EXAMPLE 2   Solve each of the given quadratic inequalities.

(a) $x^2 - x < 12$

Solution

Step 1   Rewrite the inequality so that all the terms are on the left side and 0 is on the right side.
Hence we add –12 to each side and obtain $x^2 - x - 12 < 0$.

Step 2   Find the x-intercepts by setting $y = 0$ and solving for x.
We find the x-intercepts of $y = x^2 - x - 12$ by setting $y = 0$ and solving for x:

$$x^2 - x - 12 = 0$$
$$(x + 3)(x - 4) = 0$$
$$x + 3 = 0 \quad \text{or} \quad x - 4 = 0$$
$$x = -3 \qquad x = 4.$$

Step 3   Divide the x-axis (number line) into regions using the solutions found in Step 2.
These numbers divide the x-axis into three regions, as indicated in Figure 2.54.

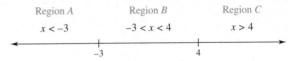

Figure 2.54

Step 4   Test a point in each region by choosing a value for x and substituting it in to the equation for y.
In each region, the graph of $y = x^2 - x - 12$ is an unbroken curve, so it will be entirely above or entirely below the axis. It can pass from above to below the x-axis only at the x-intercepts. To see whether the graph is above or below the x-axis when x is in region A, choose a value of x in region A, say, $x = -5$, and substitute it into the equation:

$$y = x^2 - x - 12 = (-5)^2 - (-5) - 12 = 18.$$

Therefore, the point $(-5, 18)$ is on the graph. Since its y-coordinate 18 is positive, this point lies above the x-axis; hence, the entire graph lies above the x-axis in region A.
Similarly, we can choose a value of x in region B, say, $x = 0$. Then

$$y = x^2 - x - 12 = 0^2 - 0 - 12 = -12,$$

so that $(0, -12)$ is on the graph. Since this point lies below the x-axis (why?), the entire graph in region B must be below the x-axis. Finally, in region C, let $x = 5$. Then $y = 5^2 - 5 - 12 = 8$, so that $(5, 8)$ is on the graph, and the entire graph in region C lies above the x-axis. We can summarize the results as follows:

Interval	$x < -3$	$-3 < x < 4$	$x > 4$
Test value in interval	-5	0	5
Value of $x^2 - x - 12$	18	-12	8
Graph	above x-axis	below x-axis	above x-axis
Conclusion	$x^2 - x - 12 > 0$	$x^2 - x - 12 < 0$	$x^2 - x - 12 > 0$

Step 5   Determine which regions satisfy the original inequality and graph the solution.
The last row shows that the only region where $x^2 - x - 12 < 0$ is region B, so the solutions of the inequality are all numbers x with $-3 < x < 4$—that is, the interval $(-3, 4)$, as shown in the number line graph in Figure 2.55.

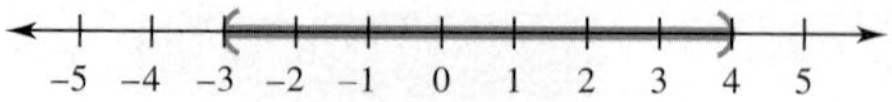

Figure 2.55

(b) $x^2 - x - 12 > 0$

Solution Use the chart in part (a). The last row shows that $x^2 - x - 12 > 0$ only when x is in region A or region C. Hence, the solutions of the inequality are all numbers x with $x < -3$ or $x > 4$—that is, all numbers in the interval $(-\infty, -3)$ or the interval $(4, \infty)$. ✔₁

✓ **Checkpoint 1**

Solve each inequality. Graph the solution on the number line.

(a) $x^2 - 2x < 15$

(b) $2x^2 - 3x - 20 < 0$

EXAMPLE 3   Solve the quadratic inequality $r^2 + 3r \geq 4$.

Solution

Step 1 First rewrite the inequality so that one side is 0:

$$r^2 + 3r \geq 4$$
$$r^2 + 3r - 4 \geq 0. \quad \text{Add } -4 \text{ to both sides.}$$

Step 2 Now solve the corresponding equation (which amounts to finding the x-intercepts of $y = r^3 + 3r - 4$):

$$r^2 + 3r - 4 = 0$$
$$(r - 1)(r + 4) = 0$$
$$r = 1 \quad \text{or} \quad r = -4.$$

Step 3 These numbers separate the number line into three regions, as shown in Figure 2.56. Test a number from each region:

Step 4 Let $r = -5$ from region A: $(-5)^2 + 3(-5) - 4 = 6 > 0$.

Let $r = 0$ from region B: $(0)^2 + 3(0) - 4 = -4 < 0$.

Let $r = 2$ from region C: $(2)^2 + 3(2) - 4 = 6 > 0$.

Step 5 We want the inequality to be positive or 0. The solution includes numbers in region A and in region C, as well as -4 and 1, the endpoints. The solution, which includes all numbers in the interval $(-\infty, -4]$ or the interval $[1, \infty)$, is graphed in Figure 2.56. ✔₂

✓ **Checkpoint 2**

Solve each inequality. Graph each solution.

(a) $k^2 + 2k - 15 \geq 0$

(b) $3m^2 + 7m \geq 6$

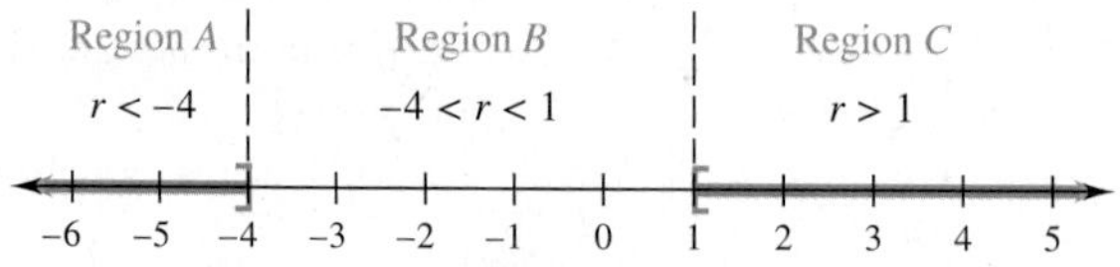

Figure 2.56

EXAMPLE 4   Solve $q^3 - 4q > 0$.

Solution

Step 1 Step 1 is already complete in the statement of the problem.

Step 2 Solve the corresponding equation by factoring:

$$q^3 - 4q = 0$$
$$q(q^2 - 4) = 0$$
$$q(q + 2)(q - 2) = 0$$
$$q = 0 \quad \text{or} \quad q + 2 = 0 \quad \text{or} \quad q - 2 = 0$$
$$q = 0 \qquad q = -2 \qquad q = 2.$$

Step 3 These three numbers separate the number line into the four regions shown in Figure 2.57.

Step 4   Test a number from each region:

A: If $q = -3, (-3)^3 - 4(-3) = -15 < 0.$

B: If $q = -1, (-1)^3 - 4(-1) = 3 > 0.$

C: If $q = 1, (1)^3 - 4(1) = -3 < 0.$

D: If $q = 3, (3)^3 - 4(3) = 15 > 0.$

Step 5   The numbers that make the polynomial positive are in the interval $(-2, 0)$ or the interval $(2, \infty)$, as graphed in Figure 2.57. ✓₃

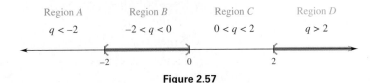

Figure 2.57

✓ **Checkpoint 3**

Solve each inequality. Graph each solution.

(a) $m^3 - 9m > 0$

(b) $2k^3 - 50k \le 0$

A graphing calculator can be used to solve inequalities without the need to evaluate at a test number in each interval. It is also useful for finding approximate solutions when the *x*-intercepts of the graph cannot be found algebraically.

EXAMPLE 5   Use a graphing calculator to solve $x^3 - 5x^2 + x + 6 > 0.$

Solution Begin by graphing $y = x^3 - 5x^2 + x + 6$ (Figure 2.58). Find the *x*-intercepts by solving $x^3 - 5x^2 + x + 6 = 0$. Since this cannot readily be done algebraically, use the graphical root finder to determine that the solutions (*x*-intercepts) are approximately $-.9254$, 1.4481, and 4.4774.

The graph is above the *x*-axis when $-.9254 < x < 1.4481$ and when $x > 4.4774$. Therefore, the approximate solutions of the inequality are all numbers in the interval $(-.9254, 1.4481)$ or the interval $(4.4774, \infty)$. ✓₄

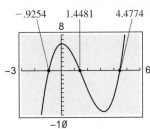

Figure 2.58

✓ **Checkpoint 4**

Use graphical methods to find approximate solutions of these inequalities.

(a) $x^2 - 6x + 2 > 0$

(b) $x^2 - 6x + 2 < 0$

EXAMPLE 6   **Break-Even Point** A company sells wholesale portable DVD players for $39 each. The variable cost of producing *x* thousand players is $5.5x - 4.9x^2$ (in thousands of dollars), and the fixed cost is $550 (in thousands). Find the values of *x* for which the company will break even or make a profit on the product.

Solution If *x* thousand DVD players are sold at $39 each, then

$$R = 39 \times x = 39x.$$

The cost expression (in thousands of dollars) is

$$\text{Cost} = \text{Fixed Costs} + \text{Variable Costs}$$
$$C = 550 + (5.5x - 4.9x^2).$$

Therefore, to break even or earn a profit, we need Revenue (*R*) greater than or equal to Cost (*C*).

$$R \ge C$$
$$39x \ge 550 + 5.5x - 4.9x^2$$
$$4.9x^2 + 33.5x - 550 \ge 0.$$

Now graph $4.9x^2 + 33.5x - 550$. Since *x* has to be positive in this situation (why?), we need only look at the graph in the right half of the plane. Here, and in other cases, you may have to try several viewing windows before you find one that shows what you need. Once you find a suitable window, such as in Figure 2.59, use the graphical root finder to determine the relevant *x*-intercept. Figure 2.59 shows the intercept is approximately $x \approx 7.71$. Hence, the company must manufacture at least $7.71 \times 1000 = 7710$ portable DVD players to make a profit.

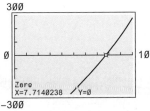

Figure 2.59

Rational Inequalities

Inequalities with quotients of algebraic expressions are called **rational inequalities**. These inequalities can be solved in much the same way as polynomial inequalities can.

Steps for Solving Inequalities Involving Rational Expressions

1. Rewrite the inequality so that all the terms are on the left side and the 0 is on the right side.
2. Write the left side as a single fraction.
3. Set the numerator and the denominator equal to 0 and solve for x.
4. Divide the x-axis (number line) into regions using the solutions found in Step 3.
5. Test a point in each region by choosing a value for x and substituting it into the equation for y.
6. Determine which regions satisfy the original inequality and graph the solution.

EXAMPLE 7   Solve the rational inequality

$$\frac{5}{x + 4} \geq 1.$$

Solution

Step 1   Write an equivalent inequality with one side equal to 0:

$$\frac{5}{x + 4} \geq 1$$

$$\frac{5}{x + 4} - 1 \geq 0.$$

Step 2   Write the left side as a single fraction:

$$\frac{5}{x + 4} - \frac{x + 4}{x + 4} \geq 0 \qquad \text{Obtain a common denominator.}$$

$$\frac{5 - (x + 4)}{x + 4} \geq 0 \qquad \text{Subtract fractions.}$$

$$\frac{5 - x - 4}{x + 4} \geq 0 \qquad \text{Distributive property}$$

$$\frac{1 - x}{x + 4} \geq 0.$$

Step 3   The quotient can change sign only at places where the denominator is 0 or the numerator is 0. (In graphical terms, these are the only places where the graph of $y = \dfrac{1 - x}{x + 4}$ can change from above the x-axis to below.) This happens when

$$1 - x = 0 \qquad \text{or} \qquad x + 4 = 0$$

$$x = 1 \qquad \text{or} \qquad x = -4.$$

Step 4   As in the earlier examples, the numbers -4 and 1 divide the x-axis into three regions:

$$x < -4, \quad -4 < x < 1, \quad x > 1.$$

Step 5 Test a number from each of these regions:

$$\text{Let } x = -5: \quad \frac{1 - (-5)}{-5 + 4} = -6 < 0.$$

$$\text{Let } x = 0: \quad \frac{1 - 0}{0 + 4} = \frac{1}{4} > 0.$$

$$\text{Let } x = 2: \quad \frac{1 - 2}{2 + 4} = -\frac{1}{6} < 0.$$

Step 6 The test shows that numbers in $(-4, 1)$ satisfy the inequality. With a quotient, the endpoints must be considered individually to make sure that no denominator is 0. In this inequality, -4 makes the denominator 0, while 1 satisfies the given inequality. Write the solution in interval notation as $(-4, 1]$ and graphically as in Figure 2.60.

Figure 2.60

CAUTION As suggested by Example 7, be very careful with the endpoints of the intervals in the solution of rational inequalities. ✓₅

✓ **Checkpoint 5**

Solve each inequality.

(a) $\dfrac{3}{x - 2} \geq 4$

(b) $\dfrac{p}{1 - p} < 3$

(c) Why is 2 excluded from the solution in part (a)?

EXAMPLE 8 Solve

$$\frac{2x - 1}{3x + 4} < 5.$$

Solution

Step 1 Write an equivalent inequality with 0 on one side by adding -5 to each side.

$$\frac{2x - 1}{3x + 4} - 5 < 0 \qquad \text{Get 0 on the right side.}$$

Step 2 Write the left side as a single fraction by using $3x + 4$ as a common denominator.

$$\frac{2x - 1 - 5(3x + 4)}{3x + 4} < 0 \qquad \text{Obtain a common denominator.}$$

$$\frac{-13x - 21}{3x + 4} < 0 \qquad \text{Distribute in the numerator and combine terms.}$$

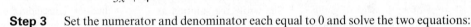

Figure 2.61

Step 3 Set the numerator and denominator each equal to 0 and solve the two equations:

$$-13x - 21 = 0 \qquad \text{or} \qquad 3x + 4 = 0$$

$$x = -\frac{21}{13} \qquad \text{or} \qquad x = -\frac{4}{3}$$

✓ **Checkpoint 6**

Solve each rational inequality.

(a) $\dfrac{3y - 2}{2y + 5} < 1$

(b) $\dfrac{3c - 4}{2 - c} \geq -5$

Step 4 The values $-21/13$ and $-4/3$ divide the x-axis into three regions:

$$x < -\frac{21}{13}, \quad -\frac{21}{13} < x < -\frac{4}{3}, \quad x > -\frac{4}{3}.$$

Steps 5 and 6 Testing points from each interval yields that the quotient is negative for numbers in the interval $(-\infty, -21/13)$ or $(-4/3, \infty)$. Neither endpoint satisfies the given inequality. ✓₆

✓ **Checkpoint 7**

(a) Solve the inequality

$$\frac{10}{x + 2} \geq 3$$

by first multiplying both sides by $x + 2$.

(b) Show that this method produces a wrong answer by testing $x = -3$.

CAUTION In problems like those in Examples 7 and 8, you should *not* begin by multiplying both sides by the denominator to simplify the inequality. Doing so will usually produce a wrong answer. For the reason, see Exercise 38. For an example, see Checkpoint 7. ✓₇

TECHNOLOGY TIP Rational inequalities can also be solved graphically. In Example 7, for instance, after rewriting the original inequality in the form $\dfrac{1 - x}{x + 4} \geq 0$, determine the

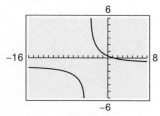

Figure 2.62

values of x that make the numerator and denominator 0 (namely, $x = 1$ and $x = -4$). Then graph $\dfrac{1 - x}{x + 4}$. Figure 2.62 shows that the graph is above the x-axis when it is between the vertical asymptote at $x = -4$ and the x-intercept at $x = 1$. So the solution of the inequality is the interval $(-4, 1]$. (When the values that make the numerator and denominator 0 cannot be found algebraically, as they were here, you can use the root finder to approximate them.)

2.5 Exercises

Solve each of these quadratic inequalities. Graph the solutions on the number line. (See Examples 2 and 3.)

1. $(x + 4)(2x - 3) \leq 0$
2. $(5y - 1)(y + 3) > 0$
3. $r^2 + 4r > -3$
4. $z^2 + 6z > -8$
5. $4m^2 + 7m - 2 \leq 0$
6. $6p^2 - 11p + 3 \leq 0$
7. $4x^2 + 3x - 1 > 0$
8. $3x^2 - 5x > 2$
9. $x^2 \leq 36$
10. $y^2 \geq 9$
11. $p^2 - 16p > 0$
12. $r^2 - 9r < 0$

Solve these inequalities. (See Example 4.)

13. $x^3 - 9x \geq 0$
14. $p^3 - 25p \leq 0$
15. $(x + 7)(x + 2)(x - 2) \geq 0$
16. $(2x + 4)(x^2 - 9) \leq 0$
17. $(x + 5)(x^2 - 2x - 3) < 0$
18. $x^3 - 2x^2 - 3x \leq 0$
19. $6k^3 - 5k^2 < 4k$
20. $2m^3 + 7m^2 > 4m$

21. A student solved the inequality $p^2 < 16$ by taking the square root of both sides to get $p < 4$. She wrote the solution as $(-\infty, 4)$. Is her solution correct?

⊞ *Use a graphing calculator to solve these inequalities. (See Example 5.)*

22. $6x + 7 < 2x^2$
23. $.5x^2 - 1.2x < .2$
24. $3.1x^2 - 7.4x + 3.2 > 0$
25. $x^3 - 2x^2 - 5x + 7 \geq 2x + 1$
26. $x^4 - 6x^3 + 2x^2 < 5x - 2$
27. $2x^4 + 3x^3 < 2x^2 + 4x - 2$
28. $x^5 + 5x^4 > 4x^3 - 3x^2 - 2$

Solve these rational inequalities. (See Examples 7 and 8.)

29. $\dfrac{r - 4}{r - 1} \geq 0$
30. $\dfrac{z + 6}{z + 4} > 1$
31. $\dfrac{a - 2}{a - 5} < -1$
32. $\dfrac{1}{3k - 5} < \dfrac{1}{3}$
33. $\dfrac{1}{p - 2} < \dfrac{1}{3}$
34. $\dfrac{7}{k + 2} \geq \dfrac{1}{k + 2}$

35. $\dfrac{5}{p + 1} > \dfrac{12}{p + 1}$
36. $\dfrac{x^2 - 4}{x} > 0$
37. $\dfrac{x^2 - x - 6}{x} < 0$

38. Determine whether $x + 4$ is positive or negative when

   (a) $x > -4$;   (b) $x < -4$.

   (c) If you multiply both sides of the inequality $\dfrac{1 - x}{x + 4} \geq 0$ by $x + 4$, should you change the direction of the inequality sign? If so, when?

   (d) Explain how you can use parts (a)–(c) to solve $\dfrac{1 - x}{x + 4} \geq 0$ correctly.

⊞ *Use a graphing calculator to solve these inequalities. You may have to approximate the roots of the numerators or denominators.*

39. $\dfrac{2x^2 + x - 1}{x^2 - 4x + 4} \leq 0$
40. $\dfrac{x^3 - 3x^2 + 5x - 29}{x^2 - 7} > 3$

41. **Profit** An analyst has found that her company's profits, in hundreds of thousands of dollars, are given by $P = 2x^2 - 12x - 32$, where x is the amount, in hundreds of dollars, spent on advertising. For what values of x does the company make a profit?

42. **Profit** The commodities market is highly unstable; money can be made or lost quickly on investments in soybeans, wheat, and so on. Suppose that an investor kept track of his total profit P at time t, in months, after he began investing, and he found that $P = 4t^2 - 30t + 14$. Find the time intervals during which he has been ahead.

43. **Profit** The manager of a 200-unit apartment complex has found that the profit is given by

$$P = x^2 + 300x - 18,000,$$

where x is the number of apartments rented. For what values of x does the complex produce a profit?

44. **Profit** A door-to-door knife salesman finds that his weekly profit can be modeled by the equation

$$P = x^2 + 5x - 530$$

where x is the number of pitches he makes in a week. For what values of x does the salesman need to make in order to earn a profit?

✓ Checkpoint Answers

1. (a) $(-3, 5)$

(b) $(-5/2, 4)$

2. (a) All numbers in $(-\infty, -5]$ or $[3, \infty)$

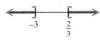

(b) All numbers in $(-\infty, -3]$ or $[2/3, \infty)$

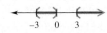

3. (a) All numbers in $(-3, 0)$ or $(3, \infty)$

(b) All numbers in $(-\infty, -5]$ or $[0, 5]$

4. (a) All numbers in $(-\infty, .3542)$ or $(5.6458, \infty)$

(b) All numbers in $(.3542, 5.6458)$

5. (a) $(2, 11/4]$

(b) All numbers in $(-\infty, 3/4)$ or $(1, \infty)$

(c) When $x = 2$, the fraction is undefined.

6. (a) $(-5/2, 7)$

(b) All numbers in $(-\infty, 2)$ or $[3, \infty)$

7. (a) $x \le \dfrac{4}{3}$

(b) $x = -3$ is a solution of $x \le \dfrac{4}{3}$, but not of the original inequality $\dfrac{10}{x + 2} \ge 3$.

CHAPTER 2 Summary and Review

Key Terms and Symbols*

2.1 Cartesian coordinate system
x-axis
y-axis
origin
ordered pair
x-coordinate
y-coordinate
quadrant
solution of an equation
graph

x-intercept
y-intercept
[viewing window]
[trace]
[graphical root finder]
[maximum and minimum finder]
graph reading

2.2 change in x
change in y

slope
slope–intercept form
parallel and perpendicular lines
point–slope form
linear equations
general form

2.3 linear models
residual
[least-squares regression line]

[linear regression]
[correlation coefficient]

2.4 linear inequality
properties of inequality
absolute-value inequality

2.5 polynomial inequality
algebraic solution methods
[graphical solution methods]
rational inequality

Chapter 2 Key Concepts

Slope of a Line | The **slope** of the line through the points (x_1, y_1) and (x_2, y_2), where $x_1 \ne x_2$, is $m = \dfrac{y_2 - y_1}{x_2 - x_1}$.

Equation of a Line | The line with equation $y = mx + b$ has slope m and y-intercept b.
The line with equation $y - y_1 = m(x - x_1)$ has slope m and goes through (x_1, y_1).
The line with equation $ax + by = c$ (with $a \ne 0, b \ne 0$) has x-intercept c/a and y-intercept c/b.
The line with equation $x = k$ is vertical, with x-intercept k, no y-intercept, and undefined slope.
The line with equation $y = k$ is horizontal, with y-intercept k, no x-intercept, and slope 0.

Parellel and Perpendicular Lines | Nonvertical **parallel lines** have the same slope, and **perpendicular lines,** if neither is vertical, have slopes with a product of -1.

*Terms in brackets deal with material in which a graphing calculator or other technology is used.

Chapter 2 Review Exercises

Which of the ordered pairs $(-2, 3)$, $(0, -5)$, $(2, -3)$, $(3, -2)$, $(4, 3)$, *and* $(7, 2)$ *are solutions of the given equation?*

1. $y = x^2 - 2x - 5$ **2.** $x - y = 5$

Sketch the graph of each equation.

3. $5x - 3y = 15$ **4.** $2x + 7y - 21 = 0$

5. $y + 3 = 0$ **6.** $y - 2x = 0$

7. $y = .25x^2 + 1$ **8.** $y = \sqrt{x + 4}$

9. The following temperature graph was recorded in Bratenahl, Ohio:

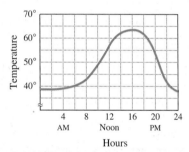

Hours

(a) At what times during the day was the temperature over 55°?

(b) When was the temperature below 40°?

10. Greenville, South Carolina, is 500 miles south of Bratenahl, Ohio, and its temperature is 7° higher all day long. (See the graph in Exercise 9.) At what time was the temperature in Greenville the same as the temperature at noon in Bratenahl?

11. In your own words, define the slope of a line.

In Exercises 12–21, find the slope of the line defined by the given conditions.

12. Through $(-1, 3)$ and $(2, 6)$

13. Through $(4, -5)$ and $(1, 4)$

14. Through $(8, -3)$ and the origin

15. Through $(8, 2)$ and $(0, 4)$

16. $3x + 5y = 25$ **17.** $6x - 2y = 7$

18. $x - 2 = 0$ **19.** $y = -4$

20. Parallel to $3x + 8y = 0$

21. Perpendicular to $x = 3y$

22. Graph the line through $(0, 5)$ with slope $m = -2/3$.

23. Graph the line through $(-4, 1)$ with $m = 3$.

24. What information is needed to determine the equation of a line?

Find an equation for each of the following lines.

25. Through $(5, -1)$, slope $2/3$

26. Through $(8, 0)$, slope $-1/4$

27. Through $(5, -2)$ and $(1, 3)$

28. Through $(2, -3)$ and $(-3, 4)$

29. Undefined slope, through $(-1, 4)$

30. Slope 0, through $(-2, 5)$

31. x-intercept -3, y-intercept 5

32. Here is a sample SAT question: Which of the following is an equation of the line that has a y-intercept of 2 and an x-intercept of 3?

(a) $-2x + 3y = 4$ (b) $-2x + 3y = 6$

(c) $2x + 3y = 4$ (d) $2x + 3y = 6$

(e) $3x + 2y = 6$

33. **Student Loans** In 2010, the total owed on federal student loans was approximately $\$.80$ trillion. In 2015, that amount was approximately $\$1.2$ trillion. (Data from: *The Wall Street Journal.*)

(a) Assuming the increase in student loan debt is linear, write an equation that gives the total amount owed in year x, with $x = 10$ corresponding to the year 2010.

(b) Is the slope of the line positive or negative? Why?

(c) Assuming the linear trend continues, estimate the amount owed in 2017.

34. **Baseball** It appears that bunting is becoming more rare for nonpitchers. In 2011, the average number of plate appearances per bunt for nonpitchers was approximately 175. In 2016, that value was 337.

(a) Assuming the trend is linear, write an equation that gives the average plate appearances per bunt in year x, with $x = 11$ corresponding to the year 2011.

(b) Is the slope of the line positive or negative? Why?

(c) Assuming the linear trend continues, estimate the average plate appearances per bunt in 2017.

35. **Median Income** The following table gives the median household income, meaning that half the households in the United States earned less than that amount and half the households earned more than that amount (see Chapter 10), for various years. (Data from: U.S. Census Bureau.)

Year	2005	2010	2014	2015
Household Income	$46,326	$49,276	$53,657	$56,516

(a) Use the data from the years 2005 and 2010 to find a linear model for these data, with $x = 5$ corresponding to the year 2005.

(b) Find the least-squares regression line for all these data.

(c) Use the models in parts (a) and (b) to estimate the median household income for the year 2015 to the nearest dollar. Compare the estimates to the actual value for 2015.

(d) Assuming the trend continues, use the model for part (b) to estimate the median household income for the year 2018 to the nearest dollar.

36. **Charitable Giving** The following table gives the total amount in charitable giving (in billions of dollars) for various years. (Data from: ProQuest Statistical Abstract of the United States: 2016.)

Year	2008	2011	2013	2014
Charitable Giving	$299.6	$298.5	$334.5	$358.4

(a) Use the data from the years 2008 and 2013 to find a linear model for these data, with $x = 8$ corresponding to the year 2008.

(b) Find the least-squares regression line for all these data.

(c) Use the models in parts (a) and (b) to estimate the total amount of charitable giving in 2014 to the nearest billion. Compare the estimates to the actual value for 2014.

(d) Assuming the trend continues, use the model for part (b) to estimate the total amount of charitable giving in 2019 to the nearest billion.

37. Two-Year School Tuition The following table shows the average annual tuition for public two-year institutions. (Data from: The College Board.)

Year	2012	2013	2014	2015	2016
Tuition	$3154	$3241	$3336	$3440	$3520

(a) Let $x = 12$ correspond to the year 2012, and find the least-squares regression line for these data.

(b) Graph the data from the years 2012 to 2016 with the least-squares regression line.

(c) Does the least-squares regression line form a good fit?

(d) What is the correlation coefficient?

38. Four-Year School Tuition The following table shows the average annual tuition for public four-year institutions. (Data from: The College Board.)

Year	2012	2013	2014	2015	2016
Tuition	$8646	$8885	$9145	$9420	$9650

(a) Let $x = 12$ correspond to the year 2012, and find the least-squares regression line for these data.

(b) Graph the data from the years 2012 to 2016 with the least-squares regression line.

(c) Does the least-squares regression line form a good fit?

(d) What is the correlation coefficient?

Solve each inequality.

39. $-6x + 3 < 2x$

40. $12z \geq 5z - 7$

41. $2(3 - 2m) \geq 8m + 3$

42. $6p - 5 > -(2p + 3)$

43. $-3 \leq 4x - 1 \leq 7$

44. $0 \leq 3 - 2a \leq 15$

45. $|b| \leq 8$

46. $|a| > 7$

47. $|2x - 7| \geq 3$

48. $|4m + 9| \leq 16$

49. $|5k + 2| - 3 \leq 4$

50. $|3z - 5| + 2 \geq 10$

51. Natural Science Here is a sample SAT question: For pumpkin carving, Mr. Sephera will not use pumpkins that weigh less than 2 pounds or more than 10 pounds. If x represents the weight of a pumpkin (in pounds) that he will *not* use, which of the following inequalities represents all possible values of x?

(a) $|x - 2| > 10$ (b) $|x - 4| > 6$

(c) $|x - 5| > 5$ (d) $|x - 6| > 4$

(e) $|x - 10| > 4$

52. Snow-Thrower Pricing Prices at several retail outlets for a new 24-inch, 2-cycle snow thrower were all within $55 of $600. Write this information as an inequality, using absolute-value notation.

53. Data Plans One cellular company offers a data plan charging $45 a month plus $8 per gigabyte of data. Another cellular company offers a flat rate of $65 per month with unlimited data usage. For what range of data use (in gigabytes) is the first plan cheaper?

54. Car Rental One car rental firm charges $125 for a weekend rental (Friday afternoon through Monday morning) and gives unlimited mileage. A second firm charges $95 plus $.20 a mile. For what range of miles driven is the second firm cheaper?

Solve each inequality.

55. $r^2 + r - 6 < 0$

56. $y^2 + 4y - 5 \geq 0$

57. $2z^2 + 7z \geq 15$

58. $3k^2 \leq k + 14$

59. $(x - 3)(x^2 + 7x + 10) \leq 0$

60. $(x + 4)(x^2 - 1) \geq 0$

61. $\dfrac{m + 2}{m} \leq 0$

62. $\dfrac{q - 4}{q + 3} > 0$

63. $\dfrac{5}{p + 1} > 2$

64. $\dfrac{6}{a - 2} \leq -3$

65. $\dfrac{2}{r + 5} \leq \dfrac{3}{r - 2}$

66. $\dfrac{1}{z - 1} > \dfrac{2}{z + 1}$

Case Study 2   Using Extrapolation and Interpolation for Prediction

One reason for developing a mathematical model is to make predictions. If your model is a least-squares line, you can predict the y-value corresponding to some new x-value by substituting that x-value into an equation of the form $\hat{y} = mx + b$. (We use $\hat{y}$ to remind us that we're getting a predicted value rather than an actual data value.) Data analysts distinguish between two very different kinds of prediction: *interpolation* and *extrapolation*. An interpolation uses a new x inside the x-range of your original data. For example, if you have inflation data at five-year intervals from 1950 to 2010, estimating the rate of inflation in 1957 is an interpolation problem. But if you use the same data to estimate what the inflation rate was in 1920, or what it will be in 2020, you are extrapolating.

In general, interpolation is much safer than extrapolation, because data that are approximately linear over a short interval may be nonlinear over a larger interval. Let us examine a case of the dangers of extrapolation. Figure 1 shows the net income (in millions of dollars) for Chipotle Mexican Grill, Inc., from the years 2010–2015. Also on Figure 1, we see the least-squares regression line. (Data from: www.morningstar.com.)

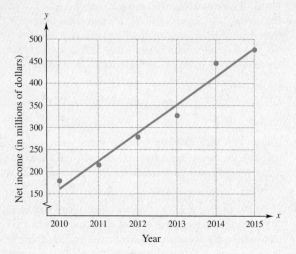

Figure 1

As we see from the graph, the fit is quite good and the linear trend seems quite clear. Additionally, the correlation coefficient is $r \approx .986$, which is very high. The least-squares regression equation is

$$y = 63.54x - 474.3,$$

where $x = 10$ corresponds to the year 2010. If we extrapolate to the next year (2016), we predict the net income for Chipotle to be

$$63.54(16) - 474.3 = \$542.34 \text{ million.}$$

The problem, however, is that Chipotle suffered from *E. coli* contamination in several states and a norovirus outbreak in the Boston, Massachusetts, area in late 2015. The net earnings in 2016 were a paltry $23 million—substantially lower than what our regression model would have predicted.

Using the past to predict the future is generally a dangerous business because we often cannot foresee events with major impact such as recessions, natural disasters, etc. Predictions into the future are often made, however, because planning has to occur and we use the best data we have available. One merely needs to realize that predictions into the future are highly likely to be inaccurate.

One way to determine if a model is likely to be inaccurate in the future is to examine the residuals. In Section 2.3, we defined a residual to be the difference between the actual value y and its predicted value $\hat{y}$.

$$\text{Residual} = y - \hat{y}.$$

Graphing these residuals on the y-axis and the predictor variable on the x-axis can be illuminating. Let's look at another historical example. Figure 2 shows the revenue (in millions of dollars) for the social media company Twitter for the years 2010 through 2015. (Data from: www.morningstar.com.)

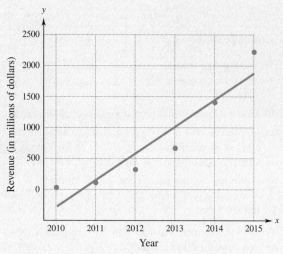

Figure 2

The regression equation for the least-squares line in Figure 2 is

$$y = 434x - 4635,$$

where $x = 10$ corresponds to the year 2010. Because the correlation coefficient is $r \approx .943$, our linear model appears to fit these data well. We notice, however, that the predictions underpredict, then overpredict, and then underpredict. We can get a better look at this pattern by plotting the residuals. To find them, we put each value of the independent variable into the regression equation, calculate the predicted value $\hat{y}$, and subtract it from the actual value of y, as in the following table.

Year	2010	2011	2012	2013	2014	2015
x	10	11	12	13	14	15
y	28	106	317	665	1403	2218
$\hat{y}$	−295.4	138.5	572.5	1006.5	1440.5	1874.4
Residual $y - \hat{y}$	323.4	−32.5	−255.5	−341.5	−37.5	343.6

The residuals in Figure 3 indicate that our data have a nonlinear U-shaped component that is not captured by a linear model. When a residual plot shows a pattern such as this (or a similar pattern with an upside down U), extrapolating from the data is probably not a good idea.

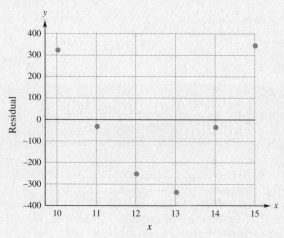

Figure 3

Exercises

Stock Price The following table gives the stock price for the large equipment manufacturer Caterpillar at the start of the given quarters. Let $x = 1$ correspond to the first quarter of the year 2015.

Date	Quarter	Stock Price	Date	Quarter	Stock Price
January 2015	1	$91.77	January 2016	5	$67.96
April 2015	2	$80.03	April 2016	6	$76.54
July 2015	3	$84.82	July 2016	7	$75.81
October 2015	4	$65.36	October 2016	8	$88.77

Year	Average Hourly Wage (Dollars)
1970	3.40
1975	4.73
1980	6.85
1985	8.74
1990	10.20
1995	11.65
2000	14.02
2005	16.13
2010	19.06
2015	21.03

1. If you have access to the appropriate technology, verify that the least-squares regression line that models these data (with rounded coefficients) is $y = -.77x + 82.3$.

2. Use the model from Exercise 1 to calculate the predicted values $\hat{y}$ for each quarter in the table.

3. Use your answers from Exercise 2 to calculate the residuals for each quarter in the table. Make a graph of the residuals similar to Figure 3.

4. Do you think the linear model of Exercise 1 is a good fit for these data? Why or why not?

Hourly Wages The following table gives the average hourly earnings (in dollars) of U.S. production workers from 1970 to 2010. (Data from: U.S. Bureau of Labor Statistics.)

5. If you have appropriate technology, verify that the least-squares regression line that models these data (with coefficients rounded) is $y = .3911x - 24.6$, where y is the average hourly wage in year x and $x = 0$ corresponds to 1900.

6. Use the model from Exercise 5 to interpolate a prediction of the hourly wage in 2012. The actual value was $14.97. How close was your prediction?

7. Use the model from Exercise 5 to extrapolate to 1960 and predict the hourly average wage. Why is this prediction nonsensical?

8. Using the model from Exercise 5, find the average hourly wage for each year in the table, and subtract it from the actual value in the second column. This gives you a table of the residuals. Plot your residuals as points on a graph.

9. What will happen if you try linear regression on the *residuals*? If you're not sure, use technology such as a graphing calculator to find the regression equation for the residuals. Why does this result make sense?

Extended Project

Go the library or use the Internet to obtain the average amount of student loan debt for the most recent years available.

1. Create a scatterplot of the data with the amount on the y-axis and the year on the x-axis and assess the trend.

2. Fit a linear model to the data and predict the loan debt amount for each year.

3. Calculate the residuals from each year in the data set. Graph the residuals.

4. Do the residuals show a U-shape?

5. Find the correlation coefficient for your data.

Functions and Graphs

3

CHAPTER OUTLINE

The modern world is overwhelmed with data—from the cost of housing, utility bills, and student loans to trends in consumer confidence, corporate revenue, smoking rates, and alcohol consumption. Examples and exercises in this chapter investigate functions and graphs that describe these and other real-world phenomena. Functions enable us to construct mathematical models that can sometimes be used to estimate outcomes. Graphs of functions allow us to visualize a situation and to detect trends more easily.

Functions are an extremely useful way of describing many real-world situations in which the value of one quantity varies with, depends on, or determines the value of another. In this chapter, we will introduce functions, learn how to use functional notation, develop skills in constructing and interpreting the graphs of functions, and learn to apply this knowledge in a variety of situations.

3.1 Functions

To understand the origin of the concept of a function, we consider some real-life situations in which one numerical quantity depends on, corresponds to, or determines another. The way in which one quantity depends on another can be described verbally, graphically, numerically using a table of data, or algebraically by means of a formula.

EXAMPLE 1   **Cell Phone Data Plan** The amount of the bill you pay for Internet data usage on your cell phone *depends on* the number of gigabytes of data you use. The way in which the usage determines the bill is described in the data plan. In this example, the way in which one numerical quantity depends on another is described verbally.

EXAMPLE 2 **Stock Exchange** The graph and table of data in Figure 3.1 show the NASDAQ Composite Index from January 2011 through January 2017. The NASDAQ index is a statistical measure of how well a portion of the market is performing. The index tracks approximately 4000 stocks that are traded on the National Association of Securities Dealers Automated Quotations (NASDAQ) System electronic stock exchange. The graph and table display the index value *corresponding to* each year. In this example, the way in which one quantity corresponds to another is described graphically and numerically. (Data from: www.morningstar.com.)

Year	NASDAQ
2011	2700.08
2012	2813.84
2013	3142.13
2014	4103.88
2015	4635.24
2016	4613.95
2017	5614.79

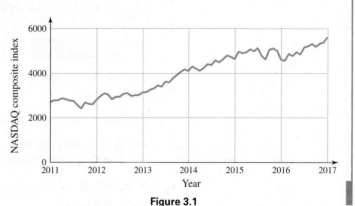

Figure 3.1

EXAMPLE 3 **Physical Science** Suppose a rock is dropped straight down from a high point. From physics, we know that the distance traveled by the rock in t seconds is $16t^2$ feet. So the distance *depends on* the time, and this relationship is described by an algebraic formula.

These examples share a couple of features. First, each involves two sets of numbers, which we can think of as inputs and outputs. Second, in each case, there is a rule by which each input determines an output, as summarized here:

	Set of Inputs	Set of Outputs	Rule
Example 1	All data usages	All bill amounts	The data plan
Example 2	Years	Index values	Graph or table
Example 3	Seconds elapsed after dropping the rock	Distances the rock travels	Distance $= 16t^2$

Each of these examples may be mentally represented by an idealized calculator that has a single operation key: A number is entered [*input*], the rule key is pushed [*rule*], and an answer is displayed [*output*]. The formal definition of a function incorporates these same common features (input–rule–output), with a slight change in terminology.

A **function** consists of a set of inputs called the **domain,** a set of outputs called the **range,** and a rule by which each input determines *exactly one* output.

✓ **Checkpoint 1**

Find the domain and range of the function in Example 3, assuming (unrealistically) that the rock can fall forever.

Answers to Checkpoint exercises are found at the end of the section.

EXAMPLE 4 Find the domains and ranges for the functions in Examples 1 and 2.

Solution In Example 1, the domain consists of all possible data usage amounts and the range consists of all possible bill amounts.

In Example 2, the domain is the set of years from 2011 to 2017, all real numbers in the interval [2011, 2017]. The range consists of the NASDAQ Composite Index values that actually occur during those years. ✓₁

Be sure that you understand the phrase "exactly one output" in the definition of the rule of a function. In Example 1, for instance, your data usage (input) determines exactly one bill amount (output)—you can't be charged two different amounts for the same data usage. However, it is quite possible for the amount of the bill (output) to be the same for using two different amounts of data (the input)—for example, using 2.9 gigabytes might cost the same as using 2.1 gigabytes of data. In other words, we can say the following:

> **In a function, each input produces a single output,**
> **but different inputs may produce the same output.**

EXAMPLE 5 Which of the following rules describe functions?

(a) Use the optical reader at the checkout counter of the grocery store to convert codes to prices.

Solution For each code, the reader produces exactly one price, so this is a function.

(b) A loan officer at Bank of America asks each applicant to provide her or his Social Security number (customers must have a social security number in order to apply).

Solution Because each applicant (input) has exactly one Social Security number (output), this is a function.

(c) A loan officer at Bank of America asks each applicant to provide her or his credit card number.

Solution Because an applicant may have more than one credit card (or no credit cards), this is not a function.

(d) Enter a number in a calculator and press the x^2 key.

Solution This is a function, because the calculator produces just one number x^2 for each number x that is entered.

(e) Assign to each number x the number y given by this table:

x	1	1	2	2	3	3
y	3	-3	-5	-5	8	-8

Solution Since $x = 1$ corresponds to more than one y-value (as does $x = 3$), this table does not define a function.

(f) Assign to each number x the number y given by the equation $y = 3x - 5$.

Solution Because the equation determines a unique value of y for each value of x, it defines a function. ✓ **2**

The equation $y = 3x - 5$ in part (f) of Example 5 defines a function, with x as input and y as output, because each value of x determines a *unique* value of y. In such a case, the equation is said to **define y as a function of x.**

EXAMPLE 6 Decide whether each of the following equations defines y as a function of x.

(a) $y = -4x + 11$

Solution For a given value of x, calculating $-4x + 11$ produces exactly one value of y. (For example, if $x = -7$, then $y = -4(-7) + 11 = 39$). Because one value of the input variable leads to exactly one value of the output variable, $y = -4x + 11$ defines y as a function of x.

✓ **Checkpoint 2**

Do the following define functions?

(a) The correspondence defined by the rule $y = x^2 + 5$, where x is the input and y is the output.

(b) The correspondence defined by entering a nonzero number in a calculator and pressing the $1/x$ key.

(c) The correspondence between a computer, x, and several users of the computer, y.

(b) $y^2 = x$

Solution Suppose $x = 36$. Then $y^2 = x$ becomes $y^2 = 36$, from which $y = 6$ or $y = -6$. Since one value of x can lead to two values of y, $y^2 = x$ does *not* define y as a function of x. ✔3

✓ **Checkpoint 3**

Do the following define y as a function of x?

(a) $y = -6x + 1$

(b) $y = x^2$

(c) $x = y^2 - 1$

(d) $y < x + 2$

Almost all the functions in this text are defined by formulas or equations, as in part (a) of Example 6. The domain of such a function is determined by the following **agreement on domains.**

> Unless otherwise stated, assume that the domain of any function defined by a formula or an equation is the largest set of real numbers (inputs) that each produces a real number as output.

EXAMPLE 7 Each of the given equations defines y as a function of x. Find the domain of each function.

(a) $y = x^4$

Solution Any number can be raised to the fourth power, so the domain is the set of all real numbers, which is sometimes written as $(-\infty, \infty)$.

(b) $y = \sqrt{x - 6}$

Solution For y to be a real number, $x - 6$ must be nonnegative. This happens only when $x - 6 \geq 0$—that is, when $x \geq 6$. So the domain is the interval $[6, \infty)$.

(c) $y = \sqrt{4 - x}$

Solution For y to be a real number here, we must have $4 - x \geq 0$, which is equivalent to $x \leq 4$. So the domain is the interval $(-\infty, 4]$.

(d) $y = \sqrt[3]{4 - x}$

Solution We can take the cube root of any number, so the domain is the set of all real numbers, $(-\infty, \infty)$.

(e) $y = \dfrac{1}{x + 3}$

Solution Because the denominator cannot be 0, $x \neq -3$, and the domain consists of all numbers in the intervals

$$(-\infty, -3) \quad \text{or} \quad (-3, \infty).$$

(f) $y = \dfrac{\sqrt{x}}{x^2 - 3x + 2}$

✓ **Checkpoint 4**

Give the domain of each function.

(a) $y = 3x + 1$

(b) $y = x^2$

(c) $y = \sqrt{-x}$

(d) $y = \dfrac{3}{x^2 - 1}$

Solution The numerator is defined only when $x \geq 0$. The domain cannot contain any numbers that make the denominator 0—that is, the numbers that are solutions of

$$x^2 - 3x + 2 = 0.$$
$$(x - 1)(x - 2) = 0 \qquad \text{Factor.}$$
$$x - 1 = 0 \quad \text{or} \quad x - 2 = 0 \qquad \text{Zero-factor property}$$
$$x = 1 \quad \text{or} \qquad x = 2.$$

Therefore, the domain consists of all nonnegative real numbers except 1 and 2. ✔4

Functional Notation

In actual practice, functions are seldom presented in the style of domain–rule–range, as they have been here. Functions are usually denoted by a letter such as f. If x is an input, then $f(x)$ denotes the output number that the function f produces from the input x. The symbol $f(x)$ is read "f of x." The rule is usually given by a formula, such as $f(x) = \sqrt{x^2 + 1}$. This formula can be thought of as a set of directions:

Name of function   Input number

$$\underline{f(x)} = \underline{\sqrt{x^2 - 1}}$$

Output number   Directions that tell you what to do with input x in order to produce the corresponding output $f(x)$—namely, "square it, subtract 1, and take the square root of the result."

For example, to find $f(3)$ (the output number produced by the input 3), simply replace x by 3 in the formula:

$$f(3) = \sqrt{3^2 - 1}$$
$$= \sqrt{8}.$$

Similarly, replacing x by -5 shows that

$$f(-5) = \sqrt{(-5)^2 - 1} = \sqrt{25 - 1} = \sqrt{24}.$$

If we try to replace x with 0, we get

$$f(0) = \sqrt{(0)^2 - 1} = \sqrt{-1}, \text{ which is not a real number.}$$

Therefore, $x = 0$ is not in the domain of the function, and we say that $f(0)$ is *not defined*.

These directions can be applied to any quantities, such as $a + b$ or c^4 (where a, b, and c are real numbers). Thus, to compute $f(a + b)$, the output corresponding to input $a + b$, we square the input [obtaining $(a + b)^2$], subtract 1 [obtaining $(a + b)^2 - 1$], and take the square root of the result:

$$f(a + b) = \sqrt{(a + b)^2 - 1}$$
$$= \sqrt{a^2 + 2ab + b^2 - 1}.$$

Similarly, the output $f(c^4)$, corresponding to the input c^4, is computed by squaring the input $[(c^4)^2]$, subtracting 1 $[(c^4)^2 - 1]$, and taking the square root of the result:

$$f(c^4) = \sqrt{(c^4)^2 - 1}$$
$$= \sqrt{c^8 - 1}.$$

EXAMPLE 8   Let $g(x) = -x^2 + 4x - 5$. Find each of the given outputs.

(a) $g(-2)$

Solution Replace x with -2:

$$g(-2) = -(-2)^2 + 4(-2) - 5$$
$$= -4 - 8 - 5$$
$$= -17.$$

(b) $g(x + h)$

Solution Replace x by the quantity $x + h$ in the rule of g:

$$g(x + h) = -(x + h)^2 + 4(x + h) - 5$$
$$= -(x^2 + 2xh + h^2) + (4x + 4h) - 5$$
$$= -x^2 - 2xh - h^2 + 4x + 4h - 5.$$

(c) $g(x + h) - g(x)$

Solution Use the result from part (b) and the rule for $g(x)$:

$$g(x + h) - g(x) = \overbrace{(-x^2 - 2xh - h^2 + 4x + 4h - 5)}^{g(x+h)} - \overbrace{(-x^2 + 4x - 5)}^{g(x)}$$
$$= -2xh - h^2 + 4h.$$

(d) $\dfrac{g(x + h) - g(x)}{h}$ (assuming that $h \neq 0$)

Solution The numerator was found in part (c). Divide it by h as follows:

$$\frac{g(x + h) - g(x)}{h} = \frac{-2xh - h^2 + 4h}{h}$$
$$= \frac{h(-2x - h + 4)}{h} \qquad \text{Factor numerator.}$$
$$= -2x - h + 4. \qquad \text{Simplify.}$$

The quotient found in Example 8(d),

$$\frac{g(x + h) - g(x)}{h},$$

is called the **difference quotient** of the function g. Difference quotients are important in calculus. ✔5

⛔ **CAUTION** Functional notation is *not* the same as ordinary algebraic notation. You cannot simplify an expression such as $f(x + h)$ by writing $f(x) + f(h)$. To see why, consider the answers to Checkpoints 5(c) and (d), which show that

$$f(1 + 3) \neq f(1) + f(3).$$

Applications

To describe some real-world situations, we need a function with a multipart rule, as the following example illustrates. Such a function is called a **piecewise-defined function**.

EXAMPLE 9 **Income Tax** If you were a single person in Connecticut in 2017 with a taxable income of x dollars and $x \leq \$50,000$, then your state income tax $T(x)$ was determined by the rule

$$T(x) = \begin{cases} .03x & \text{if } 0 \leq x \leq 10,000 \\ 300 + .05(x - 10,000) & \text{if } 10,000 < x \leq 50,000. \end{cases}$$

Find the income tax paid by a single person with the given taxable income. (Data from: www.tax-brackets.org.)

(a) $9200

Solution We must find $T(9200)$. Since 9200 is less than 10,000, the first piece of the rule applies (the first line of the piecewise-defined function):

$$T(x) = .03x$$
$$T(9200) = .03(9200) = \$276. \qquad \text{Let } x = 9200.$$

(b) $30,000

Solution Now we must find $T(30,000)$. Since 30,000 is greater than $10,000, the second piece of the rule applies:

$$T(x) = 300 + .05(x - 10,000)$$
$$T(30,000) = 300 + .05(30,000 - 10,000) \qquad \text{Let } x = 30,000.$$
$$= 300 + .05(20,000) \qquad \text{Simplify.}$$
$$= 300 + 1000 = \$1300. \text{ ✔6}$$

✔ **Checkpoint 5**

Let $f(x) = 5x^2 - 2x + 1$. Find the following.

(a) $f(1)$

(b) $f(3)$

(c) $f(1 + 3)$

(d) $f(1) + f(3)$

(e) $f(m)$

(f) $f(x + h) - f(x)$

(g) $\dfrac{f(x + h) - f(x)}{h} (h \neq 0)$

✔ **Checkpoint 6**

Use Example 9 to find the tax on each of these incomes.

(a) $48,750

(b) $7345

EXAMPLE 10 **Sales** Suppose the projected sales (in thousands of dollars) of a small company over the next 10 years are approximated by the function

$$S(x) = .07x^4 - .05x^3 + 2x^2 + 7x + 62$$

where $x = 0$ corresponds to the current year.

(a) What are the projected sales for the current year?

Solution Since the current year corresponds to $x = 0$, the sales for this year are given by $S(0)$. Substituting 0 for x in the rule for S, we see that $S(0) = 62$. So the current projected sales are $62,000.

(b) What will sales be in four years?

Solution The sales in four years from now are given by $S(4)$, which can be computed by hand or with a calculator:

$$S(x) = .07x^4 - .05x^3 + 2x^2 + 7x + 62$$
$$S(4) = .07(4)^4 - .05(4)^3 + 2(4)^2 + 7(4) + 62 \qquad \text{Let } x = 4.$$
$$= 136.72.$$

Thus, sales are projected to be $136,720.

✔ **Checkpoint 7**

A developer estimates that the total cost of building x large apartment complexes in a year is approximated by

$$A(x) = x^2 + 80x + 60,$$

where $A(x)$ represents the cost in hundred thousands of dollars. Find the cost of building

(a) 4 complexes;

(b) 10 complexes.

X	Y1	
5	184.5	
6	255.92	
7	359.92	
8	507.12	
9	709.82	
10	982	
11	1339.3	
X=5		

Figure 3.2

📱 **EXAMPLE 11** **Sales** Use the table feature of the graphing calculator to find the projected sales of the company in Example 10 for years 5 through 10.

Solution Enter the sales equation $y = .07x^4 - .05x^3 + 2x^2 + 7x + 62$ into the equation memory of the calculator (often called the Y = list). Check your instruction manual for how to set the table to start at $x = 5$ and go at least through $x = 10$. Then display the table, as in Figure 3.2. The figure shows that sales are projected to rise from $184,500 in year 5 to $982,000 in year 10.

3.1 Exercises

For each of the following rules, state whether it defines y as a function of x or not. (See Examples 5 and 6.)

1.

x	3	2	1	0	−1	−2	−3
y	9	4	1	0	1	4	9

2.

x	10	20	30	40	50	60
y	−1	−1	0	0	1	1

3.

x	9	4	1	0	1	4	9
y	3	2	1	0	−1	−2	−3

4.

x	−1	−1	0	0	1	1
y	10	20	30	40	50	60

5. $y = x^3$

6. $y = \sqrt{x - 1}$

7. $x = y^2 + 3$

8. $x = |y + 2|$

9. $y = \dfrac{-1}{x - 1}$

10. $y = \dfrac{4}{2x + 3}$

State the domain of each function. (See Example 7.)

11. $f(x) = 4x - 1$

12. $f(x) = 2x + 7$

13. $f(x) = x^4 - 1$

14. $f(x) = (2x + 5)^2$

15. $f(x) = \sqrt{5 - x}$

16. $f(x) = \sqrt{-x} + 3$

17. $f(x) = \dfrac{1}{x - 2}$

18. $g(x) = \dfrac{x}{x^2 + x - 2}$

19. $g(x) = \dfrac{x^2 + 4}{x^2 - 4}$

20. $g(x) = \dfrac{x^2 - 1}{x^2 + 1}$

21. $h(x) = \dfrac{\sqrt{x + 4}}{x^2 + x - 12}$

22. $\dfrac{\sqrt{x + 1}}{x^2 - x - 6}$

For each of the following functions, find

(a) $f(4)$; **(b)** $f(-3)$; **(c)** $f(2.7)$; **(d)** $f(-4.9)$.

(See Examples 8 and 9.)

23. $f(x) = 8$

24. $f(x) = 0$

25. $f(x) = 2x^2 + 4x$

26. $f(x) = x^2 - 2x$

27. $f(x) = \sqrt{5 - x}$

28. $f(x) = \sqrt{x + 3}$

29. $f(x) = |x^2 - 6x - 4|$ **30.** $f(x) = |x^3 - x^2 + x - 1|$

31. $f(x) = \dfrac{\sqrt{x-1}}{x^2 - 1}$ **32.** $f(x) = \sqrt{-x} + \dfrac{2}{x+1}$

33. $f(x) = \begin{cases} -2x + 4 & \text{if } x \le 1 \\ 3 & \text{if } 1 < x < 4 \\ x + 1 & \text{if } x \ge 4 \end{cases}$

34. $f(x) = \begin{cases} x^2 & \text{if } x < 2 \\ 5x - 7 & \text{if } x \ge 2 \end{cases}$

For each of the following functions, find

(a) $f(p)$;   **(b)** $f(-r)$;   **(c)** $f(m + 3)$.

(See Example 8.)

35. $f(x) = 6 - x$ **36.** $f(x) = 3x + 5$

37. $f(x) = \sqrt{4 - x}$ **38.** $f(x) = \sqrt{-2x}$

39. $x^2 - 1$ **40.** $1 - x^2$

41. $f(x) = \dfrac{3}{x - 1}$ **42.** $f(x) = \dfrac{-1}{5 + x}$

For each of the following functions, find the difference quotient

$$\frac{f(x + h) - f(x)}{h} \quad (h \ne 0).$$

(See Example 8.)

43. $f(x) = 2x - 4$ **44.** $f(x) = 2 + 4x$

45. $f(x) = x^2 + 1$ **46.** $f(x) = x^2 - x$

▦ *Use a calculator to work these exercises. (See Examples 9 and 10.)*

47. Gross Domestic Product The gross domestic product (GDP) of the United States, which measures the overall size of the economy in billions of dollars, can be approximated by the function

$$f(x) = 589x + 8858,$$

where $x = 10$ corresponds to the year 2010. Estimate the GDP (to the nearest billion dollars) in the given years. (Data from: U.S. Bureau of Economic Analysis.)

(a) 2013   **(b)** 2015   **(c)** 2016

48. Consumer Consumption Personal consumption expenditures in billions of dollars in the United States can be approximated by the function

$$f(x) = 403x + 6061,$$

where $x = 10$ corresponds to the year 2010. Estimate personal consumption (to the nearest billion dollars) in the given years. (Data from: U.S. Bureau of Economic Analysis.)

(a) 2010   **(b)** 2015   **(c)** 2016

49. Stock Prices The closing stock price of Microsoft Corporation for the years 2010 to 2016 can be approximated by the function

$$f(x) = .8x^2 - 17x + 119,$$

where $x = 10$ corresponds to the start of the year 2010. Find the following function values and explain what your answers mean in practical terms. (Data from: www.morningstar.com.)

(a) $f(12)$ **(b)** $f(14)$ **(c)** $f(16)$

50. Stock Prices The closing stock price of American Express Company for the years 2013 to 2015 can be approximated by the function

$$f(x) = -15.27x^2 + 438.5x - 3061,$$

where $x = 13$ corresponds to the start of the year 2013. Find the following function values and explain what your answers mean in practical terms. (Data from: www.morningstar.com.)

(a) $f(13)$ **(b)** $f(14)$ **(c)** $f(15)$

51. Taxes Alabama state income tax for a single person in 2017 was determined by the rule

$$T(x) = \begin{cases} .02x & \text{if } 0 \le x \le 500 \\ 10 + .04(x - 500) & \text{if } 500 < x \le 3000 \\ 110 + .05(x - 3000) & \text{if } x > 3000, \end{cases}$$

where x is the person's taxable income. Find the following function values and interpret your answers. (Data from: www.tax-brackets.org.)

(a) $T(100)$ **(b)** $T(2500)$ **(c)** $T(103,000)$

52. Taxes Mississippi state income tax for a single person in 2017 was determined by the rule

$$T(x) = \begin{cases} .03x & \text{if } 0 \le x \le 5000 \\ 150 + .04(x - 5000) & \text{if } 5000 < x \le 10,000 \\ 350 + .05(x - 10,000) & \text{if } x > 10,000, \end{cases}$$

where x is the person's taxable income. Find the following function values and interpret your answers. (Data from: www.tax-brackets.org.)

(a) $T(100)$ **(b)** $T(7000)$ **(c)** $T(110,000)$

53. Social Science The mean age of a woman in the United States when her first child is born can be approximated by the function

$$f(x) = 21x^{.0565},$$

where $x = 10$ corresponds to the year 2010. Estimate the mean age of a woman at the birth of her first child in the following years. Round your answers to one decimal place. (Data from: The General Social Survey.)

(a) 2010   **(b)** 2012   **(c)** 2014

54. Water Pollution High concentrations of zinc ions in water are lethal to rainbow trout. The function

$$f(x) = \left(\frac{x}{1960}\right)^{-.833}$$

gives the approximate average survival time (in minutes) for trout exposed to x milligrams per liter (mg/L) of zinc ions. Find the survival time (to the nearest minute) for the given concentrations of zinc ions.

(a) 110   **(b)** 525   **(c)** 1960   **(d)** 4500

55. Physical Science The distance from Chicago to Sacramento, California, is approximately 2050 miles. A plane flying directly to Sacramento passes over Chicago at noon. If the plane travels at 500 mph, find the rule of the function $f(t)$ that gives the distance of the plane from Sacramento at time t hours (with $t = 0$ corresponding to noon).

56. Physical Science The distance from Toronto, Ontario to Dallas, Texas is approximately 1200 miles. A plane flying directly to Dallas passes over Toronto at 2 p.m. If the plane travels at 550 mph, find the rule of the function $f(t)$ that gives the distance of the plane from Dallas at time t hours (with $t = 0$ corresponding to 2 p.m.).

57. Cost, Revenue, and Profit A pretzel factory has daily fixed costs of $1800. In addition, it costs 50 cents to produce each bag of pretzels. A bag of pretzels sells for $1.20.

(a) Find the rule of the cost function $c(x)$ that gives the total daily cost of producing x bags of pretzels.

(b) Find the rule of the revenue function $r(x)$ that gives the daily revenue from selling x bags of pretzels.

(c) Find the rule of the profit function $p(x)$ that gives the daily profit from x bags of pretzels. (*Hint:* Recall that profit = revenue − cost.)

58. Cost, Revenue, and Profit An aluminum can factory has daily fixed costs of $120,000 per day. In addition, it costs $.03 to produce each can. They can sell a can to a soda manufacturer for $.05 a can.

(a) Find the rule of the cost function $c(x)$ that gives the total daily cost of producing x aluminum cans.

(b) Find the rule of the revenue function $r(x)$ that gives the daily revenue from selling x aluminum cans.

(c) Find the rule of the profit function $p(x)$ that gives the daily profit from x aluminum cans.

Use the table feature of a graphing calculator to do these exercises. (See Example 11.)

59. Imports Total imports to the United States in billions of dollars for the period 2010 to 2014 can be approximated by the function

$$f(x) = 28.2x^3 - 208x^2 + 507x + 2348,$$

where $x = 0$ corresponds to the year 2010. Create a table that gives total imports for the years 2010 to 2014 (that is, $x = 0$ to $x = 4$). (Data from: U.S. Census Bureau.)

60. Inflation The inflation rate in the United States (expressed as a percentage) for the period 2010 to 2014 can be approximated by the function

$$f(x) = .28x^3 - 1.84x^2 + 2.88x + 1.68,$$

where $x = 0$ corresponds to the year 2010. Create a table that gives the inflation rate for the years 2010 to 2014 (that is, $x = 0$ to $x = 4$). (Data from: International Monetary Fund.)

✓**Checkpoint Answers**

1. The domain consists of all possible times—that is, all nonnegative real numbers. The range consists of all possible distances; thus, the range is also the set of all nonnegative real numbers.

2. (a) Yes   (b) Yes   (c) No

3. (a) Yes   (b) Yes   (c) No   (d) No

4. (a) $(-\infty, \infty)$ (b) $(-\infty, \infty)$ (c) $(-\infty, 0]$
   (d) All real numbers except 1 and −1

5. (a) 4   (b) 40   (c) 73   (d) 44
   (e) $5m^2 - 2m + 1$ (f) $10xh + 5h^2 - 2h$
   (g) $10x + 5h - 2$

6. (a) $2237.50   (b) $220.35

7. (a) $39,600,000   (b) $96,000,000

3.2 Graphs of Functions

The **graph** of a function f is defined to be the graph of the *equation* $y = f(x)$. It consists of all points $(x, f(x))$—that is, every point whose first coordinate is an input number from the domain of f and whose second coordinate is the corresponding output number.

EXAMPLE 1 The graph of the function $g(x) = .5x - 3$ is the graph of the equation $y = .5x - 3$. So the graph is a straight line with slope .5 and y-intercept −3. To graph the line, we plot just two points. Because we know the y-intercept, we already have the point $(0, -3)$. For a second point, we will find the x-intercept (although any other point would suffice). Set $y = 0$ and solve for x:

$$y = 0$$
$$.5x - 3 = 0$$
$$.5x = 3$$
$$x = 6.$$

✓**Checkpoint 1**

Graph $g(x) = 3 - .5x$.

Now we have the point $(6, 0)$, and we graph the line shown in Figure 3.3. For a refresher on graphing lines, see Section 2.2. ✓

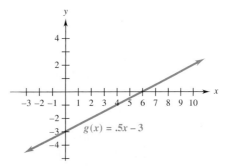

Figure 3.3

A function whose graph is a straight line, as in Example 1, is called a **linear function.** The rule of a linear function can always be put into the form

$$f(x) = ax + b$$

for some constants a and b.

Piecewise Linear Functions

We now consider functions whose graphs consist of straight-line segments. Such functions are called **piecewise linear functions** and are typically defined with different equations for different parts of the domain.

EXAMPLE 2   Graph the following function:

$$f(x) = \begin{cases} x + 1 & \text{if } x \le 2 \\ -2x + 7 & \text{if } x > 2. \end{cases}$$

Solution Consider the two parts of the rule of f. The graphs of $y = x + 1$ and $y = -2x + 7$ are straight lines. The graph of f consists of

the part of the line $y = x + 1$ with $x \le 2$ and

the part of the line $y = -2x + 7$ with $x > 2$.

Each of these line segments can be graphed by plotting two points in the appropriate interval, as shown in Figure 3.4.

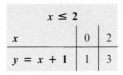

$x \le 2$		
x	0	2
$y = x + 1$	1	3

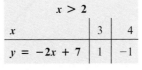

$x > 2$		
x	3	4
$y = -2x + 7$	1	-1

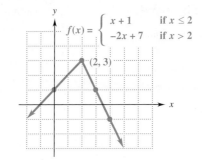

Figure 3.4

Note that the left and right parts of the graph each extend to the vertical line through $x = 2$, where the two halves of the graph meet at the point $(2, 3)$. ✓ 2

✓ **Checkpoint 2**

Graph

$$f(x) = \begin{cases} x + 2 & \text{if } x < 0 \\ 2 - x & \text{if } x \ge 0. \end{cases}$$

EXAMPLE 3   Graph the function

$$f(x) = \begin{cases} x - 2 & \text{if } x \le 3 \\ -x + 8 & \text{if } x > 3. \end{cases}$$

Solution The graph consists of parts of two lines. To find the left side of the graph, choose two values of x with $x \leq 3$, say, $x = 0$ and $x = 3$. Then find the corresponding points on $y = x - 2$, namely, $(0, -2)$ and $(3, 1)$. Use these points to draw the line segment to the left of $x = 3$, as in Figure 3.5. Next, choose two values of x with $x > 3$, say, $x = 4$ and $x = 6$, and find the corresponding points on $y = -x + 8$, namely, $(4, 4)$ and $(6, 2)$. Use these points to draw the line segment to the right of $x = 3$, as in Figure 3.5.

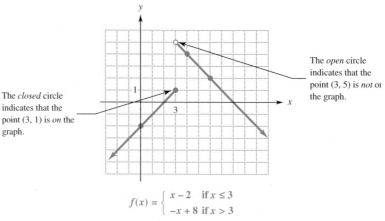

The *closed* circle indicates that the point $(3, 1)$ is *on* the graph.

The *open* circle indicates that the point $(3, 5)$ is *not* on the graph.

$$f(x) = \begin{cases} x - 2 & \text{if } x \leq 3 \\ -x + 8 & \text{if } x > 3 \end{cases}$$

Figure 3.5

Note that both line segments of the graph of f extend to the vertical line through $x = 3$. Because the two graphs do not connect at the input $x = 3$, the graph has a break, or discontinuity, at $x = 3$. ✓ 3

✓ **Checkpoint 3**

Graph

$$f(x) = \begin{cases} -2x - 3 & \text{if } x < 1 \\ x - 2 & \text{if } x \geq 1. \end{cases}$$

EXAMPLE 4 Graph the **absolute-value function,** whose rule is $f(x) = |x|$.

Solution The definition of absolute value on page 8 shows that the rule of f can be written as

$$f(x) = \begin{cases} x & \text{if } x \geq 0 \\ -x & \text{if } x < 0. \end{cases}$$

So the right half of the graph (that is, where $x \geq 0$) will consist of a portion of the line $y = x$. It can be graphed by plotting two points, say, $(0, 0)$ and $(1, 1)$. The left half of the graph (where $x < 0$) will consist of a portion of the line $y = -x$, which can be graphed by plotting $(-2, 2)$ and $(-1, 1)$, as shown in Figure 3.6. ✓ 4

✓ **Checkpoint 4**

Graph each function.

(a) $f(x) = |x - 4|$

(b) $f(x) = |2x - 4|$

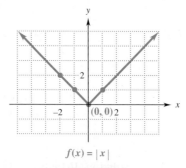

$$f(x) = |x|$$

Figure 3.6

Many real-world problems can be modeled using piecewise-defined functions. For example, every employee pays Social Security and Medicare taxes, and these formulas are piecewise-defined (see Exercises 37 and 38). The costs for many products and services are piecewise-defined due to either volume discounts or higher rates for excessive usage, as in the following example.

EXAMPLE 5   Utility Bills

The Birmingham Water Works charges its residential customers with a ⅝-inch water line a basic service charge of approximately $25 per month. The first 15 hundred cubic feet (CCF) of water cost about $2 per CCF. To encourage water conservation, excessive water usage is charged at a higher rate. It costs about $4 per CCF for each CCF above 15. (Data from: Birmingham Water Works.)

(a) What would be the amount of the water bill for a customer who used no water?

Solution The customer would just pay the $25 basic service charge.

(b) Find the water bill for a customer who used 15 CCF.

Solution The customer would pay the $25 service charge plus $2 for each CCF used, for a total of $25 + $2(15) = $25 + $30 = $55.

(c) How much would it cost to use 25 CCF of water?

Solution It would cost $55 for the first 15 CCF (as calculated above), plus $4 per CCF for each additional CCF above 15, for a total of $55 + $4(25 − 15) = $55 + $40 = $95.

(d) Find a rule for the function $f(x)$ that gives the amount of the water bill when x CCF of water are used.

Solution For up to 15 CCF, the cost is the $25 fee plus $2 per CCF, for a total of $25 + 2x$. For usage above 15 CCF, the cost is $55 plus $4 per CCF for each additional CCF above 15, for a total of $55 + 4(x − 15)$. So the amount of the water bill is the piecewise-defined function

$$f(x) = \begin{cases} 25 + 2x & \text{if } 0 \le x \le 15 \\ 55 + 4(x - 15) & \text{if } x > 15. \end{cases}$$

(e) Graph $f(x)$.

Solution For $0 \le x \le 15$, the graph is part of the line $y = 25 + 2x$. We can easily graph this line segment by plotting and connecting the points (0, 25) and (15, 55). For $x > 15$, the graph is part of the line $y = 55 + 4(x − 15)$. Use the points (15, 55) and (25, 95) to sketch the portion of this line with $x > 15$. The graph is shown in Figure 3.7. ✓ 5

✓ Checkpoint 5

In Example 5, what would be the water bill for the following amounts of water?

(a) 5 CCF

(b) 30 CCF

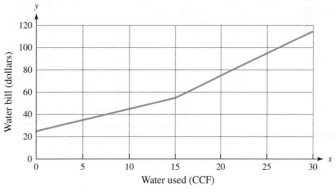

Figure 3.7

▦ **TECHNOLOGY TIP** To graph most piecewise linear functions on a graphing calculator, you must use a special syntax. For example, on TI calculators, the best way to obtain the graph in Example 3 is to graph two separate equations on the same screen:

$$y_1 = (x - 2)/(x \le 3) \quad \text{and} \quad y_2 = (-x + 8)/(x > 3).$$

The inequality symbols are in the TEST (or CHAR) menu. However, most calculators will graph absolute-value functions directly. To graph $t(x) = |x + 2|$, for instance, graph the equation $y = abs(x + 2)$. "Abs" (for absolute value) is on the keyboard or in the MATH menu.

Step Functions

The **greatest-integer function,** usually written $f(x) = [x]$, is defined by saying that $[x]$ denotes the largest integer that is less than or equal to x. For example, $[8] = 8, [7.45] = 7, [\pi] = 3$, $[-1] = -1, [-2.6] = -3$, and so on.

✓Checkpoint 6

Graph $y = [\frac{1}{2}x + 1]$.

EXAMPLE 6   Graph the greatest-integer function $f(x) = [x]$.

Solution Consider the values of the function between each two consecutive integers—for instance,

x	$-2 \leq x < -1$	$-1 \leq x < 0$	$0 \leq x < 1$	$1 \leq x < 2$	$2 \leq x < 3$
$[x]$	-2	-1	0	1	2

Thus, between $x = -2$ and $x = -1$, the value of $f(x) = [x]$ is always -2, so the graph there is a horizontal line segment, all of whose points have second coordinate -2. The rest of the graph is obtained similarly (Figure 3.8). An open circle in that figure indicates that the endpoint of the segment is *not* on the graph, whereas a closed circle indicates that the endpoint *is* on the graph. **✓6**

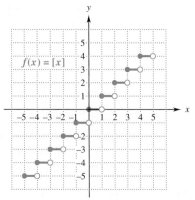

Figure 3.8

Functions whose graphs resemble the graph of the greatest-integer function are sometimes called **step functions.**

Step functions occur in many real-life situations, such as the fare for an Uber ride, the cost of parking in a parking garage, and the bill for the data plan on your cell phone (for example, the cost might jump up by $15 to buy an extra gigabyte of data). The following is another common example.

EXAMPLE 7   **Postal Rates** To ship a first-class package to Brazil in 2016, the U.S. Post Office charged a fee of $13 for the first 8 ounces, $9 for an additional 24 ounces or less, $9 for the next 24 ounces or less, and then $20 more for up to a maximum of 64 ounces. Let $D(x)$ represent the cost to send a package weighing x ounces. Graph $y = D(x)$ for x in the interval $(0, 64]$. (Data from: www.usps.com.)

Solution For x in the interval $(0, 8]$, $y = 13$. For x in the interval $(8, 32]$, $y = 13 + 9 = 22$. For x in $(32, 48]$, $y = 22 + 9 = 31$. Finally, for x in $(48, 64]$, $y = 31 + 20 = 51$. The graph is a step function and is shown in Figure 3.9.

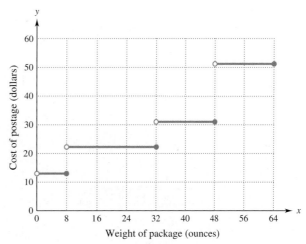

Figure 3.9

$$D(x) = \begin{cases} 13 & \text{if } 0 < x \le 8 \\ 22 & \text{if } 8 < x \le 32 \\ 31 & \text{if } 32 < x \le 48 \\ 51 & \text{if } 48 < x \le 64 \end{cases}$$

✓7

✓ **Checkpoint 7**

To mail a letter to Canada in 2016, the U.S. Post Office charged $1.20 for up to and including the first two ounces, an additional $.47 for up to and including an additional ounce, and then an additional $.49 up to a maximum weight of 3.5 ounces. Let $L(x)$ represent the cost of sending a letter weighing x ounces. Graph $L(x)$ for x in the interval (0, 3.5].

🖩 **TECHNOLOGY TIP** On most graphing calculators, the greatest-integer function is denoted INT or FLOOR. (Look on the MATH menu or its NUM submenu.) When graphing these functions, put your calculator in "dot" graphing mode rather than the usual "connected" mode to avoid erroneous vertical line segments in the graph.

Other Functions

The graphs of many functions do not consist only of straight-line segments. The most basic method for graphing functions by hand is to plot points. This method was introduced in Section 2.1 and is summarized here. In Sections 3.4–3.6, we will learn more techniques to aid with graphing. Calculus provides additional strategies.

> ### Graphing a Function by Plotting Points
>
> 1. Determine the domain of the function.
> 2. Select a few numbers in the domain of f (include both negative and positive ones when possible), and compute the corresponding values of $f(x)$.
> 3. Plot the points $(x, f(x))$ computed in Step 2. Use these points and any other information you may have about the function to make an educated guess about the shape of the entire graph.
> 4. Unless you have information to the contrary, assume that the graph is continuous (unbroken) wherever it is defined.

This method was used to find the graphs of the functions $f(x) = x^2 - 2x - 8$ and $g(x) = \sqrt{x + 2}$ in Examples 4 and 5 of Section 2.1. Here are some more examples.

EXAMPLE 8   Graph $g(x) = \sqrt{x - 1}$.

Solution Because the rule of the function is defined only when $x - 1 \ge 0$ (that is, when $x \ge 1$), the domain of g is the interval $[1, \infty)$. Use a calculator to make a table of values such as the one in Figure 3.10. Plot the corresponding points and connect them to get the graph in Figure 3.10. ✓8

✓ **Checkpoint 8**

Graph $f(x) = \sqrt{5 - 2x}$.

x	$g(x) = \sqrt{x-1}$
1	0
2	1
3	$\sqrt{2} \approx 1.414$
5	2
7	$\sqrt{6} \approx 2.449$
10	3

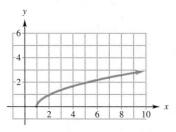

Figure 3.10

EXAMPLE 9   Graph the piecewise defined function

$$f(x) = \begin{cases} x^2 & \text{if } x \le 2 \\ \sqrt{x-1} & \text{if } x > 2. \end{cases}$$

Solution When $x \le 2$, the rule of the function is $f(x) = x^2$. Make a table of values such as the one in Figure 3.11. Plot the corresponding points and connect them to get the left half of the graph in Figure 3.11. When $x > 2$, the rule of the function is $f(x) = \sqrt{x-1}$, whose graph is shown in Figure 3.10. In Example 8, the entire graph was given, beginning at $x = 1$. Here we use only the part of the graph to the right of $x = 2$, as shown in Figure 3.11. The open circle at $(2, 1)$ indicates that this point is not part of the graph of f (why?).

x	x^2
-2	4
-1	1
0	0
1	1
2	4

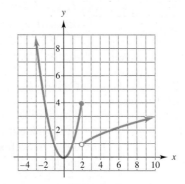

Figure 3.11

Graph Reading

Graphs are often used in business and the social sciences to present data. It is just as important to know how to *read* such graphs as it is to construct them.

EXAMPLE 10   **Housing Prices** Figure 3.12 shows the median sales prices (in thousands of dollars) of new privately owned, one-family houses by region of the United States (Northeast, Midwest, South, and West) for the years 2008 to 2015. (Data from: U.S. Department of Housing and Urban Development.)

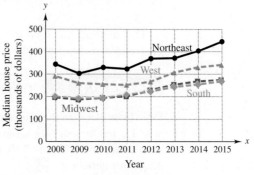

Figure 3.12

(a) How do prices in the West compare with those in the Midwest over the period shown in the graph?

Solution The prices of the houses in the West started out approximately $100,000 higher than those in the Midwest. In both regions, prices declined initially, but prices in both regions rebounded. By 2012, the gap between the two regions shrunk to less than $50,000 but widened again to about $75,000 by 2015.

(b) Which region typically had the highest prices?

Solution The Northeast had the highest median sales price for all years shown.

EXAMPLE 11   **Commodity Prices** The average price $f(x)$ of a barrel of crude oil (in dollars) in year x is shown in Figure 3.13 for the years 2005 to 2015, with projected prices shown through the year 2025. (Data from: www.worldbank.org.)

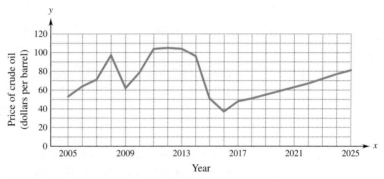

Figure 3.13

(a) Estimate the function values $f(2015)$ and $f(2025)$, and interpret your answers.

Solution The point (2015, 50) is on the graph, which means that $f(2015) = 50$. This tells us that, in the year 2015 (the input), the average price of a barrel of crude oil was $50 (the output). Similarly, we see that $f(2025) = 80$ because the point (2025, 80) is on the graph. This means that the price of crude oil is projected to be $80 per barrel in 2025.

(b) During what period did oil prices drop below $40 per barrel?

Solution Look for points on the graph whose second coordinates are 40 or below—that is, points on or below the horizontal line through 40. This occurs in the year 2016.

(c) During what period did oil prices soar above $100 per barrel?

Solution Look for points at or above the horizontal line through 100—that is, points with a second coordinate greater than or equal to 100. They occur from 2011 to 2013.

The Vertical-Line Test

The following fact distinguishes the graphs of function from other graphs.

> ## Vertical-Line Test
>
> No vertical line intersects the graph of a function $y = f(x)$ at more than one point.

Figure 3.14

In other words, if a vertical line intersects a graph at more than one point, the graph is not the graph of a function. To see why this is true, consider the graph in Figure 3.14. The vertical line $x = 3$ intersects the graph at two points. If this were the graph of a function f, it would mean that $f(3) = 2$ (because $(3, 2)$ is on the graph) and that $f(3) = -1$ (because $(3, -1)$ is on the graph). This is impossible, because a *function* can have only one value when $x = 3$ (because each input determines exactly one output). Therefore, the graph in Figure 3.14 cannot be the graph of a function. A similar argument works in the general case.

EXAMPLE 12   Use the vertical-line test to determine which of the graphs in Figure 3.15 are the graphs of functions.

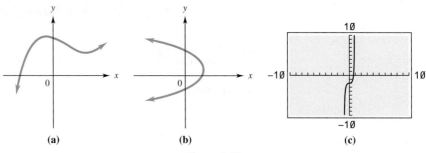

(a)   **(b)**   **(c)**

Figure 3.15

Solution To use the vertical-line test, imagine dragging a ruler held vertically across the graph from left to right. If the graph is that of a function, the edge of the ruler would hit the graph only once for every x-value. If you do this for graph (a), every vertical line intersects the graph in at most one point, so this graph is the graph of a function. Many vertical lines (including the y-axis) intersect graph (b) twice, so it is not the graph of a function.

Graph (c) appears to fail the vertical-line test near $x = 1$ and $x = -1$, indicating that it is not the graph of a function. But this appearance is misleading because of the low resolution of the calculator screen. The table in Figure 3.16 and the very narrow segment of the graph in Figure 3.17 show that the graph actually rises as it moves to the right. The same thing happens near $x = -1$. So this graph *does* pass the vertical-line test and *is* the graph of a function. (Its rule is $f(x) = 15x^{11} - 2$.) The moral of this story is that you can't always trust images produced by a graphing calculator. When in doubt, try other viewing windows or a table to see what is really going on. ✓ 9

✓ **Checkpoint 9**

Find a viewing window that indicates the actual shape of the graph of the function $f(x) = 15x^{11} - 2$ of Example 12 near the point $(-1, -17)$.

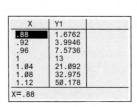

Figure 3.16   **Figure 3.17**

3.2 Exercises

Graph each function. (See Examples 1–4.)

1. $f(x) = -.5x + 2$

2. $g(x) = 3 - x$

3. $f(x) = \begin{cases} x + 3 & \text{if } x \le 1 \\ 4 & \text{if } x > 1 \end{cases}$

4. $g(x) = \begin{cases} 2x - 1 & \text{if } x < 0 \\ -1 & \text{if } x \ge 0 \end{cases}$

5. $y = \begin{cases} 4 - x & \text{if } x \le 0 \\ 3x + 4 & \text{if } x > 0 \end{cases}$

6. $y = \begin{cases} x + 5 & \text{if } x \le 1 \\ 2 - 3x & \text{if } x > 1 \end{cases}$

7. $f(x) = \begin{cases} |x| & \text{if } x < 2 \\ -2x & \text{if } x \ge 2 \end{cases}$

8. $g(x) = \begin{cases} -|x| & \text{if } x \le 1 \\ 2x & \text{if } x > 1 \end{cases}$

9. $f(x) = |x - 4|$

10. $g(x) = |4 - x|$

11. $f(x) = |3 - 3x|$

12. $g(x) = -|x|$

13. $y = -|x - 1|$

14. $f(x) = |x| - 2$

15. $y = |x - 2| + 3$

16. $|x| + |y| = 1$ (*Hint:* This is not the graph of a function, but is made up of four straight-line segments. Find them by using the definition of absolute value in these four cases: $x \geq 0$ and $y \geq 0$; $x \geq 0$ and $y < 0$; $x < 0$ and $y \geq 0$; $x < 0$ and $y < 0$).

Graph each function. (See Examples 6 and 7.)

17. $f(x) = [x - 3]$

18. $g(x) = [x + 3]$

19. $g(x) = [-x]$

20. $f(x) = [x] + [-x]$ (The graph contains horizontal segments, but is *not* a horizontal line.)

21. **Postal Rates** The accompanying table gives rates charged by the U.S. Postal Service for first-class letters in March, 2016. Graph the function $f(x)$ that gives the price of mailing a first-class letter, where x represents the weight of the letter in ounces and $0 \leq x \leq 3.5$. (Data from: www.usps.com.)

Weight Not Over	Price
1 ounce	$ .49
2 ounces	$ .71
3 ounces	$ .93
3.5 ounces	$1.15

22. **Postal Rates** The accompanying table gives rates charged by the U.S. Postal Service for first-class large flat envelopes in March, 2016. Graph the function $f(x)$ that gives the price of mailing a first-class large flat envelope, where x represents the weight of the envelope in ounces and $0 \leq x \leq 6$. (Data from: www.usps.com.)

Weight Not Over	Price
1 ounce	$ .98
2 ounces	$1.20
3 ounces	$1.42
4 ounces	$1.64
5 ounces	$1.86
6 ounces	$2.08

Graph each function. (See Examples 8–9.)

23. $g(x) = \sqrt{-x}$

24. $h(x) = \sqrt{x} - 1$

25. $f(x) = \begin{cases} x^2 & \text{if } x < 2 \\ -2x + 2 & \text{if } x \geq 2 \end{cases}$

26. $g(x) = \begin{cases} \sqrt{-x} & \text{if } x \leq -4 \\ \dfrac{x^2}{4} & \text{if } x > -4 \end{cases}$

Determine whether each graph is a graph of a function or not. (See Example 12.)

27.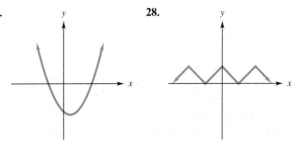

28.

29.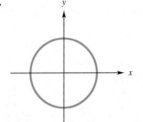

30.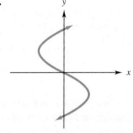

31.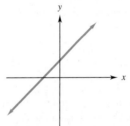

32.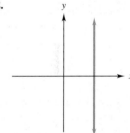

Use a graphing calculator or other technology to graph each of the given functions. If the graph has any endpoints, indicate whether they are part of the graph or not.

33. $f(x) = .2x^3 - .8x^2 - 4x + 9.6$

34. $g(x) = .1x^4 - .3x^3 - 1.3x^2 + 1.5x$

35. $g(x) = \begin{cases} 2x^2 + x & \text{if } x < 1 \\ x^3 - x - 1 & \text{if } x \geq 1 \end{cases}$

(*Hint:* See the Technology Tip on page 125.)

36. $f(x) = \begin{cases} x|x| & \text{if } x \leq 0 \\ -x^2|x| + 2 & \text{if } x > 0 \end{cases}$

Work these exercises. (See Examples 2, 3, 10, and 11.)

37. **Medicare** An employee must pay Medicare tax at a base rate plus an additional Medicare tax if wages exceed a certain threshold. For a single person in 2016, the employee's Medicare tax in dollars was determined by the piecewise linear function

$$f(x) = \begin{cases} 14.5x & \text{if } 0 \leq x \leq 200 \\ 2900 + 23.5(x - 200) & \text{if } x > 200 \end{cases},$$

where x is the employee's Medicare wages in thousands of dollars. (Data from: Internal Revenue Service.)

(a) Find the function values $f(0)$, $f(200)$, $f(400)$.

(b) Graph this function for wages up to $400,000 (that is, $0 \leq x \leq 400$).

38. Social Security An employee must pay Social Security tax up to a maximum wage that is subject to the tax. In 2016, an employee's Social Security tax in dollars was determined by the piecewise linear function

$$f(x) = \begin{cases} 62x & \text{if } 0 \le x \le 118.5 \\ 7347 & \text{if } x > 118.5 \end{cases},$$

where x is the employee's Social Security wages in thousands of dollars. (Data from: Internal Revenue Service.)

(a) Find the function values $f(0), f(118.5), f(400)$.

(b) Graph this function for wages up to $400,000 (that is, $0 \le x \le 400$).

39. Taxes Mississippi state income tax for a single person in 2017 was determined by the rule

$$T(x) = \begin{cases} .03x & \text{if } 0 \le x \le 5000 \\ 150 + .04(x - 5000) & \text{if } 5000 < x \le 10{,}000, \\ 350 + .05(x - 10{,}000) & \text{if } x > 10{,}000 \end{cases}$$

where x is the person's taxable income in dollars. Graph the function $T(x)$ for taxable incomes between 0 and $15,000. (Data from: www.tax-brackets.org.)

40. Taxes The Alabama state income tax for a single person in 2017 was determined by the rule

$$T(x) = \begin{cases} .02x & \text{if } 0 \le x \le 500 \\ 10 + .04(x - 500) & \text{if } 500 < x \le 3000 \\ 110 + .05(x - 3000) & \text{if } x > 3000 \end{cases}$$

where x is the person's taxable income in dollars. Graph the function $T(x)$ for taxable incomes between 0 and $5000.

41. Smoking Rates The following graph shows the percentage of the adult population of the United States who smoke cigarettes. (Data from: www.cdc.gov.)

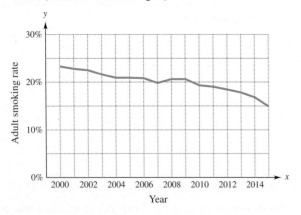

(a) According to the graph, approximately what was the adult smoking rate in 2015?

(b) Let $x = 0$ correspond to the year 2000. Approximate a linear function for the smoking rate (as a percentage) using the points $(5, 21)$ and $(15, 15)$.

(c) What does this model estimate for the smoking rate in 2000? In 2017?

(d) The Centers for Disease Control and Prevention (CDC) reports that a *Healthy People 2020* goal is that only 12% of the adult population will smoke cigarettes. If the trend continues, will that goal be met?

42. Smoking Rates Use the graph in the previous exercise to answer the following questions.

(a) What is the general trend for the adult smoking rate?

(b) Let $x = 0$ correspond to the year 2000. Approximate a linear function for the smoking rate (as a percentage) using the points $(7, 19.8)$. and $(15, 15)$.

(c) What does this model estimate for the smoking rate in 2000? In 2017?

43. Consumer Price Index The accompanying graph shows the annual percentage change in the consumer price indexes (CPIs) for various sectors of the economy. (Data from: Bureau of Labor Statistics.)

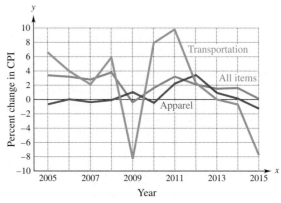

(a) During what year(s) was the change in CPI for transportation the largest?

(b) During what year(s) was the change in CPI for apparel the largest?

(c) During what year(s) was the change in CPI for all items approximately zero?

44. Consumer Price Index Use the graph in the previous exercise to answer the following questions.

(a) During what year(s) did the CPI for all items show a decrease?

(b) When was the CPI for transportation increasing at the greatest rate?

(c) During what year(s) did the CPI for apparel increase at a faster rate than the CPI for all items?

45. Apple Shareholders' Equity The following graph shows total assets and total liabilities for Apple Inc. in billions of dollars. (Data from: Apple Inc. Annual Report.)

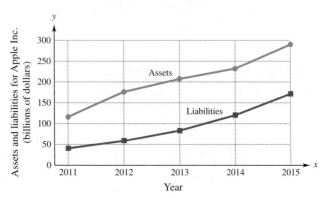

(a) Shareholders' equity is a corporation's total assets minus total liabilities. Estimate Apple's shareholders' equity in the year 2015.

(b) What is the domain of the assets function shown in the graph?

(c) Estimate the range of the liabilities function shown in the graph.

46. iPhone Sales The following graph shows iPhones sales in millions of units. (Data from: Apple Inc. Annual Report.)

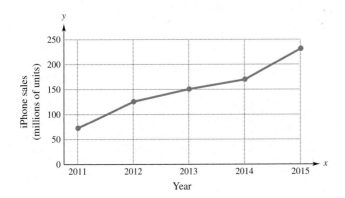

(a) Estimate and interpret the function value $f(2013)$.

(b) What is the domain shown in the graph?

(c) Estimate the range shown in the graph.

47. Postal Rates Whenever postage rates change, some newspaper publishes a graph like this one, which shows the price of a first-class stamp from 2006 to 2016:

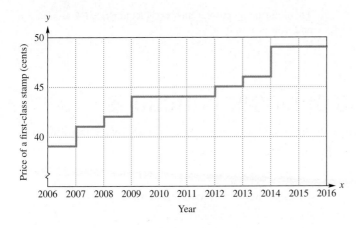

(a) Let f be the function whose rule is $f(x) = $ cost of a first-class stamp in year x. Find $f(2010)$ and $f(2015)$.

(b) Explain why the graph in the figure is not the graph of the *function f*. What must be done to the figure to make it an accurate graph of the *function f*?

48. Business A chain-saw rental firm charges $20 per day or fraction of a day to rent a saw, plus a fixed fee of $7 for resharpening the blade. Let $S(x)$ represent the cost of renting a saw for x days. Find each of the following.

(a) $S\left(\dfrac{1}{2}\right)$ **(b)** $S(1)$ **(c)** $S\left(1\dfrac{2}{3}\right)$ **(d)** $S\left(4\dfrac{3}{4}\right)$

(e) What does it cost to rent for $5\dfrac{7}{8}$ days?

(f) A portion of the graph of $y = S(x)$ is shown on the following page. Explain how the graph could be continued.

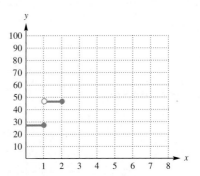

(g) What is the domain variable?

(h) What is the range variable?

(i) Write a sentence or two explaining what (c) and its answer represent.

(j) We have left $x = 0$ out of the graph. Discuss why it should or should not be included. If it were included, how would you define $S(0)$?

49. Business You need to rent a van to pick up a new couch you have purchased. The cost of the van is $19.99 for the first 75 minutes and then an additional $5 for each block of 15 minutes beyond 75. Find the cost to rent a van for

(a) 2 hours; **(b)** 1.5 hours;

(c) 3.5 hours; **(d)** 4 hours.

(e) Graph the ordered pairs (hours, cost).

50. Business A delivery company charges $25 plus 60¢ per mile or part of a mile. Find the cost for a trip of

(a) 3 miles; **(b)** 4.2 miles;

(c) 5.9 miles; **(d)** 8 miles.

(e) Graph the ordered pairs (miles, cost).

(f) Is this a function?

Work these problems.

51. Bacteria Growth A laboratory culture contains about 1 million bacteria at midnight. The culture grows very rapidly until noon, when a bactericide is introduced and the bacteria population plunges. By 4 p.m., the bacteria have adapted to the bactericide and the culture slowly increases in population until 9 p.m., when the culture is accidentally destroyed by the cleanup crew. Let $g(t)$ denote the bacteria population at time t (with $t = 0$ corresponding to midnight). Draw a plausible graph of the function g. (Many correct answers are possible.)

52. Flight Paths A plane flies from Austin, Texas, to Cleveland, Ohio, a distance of 1200 miles. Let f be the function whose rule is

$f(t) = $ distance (in miles) from Austin at time t hours,

with $t = 0$ corresponding to the 4 p.m. takeoff. In each part of this exercise, draw a plausible graph of f under the given circumstances. (There are many correct answers for each part.)

(a) The flight is nonstop and takes between 3.5 and 4 hours.

(b) Bad weather forces the plane to land in Dallas (about 200 miles from Austin) at 5 p.m., remain overnight, and leave at 8 a.m. the next day, flying nonstop to Cleveland.

(c) The plane flies nonstop, but due to heavy traffic it must fly in a holding pattern for an hour over Cincinnati (about 200 miles from Cleveland) and then go on to Cleveland.

4. (a)
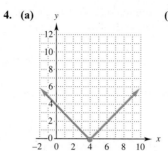
(b)
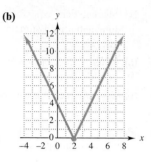

✓ Checkpoint Answers

5. (a) $35   **(b)** $115

1.
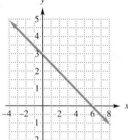

2.
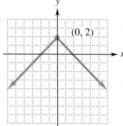

6.
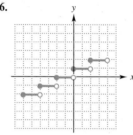

7.
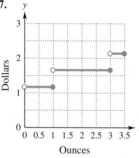

3.
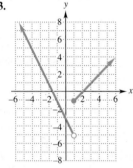

8.
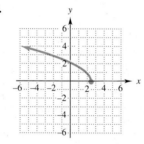

9. There are many correct answers, including, $-1.4 \le x \le -.6$ and $-30 \le y \le 0$.

3.3 Applications of Linear Functions

Most of this section deals with the basic business relationships that were introduced in Section 1.2:

> Revenue = (Price per item) × (Number of items);
> Cost = Fixed Costs + Variable Costs;
> Profit = Revenue − Cost.

The examples will use only linear functions, but the methods presented here also apply to more complicated functions.

Cost Analysis

Recall that fixed costs are for such things as buildings, machinery, real-estate taxes, and product design. Within broad limits, the fixed cost is constant for a particular product and does not change as more items are made. Variable costs are for labor, materials, shipping, and so on, and depend on the number of items made.

If $C(x)$ is the cost of making x items, then the fixed cost (the cost that occurs even when no items are produced) can be found by letting $x = 0$. For example, for the cost function $C(x) = 45x + 250,000$, the fixed cost is

$$C(0) = 45(0) + 250,000 = \$250,000.$$

In this case, the variable cost of making x items is $45x$—that is, $45 per item manufactured.

EXAMPLE 1   **Prescription Drugs** An anticlot drug can be made for $10 per unit. The total cost to produce 100 units is $1500.

(a) Assuming that the cost function is linear, find its rule.

Solution Since the cost function $C(x)$ is linear, its rule is of the form $C(x) = mx + b$. We are given that m (the cost per item) is 10, so the rule is $C(x) = 10x + b$. To find b, use the fact that it costs $1500 to produce 100 units, which means that

$$C(100) = 1500$$
$$10(100) + b = 1500 \qquad \text{\scriptsize $C(x) = 10x + b$.}$$
$$1000 + b = 1500$$
$$b = 500.$$

So the rule of the cost function is $C(x) = 10x + 500$.

(b) What are the fixed costs?

✓ **Checkpoint 1**

The total cost of producing 10 calculators is $100. The variable costs per calculator are $4. Find the rule of the linear cost function.

Solution The fixed costs are $C(0) = 10(0) + 500 = \$500.$ ✓₁

If $C(x)$ is the total cost to produce x items, then the **average cost** per item is given by

$$\overline{C}(x) = \frac{C(x)}{x}.$$

As more and more items are produced, the average cost per item typically decreases.

EXAMPLE 2   **Average Cost** Find the average cost of producing 100 and 1000 units of the anticlot drug in Example 1.

Solution The cost function is $C(x) = 10x + 500$, so the average cost of producing 100 units is

$$\overline{C}(100) = \frac{C(100)}{100} = \frac{10(100) + 500}{100} = \frac{1500}{100} = \$15.00 \text{ per unit.}$$

✓ **Checkpoint 2**

In Checkpoint 1, find the average cost per calculator when 100 are produced.

The average cost of producing 1000 units is

$$\overline{C}(1000) = \frac{C(1000)}{1000} = \frac{10(1000) + 500}{1000} = \frac{10,500}{1000} = \$10.50 \text{ per unit.} ✓₂$$

x	$f(x) = 3x + 5$
1	8
2	11
3	14
4	17
5	20

Rates of Change

The rate at which a quantity (such as revenue or profit) is changing can be quite important. For instance, if a company determines that the rate of change of its revenue is decreasing, then sales growth is slowing down, a trend that may require a response.

The rate of change of a linear function is easily determined. For example, suppose $f(x) = 3x + 5$, and consider the table of values in the margin. The table shows that each time x changes by 1, the corresponding value of $f(x)$ changes by 3. Thus, the rate of change of $f(x) = 3x + 5$ with respect to x is 3, which is the slope of the line $y = 3x + 5$. The same thing happens for any linear function:

> **The rate of change of a linear function $f(x) = mx + b$ is the slope m.**

In particular, the rate of change of a linear function is constant.

The value of a computer, or an automobile, or a machine *depreciates* (decreases) over time. **Linear depreciation** means that the value of the item at time x is given by a linear function $f(x) = mx + b$. The slope m of this line gives the rate of depreciation.

EXAMPLE 3   **Automobile Depreciation** According to Kelly Blue Book, a Honda Civic four-door sedan that is worth $24,535 today will be worth $17,020 in three years (if it is in excellent condition with low mileage).

(a) Assuming linear depreciation, find the depreciation function for this car.

Solution We know the car is worth $24,535 now ($x = 0$) and will be worth $17,020 in three years ($x = 3$). So the points $(0, 24,535)$ and $(3, 17,020)$ are on the graph of the linear depreciation function and can be used to determine its slope:

$$m = \frac{17,020 - 24,535}{3 - 0} = \frac{-7515}{3} = -2505.$$

Using the point $(0, 24,535)$, we find that the equation of the line is

$$y - y_1 = m(x - x_1) \qquad \text{Point–slope form}$$
$$y - 24,535 = -2505(x - 0) \qquad \text{Substitute}$$
$$y = -2505x + 24,535. \qquad \text{Slope–intercept form}$$

Therefore, the rule of the depreciation function is $f(x) = -2505x + 24,535$.

(b) What will the car be worth five years from now?

Solution Evaluate f when $x = 5$:

$$f(x) = -2505x + 24,535$$
$$f(5) = -2505(5) + 24,535 = -12,525 + 24,535 = \$12,010.$$

(c) At what rate is the car depreciating?

Solution The depreciation rate is given by the slope of $f(x) = -2505x + 24,535$, namely, -2505. This negative slope means that the car is decreasing in value an average of $2505 per year. ✔️ 3

✔ **Checkpoint 3**

Using the information from Example 3, determine what the car will be worth in 6 years.

Marginal Analysis

In economics, the rate of change of the cost function is called the **marginal cost.** Marginal cost is important to management in making decisions in such areas as cost control, pricing, and production planning. When the cost function is linear, say, $C(x) = mx + b$, the marginal cost is the number m (the slope of the graph of C). Marginal cost can also be thought of as the cost of producing one more item, as the next example demonstrates.

EXAMPLE 4   **Marginal Cost** An electronics company manufactures tablets. The cost function for one of its models is $C(x) = 160x + 750,000$.

(a) What are the fixed costs for this product?

Solution The fixed costs are $C(0) = 160(0) + 750,000 = \$750,000$.

(b) What is the marginal cost?

Solution The slope of $C(x) = 160x + 750,000$ is 160, so the marginal cost is $160 per item.

(c) After 50,000 units have been produced, what is the cost of producing one more?

Solution The cost of producing 50,000 is

$$C(50,000) = 160(50,000) + 750,000 = \$8,750,000.$$

The cost of 50,001 units is

$$C(50,001) = 160(50,001) + 750,000 = \$8,750,160.$$

The cost of the additional unit is the difference

$$C(50,001) - C(50,000) = 8,750,160 - 8,750,000 = \$160.$$

Thus, the cost of one more item is the marginal cost. ✓₄

Similarly, the rate of change of a revenue function is called the **marginal revenue**. When the revenue function is linear, the marginal revenue is the slope of the line, as well as the revenue from producing one more item.

✓ Checkpoint 4

The cost in dollars to produce x kilograms of chocolate candy is given by $C(x) = 3.5x + 800$. Find each of the following.

(a) The fixed cost

(b) The total cost for 12 kilograms

(c) The marginal cost of the 40th kilogram

(d) The marginal cost per kilogram

EXAMPLE 5   Marginal Revenue The energy company New York State Electric and Gas charges each residential customer a basic fee for electricity of $15.11, plus $.0333 per kilowatt hour (kWh).

(a) Assuming there are 700,000 residential customers, find the company's revenue function.

Solution The monthly revenue from the basic fee is

$$15.11(700,000) = \$10,577,000.$$

If x is the total number of kilowatt hours used by all customers, then the revenue from electricity use is $.0333x$. So the monthly revenue function is given by

$$R(x) = .0333x + 10,577,000.$$

(b) What is the marginal revenue?

Solution The marginal revenue (the rate at which revenue is changing) is given by the slope of the rate function: $0.0333 per kWh. ✓₅

Examples 4 and 5 are typical of the general case, as summarized here.

✓ Checkpoint 5

Assume that the average customer in Example 5 uses 1600 kWh in a month.

(a) What is the total number of kWh used by all customers?

(b) What is the company's monthly revenue?

Marginal cost is the rate of change of the cost function. For a **linear cost function** $C(x) = mx + b$, the marginal cost is m (the slope of the cost line) and the fixed cost is b (the y-intercept of the cost line). The marginal cost is the cost of producing one more item.

   Marginal revenue is the rate of change of the revenue function. For a **linear revenue function** $R(x) = kx + d$, the marginal revenue is k (the slope of the revenue line), which is the revenue from selling one more item.

Break-Even Analysis

A typical company must analyze its costs and the potential market for its product to determine when (or even whether) it will make a profit.

EXAMPLE 6 **Retail Profit** A company manufactures a 65-inch smart TV that sells to retailers for $550. The cost of making x of these TVs for a month is given by the cost function $C(x) = 250x + 213{,}000$.

(a) Find the function R that gives the revenue from selling x TVs.

Solution Since revenue is the product of the price per item and the number of items, $R(x) = 550x$.

(b) What is the revenue from selling 600 TVs?

Solution Evaluate the revenue function R at 600:

$$R(600) = 550(600) = \$330{,}000.$$

(c) Find the profit function P.

Solution Since Profit = Revenue − Cost,

$$P(x) = R(x) - C(x) = 550x - (250x + 213{,}000) = 300x - 213{,}000.$$

(d) What is the profit from selling 500 TVs?

Solution Evaluate the profit function at 500 to obtain

$$P(500) = 300(500) - 213{,}000 = -63{,}000,$$

that is, a loss of $63,000.

A company can make a profit only if the revenue on a product exceeds the cost of manufacturing it. The number of units at which revenue equals cost (that is, profit is 0) is the **break-even point.**

EXAMPLE 7 **Break-Even Point** Find the break-even point for the company in Example 6.

Solution The company will break even when revenue equals cost—that is, when

$$R(x) = C(x)$$
$$550x = 250x + 213{,}000$$
$$300x = 213{,}000 \qquad \text{Subtract } 250x \text{ from both sides.}$$
$$x = 710.$$

The company breaks even by selling 710 TVs. The graphs of the revenue and cost functions and the break-even point (where $x = 710$) are shown in the Figure 3.18. The company must sell more than 710 TVs ($x > 710$) in order to make a profit. ✔ 6

✔ **Checkpoint 6**

For a certain newsletter, the cost equation is $C(x) = .90x + 1500$, where x is the number of newsletters sold. The newsletter sells for $1.25 per copy. Find the break-even point.

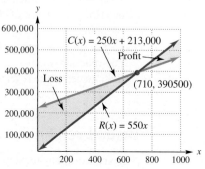

Figure 3.18

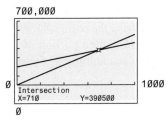

Figure 3.19

📋 **TECHNOLOGY TIP** The break-even point in Example 7 can be found on a graphing calculator by graphing the cost and revenue functions on the same screen and using the calculator's intersection finder, as shown in Figure 3.19. Depending on the calculator, the intersection finder is in the CALC or G-SOLVE menu or in the MATH or FCN submenu of the GRAPH menu.

Supply and Demand

The supply of and demand for an item are usually related to its price. Producers will supply large numbers of the item at a high price, but consumer demand will be low. As the price of the item decreases, consumer demand increases, but producers are less willing to supply large numbers of the item. The curves showing the quantity that will be supplied at a given price and the quantity that will be demanded at a given price are called **supply and demand curves,** respectively. In supply-and-demand problems, we use p for price and q for quantity. We will discuss the economic concepts of supply and demand in more detail in later chapters.

EXAMPLE 8   **Economics** Suppose that an economist has studied the supply and demand for aluminum siding and has determined that the price per unit,[*] p, and the quantity demanded, q, are related by the linear equation

$$p = 60 - \frac{3}{4}q. \qquad \text{Demand Function}$$

(a) Find the demand at a price of \$40 per unit.

Solution Let $p = 40$. Then we have

$$p = 60 - \frac{3}{4}q$$

$$40 = 60 - \frac{3}{4}q \qquad \text{Let } p = 40.$$

$$-20 = -\frac{3}{4}q \qquad \text{Add } -60 \text{ to both sides.}$$

$$\frac{80}{3} = q. \qquad \text{Multiply both sides by } -\frac{4}{3}.$$

At a price of \$40 per unit, $80/3$ (or $26\frac{2}{3}$) units will be demanded.

(b) Find the price if the demand is 32 units.

Solution Let $q = 32$. Then we have

$$p = 60 - \frac{3}{4}q$$

$$p = 60 - \frac{3}{4}(32) \qquad \text{Let } q = 32.$$

$$p = 60 - 24$$

$$p = 36.$$

With a demand of 32 units, the price is \$36.

(c) Graph $p = 60 - \frac{3}{4}q$.

Solution It is customary to use the horizontal axis for the quantity q and the vertical axis for the price p. In part (a), we saw that $80/3$ units would be demanded at a price of

---

[*]An appropriate unit here might be, for example, one thousand square feet of siding.

$40 per unit; this gives the ordered pair (80/3, 40). Part (b) shows that with a demand of 32 units, the price is $36, which gives the ordered pair (32, 36). Using the points (80/3, 40) and (32, 36) yields the demand graph depicted in Figure 3.20. Only the portion of the graph in Quadrant I is shown, because supply and demand are meaningful only for positive values of p and q. ✔7

✔ **Checkpoint 7**

Suppose price and quantity demanded are related by
$p = 100 - 4q$.

(a) Find the price if the quantity demanded is 10 units.

(b) Find the quantity demanded if the price is $80.

(c) Write the corresponding ordered pairs.

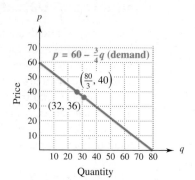

Figure 3.20

(d) From Figure 3.20, at a price of $30, what quantity is demanded?

Solution Price is located on the vertical axis. Look for 30 on the p-axis, and read across to where the line $p = 30$ crosses the demand graph. As the graph shows, this occurs where the quantity demanded is 40.

(e) At what price will 60 units be demanded?

Solution Quantity is located on the horizontal axis. Find 60 on the q-axis, and read up to where the vertical line $q = 60$ crosses the demand graph. This occurs where the price is about $15 per unit.

(f) What quantity is demanded at a price of $60 per unit?

Solution The point $(0, 60)$ on the demand graph shows that the demand is 0 at a price of $60 (that is, there is no demand at such a high price).

EXAMPLE 9 **Economics** Suppose the economist in Example 8 concludes that the supply q of siding is related to its price p by the equation

$$p = .85q. \quad \text{Supply Function}$$

(a) Find the supply if the price is $51 per unit.

Solution
$$51 = .85q \quad \text{Let } p = 51.$$
$$60 = q. \quad \text{Divide both sides by .85.}$$

If the price is $51 per unit, then 60 units will be supplied to the marketplace.

(b) Find the price per unit if the supply is 20 units.

Solution
$$p = .85(20) = 17. \quad \text{Let } q = 20.$$

If the supply is 20 units, then the price is $17 per unit.

(c) Graph the supply equation $p = .85q$.

Solution As with demand, each point on the graph has quantity q as its first coordinate and the corresponding price p as its second coordinate. Part (a) shows that the ordered pair (60, 51) is on the graph of the supply equation, and part (b) shows that (20, 17) is on the graph. Using these points, we obtain the supply graph in Figure 3.21.

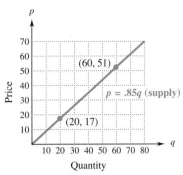

Figure 3.21

(d) Use the graph in Figure 3.21 to find the approximate price at which 35 units will be supplied. Then use algebra to find the exact price.

Solution The point on the graph with first coordinate $q = 35$ is approximately (35, 30). Therefore, 35 units will be supplied when the price is approximately \$30. To determine the exact price algebraically, substitute $q = 35$ into the supply equation:

$$p = .85q = .85(35) = \$29.75.$$

When the supply of a quantity exceeds the demand, we say there is a **surplus of supply**. When demand exceeds supply, we say there is a **shortage of supply**.

EXAMPLE 10   **Economics** The supply and demand curves of Examples 8 and 9 are shown in Figure 3.22. Determine graphically whether there is a surplus or a shortage of supply at a price of \$40 per unit.

Solution Find 40 on the vertical axis in Figure 3.22 and read across to the point where the horizontal line $p = 40$ crosses the supply and demand graphs. At a price of \$40, the quantity supplied is about 47 units, and the quantity demanded is about 27 units. Since the quantity supplied is greater than the quantity demanded, there is a surplus of supply.

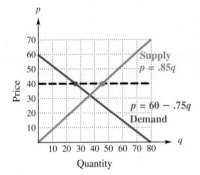

Figure 3.22

Supply and demand are equal at the point where the supply curve intersects the demand curve. This is the **equilibrium point.** Its second coordinate is the **equilibrium price,** the price at which the same quantity will be supplied as is demanded. Its first coordinate is the quantity that will be demanded and supplied at the equilibrium price; this number is called the **equilibrium quantity.**

EXAMPLE 11   **Economics** In the situation described in Examples 8–10, what is the equilibrium quantity? What is the equilibrium price?

Solution The equilibrium point is where the supply and demand curves in Figure 3.22 intersect. To find the quantity q at which the price given by the demand equation $p = 60 - .75q$ (Example 8) is the same as that given by the supply equation $p = .85q$ (Example 9), set these two expressions for p equal to each other and solve the resulting equation:

$$\text{Demand} = \text{Supply}$$
$$60 - .75q = .85q$$
$$60 = 1.6q \qquad \text{Subtract } .75q \text{ from both sides.}$$
$$37.5 = q. \qquad \text{Divide both sides by 1.6.}$$

Therefore, the equilibrium quantity is 37.5 units, the number of units for which supply will equal demand. Substituting $q = 37.5$ into either the demand or supply equation shows that

$$p = 60 - .75(37.5) = 31.875 \quad \text{or} \quad p = .85(37.5) = 31.875.$$

Checkpoint 8

The demand for a certain commodity is related to the price by $p = 80 - (2/3)q$. The supply is related to the price by $p = (4/3)q$. Find

(a) the equilibrium quantity;

(b) the equilibrium price.

So the equilibrium price is $31.875 (or $31.88, rounded). (To avoid error, it is a good idea to substitute into both equations, as we did here, to be sure that the same value of p results; if it does not, a mistake has been made.) In this case, the equilibrium point—the point whose coordinates are the equilibrium quantity and price—is (37.5, 31.875). ✓8

TECHNOLOGY TIP The equilibrium point (37.5, 31.875) can be found on a graphing calculator by graphing the supply and demand curves on the same screen and using the calculator's intersection finder to locate their point of intersection.

3.3 Exercises

Business *Write a cost function for each of the given scenarios. Identify all variables used. (See Example 1.)*

1. A chain-saw rental firm charges $25, plus $5 per hour.

2. A trailer-hauling service charges $95, plus $8 per mile.

3. A parking garage charges $8.00, plus $2.50 per hour.

4. For a 1-day rental, a car rental firm charges $65, plus 45¢ per mile.

Business *Assume that each of the given situations can be expressed as a linear cost function. Find the appropriate cost function in each case. (See Examples 1 and 4.)*

5. Fixed cost, $200; 50 items cost $2000 to produce.

6. Fixed cost, $2000; 40 items cost $5000 to produce.

7. Marginal cost, $120; 100 items cost $15,800 to produce.

8. Marginal cost, $90; 150 items cost $16,000 to produce.

Business *In Exercises 9–12, a cost function is given. Find the average cost per item when the required numbers of items are produced. (See Example 2.)*

9. $C(x) = 12x + 1800$; 50 items, 500 items, 1000 items

10. $C(x) = 80x + 12,000$; 100 items, 1000 items, 10,000 items

11. $C(x) = 6.5x + 9800$; 200 items, 2000 items, 5000 items

12. $C(x) = 8.75x + 16,500$; 1000 items, 10,000 items, 75,000 items

Business *Work these exercises. (See Example 3.)*

13. **Automobile Depreciation** The manufacturer's suggested retail price (MSRP) for a base model 2016 Toyota Prius is $25,035, and it is expected to be worth $14,115 is 4 years.

    (a) Find a linear depreciation function for this car.

    (b) Estimate the value of the car 5 years from now.

    (c) At what rate is the car depreciating?

14. A computer that cost $1250 new is expected to depreciate linearly at a rate of $250 per year.

    (a) Find the depreciation function f.

    (b) Explain why the domain of f is [0, 5].

15. A machine is now worth $120,000 and will be depreciated linearly over an 8-year period, at which time it will be worth $25,000 as scrap.

(a) Find the rule of the depreciation function f.

(b) What is the domain of f?

(c) What will the machine be worth in 6 years?

16. A house increases in value in an approximately linear fashion from $222,000 to $300,000 in 6 years.

    (a) Find the *appreciation function* that gives the value of the house in year x.

    (b) If the house continues to appreciate at this rate, what will it be worth 12 years from now?

Business *Work these problems. (See Examples 2 and 4.)*

17. The total cost (in dollars) of producing x college algebra books is $C(x) = 42.5x + 80,000$.

    (a) What are the fixed costs?

    (b) What is the marginal cost per book?

    (c) What is the total cost of producing 1000 books? 32,000 books?

    (d) What is the average cost when 1000 are produced? when 32,000 are produced?

18. The total cost (in dollars) of producing x DVDs is $C(x) = 6.80x + 450,000$.

    (a) What are the fixed costs?

    (b) What is the marginal cost per DVD?

    (c) What is the total cost of producing 50,000 DVDs? 600,000 DVDs?

    (d) What is the average cost per DVD when 50,000 are produced? when 500,000 are produced?

19. The owner of a food truck found that the cost of producing 50 tacos is $58, while the cost of producing 250 tacos is $142. Assume that the cost $C(x)$ is a linear function of x, the number of tacos produced.

    (a) Find a formula for $C(x)$.

    (b) What are the fixed costs?

    (c) Find the total cost of producing 1000 tacos.

    (d) Find the total cost of producing 1001 tacos.

    (e) Find the marginal cost of producing the 1001st taco.

    (f) Find the marginal cost of producing *any* taco.

20. In deciding whether to set up a new manufacturing plant, company analysts have determined that a linear function is a reasonable estimation for the total cost $C(x)$ in dollars of producing x items. They estimate the cost of producing 10,000 items as $547,500 and the cost of producing 50,000 items as $737,500.

(a) Find a formula for $C(x)$.

(b) Find the total cost of producing 100,000 items.

(c) Find the marginal cost of the items to be produced in this plant.

Business *Work these problems. (See Example 5.)*

21. Water Utility Revenue For the year 2016, a resident of the city of Dallas, Texas, with a ⅝-inch water meter pays $5.12 per month plus $1.87 per 1000 gallons of water used (up to 4000 gallons). If the city of Dallas has 550,000 customers that use less than 4000 gallons a month, find its monthly revenue function $R(x)$, where the total number of gallons x is measured in thousands. (Data from: dallascityhall.com/departments/waterutilities.)

22. Gas Utility Revenue The Laclede Gas Company in St. Louis, Missouri, charged customers $19.50 per month plus $.91686 per therm used (up to 30 terms) in the winter of 2017–2018. If the company has 150,000 customers that use less than 30 therms a month, find its monthly revenue function $R(x)$, where x is the total number of terms sold. (Data from: http://www.lacledegas.com.)

Business *Assume that each row of the accompanying table has a linear cost function. Find (a) the cost function; (b) the revenue function; (c) the profit function; (d) the profit on 100 items. (See Example 6.)*

	Fixed Cost	Marginal Cost per Item	Item Sells For
23.	$750	$10	$35
24.	$150	$11	$20
25.	$300	$18	$28
26.	$17,000	$30	$80

27. Profit–Volume Graph In the following profit–volume chart, *EF* and *GH* represent the profit–volume graphs of a single-product company for, 2016 and 2017, respectively. (Adapted from: Uniform CPA Examination, American Institute of Certified Public Accountants.)

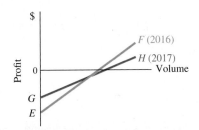

If the 2016 and 2017 unit sales prices are identical, how did the total fixed costs and unit variable costs of 2017 change compared with their values in 2016? Choose one:

	2016 Total Fixed Costs	2017 Unit Variable Costs
(a)	Decreased	Increased
(b)	Decreased	Decreased
(c)	Increased	Increased
(d)	Increased	Decreased

28. Use the figure and assumptions from the previous exercise to answer the following questions.

(a) Explain how you can tell from the graph which year had lower total fixed costs.

(b) Explain how you can tell from the graph which year had lower unit variable costs.

Use algebra to find the intersection points of the graphs of the given equations. (See Examples 7 and 11.)

29. $2x - y = 7$ and $y = 8 - 3x$

30. $6x - y = 2$ and $y = 4x + 7$

31. $y = 3x - 7$ and $y = 7x + 4$

32. $y = 3x + 5$ and $y = 12 - 2x$

Business *Work the following problems. (See Example 7.)*

33. An insurance company claims that for x thousand policies, its monthly revenue in dollars is given by $R(x) = 125x$ and its monthly cost in dollars is given by $C(x) = 100x + 5000$.

(a) Find the break-even point.

(b) Graph the revenue and cost equations on the same axes.

(c) From the graph, estimate the revenue and cost when $x = 100$ (100,000 policies).

34. The owners of a parking lot have determined that their weekly revenue and cost in dollars are given by $R(x) = 80x$ and $C(x) = 50x + 2400$, where x is the number of long-term parkers.

(a) Find the break-even point.

(b) Graph $R(x)$ and $C(x)$ on the same axes.

(c) From the graph, estimate the revenue and cost when there are 60 long-term parkers.

35. The revenue (in millions of dollars) from the sale of x units at a home supply outlet is given by $R(x) = .21x$. The profit (in millions of dollars) from the sale of x units is given by $P(x) = .084x - 1.5$.

(a) Find the cost equation.

(b) What is the cost of producing 7 units?

(c) What is the break-even point?

36. The profit (in millions of dollars) from the sale of x million units of Blue Glue is given by $P(x) = .7x - 25.5$. The cost is given by $C(x) = .9x + 25.5$.

(a) Find the revenue equation.

(b) What is the revenue from selling 10 million units?

(c) What is the break-even point?

Business *Suppose you are the manager of a firm. The accounting department has provided cost estimates, and the sales department sales estimates, on a new product. You must analyze the data they give you, determine what it will take to break even, and decide whether to go ahead with production of the new product. (See Example 7.)*

37. Cost is estimated by $C(x) = 80x + 7000$ and revenue is estimated by $R(x) = 95x$; no more than 400 units can be sold.

38. Cost is estimated by $C(x) = 37x + 9000$ and revenue is estimated by $R(x) = 53x$; no more than 500 items can be sold.

39. Cost is $C(x) = 125x + 42,000$ and revenue is $R(x) = 165.5x$; no more than 2000 units can be sold.

40. Cost is $C(x) = 1750x + 95,000$ and revenue is $R(x) = 1975x$; no more than 600 units can be sold.

41. **Female Workforce** The accompanying graph shows the percentage of the female population in the workforce for France and Spain for the years 2004 through 2013. Estimate the point at which the two countries had the same percentage of females in the workforce. (Data from: Organization for Economic Cooperation and Development.)

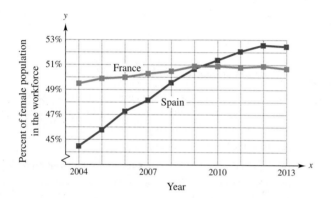

42. **Internet Users** The graph below shows the percentage of the population who are Internet users in Switzerland and the United Arab Emirates for the years 2006 through 2014. Find the point at which the two countries had the same percentage of Internet users. (Data from: www.worldbank.org.)

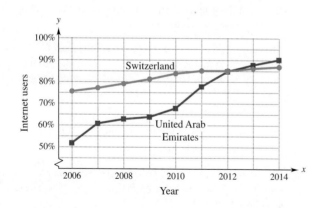

43. **Population Growth** The population (in thousands of people) of Arizona for the years 2010 through 2016 can be

approximated by the function $f(x) = 84x + 5540$. Similarly, the population (in thousands of people) of Massachusetts can be approximated by the function $g(x) = 48x + 6080$. For both of these models, let $x = 10$ correspond to the year 2010. (Data from: U.S. Census Bureau.)

(a) Graph both functions on the same coordinate axes for $x = 10$ to $x = 16$.

(b) Do the graphs intersect in this window?

(c) In what year were the populations of these two states approximately the same?

44. **Population Growth** The population (in thousands of people) of Florida for the years 2010 through 2016 can be approximated by the function $f(x) = 280x + 16,000$. Similarly, the population (in thousands of people) of New York can be approximated by the function $g(x) = 80x + 18,600$. For both of these models, let $x = 10$ correspond to the year 2010. (Data from: U.S. Census Bureau.)

(a) Graph both functions on the same coordinate axes for $x = 10$ to $x = 16$.

(b) Do the graphs intersect in this window?

(c) In what year were the populations of these two states approximately the same?

Business *Use the supply and demand curves in the accompanying graph to answer Exercises 45–48. (See Examples 8–11.)*

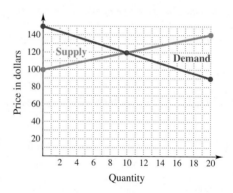

45. At what price are 20 items supplied?

46. At what price are 20 items demanded?

47. Find the equilibrium quantity.

48. Find the equilibrium price.

Economics *Work the following exercises. (See Examples 8–11.)*

49. Suppose that the demand and price for a certain brand of shampoo are related by

$$p = 16 - \frac{5}{4}q, \quad \text{(demand)}$$

where p is price in dollars and q is the quantity demanded. Find the price for a demand of

(a) 0 units;   (b) 4 units;   (c) 8 units.

Find the demand for the shampoo at a price of

(d) $6;   **(e)** $11;   **(f)** $16.

(g) Graph the demand curve $p = 16 - (5/4)q$.

Suppose the price and supply of the shampoo are related by

$$p = \frac{3}{4}q, \quad \text{(supply)}$$

where q represents the quantity supplied and p the price. Find the supply when the price is

(h) $0;   **(i)** $10;   **(j)** $20.

(k) Graph the supply curve $p = (3/4)q$ on the same axes used for part (g).

(l) Find the equilibrium quantity.

(m) Find the equilibrium price.

50. Let the supply and demand for radial tires in dollars be given by

$$\text{supply: } p = \frac{3}{2}q; \quad \text{demand: } p = 81 - \frac{3}{4}q.$$

(a) Graph these equations on the same axes.

(b) Find the equilibrium quantity.

(c) Find the equilibrium price.

51. Let the supply and demand for bananas in cents per pound be given by

$$\text{supply: } p = \frac{2}{5}q; \quad \text{demand: } p = 100 - \frac{2}{5}q.$$

(a) Graph these equations on the same axes.

(b) Find the equilibrium quantity.

(c) Find the equilibrium price.

(d) At a price of 40 cents per pound, is there a surplus or a shortage of supply?

52. Let the supply and demand for sugar be given by

$$\text{supply: } p = 1.4q - .6$$

and

$$\text{demand: } p = -2q + 3.2,$$

where p is in dollars per pound.

(a) Graph these on the same axes.

(b) Find the equilibrium quantity.

(c) Find the equilibrium price.

(d) At a price of $2 per pound, is there a surplus or a shortage of supply?

53. Explain why the graph of the (total) cost function is always above the x-axis and can never move downward as you go from left to right. Is the same thing true of the graph of the average cost function?

54. Explain why the graph of the profit function can rise or fall (as you go from left to right) and can be below the x-axis.

✓**Checkpoint Answers**

1. $C(x) = 4x + 60.$ **2.** $4.60   **3.** $9505

4. (a) $800   **(b)** $842

  (c) $3.50   **(d)** $3.50

5. (a) 1,120,000,000   **(b)** $47,873,000

6. 4286 newsletters

7. (a) $60   **(b)** 5 units   **(c)** (10, 60); (5, 80)

8. (a) 40   **(b)** $160/3 \approx $53.33

3.4 Quadratic Functions and Applications

A **quadratic function** is a function whose rule is given by a quadratic polynomial, such as

$$f(x) = x^2, \quad g(x) = 3x^2 + 30x + 67, \quad \text{and} \quad h(x) = -x^2 + 4x.$$

Thus, a quadratic function is a function whose rule can be written in the form

$$f(x) = ax^2 + bx + c$$

for some constants a, b, and c, with $a \neq 0$.

EXAMPLE 1   Graph each of these quadratic functions:

$$f(x) = x^2; \quad g(x) = 4x^2; \quad h(x) = -.2x^2.$$

Solution In each case, choose several numbers (negative, positive, and 0) for x, find the values of the function at these numbers, and plot the corresponding points. Then connect the points with a smooth curve to obtain Figure 3.23, on the next page.

$f(x) = x^2$					
x	-2	-1	0	1	2
x^2	4	1	0	1	4

$g(x) = 4x^2$					
x	-2	-1	0	1	2
$4x^2$	16	4	0	4	16

$h(x) = -.2x^2$					
x	-5	-3	-1	0	2	4
$-.2x^2$	-5	-1.8	$-.2$	0	$-.8$	-3.2

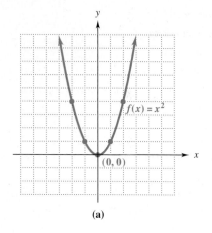

(a)

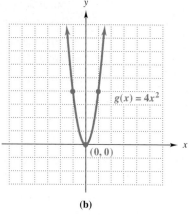

(b)

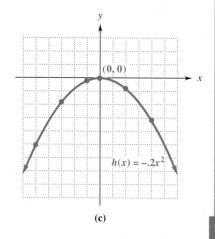

(c)

Figure 3.23

Each of the curves in Figure 3.23 is a **parabola.** It can be shown that the graph of every quadratic function is a parabola. Parabolas have many useful properties. Cross sections of radar dishes and spotlights form parabolas. Disks often visible on the sidelines of televised football games are microphones having reflectors with parabolic cross sections. These microphones are used by the television networks to pick up the signals shouted by the quarterbacks.

All parabolas have the same basic "cup" shape, although the cup may be broad or narrow and open upward or downward. The general shape of a parabola is determined by the coefficient of x^2 in its rule, as summarized here and illustrated in Example 1.

✓ **Checkpoint 1**

Graph each quadratic function.

(a) $f(x) = x^2 - 4$

(b) $g(x) = -x^2 + 4$

The graph of a quadratic function $f(x) = ax^2 + bx + c$ is a parabola.

If $a > 0$, the parabola opens upward.   [*Figure 3.23(a) and 3.23(b)*]

If $a < 0$, the parabola opens downward.   [*Figure 3.23(c)*]

If $|a| < 1$, the parabola appears wider than the graph of $y = x^2$. [*Figure 3.23(c)*]

If $|a| > 1$, the parabola appears narrower than the graph of $y = x^2$.
   [*Figure 3.23(b)*]

When a parabola opens upward [as in Figure 3.23(a), (b)], its lowest point is called the **vertex.** When a parabola opens downward [as in Figure 3.23(c)], its highest point is called the **vertex.** The vertical line through the vertex of a parabola is called the **axis of the parabola.** For example, (0, 0) is the vertex of each of the parabolas in Figure 3.23, and the axis of each parabola is the y-axis. If you were to fold the graph of a parabola along its axis, the two halves of the parabola would match exactly. This means that a parabola is *symmetric* about its axis. For this reason, the axis of the parabola is sometimes called the axis of symmetry.

In real-world problems, finding the vertex of a parabola is important because it tells us the maximum or minimum value, such as maximum profit or minimum cost. Although the vertex of a parabola can be approximated by a graphing calculator's maximum or minimum finder, its exact coordinates can be found algebraically, as in the following examples.

EXAMPLE 2 Consider the function $g(x) = 2(x - 3)^2 + 1$.

(a) Show that g is a quadratic function.

Solution Multiply out the rule of g to show that it has the required form:

$$g(x) = 2(x - 3)^2 + 1$$
$$= 2[x^2 + 2(x)(-3) + (-3)^2] + 1 \qquad \text{\small{$(a + b)^2 = a^2 + 2ab + b^2$}}$$
$$= 2(x^2 - 6x + 9) + 1$$
$$= 2x^2 - 12x + 18 + 1$$
$$g(x) = 2x^2 - 12x + 19.$$

According to the preceding box, the graph of g is a somewhat narrow, upward-opening parabola.

(b) Show that the vertex of the graph of $g(x) = 2(x - 3)^2 + 1$ is $(3, 1)$.

Solution Since $g(3) = 2(3 - 3)^2 + 1 = 0 + 1 = 1$, the point $(3, 1)$ is on the graph. The vertex of an upward-opening parabola is the lowest point on the graph, so we must show that $(3, 1)$ is the lowest point. Let x be any number except 3 (so that $x - 3 \neq 0$). Then the quantity $2(x - 3)^2$ is positive, and hence

$$g(x) = 2(x - 3)^2 + 1 = \text{(a positive number)} + 1,$$

which means that $g(x) > 1$. Therefore, every point $(x, g(x))$ on the graph, where $x \neq 3$, has second coordinate $g(x)$ greater than 1. Hence $(x, g(x))$ lies *above* $(3, 1)$. In other words, $(3, 1)$ is the lowest point on the graph—the vertex of the parabola.

(c) Graph $g(x) = 2(x - 3)^2 + 1$.

Solution Plot some points on both sides of the vertex $(3, 1)$ to obtain the graph in Figure 3.24. The vertical line $x = 3$ through the vertex is the axis of the parabola.

x	y
1	9
2	3
3	1
4	3
5	9

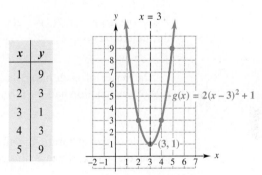

Figure 3.24

In Example 2, notice how the rule of the function g is related to the coordinates of the vertex:

$$g(x) = 2(x - 3)^2 + 1. \qquad (3, 1).$$

Arguments similar to those in Example 2 lead to the following fact.

> The graph of the quadratic function $f(x) = a(x - h)^2 + k$ is a parabola with vertex (h, k). It opens upward when $a > 0$ and downward when $a < 0$.

EXAMPLE 3 Determine algebraically whether the given parabola opens upward or downward, and find its vertex.

(a) $f(x) = -3(x - 4)^2 - 7$

Solution The rule of the function is in the form $f(x) = a(x - h)^2 + k$ (with $a = -3, h = 4$, and $k = -7$). The parabola opens downward ($a < 0$), and its vertex is $(h, k) = (4, -7)$.

(b) $g(x) = 2(x + 3)^2 + 5$

Solution Be careful here: The vertex is *not* (3, 5). To put the rule of $g(x)$ in the form $a(x - h)^2 + k$, we must rewrite it so that there is a minus sign inside the parentheses:

$$g(x) = 2(x + 3)^2 + 5$$
$$= 2(x - (-3))^2 + 5.$$

This is the required form, with $a = 2, h = -3$, and $k = 5$. The parabola opens upward, and its vertex is $(-3, 5)$. ✔2

✓ Checkpoint 2

Determine the vertex of each parabola, and graph the parabola.

(a) $f(x) = (x + 4)^2 - 3$

(b) $f(x) = -2(x - 3)^2 + 1$

EXAMPLE 4 Find the rule of a quadratic function whose graph has vertex (3, 4) and passes through the point (6, 22).

Solution The graph of $f(x) = a(x - h)^2 + k$ has vertex (h, k). We want $h = 3$ and $k = 4$, so that $f(x) = a(x - 3)^2 + 4$. Since (6, 22) is on the graph, we must have $f(6) = 22$. Therefore,

$$f(x) = a(x - 3)^2 + 4$$
$$f(6) = a(6 - 3)^2 + 4$$
$$22 = a(3)^2 + 4$$
$$22 = 9a + 4$$
$$9a = 18$$
$$a = 2.$$

Thus, the graph of $f(x) = 2(x - 3)^2 + 4$ is a parabola with vertex (3, 4) that passes through (6, 22).

The vertex of each parabola in Examples 2 and 3 was easily determined because the rule of the function had the form

$$f(x) = a(x - h)^2 + k.$$

The rule of *any* quadratic function can be put in this form by using the technique of **completing the square**, which is illustrated in the next example.

EXAMPLE 5 Determine the vertex of the graph of $f(x) = x^2 - 4x + 2$. Then graph the parabola.

Solution In order to get $f(x)$ in the form $a(x - h)^2 + k$, we use the technique of completing the square. Take *half the coefficient of* x, namely $\frac{1}{2}(-4) = -2$, and *square it:* $(-2)^2 = 4$. Then proceed as follows.

$$f(x) = x^2 - 4x + 2$$
$$= x^2 - 4x + \underline{\quad} + 2 \qquad \text{Leave space for the squared term and its negative.}$$
$$= x^2 - 4x + 4 - 4 + 2 \qquad \text{Add and subtract 4.}$$
$$= (x^2 - 4x + 4) - 4 + 2 \qquad \text{Insert parentheses.}$$
$$= (x - 2)^2 - 2 \qquad \text{Factor expression in parentheses and add.}$$

✓ Checkpoint 3

Rewrite the rule of each function by completing the square (add and subtract half the coefficient of x), and use this form to find the vertex of the graph.

(a) $f(x) = x^2 + 6x + 5$

(b) $g(x) = x^2 - 12x + 33$

Adding and subtracting 4 did not change the rule of $f(x)$, but did make it possible to have a perfect square as part of its rule: $f(x) = (x - 2)^2 - 2$. Now we can see that the graph is an upward-opening parabola, as shown in Figure 3.25, on the following page. ✔3

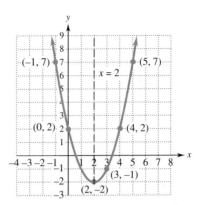

Figure 3.25

⚠️ **CAUTION** The technique of completing the square only works when the coefficient of x^2 is 1. To find the vertex of a quadratic function such as

$$f(x) = 2x^2 + 12x - 19,$$

you must first factor out the coefficient of x^2 and write the rule as

$$f(x) = 2(x^2 + 6x - \tfrac{19}{2}).$$

Now complete the square on the expression in parentheses by adding and subtracting 9 (the square of half the coefficient of x), and proceed as in Example 5.

The technique of completing the square can be used to rewrite the general equation $f(x) = ax^2 + bx + c$ in the form $f(x) = a(x - h)^2 + k$. When this is done, we obtain a formula for the coordinates of the vertex.

The graph of the quadratic function $f(x) = ax^2 + bx + c$ is a parabola with vertex (h, k), where

$$h = \frac{-b}{2a} \quad \text{and} \quad k = f(h).$$

Additionally, the fact that the vertex of a parabola is the highest or lowest point on the graph can be used in applications to find a maximum or minimum value.

EXAMPLE 6   **Microbrewery Profits** Shanese owns and operates her own microbrewery. She has hired a consultant to analyze her business operations. The consultant tells her that her daily profits from the sale of x cases of beer are given by

$$P(x) = -x^2 + 120x.$$

(a) Find the vertex, determine if it is a maximum or minimum, write the equation of the axis of the parabola, compute the x- and y-intercepts, and sketch the graph of the profit function $P(x)$.

Solution Since $P(x) = -1x^2 + 120x + 0$ is a quadratic function with $a = -1$, $b = 120$, and $c = 0$, the preceding box tells us that the vertex is (h, k) where the x-value of the vertex is

$$h = \frac{-b}{2a} = \frac{-120}{2(-1)} = 60.$$

The y-value of the vertex is

$$k = P(h) = P(60) = -(60)^2 + 120(60) = 3600.$$

The vertex is (60, 3600) and since a is negative, it is a maximum because the parabola opens downward. The axis of the parabola is the vertical line through the vertex, namely, $x = 60$. The intercepts are found by setting x and y equal to 0.

x-intercepts	**y-intercept**
Set $P(x) = y = 0$, so that	Set $x = 0$ to obtain

$$
\begin{aligned}
0 &= -x^2 + 120x \\
0 &= x(-x + 120) \\
x = 0 \quad &\text{or} \quad -x + 120 = 0 \\
&\qquad\qquad -x = -120 \\
&\qquad\qquad\;\; x = 120
\end{aligned}
$$

$$P(0) = -(0)^2 + 120(0) = 0$$

The y-intercept is 0.

The x–intercepts are 0 and 120.

Figure 3.26 shows the profit function $P(x)$.

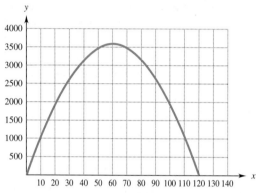

Figure 3.26

(b) Explain what the vertex tells us in this situation.

Solution Because the maximum occurs at the vertex, the *number of cases* of beer that should be sold to achieve the maximum profit is the x-value of the vertex, namely, 60 cases, and the resulting *maximum profit* is the y-value of the vertex, namely, $3600. ✔ 4

✓**Checkpoint 4**

When a company sells x units of a product, its profit is $P(x) = -2x^2 + 40x + 280$. Find

(a) the number of units that should be sold so that maximum profit is received;

(b) the maximum profit.

TECHNOLOGY TIP The maximum or minimum finder on a graphing calculator can approximate the vertex of a parabola with a high degree of accuracy. The max–min finder is in the CALC menu or in the MATH or FCN submenu of the GRAPH menu. Similarly, the calculator's graphical root finder can approximate the x-intercepts of a parabola.

 Supply and demand curves were introduced in Section 3.3. Here is a quadratic example.

EXAMPLE 7   **Economics** Suppose that the price of and demand for an item are related by

$$p = 150 - 6q^2, \qquad \text{Demand function}$$

where p is the price (in dollars) and q is the number of items demanded (in hundreds). Suppose also that the price and supply are related by

$$p = 10q^2 + 2q, \qquad \text{Supply function}$$

where p is the price (in dollars) and q is the number of items supplied (in hundreds). Find the equilibrium quantity and the equilibrium price.

Solution The graphs of both of these equations are parabolas as seen in Figure 3.27, on the following page. Only those portions of the graphs which lie in the first quadrant are included, because none of supply, demand, or price can be negative.

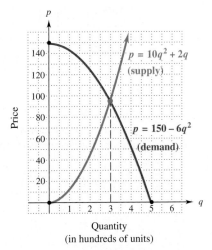

Figure 3.27

The point where the demand and supply curves intersect is the equilibrium point. Its first coordinate is the equilibrium quantity, and its second coordinate is the equilibrium price. These coordinates may be found in two ways.

Algebraic Method At the equilibrium point, the second coordinate of the demand curve must be the same as the second coordinate of the supply curve, so that

$$\text{Demand} = \text{Supply}$$
$$150 - 6q^2 = 10q^2 + 2q.$$

Write this quadratic equation in standard form as follows:

$$0 = 16q^2 + 2q - 150 \qquad \text{Add } -150 \text{ and } 6q^2 \text{ to both sides.}$$
$$0 = 8q^2 + q - 75. \qquad \text{Multiply both sides by } \tfrac{1}{2}.$$

This equation can be solved by the quadratic formula, explained in Section 1.7. Here, $a = 8$, $b = 1$, and $c = -75$:

$$q = \frac{-1 \pm \sqrt{1 - 4(8)(-75)}}{2(8)} \qquad \frac{-b \pm \sqrt{b^2 - 4ac}}{2a}$$
$$= \frac{-1 \pm \sqrt{1 + 2400}}{16} \qquad -4(8)(-75) = 2400$$
$$= \frac{-1 \pm 49}{16} \qquad \sqrt{1 + 2400} = \sqrt{2401} = 49$$
$$q = \frac{-1 + 49}{16} = \frac{48}{16} = 3 \quad \text{or} \quad q = \frac{-1 - 49}{16} = -\frac{50}{16} = -\frac{25}{8}.$$

It is not possible to make $-25/8$ units, so discard that answer and use only $q = 3$. Hence, the equilibrium quantity is 300. Find the equilibrium price by substituting 3 for q in either the supply or the demand function (and check your answer by using the other one). Using the supply function gives

$$p = 10q^2 + 2q$$
$$p = 10(3)^2 + 2(3) \qquad \text{Let } q = 3.$$
$$= 10(9) + 6$$
$$p = \$96.$$

Graphical Method Graph the two functions on a graphing calculator, and use the intersection finder to determine that the equilibrium point is (3, 96), as in Figure 3.28. ✓ 5

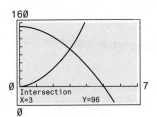

Figure 3.28

✓ **Checkpoint 5**

The price and demand for an item are related by $p = 32 - x^2$, while price and supply are related by $p = x^2$. Find

(a) the equilibrium quantity;

(b) the equilibrium price.

EXAMPLE 8 **Rental Income** The rental manager of a small apartment complex with 16 units has found from experience that each $40 increase in the monthly rent results in an empty apartment. All 16 apartments will be rented at a monthly rent of $500. How many $40 increases will produce maximum monthly income for the complex?

Solution Let x represent the number of $40 increases. Then the number of apartments rented will be $16 - x$. Also, the monthly rent per apartment will be $500 + 40x$. (There are x increases of $40, for a total increase of $40x$.) The monthly income, $I(x)$, is given by the number of apartments rented times the rent per apartment, so

$$I(x) = (16 - x)(500 + 40x)$$

FIRST OUTSIDE INSIDE LAST

$$= 8000 + 640x - 500x - 40x^2 \qquad \text{FOIL method}$$

$$= 8000 + 140x - 40x^2.$$

Since x represents the number of $40 increases and each $40 increase causes one empty apartment, x must be a whole number. Because there are only 16 apartments, $0 \leq x \leq 16$.

The graph of $I(x) = 8000 + 140x - 40x^2$ is a downward-opening parabola (why?), and the value of x that produces maximum income occurs at the vertex (see Figure 3.29). The box on page 149 shows that the vertex is $(1.75, 8122.50)$. Since x must be a whole number, evaluate $I(x)$ at $x = 1$ and $x = 2$ to see which one gives the best result:

$$\text{If } x = 1, \text{ then } I(1) = -40(1)^2 + 140(1) + 8000 = 8100.$$

$$\text{If } x = 2, \text{ then } I(2) = -40(2)^2 + 140(2) + 8000 = 8120.$$

So maximum income occurs when $x = 2$. The manager should charge a rent of $500 + 2(40) = \$580$, leaving two apartments vacant.

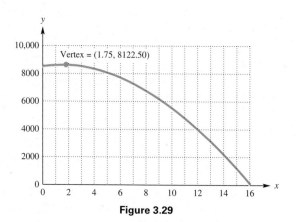

Figure 3.29

Quadratic Models

Real-world data can sometimes be used to construct a quadratic function that approximates the data. Such **quadratic models** can then be used to investigate the real-world scenarios.

EXAMPLE 9 **Price of Crude Oil** The price of Dubai crude oil, in dollars per barrel, is given in the following table for the years 2010 through 2015. (Data from: Organization of Petroleum Exporting Countries.)

Year	2010	2011	2012	2013	2014	2015
Price per barrel	$78	$106	$110	$105	$97	$51

(a) Let $x = 0$ correspond to the year 2010. Display the information graphically.

Solution Plot the points (0, 78), (1, 106), and so on, as shown in Figure 3.30.

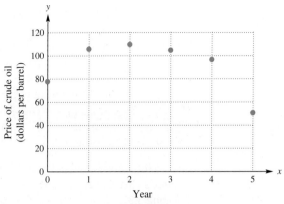

Figure 3.30

(b) The shape of the data points in Figure 3.30 resembles a downward-opening parabola. Use the year 2012 as the maximum and the year 2010 as another data point to find a quadratic model $f(x) = a(x - h)^2 + k$ for these data.

Solution Recall that when a quadratic function is written in the form $f(x) = a(x - h)^2 + k$, the vertex of the parabola is (h, k). On the basis of Figure 3.30, let $(2, 110)$ be the vertex, so that

$$f(x) = a(x - h)^2 + k$$
$$f(x) = a(x - 2)^2 + 110.$$

To find a, use the fact that the price of oil was \$78 in 2010. Because $x = 0$ corresponds to the year 2010, the point (0, 78) lies on the parabola, therefore

$$f(x) = a(x - 2)^2 + 110$$
$$78 = a(0 - 2)^2 + 110 \qquad \text{Substitute 0 for } x \text{ and 78 for } f(x).$$
$$-32 = 4a \qquad \text{Subtract 110 from both sides.}$$
$$a = -8 \qquad \text{Divide both sides by 4 and reverse the equation.}$$

Therefore, $f(x) = -8(x - 2)^2 + 110$ is a quadratic model for these data. If we expand the form into $f(x) = ax^2 + bx + c$, we obtain

$$f(x) = -8(x - 2)^2 + 110$$
$$= -8(x^2 - 4x + 4) + 110 \qquad \text{Expand inside the parentheses.}$$
$$= -8x^2 + 32x - 32 + 110 \qquad \text{Distribute the } -8.$$
$$= -8x^2 + 32x + 78. \qquad \text{Add the last two terms.}$$

Figure 3.31 shows the original data with the graph of f. The graph appears to fit the data rather well except for the last two points. We will find a better fitting model in the next example. ✔6

✓ **Checkpoint 6**

Find another quadratic model in Example 9(b) by using (2, 110) as the vertex, but use the year 2013 for the other data point.

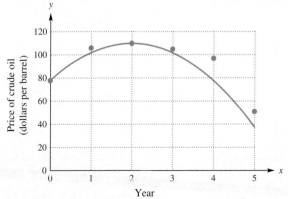

Figure 3.31

Quadratic Regression

Linear regression was used in Section 2.3 to construct a linear function that modeled a set of data points. When the data points appear to lie on a parabola rather than on a straight line (as in Example 9), a similar least-squares regression procedure is available on most graphing calculators and spreadsheet programs to construct a quadratic model for the data. Simply follow the same steps as in linear regression, with one exception: Choose quadratic, rather than linear, regression. (Both options are on the same menu.)

EXAMPLE 10 Use a graphing calculator to find a quadratic-regression model for the data in Example 9.

Solution Enter the first coordinates of the data points as list L_1 and the second coordinates as list L_2 (go to the STAT EDIT menu). Performing a quadratic regression, as in Figure 3.32 (go to the STAT CALC menu and choose QuadReg), leads to the model

$$g(x) = -7.46x^2 + 32.55x + 78.21.$$

Figure 3.33 shows the data with the quadratic regression curve.

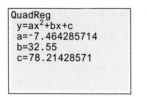

Figure 3.32   Figure 3.33

3.4 Exercises

The graph of each of the functions in Exercises 1–4 is a parabola. Without graphing, determine whether the parabola opens upward or downward. (See Example 1.)

1. $f(x) = x^2 - 3x - 12$ **2.** $g(x) = -x^2 + 5x + 15$

3. $h(x) = -3x^2 + 14x + 1$ **4.** $f(x) = 6.5x^2 - 7.2x + 4$

Without graphing, determine the vertex of the parabola that is the graph of the given function. State whether the parabola opens upward or downward. (See Examples 2 and 3.)

5. $f(x) = 2(x - 5)^2 + 7$ **6.** $g(x) = 7(x - 8)^2 - 3$

7. $h(x) = -4(x + 1)^2 - 9$ **8.** $f(x) = -8(x + 12)^2 + 9$

Match each function with its graph, which is one of those shown. (See Examples 1–3.)

9. $f(x) = x^2 + 2$ **10.** $g(x) = x^2 - 2$

11. $g(x) = (x - 2)^2$ **12.** $f(x) = -(x + 2)^2$

13. $f(x) = 2(x - 2)^2 + 2$

14. $g(x) = -2(x - 2)^2 - 2$

15. $g(x) = -2(x + 2)^2 + 2$

16. $f(x) = 2(x + 2)^2 - 2$

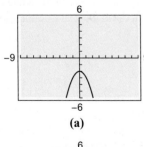

(a)

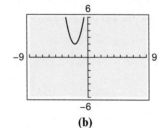

(b)

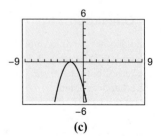

(c)

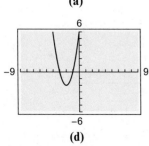

(d)

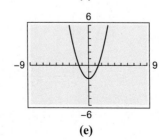

(e)

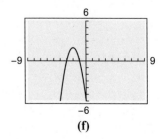

(f)

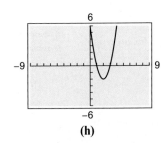

(g)

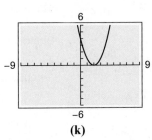

(h)

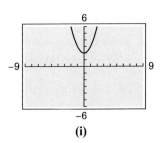

(i)

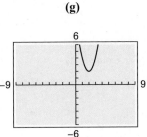

(j)

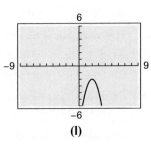

(k)

(l)

Find the rule of a quadratic function whose graph has the given vertex and passes through the given point. (See Example 4.)

17. Vertex (1, 2); point (5, 6)

18. Vertex (−3, 2); point (2, 1)

19. Vertex (−1, −2); point (1, 2)

20. Vertex (2, −4); point (5, 2)

Without graphing, find the vertex of the parabola that is the graph of the given function. (See Examples 5 and 6.)

21. $f(x) = x^2 + 6x - 3$ **22.** $g(x) = x^2 + 10x + 9$

23. $f(x) = 3x^2 - 12x + 5$ **24.** $g(x) = -4x^2 - 16x + 9$

Without graphing, determine the x- and y-intercepts of each of the given parabolas. (See Example 6.)

25. $f(x) = 3(x - 2)^2 - 3$ **26.** $g(x) = 2x^2 + 8x + 6$

27. $g(x) = x^2 - 10x + 20$ **28.** $f(x) = x^2 - 4x - 1$

Graph each parabola and find its vertex and the axis of the parabola. (See Examples 1–6.)

29. $f(x) = (x + 2)^2$ **30.** $f(x) = -(x + 5)^2$

31. $f(x) = x^2 - 4x + 6$ **32.** $f(x) = x^2 + 6x + 3$

Work these problems. (See Example 6.)

33. McDonald's Revenue The revenue for McDonald's Corporation (in billions of dollars) for the years 2010 to 2015 can be approximated by the quadratic function

$$R(x) = -\frac{1}{2}x^2 + 13x - 57,$$

where $x = 10$ corresponds to the year 2010. (Data from: www.morningstar.com.)

(a) Find the vertex.

(b) In what year during this period was revenue at a maximum?

(c) What was the maximum revenue, to the nearest billion dollars?

34. Souvenir Sales Janya owns a factory that manufactures souvenir key chains. Her weekly profit (in hundreds of dollars) is given by $P(x) = -2x^2 + 60x - 120$, where x is the number of cases of key chains sold.

(a) What is the largest number of cases she can sell and still make a profit?

(b) Explain how it is possible for her to lose money if she sells more cases than your answer in part (a).

(c) How many cases should she make and sell in order to maximize her profits?

35. Nerve Impulses A researcher in physiology has decided that a good mathematical model for the number of impulses fired after a nerve has been stimulated is given by $y = -x^2 + 20x - 60$, where y is the number of responses per millisecond and x is the number of milliseconds since the nerve was stimulated.

(a) When will the maximum firing rate be reached?

(b) What is the maximum firing rate?

36. Bullet Trajectory A bullet is fired upward from ground level. Its height above the ground (in feet) at time t seconds is given by

$$H(t) = -16t^2 + 960t.$$

(a) Find the vertex.

(b) Find the time at which the bullet reaches its maximum height.

(c) Find the maximum height.

(d) Find the time at which the bullet hits the ground (when the height is equal to zero).

37. Automobile Fatalities According to data from the National Safety Council, the fatal-accident rate per 100,000 licensed drivers can be approximated by the function $f(x) = .0328x^2 - 3.55x + 115$, where x is the age of the driver ($16 \leq x \leq 88$). At what age is the rate the lowest?

38. Population Dynamics Using data from the U.S. Census Bureau, the population (in thousands) of Detroit, Michigan can be approximated by $g(x) = -.422x^2 + 48.84x + 248$, where $x = 0$ corresponds to the year 1900. In what year, according to the given model, did Detroit have its highest population?

Work the following problems. (See Example 7.)

39. Business Suppose the supply of and demand for a certain textbook are given by

$$\text{supply: } p = \frac{1}{5}q^2 \quad \text{and} \quad \text{demand: } p = -\frac{1}{5}q^2 + 40,$$

where p is price and q is quantity. How many books are demanded at a price of

(a) 10? (b) 20? (c) 30? (d) 40?

How many books are supplied at a price of

(e) 5? (f) 10? (g) 20? (h) 30?

(i) Graph the supply and demand functions on the same axes.

40. Business Find the equilibrium quantity and the equilibrium price in Exercise 39.

41. Business Suppose the price p of widgets is related to the quantity q that is demanded by

$$p = 640 - 5q^2,$$

where q is measured in hundreds of widgets. Find the price when the number of widgets demanded is

(a) 0; (b) 5; (c) 10.

Suppose the supply function for widgets is given by $p = 5q^2$, where q is the number of widgets (in hundreds) that are supplied at price p.

(d) Graph the demand function $p = 640 - 5q^2$ and the supply function $p = 5q^2$ on the same axes.

(e) Find the equilibrium quantity.

(f) Find the equilibrium price.

42. Business The supply function for a commodity is given by $p = q^2 + 200$, and the demand function is given by $p = -10q + 3200$

(a) Graph the supply and demand functions on the same axes.

(b) Find the equilibrium point.

(c) What is the equilibrium quantity? the equilibrium price?

Business *Find the equilibrium quantity and equilibrium price for the commodity whose supply and demand functions are given.*

43. Supply: $p = 45q$; demand: $p = -q^2 + 10{,}000$.

44. Supply: $p = q^2 + q + 10$; demand: $p = -10q + 3060$.

45. Supply: $p = q^2 + 20q$; demand: $p = -2q^2 + 10q + 3000$.

46. Supply: $p = .2q + 51$; demand: $p = \dfrac{3000}{q + 5}$.

Business *The revenue function $R(x)$ and the cost function $C(x)$ for a particular product are given. These functions are valid only for the specified domain of values. Find the number of units that must be produced to break even.*

47. $R(x) = 200x - x^2$; $C(x) = 70x + 2200$; $0 \le x \le 100$

48. $R(x) = 300x - x^2$; $C(x) = 65x + 7000$; $0 \le x \le 150$

49. $R(x) = 400x - 2x^2$; $C(x) = -x^2 + 200x + 1900$; $0 \le x \le 100$

50. $R(x) = 500x - 2x^2$; $C(x) = -x^2 + 270x + 5125$; $0 \le x \le 125$

Business *A store owner finds that at a price of $p(x)$ dollars, x units of a certain item will be sold (the demand). Recall that revenue = demand × price. In Exercises 51 and 52, find the following:*

(a) *the revenue function;*

(b) *the number of items that leads to the maximum revenue;*

(c) *the maximum revenue.*

51. $p(x) = 84 - 2x$

52. $p(x) = 192 - 3x$

Business *Work each problem. (See Example 8.)*

53. A homeowner who listed her house for rent on Airbnb found that at a price of

$$p(x) = 300 - 10x$$

dollars per night, x nights will be booked in a month.

(a) Find an expression for the total revenue from renting her house for x nights. (*Hint:* Revenue = Demand × Price.)

(b) Find the number of rented nights that leads to the maximum revenue.

(c) Find the maximum revenue.

54. The manager of a bicycle shop has found that, at a price (in dollars) of $p(x) = 150 - \dfrac{x}{4}$ per bicycle, x bicycles will be sold.

(a) Find an expression for the total revenue from the sale of x bicycles. (*Hint:* Revenue = Demand × Price.)

(b) Find the number of bicycle sales that leads to maximum revenue.

(c) Find the maximum revenue.

55. A charter flight charges a fare of $200 per person, plus $4 per person for each unsold seat on the plane. If the plane holds 100 passengers and if x represents the number of unsold seats, find the following:

(a) an expression for the total revenue received for the flight. (*Hint*: Multiply the number of people flying, $100 - x$, by the price per ticket, $200 + 4x$);

(b) the graph for the expression in part (a);

(c) the number of unsold seats that will produce the maximum revenue;

(d) the maximum revenue.

56. The revenue of a charter bus company depends on the number of unsold seats. If 100 seats are sold, the price is $50 per seat. Each unsold seat increases the price per seat by $1. Let x represent the number of unsold seats.

(a) Write an expression for the number of seats that are sold.

(b) Write an expression for the price per seat.

(c) Write an expression for the revenue.

(d) Find the number of unsold seats that will produce the maximum revenue.

(e) Find the maximum revenue.

57. A farmer wants to find the best time to take her hogs to market. The current price is 88 cents per pound, and her hogs weigh an average of 90 pounds. The hogs gain 5 pounds per week, and the market price for hogs is falling each week by 2 cents per pound.

How many weeks should the farmer wait before taking her hogs to market in order to receive as much money as possible? At that time, how much money (per hog) will she get?

58. The manager of a peach orchard is trying to decide when to arrange for picking the peaches. If they are picked now, the average yield per tree will be 100 pounds, which can be sold for 40¢ per pound. Past experience shows that the yield per tree will increase about 5 pounds per week, while the price will decrease about 2¢ per pound per week.

(a) Let x represent the number of weeks that the manager waits. Find the price per pound.

(b) Find the number of pounds per tree.

(c) Find the total revenue from a tree.

(d) When should the peaches be picked in order to produce the maximum revenue?

(e) What is the maximum revenue?

Work these exercises. (See Examples 4 and 9.)

59. Federal Debt The United States federal debt (in billions of dollars) is shown in the following table for the years 2000 through 2008. (Data from: U.S. Department of the Treasury.)

Year	Debt (billions)
2000	5674
2001	5807
2002	6228
2003	6783
2004	7379
2005	7933
2006	8507
2007	9008
2008	10,025

(a) Let $x = 0$ correspond to the year 2000. Use (0, 5674) as the vertex and the data for the year 2007 ($x = 7$) to find a quadratic function in the form $f(x) = a(x - h)^2 + k$ that models these data (round coefficients to the nearest integer).

(b) Use the model to estimate the federal debt in 2008. Is the value estimated by the model reasonably close to the actual value shown in the table?

60. Federal Debt The United States federal debt (in billions of dollars) is shown in the following table for the years 2009 through 2015. (Data from: U.S. Department of the Treasury.)

Year	Debt (billions)
2009	11,910
2010	13,562
2011	14,790
2012	16,066
2013	16,738
2014	17,824
2015	18,150

(a) Let $x = 0$ correspond to the year 2000. Use (15, 18150) as the vertex and the data for the year 2011 ($x = 11$) to find a quadratic function in the form $f(x) = a(x - h)^2 + k$ that models these data.

(b) Use the model to estimate the federal debt in 2014. Is the value estimated by the model reasonably close to the actual value shown in the table?

61. Commodity Pricing The price of soybean meal (in dollars per metric ton) is shown in the following table for the years 2011 to 2015. (Data from: U.S. Department of Agriculture.)

Year	Price
2011	$398
2012	$524
2013	$545
2014	$528
2015	$395

(a) Let $x = 0$ correspond to the year 2000. Use (13, 545) as the vertex and the data for the year 2015 ($x = 15$) to find a quadratic function in the form $f(x) = a(x - h)^2 + k$ that models these data.

(b) Use the model to estimate the price of soybean meal in 2011. Is the value estimated by the model reasonably close to the actual value shown in the table?

62. Forest Fires The acreage (in millions) consumed by forest fires in the United States is given in the following table for selected years. (Data from: National Interagency Fire Center.)

Year	Acres
1985	2.9
1988	5.0
1991	3.0
1994	4.1
1997	2.9
2000	7.4
2003	4.0
2006	9.9
2009	5.9
2012	9.2
2015	10.1

(a) Let $x = 0$ correspond to the year 1985. Use (0, 2.9) as the vertex and the data for the year 2015 ($x = 30$) to find a quadratic function in the form $f(x) = a(x - h)^2 + k$ that models these data (round coefficients to three decimal places).

(b) Use the model to estimate the acreage destroyed by forest fires in 2010 ($x = 25$).

In Exercises 63–66, use quadratic regression to find a function that models these data. Plot the data and the quadratic regression curve. (See Example 10.)

63. Use the data in Exercise 59, with $x = 0$ corresponding to the year 2000. What does the graphing calculator model estimate for the year 2008? Compare your answer with that for Exercise 59.

64. Use the data in Exercise 60, with $x = 0$ corresponding to the year 2000. What does the graphing calculator model estimate for the year 2014? Compare your answer with that for Exercise 60.

65. Use the data in Exercise 61, with $x = 0$ corresponding to the year 2000. What does the graphing calculator model estimate for the year 2011? Compare your answer with that for Exercise 61.

66. Use the data in Exercise 62, with $x = 0$ corresponding to the year 1985. What does the graphing calculator model estimate for the year 2010 ($x = 25$)? Compare your answer with that for Exercise 62.

Business *Recall that profit equals revenue minus cost. In Exercises 67 and 68, find the following:*

   (a) *The break-even point (to the nearest tenth)*

   (b) *The x-value that makes profit a maximum*

   (c) *The maximum profit*

   (d) *For what x-values will a loss occur?*

   (e) *For what x-values will a profit occur?*

67. $R(x) = 400x - 2x^2$ and $C(x) = 200x + 2000$, with $0 \le x \le 100$

68. $R(x) = 900x - 3x^2$ and $C(x) = 450x + 5000$, with $20 \le x \le 150$

✓**Checkpoint Answers**

1. (a)

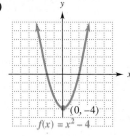

$f(x) = x^2 - 4$

(b)

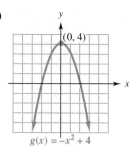

$g(x) = -x^2 + 4$

2. (a)

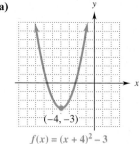

$f(x) = (x + 4)^2 - 3$

(b)

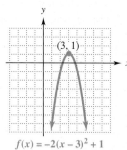

$f(x) = -2(x - 3)^2 + 1$

3. (a) $f(x) = (x + 3)^2 - 4; (-3, -4)$

   (b) $g(x) = (x - 6)^2 - 3; (6, -3)$

4. (a) 10 units   **(b)** $480

5. (a) 4   **(b)** 16

6. $f(x) = -5x^2 + 20x + 90$

3.5 Polynomial Functions

A **polynomial function of degree** n is a function whose rule is given by a polynomial of degree n.[*] For example

$$f(x) = 3x - 2 \qquad \text{polynomial function of degree 1;}$$
$$g(x) = 3x^2 + 4x - 6 \qquad \text{polynomial function of degree 2;}$$
$$h(x) = x^4 + 5x^3 - 6x^2 + x - 3 \qquad \text{polynomial function of degree 4.}$$

---

[*] The degree of a polynomial was defined on page 13.

Basic Graphs

The simplest polynomial functions are those whose rules are of the form $f(x) = ax^n$ (where a is a constant).

EXAMPLE 1 Graph $f(x) = x^3$.

Solution Make a table of data with ordered pairs belonging to the graph, as in Figure 3.34(a). Be sure to choose some negative x-values, $x = 0$, and some positive x-values in order to get representative ordered pairs. Find as many ordered pairs as you need in order to see the shape of the graph. Then plot the ordered pairs and draw a smooth curve through them to obtain the graph in Figure 3.34(b).

✓ **Checkpoint 1**

Graph $f(x) = -\dfrac{1}{2}x^3$

x	$f(x)$
2	8
1	1
0	0
−1	−1
−2	−8

Figure 3.34(a)

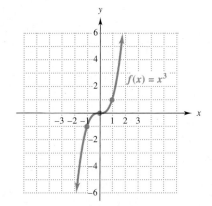

Figure 3.34(b)

EXAMPLE 2 Graph

$$f(x) = \frac{3}{2}x^4.$$

Solution The table in Figure 3.35(a) gives some typical ordered pairs and leads to the graph in Figure 3.35(b).

✓ **Checkpoint 2**

Graph $g(x) = -2x^4$.

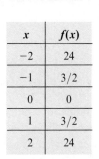

x	$f(x)$
−2	24
−1	3/2
0	0
1	3/2
2	24

Figure 3.35(a)

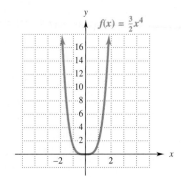

Figure 3.35(b)

The graph of $f(x) = ax^n$ has one of the four basic shapes illustrated in Examples 1 and 2 and in Checkpoints 1 and 2. The basic shapes are summarized in the following box.

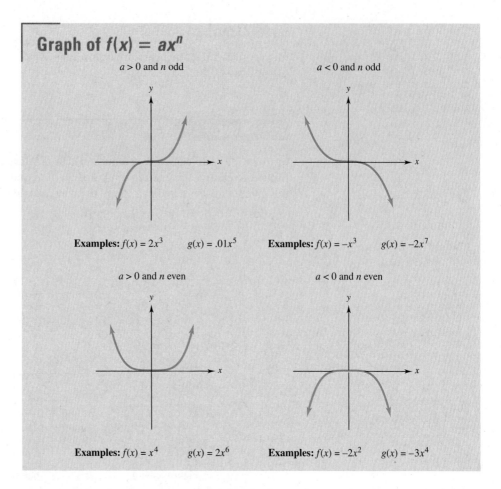

Graph of $f(x) = ax^n$

$a > 0$ and n odd

$a < 0$ and n odd

Examples: $f(x) = 2x^3$ $g(x) = .01x^5$

Examples: $f(x) = -x^3$ $g(x) = -2x^7$

$a > 0$ and n even

$a < 0$ and n even

Examples: $f(x) = x^4$ $g(x) = 2x^6$

Examples: $f(x) = -2x^2$ $g(x) = -3x^4$

Properties of Polynomial Graphs

Unlike the graphs in the preceding figures, the graphs of more complicated polynomial functions may have several "peaks" and "valleys," as illustrated in Figure 3.36, on the following page. The locations of the peaks and valleys can be accurately approximated by a maximum or minimum finder on a graphing calculator. Calculus is needed to determine their exact location.

The total number of peaks and valleys in a polynomial graph, as well as the number of x-intercepts, depends on the degree of the polynomial, as shown in Figure 3.36 and summarized here.

Polynomial	Degree	Number of peaks & valleys	Number of x-intercepts
$f(x) = x^3 - 4x + 2$	3	2	3
$f(x) = x^5 - 5x^3 + 4x$	5	4	5
$f(x) = 1.5x^4 + x^3 - 4x^2 - 3x + 4$	4	3	2
$f(x) = -x^6 + x^5 + 2x^4 + 1$	6	3	2

In each case, the number of x-intercepts is *at most* the degree of the polynomial. The total number of peaks and valleys is at most *one less than* the degree of the polynomial. The same thing is true in every case.

1. The total number of peaks and valleys on the graph of a polynomial function of degree n is at most $n - 1$.

2. The number of x-intercepts on the graph of a polynomial function of degree n is at most n.

(a) $f(x) = x^3 - 4x + 2$
1 peak
1 valley $\Big\}$ total 2
3 x-intercepts

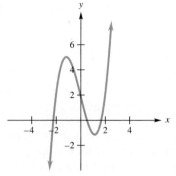

(b) $f(x) = x^5 - 5x^3 + 4x$
2 peaks
2 valleys $\Big\}$ total 4
5 x-intercepts

(c) $f(x) = 1.5x^4 + x^3 - 4x^2 - 3x + 4$
1 peak
2 valleys $\Big\}$ total 3
2 x-intercepts

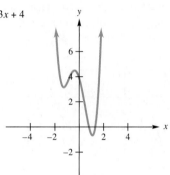

(d) $f(x) = -x^6 + x^5 + 2x^4 + 1$
2 peaks
1 (shallow) valley $\Big\}$ total 3
2 x-intercepts

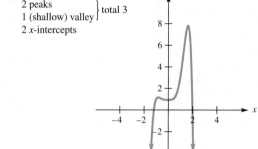

Figure 3.36

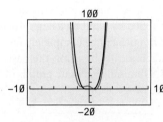

Figure 3.37

The domain of every polynomial function is the set of all real numbers, which means that its graph extends forever to the left and right. We can indicate this by the arrows on the ends of polynomial graphs, as shown in Figure 3.36.

Although there may be peaks, valleys, and bends in a polynomial graph, the far ends of the graph are easy to describe: *they look like the graph of the highest-degree term of the polynomial.* Consider, for example, $f(x) = 1.5x^4 + x^3 - 4x^2 - 3x + 4$, whose highest-degree term is $1.5x^4$ and whose graph is shown in Figure 3.36(c). The ends of the graph shoot upward, just as the graph of $y = 1.5x^4$ does in Figure 3.35. When $f(x)$ and $y = 1.5x^4$ are graphed in the same large viewing window of a graphing calculator (Figure 3.37), the graphs look almost identical, except near the origin. This is an illustration of the following facts.

> The graph of a polynomial function is a smooth, unbroken curve that extends forever to the left and right. When $|x|$ is large, the graph resembles the graph of its highest-degree term and moves sharply away from the x-axis.

EXAMPLE 3   Let $g(x) = x^3 - 11x^2 - 32x + 24$, and consider the graph in Figure 3.38.

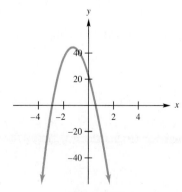

Figure 3.38

(a) Is Figure 3.38 a complete graph of $g(x)$; that is, does it show all the important features of the graph?

Solution The far ends of the graph of $g(x)$ should resemble the graph of its highest-degree term x^3. The graph of $f(x) = x^3$ in Figure 3.34 moves upward at the far right, but the graph in Figure 3.38 does not. So Figure 3.38 is *not* a complete graph.

(b) Use a graphing calculator to find a complete graph of $g(x)$.

Solution Since the graph of $g(x)$ must eventually start rising on the right side (as does the graph of x^3), a viewing window that shows a complete graph must extend beyond $x = 4$. By experimenting with various windows, we obtain Figure 3.39. This graph shows a total of two peaks and valleys and three x-intercepts (the maximum possible for a polynomial of degree 3). At the far ends, the graph of $g(x)$ resembles the graph of $f(x) = x^3$. Therefore, Figure 3.39 is a complete graph of $g(x)$. ✔₃

✓ Checkpoint 3

Find a viewing window on a graphing calculator that shows a complete graph of $f(x) = -.7x^4 + 119x^2 + 400$. (*Hint*: The graph has two x-intercepts and the maximum possible number of peaks and valleys.)

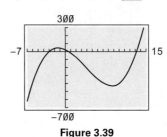

Figure 3.39

Graphing Techniques

Accurate graphs of first- and second-degree polynomial functions (lines and parabolas) are easily found algebraically, as we saw in Sections 2.2 and 3.4. All polynomial functions of degree 3, and some of higher degree, can be accurately graphed by hand by using calculus and algebra to locate the peaks and valleys. When a polynomial can be completely factored, the general shape of its graph can be determined algebraically by using the basic properties of polynomial graphs, as illustrated in Example 4. Obtaining accurate graphs of other polynomial functions generally requires the use of technology.

EXAMPLE 4   Graph $f(x) = (2x + 3)(x - 1)(x + 2)$.

Solution Note that $f(x)$ is a polynomial of degree 3. (If you don't see why, do Checkpoint 4.) Begin by finding any x-intercepts. Set $f(x) = 0$ and solve for x: ✔₄

✓ Checkpoint 4

Multiply out the expression for $f(x)$ in Example 4 and determine its degree.

$$f(x) = 0$$
$$(2x + 3)(x - 1)(x + 2) = 0.$$

Solve this equation by setting each of the three factors equal to 0:

$$2x + 3 = 0 \quad \text{or} \quad x - 1 = 0 \quad \text{or} \quad x + 2 = 0$$
$$x = -\frac{3}{2} \qquad\qquad x = 1 \qquad\qquad x = -2.$$

The three numbers $-3/2$, 1, and -2 divide the x-axis into four intervals:

$$x < -2, \quad -2 < x < -\frac{3}{2}, \quad -\frac{3}{2} < x < 1, \quad \text{and} \quad 1 < x.$$

These intervals are shown in Figure 3.40.

Figure 3.40

Since the graph is an unbroken curve, it can change from above the x-axis to below it only by passing through the x-axis. As we have seen, this occurs only at the x-intercepts: $x = -2, -3/2$, and 1. Consequently, in the interval between two intercepts (or to the left of $x = -2$ or to the right of $x = 1$), the graph of $f(x)$ must lie entirely above or entirely below the x-axis.

We can determine whether the graph lies above or below an interval by evaluating $f(x) = (2x + 3)(x - 1)(x + 2)$ at a number in that interval. For example, $x = -3$ is in the interval where $x < -2$, and

$$f(-3) = (2(-3) + 3)(-3 - 1)(-3 + 2)$$
$$= -12.$$

Therefore, $(-3, -12)$ is on the graph. Since this point lies below the x-axis, all points in this interval (that is, all points with $x < -2$) must lie below the x-axis. By testing numbers in the other intervals, we obtain the following table.

Interval	$x < -2$	$-2 < x < -3/2$	$-3/2 < x < 1$	$x > 1$
Test Number	-3	$-7/4$	0	2
Value of $f(x)$	-12	$11/32$	-6	28
Sign of $f(x)$	Negative	Positive	Negative	Positive
Graph	Below x-axis	Above x-axis	Below x-axis	Above x-axis

Since the graph intersects the x-axis at the intercepts $x = -2$ and $x = -3/2$, and is above the x-axis between these intercepts, there must be at least one peak there. Similarly, there must be at least one valley between $x = -3/2$ and $x = 1$, because the graph is below the x-axis there. A polynomial function of degree 3 can have a total of at most $3 - 1 = 2$ peaks and valleys. So there must be exactly one peak and exactly one valley on this graph.

Furthermore, when $|x|$ is large, the graph must resemble the graph of $y = 2x^3$ (the highest-degree term). The graph of $y = 2x^3$, like the graph of $y = x^3$ in Figure 3.34, moves upward to the right and downward to the left. Using these facts and plotting the x-intercepts shows that the graph must have the general shape shown in Figure 3.41. Plotting additional points leads to the reasonably accurate graph in Figure 3.42. We say "reasonably accurate" because we cannot be sure of the exact locations of the peaks and valleys on the graph without using calculus. 5

✓ **Checkpoint 5**

Graph $f(x) = .4(x - 2)(x + 3)(x - 4)$.

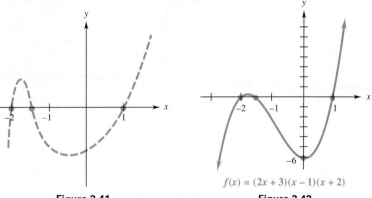

Figure 3.41   Figure 3.42

$f(x) = (2x + 3)(x - 1)(x + 2)$

Polynomial Models

Regression procedures similar to those presented for linear (degree 1) regression in Section 2.3 and quadratic (degree 2) regression in Section 3.5 can be used to find cubic (degree 3) and quartic (degree 4) polynomial models for appropriate data.

⊞ EXAMPLE 5 **Johnson & Johnson Profits** The following table shows the revenue and costs (in millions of dollars) for Johnson & Johnson. (Data from: www.morningstar.com.)

Year	2008	2009	2010	2011	2012	2013	2014	2015
Revenue	63,747	61,897	61,587	65,030	67,224	71,312	74,331	70,074
Costs	18,511	18,447	18,792	20,360	21,658	22,342	22,746	21,536

(a) Let $x = 8$ correspond to the year 2008, and use cubic regression to obtain models for the revenue data $R(x)$ and the costs $C(x)$.

Solution The procedure is the same as for linear regression, just choose cubic regression instead (go to the STAT CALC menu, and choose CubicReg). The functions obtained are:

$$R(x) = -234x^3 + 8139x^2 - 90,692x + 388,407 \text{ and}$$
$$C(x) = -73x^3 + 2461x^2 - 26,151x + 107,839.$$

(b) Plot both sets of data and graph $R(x)$ and $C(x)$ on the same screen.

Solution See Figure 3.43, which shows that the models fit these data well.

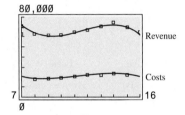

Figure 3.43

(c) Find the profit function $P(x)$ and show its graph. Were profits always increasing during this time period or did they fluctuate?

Solution The profit function is the difference between the revenue function and the cost function. We subtract the coefficients of the cost function from the respective coefficients of the revenue function:

$$\begin{aligned} P(x) &= R(x) - C(x) \\ &= -234x^3 + 8139x^2 - 90,692x + 388,407 \\ &\quad -(-73x^3 + 2461x^2 - 26,151x + 107,839) \\ &= -161x^3 + 5678x^2 - 64,541x + 280,568. \end{aligned}$$

The graph of $P(x)$ appears in Figure 3.44, which shows that profits for this company fluctuate over time.

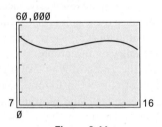

Figure 3.44

⊞ EXAMPLE 6 **Personal Savings** The table below shows gross savings in the United States in trillions of dollars. (Data from: www.worldbank.org.)

Year	Gross savings (trillions)
2000	2126
2001	2076
2002	1998
2003	1998
2004	2155
2005	2349
2006	2659
2007	2512
2008	2281
2009	2082
2010	2272
2011	2450
2012	2868
2013	2949

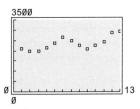

Figure 3.45

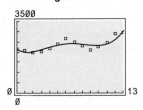

Figure 3.46

(a) Plot these data on a graphing calculator or cell phone app with $x = 0$ corresponding to the year 2000.

Solution The points plotted in Figure 3.45 suggest the general shape of a fourth-degree polynomial. A fourth-degree polynomial is called a *quartic* function (not to be confused with a *quadratic* function, which is a second-degree polynomial whose graph is a parabola).

(b) Used quartic regression to obtain a model for these data and graph the function on the same screen as the data points.

Solution The function obtained from quartic regression (look for QuartReg in the STAT CALC menu) is

$$f(x) = .686x^4 - 16.4x^3 + 121.6x^2 - 245x + 2139.$$

The graph in Figure 3.46 shows this is a reasonable model for the data.

3.5 Exercises

Graph each of the given polynomial functions. (See Examples 1 and 2.)

1. $f(x) = x^4$

2. $g(x) = -.5x^6$

3. $h(x) = -.2x^5$

4. $f(x) = x^7$

In Exercises 5–8, state whether the graph could possibly be the graph of (a) some polynomial function; (b) a polynomial function of degree 3; (c) a polynomial function of degree 4; (d) a polynomial function of degree 5. (See Example 3.)

5.

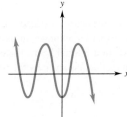

6.

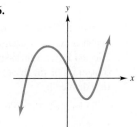

7.

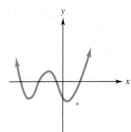

8.

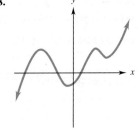

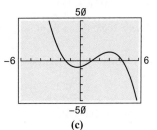

(c)

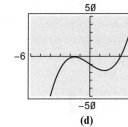

(d)

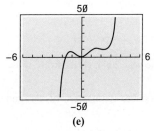

(e)

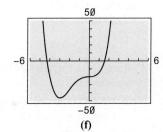

(f)

9. $f(x) = x^3 - 7x - 9$

10. $f(x) = -x^3 + 4x^2 + 3x - 8$

11. $f(x) = x^4 - 5x^2 + 7$

12. $f(x) = x^4 + 4x^3 - 20$

13. $f(x) = -x^5 + 4x^4 + x^3 - 16x^2 + 12x + 5$

14. $f(x) = .7x^5 - 2.5x^4 - x^3 + 8x^2 + x + 2$

Graph each of the given polynomial functions. (See Example 4.)

15. $f(x) = (x + 3)(x - 4)(x + 1)$

16. $f(x) = (x - 5)(x - 1)(x + 1)$

17. $f(x) = x^2(x + 3)(x - 1)$

18. $f(x) = x^2(x + 2)(x - 2)$

19. $f(x) = x^3 - x^2 - 20x$

20. $f(x) = x^3 - 4x$

In Exercises 9–14, match the given polynomial function to its graph [(a)–(f)], without using a graphing calculator. (See Example 3 and the two boxes preceding it.)

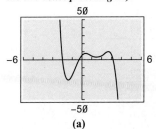

(a)

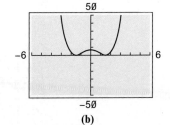

(b)

Work these exercises.

21. **Home Depot Revenue** The revenue for Home Depot Inc. (in billions of dollars) for the years 2004 to 2015 can be approximated by the function

$$R(x) = .19x^3 - 5.2x^2 + 44x - 40,$$

where $x = 4$ corresponds to the year 2004. (Data from: www.morningstar.com.)

(a) Find $R(4)$, round your answer to the nearest integer, and explain what your answer means in practical terms.

(b) Find the following (round your answers to the nearest integer): $R(6)$; $R(11)$; $R(14)$.

(c) Graph $R(x)$.

(d) Is revenue always increasing or does it fluctuate?

22. **Caterpillar Revenue** The revenue for Caterpillar Corporation (in billions of dollars) for the years 2008 to 2015 can be approximated by the function

$$R(x) = -.638x^3 + 21.1x^2 - 225x + 825,$$

where $x = 8$ corresponds to the year 2008. (Data from: www.morningstar.com.)

(a) Find the following (round your answers to the nearest integer): $R(8)$; $R(9)$; $R(13)$; $R(15)$.

(b) Graph $R(x)$.

(c) Is revenue always increasing or does it fluctuate?

23. **Home Depot Costs** The cost function for Home Depot Inc. (in billions of dollars) for the years 2004 to 2015 can be approximated by the function

$$C(x) = .13x^3 - 3.6x^2 + 31x - 34,$$

where $x = 4$ corresponds to the year 2004. (Data from: www.morningstar.com.)

(a) Find $C(4)$, round your answer to the nearest integer, and interpret your answer.

(b) Find the following (round your answers to the nearest integer): $C(6)$; $C(11)$; $C(14)$.

(c) Graph $C(x)$.

(d) Are costs always increasing or do they fluctuate?

24. **Caterpillar Costs** The cost function for Caterpillar Corporation (in billions of dollars) for the years 2008 to 2015 can be approximated by the function

$$C(x) = -.493x^3 + 16.4x^2 - 176x + 645,$$

where $x = 8$ corresponds to the year 2008. (Data from: www.morningstar.com.)

(a) Find the following (round your answers to the nearest integer): $C(8)$; $C(9)$; $C(13)$; $C(15)$.

(b) Graph $C(x)$.

(c) Are costs always increasing or do they fluctuate?

25. **Home Depot Profit** Find the profit function for Home Depot Inc. (in billions of dollars) by using the revenue function $R(x)$ from Exercise 21 and the cost function $C(x)$ from Exercise 23. Use your function to find the profit for the year 2014.

26. **Caterpillar Profit** Find the profit function for Caterpillar Corporation (in billions of dollars) by using the revenue function $R(x)$ from Exercise 22 and the cost function $C(x)$ from Exercise 24. Use your function to find the profit for the year 2015.

27. **Finance** An idealized version of the Laffer curve (originated by economist Arthur Laffer) is shown in the accompanying graph. According to this theory, decreasing the tax rate, say, from x_2 to x_1, may actually increase the total revenue to the government. The theory is that people will work harder and earn more if they are taxed at a lower rate, which means higher total tax revenues than would be the case at a higher rate. Suppose that the Laffer curve is given by the function

$$f(x) = \frac{x(x - 100)(x - 160)}{240} \quad (0 \le x \le 100),$$

where $f(x)$ is government revenue (in billions of dollars) from a tax rate of x percent. Find the revenue from the given tax rates.

(a) 20%   (b) 40%   (c) 50%   (d) 70%

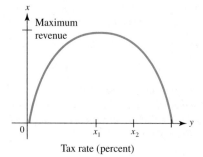

Tax rate (percent)

28. **Cardiac Output** A technique for measuring cardiac output depends on the concentration of a dye after a known amount is injected into a vein near the heart. In a normal heart, the concentration of the dye at time x (in seconds) is given by the function defined by

$$g(x) = -.006x^4 + .140x^3 - .053x^2 + 1.79x.$$

Find the following: $g(0)$; $g(1)$; $g(2)$; $g(3)$.

▦**Polynomial Models** *Use a graphing calculator to do the following cubic and quartic regression problems. (See Examples 5 and 6.)*

29. **Public Schools** Enrollment in public elementary and secondary schools (in thousands) in the United States is shown in the following table for selected years, including projected enrollment through the year 2024. (Data from: U.S. National Center for Education Statistics.)

Year	Enrollment (thousands)
2000	47,204
2005	49,113
2010	49,484
2015	50,094
2016	50,229
2020	51,547
2024	52,920

(a) Plot the data with $x = 0$ corresponding to the year 2000.

(b) Use cubic regression to find a third-order polynomial function $f(x)$ that models these data (round coefficients to the nearest integer).

(c) Graph $f(x)$ on the same screen as the data points. Does the graph appear to fit the data well?

(d) According to the model, what is the project enrollment for the year 2023?

30. **Gross National Income** The following table shows the gross national income (GNI) in dollars per capita for middle- and upper-income countries for selected years. (Data from: Organization for Economic Cooperation and Development.)

Year	2000	2002	2004	2006	2008	2010	2012	2014
GNI	1768	1842	2335	3148	4403	5472	6949	7901

(a) Plot the data with $x = 0$ corresponding to the year 2000.

(b) Use cubic regression to find a third-order polynomial function $f(x)$ that models these data (round coefficients to one decimal place).

(c) Graph $y = f(x)$ on the same screen as the data points. Does the graph appear to fit the data well?

(d) What does the model predict for the GNI in the year 2017?

31. **Consumer Confidence** The Conference Board's consumer confidence index (CCI) is listed for the years 2008 to 2015 in the table below. (Data from: The Conference Board.)

Year	2008	2009	2010	2011	2012	2013	2014	2015
CCI	88	40	57	65	62	58	79	104

(a) Plot the data with $x = 8$ corresponding to the year 2008.

(b) Use quartic regression to find a fourth-order polynomial function $f(x)$ that models these data.

(c) Graph $y = f(x)$ on the same screen as the data points. Does the graph appear to be a reasonable model for these data?

(d) What does the model estimate for the CCI in 2015? Is this close to the actual index value shown in the table?

32. **American Express Revenue** The American Express Company reported its revenue (in billions) as shown in the following table. (Data from: www.morningstar.com.)

Year	2008	2009	2010	2011	2012	2013	2014	2015
Revenue (billions)	28.4	24.5	27.8	30.0	31.6	33.0	34.3	32.8

(a) Plot the data with $x = 8$ corresponding to the year 2008.

(b) Use cubic regression to find a third-order polynomial function $f(x)$ that models these data (round coefficients to two decimal places).

(c) Graph $f(x)$ on the same screen as the data points. Does the graph appear to fit the data well?

(d) What does the model estimate for the revenue in 2011? Is this close to the actual revenue shown in the table?

Exercises 33–36 require a graphing calculator. Find a viewing window that shows a complete graph of the polynomial function (that is, a graph that includes all the peaks and valleys and that indicates how the curve moves away from the x-axis at the far left and far right). There are many correct answers. Consider your answer correct if it shows all the features that appear in the window given in the answers. (See Example 3.)

33. $g(x) = x^3 - 3x^2 - 4x - 5$

34. $f(x) = x^4 - 10x^3 + 35x^2 - 50x + 24$

35. $f(x) = 2x^5 - 3.5x^4 - 10x^3 + 5x^2 + 12x + 6$

36. $g(x) = x^5 + 8x^4 + 20x^3 + 9x^2 - 27x - 7$

✓ Checkpoint Answers

1. 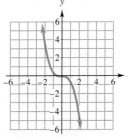   2.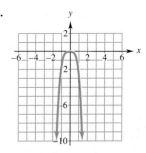

3. Many correct answers, including $-15 \le x \le 15$ and $-2000 \le y \le 6000$.

4. $f(x) = 2x^3 + 5x^2 - x - 6$; degree 3.

5.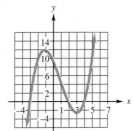

3.6 Rational Functions

A **rational function** is a function whose rule is the quotient of two polynomials, such as

$$f(x) = \frac{2}{1 + x}, \qquad g(x) = \frac{3x + 2}{2x + 4}, \qquad \text{and} \qquad h(x) = \frac{x^2 - 2x - 4}{x^3 - 2x^2 + x}.$$

Thus, a rational function is a function whose rule can be written in the form

$$f(x) = \frac{P(x)}{Q(x)},$$

where $P(x)$ and $Q(x)$ are polynomials, with $Q(x) \neq 0$. The function is undefined for any values of x that make $Q(x) = 0$, so there are breaks in the graph at these numbers.

Linear Rational Functions

We begin with rational functions in which both numerator and denominator are first-degree or constant polynomials. Such functions are sometimes called **linear rational functions.**

EXAMPLE 1   Graph the rational function defined by

$$f(x) = \frac{2}{1 + x}.$$

Solution   This function is undefined for $x = -1$, since -1 leads to a 0 denominator. For that reason, the graph of this function will not intersect the vertical line $x = -1$. Since x can take on any value except -1, the values of x can approach -1 as closely as desired from either side of -1, as shown in the following table of values.

x approaches -1

x	-1.5	-1.2	-1.1	-1.01	$-.99$	$-.9$	$-.8$	$-.5$
$1 + x$	$-.5$	$-.2$	$-.1$	$-.01$	$.01$	$.1$	$.2$	$.5$
$f(x) = \dfrac{2}{1 + x}$	-4	-10	-20	-200	200	20	10	4

$|f(x)|$ gets larger and larger

The preceding table suggests that as x gets closer and closer to -1 from either side, the demonimator, $1 + x$, gets closer and closer to 0, so the absolute value of $f(x)$ gets larger and larger. The part of the graph near $x = -1$ in Figure 3.47, on the following page, shows this behavior. The vertical line $x = -1$ that is approached by the curve is called a *vertical asymptote*. For convenience, the vertical asymptote is indicated by a dashed line in Figure 3.47, but this line is *not* part of the graph of the function. However, the graph tends toward this vertical line as x approaches -1.

As $|x|$ gets larger and larger, so does the absolute value of the denominator $1 + x$. Hence, $f(x) = 2/(1 + x)$ gets closer and closer to 0, as shown in the following table.

$|x|$ gets larger and larger

x	-101	-11	-2	0	9	99
$1 + x$	-100	-10	-1	1	10	100
$f(x) = \dfrac{2}{1 + x}$	$-.02$	$-.2$	-2	2	$.2$	$.02$

$f(x)$ approaches 0

✓ Checkpoint 1

Graph the following.

(a) $f(x) = \dfrac{3}{5 - x}$

(b) $f(x) = \dfrac{-4}{x + 4}$

The horizontal line $y = 0$ is called a *horizontal asymptote* for this graph. Using the asymptotes and plotting the intercept and other points gives the graph of Figure 3.47. ✓₁

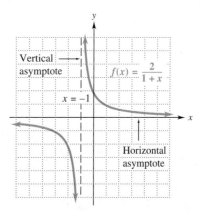

Figure 3.47

Example 1 suggests the following conclusion, which applies to all rational functions.

> If a number c makes the denominator zero, while leaving the numerator nonzero, in the expression defining a rational function, then the line $x = c$ is a **vertical asymptote** for the graph of the function. The graph of the function approaches the vertical line as x approaches c.

If the graph of a function approaches a horizontal line very closely when x is very large or very small, we say that this line is a **horizontal asymptote** of the graph. In Example 1, the horizontal asymptote was the x-axis. This is not always the case, however, as the next example illustrates.

EXAMPLE 2   Graph

$$f(x) = \frac{3x + 2}{2x + 4}.$$

Solution Find the vertical asymptote by setting the denominator equal to 0 and then solving for x:

$$2x + 4 = 0$$
$$x = -2.$$

In order to see what the graph looks like when $|x|$ is very large, we rewrite the rule of the function. When $x \neq 0$, dividing both numerator and denominator by x does not change the value of the function:

$$f(x) = \frac{3x + 2}{2x + 4} = \frac{\dfrac{3x + 2}{x}}{\dfrac{2x + 4}{x}}$$

$$= \frac{\dfrac{3x}{x} + \dfrac{2}{x}}{\dfrac{2x}{x} + \dfrac{4}{x}} = \frac{3 + \dfrac{2}{x}}{2 + \dfrac{4}{x}}.$$

✓ **Checkpoint 2**

Graph the following.

(a) $f(x) = \dfrac{2x - 5}{x - 2}$

(b) $f(x) = \dfrac{3 - x}{x + 1}$

Now, when $|x|$ is very large, the fractions $2/x$ and $4/x$ are very close to 0. (For instance, when $x = 200, 4/x = 4/200 = .02$.) Therefore, the numerator of $f(x)$ is very close to $3 + 0 = 3$, and the denominator is very close to $2 + 0 = 2$. Hence, $f(x)$ is very close to $3/2$ when $|x|$ is large, so the line $y = 3/2$ is the horizontal asymptote of the graph, as shown in Figure 3.48, on the following page. ▼2

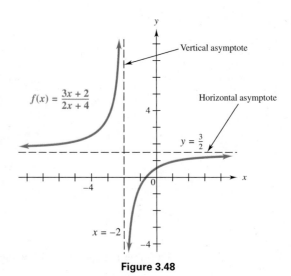

Figure 3.48

🖩 **TECHNOLOGY TIP** Depending on the viewing window, a graphing calculator may not accurately represent the graph of a rational function. For example, the graph of $f(x) = \dfrac{3x + 2}{2x + 4}$ in Figure 3.49, which should look like Figure 3.48, has an erroneous vertical line at the place where the graph has a vertical asymptote. This problem can usually be avoided by using a window that has the vertical asymptote at the center of the x-axis, as in Figure 3.50.

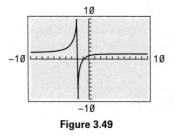

Figure 3.49

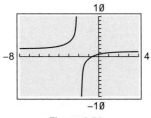

Figure 3.50

The horizontal asymptotes of a linear rational function are closely related to the coefficients of the x-terms of the numerator and denominator, as illustrated in Examples 1 and 2:

	Function	**Horizontal Asymptote**
Example 1:	$f(x) = \dfrac{2}{1 + x} = \dfrac{0x + 2}{1x + 1}$	$y = \dfrac{0}{1} = 0$ (the x-axis)
Example 2:	$f(x) = \dfrac{3x + 2}{2x + 4}$	$y = \dfrac{3}{2}$

The same pattern holds in the general case.

The graph of $f(x) = \dfrac{ax + b}{cx + d}$ (where $c \neq 0$ and $ad \neq bc$) has a vertical asymptote at the root of the denominator (the value of x that makes the denominator 0), and has horizontal asymptote $y = \dfrac{a}{c}$.

Other Rational Functions

When the numerator or denominator of a rational function has degree greater than 1, the graph of the function can be more complicated than those in Examples 1 and 2. The graph might have no horizontal asymptote, it might have several vertical asymptotes or no vertical asymptote, and it might have peaks and valleys.

EXAMPLE 3 Graph

$$f(x) = \frac{2x^2}{x^2 - 4}.$$

Solution Find the vertical asymptotes by setting the denominator equal to 0 and solving for x:

$$x^2 - 4 = 0 \qquad \text{Set denominator equal to 0.}$$
$$(x + 2)(x - 2) = 0 \qquad \text{Factor.}$$
$$x + 2 = 0 \quad \text{or} \quad x - 2 = 0 \qquad \text{Set each term equal to 0.}$$
$$x = -2 \quad \text{or} \quad x = 2. \qquad \text{Solve for } x.$$

Since neither of these numbers makes the numerator 0, the lines $x = -2$ and $x = 2$ are vertical asymptotes of the graph. The horizontal asymptote can be determined by dividing both the numerator and denominator of $f(x)$ by x^2 (the highest power of x that appears in either one):

$$f(x) = \frac{2x^2}{x^2 - 4} = \frac{\dfrac{2x^2}{x^2}}{\dfrac{x^2 - 4}{x^2}} = \frac{\dfrac{2x^2}{x^2}}{\dfrac{x^2}{x^2} - \dfrac{4}{x^2}} = \frac{2}{1 - \dfrac{4}{x^2}}.$$

When $|x|$ is very large, the fraction $4/x^2$ is very close to 0, so the denominator is very close to $1 - 0 = 1$ and $f(x)$ is very close to 2. Hence, the line $y = 2$ is the horizontal asymptote of the graph. Using this information and plotting several points in each of the three regions determined by the vertical asymptotes, we obtain Figure 3.51. ✓₃

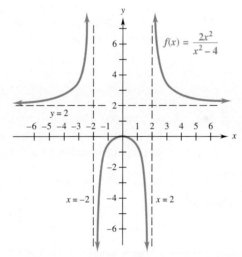

Figure 3.51

✓ **Checkpoint 3**

List the vertical and horizontal asymptotes of the given function.

(a) $f(x) = \dfrac{3x + 5}{x + 5}$

(b) $g(x) = \dfrac{2 - x^2}{x^2 - 4}$

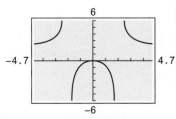

Figure 3.52

▦ **TECHNOLOGY TIP** When a function whose graph has more than one vertical asymptote (as in Example 3) is graphed on a graphing calculator, erroneous vertical lines can sometimes be avoided by using a *decimal window* (with the y-range adjusted to show the graph). On TI, use (Z)DECIMAL in the ZOOM or VIEWS menu. Figure 3.52 shows the function of Example 3 graphed in a decimal window on a TI-84+. (The x-range may be different on other calculators.)

The arguments used to find the horizontal asymptotes in Examples 1–3 above work in the general case. Example 6 at the end of this section shows a graph with no horizontal asymptote. This leads to the following conclusion.

If the numerator of a rational function $f(x)$ is of *smaller* degree than the denominator, then the x-axis (the line $y = 0$) is the horizontal asymptote of the graph. If the numerator and denominator are of the *same* degree, say, $f(x) = \dfrac{ax^n + \cdots}{cx^n + \cdots}$, then the line $y = \dfrac{a}{c}$ is the horizontal asymptote. If the numerator is of a *larger* degree than the denominator, the graph has no horizontal asymptote.*

Applications

Rational functions have a variety of applications, some of which are explored next.

> **EXAMPLE 4**   **Pollution Control** In many situations involving environmental pollution, much of the pollutant can be removed from the air or water at a fairly reasonable cost, but the small amounts of the pollutant that remain can be very expensive to remove.
>
>   Cost as a function of the percentage of pollutant removed from the environment can be calculated for various percentages of removal, with a curve fitted through the resulting data points. This curve then leads to a function that approximates the situation. Rational functions often are a good choice for these **cost–benefit functions.**
>
>   For example, suppose a cost–benefit function is given by
>
> $$f(x) = \frac{18x}{106 - x},$$
>
> where $f(x)$ is the cost (in thousands of dollars) of removing x percent of a certain pollutant. The domain of x is the set of all numbers from 0 to 100, inclusive; any amount of pollutant from 0% to 100% can be removed. To remove 100% of the pollutant here would cost
>
> $$f(100) = \frac{18(100)}{106 - 100} = 300,$$
>
> or $300,000. Check that 95% of the pollutant can be removed for about $155,000, 90% for about $101,000, and 80% for about $55,000, as shown in Figure 3.53 (in which the displayed y-coordinates are rounded to the nearest integer). ✔ 4

✓ **Checkpoint 4**

Using the function in Example 4, find the cost to remove the following percentages of pollutants.

(a) 70%

(b) 85%

(c) 98%

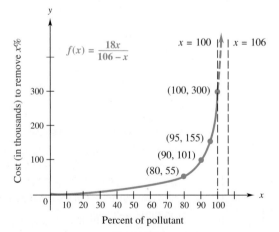

Figure 3.53

---

*In this case, the graph may have nonhorizontal lines or other curves as asymptotes; see Exercises 31 and 32 at the end of this section for examples.

In management, **product-exchange functions** give the relationship between quantities of two items that can be produced by the same machine or factory. For example, an oil refinery can produce gasoline, heating oil, or a combination of the two; a winery can produce red wine, white wine, or a combination of the two. The next example discusses a product-exchange function.

EXAMPLE 5 **Agricultural Production** The product-exchange function for the Fruits of the Earth Winery for red wine x and white wine y, in number of cases, is

$$y = \frac{150,000 - 75x}{1200 + x}.$$

Graph the function and find the maximum quantity of each kind of wine that can be produced.

Solution Only nonnegative values of x and y make sense in this situation, so we graph the function in the first quadrant (Figure 3.54). Note that the y-intercept of the graph (found by setting $x = 0$) is 125 and the x-intercept (found by setting $y = 0$ and solving for x) is 2000. Since we are interested only in the portion of the graph in Quadrant I, we can find a few more points in that quadrant and complete the graph, as shown in Figure 3.54.

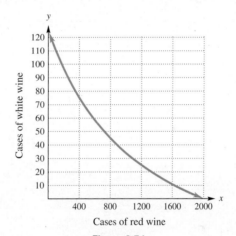

Figure 3.54

The maximum value of y occurs when $x = 0$, so the maximum amount of white wine that can be produced is 125 cases, as given by the y-intercept. The x-intercept gives the maximum amount of red wine that can be produced: 2000 cases. ✔ 5

✓ **Checkpoint 5**

Rework Example 5 with the product-exchange function

$$y = \frac{70,000 - 10x}{70 + x}$$

to find the maximum amount of each wine that can be produced.

EXAMPLE 6 **Business** A retailer buys 2500 specialty lightbulbs from a distributor each year. In addition to the cost of each bulb, there is a fee for each order, so she wants to order as few times as possible. However, storage costs are higher when there are fewer orders (and hence more bulbs per order to store). Past experience shows that the total annual cost (for the bulbs, ordering fees, and storage costs) is given by the rational function.

$$C(x) = \frac{.98x^2 + 1200x + 22,000}{x},$$

where x is the number of bulbs ordered each time. How many bulbs should be ordered each time in order to have the smallest possible cost?

Solution Graph the cost function $C(x)$ in a window with $0 \le x \le 2500$ (because the retailer cannot order a negative number of bulbs and needs only 2500 for the year).

For each point on the graph in Figure 3.55,

the x-coordinate is the number of bulbs ordered each time;
the y-coordinate is the annual cost when x bulbs are ordered each time.

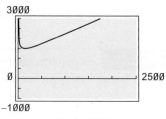

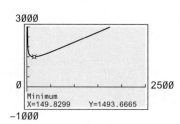

Figure 3.55   **Figure 3.56**

Use the minimum finder on a graphing calculator (look in the CALC menu) to find the point with the smallest y-coordinate, which is approximately $(149.83, 1493.67)$, as shown in Figure 3.56. Since the retailer cannot order part of a lightbulb, she should order 150 bulbs each time, for an approximate annual cost of $1494.

3.6 Exercises

Graph each function. Give the equations of the vertical and horizontal asymptotes. (See Examples 1–3.)

1. $f(x) = \dfrac{1}{x + 5}$

2. $g(x) = \dfrac{-7}{x - 6}$

3. $f(x) = \dfrac{-3}{2x + 5}$

4. $h(x) = \dfrac{-4}{2 - x}$

5. $f(x) = \dfrac{3x}{x - 1}$

6. $g(x) = \dfrac{x - 2}{x}$

7. $f(x) = \dfrac{x - 3}{x + 5}$

8. $f(x) = \dfrac{x + 1}{x - 4}$

9. $f(x) = \dfrac{2 - x}{x - 3}$

10. $g(x) = \dfrac{3x - 2}{x + 3}$

11. $f(x) = \dfrac{4x - 8}{8x + 1}$

12. $f(x) = \dfrac{3x + 2}{2x + 4}$

13. $h(x) = \dfrac{x + 1}{x^2 + 2x - 8}$

14. $g(x) = \dfrac{1}{x(x + 2)^2}$

15. $f(x) = \dfrac{x^2 + 4}{x^2 - 4}$

16. $f(x) = \dfrac{x - 1}{x^2 - 2x - 3}$

Find the equations of the vertical asymptotes of each of the given rational functions. (See Example 3.)

17. $g(x) = \dfrac{x + 2}{x^2 - 1}$

18. $f(x) = \dfrac{x - 3}{x^2 + x - 2}$

19. $g(x) = \dfrac{x^2 + 2x}{x^2 - 4x - 5}$

20. $f(x) = \dfrac{x^2 - 2x - 4}{x^3 - 2x^2 + x}$

Average Cost *For Exercises 21 and 22, recall that if the cost of producing x units is $C(x)$ then the average cost function (defined Section 3.3) is $\overline{C}(x) = \dfrac{C(x)}{x}$.*

21. An artist who makes handmade earrings has fixed costs of $100 for a table at the Portland Art Walk. The marginal cost (the cost of producing one additional pair of earrings) is $5 per pair.

(a) Find the linear cost function $C(x)$.

(b) Find the average cost function $\overline{C}(x)$.

(c) Find $\overline{C}(5)$, and interpret your answer.

(d) Find $\overline{C}(50)$, and interpret your answer.

(e) Find the horizontal asymptote of $\overline{C}(x)$, and explain what it means in practical terms.

22. A taxi driver must pay a daily fee of $75 to the cab company and has marginal costs of $.50 per mile.

(a) Find the daily cost function $C(x)$.

(d) Find the average cost function $\overline{C}(x)$.

(c) Find $\overline{C}(50)$, and interpret your answer.

(d) Find $\overline{C}(250)$, and interpret your answer.

(e) Find the horizontal asymptote of $\overline{C}(x)$, and explain what it means in practical terms.

Work these problems. (See Example 2.)

23. **Biological Modeling** The function

$$f(x) = \frac{5x}{2 + x}$$

is used in biology to give the growth rate (as a percentage) of a certain population in the presence of a quantity of x units of food (where $x \geq 0$). This is an example of the concept called Michaelis–Menten kinetics.[*]

(a) Find $f(0), f(2), f(8),$ and $f(78),$ and explain what your answers mean.

[*]W. W. Chen, M. Neipel, and P. K. Sorger, "Classic and Contemporary Approaches to Modeling Biochemical Reactions," *Genes Dev* 24, no. 17 (2010): 1861–1875.

(b) Find the horizontal asymptote.

(c) Does the graph have any vertical asymptotes on the domain $x \geq 0$?

(d) Graph the function for $x \geq 0$.

(e) What do you think the horizontal asymptote represents?

24. NASA The failure of several O-rings in field joints was the cause of the fatal crash of the *Challenger* space shuttle. NASA data from 24 successful launches prior to *Challenger* suggested that O-ring failure was related to launch temperature by a function similar to

$$N(t) = \frac{600 - 7t}{4t - 100} \quad (50 \leq t \leq 85),$$

where t is the temperature (in °F) at launch and N is the approximate number of O-rings that fail. Assume that this function accurately models the number of O-ring failures that would occur at lower launch temperatures (an assumption NASA did not make).

(a) Does $N(t)$ have a vertical asymptote? At what value of t does it occur?

(b) Without actually graphing the function, what would you conjecture that the graph would look like just to the right of the vertical asymptote? What does this suggest about the number of O-ring failures that might be expected near that temperature? (The temperature at the *Challenger* launching was 31°.)

(c) Confirm your conjecture by graphing $N(t)$ between the vertical asymptote and $t = 85$.

25. Pollution Control Suppose a cost–benefit model is given by

$$f(x) = \frac{4.3x}{100 - x},$$

where $f(x)$ is the cost, in thousands of dollars, of removing x percent of a given pollutant. Find the cost of removing each of the given percentages of pollutants.

(a) 50%   **(b)** 70%   **(c)** 80%

(d) 90%   **(e)** 95%   **(f)** 98%

(g) 99%

(h) Is it possible, according to this model, to remove *all* the pollutant?

(i) Graph the function.

26. Pollution Control Suppose a cost–benefit model is given by

$$f(x) = \frac{6.2x}{112 - x},$$

where $f(x)$ is the cost, in thousands of dollars, of removing x percent of a certain pollutant. Find the cost of removing the given percentages of pollutants.

(a) 0%   **(b)** 50%   **(c)** 80%

(d) 90%   **(e)** 95%   **(f)** 99%

(g) 100%   **(h)** Graph the function.

Business *Sketch the portion of the graph in Quadrant I of each of the functions defined in Exercises 27 and 28, and then find the x-intercept and y-intercept in order to estimate the maximum quantities of each product that can be produced. (See Example 5.)*

27. The product-exchange function for gasoline x and heating oil y, in hundreds of gallons per day, is

$$y = \frac{125,000 - 25x}{125 + 2x}.$$

28. A drug factory found that the product-exchange function for a red tranquilizer x and a blue tranquilizer y is

$$y = \frac{900,000,000 - 30,000x}{x + 90,000}.$$

Use a graphing calculator to do Exercises 29–32. (See Example 6.)

29. Finance Another model of a Laffer curve (see Exercise 27 of Section 3.5) is given by

$$f(x) = \frac{300x - 3x^2}{10x + 200},$$

where $f(x)$ is government revenue (in billions of dollars) from a tax rate of x percent. Find the revenue from the given tax rates.

(a) 16%   **(b)** 25%   **(c)** 40%

(d) 55%   **(e)** Graph $f(x)$.

(f) Use the maximum finder to find the tax rate that produces maximum revenue. What is the maximum revenue?

30. Business When no more than 110 units are produced, the cost of producing x units is given by

$$C(x) = .2x^3 - 25x^2 + 1531x + 25,000.$$

Use the minimum finder to determine the number of units that should be produced in order to have the lowest possible average cost.

31. (a) Graph $f(x) = \dfrac{x^3 + 3x^2 + x + 1}{x^2 + 2x + 1}$.

(b) Does the graph appear to have a horizontal asymptote? Does the graph appear to have some nonhorizontal straight line as an asymptote?

(c) Graph $f(x)$ and the line $y = x + 1$ on the same screen. Does this line appear to be an asymptote of the graph of $f(x)$?

32. (a) Graph $g(x) = \dfrac{x^3 - 2}{x - 1}$ in the window with $-5 \leq x \leq 5$ and $-6 \leq y \leq 12$.

(b) Graph $g(x)$ and the parabola $y = x^2 + x + 1$ on the same screen. How do the two graphs compare when $|x| \geq 2$?

✓ Checkpoint Answers

1. (a)

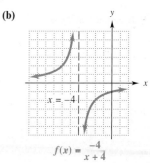

$$f(x) = \frac{3}{5 - x} \qquad\qquad f(x) = \frac{-4}{x + 4}$$

2. (a) 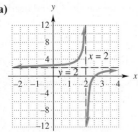 **(b)**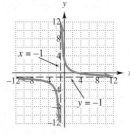

3. (a) Vertical, $x = -5$; horizontal, $y = 3$

(b) Vertical, $x = -2$ and $x = 2$; horizontal, $y = -1$

4. (a) \$35,000

(b) About \$73,000

(c) About \$221,000

5. 7000 cases of red, 1000 cases of white

CHAPTER 3 Summary and Review

Key Terms and Symbols

3.1 function
domain
range
functional notation
piecewise-defined function

3.2 graph
linear function
piecewise linear function
absolute-value function
greatest-integer function
step function

graph reading
vertical-line test
3.3 fixed costs
variable cost
average cost
linear depreciation
rate of change
marginal cost
marginal revenue
linear cost function
linear revenue function

break-even point
supply and demand curves
shortage of supply
surplus of supply
equilibrium point
equilibrium price
equilibrium quantity
3.4 quadratic function
parabola
vertex
axis of the parabola

quadratic model
3.5 polynomial function
graph of $f(x) = ax^n$
properties of polynomial
graphs
polynomial models
3.6 rational function
linear rational function
vertical asymptote
horizontal asymptote

Chapter 3 Key Concepts

Functions

A **function** consists of a set of inputs called the **domain,** a set of outputs called the **range,** and a rule by which each input determines exactly one output.

If a vertical line intersects a graph in more than one point, the graph is not that of a function.

Linear Functions

A **linear cost function** has equation $C(x) = mx + b$, where m is the **marginal cost** (the cost of producing one more item) and b is the **fixed cost.**

If $p = f(q)$ gives the price per unit when q units can be supplied and $p = g(q)$ gives the price per unit when q units are demanded, then the **equilibrium quantity** is the q-value such that $f(q) = g(q)$, and the **equilibrium price** is the corresponding p-value (on either the supply curve or demand curve because they are equal at that point). The **equilibrium point** is the point whose first coordinate is the equilibrium quantity and whose second coordinate is the equilibrium price.

Quadratic Functions

The **quadratic function** defined by $f(x) = a(x - h)^2 + k$ (with $a \neq 0$) has a graph that is a **parabola** with vertex (h, k) and axis $x = h$. The parabola opens upward if $a > 0$ and downward if $a < 0$.

If the equation is in the form $f(x) = ax^2 + bx + c$, the vertex is $\left(-\dfrac{b}{2a}, f\left(-\dfrac{b}{2a}\right)\right)$.

Polynomial Functions

When $|x|$ is large, the graph of a **polynomial function** resembles the graph of its highest-degree term ax^n. The graph of $f(x) = ax^n$ is described on page 160.

On the graph of a polynomial function of degree n,

the total number of peaks and valleys is at most $n - 1$;

the total number of x-intercepts is at most n.

Rational Functions

If a number c makes the denominator of a **rational function** 0, but the numerator nonzero, then the line $x = c$ is a **vertical asymptote** of the graph.

Whenever the values of $f(x)$ approach, but do not equal, some number k as $|x|$ gets larger and larger, the line $y = k$ is a **horizontal asymptote** of the graph.

If the numerator of a rational function is of *smaller* degree than the denominator, then the *x*-axis ($y = 0$) is the horizontal asymptote of the graph.

If the numerator and denominator of a rational function are of the *same* degree, say, $f(x) = \dfrac{ax^n + \cdots}{cx^n + \cdots}$, then the line $y = \dfrac{a}{c}$ is the horizontal asymptote of the graph.

If the numerator of a rational function is of a *larger* degree than the denominator, then the graph has no horizontal asymptote.

Chapter 3 Review Exercises

In Exercises 1–6, state whether or not the given rule defines y as a function x.

1.

x	3	2	1	0	1	2
y	8	5	2	0	-2	-5

2.

x	2	1	0	-1	-2
y	5	3	1	-1	-3

3. $y = \sqrt{x}$

4. $x = |y|$

5. $x = y^2 + 1$

6. $y = 5x - 2$

For the functions in Exercises 7–10, find

(a) $f(6)$;   (b) $f(-2)$;   (c) $f(p)$;   (d) $f(r + 1)$.

7. $f(x) = 4x - 1$

8. $f(x) = 3 - 4x$

9. $f(x) = -x^2 + 2x - 4$

10. $f(x) = 8 - x - x^2$

11. Let $f(x) = 5x - 3$ and $g(x) = -x^2 + 4x$. Find each of the following:

(a) $f(-2)$ (b) $g(3)$ (c) $g(-k)$

(d) $g(3m)$ (e) $g(k - 5)$ (f) $f(3 - p)$

12. Let $f(x) = x^2 + x + 1$. Find each of the following:

(a) $f(3)$ (b) $f(1)$ (c) $f(4)$

(d) Based on your answers in parts (a)–(c), is it true that $f(a + b) = f(a) + f(b)$ for all real numbers a and b?

Graph the functions in Exercises 13–24.

13. $f(x) = |x| - 3$

14. $f(x) = -|x| - 2$

15. $f(x) = -|x + 1| + 3$

16. $f(x) = 2|x - 3| - 4$

17. $f(x) = [x - 3]$

18. $f(x) = \left[\dfrac{1}{2}x - 2\right]$

19. $f(x) = \begin{cases} -4x + 2 & \text{if } x \le 1 \\ 3x - 5 & \text{If } x > 1 \end{cases}$

20. $f(x) = \begin{cases} 3x + 1 & \text{if } x < 2 \\ -x + 4 & \text{if } x \ge 2 \end{cases}$

21. $f(x) = \begin{cases} x^2 & \text{if } x \le 0 \\ x & \text{if } x > 0 \end{cases}$

22. $f(x) = \begin{cases} |x| & \text{if } x < 3 \\ 6 - x & \text{if } x \ge 3 \end{cases}$

23. $h(x) = \sqrt{x} + 2$

24. $f(x) = \sqrt{x^2}$

25. Business Let $f(x)$ be a function that gives the cost to rent a power washer for x hours. The cost is a flat \$45 for renting the washer, plus \$20 per day or fraction of a day for using the washer.

(a) Graph $f(x)$.   (b) Give the domain and range.

(c) You want to rent the washer, but can spend no more than \$90. What is the maximum number of days you can use it?

26. Business A tree removal service assesses a \$400 fee and then charges \$80 per hour or fraction of an hour for the time on an owner's property. Let $f(x)$ be the total cost for x hours.

(a) Is \$750 enough for 5 hours work?

(b) Graph $f(x)$.   (c) Give the domain and range.

27. Alcohol Consumption The following graph shows the percentage of twelfth graders in the United States by gender who had a drink of alcohol in the past 30 days. (Data from: Centers for Disease Control and Prevention.)

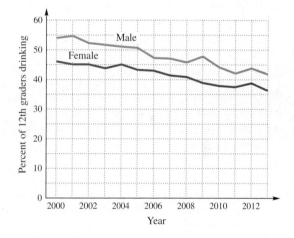

(a) What is the general trend for alcohol use by female twelfth graders during this time period?

(b) According to the graph, approximately what percentage of female twelfth graders drank alcohol in 2004?

(c) Let $x = 0$ correspond to the year 2000. Approximate a linear function for the percentage of female twelfth graders who drank alcohol using the points (4, 45) and (8, 41).

(d) What does this model estimate for this percentage in 2017?

(e) If the trend continues, when will this percentage drop to 33%?

28. Alcohol Consumption Answer the following questions by referring to the graph in Exercise 27, which shows the percentage of twelfth graders in the United States by gender who had a drink in the past 30 days.

(a) What is the general trend for alcohol use by male twelfth graders during this time period?

(b) According to the graph, approximately what percentage of male twelfth graders drank alcohol in 2001?

(c) Let $x = 0$ correspond to the year 2000. Approximate a linear function for the percentage of male twelfth graders who drank alcohol using the points $(1, 55)$ and $(12, 44)$.

(d) What does this model estimate for this percentage in 2017?

(e) If the trend continues, when will this percentage drop to 40%?

Business *In Exercises 29–32, find the following:*

(a) *the linear cost function;*

(b) *the marginal cost;*

(c) *the average cost per unit to produce 100 units.*

29. Eight units cost $300; fixed cost is $60.

30. Fixed cost is $2000; 36 units cost $8480.

31. Twelve units cost $445; 50 units cost $1585.

32. Thirty units cost $1500; 120 units cost $5640.

33. Business The cost of producing x ink cartridges for a printer is given by $C(x) = 24x + 18,000$. Each cartridge can be sold for $28.

(a) What are the fixed costs?

(b) Find the revenue function.

(c) Find the break-even point.

(d) If the company sells exactly the number of cartridges needed to break even, what is its revenue?

34. Business The cost of producing x laser printers is given by $C(x) = 325x + 2,300,000$. Each printer can be sold for $450.

(a) What are the fixed costs?

(b) Find the revenue function.

(c) Find the break-even point.

(d) If the company sells exactly the number of printers needed to break even, what is its revenue?

35. Business Suppose the demand and price for the HBO cable channel are related by $p = -.5q + 30.95$, where p is the monthly price in dollars and q is measured in millions of subscribers. If the price and supply are related by $p = .3q + 2.15$, what are the equilibrium quantity and price?

36. Business Suppose the supply and price for prescription-strength Tylenol are related by $p = .0015q + 1$, where p is the price (in dollars) of a 30-day prescription. If the demand is related to price by $p = -.0025q + 64.36$, what are the equilibrium quantity and price?

Without graphing, determine whether each of the following parabolas opens upward or downward, and find its vertex.

37. $f(x) = 3(x - 2)^2 + 6$ **38.** $f(x) = 2(x + 3)^2 - 5$

39. $g(x) = -4(x + 1)^2 + 8$ **40.** $g(x) = -5(x - 4)^2 - 6$

Graph each of the following quadratic functions, and label its vertex.

41. $f(x) = x^2 - 9$ **42.** $f(x) = 5 - 2x^2$

43. $f(x) = x^2 + 2x - 6$ **44.** $f(x) = -x^2 + 8x - 1$

45. $f(x) = -x^2 - 6x + 5$ **46.** $f(x) = 5x^2 + 20x - 2$

47. $f(x) = 2x^2 - 12x + 10$ **48.** $f(x) = -3x^2 - 12x - 2$

Determine whether the functions in Exercises 49–52 have a minimum or a maximum value, and find that value.

49. $f(x) = x^2 + 6x - 2$ **50.** $f(x) = x^2 + 4x + 5$

51. $g(x) = -4x^2 + 8x + 3$ **52.** $g(x) = -3x^2 - 6x + 3$

Solve each problem.

53. Youth Population At the start of this millennium, the youth population in the United States was rising, but later it began to fall. The function

$$f(x) = -29x^2 + 522x + 71741$$

models the number of people under 18 years old (in thousands) in year x, where $x = 0$ corresponds to the year 2000. (Data from: U.S. Census Bureau.)

(a) According to this model, what was the youth population in 2000?

(b) Find the vertex. Does a maximum or minimum occur at the vertex?

(c) According to this model, what year did the youth population peak?

(d) What was the maximum youth population with this model?

54. Population of China The declining birth rate in China has demographers believing that the population of China will soon peak. A model for the population of China (in millions) is

$$f(x) = -.2x^2 + 10.4x + 1263,$$

where $x = 0$ corresponds to the year 2000. (Data from: U.S. Census Bureau.)

(a) According to this model, what was the population of China in 2000?

(b) Find the vertex. Does a maximum or minimum occur at the vertex?

(c) According to this model, in what year does the population of China peak?

(d) What is maximum projected population?

55. Student Loans Interest rates for subsidized Stafford loans can be approximated for the years 2007 to 2014 by the function

$$f(x) = .15x^2 - 3.6x + 25,$$

where $x = 7$ corresponds to the year 2007. According to this model, during what year were interest rates lowest, and what was the lowest interest rate? (Data from: www.studentaid.ed.gov.)

56. Natural Gas Pricing The price of European natural gas (in dollars per million British thermal units [BTUs]) can be approximated for the years 2010 to 2015 by the function

$$f(x) = -.65x^2 + 16.9x - 98,$$

where $x = 10$ corresponds to the year 2010. According to this model, during what year were European natural gas prices highest, and what was the highest price? (Data from: www.worldbank.org.)

57. Netflix Revenue Netflix Inc. reported revenue (in millions of dollars) as shown in the following table. (Data from: www.morningstar.com.)

Year	Revenue
2006	997
2007	1205
2008	1365
2009	1670
2010	2163
2011	3205
2012	3609
2013	4375
2014	5505
2015	6780

(a) Let $x = 6$ correspond to the year 2006. Find a quadratic function in the form $f(x) = a(x - h)^2 + k$ that models these data using $(6, 997)$ as the vertex and the data for 2010.

(b) Estimate the revenue in 2016.

58. Netflix Costs Netflix Inc. reported costs (in millions of dollars) as shown in the following table. (Data from: www.morningstar.com.)

Year	Costs
2006	627
2007	786
2008	910
2009	1079
2010	1357
2011	2040
2012	2626
2013	3083
2014	3753
2015	4591

(a) Let $x = 6$ correspond to the year 2006. Find a quadratic function in the form $f(x) = a(x - h)^2 + k$ that models these data using $(6, 627)$ as the vertex and the data for 2010.

(b) Estimate the costs in 2016.

59. Use quadratic regression and the data from Exercise 57 to find a function $R(x)$ that models these data.

(a) Give the function (round coefficients to the nearest integer).

(b) Graph the function with the data points. Does the model fit the data well?

(c) Find the estimated revenue with this model for the year 2016.

60. Use quadratic regression and the data from Exercise 58 to find a function $C(x)$ that models these data.

(a) Give the function (round coefficients to the nearest integer).

(b) Graph the function with the data points. Does the model fit the data well?

(c) Find the estimated costs with this model for the year 2016.

Graph each of the following polynomial functions.

61. $g(x) = x^3 - 4x$

62. $f(x) = x^4 - 5$

63. $f(x) = x(x - 4)(x + 1)$

64. $f(x) = (x - 1)(x + 2)(x - 3)$

65. $f(x) = x^4 - 5x^2 - 6$

66. $f(x) = x^4 - 7x^2 - 8$

Use a graphing calculator to do Exercises 67–70.

67. Forward Industries Revenue The following table shows the revenue (in millions of dollars) for Forward Industries Inc., a company that designs protective solutions for handheld electronic devices. (Data from: www.morningstar.com.)

Year	Revenue
2006	31
2007	22
2008	20
2009	17
2010	19
2011	23
2012	29
2013	31
2014	33
2015	30

(a) Let $x = 6$ correspond to the year 2006. Use cubic regression to find a third-order polynomial function $R(x)$ that models these data (round coefficients to three decimal places).

(b) Graph $R(x)$ on the same screen as the data points. Does the graph appear to fit the data well?

(c) What does the model estimate for the revenue in 2012? Is it close to the actual revenue shown in the table?

68. Forward Industries Costs The following table shows the costs (in millions of dollars) for Forward Industries Inc. (Data from: www.morningstar.com.)

Year	Costs
2006	23
2007	17
2008	16
2009	14
2010	15
2011	18
2012	25
2013	25
2014	27
2015	24

(a) Use cubic regression to find a third-order polynomial function $C(x)$ that models these data (round coefficients to three decimal places).

(b) Graph $C(x)$ on the same screen as the data points. Does the graph appear to fit the data well?

(c) What does the model estimate for the costs in 2014? Is it close to the actual costs shown in the table?

69. Forward Industries Profit Use the revenue function $R(x)$ found in Exercise 67 and the cost function $C(x)$ found in Exercise 68 to find the profit function $P(x)$ for Forward Industries, and estimate the profit in the year 2016.

70. Netflix Profit Use the revenue function $R(x)$ found in Exercise 57 and the cost function $C(x)$ found in Exercise 58 to find the profit function $P(x)$ for Netflix, and estimate the profit in the year 2016.

List the vertical and horizontal asymptotes of each function, and sketch its graph.

71. $f(x) = \dfrac{1}{x - 3}$

72. $f(x) = \dfrac{-2}{x + 4}$

73. $f(x) = \dfrac{-3}{2x - 4}$

74. $f(x) = \dfrac{5}{3x + 7}$

75. $g(x) = \dfrac{5x - 2}{4x^2 - 4x - 3}$

76. $g(x) = \dfrac{x^2}{x^2 - 1}$

77. Business The average cost per carton of producing x cartons of cocoa is given by

$$A(x) = \frac{650}{2x + 40}.$$

Find the average cost per carton to make the given number of cartons.

(a) 10 cartons   (b) 50 cartons   (c) 70 cartons

(d) 100 cartons   (e) Graph $A(x)$.

78. Business The cost and revenue functions (in dollars) for a frozen-yogurt shop are given by

$$C(x) = \frac{400x + 400}{x + 4} \quad \text{and} \quad R(x) = 100x,$$

where x is measured in hundreds of units.

(a) Graph $C(x)$ and $R(x)$ on the same set of axes.

(b) What is the break-even point for this shop?

(c) If the profit function is given by $P(x)$, does $P(1)$ represent a profit or a loss?

(d) Does $P(4)$ represent a profit or a loss?

79. Business The supply and demand functions for the yogurt shop in Exercise 78 are

$$\text{supply: } p = \frac{q^2}{4} + 25$$

and

$$\text{demand: } p = \frac{500}{q},$$

where p is the price in dollars for q hundred units of yogurt. Graph both functions on the same axes, and from the graph, estimate the equilibrium point.

80. Business A cost–benefit curve for pollution control is given by

$$y = \frac{9.2x}{106 - x},$$

where y is the cost, in thousands of dollars, of removing x percent of a specific industrial pollutant. Find y for each of the given values of x.

(a) $x = 50$ (b) $x = 98$

(c) What percent of the pollutant can be removed for $22,000?

Case Study 3 Maximizing Profit

For many people, visits to a coin-operated laundry are a regular part of life. In 2017, this industry generated nearly $5 billion in gross revenue from about 29,500 locations nationwide, the majority of which are individually owned and operated. (Data from: Coin Laundry Association.) Suppose that one business owner raised the price of a washing machine load from $2.50 to $2.75 and found that daily demand dropped from 700 loads to 650 loads. The business owner noted that revenue actually increased but wondered about bottom-line profits. The business owner is interested in learning about both the *maximum revenue* and the *maximum profit*. We will investigate both in this case study.

As usual, we express the price, p, as a function of the quantity demanded, x. We summarize the given data in the following table.

Washing Machine Operations

Daily Demand, x	Price per Load, p
650	$2.75
700	$2.50

Assuming linear demand, we can use these data to find the slope of the demand function:

$$m = \frac{\Delta p}{\Delta x} = \frac{2.50 - 2.75}{700 - 650} = \frac{-.25}{50} = -.005.$$

Using point–slope form, we find the equation of the linear demand function:

$$y - y_1 = m(x - x_1) \qquad \text{Point–slope form using } x \text{ and } y$$
$$p - p_1 = m(x - x_1) \qquad \text{Point–slope form using } x \text{ and } p$$
$$p - 2.75 = -.005(x - 650) \qquad \text{Substitute 2.75 for } p_1, -.005$$
$$\text{for } m, \text{ and 650 for } x_1.$$
$$p - 2.75 = -.005x + 3.25 \qquad \text{Distribute the } -.005.$$
$$p = -.005x + 6 \qquad \text{Add 2.75 to both sides.}$$

We find the revenue function by multiplying the quantity demanded by the price per item:

$$R(x) = xp$$
$$= x(-.005x + 6)$$
$$= -.005x^2 + 6x.$$

The graph of the revenue function is a downward-opening parabola, so the maximum occurs at the vertex (h, k). The x-value of the vertex is

$$h = \frac{-b}{2a} = \frac{-6}{2(-.005)} = \frac{-6}{-.01} = 600.$$

The corresponding y-value of the vertex is

$$k = R(h) = -.005(600)^2 + 6(600) = 1800.$$

The vertex of the revenue function tells us that the maximum daily revenue is $1800, and it occurs when the demand for washing machines is 600 loads. The price per load corresponding to a demand of $x = 600$ is $p = -.005(600) + 6 = \$3$ per load.

   To investigate the maximum profit, we ask the business owner about operating costs. The owner reports fixed costs of $1040 per day and variable costs of $.50 per washing machine load. Let's assume that there are about an equal number of washer and dryer loads, so we will attribute half of the fixed costs

to washing machine operations. Therefore, the cost function for washing machines is

$$C(x) = 520 + .50x.$$

This leads to the profit function

$$\begin{aligned} P(x) &= R(x) - C(x) \\ &= (-.005x^2 + 6x) - (520 + .50x) \\ &= -.005x^2 + 5.50x - 520. \end{aligned}$$

Exercises

1. Find the maximum profit and the number of washing machine loads that leads to the maximum profit.

2. Is the quantity of washing machine loads the same for the maximum revenue and the maximum profit? If not, which quantity should the owner prefer?

3. Based on this information, what price should the owner charge?

Extended Project

Suppose the owner of the laundry has hired your consulting firm to determine an optimal price to charge per dryer load given the data below.

Dryer Operations

Daily Demand, x	Price per Load, p
600	$1.25
700	$1.00

Assume half of the fixed costs are associated with dryer operations, and that variable costs are $.25 per dryer load. Write a report explaining your calculations and findings. Be sure to include the following:

(a) the number of loads corresponding to the maximum profit from dryers;

(b) the corresponding maximum profit from dryers; and

(c) the price per load to charge that corresponds to the maximum dryer profit.

Exponential and Logarithmic Functions

4

CHAPTER

Population growth (of humans, fish, bacteria, etc.), compound interest, radioactive decay, and a host of other phenomena can be described by exponential functions. The growth in the number of subscribers to a new tech service such as Netflix often follows an exponential pattern. Archeologists sometimes use carbon-14 dating to determine the approximate age of an artifact such as a mummy. This procedure involves using logarithms to solve an exponential equation. The Richter scale for measuring the magnitude of an earthquake is also a logarithmic function.

Exponential and logarithmic functions play a key role in management, economics, the social and physical sciences, and engineering. We begin with exponential growth and exponential decay functions.

4.1 Exponential Functions

In polynomial functions such as $f(x) = x^2 + 5x - 4$, the variable is raised to various constant exponents. In **exponential functions,** such as

$$f(x) = 10^x, \quad g(x) = 750(1.05^x), \quad h(x) = 3^{.6x}, \quad \text{and} \quad k(x) = 2^{-x^2},$$

the variable is in the exponent and the **base** is a positive constant. We begin with the simplest type of exponential function, whose rule is of the form $f(x) = a^x$, with $a > 0$.

EXAMPLE 1   Graph $f(x) = 2^x$, and estimate the height of the graph when $x = 50$.

Solution Either use a graphing calculator or graph by hand: Make a table of values, plot the corresponding points, and join them by a smooth curve, as in Figure 4.1. The graph has y-intercept 1 and rises steeply to the right. Note that the graph gets very close to the x-axis on the left, but always lies *above* the axis (because *every* power of 2 is positive).

x	y
-3	$1/8$
-2	$1/4$
-1	$1/2$
0	1
1	2
2	4
3	8

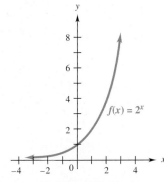

Figure 4.1

The graph illustrates **exponential growth,** which is far more explosive than polynomial growth. At $x = 50$, the graph would be 2^{50} units high. Since there are approximately 6 units to the inch in Figure 4.1, and since there are 12 inches to the foot and 5280 feet to the mile, the height of the graph at $x = 50$ would be approximately

$$\frac{2^{50}}{6 \times 12 \times 5280} \approx 2{,}961{,}647{,}482 \text{ miles!}$$

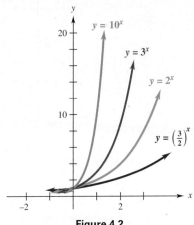

Figure 4.2

When $a > 1$, the graph of the exponential function $h(x) = a^x$ has the same basic shape as the graph of $f(x) = 2^x$, as illustrated in Figure 4.2 and summarized in the next box.

✓**Checkpoint 1**

(a) Fill in this table:

x	$g(x) = 3^x$
-3	
-2	
-1	
0	
1	
2	
3	

(b) Sketch the graph of $g(x) = 3^x$.

Answers to Checkpoint exercises are found at the end of the section.

When $a > 1$, the function $f(x) = a^x$ has the set of all real numbers as its domain. Its graph has the shape shown on the following page and all five of the properties listed below.

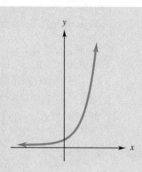

1. The graph is above the x-axis.
2. The y-intercept is 1.
3. The graph climbs steeply to the right.
4. The negative x-axis is a horizontal asymptote.
5. The larger the base a, the more steeply the graph rises to the right.

EXAMPLE 2   Consider the function $g(x) = 2^{-x}$.

(a) Rewrite the rule of g so that no minus signs appear in it.

Solution By the definition of negative exponents,

$$g(x) = 2^{-x} = \frac{1}{2^x} = \left(\frac{1}{2}\right)^x.$$

(b) Graph $g(x)$.

Solution Either use a graphing calculator or graph by hand in the usual way, as shown in Figure 4.3.

x	$y = 2^{-x}$
-3	8
-2	4
-1	2
0	1
1	$1/2$
2	$1/4$
3	$1/8$

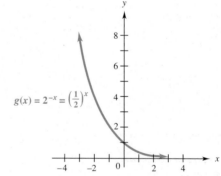

Figure 4.3

✓ Checkpoint 2

Graph $h(x) = (1/3)^x$.

The graph falls sharply to the right, but never touches the x-axis, because every power of $\frac{1}{2}$ is positive. This is an example of **exponential decay.** ✓₂

When $0 < a < 1$, the graph of $k(x) = a^x$ has the same basic shape as the graph of $g(x) = (1/2)^x$, as illustrated in Figure 4.4 and summarized in the next box, on the following page.

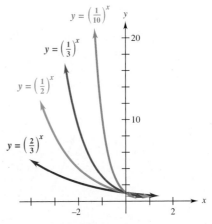

Figure 4.4

When $0 < a < 1$, the function $f(x) = a^x$ has the set of all real numbers as its domain. Its graph has the shape shown here and all five of the properties listed.

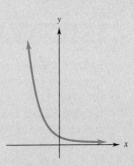

1. The graph is above the x-axis.
2. The y-intercept is 1.
3. The graph falls sharply to the right.
4. The positive x-axis is a horizontal asymptote.
5. The smaller the base a, the more steeply the graph falls to the right.

✓**Checkpoint 3** ▦

Use a graphing calculator to graph $f(x) = 4^x$ and $g(x) = \left(\frac{1}{4}\right)^x$ on the same screen.

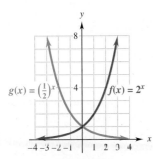

Figure 4.5

EXAMPLE 3   In each case, graph $f(x)$ and $g(x)$ on the same set of axes and explain how the graphs are related.

(a) $f(x) = 2^x$ and $g(x) = (1/2)^x$

Solution The graphs of f and g are shown in Figures 4.1 and 4.3, above. Placing them on the same set of axes, we obtain Figure 4.5. It shows that the graph of $g(x) = (1/2)^x$ is the mirror image of the graph of $f(x) = 2^x$, with the y-axis as the mirror. ✓₃

(b) $f(x) = 3^{1-x}$ and $g(x) = 3^{-x}$

Solution Choose values of x that make the exponent positive, zero, and negative, and plot the corresponding points. The graphs are shown in Figure 4.6, on the following page. The graph of $f(x) = 3^{1-x}$ has the same shape as the graph of $g(x) = 3^{-x}$, but is shifted 1 unit to the right, making the y-intercept $(0, 3)$ rather than $(0, 1)$.

(c) $f(x) = 2^{.6x}$ and $g(x) = 2^x$

Solution Comparing the graphs of $f(x) = 2^{.6x}$ and $g(x) = 2^x$ in Figure 4.7, we see that the graphs are both increasing, but the graph of $f(x)$ rises at a slower rate. This happens because of the .6 in the exponent. If the coefficient of x were greater than 1, the graph would rise at a faster rate than the graph of $g(x) = 2^x$. ✓ 4

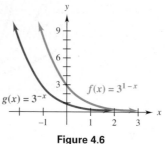

Figure 4.6

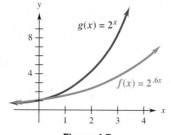

Figure 4.7

✓ **Checkpoint 4**

Graph $f(x) = 2^{x+1}$.

When the exponent involves a nonlinear expression in x, the graph of an exponential function may have a much different shape than the preceding ones have.

EXAMPLE 4   Graph $f(x) = 2^{-x^2}$.

Solution Either use a graphing calculator or plot points and connect them with a smooth curve, as in Figure 4.8. The graph is symmetric about the y-axis; that is, if the figure were folded on the y-axis, the two halves would match. This graph has the x-axis as a horizontal asymptote. The domain is still all real numbers, but here the range is $0 < y \le 1$. Graphs such as this are important in probability, where the *normal* curve has an equation similar to $f(x)$ in this example. ✓ 5

✓ **Checkpoint 5**

Graph $f(x) = \left(\frac{1}{2}\right)^{-x^2}$.

x	y
-2	$1/16$
-1.5	.21
-1	$1/2$
$-.5$	.84
0	1
.5	.84
1	$1/2$
1.5	.21
2	$1/16$

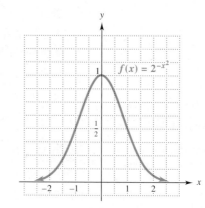

Figure 4.8

The Number *e*

A certain irrational number, denoted e, plays an important role in many mathematical and scientific contexts. To 12 decimal places,

$$e \approx 2.718281828459.$$

Perhaps the single most useful exponential function is the function defined by $f(x) = e^x$.

e^{(5)}
                    148.4131591
e^{(-1.4)}
                    .2465969639
e^{(1)}
                    2.718281828

Figure 4.9

✓ **Checkpoint 6**

Evaluate the following powers of e.

(a) $e^{.06}$

(b) $e^{-.06}$

(c) $e^{2.30}$

(d) $e^{-2.30}$

📱 **TECHNOLOGY TIP** To evaluate powers of e with a calculator, use the e^x key, as in Figure 4.9. The figure also shows how to display the decimal expansion of e by calculating e^1.

In Figure 4.10, the functions defined by

$$g(x) = 2^x, \qquad f(x) = e^x, \qquad \text{and} \qquad h(x) = 3^x$$

are graphed for comparison.

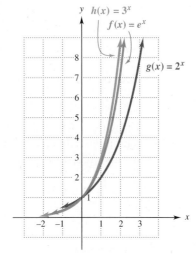

Figure 4.10

EXAMPLE 5   **Wine Consumption** The annual amount of wine (in millions of gallons) consumed in the United States can be approximated by the function

$$f(x) = 34e^{.029x},$$

where $x = 50$ corresponds to the year 1950. (Data from: www.wineinstitute.org.)

(a) How much wine was consumed in 1990?

Solution Because 1990 corresponds to $x = 90$, we evaluate $f(90)$:

$$f(90) = 34e^{.029(90)} \approx 462 \text{ million gallons.}$$

So the consumption was approximately 462 million gallons.

(b) How much wine was consumed in 2015?

Solution Because 2015 corresponds to $x = 115$, we evaluate $f(115)$:

$$f(115) = 34e^{.029(115)} \approx 955 \text{ million gallons.}$$

📱 **(c)** Use a graphing calculator to determine the first full year in which consumption will exceed 1100 million (1.1 billion) gallons.

Solution Because f measures consumption in millions of gallons, we must solve the equation $f(x) = 1100$, that is,

$$34e^{.029x} = 1100.$$

One way to do this is to find the intersection point of the graphs of $y = 34e^{.029x}$ and $y = 1100$. (The calculator's intersection finder is in the same menu as its root [or zero] finder.) Figure 4.11 shows that this point is approximately (119.9, 1100). Hence, the first full year in which consumption will exceed 1.1 billion gallons is 2020. ✓

1300

50 ⌐   ⌐ 130
   Intersection
   X=119.88638   Y=1100
-300

Figure 4.11

✓ **Checkpoint 7**

Per-person wine consumption (in gallons) can be approximated by $g(x) = .38e^{.018x}$, where $x = 50$ corresponds to the year 1950.

(a) Estimate the per-person wine consumption in 2000.

📱 (b) Determine the first full year in which the per-person wine consumption exceeded 2.75 gallons.

EXAMPLE 6   **AIG Assets** The assets (in billions of dollars) for the multinational insurance company American International Group Inc. (AIG) can be approximated by the function $A(x) = 1756e^{-.091x}$, where $x = 7$ corresponds to the year 2007. (Data from: *The Wall Street Journal.*)

(a) What were the approximate assets of AIG in 2007 (prior to the great recession)? What about in 2015?

Solution Evaluate the function at $x = 7$ and $x = 15$:

$$A(7) = 1756e^{-.091(7)} \approx \$928.7 \text{ billion};$$
$$A(15) = 1756e^{-.091(15)} \approx \$448.4 \text{ billion}.$$

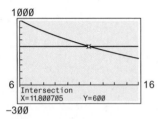

Figure 4.12

(b) What was the first full year when AIG assets fell below $600 billion?

Solution We must solve the equation

$$1756e^{-.091x} = 600.$$

Using the method of Example 5(c), we must graph $y = 1756e^{-.091x}$ and $y = 600$, and find their intersection point. Figure 4.12 shows that this point is approximately (11.8, 600). So the first full year in which AIG assets fell below $600 billion is 2012.

4.1 Exercises

Classify each function as linear, quadratic, or exponential.

1. $f(x) = 6^x$

2. $g(x) = -5x$

3. $h(x) = 4x^2 - x + 5$

4. $k(x) = 4^{x+3}$

5. $f(x) = 675(1.055^x)$

6. $g(x) = 12e^{x^2+1}$

Without graphing,

(a) *describe the shape of the graph of each function;*

(b) *find the second coordinates of the points with first coordinates 0 and 1. (See Examples 1–3.)*

7. $f(x) = .6^x$

8. $g(x) = 4^{-x}$

9. $h(x) = 2^{.5x}$

10. $k(x) = 5(3^x)$

11. $f(x) = e^{-x}$

12. $g(x) = 3(16^{x/4})$

Graph each function. (See Examples 1–3.)

13. $f(x) = 3^x$

14. $g(x) = 3^{.5x}$

15. $f(x) = 2^{x/2}$

16. $g(x) = e^{x/4}$

17. $f(x) = (1/5)^x$

18. $g(x) = 2^{3x}$

19. Graph these functions on the same axes.

   (a) $f(x) = 2^x$ **(b)** $g(x) = 2^{x+3}$ **(c)** $h(x) = 2^{x-4}$

   (d) If c is a positive constant, explain how the graphs of $y = 2^{x+c}$ and $y = 2^{x-c}$ are related to the graph of $f(x) = 2^x$.

20. Graph these functions on the same axes.

   (a) $f(x) = 3^x$ **(b)** $g(x) = 3^x + 2$ **(c)** $h(x) = 3^x - 4$

   (d) If c is a positive constant, explain how the graphs of $y = 3^x + c$ and $y = 3^x - c$ are related to the graph of $f(x) = 3^x$.

The accompanying figure shows the graphs of $y = a^x$ for $a = 1.8, 2.3, 3.2, .4, .75,$ and $.31$. They are identified by letter, but not necessarily in the same order as the values just given. Use your knowledge of how the exponential function behaves for various powers of a to match each lettered graph with the correct value of a.

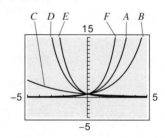

21. *A*

22. *B*

23. *C*

24. *D*

25. *E*

26. *F*

In Exercises 27 and 28, the graph of an exponential function with base a is given. Follow the directions in parts (a)–(f) in each exercise.

27.

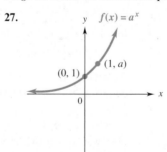

(a) Is $a > 1$ or is $0 < a < 1$?

(b) Give the domain and range of f.

(c) Sketch the graph of $g(x) = -a^x$.

(d) Give the domain and range of g.

(e) Sketch the graph of $h(x) = a^{-x}$.

(f) Give the domain and range of h.

28.

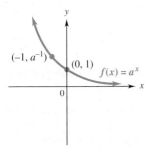

(a) Is $a > 1$ or is $0 < a < 1$?

(b) Give the domain and range of f.

(c) Sketch the graph of $g(x) = a^x + 2$.

(d) Give the domain and range of g.

(e) Sketch the graph of $h(x) = a^{x+2}$.

(f) Give the domain and range of h.

29. If $f(x) = a^x$ and $f(3) = 27$, find the following values of $f(x)$.

(a) $f(1)$ (b) $f(-1)$ (c) $f(2)$ (d) $f(0)$

30. Give a rule of the form $f(x) = a^x$ to define the exponential function whose graph contains the given point.

(a) $(3, 8)$ (b) $(-3, 64)$

Graph each function. (See Example 4.)

31. $f(x) = 2^{-x^2+2}$ **32.** $g(x) = 2^{x^2-2}$

33. $f(x) = x \cdot 2^x$ **34.** $f(x) = x^2 \cdot 2^x$

Work the following exercises.

35. Finance If \$1 is deposited into an account paying 6% per year compounded annually, then after t years the account will contain

$$y = (1 + .06)^t = (1.06)^t$$

dollars.

(a) Use a calculator to complete the following table:

t	0	1	2	3	4	5	6	7	8	9	10
y	1					1.34					1.79

(b) Graph $y = (1.06)^t$.

36. Finance If money loses value at the rate of 3% per year, the value of \$1 in t years is given by

$$y = (1 - .03)^t = (.97)^t.$$

(a) Use a calculator to complete the following table:

t	0	1	2	3	4	5	6	7	8	9	10
y	1				.86						.74

(b) Graph $y = (.97)^t$.

Work these problems. (See Example 5.)

37. Finance If money loses value, then as time passes, it takes more dollars to buy the same item. Use the results of Exercise 36(a) to answer the following questions.

(a) Suppose a house costs \$105,000 today. Estimate the cost of the same house in 10 years. (*Hint:* Solve the equation $.74t = \$105,000$.)

(b) Estimate the cost of a \$50 textbook in 8 years.

38. Natural Science Biologists have found that the oxygen consumption of yearling salmon is given by $g(x) = 100e^{.7x}$, where x is the speed in feet per second. Find each of the following.

(a) the oxygen consumption when the fish are still;

(b) the oxygen consumption when the fish are swimming at a speed of 2 feet per second.

39. Prudential Financial The assets (in billions of dollars) for Prudential Financial, Inc. can be approximated by the function $A(x) = 315e^{.056x}$, where $x = 7$ corresponds to the year 2007. (Data from: *The Wall Street Journal.*) Find the assets in each of the following years.

(a) 2012   (b) 2015   (c) 2018

40. Car Loan The monthly payment on a car loan at 12% interest per year on the unpaid balance is given by

$$f(n) = \frac{P}{\frac{1 - 1.01^{-n}}{.01}},$$

where P is the amount borrowed and n is the number of months over which the loan is paid back. Find the monthly payment for each of the following loans.

(a) \$8000 for 48 months   (b) \$8000 for 24 months

(c) \$6500 for 36 months   (d) \$6500 for 60 months

41. Natural Science The amount of plutonium remaining from 1 kilogram after x years is given by the function $W(x) = 2^{-x/24,360}$. How much will be left after

(a) 1000 years?   (b) 10,000 years?   (c) 15,000 years?

(d) Estimate how long it will take for the 1 kilogram to decay to half its original weight. Your answer may help to explain why nuclear waste disposal is a serious problem.

Business *The scrap value of a machine is the value of the machine at the end of its useful life. By one method of calculating scrap value, where it is assumed that a constant percentage of value is lost annually, the scrap value is given by*

$$S = C(1 - r)^n,$$

where C is the original cost, n is the useful life of the machine in years, and r is the constant annual percentage of value lost. Find the scrap value for each of the following machines.

42. Original cost, \$68,000; life, 10 years; annual rate of value loss, 8%

43. Original cost, \$244,000; life, 12 years; annual rate of value loss, 15%

44. Use the graphs of $f(x) = 2^x$ and $g(x) = 2^{-x}$ (not a calculator) to explain why $2^x + 2^{-x}$ is approximately equal to 2^x when x is very large.

Work the following problems. (See Examples 5 and 6.)

45. World Population There were fewer than a billion people on Earth when Thomas Jefferson died in 1826, and there are now more than 6 billion. If the world population continues to grow as expected, the population (in billions) in year t will be given by the function $P(t) = 4.834(1.01^{(t-1980)})$. (Data from: U.S. Census Bureau.) Estimate the world population in the following years.

(a) 2015   (b) 2020   (c) 2040

(d) What will the world population be when you reach 65 years old?

46. Netflix Costs The costs (in billions of dollars) related to streaming content for Netflix can be approximated by the function $C(x) = .35(1.26)^x$, where $x = 10$ corresponds to the year 2010. (Data from: *The Wall Street Journal.*)

(a) Find the costs for streaming content in 2015.

(b) Assuming the model stays accurate, what is the first full year when costs exceed $15 billion?

GDP *Use the following information to answer Exercises 47–50. The gross domestic product (GDP; in tens of trillions of U.S. dollars) for China and the United States can be approximated as follows:*

$$\text{China: } f(x) = .00011(1.102)^x,$$

$$\text{United States: } g(x) = .0132(1.067)^x,$$

where $x = 60$ corresponds to the year 1960. (Data from: www.worldbank.org.)

47. Find the projected GDP in 2015 and 2025 for China.

48. Find the projected GDP in 2015 and 2025 for the United States.

49. If the trend continues, what is the first full year in which China's GDP surpasses $25 trillion?

50. If the trend continues, what is the first full year in which the U.S. GDP surpasses $60 trillion?

51. Asset Management The amount of money (in trillions of dollars) that is invested in passively managed mutual funds can be approximated by the function

$$A(x) = .34(1.19)^x,$$

where $x = 1$ corresponds to the year 2001. (Data from: *The Wall Street Journal.*)

(a) What was the amount of money in passively managed mutual funds in 2012 and 2016?

(b) If the trend continues, what is the first full year when the amount passively managed exceeds $6.5 trillion?

52. Imports from Vietnam The value of U.S. imports from Vietnam (in billions of dollars) can be approximated by the function

$$f(x) = 2.48e^{.179x},$$

where $x = 9$ corresponds to the year 2009. (Data from: *The Wall Street Journal.*)

(a) Find the value of U.S. imports in 2014.

(b) If the trend continues, what is the first full year when imports exceed $60 billion in value?

53. Subprime Mortgages The amount of money (in billions of dollars) lent to customers with credit scores below 620 for subprime mortgages can be approximated by the function

$$g(x) = 243.5e^{-.25x},$$

where $x = 1$ corresponds to the year 2001. (Data from: *The Wall Street Journal.*)

(a) Find the value of subprime mortgage lending in 2016 for the described customer base.

(b) If the trend continues, what is the first full year in which subprime lending falls below $3 billion?

54. Subprime Mortgages The amount of money (in billions of dollars) lent to customers with a credit score below 660 for subprime mortgages can be approximated by the function

$$h(x) = 386e^{-.19x},$$

where $x = 1$ corresponds to the year 2001. (Data from: *The Wall Street Journal.*)

(a) Find the value of subprime mortgage lending in 2017 for the described customer base.

(b) If the trend continues, what is the first full year where subprime lending falls below $10 billion?

✓ Checkpoint Answers

1. **(a)** The entries in the second column are $1/27$, $1/9$, $1/3$, 1, 3, 9, 27, respectively.

(b)

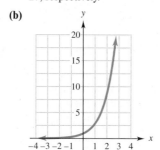

2.

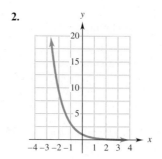

3.

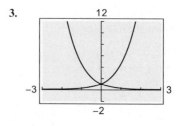

4.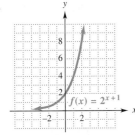

5.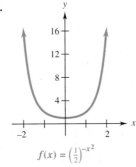

$$f(x) = \left(\tfrac{1}{2}\right)^{-x^2}$$

6. (a) 1.06184

(b) .94176

(c) 9.97418

(d) .10026

7. (a) About 2.3 gallons per person

(b) 2010 ($x \approx 109.95$)

4.2 Applications of Exponential Functions

In many situations in biology, economics, and the social sciences, a quantity changes at a rate proportional to the quantity present. For example, a country's population might be increasing at a rate of 1.3% a year. In such cases, the amount present at time t is given by an **exponential growth function.**

It is understood that growth can involve either growing larger or growing smaller.

Exponential Growth Function

Under normal conditions, growth can be described by a function of the form

$$f(t) = y_0 e^{kt} \qquad \text{or} \qquad f(t) = y_0 b^t,$$

where $f(t)$ is the amount present at time t, y_0 is the amount present at time $t = 0$, and k and b are constants that depend on the rate of growth.

When $f(t) = y_0 e^{kt}$, and $k > 0$, we describe $f(t)$ as modeling exponential growth, and when $k < 0$, we describe $f(t)$ as modeling exponential decay. When $f(t) = y_0 b^t$, and $b > 1$, we describe $f(t)$ as modeling exponential growth, and when $0 < b < 1$, we describe $f(t)$ as modeling exponential decay.

 NOTE Writing an exponential function as $f(t) = y_0 e^{kt}$ or $f(t) = y_0 b^t$ is writing the same thing in two different ways. For example, if we let $k = 2.5$, and $b = e^{2.5}$, then notice that

$$e^{kt} = (e^k)^t = (e^{2.5})^t = b^t.$$

Sometimes it is helpful to write the function using e in the formulation, and sometimes it is helpful to write the function using the formulation with $b = e^k$. Because e is a constant itself, it becomes a different-valued constant that we simply represent with a different letter (b) when e is raised to the k power.

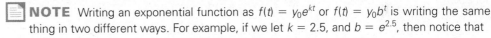

EXAMPLE 1   **Total Debt** The amount of debt across all sectors (in billions of dollars) can be approximated by the exponential function

$$f(t) = y_0 e^{.083t},$$

where t is time in years, and $t = 0$ corresponds to the year 1970. (Data from: The Federal Reserve Bank of St. Louis.)

(a) If the amount of total debt was \$1946 billion in the year 1970, find the total amount in the year 2017.

Solution Because y_0 represents the total debt when $t = 0$ (that is, in 1970), we have $y_0 = 1946$. So the growth function is $f(t) = 1946e^{.083t}$. To find the total debt in 2017, evaluate $f(t)$ at $t = 47$ (which corresponds to the year 2017):

$$f(t) = 1946e^{.083t}$$
$$f(47) = 1946e^{.083(47)} \approx 96,233.$$

Hence, the total debt in 2017 was approximately $96,233 billion (or $96.233 trillion).

(b) If the model remains accurate, what will the total debt be in the year 2021?

Solution Because 2021 corresponds to $t = 51$, evaluate $f(t)$ at $t = 51$:

$$f(51) = 1946e^{.083(51)} \approx \$134,126 \text{ billion.} \quad ✓_1$$

✓ **Checkpoint 1**

Suppose the number of bacteria in a culture at time t is

$$y = 500e^{.4t},$$

where t is measured in hours.

(a) How many bacteria are present initially?

(b) How many bacteria are present after 10 hours?

EXAMPLE 2 **Single-Family Houses** During the years from 2005 to 2011 (just before, during, and immediately after the recession years of 2007 to 2009), the number of single-family houses sold decreased at a rate that can be modeled well by an exponential function. The number of single-family houses sold (in thousands) can be approximated by the function

$$g(x) = 1483e^{-.28x},$$

where $x = 0$ corresponds to the year 2005. (Data from: *The Wall Street Journal*.)

(a) Find the number of houses sold in 2008.

Solution Evaluate $g(x)$ at $x = 3$ because 2008 corresponds to $x = 3$.

$$g(3) = 1483e^{-.28(3)} \approx 640 \text{ thousand.}$$

Thus, the number of houses sold in 2008 was approximately 640 thousand.

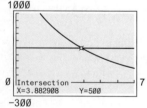

(b) What was the first full year in which the number of houses sold fell below 500 thousand?

Solution Graph $y = 1483e^{-.28x}$ and $y = 500$ on the same screen and find the x-coordinate of their intersection point. Figure 4.13 shows that $x \approx 3.88$, and we round up to the next full year of $x = 4$. Because $x = 0$ corresponds to 2005, $x = 4$ corresponds to 2009. Thus, 2009 is the first full year in which the number of houses sold fell below 500 thousand.

Figure 4.13

When a quantity is known to grow exponentially, it is sometimes possible to find a function that models its growth from a small amount of data.

EXAMPLE 3 **Finance** When money is placed in a bank account that pays compound interest, the amount in the account grows exponentially, as we shall see in Chapter 5. Suppose such an account grows from $1000 to $1316 in 7 years.

(a) Find a growth function of the form $f(t) = y_0b^t$ that gives the amount in the account at time t years.

Solution The values of the account at time $t = 0$ and $t = 7$ are given; that is, $f(0) = 1000$ and $f(7) = 1316$. Solve the first of these equations for y_0:

$$f(0) = 1000$$
$$y_0b^0 = 1000 \qquad \text{Rule of } f$$
$$y_0 = 1000. \qquad b^0 = 1$$

So the rule of f has the form $f(t) = 1000b^t$. Now solve the equation $f(7) = 1316$ for b:

$$f(7) = 1316$$
$$1000b^7 = 1316 \qquad \text{Rule of } f$$
$$b^7 = 1.316 \qquad \text{Divide both sides by 1000.}$$
$$b = (1.316)^{1/7} \approx 1.04. \qquad \text{Take the seventh root of each side.}$$

So the rule of the function is $f(t) = 1000(1.04)^t$.

(b) How much is in the account after 12 years?

Solution $f(12) = 1000(1.04)^{12} = \$1601.03$.

✓ **Checkpoint 2**

Suppose an investment grows exponentially from \$500 to \$587.12 in three years.

(a) Find a function of the form $f(t) = y_0b^t$ that gives the value of the investment after t years.

(b) How much is the investment worth after 10 years?

EXAMPLE 4 **Infant Mortality** Infant mortality rates in the United States are shown in the following table. (Data from: U.S. National Center for Health Statistics.)

Year	Rate	Year	Rate
1920	76.7	1980	12.6
1930	60.4	1990	9.2
1940	47.0	1995	7.6
1950	29.2	2000	6.9
1960	26.0	2010	6.1
1970	20.0	2014	5.8

(a) Let $t = 0$ correspond to 1920. Use the data for 1920 and 2014 to find a function of the form $f(t) = y_0b^t$ that models these data.

Solution Since the rate is 76.7 when $t = 0$, we have $y_0 = 76.7$. Hence, $f(t) = 76.7b^t$. Because 2014 corresponds to $t = 94$, we have $f(94) = 5.8$; that is,

$$76.7b^{94} = 5.8 \qquad \text{Rule of } f$$
$$b^{94} = \frac{5.8}{76.7} \qquad \text{Divide both sides by 76.7.}$$
$$b = \left(\frac{5.8}{76.7}\right)^{\frac{1}{94}} \approx .9729 \qquad \text{Take } 94^{\text{th}} \text{ roots on both sides.}$$

Therefore, the function is $f(t) = 76.7(.9729^t)$.

(b) Use exponential regression on a graphing calculator to find another model for these data.

Solution The procedure for entering the data and finding the function is the same as that for linear regression (just choose "exponential" instead of "linear"), as explained in Section 2.3. The calculator produces the equation

$$g(t) = 78.167(.9712^t).$$

This function fits the data very well, as shown in Figure 4.14.

(c) Use the preceding model, and assume it continues to remain accurate to estimate the infant mortality rate in the years 2018 and 2022.

Solution Evaluate the function $g(t)$ at $t = 98$ and $t = 102$. This yields

$$g(98) = 78.167(.9712^{98}) \approx 4.5$$
$$g(102) = 78.167(.9712^{102}) \approx 4.0.$$

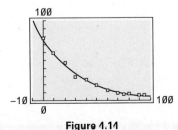

Figure 4.14

Other Exponential Models

When a quantity changes exponentially, but does not either grow very large or decrease practically to 0, as in Examples 1–3, different functions are needed.

EXAMPLE 5   **Business** Sales of a new product often grow rapidly at first and then begin to level off with time. Suppose the annual sales of an inexpensive can opener are given by

$$S(x) = 10,000(1 - e^{-.5x}),$$

where $x = 0$ corresponds to the time the can opener went on the market.

(a) What were the sales in each of the first three years?

Solution At the end of one year ($x = 1$), sales were

$$S(1) = 10,000(1 - e^{-.5(1)}) \approx 3935.$$

Sales in the next two years were

$$S(2) = 10,000(1 - e^{-.5(2)}) \approx 6321 \quad \text{and} \quad S(3) = 10,000(1 - e^{-.5(3)}) \approx 7769.$$

(b) What were the sales at the end of the 10th year?

Solution $S(10) = 10,000(1 - e^{-.5(10)}) \approx 9933.$

(c) Graph the function S. What does it suggest?

Solution The graph can be obtained by plotting points and connecting them with a smooth curve or by using a graphing calculator, as in Figure 4.15. The graph indicates that sales will level off after the 12th year, to around 10,000 can openers per year. ✓3

A variety of activities can be modeled by **logistic functions,** whose rules are of the form

$$f(x) = \frac{c}{1 + ae^{kx}}.$$

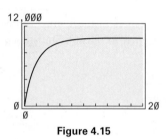

Figure 4.15

✓**Checkpoint 3**

Suppose the value of the assets (in thousands of dollars) of a certain company after t years is given by

$$V(t) = 100 - 75e^{-.2t}.$$

(a) What is the initial value of the assets?

(b) What is the limiting value of the assets?

(c) Find the value after 10 years.

(d) Graph $V(t)$.

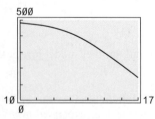

Figure 4.16

✓**Checkpoint 4**

In Example 6, find the price of scrap steel in 2013.

📇 **TECHNOLOGY TIP**

Many graphing calculators can find a logistic model for appropriate data.

EXAMPLE 6   **Price of Scrap Steel** The price (in dollars) for U.S. scrap steel per ton can be approximated by the function

$$h(t) = \frac{496.7}{1 + .000048e^{.64t}},$$

where $t = 11$ corresponds to the year 2011. (Data from: *The Wall Street Journal.*)

(a) What was the price of U.S. scrap steel in 2011?

Solution When $t = 11$,

$$h(11) = \frac{496.7}{1 + .000048e^{.64(11)}} = \$470.90.$$

So the price of scrap steel was \$470.90 a ton in 2011.

(b) What was the price in 2016?

Solution When $t = 16$,

$$h(16) = \frac{496.7}{1 + .000048e^{.64(16)}} = \$211.90.$$

So the price of scrap steel dropped to \$211.90 a ton by 2016.

(c) Draw a graph of $h(t)$ from 2010 to 2017.

Solution The TI-84 produces the graph shown in Figure 4.16. ✓4

4.2 Exercises

1. **Finance** Suppose you owe $800 on your credit card and you decide to make no new purchases and to make the minimum monthly payment on the account. Assuming that the interest rate on your card is 1% per month on the unpaid balance and that the minimum payment is 2% of the total (balance plus interest), your balance after t months is given by

$$B(t) = 800(.9898^t).$$

Find your balance at each of the given times.

(a) six months

(b) one year (remember that t is in months)

(c) five years

(d) eight years

(e) On the basis of your answers to parts (a)–(d), what advice would you give to your friends about minimum payments?

2. **Finance** Suppose you owe $1500 on your credit card and you decide to make no new purchases and to make the minimum monthly payment on the account. Assuming the interest rate on the card is 1.1% per month on the unpaid balance, and the minimum payment is 5% of the total (balance plus interest), your balance after t months is given by

$$B(t) = 1500(.9604)^t.$$

Find the balance at each of the given times.

(a) five months

(b) one year (remember that t is in months)

(c) Is the balance paid off in two years?

3. **Natural Gas Production** The annual amount of energy produced in the United States from dry natural gas (in trillion cubic feet) can be approximated by the function

$$g(t) = 14.66(1.042)^t,$$

where $t = 10$ corresponds to the year 2010. (Data from: U.S. Energy Information Administration.)

(a) Find the amount of dry natural gas energy produced in 2015.

(b) If the model continues to be accurate, project the amount of dry natural gas energy produced in 2022.

4. **Oil Production** The annual amount of U.S. crude-oil production (in millions of barrels) can be approximated by the function

$$f(t) = 871(1.091)^t,$$

where $t = 8$ corresponds to the year 2008. (Data from: U.S. Energy Information Administration.)

(a) Find the amount of production in 2010.

(b) If the trend continues, find the amount of production in 2020.

In each of the following problems, find an exponential function of the form $f(t) = y_0 b^t$ to model the given data. (See Examples 3 and 4.)

5. **Office Rent** The annual average price per square foot for office space in San Francisco was $30.48 in 2004, and it was $66.71 in 2015. (Data from: *San Francisco Business Times.*)

(a) Find a model for these data in which $t = 0$ corresponds to the year 2004.

(b) If the model remains accurate, what was the predicted rent per square foot in 2016?

(c) By experimenting with different values (or using a graphing calculator to solve an appropriate equation), estimate the first full year in which the price per square foot exceeded $45.

6. **Population Growth** The U.S. Census Bureau predicts that the African-American population will increase from 35.3 million in 2000 to 61.8 million in 2060.

(a) Find a model for these data, in which $t = 0$ corresponds to 2000.

(b) What is the projected African-American population in 2020? in 2030?

(c) By experimenting with different values of t (or by using a graphing calculator to solve an appropriate equation), estimate the first full year in which the African-American population will exceed 55 million.

7. **Personal Consumption Expenditures** The total amount of expenditures in the United States on video and audio equipment, computers, and related services was approximately $9.6 billion in 1990 and $109.2 billion in 2015. (Data from: The Federal Reserve Bank of St. Louis.)

(a) Let $t = 0$ correspond to the year 1990, and find a model for these data.

(b) According to the model, what were the expenditures in 2014?

(c) If the model remains accurate, estimate the expenditures for 2020.

8. **Medicare Expenditures** Medicare expenditures were $110 billion in 1990 and $646 billion in 2015. (Data from: Centers for Medicare and Medicaid Services.)

(a) Let $t = 0$ correspond to the year 1990, and find a model for these data.

(b) According to the model, what were Medicare expenditures in 2012?

(c) If the model remains accurate, estimate Medicare expenditures in 2025.

In the following exercises, find the exponential model as follows: If you do not have suitable technology, use the first and last data points to find a function. (See Examples 3 and 4.) If you have a graphing calculator or other suitable technology, use exponential regression to find a function. (See Example 4.)

9. **Chinese Bank Assets** The table shows the total assets (in trillion yuan) held in Chinese banks for the given years.

Year	2000	2005	2007	2009	2011	2013	2015
Assets	17	41	53	79	118	151	199

(a) Find the exponential model for these data, where $t = 0$ corresponds to the year 2000.

(b) Assume that the model remains accurate, and predict the total assets in 2018 and 2021.

(c) Use a graphing calculator (or trial and error) to determine the first full year in which the total assets exceed 450 trillion yuan.

10. Atmospheric Pressure The table shows the atmospheric pressure (in millibars) at various altitudes (in meters).

Altitude	Pressure
0	1,013
1000	899
2000	795
3000	701
4000	617
5000	541
6000	472
7000	411
8000	357
9000	308
10,000	265

(a) Find an exponential model for these data, in which t is measured in thousands. (For instance, $t = 2$ is 2000 meters.)

(b) Use the function in part (a) to estimate the atmospheric pressure at 1500 meters and 11,000 meters. Compare your results with the actual values of 846 millibars and 227 millibars, respectively.

11. Heart Disease The table shows the age-adjusted death rates per 100,000 Americans for heart disease. (Data from: U.S. Center for Health Statistics.)

Year	Death Rate
2000	257.6
2002	240.8
2004	217.0
2006	210.2
2008	186.5
2010	178.5
2012	170.5
2014	167.0

(a) Find an exponential model for these data, where $t = 0$ corresponds to the year 2000.

(b) Assuming the model remains accurate, estimate the death rate in 2019 and 2023.

(c) Use a graphing calculator (or trial and error) to determine the first full year in which the death rate will be below 100.

12. Business The table shows outstanding consumer credit (in billions of dollars) at the beginning of various years. (Data from: U.S. Federal Reserve.)

Year	Credit
1980	350
1985	524
1990	798
1995	1010
2000	1539
2005	2197
2010	2544
2015	3330

(a) Find an exponential model for these data, where $t = 0$ corresponds to the year 1980.

(b) If this model remains accurate, what will the outstanding consumer credit be in 2020 and 2025?

(c) What is the first full year where consumer credit exceeds $6500 billion?

Work the following problems. (See Example 5.)

13. Worker Productivity Assembly-line operations tend to have a high turnover of employees, forcing the companies involved to spend much time and effort in training new workers. It has been found that a worker who is new to the operation of a certain task on the assembly line will produce $P(t)$ items on day t, where

$$P(t) = 25 - 25e^{-.3t}.$$

(a) How many items will be produced on the first day?

(b) How many items will be produced on the eighth day?

(c) According to the function, what is the maximum number of items the worker can produce?

14. Words per Minute The number of words per minute that an average person can type is given by

$$W(t) = 60 - 30e^{-.5t},$$

where t is time in months after the beginning of a typing class. Find each of the following.

(a) $W(0)$ (b) $W(1)$

(c) $W(4)$ (d) $W(6)$

Natural Science *Newton's law of cooling says that the rate at which a body cools is proportional to the difference in temperature between the body and an environment into which it is introduced. Using calculus, we can find that the temperature $F(t)$ of the body at time t after being introduced into an environment having constant temperature T_0 is*

$$F(t) = T_0 + Cb^t,$$

where C and b are constants. Use this result in Exercises 15 and 16.

15. Boiling water at 100° Celsius is placed in a freezer at −18° Celsius. The temperature of the water is 50° Celsius after 24 minutes. Find the temperature of the water after 76 minutes.

16. Paisley refuses to drink coffee cooler than 95°F. She makes coffee with a temperature of 170°F in a room with a temperature of 70°F. The coffee cools to 120°F in 10 minutes. What is the longest amount of time she can let the coffee sit before she drinks it?

17. **Internet Use in China** The percentage of Chinese residents who use the Internet can be approximated by the function

$$g(t) = \frac{49.8}{1 + 52.8e^{-.47t}},$$

where $t = 0$ corresponds to the year 2000. (Data from: *The Wall Street Journal.*)

   (a) What percentage of Chinese residents used the Internet in 2006? In 2015?

   (b) According to this model, will participation reach 50% by 2050?

18. **Seat-Belt Use** Data from the National Highway Traffic Safety Administration indicate that, in year t, the approximate percentage of people in the United States who wear seat belts when driving is given by

$$g(t) = \frac{88.2}{1 + .243e^{-.18t}},$$

where $t = 0$ corresponds to the year 2000.

   (a) What percentage of drivers used seat belts in 2014?

   (b) According to this model, does seat-belt use ever reach 100%?

19. **Food Assistance** The amount of money the U.S. government spent on food and nutrition assistance (in billions of dollars) can be approximated by the function

$$f(x) = \frac{120.9}{1 + 3.42e^{-.23x}},$$

where $x = 0$ corresponds to the year 2000. (Data from: U.S. Office of Management and Budget.)

   (a) Find the amount spent in 2000 and 2015.

   (b) Use the graph to determine the first full year in which expenditures for food and nutrition assistance was higher than $100 billion.

20. **Fish Population** The population of fish in a certain lake at time t months is given by the function

$$p(t) = \frac{20,000}{1 + 24(2^{-.36t})}.$$

   (a) Graph the population function from $t = 0$ to $t = 48$ (a four-year period).

   (b) What was the population at the beginning of the period?

   (c) Use the graph to estimate the one-year period in which the population grew most rapidly.

   (d) When do you think the population will reach 25,000? What factors in nature might explain your answer?

21. **Labor Force Participation** The percentage of adults participating in the labor force can be approximated by the function

$$L(x) = \frac{67.9}{1 + .014e^{.132x}},$$

where $x = 0$ corresponds to the year 2000. (Data from: *The Wall Street Journal.*)

   (a) Estimate the percentage of labor force participation in 2005 and 2015.

   (b) Graph $L(x)$ from 2000 to 2015.

22. **Automobile Accidents** The probability $P(t)$ percent of having an automobile accident is related to the alcohol level t of the driver's blood by the function $P(t) = e^{21.459t}$.

   (a) Graph $P(t)$ in a viewing window with $0 \leq t \leq .2$ and $0 \leq P(t) \leq 100$.

   (b) At what blood alcohol level is the probability of an accident at least 50%? What is the legal blood alcohol level in your state?

✓ **Checkpoint Answers**

1. (a) 500
   (b) About 27,300

2. (a) $f(t) = 500(1.055)^t$
   (b) $854.07

3. (a) $25,000
   (b) $100,000
   (c) $89,850
   (d)

4. $414.94

4.3 Logarithmic Functions

Until the development of computers and calculators, logarithms were the only effective tool for large-scale numerical computations. They are no longer needed for this purpose, but logarithmic functions still play a crucial role in calculus and in many applications.

Common Logarithms

Logarithms are simply *a new language for old ideas*—essentially, a special case of exponents.

Definition of Common (Base 10) Logarithms

$$y = \log x \qquad \text{means} \qquad 10^y = x.$$

Log x, which is read "the logarithm of x," is the answer to the question,

> To what exponent must 10 be raised to produce x?

EXAMPLE 1   To find log 10,000, ask yourself, "To what exponent must 10 be raised to produce 10,000?" Since $10^4 = 10,000$, we see that log 10,000 = 4. Similarly,

$$\log 1 = 0 \qquad \text{because} \qquad 10^0 = 1;$$

$$\log .01 = -2 \qquad \text{because} \qquad 10^{-2} = \frac{1}{10^2} = \frac{1}{100} = .01;$$

$$\log \sqrt{10} = 1/2 \qquad \text{because} \qquad 10^{1/2} = \sqrt{10}.$$ ☑1

✓ **Checkpoint 1**

Find each common logarithm.

(a) log 100

(b) log 1000

(c) log .1

EXAMPLE 2   Log (-25) is the exponent to which 10 must be raised to produce -25. But every power of 10 is positive! So there is no exponent that will produce -25. *Logarithms of negative numbers and 0 are not defined.*

```
log(359)
              2.555094449
10^(2.5551)
              359.004589
log(.026)
             -1.585026652
```

Figure 4.17

EXAMPLE 3   **(a)** We know that log 359 must be a number between 2 and 3 because $10^2 = 100$ and $10^3 = 1000$. By using the "log" key on a calculator, we find that log 359 (to four decimal places) is 2.5551. You can verify this statement by computing $10^{2.5551}$; the result (rounded) is 359. See the first two lines in Figure 4.17.

(b) When 10 is raised to a negative exponent, the result is a number less than 1. Consequently, the logarithms of numbers between 0 and 1 are negative. For instance, log .026 = -1.5850, as shown in the third line in Figure 4.17. ☑2

✓ **Checkpoint 2**

Find each common logarithm.

(a) log 27

(b) log 1089

(c) log .00426

📄 **NOTE** On most scientific calculators, enter the number followed by the log key. On graphing calculators, press the log key followed by the number, as in Figure 4.17.

Natural Logarithms

The most widely used logarithm today is defined in terms of the number e (whose decimal expansion begins 2.71828 . . .) rather than 10. It has a special name and notation.

Definition of Natural (Base e) Logarithms

$$y = \ln x \qquad \text{means} \qquad e^y = x.$$

Thus, the number **ln** x (which is sometimes read "el-en x") is the exponent to which e must be raised to produce the number x. For instance, ln 1 = 0 because $e^0 = 1$. Although logarithms to the base e may not seem as "natural" as common logarithms, there are several reasons for using them, some of which are discussed in Section 4.4.

```
ln(85)
            4.442651256
e^4.4427
            85.0041433
ln(.38)
            -.9675840263
```

Figure 4.18

✓ **Checkpoint 3**

Find the following.

(a) ln 6.1

(b) ln 20

(c) ln .8

(d) ln .1

EXAMPLE 4   **(a)** To find ln 85, use the ‖LN‖ key of your calculator, as in Figure 4.18. The result is 4.4427. Thus, 4.4427 is the exponent (to four decimal places) to which e must be raised to produce 85. You can verify this result by computing $e^{4.4427}$; the answer (rounded) is 85. See Figure 4.18.

(b) A calculator shows that ln .38 = $-.9676$ (rounded), which means that $e^{-.9676} \approx .38$. See Figure 4.18. ✓₃

EXAMPLE 5   You don't need a calculator to find ln e^8. Just ask yourself, "To what exponent must e be raised to produce e^8?" The answer, obviously, is 8. So ln $e^8 = 8$.

Other Logarithms

The procedure used to define common and natural logarithms can be carried out with any positive number $a \neq 1$ as the base in place of 10 or e.

Definition of Logarithms to the Base a

$$y = \log_a x \qquad \text{means} \qquad a^y = x.$$

Read $y = \log_a x$ as "y is the logarithm of x to the base a." As was the case with common and natural logarithms, $\log_a x$ is an *exponent*; it is the answer to the question,

To what power must a be raised to produce x?

For example, suppose $a = 2$ and $x = 16$. Then $\log_2 16$ is the answer to the question,

To what power must 2 be raised to produce 16?

It is easy to see that $2^4 = 16$, so $\log_2 16 = 4$. In other words, the exponential statement $2^4 = 16$ is equivalent to the logarithmic statement $4 = \log_2 16$.

In the definition of a logarithm to base a, note carefully the relationship between the base and the exponent:

Logarithmic form: $y = \log_a x$

Exponential form: $a^y = x$

Common and natural logarithms are the special cases when $a = 10$ and when $a = e$, respectively. Both $\log u$ and $\log_{10} u$ mean the same thing. Similarly, ln u and $\log_e u$ mean the same thing.

Graphing Logarithmic Functions

The graph of the logarithmic function $h(x) = \ln x$ has the same basic shape as the graph of $f(x) = \log_a x$, as illustrated in Figure 4.19 and summarized in the next box.

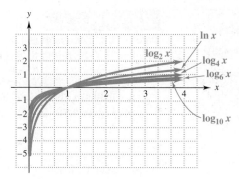

Figure 4.19

When $a > 0$ and $a \neq 1$, the function $f(x) = \log_a x$ has the set of nonnegative real numbers as its domain. Its graph has the shape shown below and the properties listed.

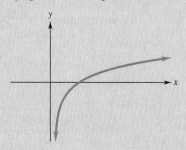

1. The graph is below the x-axis when $0 < x < 1$ and above the x-axis when $x > 1$.
2. The x-intercept is 1. In other words, $\log_a 1 = 0$ for all a such that $a > 0$ and $a \neq 1$.
3. The graph climbs to the right and then grows more slowly.
4. The y-axis is a vertical asymptote.
5. The larger the base a, the less steeply the graph rises to the right.

EXAMPLE 6   This example shows several statements written in both exponential and logarithmic form.

Exponential Form	Logarithmic Form
(a) $\ 3^2 = 9$	$\log_3 9 = 2$
(b) $\ (1/5)^{-2} = 25$	$\log_{1/5} 25 = -2$
(c) $\ 10^5 = 100{,}000$	$\log_{10} 100{,}000 \ (\text{or } \log 100{,}000) = 5$
(d) $\ 4^{-3} = 1/64$	$\log_4 (1/64) = -3$
(e) $\ 2^{-4} = 1/16$	$\log_2 (1/16) = -4$
(f) $\ e^0 = 1$	$\log_e 1 \ (\text{or } \ln 1) = 0$

✓ **Checkpoint 4**

Write the logarithmic form of
(a) $5^3 = 125$;
(b) $3^{-4} = 1/81$;
(c) $8^{2/3} = 4$.

✓ **Checkpoint 5**

Write the exponential form of
(a) $\log_{16} 4 = 1/2$;
(b) $\log_3 (1/9) = -2$;
(c) $\log_{16} 8 = 3/4$.

Properties of Logarithms

Some of the important properties of logarithms arise directly from their definition.

Let x and a be any positive real numbers, with $a \neq 1$, and r be any real number. Then

(a) $\log_a 1 = 0$;   (b) $\log_a a = 1$;
(c) $\log_a a^r = r$;   (d) $a^{\log_a x} = x$.

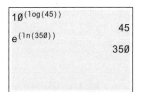

Figure 4.20

Property (a) was discussed in Example 1 (with $a = 10$). Property (c) was illustrated in Example 5 (with $a = e$ and $r = 8$). Property (b) is property (c) with $r = 1$. To understand property (d), recall that $\log_a x$ is the exponent to which a must be raised to produce x. Consequently, when you raise a to this exponent, the result is x, as illustrated for common and natural logarithms in Figure 4.20.

The following properties are part of the reason that logarithms are so useful. They will be used in the next section to solve exponential and logarithmic equations.

The Product, Quotient, and Power Properties

Let x, y, and a be any positive real numbers, with $a \neq 1$. Let r be any real number. Then

$$\log_a xy = \log_a x + \log_a y \qquad \text{Product property}$$

$$\log_a \frac{x}{y} = \log_a x - \log_a y \qquad \text{Quotient property}$$

$$\log_a x^r = r \log_a x \qquad \text{Power property}$$

Each of these three properties is illustrated on a calculator in Figure 4.21.

```
log(7*8)
        1.748188027
log(7)+log(8)
        1.748188027
```
```
ln(50/9)
        1.714798428
ln(50)-ln(9)
        1.714798428
```
```
log(12⁹)
        9.712631214
9*log(12)
        9.712631214
```

Figure 4.21

To prove the product property, let

$$m = \log_a x \qquad \text{and} \qquad n = \log_a y.$$

Then, by the definition of logarithm,

$$a^m = x \qquad \text{and} \qquad a^n = y.$$

Multiply to get

$$a^m \cdot a^n = x \cdot y,$$

or, by a property of exponents,

$$a^{m+n} = xy.$$

Use the definition of logarithm to rewrite this last statement as

$$\log_a xy = m + n.$$

Replace m with $\log_a x$ and n with $\log_a y$ to get

$$\log_a xy = \log_a x + \log_a y.$$

The quotient and power properties can be proved similarly.

EXAMPLE 7   Using the properties of logarithms, we can write each of the following as a single logarithm:[*]

(a) $\log_a x + \log_a (x - 1) = \log_a x(x - 1);$ Product property

(b) $\log_a (x^2 + 4) - \log_a (x + 6) = \log_a \dfrac{x^2 + 4}{x + 6};$ Quotient property

(c) $\log_a 9 + 5 \log_a x = \log_a 9 + \log_a x^5 = \log_a 9x^5.$ Product and power properties

---

*Here and elsewhere, we assume that variable expressions represent positive numbers and that the base a is positive, with $a \neq 1$.

✓ **Checkpoint 6**

Write each expression as a single logarithm, using the properties of logarithms.

(a) $\log_a 5x + \log_a 3x^4$

(b) $\log_a 3p - \log_a 5q$

(c) $4 \log_a k - 3 \log_a m$

⚠️ **CAUTION** There is no logarithm property that allows you to simplify the logarithm of a sum, such as $\log_a (x^2 + 4)$. In particular, $\log_a (x^2 + 4)$ is *not* equal to $\log_a x^2 + \log_a 4$. The product property of logarithms shows that $\log_a x^2 + \log_a 4 = \log_a 4x^2$.

EXAMPLE 8 Assume that $\log_6 7 \approx 1.09$ and $\log_6 5 \approx .90$. Use the properties of logarithms to find each of the following:

(a) $\log_6 35 = \log_6 (7 \cdot 5) = \log_6 7 + \log_6 5 \approx 1.09 + .90 = 1.99$;

(b) $\log_6 5/7 = \log_6 5 - \log_6 7 \approx .90 - 1.09 = -.19$;

(c) $\log_6 5^3 = 3 \log_6 5 \approx 3(.90) = 2.70$;

(d) $\log_6 6 = 1$;

(e) $\log_6 1 = 0$. ✓₇

In Example 8, several logarithms to the base 6 were given. However, they could have been found by using a calculator and the following formula.

Change-of-Base Theorem

For any positive numbers a and x (with $a \neq 1$),

$$\log_a x = \frac{\ln x}{\ln a}.$$

EXAMPLE 9 To find $\log_7 3$, use the theorem with $a = 7$ and $x = 3$:

$$\log_7 3 = \frac{\ln 3}{\ln 7} \approx \frac{1.0986}{1.9459} \approx .5646.$$

You can check this result on your calculator by verifying that $7^{.5646} \approx 3$.

Logarithmic Functions

For a given *positive* value of x, the definition of logarithm leads to exactly one value of y, so that $y = \log_a x$ defines a function.

If $a > 0$ and $a \neq 1$, the **logarithmic function** with base a is defined as

$$f(x) = \log_a x.$$

The most important logarithmic function is the natural logarithmic function.

EXAMPLE 10 Graph $f(x) = \ln x$ and $g(x) = e^x$ on the same axes.

Solution For each function, use a calculator to compute some ordered pairs. Then plot the corresponding points and connect them with a curve to obtain the graphs in Figure 4.22.

The dashed line in Figure 4.22 is the graph of $y = x$. Observe that the graph of $f(x) = \ln x$ is the mirror image of the graph of $g(x) = e^x$, with the line $y = x$ being the mirror. A pair of functions whose graphs are related in this way are said to be **inverses** of each other. A more complete discussion of inverse functions is given in many college algebra books. ✓₈

✓ Checkpoint 7

Use the properties of logarithms to rewrite and evaluate each expression, given that $\log_3 7 \approx 1.77$ and $\log_3 5 \approx 1.46$.

(a) $\log_3 35$

(b) $\log_3 7/5$

(c) $\log_3 25$

(d) $\log_3 3$

(e) $\log_3 1$

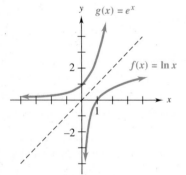

Figure 4.22

✓ Checkpoint 8

Verify that $f(x) = \log x$ and $g(x) = 10^x$ are inverses of each other by graphing $f(x)$ and $g(x)$ on the same axes.

When the base $a > 1$, the graph of $f(x) = \log_a x$ has the same basic shape as the graph of the natural logarithmic function in Figure 4.22.

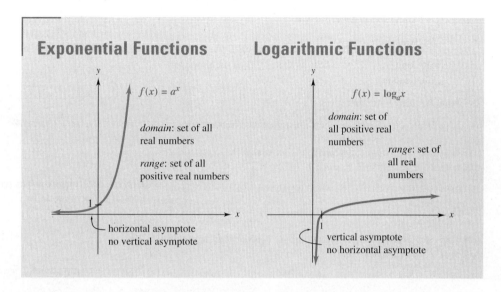

Exponential Functions

$f(x) = a^x$

domain: set of all real numbers

range: set of all positive real numbers

horizontal asymptote
no vertical asymptote

Logarithmic Functions

$f(x) = \log_a x$

domain: set of all positive real numbers

range: set of all real numbers

vertical asymptote
no horizontal asymptote

As the information in the box suggests, the functions $f(x) = \log_a x$ and $g(x) = a^x$ are inverses of each other. (Their graphs are mirror images of each other, with the line $y = x$ being the mirror.)

Applications

Logarithmic functions are useful for, among other things, describing quantities that grow, but do so at a slower rate as time goes on.

EXAMPLE 11 **Wind Energy** Annual U.S. power generation from wind (in million megawatt-hours) can be approximated by the function

$$f(x) = -335.5 + 193.1 \ln x,$$

where $x = 6$ corresponds to the year 2006. (Data from: *The Wall Street Journal.*)

(a) Find the wind power generated in 2008, 2012, and 2015.

Solution These years correspond to $x = 8$, $x = 12$, and $x = 15$, respectively, so you can use a calculator to evaluate $f(x)$ at these numbers:

$$f(8) = -335.5 + 193.1 \ln(8) \approx 66.0;$$
$$f(12) = -335.5 + 193.1 \ln(12) \approx 144.3;$$
$$f(15) = -335.5 + 193.1 \ln(15) \approx 187.4.$$

So U.S. wind power generation was about 66.0 million megawatt-hours in 2008, 144.3 million megawatt-hours in 2012, and 187.4 million megawatt-hours in 2015.

(b) If this function remains accurate, what will be the first full year in which wind power production exceeds 225 million megawatt-hours?

Solution We must solve the equation $f(x) = 225$, that is,

$$-335.5 + 193.1 \ln x = 225.$$

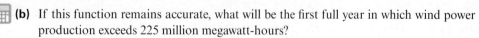

In the next section, we shall see how to do this algebraically. For now, we solve the equation graphically by graphing $f(x)$ and $y = 225$ on the same screen and finding their intersection point. Figure 4.23 shows that the x-coordinate of this intersection point is $x \approx 18.22$. Thus, the first full year in which wind power will exceed 225 million megawatt-hours is 2019.

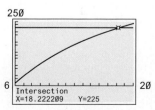

250

6   20

Intersection
X=18.222209   Y=225

Figure 4.23

4.3 Exercises

Complete each statement in Exercises 1–4.

1. $y = \log_a x$ means $x = $ _____.

2. The statement $\log_5 125 = 3$ tells us that _____ is the power of _____ that equals _____.

3. What is wrong with the expression $y = \log_b$?

4. Logarithms of negative numbers are not defined because _____.

Translate each logarithmic statement into an equivalent exponential statement. (See Examples 1, 5, and 6.)

5. $\log 100{,}000 = 5$ **6.** $\log .001 = -3$

7. $\log_9 81 = 2$ **8.** $\log_2 (1/8) = -3$

Translate each exponential statement into an equivalent logarithmic statement. (See Examples 5 and 6.)

9. $10^{1.9823} = 96$ **10.** $e^{3.2189} = 25$

11. $3^{-2} = 1/9$ **12.** $16^{1/2} = 4$

Without using a calculator, evaluate each of the given expressions. (See Examples 1, 5, and 6.)

13. $\log 1000$ **14.** $\log .0001$

15. $\log_6 36$ **16.** $\log_3 81$

17. $\log_4 64$ **18.** $\log_5 125$

19. $\log_2 \dfrac{1}{4}$ **20.** $\log_3 \dfrac{1}{27}$

21. $\ln \sqrt{e}$ **22.** $\ln(1/e)$

23. $\ln e^{8.77}$ **24.** $\log 10^{74.3}$

Use a calculator to evaluate each logarithm to three decimal places. (See Examples 3 and 4.)

25. $\log 53$ **26.** $\log .005$

27. $\ln .0068$ **28.** $\ln 354$

29. Why does $\log_a 1$ always equal 0 for any valid base a?

Write each expression as the logarithm of a single number or expression. Assume that all variables represent positive numbers. (See Example 7.)

30. $\log 20 - \log 5$ **31.** $\log 6 + \log 8 - \log 2$

32. $3 \ln 2 + 2 \ln 3$ **33.** $2 \ln 5 - \frac{1}{2} \ln 25$

34. $5 \log x - 2 \log y$ **35.** $2 \log u + 3 \log w - 6 \log v$

36. $\ln(3x + 2) + \ln(x + 4)$ **37.** $2 \ln(x + 2) - \ln(x + 3)$

Write each expression as a sum and/or a difference of logarithms, with all variables to the first degree.

38. $\log 5x^2 y^3$ **39.** $\ln \sqrt{6m^4 n^2}$

40. $\ln \dfrac{3x}{5y}$ **41.** $\log \dfrac{\sqrt{xz}}{z^3}$

42. The calculator-generated table in the figure is for $y_1 = \log(4 - x)$. Why do the values in the y_1 column show ERROR for $x \geq 4$?

X	Y1	
0	.60206	
1	.47712	
2	.30103	
3	0	
■	ERROR	
5	ERROR	
6	ERROR	

X=4

Express each expression in terms of u and v, where $u = \ln x$ and $v = \ln y$. For example, $\ln x^3 = 3(\ln x) = 3u$.

43. $\ln(x^2 y^5)$ **44.** $\ln(\sqrt{x} \cdot y^2)$

45. $\ln(x^3 / y^2)$ **46.** $\ln(\sqrt{x}/y)$

Evaluate each expression. (See Example 9.)

47. $\log_6 384$ **48.** $\log_{30} 78$

49. $\log_{35} 5646$ **50.** $\log_6 60 - \log_{60} 6$

Find numerical values for b and c for which the given statement is FALSE.

51. $\log(b + c) = \log b + \log c$ **52.** $\dfrac{\ln b}{\ln c} = \ln\left(\dfrac{b}{c}\right)$

Graph each function. (See Example 11.)

53. $y = \ln(x + 2)$ **54.** $y = \ln x + 2$

55. $y = \log(x - 3)$ **56.** $y = \log x - 3$

57. Graph $f(x) = \log x$ and $g(x) = \log(x/4)$ for $-2 \leq x \leq 8$. How are these graphs related? How does the quotient rule support your answer?

In Exercises 58 and 59, the coordinates of a point on the graph of the indicated function are displayed at the bottom of the screen. Write the logarithmic and exponential equations associated with the display.

58.

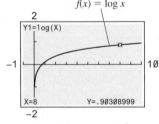

59.

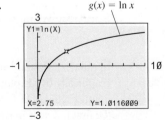

60. Match each equation with its graph. Each tick mark represents one unit.

(a) $y = \log x$　　　　**(b)** $y = 10^x$

(c) $y = \ln x$　　　　**(d)** $y = e^x$

(A)

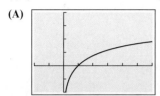

(B)

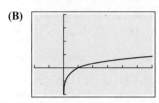

(C)

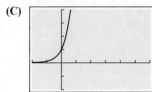

(D)

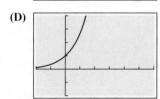

61. Finance The doubling function

$$D(r) = \frac{\ln 2}{\ln(1 + r)}$$

gives the number of years required to double your money when it is invested at interest rate r (expressed as a decimal), compounded annually. How long does it take to double your money at each of the following rates?

(a) 4%　　**(b)** 8%　　**(c)** 18%　　**(d)** 36%

(e) Round each of your answers in parts (a)–(d) to the nearest year, and compare them with these numbers: 72/4, 72/8, 72/18, and 72/36. Use this evidence to state a "rule of thumb" for determining approximate doubling time without employing the function D. This rule, which has long been used by bankers, is called the *rule of 72.*

62. Flu Virus Two people with the flu visited a college campus. The number of days, T, that it took for the flu virus to infect n people is given by

$$T = -1.43 \ln\left(\frac{10,000 - n}{4998n}\right).$$

How many days will it take for the virus to infect

(a) 500 people?　　　　**(b)** 5000 people?

63. Health Insurance Costs The average annual cost (in dollars) for health insurance in the United States can be approximated by the function

$$g(x) = -1736.6 + 1661.8 \ln x,$$

where $x = 6$ corresponds to the year 2006. (Data from: U.S. Bureau of Labor Statistics.)

(a) Estimate the average cost in 2010.

(b) Graph the function g for the period 2006 to 2015.

(c) Assuming that the graph remains accurate, what does the shape of the graph suggest regarding the average cost of health insurance?

64. Physical Science The barometric pressure p (in inches of mercury) is related to the height h above sea level (in miles) by the equation

$$h = -5 \ln\left(\frac{p}{29.92}\right).$$

The pressure readings given in parts (a)–(c) were made by a weather balloon. At what heights were they made?

(a) 29.92 in.　　**(b)** 20.05 in.　　**(c)** 11.92 in.

(d) Use a graphing calculator to determine the pressure at a height of 3 miles.

65. Dairy Expenditures The average annual expenditures for the U.S. consumer on dairy products can be approximated by the function

$$f(x) = 310.4 + 40.7 \ln x,$$

where $x = 6$ corresponds to the year 2006. (Data from: U.S. Bureau of Labor Statistics.)

(a) Estimate the average expenditures for 2009 and 2014.

(b) Assuming that the model remains accurate, what is the first full year in which expenditures on dairy products exceed $430?

66. Credit Union Assets The total assets (in billions of dollars) held by U.S. credit unions can be approximated by the function

$$f(x) = -689.9 + 695.8 \ln x,$$

where $x = 10$ corresponds to the year 2010. (Data from: Credit Union National Association.)

(a) What were the total assets in 2011 and 2015?

(b) If the model remains accurate, what is the first full year in which the total assets in credit unions exceed $1450 billion?

67. Border Patrol Budget The amount (in billions) that the United States spent on border patrol can be approximated by the function

$$g(x) = .185 + 1.209 \ln x,$$

where $x = 2$ corresponds to the year 2002. (Data from: *USA Today*.)

(a) What was the amount spent in 2005 and 2016 on border patrol?

(b) If the model remains accurate, what is the first full year in which the amount spent on border patrol exceeds $4.0 billion?

68. Opioid Deaths The number of deaths from opioids in the United States can be approximated by the function

$$f(x) = 4760 + 4705 \ln x,$$

where $x = 1$ corresponds to the year 2001. (Data from: *The Wall Street Journal*.)

(a) What was the number of deaths from opioids in 2007 and 2013?

(b) If the model remains accurate, what is the first full year in which the number of opioid deaths exceeds 18,000?

69. Apple iPhone Sales The worldwide number (in millions) of iPhones sold can be approximated by the function

$$g(x) = -875 + 399 \ln x,$$

where $x = 11$ corresponds to the year 2011. (Data from: www.statistica.com.)

(a) What was the number of worldwide sales for iPhones in 2015?

(b) If the model continues to be accurate, what was the first full year in which iPhone sales exceed 250 million?

70. Miles Traveled The total number of miles traveled by vehicles on all roads and streets in the United States (in trillions of miles) can be approximated by the function

$$h(x) = 2.8 + .091 \ln x,$$

where $x = 1$ corresponds to the year 2001. (Data from: U.S. Department of Transportation.)

(a) What was the number of total vehicle miles traveled in 2015?

(b) According to this model, what was the first full year in which the total number of vehicle miles exceeded 2.95 trillion miles?

✓ Checkpoint Answers

1. (a) 2   **(b)** 3   **(c)** −1

2. (a) 1.4314   **(b)** 3.0370   **(c)** −2.3706

3. (a) 1.8083   **(b)** 2.9957   **(c)** −.2231
   (d) −2.3026

4. (a) $\log_5 125 = 3$ **(b)** $\log_3 (1/81) = -4$ **(c)** $\log_8 4 = 2/3$

5. (a) $16^{1/2} = 4$ **(b)** $3^{-2} = 1/9$ **(c)** $16^{3/4} = 8$

6. (a) $\log_a 15x^5$ **(b)** $\log_a (3p/5q)$ **(c)** $\log_a (k^4/m^3)$

7. (a) 3.23   **(b)** .31   **(c)** 2.92
   (d) 1   **(e)** 0

8.

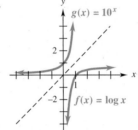

4.4 Logarithmic and Exponential Equations

Many applications involve solving logarithmic and exponential equations, so we begin with solution methods for such equations.

Logarithmic Equations

When an equation involves only logarithmic terms, use the logarithm properties to write each side as a single logarithm. Then use the following fact.

> Let u, v, and a be positive real numbers, with $a \neq 1$.
>
> If $\log_a u = \log_a v$, then $u = v$.

EXAMPLE 1   Solve $\log x = \log(x + 3) - \log(x - 1)$.

Solution First, use the quotient property of logarithms to write the right side as a single logarithm:

$$\log x = \log(x + 3) - \log(x - 1)$$

$$\log x = \log\left(\frac{x + 3}{x - 1}\right). \qquad \text{Quotient property of logarithms}$$

The fact in the preceding box now shows that

$$x = \frac{x + 3}{x - 1}$$

$$x(x - 1) = x + 3 \qquad \text{Cross multiply.}$$

$$x^2 - x = x + 3 \qquad \text{Distributive property}$$

$$x^2 - 2x - 3 = 0 \qquad \text{Add } -x - 3 \text{ to each side and combine like terms.}$$

$$(x - 3)(x + 1) = 0 \qquad \text{Factor.}$$

$$x = 3 \quad \text{or} \quad x = -1. \qquad \text{Set each factor to 0 and solve for } x.$$

Since log x is not defined when $x = -1$, the only possible solution is $x = 3$. We indicate this by crossing the value that is not a solution

$$x = 3 \quad \text{or} \quad x \cancel{=} -1.$$

Use a calculator to verify that 3 is actually a solution. ✓1

When an equation involves constants and logarithmic terms, use algebra and the logarithm properties to write one side as a single logarithm and the other as a constant. Then use the following property of logarithms, which was discussed on pages 200–201.

✓ **Checkpoint 1**

Solve each equation.

(a) $\log_2 (p + 9) - \log_2 p = \log_2 (p + 1)$

(b) $\log_3 (m + 1) - \log_3 (m - 1) = \log_3 m$

If a and u are positive real numbers, with $a \neq 1$, then

$$a^{\log_a u} = u.$$

EXAMPLE 2 Solve each equation.

(a) $\log_5 (2x - 3) = 2$

Solution Since the base of the logarithm is 5, raise 5 to the exponents given by the equation:

$$5^{\log_5 (2x-3)} = 5^2$$

On the left side, use the fact in the preceding box (with $a = 5$ and $u = 2x - 3$) to conclude that

$$2x - 3 = 25$$
$$2x = 28$$
$$x = 14.$$

Verify that 14 is a solution of the original equation.

(b) $\log(x - 16) = 2 - \log(x - 1)$

Solution First rearrange the terms to obtain a single logarithm on the left side:

$$\log(x - 16) + \log(x - 1) = 2$$
$$\log[(x - 16)(x - 1)] = 2 \qquad \text{Product property of logarithms}$$
$$\log(x^2 - 17x + 16) = 2.$$

Since the base of the logarithm is 10, raise 10 to the given powers:

$$10^{\log(x^2 - 17x + 16)} = 10^2.$$

On the left side, apply the logarithm property in the preceding box (with $a = 10$ and $u = x^2 - 17x + 16$):

$$x^2 - 17x + 16 = 100$$
$$x^2 - 17x - 84 = 0 \qquad \text{Add } -100 \text{ to each side.}$$
$$(x + 4)(x - 21) = 0 \qquad \text{Factor.}$$
$$x = -4 \quad \text{or} \quad x = 21. \qquad \text{Set each factor} = 0 \text{ and solve for } x.$$

In the original equation, when $x = -4$, $\log(x - 16) = \log(-20)$, which is not defined. So -4 is not a solution. We indicate this by crossing the value that is not a solution

$$x \cancel{= -4} \quad \text{or} \quad x = 21.$$

You should verify that 21 is a solution of the original equation.

(c) $\log_2 x - \log_2 (x - 1) = 1$

Solution Proceed as before, with 2 as the base of the logarithm:

$$\log_2 \frac{x}{x - 1} = 1 \qquad \text{Quotient property of logarithms}$$

$$2^{\log_2 x/(x-1)} = 2^1 \qquad \text{Exponentiate to the base 2.}$$

$$\frac{x}{x - 1} = 2 \qquad \text{Use the fact in the preceding box.}$$

$$x = 2(x - 1) \qquad \text{Multiply both sides by } x - 1.$$

$$x = 2x - 2 \qquad \text{Distributive property}$$

$$-x = -2 \qquad \text{Multiply both sides by } -1.$$

$$x = 2.$$

Verify that 2 is a solution of the original equation. ✓₂

✓ Checkpoint 2

Solve each equation.

(a) $\log_5 x + 2 \log_5 x = 3$

(b) $\log_6 (a + 2) - \log_6 \dfrac{a - 7}{5} = 1$

Exponential Equations

An equation in which all the variables are exponents is called an exponential equation. When such an equation can be written as two powers of the same base, it can be solved by using the following fact.

> Let a be a positive real number, with $a \neq 1$.
>
> $$\text{If } a^u = a^v, \quad \text{then} \quad u = v.$$

EXAMPLE 3   Solve $9^x = 27$.

Solution First, write both sides as powers of the same base. Since $9 = 3^2$ and $27 = 3^3$, we have

$$9^x = 27$$

$$(3^2)^x = 3^3 \qquad \text{Write 9 and 27 as powers of 3.}$$

$$3^{2x} = 3^3. \qquad \text{Power of a power}$$

Apply the fact in the preceding box (with $a = 3$, $u = 2x$, and $v = 3$):

$$2x = 3$$

$$x = \frac{3}{2}.$$

Verify that $x = 3/2$ is a solution of the original equation. ✓₃

✓ Checkpoint 3

Solve each equation.

(a) $8^{2x} = 4$

(b) $5^{3x} = 25^4$

(c) $36^{-2x} = 6$

Exponential equations involving different bases can often be solved by using the power property of logarithms, which is repeated here for natural logarithms.

> If u and r are real numbers, with u positive, then
>
> $$\ln u^r = r \ln u.$$

Although natural logarithms are used in the following examples, logarithms to any base will produce the same solutions.

EXAMPLE 4   Solve $3^x = 5$.

Solution Take natural logarithms on both sides:

$$\ln 3^x = \ln 5.$$

Apply the power property of logarithms in the preceding box (with $r = x$) to the left side:

$$x \ln 3 = \ln 5$$

$$x = \frac{\ln 5}{\ln 3} \approx 1.465. \qquad \text{Divide both sides by the constant ln 3.}$$

To check, evaluate $3^{1.465}$; the answer should be approximately 5, which verifies that the solution of the given equation is 1.465 (to the nearest thousandth). ✓ 4

Checkpoint 4

Solve each equation. Round solutions to the nearest thousandth.

(a) $2^x = 7$

(b) $5^m = 50$

(c) $3^y = 17$

⊘ **CAUTION** Be careful: $\dfrac{\ln 5}{\ln 3}$ is *not* equal to $\ln\left(\dfrac{5}{3}\right)$ or $\ln 5 - \ln 3$.

EXAMPLE 5   Solve $3^{2x-1} = 4^{x+2}$.

Solution Taking natural logarithms on both sides gives

$$\ln 3^{2x-1} = \ln 4^{x+2}.$$

Now use the power property of logarithms and the fact that $\ln 3$ and $\ln 4$ are constants to rewrite the equation:

$$(2x - 1)(\ln 3) = (x + 2)(\ln 4) \qquad \text{Power property}$$

$$2x(\ln 3) - 1(\ln 3) = x(\ln 4) + 2(\ln 4) \qquad \text{Distributive property}$$

$$2x(\ln 3) - x(\ln 4) = 2(\ln 4) + 1(\ln 3). \qquad \text{Collect terms with } x \text{ on one side.}$$

Factor out x on the left side to get

$$[2(\ln 3) - \ln 4]x = 2(\ln 4) + \ln 3.$$

Divide both sides by the coefficient of x:

$$x = \frac{2(\ln 4) + \ln 3}{2(\ln 3) - \ln 4}.$$

Using a calculator to evaluate this last expression, we find that

Checkpoint 5

Solve each equation. Round solutions to the nearest thousandth.

(a) $6^m = 3^{2m-1}$

(b) $5^{6a-3} = 2^{4a+1}$

$$x = \frac{2 \ln 4 + \ln 3}{2 \ln 3 - \ln 4} \approx 4.774. \; ✓ 5$$

Recall that $\ln e = 1$ (because 1 is the exponent to which e must be raised to produce e). This fact simplifies the solution of equations involving powers of e.

EXAMPLE 6 Solve $3e^{x^2} = 600$.

Solution First divide each side by 3 to get

$$e^{x^2} = 200.$$

Now take natural logarithms on both sides; then use the power property of logarithms:

$$e^{x^2} = 200$$
$$\ln e^{x^2} = \ln 200 \qquad \text{Take natural logarithms on both sides.}$$
$$x^2 \ln e = \ln 200 \qquad \text{Power property}$$
$$x^2 = \ln 200 \qquad \text{In } e = 1.$$
$$x = \pm\sqrt{\ln 200} \qquad \text{Take the square root of each side.}$$
$$x \approx \pm 2.302.$$

Verify that the solutions are ± 2.302, rounded to the nearest thousandth. (The symbol $\pm$ is used as a shortcut for writing the two solutions 2.302 and -2.302.) ✓6

✓**Checkpoint 6**

Solve each equation. Round solutions to the nearest thousandth.

(a) $e^{1x} = 11$

(b) $e^{3+x} = .893$

(c) $e^{2x^2-3} = 9$

▦ **TECHNOLOGY TIP**

Logarithmic and exponential equations can be solved on a graphing calculator. To solve $3^x = 5^{2x-1}$, for example, graph $y = 3^x$ and $y = 5^{2x-1}$ on the same screen. Then use the intersection finder to determine the x-coordinates of their intersection points. Alternatively, graph $y = 3^x - 5^{2x-1}$ and use the root finder to determine the x-intercepts of the graph.

Applications

Some of the most important applications of exponential and logarithmic functions arise in banking and finance. They will be thoroughly discussed in Chapter 5. The applications here are from other fields.

EXAMPLE 7 **New Jobs** The number of nonfarm jobs (in millions) added to the U.S. economy by quarter can be approximated by the function

$$f(x) = 12.1 + 2.03 \ln x,$$

where $x = 1$ corresponds to the first quarter of the year 2014. Find the first full quarter in which the number of jobs added to the economy exceeded 16 million. (Data from: The Bureau of Labor Statistics.)

Solution We need to find the value for x for which $f(x) = 16$. Thus, we need to solve the following equation for x:

$$12.1 + 2.03 \ln x = 16$$
$$2.03 \ln x = 3.9 \qquad \text{Subtract 12.1 from each side.}$$
$$\ln x = \frac{3.9}{2.03} \qquad \text{Divide each side by 2.03.}$$
$$e^{\ln x} = e^{3.9/2.03} \qquad \text{Exponentiate each side.}$$
$$x \approx 6.83.$$

Because x is fractional and the question asks for the "first full quarter," we round up to $x = 7$, which would be the third quarter of 2015. ✓7

✓**Checkpoint 7**

Use the function in Example 7 to determine the first full quarter in which the number of jobs added to the economy exceeded 17 million.

EXAMPLE 8 **Digital Groceries** The amount of money U.S. consumers spend online for food (in billions) can be approximated by the function.

$$f(x) = .174(1.318)^x,$$

where $x = 11$ corresponds to the year 2011. Assuming that the model remains accurate, determine the first full year in which online grocery sales exceed $50 billion. (Data from: The Wall Street Journal.)

Solution We need to find the value of x for which $f(x) = 50$—that is, to solve the following equation:

$$.174(1.318)^x = 50$$

$$1.318^x = \frac{50}{.174} \qquad \text{Divide both sides by .174.}$$

$$\ln(1.318^x) = \ln\left(\frac{50}{.174}\right) \qquad \text{Take logarithms on both sides.}$$

$$x \ln(1.318) = \ln\left(\frac{50}{.174}\right) \qquad \text{Power property of logarithms}$$

$$x = \frac{\ln(50/.174)}{\ln(1.318)} \qquad \text{Divide both sides by ln(1.318).}$$

$$x \approx 20.5.$$

Thus, the first full year in which this model predicts that sales will exceed $50 billion is 2021. ✓ 8

The **half-life** of a radioactive substance is the time it takes for a given quantity of the substance to decay to one-half its original mass. The half-life depends only on the substance, not on the size of the sample. It can be shown that the amount of a radioactive substance at time t is given by the function

$$f(t) = y_0\left(\frac{1}{2}\right)^{t/h} = y_0(.5^{t/h}),$$

where y_0 is the initial amount (at time $t = 0$) and h is the half-life of the substance.

Radioactive carbon-14 is found in every living plant and animal. After the plant or animal dies, its carbon-14 decays exponentially, with a half-life of 5730 years. This fact is the basis for a technique called *carbon dating* for determining the age of fossils.

EXAMPLE 9   **Carbon Dating** A round wooden table hanging in Winchester Castle (England) was alleged to have belonged to King Arthur, who lived in the fifth century. A recent chemical analysis showed that the wood had lost 9% of its carbon-14.[*] How old is the table?

Solution The decay function for carbon-14 is

$$f(t) = y_0(.5^{t/5730}),$$

where $t = 0$ corresponds to the time the wood was cut to make the table. (That is when the tree died.) Since the wood has lost 9% of its carbon-14, the amount in the table now is 91% of the initial amount y_0 (that is, $.91y_0$). We must find the value of t for which $f(t) = .91y_0$. So we must solve the equation

$$y_0(.5^{t/5730}) = .91y_0 \qquad \text{Definition of } f(t)$$

$$.5^{t/5730} = .91 \qquad \text{Divide both sides by } y_0.$$

$$\ln .5^{t/5730} = \ln .91 \qquad \text{Take logarithms on both sides.}$$

$$\left(\frac{t}{5730}\right)\ln .5 = \ln .91 \qquad \text{Power property of logarithms}$$

$$t \ln .5 = 5730 \ln .91 \qquad \text{Multiply both sides by 5730.}$$

$$t = \frac{5730 \ln .91}{\ln .5} \approx 779.63. \qquad \text{Divide both sides by ln .5.}$$

The table is about 780 years old and therefore could not have belonged to King Arthur. ✓ 9

 **Checkpoint 8**

Use the function in Example 8 to determine the first full year in which online grocery sales exceeded $20 billion.

Checkpoint 9

How old is a skeleton that has lost 65% of its carbon-14?

[*]This is done by measuring the ratio of carbon-14 to nonradioactive carbon-12 in the table (a ratio that is approximately constant over long periods) and comparing it with the ratio in living wood.

Earthquakes are often in the news. The standard method of measuring their size, the **Richter scale,** is a logarithmic function (base 10).

EXAMPLE 10   **Richter Scale** The intensity $R(i)$ of an earthquake, measured on the Richter scale, is given by

$$R(i) = \log\left(\frac{i}{i_0}\right),$$

where i is the intensity of the ground motion of the earthquake and i_0 is the intensity of the ground motion of the so-called *zero earthquake* (the smallest detectable earthquake, against which others are measured). The underwater earthquake that caused the disastrous 2004 tsunami in Southeast Asia measured 9.1 on the Richter scale.

(a) How did the ground motion of this tsunami compare with that of the zero earthquake?

Solution In this case,

$$R(i) = 9.1$$

$$\log\left(\frac{i}{i_0}\right) = 9.1.$$

By the definition of logarithms, 9.1 is the exponent to which 10 must be raised to produce i/i_0, which means that

$$10^{9.1} = \frac{i}{i_0}, \quad \text{or equivalently,} \quad i = 10^{9.1}i_0.$$

So the earthquake that produced the tsunami had $10^{9.1}$ (about 1.26 *billion*) times more ground motion than the zero earthquake.

(b) What is the Richter-scale intensity of an earthquake with 10 times as much ground motion as the 2004 tsunami earthquake?

Solution From (a), the ground motion of the tsunami quake was $10^{9.1}i_0$. So a quake with 10 times that motion would satisfy

$$i = 10(10^{9.1}i_0) = 10^1 \cdot 10^{9.1}i_0 = 10^{10.1}i_0.$$

Therefore, its Richter scale measure would be

$$R(i) = \log\left(\frac{i}{i_0}\right) = \log\left(\frac{10^{10.1}i_0}{i_0}\right) = \log 10^{10.1} = 10.1.$$

Thus, a tenfold increase in ground motion increases the Richter scale measure by only 1. ✔10

✓**Checkpoint 10**

Find the Richter-scale intensity of an earthquake whose ground motion is 100 times greater than the ground motion of the 2004 tsunami earthquake discussed in Example 10.

4.4 Exercises

Solve each logarithmic equation. (See Example 1.)

1. $\ln(x + 3) = \ln(2x - 5)$

2. $\ln(8k - 7) - \ln(3 + 4k) = \ln(9/11)$

3. $\ln(3x + 1) - \ln(5 + x) = \ln 2$

4. $\ln(5x + 2) = \ln 4 + \ln(x + 3)$

5. $2 \ln(x - 3) = \ln(x + 5) + \ln 4$

6. $\ln(k + 5) + \ln(k + 2) = \ln 18k$

Solve each logarithmic equation. (See Example 2.)

7. $\log_3 (6x - 2) = 2$

8. $\log_5 (3x - 4) = 1$

9. $\log x - \log(x + 4) = -1$

10. $\log m - \log(m + 4) = -2$

11. $\log_3 (y + 2) = \log_3 (y - 7) + \log_3 4$

12. $\log_8 (z - 6) = 2 - \log_8 (z + 15)$

13. $\ln(x + 9) - \ln x = 1$

14. $\ln(2x + 1) - 1 = \ln(x - 2)$

15. $\log x + \log(x - 9) = 1$

16. $\log(x - 1) + \log(x + 2) = 1$

Solve each equation for c.

17. $\log(3 + b) = \log(4c - 1)$ **18.** $\ln(b + 7) = \ln(6c + 8)$

19. $2 - b = \log(6c + 5)$ **20.** $8b + 6 = \ln(2c) + \ln c$

21. Suppose you overhear the following statement: "I must reject any negative answer when I solve an equation involving logarithms." Is this correct? Write an explanation of why it is or is not correct.

22. What values of x cannot be solutions of the following equation?

$$\log_a (4x - 7) + \log_a (x^2 + 4) = 0.$$

Solve these exponential equations without using logarithms. (See Example 3.)

23. $2^{x-1} = 8$ **24.** $16^{-x+2} = 8$

25. $25^{-3x} = 3125$ **26.** $81^{-2x} = 3^{x-1}$

27. $6^{-x} = 36^{x+6}$ **28.** $16^x = 64$

29. $\left(\dfrac{3}{4}\right)^x = \dfrac{16}{9}$ **30.** $2^{x^2 - 4x} = \dfrac{1}{16}$

Use logarithms to solve these exponential equations. (See Examples 4–6.)

31. $2^x = 5$ **32.** $5^x = 8$

33. $2^x = 3^{x-1}$ **34.** $4^{x+2} = 5^{x-1}$

35. $3^{1-2x} = 5^{x+5}$ **36.** $4^{3x-1} = 3^{x-2}$

37. $e^{3x} = 6$ **38.** $e^{-3x} = 5$

39. $2e^{5a+2} = 8$ **40.** $10e^{3z-7} = 5$

Solve each equation for c.

41. $10^{4c-3} = d$ **42.** $3 \cdot 10^{2c+1} = 4d$

43. $e^{2c-1} = b$ **44.** $3e^{5c-7} = b$

Solve these equations. (See Examples 1–6.)

45. $\log_7 (r + 3) + \log_7 (r - 3) = 1$

46. $\log_4 (z + 3) + \log_4 (z - 3) = 2$

47. $\log_3 (a - 3) = 1 + \log_3 (a + 1)$

48. $\log w + \log(3w - 13) = 1$

49. $\log_2 \sqrt{2y^2} - 1 = 3/2$ **50.** $\log_2 (\log_2 x) = 1$

51. $\log_2 (\log_3 x) = 1$ **52.** $\dfrac{\ln(2x + 1)}{\ln(3x - 1)} = 2$

53. $5^{-2x} = \dfrac{1}{25}$ **54.** $5^{x^2+x} = 1$

55. $2^{|x|} = 16$ **56.** $5^{-|x|} = \dfrac{1}{25}$

57. $2^{x^2-1} = 10$ **58.** $3^{2-x^2} = 8$

59. $2(e^x + 1) = 10$ **60.** $5(e^{2x} - 2) = 15$

61. Explain why the equation $4^{x^2+1} = 2$ has no solutions.

62. Explain why the equation $\log(-x) = -4$ does have a solution, and find that solution.

Work these problems. (See Examples 6, 7, and 8.)

63. **Foreign Earnings** The amount of foreign earnings (in trillions of dollars) kept overseas by S&P 500 companies can be approximated by the function

$$f(x) = .227(1.17)^x,$$

where $x = 5$ corresponds to the year 2005. Find the first full year in which the foreign earnings kept overseas exceeds the following amounts. (Data from: *The Wall Street Journal.*)

   (a) $1.5 trillion   **(b)** $2.5 trillion

64. **Nursing Degrees** The number of registered nurses (in thousands) enrolled in bachelor of nursing (BSN) degree completion programs in the United States can be approximated by the function

$$g(x) = 21.8(1.14)^x,$$

where $x = 4$ corresponds to the year 2004. Find the first full year in which the number of enrollees first exceeded the following values. (Data from: *The Wall Street Journal.*)

   (a) 100,000   **(b)** 140,000

65. **Veterans Benefits** The total expenditures on benefits for U.S. veterans (in billions of dollars) can be approximated by the function

$$h(x) = 23.5(1.08)^x,$$

where $x = 5$ corresponds to the year 1995. (Data from: U.S. Department of Veterans Affairs.)

   (a) What were the amount for total expenditures in 2013?

   (b) What was the first full year in which expenditures exceeded $110 billion?

66. **Snapchat Users** The total number of global daily users (in millions) of Snapchat can be approximated by the function

$$g(x) = 44.2(1.12)^x,$$

where $x = 1$ corresponds to the first quarter of the year 2014. (Data from: *The Wall Street Journal.*)

   (a) Give the number of global daily users for the third quarter of 2015.

   (b) What was the first full quarter in which the number of daily global users exceeded 120 million?

67. **Wind Energy** As we saw in Example 12 of Section 4.3, annual U.S. power generation from wind (in million megawatt-hours) can be approximated by the function

$$f(x) = -335.5 + 193.1 \ln x,$$

where $x = 6$ corresponds to the year 2006. (Data from: *The Wall Street Journal.*) Assuming that the model remains accurate, find the first full year in which the wind power exceeds the following values.

   (a) 250 million megawatt-hours

   (b) 300 million megawatt-hours

68. Japanese Messaging The number of daily users for Japan's Line messaging service (in millions) can be approximated by the function

$$g(x) = 53.1 + 63.1 \ln(x),$$

where $x = 1$ corresponds to the first quarter of 2013. Numbering quarters from that date, and assuming the model remains accurate, find the first full quarter in which the number of users exceeded 220 million.

69. CVS Health Earnings per Share The earnings per share for CVS Health (in dollars) can be approximated by the function

$$g(x) = -5.3 + 3.54 \ln x,$$

where $x = 7$ corresponds to the year 2007. (Data from: www.morningstar.com.)

(a) Find the earnings per share for the year 2010.

(b) Find the first full year in which the earnings per share exceeded $3.25.

70. Number of Teachers The total number of teachers in the United States (in millions) can be approximated by the function

$$h(x) = .5 + .84 \ln x,$$

where $x = 10$ corresponds to the year 1990. (Data from: U.S. National Center for Education Statistics.)

(a) Find the number of teachers in the year 2013.

(b) Find the first full year in which the number of teachers exceeded 3.0 million.

71. CVS Health Revenue The total revenue for CVS Health (in billions of dollars) can be approximated by the function

$$f(x) = 41.6e^{.088x},$$

where $x = 7$ corresponds to the year 2007. (Data from: www.morningstar.com.)

(a) What was the total revenue in 2011?

(b) What was the first full year in which revenue exceeded $140 billion?

72. Best Buy Revenue The total revenue for Best Buy Co. Inc., (in billions of dollars) can be approximated by the function

$$h(x) = 94.6e^{-.054x},$$

where $x = 12$ corresponds to the year 2012. (Data from: www.morningstar.com.)

(a) What was the total revenue in 2017?

(b) What was the first full year in which revenue was below $40 billion?

73. Twitter Stock The stock price to earnings ratio for Twitter can be approximated by the function

$$g(x) = 22e^{-.23x},$$

where $x = 1$ corresponds to the first quarter of the year 2015. What was the first full quarter where the ratio was lower than 5.0? (Data from: www.morningstar.com.)

74. Outstanding Loans The percentage of loans outstanding as a share of deposits in U.S. banks can be approximated by the function

$$h(x) = 120.7e^{-.04x},$$

where $x = 7$ corresponds to the year 2007. What was the first full year in which the percentage of loans outstanding was first below 75%? (Data from: Federal Deposit Insurance Corp.)

Work these exercises. (See Example 9.)

75. Natural Science The amount of cobalt-60 (in grams) in a storage facility at time t is given by

$$C(t) = 25e^{-.14t},$$

where time is measured in years.

(a) How much cobalt-60 was present initially?

(b) What is the half-life of cobalt-60? (*Hint:* For what value of t is $C(t) = 12.5$?)

76. Carbon Dating A Native American mummy was found recently. It had 73.6% of the amount of radiocarbon present in living beings. Approximately how long ago did this person die?

77. Carbon Dating How old is a piece of ivory that has lost 36% of its radiocarbon?

78. Carbon Dating A sample from a refuse deposit near the Strait of Magellan had 60% of the carbon-14 of a contemporary living sample. How old was the sample?

Work these problems. (See Example 10.)

79. Richter Scale In May 2008, Sichuan province, China suffered an earthquake that measured 7.9 on the Richter scale.

(a) Express the intensity of this earthquake in terms of i_0.

(b) In July of the same year, a quake measuring 5.4 on the Richter scale struck Los Angeles. Express the intensity of this earthquake in terms of i_0.

(c) How many times more intense was the China earthquake than the one in Los Angeles?

80. Richter Scale Find the Richter-scale intensity of earthquakes whose ground motion is

(a) $1000i_0$; (b) $100,000i_0$;

(c) $10,000,000i_0$.

(d) Fill in the blank in this statement: Increasing the ground motion by a factor of 10^k increases the Richter intensity by _____ units.

81. Sound The loudness of sound is measured in units called decibels. The decibel rating of a sound is given by

$$D(i) = 10 \cdot \log\left(\frac{i}{i_0}\right),$$

where i is the intensity of the sound and i_0 is the minimum intensity detectable by the human ear (the so-called *threshold sound*). Find the decibel rating of each of the sounds with the given intensities. Round answers to the nearest whole number.

(a) Whisper, $115i_0$

(b) Average sound level in the movie *Godzilla*, $10^{10}i_0$

(c) Jackhammer, $31{,}600{,}000{,}000i_0$

(d) Rock music, $895{,}000{,}000{,}000i_0$

(e) Jetliner at takeoff, $109{,}000{,}000{,}000{,}000i_0$

82. Sound Using information from Exercise 81, answer the following.

(a) How much more intense is a sound that measures 100 decibels than the threshold sound?

(b) How much more intense is a sound that measures 50 decibels than the threshold sound?

(c) How much more intense is a sound measuring 100 decibels than one measuring 50 decibels?

✓ **Checkpoint Answers**

1. (a) 3 (b) $1 + \sqrt{2} \approx 2.414$
2. (a) 5 (b) 52
3. (a) 1/3 (b) 8/3 (c) −1/4
4. (a) 2.807 (b) 2.431 (c) 2.579
5. (a) 2.710 (b) .802
6. (a) 23.979 (b) −3.113 (c) ±1.612
7. Fourth quarter of 2016 ($x \approx 11.18$)
8. 2018 ($x \approx 17.18$)
9. About 8679 years
10. 11.1

CHAPTER 4 Summary and Review

Key Terms and Symbols

4.1 exponential function
exponential growth and
decay
the number
$e \approx 2.71828 \ldots$

4.2 exponential growth function
logistic function

4.3 $\log x$ common (base-10)
logarithm of x
$\ln x$ natural (base-e)
logarithm of x
$\log_a x$ base-a logarithm
of x

product, quotient, and
power properties of
logarithms
change-of-base theorem
Inverse logarithmic
functions

4.4 logarithmic equations
exponential equations
half-life
Richter scale

Chapter 4 Key Concepts

Exponentialic Functions An important application of exponents is the **exponential growth function,** defined as $f(t) = y_0 e^{kt}$ or $f(t) = y_0 b^t$, where y_0 is the amount of a quantity present at time $t = 0$, $e \approx 2.71828$, and k and b are constants.

Logarithmic Functions The **logarithm** of x to the base a is defined as follows: For $a > 0$ and $a \neq 1$, $y = \log_a x$ means $a^y = x$. Thus, $\log_a x$ is an *exponent*, the power to which a must be raised to produce x.

Properties of Logarithms Let x, y, and a be positive real numbers, with $a \neq 1$, and let r be any real number. Then

$$\log_a 1 = 0; \quad \log_a a = 1;$$
$$\log_a a^r = r; \quad a^{\log_a x} = x.$$

Product property $\log_a xy = \log_a x + \log_a y$

Quotient property $\log_a \dfrac{x}{y} = \log_a x - \log_a y$

Power property $\log_a x^r = r \log_a x$

**Solving Exponential and
Logarithmic Equations**
Let $a > 0$, with $a \neq 1$.
If $\log_a u = \log_a v$, then $u = v$.
If $a^u = a^v$, then $u = v$.

Chapter 4 Review Exercises

Match each equation with the letter of the graph that most closely resembles the graph of the equation. Assume that a > 1.

1. $y = a^{x+2}$

2. $y = a^x + 2$

3. $y = -a^x + 2$

4. $y = a^{-x} + 2$

(a)

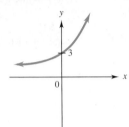

(b)

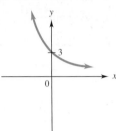

(c)

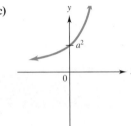

(d)

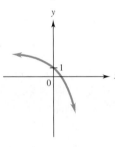

Consider the exponential function $y = f(x) = a^x$ graphed here. Answer each question on the basis of the graph.

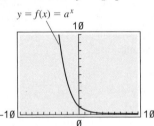

$y = f(x) = a^x$

5. What is true about the value of a in comparison to 1?

6. What is the domain of f?

7. What is the range of f?

8. What is the value of $f(0)$?

Graph each function.

9. $f(x) = 4^x$

10. $g(x) = 4^{-x}$

11. $f(x) = \ln x + 5$

12. $g(x) = \log x - 3$

Work these problems.

13. Exports to Mexico The amount of U.S. exports to Mexico (in billions of dollars) can be approximated by the function

$$g(x) = 27.1(1.10)^x,$$

where $x = 0$ corresponds to the year 1990. (Data from: *The Wall Street Journal.*)

(a) Estimate the value of U.S. exports to Mexico in 2015.

(b) What was the first full year in which exports exceeded $150 billion in value according to this model?

14. Royal Caribbean Share Price The percent gain in share price for the cruise line Royal Caribbean can be approximated by the function

$$f(x) = .0007(2.29)^x,$$

where $x = 12$ corresponds to the year 2012. (Data from: *The Wall Street Journal.*)

(a) Find the percent gain in share price for the year 2013.

(b) According to this model, what was the first full year to have the share price percentage gain be greater than 100%?

Translate each exponential statement into an equivalent logarithmic one.

15. $10^{2.53148} = 340$

16. $5^4 = 625$

17. $e^{3.8067} = 45$

18. $7^{1/2} = \sqrt{7}$

Translate each logarithmic statement into an equivalent exponential one.

19. $\log 10{,}000 = 4$

20. $\log 26.3 = 1.4200$

21. $\ln 81.1 = 4.3957$

22. $\log_2 4096 = 12$

Evaluate these expressions without using a calculator.

23. $\ln e^5$

24. $\log \sqrt[3]{10}$

25. $10^{\log 8.9}$

26. $\ln e^{3t^2}$

27. $\log_8 2$

28. $\log_8 32$

Write these expressions as a single logarithm. Assume all variables represent positive quantities.

29. $\log 4x + \log 5x^5$

30. $4 \log u - 5 \log u^6$

31. $3 \log b - 2 \log c$

32. $7 \ln x - 3(\ln x^3 + 5 \ln x)$

Solve each equation. Round to the nearest thousandth.

33. $\ln(m + 8) - \ln m = \ln 3$

34. $2 \ln(y + 1) = \ln(y^2 - 1) + \ln 5$

35. $\log(m + 3) = 2$

36. $\log x^3 = 2$

37. $\log_2 (3k + 1) = 4$

38. $\log_5 \left(\dfrac{5z}{z - 2} \right) = 2$

39. $\log x + \log(x - 3) = 1$

40. $\log_2 r + \log_2 (r - 2) = 3$

41. $2^{3x} = \dfrac{1}{64}$

42. $\left(\dfrac{9}{16}\right)^x = \dfrac{3}{4}$

43. $9^{2y+1} = 27^y$

44. $\dfrac{1}{2} = \left(\dfrac{b}{4}\right)^{1/4}$

45. $8^p = 19$

46. $3^z = 11$

47. $5 \cdot 2^{-m} = 35$

48. $2 \cdot 15^{-k} = 18$

49. $e^{-5-2x} = 5$

50. $e^{3x-1} = 12$

51. $6^{2-m} = 2^{3m+1}$

52. $5^{3r-1} = 6^{2r+5}$

53. $(1 + .003)^k = 1.089$

54. $(1 + .094)^z = 2.387$

Work these problems.

55. Population Growth A population is increasing according to the growth law $y = 4e^{.03t}$, where y is in thousands and t is in months. Match each of the questions (a), (b), (c), and (d) with one of the solution (A), (B), (C), or (D).

(a) How long will it take for the population to double?

(A) Solve $4e^{.03t} = 12$ for t.

(b) When will the population reach 12 thousand?

(B) Evaluate $4e^{.03(72)}$.

(c) How large will the population be in 6 months?

(C) Solve $4e^{.03t} = 2 \cdot 4$ for t.

(d) How large will the population be in 6 years?

(D) Evaluate $4e^{.03(6)}$.

56. Population Growth A population is increasing according to the growth law $y = 2e^{.02t}$, where y is in millions and t is in years. Match each of the questions (a), (b), (c), and (d) with one of the solutions (A), (B), (C), or (D).

(a) How long will it take for the population to triple?

(A) Evaluate $2e^{.02(1/3)}$.

(b) When will the population reach 3 million?

(B) Solve $2e^{.02t} = 3 \cdot 2$ for t.

(c) How large will the population be in 3 years?

(C) Evaluate $2e^{.02(3)}$.

(d) How large will the population be in 4 months?

(D) Solve $2e^{.02t} = 3$ for t.

57. Murders in Chicago The number of murders in Chicago can be approximated by the function

$$g(x) = 846e^{-.034x},$$

where $x = 0$ corresponds to the year 1990. (Data from: Federal Bureau of Investigation.)

(a) What was the number of deaths according to this model in 2012?

(b) According to this model, what was the first full year in which the number of murders fell below 600?

58. Natural Science The amount of polonium (in grams) present after t days is given by

$$A(t) = 10e^{-.00495t}.$$

(a) How much polonium was present initially?

(b) What is the half-life of polonium?

(c) How long will it take for the polonium to decay to 3 grams?

59. Crude Oil Futures The price per barrel (in dollars) for NYMEX-traded crude oil futures can be approximated by the function

$$g(x) = \frac{124}{1 + .118e^{.371x}},$$

where $x = 1$ corresponds to the first quarter of the year 2014. (Data from: *The Wall Street Journal.*)

(a) Find the futures price of crude oil for the first quarter of 2015.

(b) What quarter was the first full quarter in which the price of crude oil dropped below $50?

60. Coal Futures The price of a metric ton of coal on the Newcastle futures market can be approximated by the function

$$h(x) = \frac{170}{1 + .009e^{.342x}},$$

where $x = 11$ corresponds to the year 2011. (Data from: *The Wall Street Journal.*)

(a) Find the futures price of coal for the year 2016.

(b) What was the first full year in which the price of coal fell below $90 for a metric ton?

61. Richter Scale In April 2016, Ecuador suffered an earthquake that measured 7.8 on the Richter scale.

(a) Express the intensity of this earthquake in terms of i_0.

(b) That same month, a magnitude 6.0 earthquake struck offshore Mexico. Express the intensity of this earthquake in terms of i_0.

(c) How many times more intense was the earthquake in Ecuador than the one that struck offshore Mexico?

62. Natural Science One earthquake measures 4.6 on the Richter scale. A second earthquake has ground motion 1000 times greater than the first. What does the second one measure on the Richter scale?

Natural Science *Another form of Newton's law of cooling (see Section 4.2, Exercises 15 and 16) is $F(t) = T_0 + Ce^{-kt}$, where C and k are constants.*

63. A piece of metal is heated to 300° Celsius and then placed in a cooling liquid at 50° Celsius. After 4 minutes, the metal has cooled to 175° Celsius. Find its temperature after 12 minutes.

64. A frozen pizza has a temperature of 3.4° Celsius when it is taken from the freezer and left out in a room at 18° Celsius. After half an hour, its temperature is 7.2° Celsius. How long will it take for the pizza to thaw to 10° Celsius?

In Exercises 65–68, do part (a) and skip part (b) if you do not have a graphing calculator. If you have a graphing calculator, then skip part (a) and do part (b).

65. FedEx Profit The gross profit for FedEx (in billions of dollars) is given in the following table for the years 2011 to 2016. (Data from: *The Wall Street Journal.*)

Year	Revenue (billions)
2011	$25.0
2012	$26.9
2013	$27.8
2014	$28.5
2015	$30.5
2016	$33.0

(a) Let $x = 0$ correspond to the year 2011. Use the data points from 2011 and 2016 to find a function of the form $f(x) = a(b^x)$ that models these data.

(b) Use exponential regression to find a function g that models these data, with $x = 0$ corresponding to the year 2011.

(c) Assume the model remains accurate and estimate the gross profit for the year 2017.

(d) Assume the model remains accurate and find the first full year in which the gross profit will be higher than $40 billion.

66. Starbucks App The number of unique visitors (in millions) using the Starbucks app is given in the following table for 12 months starting in June 2014. (Data from: www.comscore.com.)

Month	Average minutes
Jun 2014	10.0
Jul 2014	10.4
Aug 2014	10.2
Sep 2014	11.2
Oct 2014	10.1
Nov 2014	11.6
Dec 2014	11.8
Jan 2015	13.1
Feb 2015	14.4
Mar 2015	15.7
Apr 2015	14.9
May 2015	15.9

(a) Let $x = 0$ correspond to June 2014. Use the data points from June 2014 and May 2015 to find a function of the form $f(x) = a(b^x)$ that models these data.

(b) Use exponential regression to find a function h that models these data, with $x = 0$ corresponding to June 2014.

(c) Assume the model remains accurate and estimate the number of unique visitors using the app in June 2015.

(d) Assume the model remains accurate and find the first full month in which the number of unique visitors was higher than 18 million.

67. Dunkin' Donuts App The number of unique visitors (in millions) using the Dunkin' Donuts app is given in the following table for 12 months starting in June 2014. (Data from: www.comscore.com.)

Month	Unique Visitors (millions)
Jun 2014	4.9
Jul 2014	4.9
Aug 2014	5.6
Sep 2014	7.1
Oct 2014	6.9
Nov 2014	6.0
Dec 2014	6.8
Jan 2015	6.7
Feb 2015	6.9
Mar 2015	6.2
Apr 2015	6.0
May 2015	6.6

(a) Let $x = 1$ correspond to June 2014. Use the data points from June 2014 and May 2015 to find a function of the form $f(x) = a + b \ln x$ that models these data. [*Hint:* Use (1, 4.9) to find a; then use (12, 6.6) to find b.]

(b) Use logarithmic regression to find a function g that models these data, with $x = 1$ corresponding to June 2014.

(c) Assume the model remains accurate and estimate the number of unique visitors using the app in June 2015.

(d) Assume the model remains accurate and find the first full month in which the number of unique visitors was higher than 7.0 million.

68. Bank Capital The common equity capital ratio is given in the following table for the largest banks in the United States as a percentage of risk-weighted assets. (Data from: The Federal Reserve.)

Year	Percent
2009	5.8
2010	8.1
2011	9.8
2012	10.2
2013	11.4
2014	12.1
2015	12.2

(a) Let $x = 1$ correspond to the year 2009. Use the data points from 2009 and 2015 to find a function of the form $f(x) = a + b \ln x$ that models these data. [*Hint:* See Exercise 67(a).]

(b) Use logarithmic regression to find a function h that models these data, with $x = 9$ corresponding to the year 2009.

(c) Assume the model remains accurate and estimate the common equity capital ratio for the year 2016.

(d) Assume the model remains accurate and find the first full year in which the common equity capital ratio exceeds 13%.

Case Study 4 Gapminder.org

The website www.gapminder.org is committed to displaying economic and health data for countries around the world in a visually compelling way. The goal is to make publicly available data easy for citizens of the world to understand so individuals and policy makers can comprehend trends and make data-driven conclusions. For example, the website uses the size of the bubble on a plot to indicate the population of the country, and the color indicates the region in which the country lies:

Yellow	America
Orange	Europe and Central Asia
Green	Middle East and Northern Africa
Dark Blue	Sub-Saharan Africa
Red	East Asia and the Pacific

Additionally, the website also uses animation to show how variables change over time.

Let's look at an example of two variables of interest as best we can on a printed page. We will use income per-person (in U.S. dollars and measured as gross domestic product per capita) as a measure of a country's wealth as the x-axis variable and life expectancy (in years) at birth as a measure of the health of a country on the y-axis. In Figure 4.24 we see a bubble for each country in the year 1800.

We see that all the countries are bunched together with quite low income per-person and average life expectancy ranging from 25–40 years. The two large bubbles you see are for the two most populous countries: China (red) and India (light blue). When you visit the website (please do so, it is very cool!) and press the "Play" button at the bottom left, you can see how income per-person and life expectancy change over time for each country. At its conclusion, you obtain Figure 4.25 on the following page, which shows the relationship between income per-person and average life expectancy in the year 2015.

We can see that a great deal has changed in 200 years! One great thing is that all countries now have average life expectancy

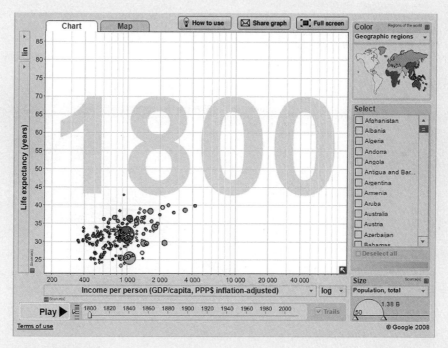

Figure 4.24 Relationship Between Income per-Person and Average Life Expectancy in 1800 from Gapminder Foundation. Reproduced by permission of Gapminder Foundation.

that is over 45 years. This is simply incredible. Also, we see there is a great range of income per-person values for a variety of countries. The country to the far right (the little orange dot) is the European country of Luxembourg. We see that many of the Sub-Sahara countries on the left (in dark blue) have the lowest per-person income and the lowest average life expectancy. The wealthier countries—

for example, countries whose per-person income is greater than $20,000—all have life expectancy over 65 years.

We can actually use the per-person income x to predict the average life expectancy $f(x)$. Note that Gapminder.org has its default showing the income per-person on a logarithmic scale. When we see a linear pattern such as this where the x-axis is on a

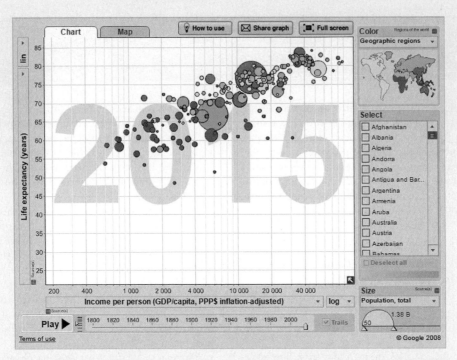

Figure 4.25   Relationship Between Income per-Person and Average Life Expectancy in 2015 from Gapminder Foundation. Reproduced by permission of Gapminder Foundation.

logarithmic scale and the y-axis is on the regular scale, a model of the form $f(x) = a + b \ln x$ often fits the data well. Since Gapminder.org makes all the data for their graphs available, we can fit a model to find values for a and b to use 2015 income per-person (x = GDP per capita) to predict life expectancy (y = years). Using regression, we obtain the following model:

$$f(x) = 23.97 + 5.221 \ln x. \tag{1}$$

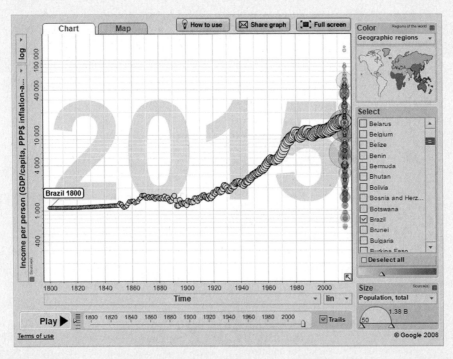

Figure 4.26   Relationship Between Time and per Capita GDP in 2015 from Gapminder Foundation. Reproduced by permission of Gapminder Foundation.

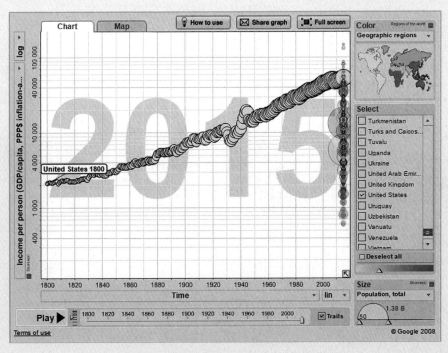

Figure 4.27   Trajectory of United States from Gapminder Foundation. Reproduced by permission of Gapminder Foundation.

Gapminder.org will also show the economic development for an individual country. Figure 4.26 shows per capita GDP on the y-axis for the country of Brazil over time. We can see that economic gains began close to the year 1900. The graph shows per-person income on the y-axis (again on a logarithmic scale) and the trend seems approximately linear since 1900. When this is the case, a model of the form $a(b^x)$ usually fits the data well. Letting $x = 0$ correspond to the year 1900, we obtain the following:

$$g(x) = 1037(1.025)^x \qquad (2)$$

The trajectory of the United States (Figure 4.27) is a bit different, with rising incomes starting in the early 1800s and continuing to the present. Notice the dips that occur in the mid-1930s (the Great Depression) and the early 1940s (World War II). Fitting a model to the data for the United States (again with $x = 0$ corresponding to the year 1900) yields the model:

$$h(x) = 6260(1.019)^x \qquad (3)$$

Exercises

For Exercises 1–6, use Equation **(1)** that provides a model using per-person income as the x-variable to predict life expectancy as the y-variable.

1. Create a graph of the function given in Equation **(1)** from $x = 1$ to 100,000.

2. Describe the shape of the graph.

3. The country furthest to the left in Figure 4.25 is the Central African Republic. In 2015, its income per-person was $599, and its life expectancy was 54. Suppose that economic development improved so that the Central African Republic has per-person income level of $4000. What does the model of Equation **(1)** predict for life expectancy?

4. In 2015, Indonesia had per-person income of $10,500 and a life expectancy of 71 years. What does the model predict for life expectancy if the per-person income improved to $20,000?

5. According to the model, what per-person income level predicts a life expectancy of 60 years?

6. According to the model, what per-person income level predicts a life expectancy of 75 years?

For Exercises 7–10, use the model in Equation **(2)** that uses year (with $x = 0$ corresponding to the year 1900) to predict the per-person income for Brazil.

7. Graph the model of Equation **(2)** from $x = 0$ to 115.

8. Describe the shape of the model.

9. According to the model, what was the per-person income for Brazil in the year 1950? In the year 2000?

10. What is the first full year in which the model indicates that per-person income exceeded $4000?

For Exercises 11–14, use the model in Equation **(3)** that uses year (with $x = 0$ corresponding to the year 1900) to predict the per-person income for the United States.

11. Graph the model of Equation **(3)** from $x = 0$ to 115.

12. Describe the shape of the model.

13. According to the model, what was the per-person income for the United States in the year 1960? In the year 2010?

14. What is the first full year in which the model indicates that per-person income exceeded $30,000?

▦ Extended Project

Go to the website www.gapminder.org and explore different relationships. Examine the relationship between a variety of variables that look at relationships among health, climate, disasters, economy, education, family, global trends, HIV, and poverty. (*Hint:* When you have the graphic display showing, click on the button for "Open Graph Menu" to see a list of options in these areas all ready to be viewed.) When observing the different graphs, notice when the x-axis, the y-axis, or both are on a logarithmic scale.

1. Find a relationship that appears linear when the x-axis is on the logarithmic scale and the y-axis is not. Now click on the button to the right of the x-axis and observe the graph on a linear scale (rather than the logarithmic scale). Describe the shape.

2. Find a relationship that appears linear when the x-axis is not on the logarithmic scale, but the y-axis is on the logarithmic scale. Now click on the button above the y-axis and observe the graph on a linear scale (rather than the logarithmic scale). Describe the shape.

3. Find a relationship that appears linear when the x-axis is on the logarithmic scale, and the y-axis is on the logarithmic scale. Now click on the button to the right of the x-axis and click on the button above the y-axis and observe the graph on a linear scale (rather than the logarithmic scale) for both axes. Describe the shape.

Mathematics of Finance

5

CHAPTER OUTLINE

We should all plan for eventual retirement, and such savings can include savings accounts and annuities. Most people also take out a loan for a big purchase, such as a car, a major appliance, or a house. People who carry a balance on their credit cards are in effect borrowing money. Loan payments must be calculated accurately, and it may take some mathematical work to find the best deal. In this chapter we examine the mathematics of interest and compounding, with examples in the savings and bond markets. We explore annuities where payments are made into and out of different kinds of financial tools. Last, we cover loan amortization in the context of auto loans and home mortgages.

It is important for both businesspersons and consumers to understand the mathematics of finance in order to make sound financial decisions. Interest formulas for borrowing and investing money are introduced in this chapter.

📝 **NOTE** We try to present realistic, up-to-date applications in this text. Because interest rates change so frequently, however, it is very unlikely that the rates in effect when this chapter was written are the same as the rates today when you are reading it. Fortunately, the mathematics of finance is the same regardless of the level of interest rates. So we have used a variety of rates in the examples and exercises. Some will be realistic and some won't by the time you see them—but all of them have occurred in the past several decades.

5.1 Simple Interest and Discount

Interest is the fee paid to use someone else's money. Interest on loans of a year or less is frequently calculated as **simple interest,** which is paid only on the amount borrowed or invested and not on past interest. The amount borrowed or deposited is called the **principal.** We can also call the principal the present value of the amount borrowed or deposited. Throughout this chapter we will use PV to represent present value. The **rate** of interest is given as a percent per year, expressed as a decimal. For example, $6\% = .06$ and $11\frac{1}{2}\% = .115$. The **time** during which the money is accruing interest is calculated in years. Simple interest is the product of the principal (present value), rate, and time.

Simple Interest

The simple interest I on PV dollars at a rate of interest r per year for t years is

$$I = PVrt.$$

It is customary in financial problems to round interest to the nearest cent.

EXAMPLE 1   **Personal Loan** To furnish her new apartment, Maggie Chan borrowed $4000 at 3% interest from her parents for 9 months. How much interest will she pay?

Solution Use the formula $I = PVrt$, with $PV = 4000, r = 0.03$, and $t = 9/12 = 3/4$ years:

$$I = PVrt$$

$$I = 4000 * .03 * \frac{3}{4} = 90.$$

Thus, Maggie pays a total of $90 in interest. ✓₁

✓**Checkpoint 1**

Find the simple interest for each loan.

(a) $2000 at 8.5% for 10 months

(b) $3500 at 10.5% for $1\frac{1}{2}$ years

Answers to Checkpoint exercises are found at the end of the section.

Simple interest is normally used for loans with a term of one year or less. A significant exception is the case of **corporate bonds** and similar financial instruments. When an investor buys a corporate bond, she is essentially lending money to the corporation that issues the bond. The corporation promises to return the principal back to the investor at a specified maturity date *and* pay the investor a specified interest rate, usually semiannually (twice a year).

To determine the amount of money paid to the investor twice a year, we use the interest formula $I = PVrt$, with PV as the amount invested in the bond (principal or present value), r as the interest rate as a decimal, and $t = \frac{1}{2}$ because the bond pays twice a year.

EXAMPLE 2   **Corporate Bond** Bank of America issued 10-year bonds with a maturity date in 2022 and an annual simple interest rate of 5.7%. The interest is paid semiannually. Santiago buys a $10,000 bond. Determine the following. (Data from: www.finra.org.)

(a) How much interest will Santiago earn every six months?

Solution Use the simple interest formula $I = PVrt$ with $PV = 10,000, r = .057$, and $t = \frac{1}{2}$.

$$I = PVrt = (10,000)(.057)\left(\frac{1}{2}\right) = \$285.$$

(b) How much interest will he earn over the 10-year life of the bond?

Solution Either use the simple interest formula with $t = 10$, that is,

$$I = PVrt = (10{,}000)(.057)(10) = \$5700,$$

or take the answer in part (a), which will be paid out every six months for 10 years for a total of 20 times. Thus, Santiago would obtain $\$285(20) = \$5700.$

Future Value

When you make a deposit at simple interest rate r for t years, the **future value** (or **maturity value**) is the sum of the deposit and the interest it has earned. In other words, the future value (FV) is the sum of the deposit (present value PV) and the interest I:

$$FV = PV + I$$
$$= PV + PVrt \qquad I = PVrt.$$
$$= PV(1 + rt). \qquad \text{Factor out } PV.$$

The following box summarizes this result.

> ### Future Value (or Maturity Value) for Simple Interest
>
> The future value (maturity value) FV of PV dollars for t years at interest rate r per year is
>
> $$FV = PV + I, \qquad \text{or} \qquad FV = PV(1 + rt).$$

EXAMPLE 3   **Personal Loan** Find each maturity value and the amount of interest paid.

(a) Hai borrows $20,000 from his parents at 5.25% to add a room onto his house. He plans to repay the loan in 9 months with a bonus he expects to receive at that time.

Solution The loan is for 9 months, or 9/12 of a year, so $t = .75$, $PV = 20{,}000$, and $r = .0525$. Use the formula to obtain

$$FV = PV(1 + rt)$$
$$= 20{,}000[1 + .0525(.75)]$$
$$= 20{,}787.50, \qquad \text{Use a calculator.}$$

or $20,787.50. The maturity value FV is the sum of the principal PV and the interest I, that is, $FV = PV + I$. To find the amount of interest paid, rearrange this equation:

$$I = FV - PV$$
$$I = \$20{,}787.50 - \$20{,}000 = \$787.50.$$

(b) A loan of $11,280 for 85 days at 9% interest.

Solution Use the formula $FV = PV(1 + rt)$, with $PV = 11{,}280$ and $r = .09$. Unless stated otherwise, we assume a 365-day year, so the period in years is $t = 85/365$. The maturity value is

$$FV = PV(1 + rt)$$
$$FV = 11{,}280\left(1 + .09 * \frac{85}{365}\right)$$
$$\approx 11{,}280(1.020958904) \approx \$11{,}516.42.$$

As in part (a), the interest is

$$I = FV - PV = \$11{,}516.42 - \$11{,}280 = \$236.42.$$ 3

✓ **Checkpoint 2**

For the given bonds, find the semiannual interest payment and the total interest paid over the life of the bond.

(a) $7500 Time Warner Cable, Inc. 30-year bond at 7.3% annual interest.

(b) $15,000 Clear Channel Communications 10-year bond at 9.0% annual interest.

✓ **Checkpoint 3**

Find each future value.

(a) $1000 at 4.6% for 6 months

(b) $8970 at 11% for 9 months

(c) $95,106 at 9.8% for 76 days

EXAMPLE 4   **Personal Loan** Sally borrows $15,000 and is required to pay $15,315 in 4 months to pay off the loan and interest. What is the simple interest rate?

Solution One way to find the rate is to solve for r in the future-value formula when $PV = 15,000$, $FV = 15,315$, and $t = 4/12 = 1/3$:

$$PV(1 + rt) = FV$$

$$15,000\left(1 + r * \frac{1}{3}\right) = 15,315$$

$$15,000 + \frac{15,000r}{3} = 15,315 \qquad \text{Multiply out left side.}$$

$$\frac{15,000r}{3} = 315 \qquad \text{Subtract 15,000 from both sides.}$$

$$15,000r = 945 \qquad \text{Multiply both sides by 3.}$$

$$r = \frac{945}{15,000} = .063. \qquad \text{Divide both sides by 15,000.}$$

Therefore, the interest rate is 6.3%. ☑4

✓ **Checkpoint 4**

You lend a friend $500. She agrees to pay you $520 in 6 months. What is the interest rate?

Present Value

In interest problems, PV always represents the amount at the beginning of the period, and FV always represents the amount at the end of the period. To find a formula for PV, we begin with the future-value formula:

$$FV = PV(1 + rt).$$

Dividing each side by $1 + rt$ gives the following formula for the present value.

Present Value for Simple Interest

The **present value** PV of a future amount of FV dollars at a simple interest rate r for t years is

$$PV = \frac{FV}{1 + rt}.$$

EXAMPLE 5   **Present Value** Find the present value of $32,000 in 4 months at 9% interest.

Solution

$$PV = \frac{FV}{1 + rt} = \frac{32,000}{1 + (.09)\left(\dfrac{4}{12}\right)} = \frac{32,000}{1.03} \approx 31,067.96.$$

A deposit of $31,067.96 today at 9% interest would produce $32,000 in 4 months. These two sums, $31,067.96 today and $32,000.00 in 4 months, are equivalent (at 9%) because the first amount becomes the second amount in 4 months. ☑5

✓ **Checkpoint 5**

Find the present value of the given future amounts. Assume 6% interest.

(a) $7500 in 1 year

(b) $89,000 in 5 months

(c) $164,200 in 125 days

EXAMPLE 6   **Court Settlement** Because of a court settlement, Jalinda owes $5000 to Madison. The money must be paid in 10 months, with no interest. Suppose Jalinda wants to pay the money today and that Madison can invest it at an annual rate of 5%. What amount should Madison be willing to accept to settle the debt?

Solution The $5000 is the future value in 10 months. So Madison should be willing to accept an amount that will grow to $5000 in 10 months at 5% interest. In other words, she should accept an amount that is the present value of $5000 under these circumstances. Use the present-value formula with $FV = 5000$, $r = .05$, and $t = 10/12 = 5/6$:

$$PV = \frac{FV}{1 + rt} = \frac{5000}{1 + .05 * \dfrac{5}{6}} = 4800.$$

Madison should be willing to accept $4800 today in settlement of the debt.

EXAMPLE 7 **Selling a Note** Larry Parks owes $6500 to Virginia Donovan. The loan is payable in one year at 6% interest. Virginia needs cash to pay medical bills, so four months before the loan is due, she sells the note (loan) to the bank. If the bank wants a return of 9% on its investment, how much should it pay Virginia for the note?

Solution First find the maturity value of the loan—the amount (with interest) that Larry must pay Virginia:

$$
\begin{aligned}
FV &= PV(1 + rt) &&\text{Maturity-value formula}\\
&= 6500(1 + .06 * 1) &&\text{Let } PV = 6500, r = .06, \text{ and } t = 1.\\
&= 6500(1.06) = \$6890.
\end{aligned}
$$

In four months, the bank will receive $6890. Since the bank wants a 9% return, compute the present value of this amount at 9% for four months:

$$
\begin{aligned}
PV &= \frac{FV}{1 + rt} &&\text{Present-value formula}\\[2mm]
&= \frac{6890}{1 + .09\left(\dfrac{4}{12}\right)} = \$6689.32. &&\text{Let } FV = 6890, r = .09, \text{ and } t = 4/12.
\end{aligned}
$$

The bank pays Virginia $6689.32 and four months later it is worth $6890.

Discount

The preceding examples dealt with loans in which money is borrowed and simple interest is charged. For most simple interest loans, both the principal (amount borrowed) and the interest are paid at the end of the loan period. With a corporate bond (which is a loan to a company by the investor who buys the bond), interest is paid during the life of the bond and the principal is paid back at maturity. In both cases,

the borrower receives the principal,

but pays back the principal *plus* the interest.

In a **simple discount loan,** however, the interest is deducted in advance from the amount of the loan and the *balance* is given to the borrower. The *full value* of the loan must be paid back at maturity. Thus,

the borrower receives the principal *less* the interest,

but pays back the principal.

The most common examples of simple discount loans are U.S. Treasury bills (T-bills), which are essentially short-term loans to the U.S. government by investors. T-bills are sold at a **discount** from their face value and the Treasury pays back the face value of the T-bill at maturity. The discount amount is the interest deducted in advance from the face value. The Treasury receives the face value less the discount, but pays back the full face value. If you are interested in more information on purchasing T-bills, visit www.treasurydirect.gov.

✓ **Checkpoint 6**

Jerrell Davis is owed $19,500 by Christine O'Brien. The money will be paid in 11 months, with no interest. If the current interest rate is 10%, how much should Davis be willing to accept today in settlement of the debt?

✓ **Checkpoint 7**

A firm accepts a $21,000 note due in 8 months, with interest of 10.5%. Two months before it is due, the firm sells the note to a broker. If the broker wants a 12.5% return on his investment, how much should he pay for the note?

> **EXAMPLE 8**   **Treasury Bills** An investor bought a 6-month $10,000 U.S. Treasury bill on July 14, 2016, that sold at a discount rate of .39%. What is the amount of the discount? What is the price of the T-bill? (Data from: www.treasurydirect.gov.)
>
> **Solution** The discount rate on a T-bill is always a simple annual interest rate. Consequently, the discount (interest) is found with the simple interest formula, using $PV = 8000$ (face value), $r = .0039$ (discount rate), and $t = .5$ (because 6 months is half a year):
>
> $$\text{Discount} = PVrt = 10,000 * .0039 * .5 = \$19.50.$$
>
> So the price of the T-bill is:
>
> $$\text{Face Value} - \text{Discount} = 10,000 - 19.50 = \$9980.50. \quad ✓_8$$

✓ **Checkpoint 8**

The maturity times and discount rates for $15,000 T-bills sold on June 23, 2016, are given. Find the discount amount and the price of each T-bill.

(a) one year; .56%

(b) six months; .34%

(c) three months; .26%

In a simple discount loan such as a T-bill, the discount rate is not the actual interest rate that the borrower pays. In Example 8, the discount rate of .39% was applied to the face value of $10,000 rather than the $9980.50 that the Treasury (the borrower) received.

> **EXAMPLE 9**   **Treasury Bills** Find the actual interest rate paid by the Treasury in Example 8.
>
> **Solution** Use the formula for simple interest, $I = PVrt$, with r as the unknown. Here, $PV = 9980.50$ (the amount the Treasury received) and $I = 19.50$ (the discount amount). Because this is a 6-month T-bill, $t = .5$, and we have
>
> $$I = PVrt$$
> $$19.50 = 9980.50(r)(.5)$$
> $$19.50 = 4990.25r \qquad \text{Multiply the terms on the right side.}$$
> $$r = \frac{19.5}{4990.25} \approx .0039076. \qquad \text{Divide both sides by 4990.25.}$$
>
> The actual interest rate is then .39076%. $\quad ✓_9$

✓ **Checkpoint 9**

Find the actual interest rate paid by the Treasury for each T-bill in Checkpoint 8.

5.1 Exercises

Unless stated otherwise, "interest" means simple interest, and "interest rate" and "discount rate" refer to annual rates. Assume 365 days in a year.

1. What factors determine the amount of interest earned on a fixed principal?

Find the interest on each of these loans. (See Example 1.)

2. $35,000 at 6% for 9 months

3. $2850 at 7% for 8 months

4. $1875 at 5.3% for 7 months

5. $3650 at 6.5% for 11 months

6. $5160 at 7.1% for 58 days

7. $2830 at 8.9% for 125 days

8. $8940 at 9%; loan made on May 7 and due September 19

9. $5328 at 8%; loan made on August 16 and due December 30

10. $7900 at 7%; loan made on July 7 and due October 25

Corporate Bonds *For each of the given corporate bonds, whose interest rates are provided, find the semiannual interest payment and the total interest earned over the life of the bond. (See Example 2, Data from: www.finra.org.)*

11. $5000 IBM, 30-year bond; 6.220%

12. $9000 Barrick Gold Corp., 10-year bond; 6.800%

13. $12,500 Morgan Stanley, 10-year bond; 4.100%

14. $4500 Goldman Sachs, 7-year bond; 2.134%

15. $6500 Amazon.com Corp, 5-year bond; 1.200%

16. $10,000 Wells Fargo, 12-year bond; 4.000%

Find the future value of each of these loans. (See Example 3.)

17. $12,000 loan at 3.5% for 3 months

18. $3475 loan at 7.5% for 6 months

19. $6500 loan at 5.25% for 8 months

20. $24,500 loan at 9.6% for 10 months

◥**21.** What is meant by the *present value* of money?

◥**22.** In your own words, describe the *maturity value* of a loan.

Find the present value of each future amount. (See Examples 5 and 6.)

23. $15,000 for 9 months; money earns 6%

24. $48,000 for 8 months; money earns 5%

25. $15,402 for 120 days; money earns 6.3%

26. $29,764 for 310 days; money earns 7.2%

Treasury Bills *The given U.S. Treasury bills were sold on July 1, 2016. Find (a) the price of the T-bill, and (b) the actual interest rate paid by the Treasury. (See Examples 8 and 9. Data from: www.treasury.gov.)*

27. Three-month $20,000 T-bill with discount rate of .27%

28. One-month $12,750 T-bill with discount rate of .23%

29. Six-month $15,500 T-bill with discount rate of .36%

30. One-year $7000 T-bill with discount rate of .44%

Treasury Bills *Historically, treasury bills offered higher rates. On January 2, 2007, the discount rates were substantially higher than in July 2016. For the following treasury bills bought in 2007, find (a) the price of the T-bill, and (b) the actual interest rate paid by the Treasury. (See Examples 8 and 9. Data from: www.treasury.gov.)*

31. Three-month $20,000 T-bill with discount rate of 5.07%

32. One-month $12,750 T-bill with discount rate of 4.74%

33. Six-month $15,500 T-bill with discount rate of 5.10%

34. Six-month $9000 T-bill with discount rate of 5.10%

Work the following applied problems.

35. Tax Penalty An accountant for a corporation forgot to pay the firm's income tax of $650,000 on time. The government charged a penalty of 5.2% simple interest for the 30 days that the money was late. Find the total amount (tax and penalty) that was paid.

36. Family Loan Wilma borrowed $12,000 from her uncle to buy a used car. Her uncle charged simple interest at 3.5% for 5 years. Find the total amount (principal and interest) that Wilma paid to her uncle.

37. Savings Gabriel Branson invested his summer earnings of $3000 in a savings account for college. The account pays 1.5% simple interest. What will be the value of this account after 2 years?

38. Savings Jada received an inheritance from her grandmother of $7500. She places the money in savings account paying simple interest at 1.8%. What will be the value of the account after 3.5 years?

39. Money Market Account An account invested in a money market fund grew from $67,081.20 to $67,359.39 in a month. What was the interest rate, to the nearest tenth?

40. CD A $100,000 certificate of deposit held for 60 days is worth $100,560.00. Assuming simple interest, what interest rate, to the nearest tenth of a percent, was earned?

41. Student Loan Danielle borrowed $7500 from her grandfather with simple interest of 7%. She eventually repaid $7675 (principal and interest). What was the time period of the loan?

42. Auto Loan Kiara received a $10,000 loan from her aunt to buy a car. Her aunt charged her 2.5% interest. How long was the term of the loan if the interest accrued was $125?

43. Construction Payment John Sun Yee will need $6000 to pay for remodeling work on his house. A contractor agrees to do the work in 10 months. What is the present value of $6000 at 3.6% simple interest to accumulate the $6000 in 10 months?

44. Tuition Payment Mariana will need to pay a tuition bill of $7500 for her daughter in 10 months. What dollar amount can she deposit today at 1.25% simple interest to have enough to pay the tuition bill?

45. Computing Hardware Payment An accounting firm has ordered 50 new computers at a cost of $1704 each. The machines will not be delivered for 7 months. What amount could the firm deposit today in an account paying 3.42% simple interest to have enough to pay for the machines in 7 months?

46. Car Dealership A car dealership has ordered 12 new vehicles from the auto manufacturer. The total amount of the purchase is $420,000. The vehicles will be delivered in three months. What amount could the dealership deposit today in an account paying 4.1% simple interest to have enough money to pay for the vehicles in three months?

47. Treasury Bill A six-month $4000 T-bill sold for $3930. What was the discount rate?

48. Treasury Bill A one-year $9600 T-bill sold for $9403. What was the discount rate?

49. Treasury Bill A three-month $7600 T-bill carries a discount of $80.75. What is the discount rate for this T-bill?

50. Treasury Bill A six-month $12,500 T-bill carries a discount of $120.25. What is the discount rate for this T-bill?

Interest Rate *Work the next set of problems, in which you are to find the annual simple interest rate. Consider any fees, dividends, or profits as part of the total interest.*

51. A stock that sold for $22 at the beginning of the year was selling for $24 at the end of the year. If the stock paid a dividend of $.50 per share, what is the simple interest rate on an investment in this stock? (*Hint:* Consider the interest to be the increase in value plus the dividend.)

52. Loan Jerry Ryan borrowed $8000 for nine months at an interest rate of 7%. The bank also charges a $100 processing fee. What is the actual interest rate for this loan?

53. Refund Loan You are due a tax refund of $760. Your tax preparer offers you a no-interest loan to be repaid by your refund check, which will arrive in four weeks. She charges a $60 fee for this service. What actual interest rate will you pay for this loan? (*Hint:* The time period of this loan is not 4/52, because a 365-day year is 52 weeks and 1 day. So use days in your computations.)

54. Refund Loan Your cousin is due a tax refund of $400 in six weeks. His tax preparer has an arrangement with a bank to get him the $400 now. The bank charges an administrative fee of $29 plus interest at 6.5%. What is the actual interest rate for this loan? (See the hint for Exercise 53.)

Finance *Work these problems. (See Example 7.)*

55. Selling a Note A building contractor gives a $13,500 promissory note to a plumber who has loaned him $13,500. The note is due in nine months with interest at 9%. Three months after the note is signed, the plumber sells it to a bank. If the bank gets a 10% return on its investment, how much will the plumber receive? Will it be enough to pay a bill for $13,650?

56. Selling a Note Shalia Johnson owes $7200 to the Eastside Music Shop. She has agreed to pay the amount in seven months at an interest rate of 10%. Two months before the loan is due, the store needs $7550 to pay a wholesaler's bill. The bank will buy the note, provided that its return on the investment is 11%. How much will the store receive? Is it enough to pay the bill?

57. Future Value Let y_1 be the future value after t years of $100 invested at 8% annual simple interest. Let y_2 be the future value after t years of $200 invested at 3% annual simple interest.

(a) Think of y_1 and y_2 as functions of t and write the rules of these functions.

(b) Without graphing, describe the graphs of y_1 and y_2.

(c) Verify your answer to part (b) by graphing y_1 and y_2 in the first quadrant.

(d) What do the slopes and y-intercepts of the graphs represent (in terms of the investment situation that they describe)?

58. If $y = 16.25t + 250$ and y is the future value after t years of P dollars at interest rate r, what are P and r? (*Hint:* See Exercise 57.)

✓**Checkpoint Answers**

1. **(a)** $141.67   **(b)** $551.25

2. **(a)** $273.75; $16,425   **(b)** $675; $13,500

3. **(a)** $1023   **(b)** $9710.03   **(c)** $97,046.68

4. 8%

5. **(a)** $7075.47   **(b)** $86,829.27   **(c)** $160,893.96

6. $17,862.60

7. $22,011.43

8. **(a)** $84; $14,916   **(b)** $25.50; $14,974.50
   (c) $9.75; $14,990.25

9. **(a)** About .56314%   **(b)** About .34058%
   (c) About .26017%

5.2 Compound Interest

With annual simple interest, you earn interest each year on your original investment. With annual **compound interest,** however, you earn interest both on your original investment *and* on any previously earned interest. To see how this process works, suppose you deposit $1000 at 5% annual interest. The following chart shows how your account would grow with both simple and compound interest:

End of Year	SIMPLE INTEREST		COMPOUND INTEREST	
	Interest Earned	Balance	Interest Earned	Balance
	Original Investment: $1000		*Original Investment:* $1000	
1	1000(.05) = $50	$1050	1000(.05) = $50	$1050
2	1000(.05) = $50	$1100	1050(.05) = $52.50	$1102.50
3	1000(.05) = $50	$1150	1102.50(.05) = $55.13*	$1157.63

As the chart shows, simple interest is computed each year on the original investment, but compound interest is computed on the entire balance at the end of the preceding year. So simple interest always produces $50 per year in interest, whereas compound interest produces $50 interest in the first year and increasingly larger amounts in later years (because you earn interest on your interest). ✓₁

✓**Checkpoint 1**

Extend the chart in the text by finding the interest earned and the balance at the end of years 4 and 5 for **(a)** simple interest and **(b)** compound interest.

---
*Rounded to the nearest cent.

EXAMPLE 1   **Savings Account** If $7000 is deposited in a savings account that pays 4% interest compounded annually, how much money is in the account after nine years?

Solution After one year, the account balance is

$$7000 + 4\% \text{ of } 7000 = 7000 + (.04)7000$$
$$= 7000(1 + .04) \qquad \text{Distributive property}$$
$$= 7000(1.04) = \$7280.$$

The initial balance has grown by a factor of 1.04. At the end of the second year, the balance is

$$7280 + 4\% \text{ of } 7280 = 7280 + (.04)7280$$
$$= 7280(1 + .04) \qquad \text{Distributive property}$$
$$= 7280(1.04) = 7571.20.$$

Once again, the balance at the beginning of the year has grown by a factor of 1.04. This is true in general: If the balance at the beginning of a year is PV dollars, then the balance at the end of the year is

$$PV + 4\% \text{ of } PV = PV + .04PV$$
$$= PV(1 + .04)$$
$$= PV(1.04).$$

So the account balance grows like this:

$$\textbf{Year 1} \qquad\qquad \textbf{Year 2} \qquad\qquad\qquad \textbf{Year 3}$$
$$7000 \to 7000(1.04) \to \underbrace{[7000(1.04)](1.04)}_{7000(1.04)^2} \to \underbrace{[7000(1.04)(1.040)](1.04)}_{7000(1.04)^3} \to \cdots.$$

At the end of nine years, the balance is

$$7000(1.04)^9 = \$9963.18 \qquad \text{(rounded to the nearest penny)}.$$

The argument used in Example 1 applies in the general case and leads to this conclusion.

Compound Interest

If PV dollars are invested at interest rate i per period, then the **compound amount** (future value) FV after n compounding periods is

$$FV = PV(1 + i)^n.$$

In Example 1, for instance, we had $PV = 7000, n = 9$, and $i = .04$ (so that $1 + i = 1 + .04 = 1.04$).

NOTE Compare this future value formula for compound interest with the one for simple interest from the previous section, using t as the number of years:

Compound interest	$FV = PV(1 + r)^t$;
Simple interest	$FV = PV(1 + rt)$.

The important distinction between the two formulas is that, in the compound interest formula, the number of years, t, is an *exponent*, so that money grows much more rapidly when interest is compounded.

> **EXAMPLE 2** **Savings Account** Suppose $1000 is deposited for six years in a savings account paying 8.31% per year compounded annually.
>
> **(a)** Find the compound amount.
>
> **Solution** In the formula above, $PV = 1000, i = .0831$, and $n = 6$. The compound amount is
>
> $$FV = PV(1 + i)^n$$
> $$FV = 1000(1.0831)^6$$
> $$FV = \$1614.40.$$
>
> **(b)** Find the amount of interest earned.
>
> **Solution** Subtract the initial deposit from the compound amount:
>
> $$\text{Amount of interest} = \$1614.40 - \$1000 = \$614.40. \quad \boxed{\checkmark}_2$$

✓ **Checkpoint 2**

Suppose $17,000 is deposited at 4% compounded annually for 11 years.

(a) Find the compound amount.

(b) Find the amount of interest earned.

📋 **TECHNOLOGY TIP**

Spreadsheets are ideal for performing financial calculations. Figure 5.1 shows a Microsoft Excel spreadsheet with the results when $1000 is invested at an annual rate of 10% using compound and simple interest for 20 years in Columns B and C, respectively. The blue balloons show the Excel 2016 formulas for generating the results in Columns B and C that would actually be typed into Columns B and C. Notice how rapidly the compound amount increases compared to the maturity value with simple interest. For more details on the use of spreadsheets in the mathematics of finance, see the *Spreadsheet Manual* that is available with this text.

	A	B	C
1	Period	Compound	Simple
2	0	$1000.00	$1000.00
3	1	$1100.00	$1100.00
4	2	$1210.00	$1200.00
5	3	$1331.00	$1300.00
6	4	$1464.10	$1400.00
7	5	$1610.51	$1500.00
8	6	$1771.56	$1600.00
9	7	$1948.72	$1700.00
10	8	$2143.59	$1800.00
11	9	$2357.95	$1900.00
12	10	$2593.74	$2000.00
13	11	$2853.12	$2100.00
14	12	$3138.43	$2200.00
15	13	$3452.27	$2300.00
16	14	$3797.50	$2400.00
17	15	$4177.25	$2500.00
18	16	$4594.97	$2600.00
19	17	$5054.47	$2700.00
20	18	$5559.92	$2800.00
21	19	$6115.91	$2900.00
22	20	$6727.50	$3000.00

=B2+.10*B2

=C2+100

Figure 5.1

> **EXAMPLE 3** **Savings Account** If a $16,000 savings account grows to $50,000 in 18 years, what is the interest rate (assuming annual compounding)?
>
> **Solution** Use the compound interest formula, with $PV = 16{,}000, FV = 50{,}000$, and $n = 18$, and solve for i:
>
> $$PV(1 + i)^n = FV$$
> $$16{,}000(1 + i)^{18} = 50{,}000$$
> $$(1 + i)^{18} = \frac{50{,}000}{16{,}000} = 3.125 \qquad \text{Divide both sides by 16,000.}$$
> $$\sqrt[18]{(1 + i)^{18}} = \sqrt[18]{3.125} \qquad \text{Take 18th roots on both sides.}$$
> $$1 + i = \sqrt[18]{3.125}$$
> $$i = \sqrt[18]{3.125} - 1 \approx .06535. \qquad \text{Subtract 1 from both sides.}$$
>
> So the interest rate is about 6.535%.

Interest can be compounded more than once a year. Common **compounding periods** include

semiannually (2 periods per year),

quarterly (4 periods per year),

monthly (12 periods per year), and

daily (usually 365 periods per year).

When the annual interest i is compounded m times per year, the interest rate per period is understood to be i/m.

EXAMPLE 4   **Money Market Accounts** In July 2016, bankrate.com advertised a money market account at Ally Bank at an interest rate of .85%. Find the value of the account if $10,000 is invested for one year and interest is compounded according to the periods given below.

(a) Annually

Solution Apply the formula $FV = PV(1 + i)^n$ with $PV = 10{,}000$, $i = .0085$, and $n = 1$:

$$FV = PV(1 + i)^n = 10{,}000(1 + .0085)^1 = 10{,}000(1.0085) = \$10{,}085.$$

(b) Semiannually

Solution Use the same formula and value of PV. Here, interest is compounded twice a year, so the number of periods is $n = 2$, and the interest rate per period is $i = \dfrac{.0085}{2}$:

$$FV = PV(1 + i)^n = 10{,}000\left(1 + \frac{.0085}{2}\right)^2 = \$10{,}085.18.$$

(c) Quarterly

Solution Proceed as in part (b), but the new interest is compounded every 3 months for a total of 4 times a year. Thus, $n = 4$, and the interest rate per period is $i = \dfrac{.0085}{4}$:

$$FV = PV(1 + i)^n = 10{,}000\left(1 + \frac{.0085}{4}\right)^4 = \$10{,}085.27.$$

(d) Monthly

Solution Interest is compounded 12 times a year, so $n = 12$ and $i = \dfrac{.0085}{12}$:

$$FV = PV(1 + i)^n = 10{,}000\left(1 + \frac{.0085}{12}\right)^{12} = \$10{,}085.33.$$

(e) Daily

Solution Interest is compounded 365 times a year, so $n = 365$ and $i = \dfrac{.0085}{365}$:

$$FV = PV(1 + i)^n = 10{,}000\left(1 + \frac{.0085}{365}\right)^{365} = \$10{,}085.36.$$

✓ **Checkpoint 3**

Find the future value for these money market accounts.

(a) Sallie Mae: $1000 at 1.60% compounded monthly for 3 years.

(b) Discover Bank: $2500 at .8% compounded daily for 9 months (assume 30 days in each month).

Example 4 shows that the more often interest is compounded, the larger is the amount of interest earned. Since interest is rounded to the nearest penny, however, there is a limit on how much can be earned. In Example 4, part (e), for instance, that limit of $10,1085.36 has been reached. Nevertheless, the idea of compounding more and more frequently leads to a method of computing interest called **continuous compounding** that is used in certain financial situations. The formula for continuous compounding is given in the following box where $e = 2.7182818\ldots$, which was introduced in Chapter 4.

> ## Continuous Compound Interest
>
> The compound amount FV for a deposit of PV dollars at an interest rate r per year compounded continuously for t years is given by
>
> $$FV = PVe^{rt}.$$

EXAMPLE 5   **Savings Account** Suppose that $5000 is invested in a savings account at an annual interest rate of 3.1% compounded continuously for 4 years. Find the compound amount.

Solution In the formula for continuous compounding, let $PV = 5000, r = .031,$ and $t = 4$. Then a calculator with an e^x key shows that

$$FV = PVe^{rt} = 5000e^{.031(4)} = \$5660.08.$$

 TECHNOLOGY TIP TI-84+ has a "TVM solver" for financial computations (in the TI APPS/financial menu); a similar one can be downloaded for TI-89. Figure 5.2 shows the solution of Example 4(e) on such a solver. The use of these solvers is explained in the next section. Most of the problems in this section can be solved just as quickly with an ordinary calculator.

Ordinary corporate or municipal bonds usually make semiannual simple interest payments. With a **zero-coupon bond,** however, there are no interest payments during the life of the bond. The investor receives a single payment when the bond matures, consisting of his original investment and the interest (compounded semiannually) that it has earned. Zero-coupon bonds are sold at a substantial discount from their face value, and the buyer receives the face value of the bond when it matures. The difference between the face value and the price of the bond is the interest earned.

EXAMPLE 6   **Bonds** Doug bought a 15-year zero-coupon bond paying 4.5% interest (compounded semiannually) for $12,824.50. What is the face value of the bond?

Solution Use the compound interest formula with $PV = 12,824.50$. Interest is paid twice a year, so the rate per period is $i = .045/2$, and the number of periods in 15 years is $n = 30$. The compound amount will be the face value:

$$FV = PV(1 + i)^n = 12,824.50(1 + .045/2)^{30} = 24,999.99618.$$

Rounding to the nearest cent, we see that the face value of the bond in $25,000.

EXAMPLE 7   **Inflation** Suppose that the inflation rate is 3.5% (which means that the overall level of prices is rising 3.5% a year). How many years will it take for the overall level of prices to double?

Solution We want to find the number of years it will take for $1 worth of goods or services to cost $2. Think of $1 as the present value and $2 as the future value, with an interest rate of 3.5%, compounded annually. Then the compound amount formula becomes

$$PV(1 + i)^n = FV$$
$$1(1 + .035)^n = 2,$$

which simplifies as

$$1.035^n = 2.$$

We must solve this equation for n. There are several ways to do this.

Graphical Use a graphing calculator (with x in place of n) to find the intersection point of the graphs of $y_1 = 1.035^x$ and $y_2 = 2$. Figure 5.3 shows that the intersection point has (approximate) x-coordinate 20.14879. So it will take about 20.15 years for prices to double.

Algebraic The same answer can be obtained by using natural logarithms, as in Section 4.4:

$$1.035^n = 2$$
$$\ln 1.035^n = \ln 2 \qquad \text{Take the logarithm of each side.}$$
$$n \ln 1.035 = \ln 2 \qquad \text{Power property of logarithms}$$
$$n = \frac{\ln 2}{\ln 1.035} \qquad \text{Divide both sides by } \ln 1.035.$$
$$n \approx 20.14879. \qquad \text{Use a calculator.}$$

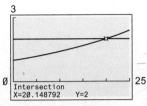

Figure 5.3

Intersection
X=20.148792   Y=2

✓ **Checkpoint 6**

Using a calculator, find the number of years it will take for $500 to increase to $750 in an account paying 6% interest compounded semiannually.

Effective Rate (Annual Percentage Yield)

If you invest $100 at 9%, compounded monthly, then your balance at the end of one year is

$$FV = PV(1 + i)^n = 100\left(1 + \frac{.09}{12}\right)^{12} = \$109.38.$$

You have earned $9.38 in interest, which is 9.38% of your original $100. In other words, $100 invested at 9.38% compounded *annually* will produce the same amount of interest (namely, $100 * .0938 = $9.38) as does 9% compounded monthly. In this situation, 9% is called the **nominal** or **stated rate**, while 9.38% is called the **effective rate** or **annual percentage yield (APY)**.

In the discussion that follows, the nominal rate is denoted r and the APY (effective rate) is denoted r_E.

Effective Rate (r_E) or Annual Percentage Yield (APY)

The *APY* r_E is the simple interest rate needed to produce the same amount of interest in one year as the nominal rate does with more frequent compounding over one year.

EXAMPLE 8   **Money Market Accounts** In July 2016, American Express offered its customers a money market savings account at a rate of .90%. Assume the account compounded interest daily and the principal is $100,000. Find the APY. (Data from bankrate.com.)

Solution The definition given previously means that we must have the following:

$$\begin{array}{c} \$100,000 \text{ at rate } r_E \\ \text{simple interest} \end{array} = \begin{array}{c} \$100,000 \text{ at .90,} \\ \text{compounded daily} \end{array}$$

$$100,000(1 + r_E) = 100,000\left(1 + \frac{.0090}{365}\right)^{365} \qquad \text{Interest formulas}$$

$$(1 + r_E) = \left(1 + \frac{.0090}{365}\right)^{365} \qquad \text{Divide both sides by 100,000.}$$

$$r_E = \left(1 + \frac{.0090}{365}\right)^{365} - 1 \qquad \text{Subtract 1 from both sides.}$$

$$r_E \approx .00904.$$

So the APY is about .904%.

The argument in Example 8 can be carried out with 100,000 replaced by PV, .0090 by r, and 365 by m. The result is the effective rate formula.

Effective Rate (APY)

The effective rate (APY) corresponding to a stated rate of annual interest r compounded m times per year is

$$r_E = \left(1 + \frac{r}{m}\right)^m - 1.$$

EXAMPLE 9 **Money Market Accounts** When interest rates are low (as they were when this text went to press), the interest rate and the APY are insignificantly different. To see when the difference is more pronounced, we will find the APY for each of the given money market checking accounts, which were advertised in October 2008 when offered rates were higher.

(a) Imperial Capital Bank: 3.35% compounded monthly.

Solution Use the effective-rate formula with $r = .0335$ and $m = 12$:

$$r_E = \left(1 + \frac{r}{m}\right)^m - 1 = \left(1 + \frac{.0335}{12}\right)^{12} - 1 \approx .034019.$$

So the APY is about 3.40%, a slight increase over the nominal rate of 3.35%.

(b) U.S. Bank: 2.33% compounded daily.

Solution Use the formula with $r = .0233$ and $m = 365$:

$$r_E = \left(1 + \frac{r}{m}\right)^m - 1 = \left(1 + \frac{.0233}{365}\right)^{365} - 1 \approx .023572.$$

The APY is about 2.36%. ✓ 7

✓ **Checkpoint 7**

Find the APY corresponding to a nominal rate of

(a) 12% compounded monthly;

(b) 8% compounded quarterly.

⊞ TECHNOLOGY TIP

Effective rates (APYs) can be computed on TI-84+ by using "Eff" in the APPS financial menu, as shown in Figure 5.4 for Example 10.

```
▸Eff(1.8,1)
                    1.8
▸Eff(1.77,4)
            1.781783071
▸Eff(1.75,12)
            1.764104916
■
```

Figure 5.4

✓ **Checkpoint 8**

Find the APY corresponding to a nominal rate of

(a) 4% compounded quarterly;

(b) 7.9% compounded daily.

EXAMPLE 10 **Money Market Accounts** Bank A offers a 3-year money market account with a rate of 1.80% compounded annually. Bank B offers a similar money market account with a rate of 1.77% compounded quarterly, and Bank C offers 1.75% compounded monthly. Which bank offers the highest APY?

Solution Compare the APYs:

$$\text{Bank A:} \quad \left(1 + \frac{.0180}{1}\right)^1 - 1 = .0180 = 1.80;$$

$$\text{Bank B:} \quad \left(1 + \frac{.0177}{4}\right)^4 - 1 \approx .0178 = 1.78;$$

$$\text{Bank C:} \quad \left(1 + \frac{.0175}{12}\right)^{12} - 1 \approx .0176 = 1.76.$$

Bank A offers the highest APY. ✓ 8

📄 NOTE Although you can find both the stated interest rate and the APY for most certificates of deposit and other interest-bearing accounts, most bank advertisements mention only the APY.

Present Value for Compound Interest

The formula for compound interest, $FV = PV(1 + i)^n$, has four variables: FV, PV, i, and n. Given the values of any three of these variables, the value of the fourth can be found. In particular, if FV (the future amount), i, and n are known, then PV can be found. Here, PV is the amount that should be deposited today to produce FV dollars in n periods.

EXAMPLE 11 **Lump Sum Payment** Keisha Jones must pay a lump sum of $6000 in 5 years. What amount deposited today at 6.2% compounded annually will amount to $6000 in 5 years?

Solution Here, $FV = 6000$, $i = .062$, $n = 5$, and PV is unknown. Substituting these values into the formula for the compound amount gives

$$6000 = PV(1.062)^5$$

$$PV = \frac{6000}{(1.062)^5} = 4441.49,$$

or $4441.49. If Keisha leaves $4441.49 for 5 years in an account paying 6.2% compounded annually, she will have $6000 when she needs it. To check your work, use the compound interest formula with $PV = \$4441.49$, $i = .062$, and $n = 5$. You should get $FV = \$6000.00$. ✓₉

As Example 11 shows, $6000 in 5 years is (approximately) the same as $4441.49 today (if money can be deposited at 6.2% annual interest). An amount that can be deposited today to yield a given amount in the future is called the *present value* of the future amount. By solving $FV = PV(1 + i)^n$ for PV, we get the following general formula for present value.

Present Value for Compound Interest

The **present value** of FV dollars compounded at an interest rate i per period for n periods is

$$PV = \frac{FV}{(1 + i)^n}, \quad \text{or} \quad PV = FV(1 + i)^{-n}.$$

EXAMPLE 12 **Bonds** A zero-coupon bond with face value $15,000 and a 6% interest rate (compounded semiannually) will mature in 9 years. What is a fair price to pay for the bond today?

Solution Think of the bond as a 9-year investment paying 6%, compounded semiannually, whose future value is $15,000. Its present value (what it is worth today) would be a fair price. So use the present-value formula with $FV = 15,000$. Since interest is compounded twice a year, the interest rate per period is $i = .06/2 = .03$ and the number of periods in 9 years is $n = 9(2) = 18$. Hence,

$$PV = \frac{FV}{(1 + i)^n} = \frac{15,000}{(1 + .03)^{18}} \approx 8810.919114.$$

So a fair price would be the present value of $8810.92. ✓₁₀

✓**Checkpoint 9**

Find PV in Example 11 if the interest rate is

(a) 6%;

(b) 10%.

✓**Checkpoint 10**

Find the fair price (present value) in Example 12 if the interest rate is 7.5%.

EXAMPLE 13 **Inflation** The average annual inflation rate for the years 2012–2014 was 1.72%. How much did an item that sells for $1000 in early 2015 cost three years before? (Data from: inflationdata.com.)

Solution Think of the price three years prior as the present value PV and $1000 as the future value FV. Then $i = .0172$, $n = 3$, and the present value is

$$PV = \frac{FV}{(1 + i)^n} = \frac{1000}{(1 + .0172)^3} = \$950.12.$$

The item cost $950.12 three years earlier. ✓₁₁

✓ **Checkpoint 11**

What did a $1000 item sell for 5 years prior if the annual inflation rate has been 3.2%?

Summary

At this point, it seems helpful to summarize the notation and the most important formulas for simple and compound interest. We use the following variables:

$PV =$ principal or present value;

$FV =$ future or maturity value;

$r =$ annual (stated or nominal) interest rate;

$t =$ number of years;

$m =$ number of compounding periods per year;

$i =$ interest rate per period;

$n =$ total number of compounding periods;

$r_E =$ effective rate (APY).

Simple Interest	**Compound Interest**	**Continuous Compounding**
$FV = PV(1 + rt)$	$FV = PV(1 + r)^n$	$FV = PVe^{rt}$
$PV = \dfrac{FV}{1 + rt}$	$PV = \dfrac{FV}{(1 + i)^n} = FV(1 + i)^{-n}$	$PV = \dfrac{FV}{e^{rt}}$
	$r_E = \left(1 + \dfrac{r}{m}\right)^m - 1$	

5.2 Exercises

Bonds *Interest on the zero-coupon bonds here is compounded semiannually.*

◼ **1.** In the preceding summary what is the difference between r and i? between t and n?

◼ **2.** Explain the difference between simple interest and compound interest.

3. What factors determine the amount of interest earned on a fixed principal?

◼ **4.** In your own words, describe the *maturity value* of a loan.

◼ **5.** What is meant by the *present value* of money?

◼ **6.** If interest is compounded more than once per year, which rate is higher, the stated rate or the effective rate?

Savings Accounts *Find the compound amount and the interest earned for each of the following deposits. (See Examples 1, 2, and 4.)*

7. $470 at 8% compounded monthly for 12 years

8. $15,000 at 4.6% compounded monthly for 11 years

9. $6500 at 4.5% compounded quarterly for 8 years

10. $9100 at 6.1% compounded quarterly for 4 years

Money Market Accounts *The following money market account rates were available at bankrate.com in July 2016. Find the compound amount and the interest earned for each of the following. (See Example 4.)*

11. Everbank: $10,000 at 1.11% compounded daily for one year

12. Alostar Bank of Commerce: $5000 at 1.05% compounded daily for one year

13. Goldman Sachs Bank: $15,000 at 1.00% compounded monthly for three years

14. Discover Bank: $60,000 at .95% compounded monthly for two years

15. Pacific National Bank: $75,000 at .90% compounded quarterly for four years

16. Colorado Federal: $80,000 at .85% compounded quarterly for 3 years

Find the interest rate (with annual compounding) that makes the statement true. (See Example 3.)

17. $3000 grows to $3606 in 5 years

18. $2550 grows to $3905 in 11 years

19. $8500 grows to $12,161 in 7 years

20. $9000 grows to $17,118 in 16 years

Find the compound amount and the interest earned when the following investments have continuous compounding. (See Example 5.)

21. $20,000 at 3.5% for 5 years

22. $15,000 at 2.9% for 10 years

23. $30,000 at 1.8% for 3 years

24. $100,000 at 5.1% for 20 years

Bonds *Find the face value (to the nearest dollar) of the zero-coupon bond. (See Example 6.)*

25. 15-year bond at 5.2%; price $4630

26. 10-year bond at 4.1%; price $13,328

27. 20-year bond at 3.5%; price $9992

28. How do the nominal, or stated, interest rate and the effective interest rate (APY) differ?

APY *Find the APY corresponding to the given nominal rates. (See Examples 8–10).*

29. 4% compounded semiannually

30. 6% compounded quarterly

31. 5% compounded quarterly

32. 4.7% compounded semiannually

Present Value *Find the present value of the given future amounts. (See Example 11.)*

33. $12,000 at 5% compounded annually for 6 years

34. $8500 at 6% compounded annually for 9 years

35. $17,230 at 4% compounded quarterly for 10 years

36. $5240 at 6% compounded quarterly for 8 years

Bonds *What price should you be willing to pay for each of these zero-coupon bonds? (See Example 12.)*

37. 5-year $5000 bond; interest at 3.5%

38. 10-year $10,000 bond; interest at 4%

39. 15-year $20,000 bond; interest at 4.7%

40. 20-year $15,000 bond; interest at 5.3%

Inflation *For Exercises 41 and 42, assume an annual inflation rate of 1.62% (the annual inflation rate of 2014 according to www.InflationData.com). Find the previous price of the following items. (See Example 13.)*

41. How much did an item that costs $5000 now cost 4 years prior?

42. How much did an item that costs $7500 now cost 5 years prior?

43. If the annual inflation rate is 3.6%, how much did an item that costs $500 now cost 2 years prior?

44. If the annual inflation rate is 1.18%, how much did an item that costs $1250 now cost 6 years prior?

45. If money can be invested at 8% compounded quarterly, which is larger, $1000 now or $1210 in 5 years? Use present value to decide.

46. If money can be invested at 6% compounded annually, which is larger, $10,000 now or $15,000 in 6 years? Use present value to decide.

Work the following applied problems.

47. Investing Lora Reilly has inherited $10,000 from her uncle's estate. She will invest the money for 2 years. She is considering two investments: a money market fund that pays a guaranteed 5.8% interest compounded daily and a 2-year Treasury note at 6% annual interest. Which investment pays the most interest over the 2-year period?

48. Investing As the prize in a contest, you are offered $1000 now or $1210 in 5 years. If money can be invested at 6% compounded annually, which is larger?

49. Bonds Which of these 20-year zero-coupon bonds will be worth more at maturity: one that sells for $4510, with a 6.1% interest rate, or one that sells for $5809, with a 4.8% interest rate?

50. Bonds Which of these 15-year zero-coupon bonds will be worth more at maturity: one that sells for $3890, with a 3.7% interest rate, or one that sells for $4510, with a 2.9% interest rate?

51. Investing A company has agreed to pay $2.9 million in 5 years to settle a lawsuit. How much must it invest now in an account paying 5% interest compounded monthly to have that amount when it is due?

52. Investing Bill Poole wants to have $20,000 available in 5 years for a down payment on a house. He has inherited $16,000. How much of the inheritance should he invest now to accumulate the $20,000 if he can get an interest rate of 5.5% compounded quarterly?

53. Inflation If inflation has been running at 3.75% per year and a new car costs $23,500 today, what would it have cost three years ago?

54. Inflation If inflation is 2.4% per year and a washing machine costs $345 today, what did a similar model cost five years ago?

Economics *Use the approach in Example 7 to find the time it would take for the general level of prices in the economy to double at the average annual inflation rates in Exercises 55–58.*

55. 3%   **56.** 4%   **57.** 5%   **58.** 5.5%

59. Electricity Demand The consumption of electricity has increased historically at 6% per year. If it continues to increase at this rate indefinitely, find the number of years before the electric utility companies will need to double their generating capacity.

60. Electricity Demand Suppose a conservation campaign coupled with higher rates causes the demand for electricity to increase at only 2% per year, as it has recently. Find the number of years before the utility companies will need to double their generating capacity.

61. Investing You decide to invest a $16,000 bonus in a money market fund that guarantees a 5.5% annual interest rate compounded monthly for 7 years. A one-time fee of $30 is charged to set up the account. In addition, there is an annual administrative charge of 1.25% of the balance in the account at the end of each year.

(a) How much is in the account at the end of the first year?

(b) How much is in the account at the end of the seventh year?

62. Investing Joe decides to invest $12,000 in a money market fund that guarantees a 4.6% annual interest rate compounded daily for 6 years. A one-time fee of $25 is charged to set up the account. In addition, there is an annual administration charge of .9% of the balance in the account at the end of each year.

(a) How much is in the account at the end of the first year?

(b) How much is in the account at the end of the sixth year?

The following exercises are similar to professional examination questions.

63. On January 1, 2015, Rhom Company exchanged equipment for a $200,000 non-interest-bearing note due on January 1, 2018. The prevailing rate of interest for a note of this type on January 1, 2015, was 10%. The present value of $1 at 10% for three periods is 0.75. What amount of interest revenue should be included in Rhom's 2016 income statement? (Adapted from the Uniform CPA Examination, American Institute of Certified Public Accountants.)

64. On January 1, 2015, Bartell Company exchanged property for a $450,000 non-interest-bearing note due on January 1, 2018. The prevailing rate of interest for a note of this type on January 1, 2015, was 7.7%. The present value of $1 at 7.7% for three periods is 0.80. What amount of interest revenue should be included in Bartell's 2016 income statement? (Adapted from the Uniform CPA Examination, American Institute of Certified Public Accountants.)

✓ Checkpoint Answers

1. (a)

Year	Interest	Balance
4	$50	$1200
5	$50	$1250

(b)

Year	Interest	Balance
4	$57.88	$1215.51
5	$60.78	$1276.29

2. (a) $26,170.72 (b) $9170.72

3. (a) $1049.14 (b) $2514.84

4. $7980.52

5. (a) $15,000 (b) $35,000

6. About 7 years ($n = 6.86$)

7. (a) 12.68% (b) 8.24%

8. (a) 4.06% (b) 8.220%

9. (a) $4483.55 (b) $3725.53

10. $7732.24 11. $854.28

5.3 Annuities, Future Value, and Sinking Funds

So far in this chapter, only lump-sum deposits and payments have been discussed. Many financial situations, however, involve a sequence of payments at regular intervals, such as weekly deposits into a savings account or monthly payments on a mortgage or car loan. Such periodic payments are the subject of this section and the next.

The analysis of periodic payments will require an algebraic technique that we now develop. Suppose x is a real number. For reasons that will become clear later, we want to find the product

$$(x - 1)(1 + x + x^2 + x^3 + \cdots + x^{11}).$$

Using the distributive property to multiply this expression out, we see that all but two of the terms cancel:

$$x(1 + x + x^2 + x^3 + \cdots + x^{11}) - 1(1 + x + x^2 + x^3 + \cdots + x^{11})$$
$$= (x + x^2 + x^3 + \cdots + x^{11} + x^{12}) - 1 - x - x^2 - x^3 - \cdots - x^{11}$$
$$= x^{12} - 1.$$

Hence, $(x - 1)(1 + x + x^2 + x^3 + \cdots + x^{11}) = x^{12} - 1$. Dividing both sides by $x - 1$, we have

$$1 + x + x^2 + x^3 + \cdots + x^{11} = \frac{x^{12} - 1}{x - 1}.$$

The same argument, with any positive integer n in place of 12 and $n - 1$ in place of 11, produces the following result:

> If x is a real number and n is a positive integer, then
> $$1 + x + x^2 + x^3 + \cdots + x^{n-1} = \frac{x^n - 1}{x - 1}.$$

The above formula is a special case of a finite geometric series. For example, when $x = 5$ and $n = 7$, we see that

$$1 + 5 + 5^2 + 5^3 + 5^4 + 5^5 + 5^6 = \frac{5^7 - 1}{5 - 1} = \frac{78{,}124}{4} = 19{,}531.$$

A calculator can easily add up the terms on the left side, but it is faster to use the formula (Figure 5.5).

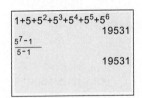

Figure 5.5

Ordinary Annuities

A sequence of equal payments made at equal periods of time is called an **annuity.** The time between payments is the **payment period,** and the time from the beginning of the first payment period to the end of the last period is called the **term of the annuity.** Annuities can be used to accumulate funds—for example, when you make regular deposits into a savings account. Or they can be used to pay out funds—as when you receive regular payments from a pension plan after you retire.

Annuities that pay out funds are considered in the next section. This section deals with annuities in which funds are accumulated by regular payments into an account or investment that earns compound interest. The **future value** of such an annuity is the final sum on deposit—that is, the total amount of all deposits and all interest earned by them.

We begin with **ordinary annuities**—ones where the payments are made at the *end* of each period and the frequency of payments is the same as the frequency of compounding the interest.

EXAMPLE 1 **Saving Accounts** $1500 is deposited at the end of each year for the next 6 years in a savings account paying 8% interest compounded annually. Find the future value of this annuity.

Solution Figure 5.6 shows the situation schematically.

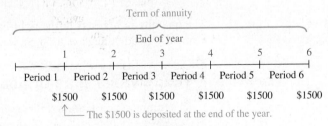

Figure 5.6

To find the future value of this annuity, look separately at each of the $1500 payments. The first $1500 is deposited at the end of period 1 and earns interest for the remaining 5 periods. From the formula in the box on page 231, the compound amount produced by this payment is

$$1500(1 + .08)^5 = 1500(1.08)^5.$$

The second $1500 payment is deposited at the end of period 2 and earns interest for the remaining 4 periods. So the compound amount produced by the second payment is

$$1500(1 + .08)^4 = 1500(1.08)^4.$$

Continue to compute the compound amount for each subsequent payment, as shown in Figure 5.7. Note that the last payment earns no interest.

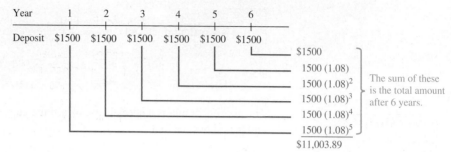

Figure 5.7

The last column of Figure 5.7 shows that the total amount after 6 years is the sum

$$1500 + 1500 \cdot 1.08 + 1500 \cdot 1.08^2 + 1500 \cdot 1.08^3 + 1500 \cdot 1.08^4 + 1500 \cdot 1.08^5$$
$$= 1500(1 + 1.08 + 1.08^2 + 1.08^3 + 1.08^4 + 1.08^5)$$
$$= \$11{,}003.89. \tag{1}$$

Now apply the algebraic fact in the box on page 241 to the expression in parentheses (with $x = 1.08$ and $n = 6$). It shows that the sum (the future value of the annuity) is

$$1500 \cdot \frac{1.08^6 - 1}{1.08 - 1} = \$11{,}003.89. \quad \boxed{1}$$

Example 1 is the model for finding a formula for the future value of any annuity. Suppose that a payment of PMT dollars is deposited at the end of each period for n periods, at an interest rate of i per period. Then the future value of this annuity can be found by using the procedure in Example 1, with these replacements:

1500	.08	1.08	6	5
↓	↓	↓	↓	↓
PMT	i	$1 + i$	n	$n - 1$

The future value FV in Example 1 is the sum **(1)**, which now becomes

$$FV = PMT[1 + (1 + i) + (1 + i)^2 + \cdots + (1 + i)^{n-2} + (1 + i)^{n-1}].$$

Apply the algebraic fact in the box on page 241 to the expression in brackets (with $x = 1 + i$). Then we have

$$FV = PMT\left[\frac{(1 + i)^n - 1}{(1 + i) - 1}\right] = PMT\left[\frac{(1 + i)^n - 1}{i}\right].$$

The quantity in brackets in the right-hand part of the preceding equation is sometimes written $s_{\overline{n}|i}$ (read "*s*-angle-*n* at *i*"). So we can summarize as follows.

Future Value of an Ordinary Annuity

The future value FV of an ordinary annuity used to accumulate funds is given by

$$FV = PMT\left[\frac{(1 + i)^n - 1}{i}\right], \quad \text{or} \quad FV = PMT \cdot s_{\overline{n}|i},$$

where

PMT is the payment at the end of each period;

$i = r/m$ is the interest rate per period, r is the annual interest rate, m is the number of periods per year; and

n is the total number of periods.

NOTE In Example 1, 8% was the annual interest rate, and thus one payment was made per year. In cases like this, $m = 1$, and so $i = r/m = r/1 = r$.

✓Checkpoint 1

Complete these steps for an annuity of $2000 at the end of each year for 3 years. Assume interest of 6% compounded annually.

(a) The first deposit of $2000 produces a total of _____.

(b) The second deposit becomes _____.

(c) No interest is earned on the third deposit, so the total in the account is _____.

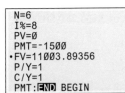

```
N=6
I%=8
PV=0
PMT=-1500
•FV=11003.89356
P/Y=1
C/Y=1
PMT:END BEGIN
```

Figure 5.8

TECHNOLOGY TIP Most computations with annuities can be done quickly with a spreadsheet program or a graphing calculator. On a calculator, use the TVM solver if there is one (see the Technology Tip on page 234).

Figure 5.8 shows how to do Example 1 on a TI-84+ TVM solver. First, enter the known quantities: N = number of payments, I% = interest rate, PV = present value, PMT = payment per period (entered as a negative amount), P/Y = number of payments per year, and C/Y = number of compoundings per year. At the bottom of the screen, set PMT: to "END" for ordinary annuities. Then put the cursor next to the unknown amount FV (future value), and press SOLVE.

Note: P/Y and C/Y should always be the same for problems in this text. If you use the solver for ordinary compound interest problems, set PMT = 0 and enter either PV or FV (whichever is known) as a negative amount.

```
N=7
I%=4.1
PV=0
PMT=350000
•FV=-2772807.587
P/Y=1
C/Y=1
PMT:END BEGIN
```

Figure 5.9

EXAMPLE 2 **Money Market Account** A rookie player in the National Football League just signed his first 7-year contract. To prepare for his future, he deposits $350,000 at the end of each year for 7 years in an account paying 4.1% compounded annually. How much will he have on deposit after 7 years?

Solution His payments form an ordinary annuity with $PMT = 350,000, n = 7$, and $i = .041$. The future value of this annuity (by the previous formula) is

$$FV = 350,000\left[\frac{(1.041)^7 - 1}{.041}\right] = \$2,772,807.59.$$

Figure 5.9 shows the result with the TVM solver. ✓₂

✓ **Checkpoint 2**

Johnson Building Materials deposits $2500 at the end of each year into an account paying 8% per year compounded annually. Find the total amount on deposit after

(a) 6 years;

(b) 10 years.

EXAMPLE 3 **Investing** Allyson opens a money market account at Silvergate Bank that in July 2016 offered a rate of 1.11%. Assume Allyson contributes $950 a month to the account and that the interest rate stays at 1.11% compounded monthly for the next 10 years. (Data from: www.bankrate.com.)

(a) What will be the value of the account in 10 years?

Solution Allyson's payments into the account form an annual annuity, with monthly payment $PMT = 950$. The interest per month is $i = \dfrac{r}{m} = \dfrac{.0111}{12}$, and the number of months in 10 years is $n = 10*12 = 120$. The future value of this annuity is

$$FV = PMT\left[\frac{(1 + i)^n - 1}{i}\right] = 950\left[\frac{(1 + .0111/12)^{120} - 1}{.0111/12}\right] = \$120,508.87.$$

```
N=120
I%=1.11
PV=0
PMT=-950
•FV=120508.8653
P/Y=12
C/Y=12
PMT:END BEGIN
```

Figure 5.10(a)

Figure 5.10(a) shows the result with the TVM solver.

(b) Suppose that, after the 10 years, Allyson was able to invest the $120,508.87 in a money market account that paid 1.24% interest compounded monthly. If she continues to deposit $950 a month into this new account, how much would be the total after an additional five years?

Solution Deal separately with the two parts of her account (the $950 contributions in the future and the $120,508.87 already in the account). The contributions form an ordinary annuity, as they did in part (a). Now we have $PMT = 950, i = \dfrac{.0124}{12}$, and $n = 5*12 = 60$. The future value of this annuity is

$$FV = PMT\left[\frac{(1 + i)^n - 1}{i}\right] = 950\left[\frac{(1 + .0124/12)^{60} - 1}{.0124/12}\right] = \$58,772.78.$$

```
N=60
I%=1.24
PV=-120508.87
PMT=-950
•FV=186985.5751
P/Y=12
C/Y=12
PMT:END BEGIN
```

Figure 5.10(b)

✓**Checkpoint 3**

Find the total value of the account in part (b) of Example 3 if the fund's return for the last 5 years is 1.42%, compounded monthly.

Meanwhile, the $120,508.87 from the first 10 years is also earning interest at 1.24% compounded monthly. By the compound amount formula (Section 5.2), the future value of this money is

$$FV = (1 + i/12)^{n*12} = 120,508.87(1 + .0124/12)^{60} = \$128,212.80.$$

So the total amount in Allyson's account after 15 years is the sum

$$\$58,772.78 + \$128,212.80 = \$186,985.58.$$

This calculation can also be done in one step using the TVM solver. Figure 5.10(b) shows how the information is entered. Note that we enter the amount from part (a) as a negative value in the PV line. ✓₃

Sinking Funds

A **sinking fund** is a fund set up to receive periodic payments. Corporations and municipalities use sinking funds to repay bond issues, to retire preferred stock, to provide for replacement of fixed assets, and for other purposes. If the payments are equal and are made at the end of regular periods, they form an ordinary annuity.

EXAMPLE 4 **Investing** A business sets up a sinking fund so that it will be able to pay off bonds it has issued when they mature. If it deposits $12,000 at the end of each quarter in an account that earns 5.2% interest, compounded quarterly, how much will be in the sinking fund after 10 years?

Solution The sinking fund is an annuity, with $PMT = 12,000$, $i = r/m = .052/4$, and $n = 4(10) = 40$. The future value is

$$FV = PMT\left[\frac{(1 + i)^n - 1}{i}\right] = 12,000\left[\frac{(1 + .052/4)^{40} - 1}{.052/4}\right] = \$624,369.81.$$

So there will be about $624,370 in the sinking fund. Figure 5.11 shows the entries in the TVM solver.

```
N=40
I%=5.2
PV=0
PMT=-12000
•FV=624369.8089
P/Y=4
C/Y=4
PMT:END BEGIN
```

Figure 5.11

EXAMPLE 5 **Investing** A firm borrows $6 million to build a small factory. The bank requires it to set up a $200,000 sinking fund to replace the roof after 15 years. If the firm's deposits earn 6% interest, compounded annually, find the payment it should make at the end of each year into the sinking fund.

Solution This situation is an annuity with future value $FV = 200,000$, interest rate $i = .06$, and $n = 15$. Solve the future-value formula for PMT:

$$FV = PMT\left[\frac{(1 + i)^n - 1}{i}\right]$$

$$200,000 = PMT\left[\frac{(1 + .06)^{15} - 1}{.06}\right] \quad \text{Let } FV = 200,000, i = .06, \text{ and } n = 15.$$

$$200,000 \approx PMT[23.27597] \quad \text{Compute the quantity in brackets.}$$

$$PMT = \frac{200,000}{23.27597} = \$8592.55. \quad \text{Divide both sides by 23.27597.}$$

So the annual payment is about $8593. Figure 5.12 shows the entries in the TVM solver. ✓₄

```
N=15
I%=6
PV=0
•PMT=-8592.5527...
FV=200000
P/Y=1
C/Y=1
PMT:END BEGIN
```

Figure 5.12

✓**Checkpoint 4**

Francisco Arce needs $8000 in 6 years so that he can go on an archaeological dig. He wants to deposit equal payments at the end of each quarter so that he will have enough to go on the dig. Find the amount of each payment if the bank pays

(a) 12% interest compounded quarterly;

(b) 8% interest compounded quarterly.

EXAMPLE 6 **Incentive Bonus** As an incentive for a valued employee to remain on the job, a company plans to offer her a $100,000 bonus, payable when she retires in 20 years. If the company deposits $200 a month in a sinking fund, what interest rate must it earn, with monthly compounding, in order to guarantee that the fund will be worth $100,000 in 20 years?

Solution The sinking fund is an annuity with $PMT = 200$, $n = 12(20) = 240$, and future value $FV = 100{,}000$. We must find the interest rate. If x is the annual interest rate in decimal form, then the interest rate per month is $i = x/12$. Inserting these values into the future-value formula, we have

$$PMT\left[\frac{(1 + i)^n - 1}{i}\right] = FV$$

$$200\left[\frac{(1 + x/12)^{240} - 1}{x/12}\right] = 100{,}000.$$

It is not possible to solve this equation algebraically. You can get a rough approximation by using a calculator and trying different values for x. With a graphing calculator, you can get an accurate solution by graphing

$$y_1 = 200\left[\frac{(1 + x/12)^{240} - 1}{x/12}\right] \quad \text{and} \quad y_2 = 100{,}000$$

and finding the x-coordinate of the point where the graphs intersect. Figure 5.13 shows that the company needs an interest rate of about 6.661%. The same answer can be obtained on a TVM solver (Figure 5.14). ✔5

Pete's Pizza deposits $5800 at the end of each quarter for 4 years.

(a) Find the final amount on deposit if the money earns 6.4% compounded quarterly.

(b) Pete wants to accumulate $110,000 in the 4-year period. What interest rate (to the nearest tenth) will be required?

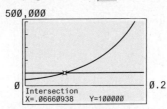

Figure 5.13

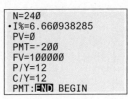

Figure 5.14

Annuities Due

The formula developed previously is for *ordinary annuities*—annuities with payments at the *end* of each period. The results can be modified slightly to apply to **annuities due**— annuities where payments are made at the *beginning* of each period.

An example will illustrate how this is done. Consider an annuity due in which payments of $100 are made for 3 years, and an ordinary annuity in which payments of $100 are made for 4 years, both with 5% interest, compounded annually. Figure 5.15 computes the growth of each payment separately (as was done in Example 1).

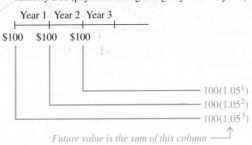

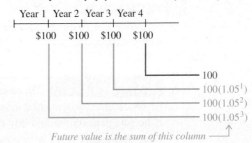

Figure 5.15

Figure 5.15 shows that the future values are the same, *except* for one $100 payment on the ordinary annuity (shown in red). So we can use the formula on page 242 to find the future value of the 4-year ordinary annuity and then subtract one $100 payment to get the future value of the 3-year annuity due:

Future value of	=	Future value of	−	One payment
3-year annuity due		4-year ordinary annuity		

$$FV = 100\left[\frac{1.05^4 - 1}{.05}\right] - 100 = \$331.01.$$

Essentially the same argument works in the general case.

Future Value of an Annuity Due

The future value FV of an annuity due used to accumulate funds is given by

$$FV = PMT\left[\frac{(1 + i)^{n+1} - 1}{i}\right] - PMT$$

$$FV = \text{an ordinary annuity} - \text{One payment},$$
of $n + 1$ payments

where

PMT is the payment at the beginning of each period;

$i = r/m$ is the interest rate per period, r is the annual interest rate, m is the number of periods per year; and

n is the total number of periods.

✓**Checkpoint 6**

(a) Ms. Black deposits $800 at the beginning of each 6-month period for 5 years. Find the final amount if the account pays 6% compounded semiannually.

(b) Find the final amount if this account were an ordinary annuity.

```
N=28
I%=1.08
PV=0
PMT=-500
•FV=14561.65577
P/Y=4
C/Y=4
PMT:END BEGIN
```
Figure 5.16

EXAMPLE 7 **Investing** Jillian will make payments of $500 at the beginning of each quarter for 7 years into a money market account at East Boston Savings Bank, which in July 2016 offered a rate of 1.08%. If interest is compounded quarterly, find the future value of this annuity due. (Data from: www.bankrate.com.)

Solution In 7 years, there are $n = 28$ quarterly periods. For an annuity due, add one period to get $n + 1 = 29$, and use the formula with $i = r/m = .0108/4 = .0027$:

$$FV = PMT\left[\frac{(1 + i)^{n+1} - 1}{i}\right] - PMT = 500\left[\frac{(1 + .0027)^{29} - 1}{.0027}\right] - 500 = \$14,561.66.$$

After 7 years, the account balance will be $14,561.66. ✓6

▦ **TECHNOLOGY TIP** When a TVM solver is used for annuities due, PMT: END at the bottom of the screen should be BEGIN. See Figure 5.16. Note that we use N = 28 periods with the calculator rather than 29 in the formula.

EXAMPLE 8 **Investing** Shen Wong plans to have a fixed amount from his paycheck directly deposited into an account that pays 5.5% interest, compounded monthly. If he gets paid on the first day of the month and wants to accumulate $13,000 in the next three-and-a-half years, how much should he deposit each month?

Solution Shen's deposits form an annuity due whose future value is $FV = 13{,}000$. The interest rate is $i = r/m = .055/12$. There are 42 months in three-and-a-half years. Since this is an annuity due, add one period, so that $n + 1 = 43$. Then solve the future-value formula for the payment PMT:

$$PMT\left[\frac{(1+i)^{n+1} - 1}{i}\right] - PMT = FV$$

$$PMT\left[\frac{(1 + .055/12)^{43} - 1}{.055/12}\right] - PMT = 13{,}000 \qquad \text{Let } i = .055/12, n = 43, \text{ and } FV = 13{,}000.$$

$$PMT\left(\left[\frac{(1 + .055/12)^{43} - 1}{.055/12}\right] - 1\right) = 13{,}000 \qquad \text{Factor out } PMT \text{ on left side.}$$

$$PMT(46.4103) = 13{,}000 \qquad \text{Compute left side.}$$

$$PMT = \frac{13{,}000}{46.4103} = 280.110. \qquad \text{Divide both sides by 46.4103.}$$

```
N=42
I%=5.5
PV=0
•PMT=-280.110136
FV=13000
P/Y=12
C/Y=12
PMT:END BEGIN
```

Figure 5.17

Shen should have $280.11 deposited from each paycheck. The TVM solver entries are shown in Figure 5.17.

We have seen in this section that sinking funds and annuities due are simply special cases of annuities. Recall that an annuity is a sequence of equal payments made at equal periods of time. When the payment is made at the *end* of the payment period, we call it a sinking fund or ordinary annuity. If the payment is made at the *beginning* of the payment period, then it is called an annuity due.

5.3 Exercises

Note: Unless stated otherwise, all payments are made at the end of the period.

Find each of these sums (to 4 decimal places).

1. $1 + 1.05 + 1.05^2 + 1.05^3 + \cdots + 1.05^{14}$

2. $1 + 1.046 + 1.046^2 + 1.046^3 + \cdots + 1.046^{21}$

Investing *As of July 2016, bankrate.com advertised 3 money market rates of 1.10%, 1.05%, and 1.00% at 3 different banks. Find the future value for the ordinary annuities with the given payments and interest rates. (See Examples 1, 2, 3(a), and 4.)*

3. $PMT = \$2000$, 1.10% compounded monthly for 3 years.

4. $PMT = \$500$, 1.00% compounded quarterly for 3 years.

5. $PMT = \$750$, 1.05% compounded semiannually for 5 years.

6. $PMT = \$900$, 1.10% compounded monthly for 5 years.

7. $PMT = \$2500$, 1.00% compounded quarterly for 10 years.

8. $PMT = \$1100$, 1.05% compounded semiannually for 10 years.

Investing *Find the final amount (rounded to the nearest dollar) in each of these retirement accounts, in which the rate of return on the account and the regular contribution change over time. (See Example 3.)*

9. $400 per month invested at 4%, compounded monthly, for 10 years; then $600 per month invested at 6%, compounded monthly, for 10 years.

10. $500 per month invested at 5%, compounded monthly, for 20 years; then $1000 per month invested at 8%, compounded monthly, for 20 years.

11. $1000 per quarter invested at 4.2%, compounded quarterly, for 10 years; then $1500 per quarter invested at 7.4%, compounded quarterly, for 15 years.

12. $1500 per quarter invested at 7.4%, compounded quarterly, for 15 years; then $1000 per quarter invested at 4.2%, compounded quarterly, for 10 years. (Compare with Exercise 11.)

Sinking Funds *Find the amount of each payment to be made into a sinking fund to accumulate the given amounts. Payments are made at the end of each period. (See Example 5.)*

13. $11,000; money earns 5% compounded semiannually for 6 years

14. $65,000; money earns 6% compounded semiannually for $4\frac{1}{2}$ years

15. $50,000; money earns 8% compounded quarterly for $2\frac{1}{2}$ years

16. $25,000; money earns 9% compounded quarterly for $3\frac{1}{2}$ years

17. $6000; money earns 6% compounded monthly for 3 years

18. $9000; money earns 7% compounded monthly for $2\frac{1}{2}$ years

Sinking Fund *Find the interest rate needed for the sinking fund to reach the required amount. Assume that the compounding period is the same as the payment period. (See Example 6.)*

19. $50,000 to be accumulated in 10 years; annual payments of $3940.

20. $100,000 to be accumulated in 15 years; quarterly payments of $1200.

21. $38,000 to be accumulated in 5 years; quarterly payments of $1675.

22. $77,000 to be accumulated in 20 years; monthly payments of $195.

23. What is meant by a sinking fund? List some reasons for establishing a sinking fund.

24. Explain the difference between an ordinary annuity and an annuity due.

Investing *In 2009, the average savings account rate was .20%, which is a much higher rate than current savings rates when this book went to press. Find the future value of each annuity due with the given rate close to the 2009 average. (See Example 7. Data from: fred.stlouisfed.org.)*

25. Payments of $500 for 10 years at .25% compounded annually

26. Payments of $1050 for 8 years at .21% compounded annually

27. Payments of $16,000 for 11 years at .18% compounded monthly

28. Payments of $25,000 for 12 years at .16% compounded monthly

29. Payments of $1000 for 9 years at .15% compounded semiannually

30. Payments of $750 for 15 years at .27% compounded semiannually

31. Payments of $200 for 7 years at .30% compounded quarterly

32. Payments of $2800 for 20 years at .32% compounded quarterly

Annuity Due *Find the payment that should be used for the annuity due whose future value is given. Assume that the compounding period is the same as the payment period. (See Example 8.)*

33. $8000; quarterly payments for 3 years; interest rate 4.4%

34. $12,000; annual payments for 6 years; interest rate 5.1%

35. $55,000; monthly payments for 12 years; interest rate 5.7%

36. $125,000; monthly payments for 9 years; interest rate 6%

37. **Cigarette Costs** The website www.theawl.com estimates that a pack-a-day smoker in New York State spends $400 a month on cigarettes (the highest cost of the 50 states). Suppose the smoker invests that amount at the end of each month in a money market account that pays .85% interest compounded monthly. What would the account be worth after 30 years?

38. **Cigarette Costs** The website www.theawl.com estimates that a pack-a-day smoker in Virginia spends $150 a month on cigarettes (the lowest cost of the 50 states). Suppose the smoker invests that amount at the end of each month in a money market account that pays .85% interest compounded monthly. What would the account be worth after 30 years?

39. **Investing** Mary Dodge deposits $12,000 at the end of each year for 9 years in an investment account with a guaranteed interest rate of 6% compounded annually.

    (a) Find the value of the account at the end of the 9 years.

    (b) Mary's sister works for an investment firm that pays 5% compounded annually. If Mary deposits money with this firm instead of the one in part (a), how much will she have in her account at the end of 9 years?

    (c) How much would Mary lose or gain over 9 years by investing in her sister's firm?

40. **Investing** Peter Healy deposits $800 at the end of each month for 5 years in an investment account with a guaranteed interest rate of 4.12% compounded monthly.

    (a) Find the value of the account at the end of 5 years.

    (b) A rival financial planner offers Peter an investment strategy of depositing $700 a month for 5 years with a guaranteed interest rate of 6.15% compounded monthly. What is the value of this investment strategy at the end of 5 years?

    (c) How much more money is gained by investing in the better strategy described in part (a) or in part (b)?

41. **Investing** At the end of each quarter, a 50-year-old woman puts $1200 in a retirement account that pays 7.2% interest compounded quarterly.

    (a) When she reaches age 60, what is the value of the account?

    (b) If no further deposits or withdrawals are made to the account, what is the value of the account when she reaches age 65?

42. **Retirement Planning** Miguel Handley is 45 years old. At the end of each month, he deposits $300 in a retirement account that pays 5.15% interest compounded monthly.

    (a) After ten years, what is the value of the account?

    (b) If no further deposits or withdrawals are made to the account, what is the value of the account when Miguel reaches age 65?

43. **Investing** A mother opened an investment account for her son on the day he was born, investing $1000. Each year on his birthday, she deposits another $1000, making the last deposit on his 18th birthday. If the account paid a return rate of 5.6% compounded annually, how much is in the account at the end of the day on the son's 18th birthday?

44. **Investing** A grandmother opens an investment account for her only granddaughter on the day she was born, investing $500. Each year on her birthday, she deposits another $500, making the last deposit on her 25th birthday. If the account paid a return rate of 6.2% compounded annually, how much is in the account at the end of the day on the granddaughter's 25th birthday?

45. **Investing** Chuck Hickman deposits $10,000 at the beginning of each year for 12 years in an account paying 5% compounded annually. He then puts the total amount on deposit in another account paying 6% compounded semiannually for another 9 years. Find the final amount on deposit after the entire 21-year period.

46. **Investing** Suppose that the best rate that the company in Example 6 can find is 6.3%, compounded monthly (rather than the 6.661% it wants). Then the company must deposit more in the sinking fund each month. What monthly deposit will guarantee that the fund will be worth $100,000 in 20 years?

47. **Investing** Diego Sanchez needs $10,000 in 8 years.

    (a) What amount should he deposit at the end of each quarter at 5% compounded quarterly so that he will have his $10,000?

    (b) Find Diego's quarterly deposit if the money is deposited at 5.8% compounded quarterly.

⟨C⟩ 48. Equipment Purchase Harv's Meats knows that it must buy a new machine in 4 years. The machine costs $12,000. In order to accumulate enough money to pay for the machine, Harv decides to deposit a sum of money at the end of each 6 months in an account paying 6% compounded semiannually. How much should each payment be?

49. Auto Purchase Barbara Margolius wants to buy a $24,000 car in 6 years. How much money must she deposit at the end of each quarter in an account paying 5% compounded quarterly so that she will have enough to pay for her car?

50. Museum Purchase The Chinns agree to sell an antique vase to a local museum for $19,000. They want to defer the receipt of this money until they retire in 5 years (and are in a lower tax bracket). If the museum can earn 5.8%, compounded annually, find the amount of each annual payment it should make into a sinking fund so that it will have the necessary $19,000 in 5 years.

51. Real Estate Purchase Diane sells some land in Nevada. She will be paid a lump sum of $60,000 in 7 years. Until then, the buyer pays 8% simple interest quarterly.

(a) Find the amount of each quarterly interest payment.

(b) The buyer sets up a sinking fund so that enough money will be present to pay off the $60,000. The buyer wants to make semiannual payments into the sinking fund; the account pays 6% compounded semiannually. Find the amount of each payment into the fund.

52. Stamp Purchase Joe Seniw bought a rare stamp for his collection. He agreed to pay a lump sum of $4000 after 5 years. Until then, he pays 6% simple interest semiannually.

(a) Find the amount of each semiannual interest payment.

(b) Joe sets up a sinking fund so that enough money will be present to pay off the $4000. He wants to make annual payments into the fund. The account pays 8% compounded annually. Find the amount of each payment.

53. Retirement Planning To save for retirement, Karla Harby put $300 each month into an ordinary annuity for 20 years. Interest was compounded monthly. At the end of the 20 years, the annuity was worth $147,126. What annual interest rate did she receive?

54. Real Estate Destiny Green made payments of $250 per month at the end of each month to purchase a piece of property. After 30 years, she owned the property, which she sold for $330,000. What annual interest rate would she need to earn on an ordinary annuity for a comparable rate of return?

55. Retirement Planning When Joe and Sarah graduate from college, each expects to work a total of 45 years. Joe begins saving for retirement immediately. He plans to deposit $600 at the end of each quarter into an account paying 8.1% interest, compounded quarterly, for 10 years. He will then leave his balance in the account, earning the same interest rate, but make no further deposits for 35 years. Sarah plans to save nothing during the first 10 years and then begin depositing $600 at the end of each quarter in an account paying 8.1% interest, compounded quarterly, for 35 years.

(a) Without doing any calculations, predict which one will have the most in his or her retirement account after 45 years. Then test your prediction by answering the following questions (calculation required to the nearest dollar).

(b) How much will Joe contribute to his retirement account?

(c) How much will be in Joe's account after 45 years?

(d) How much will Sarah contribute to her retirement account?

(e) How much will be in Sarah's account after 45 years?

56. In a 1992 Virginia lottery, the jackpot was $27 million. An Australian investment firm tried to buy all possible combinations of numbers, which would have cost $7 million. In fact, the firm ran out of time and was unable to buy all combinations, but ended up with the only winning ticket anyway. The firm received the jackpot in 20 equal annual payments of $1.35 million. Assume these payments meet the conditions of an ordinary annuity. (Data from: *Washington Post*, March 10, 1992, p. A1.)

(a) Suppose the firm can invest money at 8% interest compounded annually. How many years would it take until the investors would be further ahead than if they had simply invested the $7 million at the same rate? (*Hint:* Experiment with different values of *n*, the number of years, or use a graphing calculator to plot the value of both investments as a function of the number of years.)

(b) How many years would it take in part (a) at an interest rate of 12%?

✓ **Checkpoint Answers**

1. (a) $2247.20 (b) $2120.00 (c) $6367.20

2. (a) $18,339.82 (b) $36,216.41

3. $188,406.70

4. (a) $232.38 (b) $262.97

5. (a) $104,812.44 (b) 8.9%

6. (a) $9446.24 (b) $9171.10

5.4 Annuities, Present Value, and Amortization

In the annuities studied previously, regular deposits were made into an interest-bearing account and the value of the annuity increased from 0 at the beginning to some larger amount at the end (the future value). Now we expand the discussion to include annuities that begin with an amount of money and make regular payments out of the account each period until the value of the annuity decreases to 0. Examples of such annuities are lottery jackpots, structured settlements imposed by a court in which the party at fault (or his or her insurance company) makes regular payments to the injured party, and trust funds that pay the recipients a fixed amount at regular intervals.

In order to develop the essential formula for dealing with "payout annuities," we need another useful algebraic fact. If x is a nonzero number and n is a positive integer, verify the following equality by multiplying out the right-hand side:*

$$x^{-1} + x^{-2} + x^{-3} + \cdots + x^{-(n-1)} + x^{-n} = x^{-n}(x^{n-1} + x^{n-2} + x^{n-3} + \cdots + x^1 + 1).$$

Now use the sum formula in the box on page 241 to rewrite the expression in parentheses on the right-hand side:

$$x^{-1} + x^{-2} + x^{-3} + \cdots + x^{-(n-1)} + x^{-n} = x^{-n}\left(\frac{x^n - 1}{x - 1}\right)$$

$$= \frac{x^{-n}(x^n - 1)}{x - 1} = \frac{x^0 - x^{-n}}{x - 1} = \frac{1 - x^{-n}}{x - 1}.$$

We have proved the following result:

If x is a nonzero real number and n is a positive integer, then

$$x^{-1} + x^{-2} + x^{-3} + \cdots + x^{-n} = \frac{1 - x^{-n}}{x - 1}.$$

Present Value

 In Section 5.2, we saw that the present value of FV dollars at interest rate i per period for n periods is the amount that must be deposited today (at the same interest rate) in order to produce FV dollars in n periods. Similarly, the **present value of an annuity** is the amount that must be deposited today (at the same compound interest rate as the annuity) to provide all the payments for the term of the annuity. It does not matter whether the payments are invested to accumulate funds or are paid out to disperse funds; the amount needed to provide the payments is the same in either case. We begin with ordinary annuities.

A relative has funded an annuity for Timothy Carlo that pays him $1500 at the end of each year for six years. Let us assume that the interest rate is 8% compounded annually. We will now investigate the present value of this annuity. To do this, we look separately at each payment that Tim receives. We will then find the present value of each payment—the amount needed now in order to make the payment in the future. The sum of these present values will be the present value of the annuity because it will provide all the payments.

To find the first $1500 payment (due in one year), the present value of $1500 at 8% annual interest is needed now. According to the present-value formula for compound interest in the Summary of Section 5.2 (with $FV = 1500$, $i = .08$, and $n = 1$), this present value is

$$\frac{1500}{1 + .08} = \frac{1500}{1.08} = 1500(1.08^{-1}) \approx \$1388.89.$$

This amount will grow to $1500 in one year.

For the second $1500 payment (due in two years), we need the present value of $1500 at 8% interest, compounded annually for two years. The present-value formula for compound interest (with $FV = 1500$, $i = .08$, and $n = 2$) shows that this present value is

$$\frac{1500}{(1 + .08)^2} = \frac{1500}{1.08^2} = 1500(1.08^{-2}) \approx \$1286.01.$$

---

*Remember that powers of x are multiplied by *adding* exponents and that $x^n x^{-n} = x^{n-n} = x^0 = 1$.

Less money is needed for the second payment because it will grow over two years instead of one.

A similar calculation shows that the third payment (due in three years) has present value $1500(1.08^{-3})$. Continue in this manner to find the present value of each of the remaining payments, as summarized in Figure 5.18.

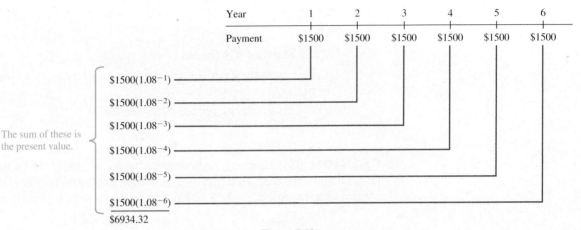

Figure 5.18

The left-hand column of Figure 5.18 shows that the present value is

$$1500 \cdot 1.08^{-1} + 1500 \cdot 1.08^{-2} + 1500 \cdot 1.08^{-3} + 1500 \cdot 1.08^{-4}$$
$$+ 1500 \cdot 1.08^{-5} + 1500 \cdot 1.08^{-6}$$
$$= 1500(1.08^{-1} + 1.08^{-2} + 1.08^{-3} + 1.08^{-4} + 1.08^{-5} + 1.08^{-6})$$
$$= \$6934.32. \tag{1}$$

Now apply the algebraic fact in the box on page 250 to the expression in parentheses (with $x = 1.08$ and $n = 6$). It shows that the sum (the present value of the annuity) is

$$1500\left[\frac{1 - 1.08^{-6}}{1.08 - 1}\right] = 1500\left[\frac{1 - 1.08^{-6}}{.08}\right] = \$6934.32.$$

This amount will provide for all six payments and leave a zero balance at the end of six years (give or take a few cents due to rounding to the nearest penny at each step).

We have just shown the model for finding a formula for the present value of any ordinary annuity. Suppose that a payment of PMT dollars is made at the end of each period for n periods, at interest rate i per period. Then the present value of this annuity can be found by using the procedure, with these replacements:

$$
\begin{array}{cccc}
1500 & .08 & 1.08 & 6 \\
\downarrow & \downarrow & \downarrow & \downarrow \\
PMT & i & 1 + i & n
\end{array}
$$

The present value is the sum in equation **(1)**, which now becomes

$$PV = PMT[(1 + i)^{-1} + (1 + i)^{-2} + (1 + i)^{-3} + \cdots + (1 + i)^{-n}].$$

Apply the algebraic fact in the box on page 250 to the expression in brackets (with $x = 1 + i$). Then we have

$$PV = PMT\left[\frac{1 - (1 + i)^{-n}}{(1 + i) - 1}\right] = PMT\left[\frac{1 - (1 + i)^{-n}}{i}\right].$$

The quantity in brackets in the right-hand part of the preceding equation is sometimes written $a_{\overline{n}i}$ (read "a-angle-n at i"). So we can summarize as follows.

Present Value of an Ordinary Annuity

The present value PV of an ordinary annuity is given by

$$PV = PMT\left[\frac{1 - (1 + i)^{-n}}{i}\right], \quad \text{or} \quad PV = PMT \cdot a_{\overline{n}|i},$$

where

PMT is the payment at the end of each period;

$i = r/m$ is the interest rate per period, r is the annual interest rate, m is the number of periods per year; and

n is the total number of periods.

⚠ CAUTION Do not confuse the formula for the present value of an annuity with the one for the future value of an annuity. Notice the difference: The numerator of the fraction in the present-value formula is $1 - (1 + i)^{-n}$, but in the future-value formula, it is $(1 + i)^n - 1$.

```
N=84
I%=4.7
•PV=142924.7907
PMT=-2000
FV=0
P/Y=12
C/Y=12
PMT:END BEGIN
```
Figure 5.19

EXAMPLE 1 **Insurance Sattlement** Camila was in an auto accident. She sued the person at fault and was awarded a structured settlement in which an insurance company will pay her $2000 at the end of each month for the next seven years. How much money should the insurance company invest now at 4.7%, compounded monthly, to guarantee that all the payments can be made?

Solution The payments form an ordinary annuity. The amount needed to fund all the payments is the present value of the annuity. Apply the present-value formula with $PMT = 2000$, $n = 7 \cdot 12 = 84$, and $i = r/m = .047/12$ (the interest rate per month). The insurance company should invest

$$PV = PMT\left[\frac{1 - (1 + i)^{-n}}{i}\right] = 2000\left[\frac{1 - (1 + .047/12)^{-84}}{.047/12}\right] = \$142{,}924.79.$$

The entries in the TVM solver appear in Figure 5.19. ✓₁

✓ **Checkpoint 1**

An insurance company offers to pay Jane Parks an ordinary annuity of $3000 per quarter for five years *or* the present value of the annuity now. If the interest rate is 6%, find the present value.

```
N=240
I%=5.2
PV=-400000
•PMT=2684.216207
FV=0
P/Y=12
C/Y=12
PMT:END BEGIN
```
Figure 5.20

EXAMPLE 2 **Retirement Planning** To supplement his pension in the early years of his retirement, Hong plans to use $400,000 of his savings as an ordinary annuity that will make monthly payments to him for 20 years. If the interest rate is 5.2%, how much will each payment be?

Solution The present value of the annuity is $PV = \$400{,}000$, the monthly interest rate is $i = r/m = .052/12$, and $n = 12 \cdot 20 = 240$ (the number of months in 20 years). Solve the present-value formula for the monthly payment PMT:

$$PV = PMT\left[\frac{1 - (1 + i)^{-n}}{i}\right]$$

$$400{,}000 = PMT\left[\frac{1 - (1 + .052/12)^{-240}}{.052/12}\right]$$

$$PMT = \frac{400{,}000}{\left[\dfrac{1 - (1 + .052/12)^{-240}}{.052/12}\right]} = \$2684.22.$$

✓ **Checkpoint 2**

Carl Dehne has $80,000 in an account paying 4.8% interest, compounded monthly. He plans to use up all the money by making equal monthly withdrawals for 15 years. If the interest rate is 4.8%, find the amount of each withdrawal.

Hong will receive $2684.22 a month (about $51,382 per year) for 20 years. The entries in the TMV solver appear in Figure 5.20. ✓₂

Loans and Amortization

If you take out a car loan or a home mortgage, you repay it by making regular payments to the bank. From the bank's point of view, your payments are an annuity that is paying the bank a fixed amount each month. The present value of this annuity is the amount you borrowed.

We can use the present value of an ordinary annuity formula, as we did in Example 2, to determine the monthly payment needed for common loans such as auto loans, student loans, and mortgages. In these cases, PV is the amount borrowed today, i is the monthly interest rate (most loans consist of monthly payments), and n is the number of payments.

EXAMPLE 3 **Auto Loan** In October 2016, Chase Bank advertised a used-car auto loan rate of 2.84% for a 5-year loan. Shelley wants to buy a used car for $18,700. Find the amount of each payment. (Data from: www.chase.com.)

Solution Use the present value formula of an ordinary annuity, with $PV = 18{,}700$, $n = 12*5 = 60$, $i = r/m = .0284/12$ (the monthly interest rate). Then solve for payment PMT.

$$PV = PMT\left[\frac{1 - (1 + i)^{-n}}{i}\right]$$

$$18{,}700 = PMT\left[\frac{1 - (1 + .0284/12)^{-60}}{.0284/12}\right]$$

$$PMT = \frac{18{,}700}{\left[\dfrac{1 - (1 + .0284/12)^{-60}}{.0284/12}\right]} \qquad \text{Solve for } PMT.$$

$$PMT = \$334.69$$

The entries in the TVM solver appear in Figure 5.21. ☑3

A loan is **amortized** if both the principal and interest are paid by a sequence of equal periodic payments. The periodic payment needed to amortize a loan may be found, as in Example 3, by solving the present-value formula for PMT.

```
N=60
I%=2.84
PV=18700
•PMT=-334.68657...
FV=0
P/Y=12
C/Y=12
PMT:END BEGIN
```

Figure 5.21

✓ **Checkpoint 3**

Yulanda uses a Chase Bank auto loan to purchase a new car priced at $35,650. Her 4-year loan has an interest rate of 2.54%. What is the monthly payment?

Amortization Payments

A loan of PV dollars at interest rate i per period may be amortized in n equal periodic payments of PMT dollars made at the end of each period, where

$$PMT = \frac{PV}{\left[\dfrac{1 - (1 + i)^{-n}}{i}\right]} = \frac{PVi}{1 - (1 + i)^{-n}}.$$

EXAMPLE 4 **Mortgage Loan** On October 10, 2016, the average rate for a 30-year fixed mortgage was 3.37%. Assume a down payment of 20% on a home purchase of $285,000. (Data from: Zillow.com.)

(a) Find the monthly payment needed to amortize this loan.

Solution The down payment is $.20(285{,}000) = \$57{,}000$. Thus, the loan amount PV is $\$285{,}000 - \$57{,}000 = \$228{,}000$. We can now apply the formula in the preceding box, with $n = 12(30) = 360$ (the number of monthly payments in 30 years), and monthly interest rate $i = r/m = .0337/12.^*$

$$PMT = \frac{PVi}{1 - (1 + i)^{-n}} = \frac{228{,}000(.0337/12)}{1 - (1 + .0337/12)^{-360}} = \$1007.35$$

---

*Mortgage rates are quoted in terms of annual interest, but it is always understood that the monthly rate is $\frac{1}{12}$ of the annual interest rate.

```
N=360
I%=3.37
PV=228000
•PMT=-1007.34843
FV=0
P/Y=12
C/Y=12
PMT:END BEGIN
```

Figure 5.22(a)

```
N=240
I%=3.37
PV=175710.2335
•PMT=-1007.35
FV=0
P/Y=12
C/Y=12
PMT:END BEGIN
```

Figure 5.22(b)

✓ **Checkpoint 4**

Find the remaining balance after 15 years.

Monthly payments of \$1007.34 are required to amortize the loan. The entries in the TVM solver appear in Figure 5.22(a).

(b) After 10, years approximately how much is owed on the mortgage?

Solution You may be tempted to say that, after 10 years of payments on a 30-year mortgage, the balance will be reduced by one-third. However, in the early months of repaying a loan, a significant portion of each payment goes to pay interest. Thus, much less than one-third of the mortgage is paid off in the first 10 years, as we now see.

After 10 years (120 payments), the 240 remaining payments can be thought of as an annuity. The present value for this annuity is the (approximate) remaining balance on the mortgage. Hence, we use the present value formula with $PMT = 1007.34$, $i = r/m = .0337/12$, and $n = 240$:

$$PV = 1007.35\left[\frac{1 - (1 + .0337/12)^{-240}}{(.0337/12)}\right] = \$175,710.23$$

So the remaining balance is about \$175,710.23. The actual balance probably differs slightly from this amount because payments and interest amounts are rounded to the nearest penny. The entries in the TVM solver appear in Figure 5.22(b). ✓4

Example 4(b) illustrates an important fact: Even though equal *payments* are made to amortize a loan, the loan *balance* does not decrease in equal steps. The method used to estimate the remaining balance in Example 4(b) works in the general case. If n payments are needed to amortize a loan and x payments have been made, then the remaining payments form an annuity of $n - x$ payments. So we apply the present-value formula with $n - x$ in place of n to obtain this result.

Remaining Balance

If a loan can be amortized by n payments of PMT dollars each at an interest rate i per period, then the *approximate* remaining balance B after x payments is

$$B = PMT\left[\frac{1 - (1 + i)^{-(n-x)}}{i}\right].$$

Amortization Schedules

The remaining-balance formula is a quick and convenient way to get a reasonable estimate of the remaining balance on a loan, but it is not accurate enough for a bank or business, which must keep its books exactly. To determine the exact remaining balance after each loan payment, financial institutions normally use an **amortization schedule,** which lists how much of each payment is interest, how much goes to reduce the balance, and how much is still owed after each payment.

EXAMPLE 5 **Personal Loan** Beth Hill borrows \$1000 for one year at 12% annual interest, compounded monthly.

(a) Find her monthly payment.

Solution Apply the amortization payment formula with $PV = 1000, n = 12$, and monthly interest rate $i = r/m = .12/12 = .01$. Her payment is

$$PMT = \frac{PVi}{1 - (1 + i)^{-n}} = \frac{1000(.01)}{1 - (1 + .01)^{-12}} = \$88.85.$$

(b) After making five payments, Beth decides to pay off the remaining balance. Approximately how much must she pay?

Solution Apply the remaining-balance formula just given, with $PMT = 88.85$, $i = .01$, and $n - x = 12 - 5 = 7$. Her approximate remaining balance is

$$B = PMT\left[\frac{1 - (1 + i)^{-(n-x)}}{i}\right] = 88.85\left[\frac{1 - (1 + .01)^{-7}}{.01}\right] = \$597.80.$$

(c) Construct an amortization schedule for Beth's loan.

Solution An amortization schedule for the loan is shown in the following table. It was obtained as follows: The annual interest rate is 12% compounded monthly, so the interest rate per month is $12\%/12 = 1\% = .01$. When the first payment is made, one month's interest, namely, $.01(1000) = \$10$, is owed. Subtracting this from the $88.85 payment leaves $78.85 to be applied to repayment. Hence, the principal at the end of the first payment period is $1000 - 78.85 = \$921.15$, as shown in the "payment 1" line of the table.

When payment 2 is made, one month's interest on the new balance of $921.15 is owed, namely, $.01(921.15) = \$9.21$. Continue as in the preceding paragraph to compute the entries in this line of the table. The remaining lines of the table are found in a similar fashion.

Payment Number	Amount of Payment	Interest for Period	Portion to Principal	Principal at End of Period
0	—	—	—	$1000.00
1	$88.85	$10.00	$78.85	921.15
2	88.85	9.21	79.64	841.51
3	88.85	8.42	80.43	761.08
4	88.85	7.61	81.24	679.84
5	88.85	6.80	82.05	597.79
6	88.85	5.98	82.87	514.92
7	88.85	5.15	83.70	431.22
8	88.85	4.31	84.54	346.68
9	88.85	3.47	85.38	261.30
10	88.85	2.61	86.24	175.06
11	88.85	1.75	87.10	87.96
12	88.84	.88	87.96	0

Note that Beth's remaining balance after five payments differs slightly from the estimate made in part (b).

The final payment in the amortization schedule in Example 5(c) differs from the other payments. It often happens that the last payment needed to amortize a loan must be adjusted to account for rounding earlier and to ensure that the final balance will be exactly 0.

TECHNOLOGY TIP To create an amortization table with the TI-84 calculator, use the amortization table program. Spreadsheets are another useful tool for creating amortization tables. Microsoft Excel (Microsoft Corporation Excel © 2016) has a built-in feature for calculating monthly payments. Figure 5.23 shows an Excel amortization table for Example 5. For more details, see the *Spreadsheet Manual*, also available with this text.

	A	B	C	D	E	F
1	Prnt#	Payment	Interest	Principal	End Principal	
2	0				1000	
3	1	88.85	10.00	78.85	921.15	
4	2	88.85	9.21	79.64	841.51	
5	3	88.85	8.42	80.43	761.08	
6	4	88.85	7.61	81.24	679.84	
7	5	88.85	6.80	82.05	597.79	
8	6	88.85	5.98	82.87	514.92	
9	7	88.85	5.15	83.70	431.22	
10	8	88.85	4.31	84.54	346.68	
11	9	88.85	3.47	85.38	261.30	
12	10	88.85	2.61	86.24	175.06	
13	11	88.85	1.75	87.10	87.96	
14	12	88.85	0.88	87.97	-0.01	

Figure 5.23

When doing financial planning, it can be helpful to determine the amount of interest paid over the course of a loan. For example, in Example 4, borrowing $228,000 at 3.37% annual interest for a 30-year fixed mortgage yields a payment of $1007.35 per month. Over the lifetime of the loan, that yields

$$(30)(12)(1007.35) = \$362,646.00$$

in total payments to the bank. Thus, $362,646 - 228,000 = 134,646$ is paid in total interest to the bank.

Many financial planners recommend making extra payments toward the principal of the loan each month. One way to make an extra payment each year is by adding $\frac{1}{12}$ of the regular monthly payment to each monthly payment. This extra amount is then applied to the principal of the loan and thus reduces the amount of future interest payments. We can use the TVM solver to find out how much more quickly (and how much less interest we pay) with this approach. We simply enter the payment amount as a negative number and then solve for the number of payments. This method is described in Example 6.

▦ EXAMPLE 6
Let us revisit the 30-year fixed mortgage loan described in Example 4. The loan amount was $228,000 at 3.37% annual interest. Each month, an extra amount of money equal to $\frac{1}{12}$ the monthly payment was added to the principal.

(a) How much quicker is the loan paid off?

Solution Paying $\frac{1}{12}$ of the required monthly payment extra each month is equivalent to making one extra mortgage payment in that year. In this case, $\frac{1}{12}$ of the monthly payment is:

$$\left(\frac{1}{12}\right)(1007.35) = \$83.95.$$

So now the monthly payment amount is $1007.35 + 83.95 = \$1091.30$. In the TVM solver, enter the information as in Figure 5.24(a). Put the curser in the N = section and solve for the number of payments. The solution appears in Figure 5.24(b).

The TVM solver tells us that it will take 316 (we round up) payments to amortize the loan. This represents 44 $(360 - 316)$ fewer payments. Because 44 payments is equivalent to 3 years and 8 months of payments, that is how much more quickly the loan will be paid off.

(b) How much money in interest payments is saved?

Solution This problem is a little trickier. Notice that in part (a), the TVM solver states that 315.0971424 is the number of payments. This means that it takes .0971424 of a payment to pay off the remaining balance for the 316th payment. However, we can find this amount by multiplying the payment amount by this fraction of a payment, which is

$$(\$1091.30)(.0971424) = \$106.01 \text{ (rounded to the nearest cent).}$$

```
N=
I%=3.37
PV=228000
PMT=-1091.3
FV=0
P/Y=12
C/Y=12
PMT:END BEGIN
```
Figure 5.24(a)

```
•N=315.0971424
 I%=3.37
 PV= 228000
 PMT=-1091.3
 FV=0
 P/Y=12
 C/Y=12
 PMT:END BEGIN
```
Figure 5.24(b)

Now we can find the total amount of money paid over the loan by multiplying the full monthly payment amount ($1091.30) by 315 and then adding this extra amount for the last payment:

$$(\$1091.30)(315) + \$106.01 = \$343,865.51$$

We saw earlier that making 360 equal payments over 30 years yielded a total loan payment amount of $362,646.00. By adding an extra $83.95 per month to our payment, the borrower would save, in interest paid to the bank,

$$\$362,646.00 - 343,865.51 = \$18,780.49. \quad ✔_5$$

Annuities Due

We want to find the present value of an annuity due in which 6 payments of PMT dollars are made at the *beginning* of each period, with interest rate i per period, as shown schematically in Figure 5.25.

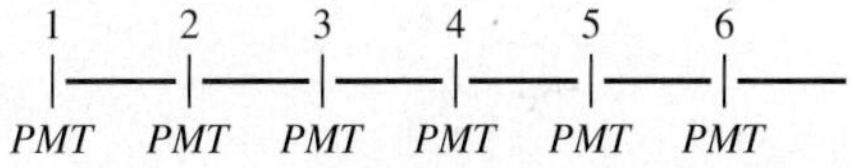

Figure 5.25

The present value is the amount needed to fund all 6 payments. Since the first payment earns no interest, PMT dollars are needed to fund it. Now look at the last 5 payments by themselves in Figure 5.26.

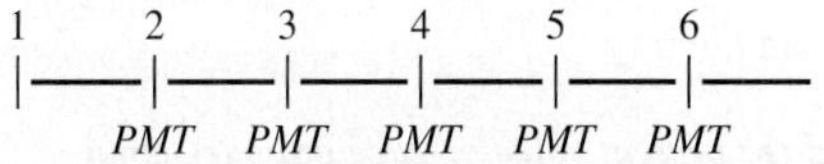

Figure 5.26

If you think of these 5 payments as being made at the end of each period, you see that they form an ordinary annuity. The money needed to fund them is the present value of this ordinary annuity. So the present value of the annuity due is given by

$$PMT + \begin{array}{c} \text{Present value of the ordinary} \\ \text{annuity of 5 payments} \end{array}$$

$$PMT + PMT\left[\frac{1 - (1 + i)^{-5}}{i}\right].$$

Replacing 6 by n and 5 by $n - 1$, and using the argument just given, produces the general result that follows.

Present Value of an Annuity Due

The present value PV of an annuity due is given by

$$PV = PMT + PMT\left[\frac{1 - (1 + i)^{-(n-1)}}{i}\right],$$

$$PV = \underset{\text{payment}}{\underset{\text{One}}{\underbrace{}}} + \underset{\text{of } n - 1 \text{ payments}}{\underset{\text{an ordinary annuity}}{\underbrace{\text{Present value of}}}}$$

where

PMT is the payment at the beginning of each period;

$i = r/m$ is the interest rate per period, r is the annual interest rate, m is the number of periods per year;

n is the total number of periods.

✓ **Checkpoint 5**

The same mortgage amount as Example 6 is paid off in 15 years at 2.75%.

(a) Find the monthly payment.

(b) If an extra $100 is paid each month toward the principal, how many payments are needed to pay off the loan?

(c) How much interest is saved by paying an extra $100 a month?

```
N=20
I%=4.22
•PV=694593.4902
PMT=-50000
FV=0
P/Y=1
C/Y=1
PMT:END BEGIN
```
Figure 5.27

✓ **Checkpoint 6**

If an Illinois player wins a Lotto jackpot of $60 million that would be paid out in 30 annual payments, and the state can earn 5.25% annual interest, what is the cash value of the jackpot?

EXAMPLE 7 **Lottery Winnings** The Illinois *Lottery Winners' Handbook* discusses the options of how to receive the winnings for a $1 million scratch ticket. One option is to take 20 annual payments of approximately $50,000, which is $1 million divided into 20 equal payments. The other option is to take a lump-sum payment (which is often called the cash value). If the Illinois Lottery Commission can earn 4.22% annual interest, how much is the cash value?

Solution The yearly payments form a 20-payment annuity due. An equivalent amount now is the present value of this annuity. Apply the present-value formula with $PMT = 50,000$, $i = .0422$, and $n = 20$:

$$PV = PMT + PMT\left[\frac{1 - (1 + i)^{-(n-1)}}{i}\right] = 50,000 + 50,000\left[\frac{1 - (1 + .0422)^{-19}}{.0422}\right]$$

$$= \$694,593.49.$$

The cash value is $694,593.49. The entries in the TVM solver appear in Figure 5.27.

5.4 Exercises

Unless noted otherwise, all payments and withdrawals are made at the end of the period.

1. Explain the difference between the present value of an annuity and the future value of an annuity.

Ordinary Annuities *Find the present value of each ordinary annuity. (See Examples 1 and 2.)*

2. Payments of $890 each year for 16 years at 6% compounded annually

3. Payments of $1400 each year for 8 years at 6% compounded annually

4. Payments of $10,000 semiannually for 15 years at 7.5% compounded semiannually

5. Payments of $50,000 quarterly for 10 years at 5% compounded quarterly

6. Payments of $15,806 quarterly for 3 years at 6.8% compounded quarterly

Ordinary Annuities *Find the amount necessary to fund the given withdrawals. (See Example 1.)*

7. Quarterly withdrawals of $650 for 5 years; interest rate is 4.9%, compounded quarterly.

8. Yearly withdrawals of $1200 for 14 years; interest rate is 5.6%, compounded annually.

9. Monthly withdrawals of $425 for 10 years; interest rate is 6.1%, compounded monthly.

10. Semiannual withdrawals of $3500 for 7 years; interest rate is 5.2%, compounded semiannually.

Ordinary Annuities *Find the payment made by the ordinary annuity with the given present value. (See Example 2.)*

11. $90,000; monthly payments for 22 years; interest rate is 4.9%, compounded monthly.

12. $45,000; monthly payments for 11 years; interest rate is 5.3%, compounded monthly.

13. $275,000; quarterly payments for 18 years; interest rate is 6%, compounded quarterly.

14. $330,000; quarterly payments for 30 years; interest rate is 6.1% compounded quarterly.

Amortization *Find the payment necessary to amortize each of the given loans. (See Example 4.)*

15. $2500; 8% interest; 6 quarterly payments

16. $41,000; 9% interest; 10 semiannual payments

17. $90,000; 7% interest; 12 annual payments

18. $140,000; 12% interest; 15 quarterly payments

19. $7400; 8.2% interest; 18 semiannual payments

20. $5500; 9.5% interest; 24 monthly payments

Mortgages *In October 2016, the mortgage interest rates listed in Exercises 21–24 for the given companies were listed at www.hsh.com. Find the monthly payment necessary to amortize the given loans. (See Example 4.)*

21. $225,000 at 3.68% for 30 years from loanDepot

22. $330,000 at 3.50% for 30 years from Quicken Loans

23. $140,000 at 2.63% for 15 years from CloseYourOwnLoan

24. $180,000 at 2.50% for 10 years from CloseYourOwnLoan

Loans *Find the monthly payment and estimate the remaining balance (to the nearest dollar). Assume interest is on the unpaid balance. The interest rates are from national averages from www.bankrate.com in October 2016. (See Examples 4 and 5.)*

25. Six-year new car loan for $26,799 at 3.74%; remaining balance after 2 years

26. Three-year used car loan for $15,875 at 2.89%; remaining balance after 1 year

27. Thirty-year mortgage for $210,000 at 3.63%; remaining balance after 12 years

28. Fifteen-year mortgage for $195,000 at 2.75%; remaining balance after 4.5 years

Amortization *Use the amortization table in Example 5(c) to answer the questions in Exercises 29–32.*

29. How much of the 5th payment is interest?

30. How much of the 10th payment is used to reduce the debt?

31. How much interest is paid in the first 5 months of the loan?

32. How much interest is paid in the last 5 months of the loan?

Lottery *Find the cash value of the lottery jackpot (to the nearest dollar). Yearly jackpot payments begin immediately (30 for Mega Millions and 30 for Powerball). Assume the lottery can invest at the given interest rate. (See Example 7.)*

33. Powerball: $57.6 million; 5.1% interest

34. Powerball: $207 million; 5.78% interest

35. Mega Millions: $41.6 million; 4.735% interest

36. Mega Millions: $23.4 million; 4.23% interest

Work the following applied problems.

37. **Financing** An auto stereo dealer sells a stereo system for $600 down and monthly payments of $30 for the next 3 years. If the interest rate is 1.25% *per month* on the unpaid balance, find

  (a) the cost of the stereo system;

  (b) the total amount of interest paid.

38. **Auto Loan** John Kushida buys a used car costing $16,000. He agrees to make payments at the end of each monthly period for 4 years. He pays 4.35% interest, in monthly payments.

  (a) What is the amount of each payment?

  (b) Find the total amount of interest Kushida will pay.

39. **Land Mortgage** A speculator agrees to pay $15,000 for a parcel of land; this amount, with interest, will be paid over 4 years with semiannual payments at an interest rate of 10%. Find the amount of each payment.

40. **Auto Loan** Alan Stasa buys a new car costing $26,750. What is the monthly payment if the interest rate is 4.2%, and the loan is for 60 months? Find the total amount of interest Alan will pay.

Student Loan *A student education loan has two repayment options. The standard plan repays the loan in 10 years with equal monthly payments. The extended plan allows from 12 to 30 years to repay the loan. A student borrows $35,000 at 7.43% interest.*

41. Find the monthly payment and total interest paid under the standard plan.

42. Find the monthly payment and total interest paid under the extended plan with 20 years to pay off the loan.

Use the formula for the approximate remaining balance to work each problem. (See Examples 4(b) and 5(b).)

43. **Business Loan** When Teresa Flores opened her law office, she bought $14,000 worth of law books and $7200 worth of office furniture. She paid $1200 down and agreed to amortize the balance with semiannual payments for 5 years at 12% interest.

  (a) Find the amount of each payment.

  (b) When her loan had been reduced below $5000, Flores received a large tax refund and decided to pay off the loan. How many payments were left at this time?

44. **Mortgage** Kareem Adams buys a house for $285,000. He pays $60,000 down and takes out a mortgage at 6.9% on the balance. Find his monthly payment and the total amount of interest he will pay if the length of the mortgage is

  (a) 15 years;

  (b) 20 years;

  (c) 25 years.

  (d) When will half the 20-year loan be paid off?

45. **Mortgage** Susan Carver will purchase a home for $257,000. She will use a down payment of 20% and finance the remaining portion at 3.9% interest for 30 years.

  (a) What will be the monthly payment?

  (b) How much will remain on the loan after making payments for 5 years?

  (c) How much interest will be paid on the total amount of the loan over the course of 30 years?

46. **Mortgage** Mohsen Manouchehri will purchase a $230,000 home with a 20-year mortgage. If he makes a down payment of 20% and the interest rate is 3.3%,

  (a) what will the monthly payment be?

  (b) how much will he owe after making payments for 8 years?

  (c) how much in total interest will he pay over the course of the 20-year loan?

Work each problem.

47. Retirement Planning Erin Pupillo and her employer contribute $400 at the end of each month to her retirement account, which earns 7% interest, compounded monthly. When she retires after 45 years, she plans to make monthly withdrawals for 30 years. If her account earns 5% interest, compounded monthly, then when she retires, what is her maximum possible monthly withdrawal (without running out of money)?

48. Financial Planning Jim Milliken won a $15,000 prize. On March 1, he deposited it in an account earning 5.2% interest, compounded monthly. On March 1 one year later, he begins to withdraw the same amount at the beginning of each month for a year. Assuming that he uses up all the money in the account, find the amount of each monthly withdrawal.

49. Retirement Planning Catherine Dohanyos plans to retire in 20 years. She will make 20 years of monthly contributions to her retirement account. One month after her last contribution, she will begin the first of 10 years of withdrawals. She wants to withdraw $2500 per month. How large must her monthly contributions be in order to accomplish her goal if the account earns interest of 7.1% compounded monthly for the duration of her contributions and the 120 months of withdrawals?

50. Retirement Planning Mateo plans to retire in 25 years. He will make 25 years of monthly contributions to his retirement account. One month after his last contribution, he will begin the first of 10 years of withdrawals. He wants to withdraw $3000 per month. How large must his monthly contributions be in order to accomplish his goal if the account earns interest of 6.8% compounded monthly for the duration of his contributions and the 120 months of withdrawals?

51. Retirement Planning Marcus plans to retire in 20 years. Marcus will make 10 years (120 months) of equal monthly payments into his account. Ten years after his last contribution, he will begin the first of 120 monthly withdrawals of $3400 per month. Assume that the retirement account earns interest of 8.2% compounded monthly for the duration of his contributions, the 10 years in between his contributions and the beginning of his withdrawals, and the 10 years of withdrawals. How large must Marcus's monthly contributions be in order to accomplish his goal?

52. Retirement Planning Joshua plans to retire in 25 years. He will make 15 years (180 months) of equal monthly payments into his account. Ten years after his last contribution, he will begin the first of 120 monthly withdrawals of $2900 per month. Assume that the retirement account earns interest of 5.4% compounded monthly for the duration of his contributions, the 10 years in between his contributions and the beginning of his withdrawals, and the 10 years of withdrawals. How large must Joshua's monthly contributions be in order to accomplish his goal?

53. Auto Loan Jessica Walters takes out a 6-year fixed-rate auto loan for $35,000 at 5.2% annual interest.

(a) Find the monthly payment for the loan.

(b) Find the total interest paid over the course of the loan.

(c) If Jessica pays an additional $100 per month toward the principal, how many payments will it take to pay off the loan?

(d) How much is the total interest paid on the loan when the additional $100 is paid each month?

54. Mortgage Ling Wu takes out a 30-year fixed rate mortgage for $325,000 at 3.9% annual interest.

(a) Find the monthly payment for the loan.

(b) Find the total interest paid over the course of the loan.

(c) If Ling pays an additional $150 per month toward the principal, how many payments will it take to pay off the loan?

(d) How much is the total interest of the loan when the additional $150 is paid each month?

In Exercises 55–58, prepare an amortization schedule showing the first four payments for each loan. (See Example 5(c).)

55. Insurance Payment An insurance firm pays $4000 for a new printer for its computer. It amortizes the loan for the printer in 4 annual payments at 8% interest.

56. Commercial Auto Loan Large semitrailer trucks cost $72,000 each. Ace Trucking buys such a truck and agrees to pay for it by a loan that will be amortized with 9 semiannual payments at 6% interest.

57. Business Loan One retailer charges $1048 for a certain computer. A firm of tax accountants buys 8 of these computers. It makes a down payment of $1200 and agrees to amortize the balance with monthly payments at 12% interest for 4 years.

58. Business Loan Joan Varozza plans to borrow $20,000 to stock her small boutique. She will repay the loan with semiannual payments for 5 years at 7% interest.

✓**Checkpoint Answers**

1. $51,505.92

2. $624.33

3. $781.86

4. $142,177.53

5. **(a)** $1547.26

   (b) 167

   (c) $3937.45

6. $31,456,910.67

CHAPTER 5 Summary and Review

Key Terms and Symbols

5.1 simple interest
principal
rate
time
future value (maturity
value)
present value
discount and
T-bills

5.2 compound interest
compound amount
compounding period
nominal rate (stated rate)
effective rate (APY)
present value

5.3 annuity
payment period

term of an annuity
ordinary annuity
future value of an
ordinary annuity
sinking fund
annuity due
future value of an
annuity due

5.4 present value of an
ordinary annuity
amortization payments
remaining balance
amortization schedule
present value of an
annuity due

Chapter 5 Key Concepts

**A Strategy for Solving
Finance Problems**

We have presented a lot of new formulas in this chapter. By answering the following questions, you can decide which formula to use for a particular problem.

1. If simple interest is being used, what is being sought—interest amount, future value, present value, or discount?

2. If compound interest is being used, does it involve a lump sum (single payment) or an annuity (sequence of payments)?

   (a) For a lump sum,

   (i) is ordinary compound interest involved?

   (ii) what is being sought—present value, future value, number of periods, or effective rate (APY)?

   (b) For an annuity,

   (i) is it an ordinary annuity (payment at the end of each period) or an annuity due (payment at the beginning of each period)?

   (ii) what is being sought—present value, future value, or payment amount?

Once you have answered these questions, choose the appropriate formula and work the problem. As a final step, consider whether the answer you get makes sense. For instance, the amount of interest or the payments in an annuity should be fairly small compared with the total future value.

Key Formulas

List of Variables

r is the annual interest rate.
m is the number of periods per year.
i is the interest rate per period. $\quad i = \dfrac{r}{m}$
t is the number of years.
n is the number of periods. $\quad n = tm$
x is the number of remaining payments.
PV is the principal or present value.
FV is the future value of a lump sum.
FV is the future value of an annuity.
PMT is the periodic payment in an annuity.
B is the remaining balance on a loan.

Interest	Simple Interest	Compound Interest
Interest	$I = PVrt$	$I = FV - PV$
Future value	$FV = PV(1 + rt)$	$FV = PV(1 + i)^n$
Present value	$PV = \dfrac{FV}{1 + rt}$	$PV = \dfrac{FV}{(1 + i)^n} = FV(1 + i)^{-n}$
		Effective rate (or APY) $r_E = \left(1 + \dfrac{r}{m}\right)^m - 1$

Discount   If D is the **discount** on a T-bill with face value PV at simple interest rate r for t years, then $D = PVrt$.

Continuous Interest   If PV dollars are deposited for t years at interest rate r per year, compounded continuously, the **compound amount (future value)** is $FV = PVe^{rt}$.

The **present value** PV of FV dollars at interest rate r per year compounded continuously for t years is

$$PV = \frac{FV}{e^{rt}}.$$

Remaining Balance (Approximate)   If a loan can be amortized by n payments of PMT dollars, each at an interest rate i per period, then the approximate remaining balance B after x payments is

$$B = PMT\left[\frac{1 - (1 + i)^{-(n-x)}}{i}\right].$$

Annuities

Ordinary annuity   Future value $FV = PMT\left[\dfrac{(1 + i)^n - 1}{i}\right] = PMT \cdot s_{\overline{n}|i}$

Present value $PV = PMT\left[\dfrac{1 - (1 + i)^{-n}}{i}\right] = PMT \cdot a_{\overline{n}|i}$

Annuity due   Future value $FV = PMT\left[\dfrac{(1 + i)^{n+1} - 1}{i}\right] - PMT$

Present value $PV = PMT + PMT\left[\dfrac{1 - (1 + i)^{-(n-1)}}{i}\right]$

Chapter 5 Review Exercises

Loans *Find the simple interest for the following loans.*

1. $4902 at 6.5% for 11 months

2. $42,368 at 9.22% for 5 months

3. $3478 at 7.4% for 88 days

4. $2390 at 8.7% from May 3 to July 28

Bonds *Find the semiannual (simple) interest payment and the total interest earned over the life of the bond.*

5. $12,000 Merck Company 6-year bond at 4.75% annual interest

6. $20,000 General Electric 9-year bond at 5.25% annual interest

Loans *Find the maturity value for each simple interest loan.*

7. $7750 at 6.8% for 4 months

8. $15,600 at 8.2% for 9 months

Present Value *Find the present value of the given future amounts; use simple interest.*

9. $3250 in 11 months; money earns 5.87%

10. $459.57 in 7 months; money earns 5.5%

11. $80,612 in 128 days; money earns 6.77%

12. A 9-month $7000 Treasury bill sells at a discount rate of 3.5%. Find the amount of the discount and the price of the T-bill.

13. A 6-month $10,000 T-bill sold at a 4% discount. Find the actual rate of interest paid by the Treasury.

14. For a given amount of money at a given interest rate for a given period greater than 1 year, does simple interest or compound interest produce more interest? Explain.

Investing *Find the compound amount and the amount of interest earned in each of the given scenarios.*

15. $2800 at 6% compounded annually for 12 years

16. $57,809.34 at 4% compounded quarterly for 6 years

17. $12,903.45 at 6.37% compounded quarterly for 29 quarters

18. $4677.23 at 4.57% compounded monthly for 32 months

Investing *Find the amount of compound interest earned by each deposit.*

19. $22,000 at 5.5%, compounded quarterly for 6 years

20. $2975 at 4.7%, compounded monthly for 4 years

Bonds *Find the face value (to the nearest dollar) of the zero-coupon bond.*

21. 5-year bond at 3.9%; price $12,366

22. 15-year bond at 5.2%; price $11,575

APY *Find the APY corresponding to the given nominal rate.*

23. 5% compounded semiannually

24. 6.5% compounded daily

Present Value *Find the present value of the given amounts at the given interest rate.*

25. $42,000 in 7 years; 12% compounded monthly

26. $17,650 in 4 years; 8% compounded quarterly

27. $1347.89 in 3.5 years; 6.2% compounded semiannually

28. $2388.90 in 44 months; 5.75% compounded monthly

Bonds *Find the price that a purchaser should be willing to pay for these zero-coupon bonds.*

29. 10-year $15,000 bond; interest at 4.4%

30. 25-year $30,000 bond; interest at 6.2%

31. What is meant by the future value of an annuity?

Annuities *Find the future value of each annuity.*

32. $1288 deposited at the end of each year for 14 years; money earns 7% compounded annually

33. $4000 deposited at the end of each quarter for 8 years; money earns 6% compounded quarterly

34. $233 deposited at the end of each month for 4 years; money earns 6% compounded monthly

35. $672 deposited at the beginning of each quarter for 7 years; money earns 5% compounded quarterly

36. $11,900 deposited at the beginning of each month for 13 months; money earns 7% compounded monthly

37. What is the purpose of a sinking fund?

Sinking Funds *Find the amount of each payment that must be made into a sinking fund to accumulate the given amounts. Assume payments are made at the end of each period.*

38. $6500; money earns 5% compounded annually; 6 annual payments

39. $57,000; money earns 6% compounded semiannually for $8\frac{1}{2}$ years

40. $233,188; money earns 5.7% compounded quarterly for $7\frac{3}{4}$ years

41. $56,788; money earns 6.12% compounded monthly for $4\frac{1}{2}$ years

Annuities *Find the present value of each ordinary annuity.*

42. Payments of $850 annually for 4 years at 5% compounded annually

43. Payments of $1500 quarterly for 7 years at 8% compounded quarterly

44. Payments of $4210 semiannually for 8 years at 5.6% compounded semiannually

45. Payments of $877.34 monthly for 17 months at 6.4% compounded monthly

Annuities *Find the amount necessary to fund the given withdrawals (which are made at the end of each period).*

46. Quarterly withdrawals of $800 for 4 years with interest rate 4.6%, compounded quarterly

47. Monthly withdrawals of $1500 for 10 years with interest rate 5.8%, compounded monthly

48. Yearly withdrawals of $3000 for 15 years with interest rate 6.2%, compounded annually

Annuities *Find the payment for the ordinary annuity with the given present value.*

49. $150,000; monthly payments for 15 years, with interest rate 5.1%, compounded monthly

50. $25,000; quarterly payments for 8 years, with interest rate 4.9%, compounded quarterly

51. Find the lump-sum deposit today that will produce the same total amount as payments of $4200 at the end of each year for 12 years. The interest rate in both cases is 4.5%, compounded annually.

52. If the current interest rate is 6.5%, find the price (to the nearest dollar) that a purchaser should be willing to pay for a $24,000 bond with coupon rate 5% that matures in 6 years.

Amortization *Find the amount of the payment necessary to amortize each of the given loans.*

53. $32,000 at 8.4% interest; 10 quarterly payments

54. $5607 at 7.6% interest; 32 monthly payments

Mortgages *Find the monthly house payments for the given mortgages.*

55. $95,000 at 3.67% for 30 years

56. $167,000 at 2.91% for 15 years

57. Find the approximate remaining balance after 5 years of payments on the loan in Exercise 55.

58. Find the approximate remaining balance after 7.5 years of payments on the loan in Exercise 56.

Student Loans *According to www.studentaid.ed.gov, in 2016 the interest rate for a direct unsubsidized student loan was 3.76%. A* *portion of an amortization table is given here for a $15,000 direct unsubsidized student loan to be paid back in 10 years. Use the table to answer Exercises 59–62.*

Payment Number	Amount of Payment	Portion to Principal	Interest for Period	Principal at End of Period
0				15000.00
1	150.16	103.16	47.00	14896.84
2	150.16	103.49	46.68	14793.35
3	150.16	103.81	46.35	14689.54
4	150.16	104.14	46.03	14585.41
5	150.16	104.46	45.70	14480.94
6	150.16	104.79	45.37	14376.16
7	150.16	105.12	45.04	14271.04
8	150.16	105.45	44.72	14165.59
9	150.16	105.78	44.39	14059.81
10	150.16	106.11	44.05	13953.71

59. How much of the seventh payment is interest?

60. How much of the fourth payment is used to reduce the debt?

61. How much interest is paid in the first 6 months of the loan?

62. How much has the debt been reduced at the end of the first 8 months?

Work the following applied problems.

63. Lottery In May 2016, a New Jersey family won a Powerball lottery prize of $429,600,000. (Data from: www.powerball.com.)

   (a) If the family had chosen to receive the money in 30 yearly payments, beginning immediately, what would have been the amount of each payment?

   (b) The family chose the one-time lump-sum cash option. If the interest rate is 3.58%, approximately how much did the family receive?

64. Business Loan A firm of attorneys deposits $15,000 of profit-sharing money in an account at 6% compounded semiannually for $7\frac{1}{2}$ years. Find the amount of interest earned.

65. Retirement Planning Tom, a graduate student, is considering investing $500 now, when he is 23, or waiting until he is 40 to invest $500. How much more money will he have at age 65 if he invests now, given that he can earn 5% interest compounded quarterly?

66. Historical CD Savings According to a financial Web site, on June 15, 2005, Frontenac Bank of Earth City, Missouri, paid 3.94% interest, compounded quarterly, on a 2-year CD, while E*TRADE Bank of Arlington, Virginia, paid 3.93% compounded daily. What was the effective rate for the two CDs, and which bank paid a higher effective rate? (Data from: www.bankrate.com.)

67. Investing Chalon Bridges deposits semiannual payments of $3200, received in payment of a debt, in an ordinary annuity at 6.8% compounded semiannually. Find the final amount in the account and the interest earned at the end of 3.5 years.

68. Retirement Benefits Each year, a firm must set aside enough funds to provide employee retirement benefits of $625,000,000 in 20 years. If the firm can invest money at 7.5% compounded monthly, what amount must be invested at the end of each month for this purpose?

69. Charitable Giving A benefactor wants to be able to leave a bequest to the college she attended. If she wants to make a donation of $2,000,000 in 10 years, how much each month does she need to place in an investment account that pays an interest rate of 5.5%, compounded monthly?

70. Retirement Planning Suppose you have built up a pension with $18,000 annual payments by working 10 years for a company when you leave to accept a better job. The company gives you the option of collecting half the full pension when you reach age 55 or the full pension at age 65. Assume an interest rate of 8% compounded annually. By age 75, how much will each plan produce? Which plan would produce the larger amount?

71. Charitable Giving In 3 years, Ms. Nguyen must pay a pledge of $7500 to her favorite charity. What lump sum can she deposit today at 10% compounded semiannually so that she will have enough to pay the pledge?

72. Home Improvement Loan To finance the $15,000 cost of their kitchen remodeling, the Chews will make equal payments at the end of each month for 36 months. They will pay interest at the rate of 7.2%. Find the amount of each payment.

73. Business Loan To expand her business, the owner of a small restaurant borrows $40,000. She will repay the money in equal payments at the end of each semiannual period for 8 years at 9% interest. What payments must she make?

74. Mortgage The Keller family bought a house for $210,000. They paid $42,000 down and took out a 30-year mortgage for the balance at 3.75%.

   (a) Find the monthly payment.

   (b) How much of the first payment is interest?

   After 15 years, the family sold their house for $255,000.

   (c) Estimate the current mortgage balance at the time of the sale.

   (d) Find the amount of money they received from the sale after paying off the mortgage.

75. Investing Miranda Williams invests $325 at the end of each month in a savings account earning 6.8% annual interest. If interest is compounded monthly, what is the value of the account in 20 years?

76. Retirement Planning Dan Hook deposits $400 a month to a retirement account that has an interest rate of 3.1%, compounded monthly. After making 60 deposits, Dan changes his job and stops making payments for 3 years during which time he earns 3.1% interest, compounded monthly. After 3 years, he starts making deposits again, but now he deposits $525 per month. What will the value of the retirement account be after Dan makes his $525 monthly deposits for 5 years?

C 77. Inheritance The proceeds of a $50,000 inheritance are left on deposit with an insurance company for 10 years at an interest rate of 4.2%, compounded annually. The balance at the end of 10 years is paid to the beneficiary in 120 equal monthly payments, with the first payment made immediately. During the payout period, interest is credited at an annual interest rate of 2.5%. What is the amount of the monthly payments?

C 78. Retirement Planning Eileen Gianiodis wants to retire on $75,000 per year for her life expectancy of 20 years after she retires. She estimates that she will be able to earn an interest rate of 10.1%, compounded annually, throughout her lifetime. To reach her retirement goal, Eileen will make annual contributions to her account for the next 30 years. One year after making her last deposit, she will receive her first retirement check. How large must her yearly contributions be?

Case Study 5 Investing in Stocks and Using the Rule of 72

The future value formulas that we used throughout Chapter 5 are based on fixed interest rates that exist through the savings or investment period. Many people invest in the stock market, however, and wish to project the value of their money several years into the future. The challenge here is that individual stock returns and mutual fund returns are not fixed and are often unpredictable. Thus, projecting a rate of return into the future always requires assumptions and risks.

With these caveats in mind, financial planners and individual investors can still make projections on investment income in the future using formulas that we have already learned. For example, according to vanguard.com, the average return on equities (stocks) from 1930 through 2010 was 9.2%. Many index funds follow the return on the market quite closely. So let us assume that we have $5000 to invest in a market index fund. We will assume an annual return of 9.2%. What is the value of the fund in 20 years?

To answer this question, we can use the future value formula

$$FV = PV(1 + i)^n \qquad (1)$$

with $PV = \$5000$, $i = .092$, and $n = 20$. Thus,

$$FV = 5000(1 + .092)^{20} = 29{,}068.51.$$

Let us look more closely at the growth of the account value in this fund over time. The table below shows the value of the fund over the course of the first 10 years.

Year	Account Value
0	$5000.00
1	5460.00
2	5962.32
3	6510.85
4	7109.85
5	7763.96
6	8478.24
7	9258.24
8	10,110.00
9	11,040.12
10	12,055.81

Notice that after 8 years, the fund's value has more than doubled from the original $5000 to $10,110.00. This is actually not surprising because investors often use an informal rule knows as the Rule of 72 to approximate how long it takes for a fund to double in value.

The Rule of 72 gives a rough estimate of how long it takes for an investment to double in value. The way it works is to take the number 72 and divide by the annual rate of return (let's round our return to 9%). So we take $72/9 = 8$, and 8 years is a rough approximation for the number of years it takes for an investment to double. The table below shows how close this rule actually works by comparing its estimated time to doubling with what is predicted by the future value formula in Equation **(1)**.

Rate of Return	Rule of 72	Actual Number of Years
4%	18.0	17.7
6%	12.0	11.9
8%	9.0	9.0
10%	7.2	7.3
12%	6.0	6.1
14%	5.1	5.3
16%	4.5	4.7
18%	4.0	4.2
20%	3.6	3.8

As we can see, the Rule of 72 is most accurate when returns are between 6% and 12%. This is generally a good range of expectations from the stock market.

The power of doubling (or exponential growth) is the reason that many financial planners urge college graduates to begin saving for retirement as soon as they can. We can use the future value of an ordinary annuity formula from Section 5.3 for projecting the value of an account when regular deposits are made at the end of each time period. Even saving $100 a month can reap long-term benefits, as the following example shows.

EXAMPLE 1 **Investing** Shanda graduates from college at age 22 and lands her first job. She manages to deposit $100 at the end of each month in an individual retirement account (IRA). Assume that the account earns an annual return of 8%. She makes these monthly deposits for 20 years. Determine the future value of her IRA.

Solution The future value of an ordinary annuity formula is

$$FV = PMT\left[\frac{(1 + i)^n - 1}{i}\right]$$

where $PMT = \$100$, $i = r/m = .08/12$, and $n = (20)(12) = 240$. Thus, we have

$$FV = PMT\left[\frac{(1 + i)^n - 1}{i}\right]$$

$$= 100\left[\frac{(1 + .08/12)^{240} - 1}{.08/12}\right] = \$58,902.04.$$

EXAMPLE 2 **Investing** Miguel graduates at the same time as Shanda (see Example 1). However, he waits 10 years before he begins making deposits to an IRA account. When he does, he invests $200 a month at the end of each month. If we assume Miguel's IRA earns an average annual return of 8% as well, what is the value of his account in 10 years?

Solution Here we use the same future value of an ordinary annuity formula, but with $PMT = \$200$, $i = .08/12$, and $n = 10*12 = 120$. Thus, we have

$$FV = PMT\left[\frac{(1 + i)^n - 1}{i}\right]$$

$$= 200\left[\frac{(1 + .08/12)^{120} - 1}{.08/12}\right] = \$36,589.21.$$

So even though Miguel and Shanda invest the same amount of money in monthly payments (Sandra invests $100(20)(12) = \$24,000$ and Miguel invests $200(10)(12) = \$24,000$), Shanda has a much higher account value at the end of the 20 years than Miguel does. In fact, we can use the Rule of 72 to approximate the value of each account when they are ready to retire, as we see in the next example.

EXAMPLE 3 Shanda and Miguel stop making monthly investments into their respective IRA accounts but continue to hold the money in an account earning 8%. Use the Rule of 72 to approximate the value of each account when both Shanda and Miguel turn 69 years old.

Solution In Examples 1 and 2, Shanda invested for 20 years starting age 22, and Miguel invested for 10 years starting at age 32. Thus, they are both 42 years old when they stop making monthly investments. The Rule of 72 tells us that, if we assume an 8% annual return,

then every 9 years, the value of the account will double. So the value of each of their accounts will double 3 times in the course of the 27 years when they reach age 69.

The table gives the amounts that Shanda and Miguel will earn after each 9-year interval.

Time	Shanda	Miguel
First doubling	(2)($58,902.04) = $117,804.08	(2)(36,589.21) = $73,178.42
Second doubling	(2)(117,804.08) = $235,608.16.	(2)(73,178.42) = $146,356.84
Third doubling	(2)(235,608.16) = $471,216.32	(2)(146,356.84) = $292,713.68

We could have also multiplied each of the account values at age 42 by 8 [which is (2)(2)(2)] to obtain the value of three doublings, but the table shows in dramatic fashion how much more money Shanda will have in her account when she turns 69 years old than Miguel will.

Exercises

1. **Investing** Shelia graduates from college at age 25 and obtains her first job. She manages to deposit $150 at the end of each month in an individual retirement account (IRA). Assume that the account earns an annual return of 6%. Shelia makes these monthly deposits for 30 years. Determine the future value of her IRA.

2. **Investing** Hector manages to deposit $300 at the end of each month in an individual retirement account (IRA) at age 30. Assume that the account earns an annual return of 12%. Hector makes these monthly deposits for 25 years. Determine the future value of his IRA.

3. **Investing** Suppose Shelia in Exercise 1 stops adding money to her account when she turns 55, but the account continues to earn a return of 6% per year.

   (a) If we use the Rule of 72, how old will Shelia be when the balance has doubled?

   (b) What will the balance be when it has doubled?

4. **Investing** Suppose Hector in Exercise 2 stops adding money to his account when he turns 55, but the account continues to earn a return of 12% per year.

   (a) If we use the Rule of 72, how old will Hector be when the balance has doubled?

   (b) What will the balance be when it has doubled?

5. **Investing** The rate of return can make a huge difference in how quickly money doubles. If $10,000 is invested for 36 years at 6%, how many times will the value double according to the Rule of 72? What will the value be at the end of 36 years?

6. **Investing** If $10,000 is invested for 36 years at 8%, how many times will the value double according to the Rule of 72? What will the value be at the end of 36 years?

Extended Project

1. Investigate the interest rates for the subsidized and unsubsidized student loans. If you have taken out student loans or plan to take out student loans before graduating, calculate your own monthly payment and how much interest you will pay over the course of the repayment period. If you have not taken out, and you do not plan to take out a student loan, contact the financial aid office of your college or university to determine the median amount borrowed with student loans at your institution. Determine the monthly payment and how much interest is paid during repayment for the typical borrowing student.

2. Determine the best interest rate for a new car purchase for a 48-month loan at a bank near you. If you finance $25,999 with such a loan, determine the payment and the total interest paid over the course of the loan. Also, determine the best interest rate for a new car purchase for a 48-month loan at a credit union near you. Determine the monthly payment and total interest paid if the same auto loan is financed through the credit union. Is it true that the credit union would save you money?

3. Suppose an investor has $1,000,000 to invest for 20 years at an annual interest rate of 2.5%. Find the value of the account if the interest is compounded monthly and if it is compounded continuously. Is there an appreciable difference in the value? If the investor invested $100,000,000, would there be an appreciable difference in value then? What if the interest rate were only .85%?

Systems of Linear Equations and Matrices

6

CHAPTER

CHAPTER OUTLINE

6.1 Systems of Two Linear Equations in Two Variables
6.2 Larger Systems of Linear Equations
6.3 Applications of Systems of Linear Equations
6.4 Basic Matrix Operations
6.5 Matrix Products and Inverses
6.6 Applications of Matrices

CASE STUDY 6
Airline Route Maps

A variety of resource allocation problems involving many variables can be handled by solving an appropriate system of linear equations. Technology (such as graphing calculators, WolframAlpha.com, computer algebra systems, and smart phone apps) is very helpful for handling large systems. Smaller ones can easily be solved by hand.

This chapter deals with **linear** (or **first-degree**) **equations** such as

$$2x + 3y = 14 \qquad \text{linear equation in two variables,}$$
$$4x - 2y + 5z = 8 \qquad \text{linear equation in three variables,}$$

and so on. A **solution** of such an equation is an ordered list of numbers that, when substituted for the variables in the order they appear, produces a true statement. For instance, $(1, 4)$ is a solution of the equation $2x + 3y = 14$ because substituting $x = 1$ and $y = 4$ produces the true statement $2(1) + 3(4) = 14$.

Many applications involve **systems of linear equations,** such as these two:

Two equations in two variables	Three equations in four variables
	$2x + y + z \qquad = 3$
$5x - 3y = 7$	$x + y + z + w = 5$
$2x + 4y = 8$	$-4x + \qquad z + w = 0.$

A **solution of a system** is a solution that satisfies *all* the equations in the system. For instance, in the right-hand system of equations on page 268, $(1, 0, 1, 3)$ is a solution of all three equations (check it) and hence is a solution of the system. By contrast, $(1, 1, 0, 3)$ is a solution of the first two equations, but not the third. Hence, $(1, 1, 0, 3)$ is not a solution of the system.

This chapter presents methods for solving such systems, including matrix methods. Matrix algebra and other applications of matrices are also discussed.

6.1   Systems of Two Linear Equations in Two Variables

The graph of a linear equation in two variables is a straight line. The coordinates of each point on the graph represent a solution of the equation (Section 2.1). Thus, the solution of a system of two such equations is represented by the point or points where the two lines intersect. There are exactly three geometric possibilities for two lines: They intersect at a single point, or they coincide, or they are distinct and parallel. As illustrated in Figure 6.1, each of these geometric possibilities leads to a different number of solutions for the system.

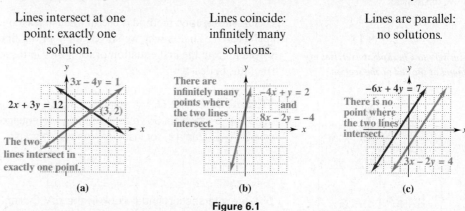

| Lines intersect at one point: exactly one solution. | Lines coincide: infinitely many solutions. | Lines are parallel: no solutions. |

Figure 6.1

The Substitution Method

Example 1 illustrates the substitution method for solving a system of two equations in two variables.

EXAMPLE 1   Solve the system

$$2x - y = 1$$
$$3x + 2y = 4.$$

Solution   Begin by solving the first equation for y:

$$2x - y = 1$$
$$-y = -2x + 1 \qquad \text{Subtract } 2x \text{ from both sides.}$$
$$y = 2x - 1. \qquad \text{Multiply both sides by } -1.$$

Substitute this expression for y in the second equation and solve for x:

$$3x + 2y = 4$$
$$3x + 2(2x - 1) = 4 \qquad \text{Substitute } 2x - 1 \text{ for } y.$$
$$3x + 4x - 2 = 4 \qquad \text{Multiply out the left side.}$$
$$7x - 2 = 4 \qquad \text{Combine like terms.}$$
$$7x = 6 \qquad \text{Add 2 to both sides.}$$
$$x = 6/7. \qquad \text{Divide both sides by 7.}$$

Therefore, every solution of the system must have $x = 6/7$. To find the corresponding solution for y, substitute $x = 6/7$ in one of the two original equations and solve for y. We will use the first equation.

$$2x - y = 1$$

$$2\left(\frac{6}{7}\right) - y = 1 \qquad \text{Substitute 6/7 for } x.$$

$$\frac{12}{7} - y = 1 \qquad \text{Multiply out the left side.}$$

$$-y = -\frac{12}{7} + 1 \qquad \text{Subtract 12/7 from both sides.}$$

$$y = \frac{12}{7} - 1 = \frac{5}{7}. \qquad \text{Multiply both sides by } -1.$$

Hence, the solution of the original system is $x = 6/7$ and $y = 5/7$. We would have obtained the same solution if we had substituted $x = 6/7$ in the second equation of the original system, as you can easily verify. ✓ 1

The substitution method is useful when at least one of the equations has a variable with coefficient 1. That is why we solved for y in the first equation of Example 1. If we had solved for x in the first equation or for x or y in the second, we would have had a lot more fractions to deal with.

▦ **TECHNOLOGY TIP** A graphing calculator can be used to solve systems of two equations in two variables. Solve each of the given equations for y and graph them on the same screen. In Example 1, for instance, we would graph

$$y_1 = 2x - 1 \qquad \text{and} \qquad y_2 = \frac{-3x + 4}{2}.$$

Use the intersection finder (look in the CALC menu) to determine the point where the graphs intersect (the solution of the system), as shown in Figure 6.2. Note that this is an *approximate* solution, rather than the exact solution found algebraically in Example 1.

The Elimination Method

The elimination method of solving systems of linear equations is often more convenient than substitution, as illustrated in the next four examples.

EXAMPLE 2 Solve the system

$$5x + y = 4$$
$$3x + 2y = 1.$$

Solution Multiply the first equation by -2, so that the coefficients of y in the two equations are negatives of each other:

$$-10x - 2y = -8 \qquad \text{First equation multiplied by } -2.$$
$$3x + 2y = 1. \qquad \text{Second equation as given.}$$

Any solution of this system of equations must also be a solution of the sum of the two equations:

$$-10x - 2y = -8$$
$$\underline{3x + 2y = 1}$$
$$-7x = -7 \qquad \text{Sum; variable } y \text{ is eliminated.}$$
$$x = 1. \qquad \text{Divide both sides by } -7.$$

✓ **Checkpoint 1**

Use the substitution method to solve this system:

$$x - 2y = 3$$
$$2x + 3y = 13.$$

Answers to Checkpoint exercises are found at the end of the section.

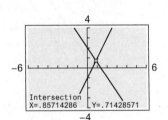

Intersection
X=.85714286   Y=.71428571

Figure 6.2

To find the corresponding value of y, substitute $x = 1$ in one of the original equations, say the first one:

$$5x + y = 4$$
$$5(1) + y = 4 \qquad \text{Substitute 1 for } x.$$
$$y = -1. \qquad \text{Subtract 5 from both sides.}$$

Therefore, the solution of the original system is $(1, -1)$. ✓₂

✓ **Checkpoint 2**

Use the elimination method to solve this system:

$$x + 2y = 4$$
$$3x - 4y = -8.$$

EXAMPLE 3   Solve the system

$$3x - 4y = 1$$
$$2x + 3y = 12.$$

Solution Multiply the first equation by 2 and the second equation by -3 to get

$$6x - 8y = 2 \qquad \text{First equation multiplied by 2.}$$
$$-6x - 9y = -36. \qquad \text{Second equation multiplied by } -3.$$

The multipliers 2 and -3 were chosen so that the coefficients of x in the two equations would be negatives of each other. Any solution of both these equations must also be a solution of their sum:

$$\begin{aligned} 6x - 8y &= 2 \\ \underline{-6x - 9y} &= \underline{-36} \\ -17y &= -34 \qquad \text{Sum; variable } x \text{ is eliminated.} \\ y &= 2. \qquad \text{Divide both sides by } -17. \end{aligned}$$

To find the corresponding value of x, substitute 2 for y in either of the original equations. We choose the first equation:

$$3x - 4y = 1$$
$$3x - 4(2) = 1 \qquad \text{Substitute 2 for } y.$$
$$3x - 8 = 1 \qquad \text{Simplify.}$$
$$3x = 9 \qquad \text{Add 8 to both sides.}$$
$$x = 3. \qquad \text{Divide both sides by 3.}$$

Therefore, the solution of the system is $(3, 2)$. The graphs of both equations of the system are shown in Figure 6.1(a) at the beginning of this section. They intersect at the point $(3, 2)$, the solution of the system. ✓₃

✓ **Checkpoint 3**

Solve the system of equations

$$3x + 2y = -1$$
$$5x - 3y = 11.$$

Draw the graph of each equation on the same set of axes.

Dependent and Inconsistent Systems

A system of equations that has a unique solution, such as those in Examples 1–3, is called an **independent system.** Now we consider systems that have infinitely many solutions or no solutions at all.

EXAMPLE 4   Solve the system

$$-4x + y = 2$$
$$8x - 2y = -4.$$

Solution If you solve each equation in the system for y, you see that the two equations are actually the same:

$$\begin{aligned} -4x + y &= 2 & 8x - 2y &= -4 \\ y &= 4x + 2 & -2y &= -8x - 4 \\ & & y &= 4x + 2. \end{aligned}$$

So the two equations have the same graph, as shown in Figure 6.1(b), and the system has infinitely many solutions (namely, all solutions of $y = 4x + 2$). A system such as this is said to be **dependent.**

You do not have to analyze a system as was just done in order to find out that the system is dependent. The elimination method will warn you. For instance, if you attempt to eliminate x in the original system by multiplying both sides of the first equation by 2 and adding the results to the second equation, you obtain

$$-8x + 2y = 4 \qquad \text{First equation multiplied by 2.}$$
$$\underline{8x - 2y = -4} \qquad \text{Second equation as given.}$$
$$ 0 = 0. \qquad \text{Both variables are eliminated.}$$

The equation "$0 = 0$" is an algebraic indication that the system is dependent and has infinitely many solutions. ✓ 4

✓ **Checkpoint 4**

Solve the following system:

$$\begin{aligned} 3x - 4y &= 13 \\ 12x - 16y &= 52. \end{aligned}$$

EXAMPLE 5 Solve the system

$$\begin{aligned} 3x - 2y &= 4 \\ -6x + 4y &= 7. \end{aligned}$$

Solution The graphs of these equations are parallel lines (each has slope $3/2$), as shown in Figure 6.1(c). Therefore, the system has no solution. However, you do not need the graphs to discover this fact. If you try to solve the system algebraically by multiplying both sides of the first equation by 2 and adding the results to the second equation, you obtain

$$6x - 4y = 8 \qquad \text{First equation multiplied by 2.}$$
$$\underline{-6x + 4y = 7} \qquad \text{Second equation as given.}$$
$$ 0 = 15. \qquad \text{False statement.}$$

The false statement "$0 = 15$" is the algebraic signal that the system has no solution. A system with no solutions is said to be **inconsistent.** ✓ 5

✓ **Checkpoint 5**

Solve the system

$$\begin{aligned} x - y &= 4 \\ 2x - 2y &= 3. \end{aligned}$$

Draw the graph of each equation on the same set of axes.

Applications

In many applications, the answer is the solution of a system of equations. Solving the system is the easy part—*finding* the system, however, may take some thought.

EXAMPLE 6 **Basketball Tickets** Eight hundred people attend a basketball game, and total ticket sales are $3102. If adult tickets are $6 and student tickets are $3, how many adults and how many students attended the game?

Solution Let x be the number of adults and y the number of students. Then

Number of adults + Number of students = Total attendance.

$$x + y = 800.$$

A second equation can be found by considering ticket sales:

Adult ticket sales + Student ticket sales = Total ticket sales

$$\left(\begin{array}{c}\text{Price}\\\text{per}\\\text{ticket}\end{array}\right) \times \left(\begin{array}{c}\text{Number}\\\text{of}\\\text{adults}\end{array}\right) + \left(\begin{array}{c}\text{Price}\\\text{per}\\\text{ticket}\end{array}\right) \times \left(\begin{array}{c}\text{Number}\\\text{of}\\\text{students}\end{array}\right) = 3102$$

$$6x + 3y = 3102.$$

✓ **Checkpoint 6**

(a) Use substitution or elimination to solve the system of equations in Example 6.

 (b) Use technology to solve the system. (*Hint:* Solve each equation for y, graph both equations on the same screen, and use the intersection finder.)

To find x and y we must solve this system of equations:

$$x + y = 800$$
$$6x + 3y = 3102.$$

The system can readily be solved by hand or by using technology. See Checkpoint 6, which shows that 234 adults and 566 students attended the game. ✓ 6

6.1 Exercises

Determine whether the given ordered list of numbers is a solution of the system of equations.

1. $(-1, 3)$
$$2x + y = 1$$
$$-3x + 2y = 9$$

2. $(2, -.5)$
$$.5x + 8y = -3$$
$$x + 5y = -5$$

Use substitution to solve each system. (See Example 1.)

3. $3x - y = 1$
$$x + 2y = -9$$

4. $x + y = 7$
$$x - 2y = -5$$

5. $3x - 2y = 4$
$$2x + y = -1$$

6. $5x - 3y = -2$
$$-x - 2y = 3$$

Use elimination to solve each system. (See Examples 2–5.)

7. $x - 2y = 5$
$$2x + y = 3$$

8. $3x - y = 1$
$$-x + 2y = 4$$

9. $2x - 2y = 12$
$$-2x + 3y = 10$$

10. $3x + 2y = -4$
$$4x - 2y = -10$$

11. $x + 3y = -1$
$$2x - y = 5$$

12. $4x - 3y = -1$
$$x + 2y = 19$$

13. $2x + 3y = 15$
$$8x + 12y = 40$$

14. $2x + 5y = 8$
$$6x + 15y = 18$$

15. $2x - 8y = 2$
$$3x - 12y = 3$$

16. $3x - 2y = 4$
$$6x - 4y = 8$$

In Exercises 17 and 18, multiply both sides of each equation by a common denominator to eliminate the fractions. Then solve the system.

17. $\dfrac{x}{2} + \dfrac{y}{3} = 8$

$$\dfrac{2x}{3} + \dfrac{3y}{2} = 17$$

18. $\dfrac{x}{5} + 3y = 31$

$$2x - \dfrac{y}{5} = 8$$

19. Millennials The number of baby boomers has been decreasing, and the number of millennials has been increasing. Now millennials have surpassed boomers in number. The following equations approximate the population y (in millions) of the given generation in year x, where $x = 14$ corresponds to the year 2014. (Data from: Pew Research Center.)

$$\text{Boomers:}\quad 10x + 13y = 1125$$
$$\text{Millennials:}\quad -2x + 7y = 495$$

Solve this system of equations, and interpret your answer.

20. Generation X The size of generation X has been decreasing, but the size of the baby boomer generation has been decreasing even more rapidly, so the number of generation Xers is projected to overtake the number of baby boomers. The population y (in millions) of the given generation in year x is approximated by

$$\text{Boomers:}\quad 10x + 13y = 1125$$
$$\text{Gen X:}\quad x + 13y = 873,$$

where $x = 14$ corresponds to the year 2014. Solve this system of equations, and interpret your answer. (Data from: Pew Research Center.)

21. Slow Midwestern Growth According to U.S. Census Bureau projections through the year 2030, the population y (in millions) of the given state in year x is approximated by

$$\text{Ohio:}\quad -x + 150y = 1710$$
$$\text{North Carolina:}\quad -41x + 300y = 2430,$$

where $x = 0$ corresponds to the year 2000. In what year are the two states projected to have the same population? Round your answer to the nearest year.

22. Booming Florida At the start of the millennium, New York was the third most populous state in the country, followed by Florida. Since that time, Florida has experienced faster growth. The population y (in millions) of the given state in year x is approximated by

$$\text{Florida:}\quad 7y - 2x = 112$$
$$\text{New York:}\quad 14y - x = 266,$$

where $x = 0$ corresponds to the year 2000. In what year did Florida overtake New York in population? To the nearest million, what was the population of these states at that time? (Data from: U.S. Census Bureau.)

23. Google Trends According to Google Trends, popular interest in light-emitting diode (LED) lightbulbs has been soaring, while interest in compact florescent lightbulbs (CFLs) has been dropping. The following equations approximate the Google Trends rating (on a scale from 0 to 100) in year x, where $x = 10$ corresponds to the year 2010. (Data from: www.google.com/trends.)

$$\text{LED:} \quad -25x + 6y = 20$$
$$\text{CFL:} \quad 15x + 2y = 322$$

When did interest in LED lighting surpass CFL lighting? Round your answer to the nearest year.

24. Heart Disease and Cancer Deaths The number of deaths y (in thousands) in year x for diseases of the heart and malignant neoplasms (cancer) can be approximated by

$$\text{Heart disease:} \quad 100x + 13y = 9243$$
$$\text{Cancer:} \quad -64x + 26y = 14{,}379,$$

where $x = 0$ corresponds to the year 2000. According to these models, during what year did deaths from heart disease equal cancer deaths? (Data from: Centers for Disease Control and Prevention.)

25. Workforce Participation for Women and Men On the basis of projections for the year 2022, the number of women and men in the workforce (in millions) can be estimated by

$$\text{Women:} \quad -7x + 16y = 1070$$
$$\text{Men:} \quad -5x + 10y = 759,$$

where $x = 14$ corresponds to the year 2014. According to these models, will the number of women in the workforce equal the number of men during the time period from 2014 to 2022 (that is, $14 \le x \le 22$)? (Data from: U.S. Department of Labor.)

26. Weekly Earnings for Women and Men On the basis of data from the years 2000 through 2015, seasonally adjusted median weekly earnings y in year x can be approximated for women and men by

$$\text{Women:} \quad -791x + 50y = 24{,}850$$
$$\text{Men:} \quad -84x + 5y = 3210,$$

where $x = 0$ corresponds to the year 2000 and y is in current dollars. If these equations remain valid in the future, will earnings for women catch up with men within the next decade? (Data from: Bureau of Labor Statistics.)

27. Theater Tickets A 200-seat theater charges $8 for adults and $5 for children. If all seats were filled and the total ticket income was $1435, how many adults and how many children were in the audience?

28. Concession Sales During intermission, the cash bar at a theater sold 72 drinks and took in a total of $663. If the theater sells beer for $7 each and wine for $10 a glass, how many servings of each beverage were sold?

29. Physical Science A plane flies 3000 miles from San Francisco to Boston in 5 hours, with a tailwind all the way. The return trip on the same route, now with a headwind, takes 6 hours. Assuming both remain constant, find the speed of the plane and the speed of the wind. [*Hint:* If x is the plane's speed and y the wind speed (in mph), then the

plane travels to Boston at $x + y$ mph because the plane and the wind go in the same direction; on the return trip, the plane travels at $x - y$ mph. (Why?) Then use the equation distance = speed $\cdot$ time.]

30. Physical Science A plane flying into a headwind travels 2200 miles in 5 hours. The return flight along the same route with a tailwind takes 4 hours. Find the wind speed and the plane's speed (assuming both are constant). (See the hint for Exercise 29.)

C 31. **(a)** Find the equation of the straight line through $(1, 2)$ and $(3, 4)$.

(b) Find the equation of the line through $(-1, 1)$ with slope 3.

(c) Find a point that lies on both of the lines in (a) and (b).

C 32. **(a)** Find an equation of the straight line through $(0, 9)$ and $(2, 1)$.

(b) Find an equation of the straight line through $(1, 5)$ with slope 2.

(c) Find a point that lies on both of the lines in (a) and (b).

✓ Checkpoint Answers

1. $(5, 1)$ **2.** $(0, 2)$

3. $(1, -2)$

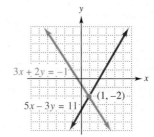

4. All ordered pairs that satisfy the equation $3x - 4y = 13$ (or $12x - 16y = 52$)

5. No solution

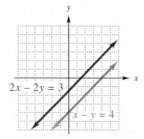

6. (a) $x = 234$ and $y = 566$

(b)

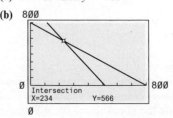

6.2 Larger Systems of Linear Equations

Two systems of equations are said to be **equivalent** if they have the same solutions. The basic procedure for solving a large system of equations is to transform the system into a simpler, equivalent system and then solve this simpler system.

Three operations, called **elementary operations,** are used to transform a system into an equivalent one:

1. **Interchange any two equations in the system.**
   Changing the order of the equations obviously does not affect the solutions of the equations or the system.

2. **Multiply an equation in the system by a nonzero constant.**
   Multiplying an equation by a nonzero constant does not change its solutions. So it does not change the solutions of the system.

3. **Replace an equation in the system by the sum of itself and a nonzero constant multiple of another equation.**
   We are not saying here that the replacement equation has the same solutions as the one it replaces—it doesn't—only that the new *system* has the same solutions as the original system (that is, the systems are equivalent).[*]

Example 1 shows how to use elementary operations on a system to eliminate certain variables and produce an equivalent system that is easily solved. The italicized statements provide an outline of the procedure. Here and later on, we use R_1 to denote the first equation in a system, R_2 the second equation, and so on.

EXAMPLE 1 Solve the system

$$2x + y - z = 2$$
$$x + 3y + 2z = 1$$
$$x + y + z = 2.$$

Solution First, *use elementary operations to produce an equivalent system in which 1 is the coefficient of x in the first equation.* One way to do this is to interchange the first two equations (another would be to multiply both sides of the first equation by $\frac{1}{2}$):

$$x + 3y + 2z = 1 \qquad \text{Interchange } R_1 \text{ and } R_2.$$
$$2x + y - z = 2$$
$$x + y + z = 2.$$

Next, *use elementary operations to produce an equivalent system in which the x-term has been eliminated from the second and third equations.* To eliminate the x-term from the second equation, replace the second equation by the sum of itself and -2 times the first equation:

$$
\begin{array}{ll}
-2R_1 & -2x - 6y - 4z = -2 \\
R_2 & 2x + y - z = 2 \\
\hline
-2R_1 + R_2 & -5y - 5z = 0
\end{array}
$$

$$x + 3y + 2z = 1$$
$$-5y - 5z = 0 \qquad -2R_1 + R_2$$
$$x + y + z = 2.$$

To eliminate the x-term from the third equation of this last system, replace the third equation by the sum of itself and -1 times the first equation:

$$
\begin{array}{ll}
-1R_1 & -x - 3y - 2z = -1 \\
R_3 & x + y + z = 2 \\
\hline
-1R_1 + R_3 & -2y - z = 1
\end{array}
$$

$$x + 3y + 2z = 1$$
$$-5y - 5z = 0$$
$$-2y - z = 1. \qquad -1R_1 + R_3$$

---

[*]This operation was used in a slightly different form in Section 6.1, when the sum of an equation and a constant multiple of another equation was found in order to eliminate one of the variables. See, for instance, Examples 2 and 3 of that section.

Now that x has been eliminated from all but the first equation, we ignore the first equation and work on the remaining ones. *Use elementary operations to produce an equivalent system in which 1 is the coefficient of y in the second equation.* This can be done by multiplying the second equation in the system by $-\frac{1}{5}$:

$$
\begin{aligned}
x + 3y + 2z &= 1 \\
y + z &= 0 \qquad -\tfrac{1}{5}R_2 \\
-2y - z &= 1.
\end{aligned}
$$

Then *use elementary operations to obtain an equivalent system in which y has been eliminated from the third equation.* Replace the third equation by the sum of itself and 2 times the second equation:

$$
\begin{aligned}
x + 3y + 2z &= 1 \\
y + z &= 0 \\
z &= 1. \qquad 2R_2 + R_3
\end{aligned}
$$

The solution of the third equation is obvious: $z = 1$. Now work backward in the system. Substitute 1 for z in the second equation and solve for y, obtaining $y = -1$. Finally, substitute 1 for z and -1 for y in the first equation and solve for x, obtaining $x = 2$. This process is known as **back substitution.** When it is finished, we have the solution of the original system, namely, $(2, -1, 1)$. It is always wise to check the solution by substituting the values for x, y, and z in *all* equations of the original system.

The procedure used in Example 1 to eliminate variables and produce a system in which back substitution works can be carried out with any system, as summarized below. In this summary, the first variable that appears in an equation with a nonzero coefficient is called the **leading variable** of that equation, and its nonzero coefficient is called the **leading coefficient.**

The Elimination Method for Solving Large Systems of Linear Equations

Use elementary operations to transform the given system into an equivalent one as follows:

1. Make the leading coefficient of the first equation 1 either by interchanging equations or by multiplying the first equation by a suitable constant.

2. Eliminate the leading variable of the first equation from each later equation by replacing the later equation by the sum of itself and a suitable multiple of the first equation.

3. Repeat Steps 1 and 2 for the second equation: Make its leading coefficient 1 and eliminate its leading variable from each later equation by replacing the later equation by the sum of itself and a suitable multiple of the second equation.

4. Repeat Steps 1 and 2 for the third equation, fourth equation, and so on, until it is not possible to go any further.

Then solve the resulting system by back substitution.

✓ **Checkpoint 1**

Use the elimination method to solve each system.

(a) $\begin{aligned} 2x + y &= -1 \\ x + 3y &= 2 \end{aligned}$

(b) $\begin{aligned} 2x - y + 3z &= 2 \\ x + 2y - z &= 6 \\ -x - y + z &= -5 \end{aligned}$

At various stages in the elimination process, you may have a choice of elementary operations that can be used. As long as the final result is a system in which back substitution can be used, the choice does not matter. To avoid unnecessary errors, choose elementary operations that minimize the amount of computation and, as far as possible, avoid complicated fractions. ✓₁

Matrix Methods

You may have noticed that the variables in a system of equations remain unchanged during the solution process. So we need to keep track of only the coefficients and the constants. For instance, consider the system in Example 1:

$$2x + y - z = 2$$
$$x + 3y + 2z = 1$$
$$x + y + z = 2.$$

This system can be written in an abbreviated form without listing the variables as

$$\begin{bmatrix} 2 & 1 & -1 & 2 \\ 1 & 3 & 2 & 1 \\ 1 & 1 & 1 & 2 \end{bmatrix}.$$

Such a rectangular array of numbers, consisting of horizontal **rows** and vertical **columns,** is called a **matrix** (plural, **matrices**). Each number in the array is an **element,** or **entry.** To separate the constants in the last column of the matrix from the coefficients of the variables, we sometimes use a vertical line, producing the following **augmented matrix:**

$$\left[\begin{array}{ccc|c} 2 & 1 & -1 & 2 \\ 1 & 3 & 2 & 1 \\ 1 & 1 & 1 & 2 \end{array}\right]. \quad ✓_2$$

The rows of the augmented matrix can be transformed in the same way as the equations of the system, since the matrix is just a shortened form of the system. The following **row operations** on the augmented matrix correspond to the elementary operations used on systems of equations.

Performing any one of the following **row operations** on the augmented matrix of a system of linear equations produces the augmented matrix of an equivalent system:

1. Interchange any two rows.
2. Multiply each element of a row by a nonzero constant.
3. Replace a row by the sum of itself and a nonzero constant multiple of another row of the matrix.

Row operations on a matrix are indicated by the same notation we used for elementary operations on a system of equations. For example, $2R_3 + R_1$ indicates the sum of 2 times row 3 and row 1. ✓₃

EXAMPLE 2   Use matrices to solve the system

$$x - 2y = 6 - 4z$$
$$x + 13z = 6 - y$$
$$-2x + 6y - z = -10.$$

Solution First, put the system in the required form, with the constants on the right side of the equals sign and the terms with variables *in the same order* in each equation on the left side of the equals sign. Then write the augmented matrix of the system.

$$\begin{array}{rcrcrcr} x & - & 2y & + & 4z & = & 6 \\ x & + & y & + & 13z & = & 6 \\ -2x & + & 6y & - & z & = & -10 \end{array} \qquad \left[\begin{array}{ccc|c} 1 & -2 & 4 & 6 \\ 1 & 1 & 13 & 6 \\ -2 & 6 & -1 & -10 \end{array}\right].$$

✓ Checkpoint 2

(a) Write the augmented matrix of the following system:

$$4x - 2y + 3z = 4$$
$$3x + 5y + z = -7$$
$$5x - y + 4z = 6.$$

(b) Write the system of equations associated with the following augmented matrix:

$$\left[\begin{array}{cc|c} 2 & -2 & -2 \\ 1 & 1 & 4 \\ 3 & 5 & 8 \end{array}\right].$$

✓ Checkpoint 3

Perform the given row operations on the matrix

$$\begin{bmatrix} -1 & 5 \\ 3 & -2 \end{bmatrix}.$$

(a) Interchange R_1 and R_2.
(b) Replace R_1 by $2R_1$.
(c) Replace R_2 by $-3R_1 + R_2$.
(d) Replace R_1 by $2R_2 + R_1$.

The matrix method is the same as the elimination method, except that row operations are used on the augmented matrix instead of elementary operations on the corresponding system of equations, as shown in the following side-by-side comparison:

Equation Method	**Matrix Method**

Replace the second equation by the sum of itself and -1 times the first equation:

$$\begin{aligned} x - 2y + 4z &= 6 \\ 3y + 9z &= 0 \\ -2x + 6y - z &= -10. \end{aligned}$$

Replace the second row by the sum of itself and -1 times the first row:

$$\leftarrow -1R_1 + R_2 \rightarrow \qquad \left[\begin{array}{ccc|c} 1 & -2 & 4 & 6 \\ 0 & 3 & 9 & 0 \\ -2 & 6 & -1 & -10 \end{array}\right].$$

Replace the third equation by the sum of itself and 2 times the first equation:

$$\begin{aligned} x - 2y + 4z &= 6 \\ 3y + 9z &= 0 \\ 2y + 7z &= 2. \end{aligned}$$

Replace the third row by the sum of itself and 2 times the first row:

$$\leftarrow 2R_1 + R_3 \rightarrow \qquad \left[\begin{array}{ccc|c} 1 & -2 & 4 & 6 \\ 0 & 3 & 9 & 0 \\ 0 & 2 & 7 & 2 \end{array}\right].$$

Multiply both sides of the second equation by $\frac{1}{3}$:

$$\begin{aligned} x - 2y + 4z &= 6 \\ y + 3z &= 0 \\ 2y + 7z &= 2. \end{aligned}$$

Multiply each element of row 2 by $\frac{1}{3}$:

$$\leftarrow \tfrac{1}{3}R_2 \rightarrow \qquad \left[\begin{array}{ccc|c} 1 & -2 & 4 & 6 \\ 0 & 1 & 3 & 0 \\ 0 & 2 & 7 & 2 \end{array}\right].$$

Replace the third equation by the sum of itself and -2 times the second equation:

$$\begin{aligned} x - 2y + 4z &= 6 \\ y + 3z &= 0 \\ z &= 2. \end{aligned}$$

Replace the third row by the sum of itself and -2 times the second row:

$$\leftarrow -2R_2 + R_3 \rightarrow \qquad \left[\begin{array}{ccc|c} 1 & -2 & 4 & 6 \\ 0 & 1 & 3 & 0 \\ 0 & 0 & 1 & 2 \end{array}\right].$$

Now use back substitution:

$$z = 2 \qquad \begin{aligned} y + 3(2) &= 0 \\ y &= -6 \end{aligned} \qquad \begin{aligned} x - 2(-6) + 4(2) &= 6 \\ x + 20 &= 6 \\ x &= -14. \end{aligned}$$

The solution of the system is $(-14, -6, 2)$.

✓**Checkpoint 4**

Complete the matrix solution of the system with this augmented matrix:

$$\left[\begin{array}{ccc|c} 1 & 1 & 1 & 2 \\ 1 & -2 & 1 & -1 \\ 0 & 3 & 1 & 5 \end{array}\right].$$

A matrix, such as the last one in Example 2, is said to be in **row echelon form** when

all rows consisting entirely of zeros (if any) are at the bottom;

the first nonzero entry in each row is 1 (called a *leading* 1); and

each leading 1 appears to the right of the leading 1s in any preceding rows.

When a matrix in row echelon form is the augmented matrix of a system of equations, as in Example 2, the system can readily be solved by back substitution. So the matrix solution method amounts to transforming the augmented matrix of a system of equations into a matrix in row echelon form.

The Gauss–Jordan Method

The **Gauss–Jordan method** is a variation on the matrix elimination method used in Example 2. It replaces the back substitution used there with additional elimination of variables, as illustrated in the next example.

EXAMPLE 3 Use the Gauss–Jordan method to solve the system in Example 2.

Solution First, set up the augmented matrix. Then apply the same row operations used in Example 2 until you reach the final row echelon matrix, which is

$$\begin{bmatrix} 1 & -2 & 4 & | & 6 \\ 0 & 1 & 3 & | & 0 \\ 0 & 0 & 1 & | & 2 \end{bmatrix}.$$

Now use additional row operations to make the entries (other than the 1s) in columns two and three into zeros. Make the first two entries in column three 0 as follows:

$$\begin{bmatrix} 1 & -2 & 4 & | & 6 \\ 0 & 1 & \mathbf{0} & | & -6 \\ 0 & 0 & 1 & | & 2 \end{bmatrix} \quad -3R_3 + R_2$$

$$\begin{bmatrix} 1 & -2 & \mathbf{0} & | & -2 \\ 0 & 1 & 0 & | & -6 \\ 0 & 0 & 1 & | & 2 \end{bmatrix} \quad -4R_3 + R_1$$

Now make the first entry in column two 0.

$$\begin{bmatrix} 1 & \mathbf{0} & \mathbf{0} & | & -14 \\ 0 & 1 & 0 & | & -6 \\ 0 & 0 & 1 & | & 2 \end{bmatrix}. \quad 2R_2 + R_1$$

The last matrix corresponds to the system $x = -14$, $y = -6$, $z = 2$. So the solution of the original system is $(-14, -6, 2)$. This solution agrees with what we found in Example 2.

The final matrix in the Gauss–Jordan method is said to be in **reduced row echelon form,** meaning that it is in row echelon form *and* every column containing a leading 1 has zeros in all its other entries, as in Example 3. As we saw there, the solution of the system can be read directly from the reduced row echelon matrix. The *reduced* row echelon form of a matrix is unique. This is not true of a matrix that is simply in row echelon form. This is why a row echelon form produced by hand may differ from a row echelon form provided by a calculator, but each will lead to the same solution of the system of equations.

In the Gauss–Jordan method, row operations may be performed in any order, but it is best to transform the matrix systematically. Either follow the procedure in Example 3 (which first puts the system into a form in which back substitution can be used and then eliminates additional variables) or work column by column from left to right, as in Checkpoint 5. ✓ 5

Technology and Systems of Equations

When working by hand, it is usually better to use the matrix elimination method (Example 2), because errors with back substitution are less likely to occur than errors in performing the additional row operations needed in the Gauss–Jordan method. When you are using technology, however, the Gauss–Jordan method (Example 3) is more efficient: The solution can be read directly from the final reduced row echelon matrix without any "hand work," as illustrated in the next example.

▦ **TECHNOLOGY TIP** To enter a matrix on a TI graphing calculator, go to the MATRIX menu and then to the EDIT submenu, and select the desired name of the matrix ([A], [B], and so on). Enter the number of rows and columns for the matrix. For example, the augmented matrix for the system in Example 2 is a 3×4 matrix because it has 3 rows and 4 columns. Next enter each element of the matrix. As usual, use QUIT to return to the home screen. To use a matrix in calculations, go to the MATRIX menu and then the NAMES submenu, and select the matrix of choice. For example, Figure 6.3, on the following page, shows the augmented matrix from Example 2 as matrix [A] on a TI-84+.

✓ **Checkpoint 5**

Use the Gauss–Jordan method to solve the system

$$\begin{aligned} x + 2y &= 11 \\ -4x + y &= -8 \\ 5x + y &= 19 \end{aligned}$$

as instructed in parts (a)–(g). Give the shorthand notation for the required row operations in parts (b)–(f) and the new matrix in each step.

(a) Set up the augmented matrix.

(b) Get 0 in row 2, column 1.

(c) Get 0 in row 3, column 1.

(d) Get 1 in row 2, column 2.

(e) Get 0 in row 1, column 2.

(f) Get 0 in row 3, column 2.

(g) List the solution of the system.

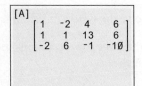

Figure 6.3

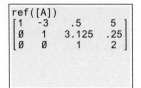

Figure 6.4

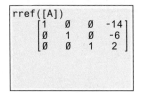

Figure 6.5

EXAMPLE 4   Use a graphing calculator to solve the system in Example 2.

Solution There are three ways to proceed.

Matrix Elimination Method Enter the augmented matrix of the system into the calculator (Figure 6.3). Use REF (in the MATH or OPS submenu of the MATRIX menu) to put this matrix in row echelon form (Figure 6.4). The system corresponding to this matrix is

$$x - 3y + \quad .5z = 5$$
$$y + 3.125z = .25$$
$$z = 2.$$

Because the calculator used a different sequence of row operations than was used in Example 2, it produced a different row echelon matrix (and corresponding system). However, back substitution shows that the solutions of this system are the same ones found in Example 2:

$$z = 2 \qquad y + 3.125(2) = .25 \qquad x - 3(-6) + .5(2) = 5$$
$$y + 6.25 = .25 \qquad x + 18 + 1 = 5$$
$$y = -6 \qquad x = -14.$$

As you can see, a significant amount of hand work is involved in this method, so it is not recommended.

Gauss–Jordan Method Enter the augmented matrix (Figure 6.3). Use RREF (in same menu as REF) to produce the same reduced row echelon matrix obtained in Example 3 (Figure 6.5). The solution of the system can be read directly from the matrix: $x = -14$, $y = -6, z = 2$.

System Solver Method This is essentially the same as the Gauss–Jordan method, with an extra step at the beginning. Call up the solver (see the Technology Tip below) and enter the number of variables and equations. The solver will display an appropriately sized matrix, with all entries 0. Change the entries so that the matrix becomes the augmented matrix of the system (Figure 6.6). Press SOLVE and the solution will appear (Figure 6.7). The calculator uses the variables x_1, x_2, and x_3 instead of x, y, and z. ✓ 6

✓ **Checkpoint 6**

On a graphing calculator, use the Gauss–Jordan method or the system solver to solve

$$x - \quad y + 5z = \quad -6$$
$$4x + 2y + 4z = \quad 4$$
$$x - 5y + 8z = -17$$
$$x + 3y + 2z = \quad 5.$$

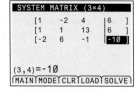

Figure 6.6

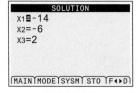

Figure 6.7

TECHNOLOGY TIP Many graphing calculators have system solvers. Select PLYSMLT2 in the APPS menu of TI-84+ or SIMULT on the TI-86 keyboard. *Note:* The TI-86 solver works only for systems that have the same number of variables as equations *and* have a unique solution (so they cannot be used in Checkpoint 6 or in Examples 5–8).

EXAMPLE 5   Use a graphing calculator to solve the system

$$x + 2y - \quad z = \quad 0$$
$$3x - \quad y + \quad z = \quad 6$$
$$7x + 28y - 8z = -5$$
$$5x + 3y - \quad z = \quad 6.$$

Solution Enter the augmented matrix into the calculator (Figure 6.8). Then use RREF to put this matrix in reduced row echelon form (Figure 6.9). In Figure 6.9, you must use the arrow key to scroll over to see the full decimal expansions in the right-hand column. This inconvenience can be avoided for many matrices on TI calculators by using FRAC (in the MATH menu), as in Figure 6.10.

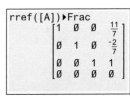

```
[A]
┌ 1    2   -1    0 ┐
│ 3   -1    1    6 │
│ 7   28   -8   -5 │
└ 5    3   -1    6 ┘
```

Figure 6.8

```
rref([A])
┌ 1   0   0    1.571428571 ┐
│ 0   1   0   -.2857142857 │
│ 0   0   1             1  │
└ 0   0   0             0  ┘
```

Figure 6.9

```
rref([A])▶Frac
┌ 1   0   0    11/7 ┐
│ 0   1   0    -2/7 │
│ 0   0   1      1  │
└ 0   0   0      0  ┘
```

Figure 6.10

The answers can now be read from the last column of the matrix in Figure 6.10: $x = 11/7$, $y = -2/7$, $z = 1$. ✓ 7

Dependent and Inconsistent Systems

Recall that a system of equations in two variables may have exactly one solution, an infinite number of solutions, or no solutions at all. This fact was illustrated geometrically in Figure 6.1. The same thing is true for systems with three or more variables (and the same terminology is used): A system has exactly one solution (an **independent system**), an infinite number of solutions (a **dependent system**), or no solutions at all (an **inconsistent system**).

EXAMPLE 6   Solve the system

$$2x + 4y = 4$$
$$3x + 6y = 8$$
$$2x + y = 7.$$

Solution   There are several possible ways to proceed.

Manual Method   Write the augmented matrix and perform row operations to obtain a first column whose entries (from top to bottom) are 1, 0, 0:

$$\begin{bmatrix} 2 & 4 & | & 4 \\ 3 & 6 & | & 8 \\ 2 & 1 & | & 7 \end{bmatrix}$$

$$\begin{bmatrix} 1 & 2 & | & 2 \\ 3 & 6 & | & 8 \\ 2 & 1 & | & 7 \end{bmatrix} \quad \tfrac{1}{2} R_1$$

$$\begin{bmatrix} 1 & 2 & | & 2 \\ 0 & 0 & | & 2 \\ 2 & 1 & | & 7 \end{bmatrix}. \quad -3R_1 + R_2$$

Stop! The second row of the matrix denotes the equation $0x + 0y = 2$. Since the left side of this equation is always 0 and the right side is 2, it has no solution. Therefore, the original system has no solution.

Calculator Method   Enter the augmented matrix into a graphing calculator and use RREF to put it into reduced row echelon form, as in Figure 6.11. The last row of that matrix corresponds to $0x + 0y = 1$, which has no solution. Hence, the original system has no solution. ✓ 8

Whenever the solution process produces a row whose elements are all 0 *except* the last one, as in Example 6, the system is inconsistent and has no solutions. However, if a row with a 0 for *every* entry is produced, it corresponds to an equation such as $0x + 0y + 0z = 0$, which has infinitely many solutions. So the system may have solutions, as in the next example.

✓ **Checkpoint 7** 🖩

Use a graphing calculator to solve the following system:

$$x + 3y = 4$$
$$4x + 8y = 4$$
$$6x + 12y = 6.$$

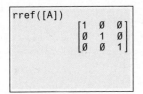

```
rref([A])
┌ 1   0   0 ┐
│ 0   1   0 │
└ 0   0   1 ┘
```

Figure 6.11

✓ **Checkpoint 8**

Solve each system.

(a) $x - y = 4$
$$-2x + 2y = 1$$

(b) $3x - 4y = 0$
$$2x + y = 0$$

EXAMPLE 7   Solve the system

$$2x - 3y + 4z = 6$$
$$x - 2y + z = 9$$
$$y + 2z = -12.$$

Solution Use matrix elimination as far as possible, beginning with the augmented matrix of the system:

$$\begin{bmatrix} 2 & -3 & 4 & | & 6 \\ 1 & -2 & 1 & | & 9 \\ 0 & 1 & 2 & | & -12 \end{bmatrix}$$

$$\begin{bmatrix} 1 & -2 & 1 & | & 9 \\ 2 & -3 & 4 & | & 6 \\ 0 & 1 & 2 & | & -12 \end{bmatrix} \quad \text{Interchange } R_1 \text{ and } R_2.$$

$$\begin{bmatrix} 1 & -2 & 1 & | & 9 \\ 0 & 1 & 2 & | & -12 \\ 0 & 1 & 2 & | & -12 \end{bmatrix} \quad -2R_1 + R_2$$

$$\begin{bmatrix} 1 & -2 & 1 & | & 9 \\ 0 & 1 & 2 & | & -12 \\ 0 & 0 & 0 & | & 0 \end{bmatrix}. \qquad \begin{aligned} x - 2y + z &= 9 \\ y + 2z &= -12. \end{aligned}$$
$$-R_2 + R_3$$

The last augmented matrix above represents the system shown to its right. Since there are only two nonzero rows in the matrix, it is not possible to continue the process. The fact that the corresponding system has one variable (namely, z) that is not the leading variable of an equation indicates a dependent system. To find its solutions, first solve the second equation for y:

$$y = -2z - 12.$$

Now substitute the result for y in the first equation and solve for x:

$$x - 2y + z = 9$$
$$x - 2(-2z - 12) + z = 9$$
$$x + 4z + 24 + z = 9$$
$$x + 5z = -15$$
$$x = -5z - 15.$$

Each choice of a value for z leads to values for x and y. For example,

$$\text{if } z = 1, \quad \text{then } x = -20 \quad \text{and} \quad y = -14;$$
$$\text{if } z = -6, \quad \text{then } x = 15 \quad \text{and} \quad y = 0;$$
$$\text{if } z = 0, \quad \text{then } x = -15 \quad \text{and} \quad y = -12.$$

There are infinitely many solutions of the original system, since z can take on infinitely many values. The solutions are all ordered triples of the form

$$(-5z - 15, -2z - 12, z),$$

where z is any real number. ✔9

✓ **Checkpoint 9**

Use the following values of z to find additional solutions for the system of Example 7.

(a) $z = 7$

(b) $z = -14$

(c) $z = 5$

Since both x and y in Example 7 were expressed in terms of z, the variable z is called a **parameter.** If we had solved the system in a different way, x or y could have been the parameter. The final system in Example 7 had one more variable than equations. If there are two more variables than equations, there usually will be two parameters, and so on.

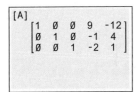

Figure 6.12

EXAMPLE 8

Row operations were used to reduce the augmented matrix of a system of three equations in four variables $(x, y, z,$ and $w)$ to the reduced row echelon matrix in Figure 6.12. Solve the system.

Solution First write out the system represented by the matrix:

$$x + \qquad\qquad 9w = -12$$
$$y - \qquad w = \quad 4$$
$$z - 2w = \quad 1.$$

Let w be the parameter. Solve the first equation for x, the second for y, and the third for z:

$$x = -9w - 12; \qquad y = w + 4; \qquad z = 2w + 1.$$

The solutions are given by $(-9w - 12, w + 4, 2w + 1, w)$, where w is any real number.

TECHNOLOGY TIP When the system solver on a TI-86 produces an error message, the system might be inconsistent (no solutions) or dependent (infinitely many solutions). You must use RREF or manual methods to solve the system.

The TI-84+ solver usually solves dependent systems directly; Figure 6.13 shows its solutions for Example 7. When the message "no solution found" is displayed, as in Figure 6.14, select RREF at the bottom of the screen to display the reduced row echelon matrix of the system. From that, you can determine the solutions (if there are any) or that no solutions are possible.

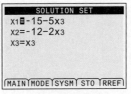

Figure 6.13

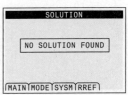

Figure 6.14

6.2 Exercises

Obtain an equivalent system by performing the stated elementary operation on the system. (See Example 1.)

1. Interchange equations 1 and 2.

$$2x - 4y + 5z = 1$$
$$x \qquad - 3z = 2$$
$$5x - 8y + 7z = 6$$
$$3x - 4y + 2z = 3$$

2. Interchange equations 1 and 3.

$$2x - 2y + \quad z = -6$$
$$3x + \quad y + 2z = \quad 2$$
$$x + \quad y - 2z = \quad 0$$

3. Multiply the third equation by $1/2$.

$$x + 2y + 4z = 3$$
$$x \qquad + 2z = 0$$
$$2x + 4y + \quad z = 3$$

4. Multiply the first equation by $-\dfrac{2}{3}$.

$$-\frac{3}{2}x + 6y - 5z = 12$$
$$9x \qquad + 7z = \quad 4$$

5. Replace the second equation by the sum of itself and -2 times the first equation.

$$x + y + 2z + 3w = 1$$
$$2x + y + 3z + 4w = 1$$
$$3x + y + 4z + 5w = 2$$

6. Replace the third equation by the sum of itself and -1 times the first equation.

$$x + 2y + 4z = \quad 6$$
$$y + \quad z = \quad 1$$
$$x + 3y + 5z = 10$$

7. Replace the third equation by the sum of itself and -2 times the second equation.

$$
\begin{aligned}
x + 12y - 3z + 4w &= 10 \\
2y + 3z + \ w &= 4 \\
4y + 5z + 2w &= 1 \\
6y - 2z - 3w &= 0
\end{aligned}
$$

8. Replace the third equation by the sum of itself and 3 times the second equation.

$$
\begin{aligned}
2x + 2y - 4z + \ w &= -5 \\
2y + 4z - \ w &= 2 \\
-6y - 4z + 2w &= 6 \\
2y + 5z - 3w &= 7
\end{aligned}
$$

Solve the system by back substitution. (See Example 1.)

9.
$$
\begin{aligned}
x + 3y - 4z + 2w &= 1 \\
y + \ z - \ w &= 4 \\
2z + 2w &= -6 \\
3w &= 9
\end{aligned}
$$

10.
$$
\begin{aligned}
x + \quad 5z + 6w &= 10 \\
y + 3z - 2w &= 4 \\
z - 4w &= -6 \\
2w &= 4
\end{aligned}
$$

11.
$$
\begin{aligned}
2x + 2y - 4z + \ w &= -5 \\
3y + 4z - \ w &= 0 \\
2z - 7w &= -6 \\
5w &= 15
\end{aligned}
$$

12.
$$
\begin{aligned}
3x - 2y - 4z + 2w &= 6 \\
2y + 5z - 3w &= 7 \\
3z + 4w &= 0 \\
3w &= 15
\end{aligned}
$$

Write the augmented matrix of each of the given systems. Do not solve the systems. (See Example 2.)

13.
$$
\begin{aligned}
2x + \ y + \ z &= 3 \\
3x - 4y + 2z &= -5 \\
x + \ y + \ z &= 2
\end{aligned}
$$

14.
$$
\begin{aligned}
3x + 4y - 2z - 3w &= 0 \\
x - 3y + 7z + 4w &= 9 \\
2x \quad + 5z - 6w &= 0
\end{aligned}
$$

Write the system of equations associated with each of the given augmented matrices. Do not solve the systems.

15. $\begin{bmatrix} 2 & 3 & 8 & | & 20 \\ 1 & 4 & 6 & | & 12 \\ 0 & 3 & 5 & | & 10 \end{bmatrix}$

16. $\begin{bmatrix} 3 & 2 & 6 & | & 18 \\ 2 & -2 & 5 & | & 7 \\ 1 & 0 & 5 & | & 20 \end{bmatrix}$

Use the indicated row operation to transform each matrix. (See Examples 2 and 3.)

17. Interchange R_1 and R_3.

$$\begin{bmatrix} 5 & -1 & 4 & | & 8 \\ 2 & 2 & 9 & | & 3 \\ 1 & 7 & -3 & | & 5 \end{bmatrix}$$

18. Interchange R_2 and R_3.

$$\begin{bmatrix} 1 & 8 & 3 & | & -2 \\ 0 & 7 & -9 & | & 3 \\ 0 & 1 & 8 & | & -4 \end{bmatrix}$$

19. Replace R_3 by $\dfrac{1}{4}R_3$.

$$\begin{bmatrix} 2 & 5 & 1 & | & -1 \\ -4 & 0 & 4 & | & 6 \\ 6 & 0 & 8 & | & -4 \end{bmatrix}$$

20. Replace R_1 with $-\dfrac{1}{6}R_1$.

$$\begin{bmatrix} -6 & 3 & -8 & | & 12 \\ 7 & 4 & -5 & | & 25 \end{bmatrix}$$

21. Replace R_2 by $2R_1 + R_2$.

$$\begin{bmatrix} -4 & -3 & 1 & -1 & | & 2 \\ 8 & 2 & 5 & 0 & | & 6 \\ 0 & -2 & 9 & 4 & | & 5 \end{bmatrix}$$

22. Replace R_3 by $-3R_1 + R_3$.

$$\begin{bmatrix} 1 & 5 & 2 & 0 & | & -1 \\ 8 & 5 & 4 & 6 & | & 6 \\ 3 & 0 & 7 & 1 & | & -4 \end{bmatrix}$$

In Exercises 23–28, the reduced row echelon form of the augmented matrix of a system of equations is given. Find the solutions of the system. (See Example 3.)

23. $\begin{bmatrix} 1 & 0 & 0 & 0 & | & 3/2 \\ 0 & 1 & 0 & 0 & | & 17 \\ 0 & 0 & 1 & 0 & | & -5 \\ 0 & 0 & 0 & 1 & | & 0 \end{bmatrix}$

24. $\begin{bmatrix} 1 & 0 & 0 & 0 & | & 7 \\ 0 & 1 & 0 & 0 & | & 2 \\ 0 & 0 & 1 & 0 & | & -5 \\ 0 & 0 & 0 & 1 & | & 3 \\ 0 & 0 & 0 & 0 & | & 0 \\ 0 & 0 & 0 & 0 & | & 0 \end{bmatrix}$

25. $\begin{bmatrix} 1 & 0 & 0 & | & -5 \\ 0 & 1 & 0 & | & 9 \\ 0 & 0 & 0 & | & 1 \end{bmatrix}$

26. $\begin{bmatrix} 1 & 0 & 0 & 0 & 0 & | & 6 \\ 0 & 1 & 0 & 0 & 0 & | & 4 \\ 0 & 0 & 1 & 0 & 0 & | & 5 \\ 0 & 0 & 0 & 0 & 1 & | & 2 \\ 0 & 0 & 0 & 0 & 0 & | & 1 \end{bmatrix}$

27. $\begin{bmatrix} 1 & 0 & 0 & 1 & | & 12 \\ 0 & 1 & 0 & 2 & | & -3 \\ 0 & 0 & 1 & 0 & | & -5 \\ 0 & 0 & 0 & 0 & | & 0 \end{bmatrix}$

28. $\begin{bmatrix} 1 & 0 & -7 & | & 10 \\ 0 & 1 & 3 & | & 25 \end{bmatrix}$

In Exercises 29–34, perform row operations on the augmented matrix as far as necessary to determine whether the system is independent, dependent, or inconsistent. (See Examples 6–8.)

29.
$$x + 2y \qquad = 0$$
$$y - z = 2$$
$$x + y + z = -2$$

30.
$$x + 2y + z = 0$$
$$y + 2z = 0$$
$$x + y - z = 0$$

31.
$$x + 2y + 4z = 6$$
$$y + z = 1$$
$$x + 3y + 5z = 10$$

32.
$$x + y + 2z + 3w = 1$$
$$2x + y + 3z + 4w = 1$$
$$3x + y + 4z + 5w = 2$$

33.
$$a - 3b - 2c = -3$$
$$3a + 2b - c = 12$$
$$-a - b + 4c = 3$$

34.
$$2x + 2y + 2z = 6$$
$$3x - 3y - 4z = -1$$
$$x + y + 3z = 11$$

Write the augmented matrix of the system and use the matrix method to solve the system. (See Examples 2 and 4.)

35.
$$-x + 3y + 2z = 0$$
$$2x - y - z = 3$$
$$x + 2y + 3z = 0$$

36.
$$3x + 7y + 9z = 0$$
$$x + 2y + 3z = 2$$
$$x + 4y + z = 2$$

37.
$$x - 2y + 4z = 6$$
$$x + 2y + 13z = 6$$
$$-2x + 6y - z = -10$$

38.
$$x - 2y + 5z = -6$$
$$x + 2y + 3z = 0$$
$$x + 3y + 2z = 5$$

39.
$$x + y + z = 200$$
$$x - 2y = 0$$
$$2x + 3y + 5z = 600$$
$$2x - y + z = 200$$

40.
$$2x - y + 2z = 3$$
$$-x + 2y - z = 0$$
$$3y - 2z = 1$$
$$x + y - z = 1$$

41.
$$x + y + z = 5$$
$$2x + y - z = 2$$
$$x - y + z = -2$$

42.
$$2x + y + 3z = 9$$
$$-x - y + z = 1$$
$$3x - y + z = 9$$

Use the Gauss–Jordan method to solve each of the given systems of equations. (See Examples 3–5.)

43.
$$x + 2y + z = 5$$
$$2x + y - 3z = -2$$
$$3x + y + 4z = -5$$

44.
$$3x - 2y + z = 6$$
$$3x + y - z = -4$$
$$-x + 2y - 2z = -8$$

45.
$$x + 3y - 6z = 7$$
$$2x - y + 2z = 0$$
$$x + y + 2z = -1$$

46.
$$x = 1 - y$$
$$2x = z$$
$$2z = -2 - y$$

47.
$$x - 2y + 4z = 9$$
$$x + y + 13z = 6$$
$$-2x + 6y - z = -10$$

48.
$$x - y + 5z = -6$$
$$3x + 3y - z = 10$$
$$x + 2y + 3z = 5$$

Solve the system by any method.

49.
$$x + 3y + 4z = 14$$
$$2x - 3y + 2z = 10$$
$$3x - y + z = 9$$
$$4x + 2y + 5z = 9$$

50.
$$x - y = 2$$
$$x + y = 4$$
$$2x + 3y = 9$$
$$3x - 2y = 6$$

51.
$$4x - 4y + 12z = 28$$
$$3x - 2y - z = 36$$

52.
$$3x + y - z = 0$$
$$2x - y + 3z = -7$$

53.
$$-8x - 9y = 11$$
$$24x + 34y = 2$$
$$16x + 11y = -57$$

54.
$$x + 2y = 3$$
$$2x + 3y = 4$$
$$3x + 4y = 5$$
$$4x + 5y = 6$$

55.
$$x - 2y + z = 5$$
$$2x + y - z = 2$$
$$-2x + 4y - 2z = 2$$

56.
$$2x + y = 7$$
$$x - y = 3$$
$$x + 3y = 4$$

57.
$$x + y - z = -20$$
$$2x - y + z = 11$$

58.
$$4x + 3y + z = 1$$
$$-2x - y + 2z = 0$$

59.
$$2x + y + 3z - 2w = -6$$
$$4x + 3y + z - w = -2$$
$$x + y + z + w = -5$$
$$-2x - 2y - 2z + 2w = -10$$

60.
$$x + y + z + w = -1$$
$$-x + 4y + z - w = 0$$
$$x - 2y + z - 2w = 11$$
$$-x - 2y + z + 2w = -9$$

61.
$$x + 2y - z = 3$$
$$3x + y + w = 4$$
$$2x - y + z + w = 2$$

62.
$$3x - 2y - 8z = 1$$
$$9x - 6y - 24z = -2$$
$$x - y + z = 1$$

63.
$$2x + 3y + z = 9$$
$$4x + y - 3z = -7$$
$$6x + 2y - 4z = -8$$

64.
$$x - 2y - z - 3w = -3$$
$$-x + y + z = 2$$
$$4y + 3z - 6w = -2$$

Work these exercises.

65. Population Growth The population y in year x of the county listed is approximated by:

Calhoun County, IL:	$x + 20y = 112$
Martin County, TX:	$-x + 10y = 38$
Schley County, GA:	$y = 5,$

where $x = 10$ corresponds to the year 2010 and y is in thousands. Solve this system of equations, and interpret your answer. (Data from: U.S. Census Bureau.)

66. Population Growth The population y in year x of the greater metropolitan area listed is approximated by the given equation in which $x = 10$ corresponds to the year 2010 and y is in thousands:

Portland, OR:	$-32x + y = 1906$
Cambridge, MA:	$-22x + y = 2026$
Charlotte, NC:	$-73x + 2y = 3704.$

Solve this system of equations, and interpret your answer. (Data from: U.S. Census Bureau.)

Use a graphing calculator to solve these problems.

67. Concert Tickets A band concert is attended by x adults, y teenagers, and z preteen children. These numbers satisfy the following equations:

$$x + 1.25y + .25z = 457.5$$
$$x + .6y + .4z = 390$$
$$3.16x + 3.48y + .4z = 1297.2.$$

How many adults, teenagers, and children were present?

68. Credit Card Debt The owner of a small business borrows money on three separate credit cards: x dollars on Mastercard, y dollars on Visa, and z dollars on American Express. These amounts satisfy the following equations:

$$1.18x + 1.15y + 1.09z = 11{,}244.25$$
$$3.54x - .55y + .27z = 3{,}732.75$$
$$.06x + .05y + .03z = 414.75.$$

How much did the owner borrow on each card?

Work these problems.

69. Graph the equations in the given system. Then explain why the graphs show that the system is inconsistent:

$$2x + 3y = 8$$
$$x - y = 4$$
$$5x + y = 7.$$

70. Explain why the graphs of the equations in the given system suggest that the system is independent:

$$-x + y = 1$$
$$-2x + y = 0$$
$$x + y = 3.$$

71. Find constants a, b, and c such that the points $(2, 3)$, $(-1, 0)$, and $(-2, 2)$ lie on the graph of the equation $y = ax^2 + bx + c$. (*Hint:* Since $(2, 3)$ is on the graph, we must have $3 = a(2^2) + b(2) + c$; that is, $4a + 2b + c = 3$. Similarly, the other two points lead to two more equations. Solve the resulting system for a, b, and c.)

72. Explain why a system with more variables than equations cannot have a unique solution (that is, be an independent system). (*Hint:* When you apply the elimination method to such a system, what must happen?)

✓ **Checkpoint Answers**

1. (a) $(-1, 1)$ **(b)** $(3, 1, -1)$

2. (a) $\begin{bmatrix} 4 & -2 & 3 & | & 4 \\ 3 & 5 & 1 & | & -7 \\ 5 & -1 & 4 & | & 6 \end{bmatrix}$

(b) $\begin{aligned} 2x - 2y &= -2 \\ x + y &= 4 \\ 3x + 5y &= 8 \end{aligned}$

3. (a) $\begin{bmatrix} 3 & -2 \\ -1 & 5 \end{bmatrix}$

(b) $\begin{bmatrix} -2 & 10 \\ 3 & -2 \end{bmatrix}$

(c) $\begin{bmatrix} -1 & 5 \\ 6 & -17 \end{bmatrix}$

(d) $\begin{bmatrix} 5 & 1 \\ 3 & -2 \end{bmatrix}$

4. $(-1, 1, 2)$

5. (a) $\begin{bmatrix} 1 & 2 & | & 11 \\ -4 & 1 & | & -8 \\ 5 & 1 & | & 19 \end{bmatrix}$

(b) $4R_1 + R_2 \begin{bmatrix} 1 & 2 & | & 11 \\ 0 & 9 & | & 36 \\ 5 & 1 & | & 19 \end{bmatrix}$

(c) $-5R_1 + R_3 \begin{bmatrix} 1 & 2 & | & 11 \\ 0 & 9 & | & 36 \\ 0 & -9 & | & -36 \end{bmatrix}$

(d) $\dfrac{1}{9} R_2 \begin{bmatrix} 1 & 2 & | & 11 \\ 0 & 1 & | & 4 \\ 0 & -9 & | & -36 \end{bmatrix}$

(e) $-2R_2 + R_1 \begin{bmatrix} 1 & 0 & | & 3 \\ 0 & 1 & | & 4 \\ 0 & -9 & | & -36 \end{bmatrix}$

(f) $9R_2 + R_3 \begin{bmatrix} 1 & 0 & | & 3 \\ 0 & 1 & | & 4 \\ 0 & 0 & | & 0 \end{bmatrix}$

(g) $(3, 4)$

6. $(1, 2, -1)$

7. $(-5, 3)$

8. (a) No solution

(b) $(0, 0)$

9. (a) $(-50, -26, 7)$

(b) $(55, 16, -14)$

(c) $(-40, -22, 5)$

6.3 Applications of Systems of Linear Equations

There are no hard and fast rules for solving applied problems, but it is usually best to begin by identifying the unknown quantities and letting each be represented by a variable. Then look at the given data to find one or more relationships among the unknown quantities that lead to equations. If the equations are all linear, use the techniques of the preceding sections to solve the system.

| EXAMPLE 1 | **Truck Rentals** A rent-a-truck company plans to spend $5 million |

EXAMPLE 1 **Truck Rentals** A rent-a-truck company plans to spend $5 million on 200 new vehicles. Each van will cost $20,000, each small truck $25,000, and each large truck $35,000. Past experience shows that the company needs twice as many vans as small trucks. How many of each kind of vehicle can the company buy?

Solution Let x be the number of vans, y the number of small trucks, and z the number of large trucks. Then

$$\left(\begin{array}{c}\text{Number of}\\ \text{vans}\end{array}\right) + \left(\begin{array}{c}\text{Number of}\\ \text{small trucks}\end{array}\right) + \left(\begin{array}{c}\text{Number of}\\ \text{large trucks}\end{array}\right) = \text{Total number of vehicles}$$

$$x + y + z = 200. \tag{1}$$

Similarly,

$$\left(\begin{array}{c}\text{Cost of }x\\ \text{vans}\end{array}\right) + \left(\begin{array}{c}\text{Cost of }y\\ \text{small trucks}\end{array}\right) + \left(\begin{array}{c}\text{Cost of }z\\ \text{large trucks}\end{array}\right) = \text{Total cost}$$

$$20{,}000x + 25{,}000y + 35{,}000z = 5{,}000{,}000.$$

Dividing both sides by 5000 produces the equivalent equation

$$4x + 5y + 7z = 1000. \tag{2}$$

Finally, the number of vans is twice the number of small trucks; that is, $x = 2y$, or equivalently,

$$x - 2y = 0. \tag{3}$$

We must solve the system given by equations (1)–(3):

$$\begin{aligned} x + y + z &= 200 \\ 4x + 5y + 7z &= 1000 \\ x - 2y \quad\;\; &= 0. \end{aligned}$$

Manual Method Form the augmented matrix and transform it into row echelon form:

$$\left[\begin{array}{ccc|c} 1 & 1 & 1 & 200 \\ 4 & 5 & 7 & 1000 \\ 1 & -2 & 0 & 0 \end{array}\right]$$

$$\left[\begin{array}{ccc|c} 1 & 1 & 1 & 200 \\ 0 & 1 & 3 & 200 \\ 0 & -3 & -1 & -200 \end{array}\right] \quad \begin{array}{l} -4R_1 + R_2 \\ -R_1 + R_3 \end{array}$$

$$\left[\begin{array}{ccc|c} 1 & 1 & 1 & 200 \\ 0 & 1 & 3 & 200 \\ 0 & 0 & 8 & 400 \end{array}\right] \quad 3R_2 + R_3$$

$$\left[\begin{array}{ccc|c} 1 & 1 & 1 & 200 \\ 0 & 1 & 3 & 200 \\ 0 & 0 & 1 & 50 \end{array}\right]. \quad \frac{1}{8}R_3$$

This row echelon matrix corresponds to the system

$$\begin{aligned} x + y + z &= 200 \\ y + 3z &= 200 \\ z &= 50. \end{aligned}$$

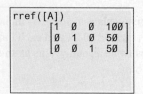

Figure 6.15

✓ **Checkpoint 1**

In Example 1, suppose that the company can spend only $2 million on 150 new vehicles and the company needs three times as many vans as small trucks. Write a system of equations to express these conditions.

Use back substitution to solve this system:

$$z = 50 \qquad y + 3z \quad = 200 \qquad x + y + z \quad = 200$$
$$y + 3(50) = 200 \qquad x + 50 + 50 = 200$$
$$y + 150 \quad = 200 \qquad x \quad + 100 = 200$$
$$y = 50 \qquad\qquad x = 100.$$

Therefore, the company should buy 100 vans, 50 small trucks, and 50 large trucks.

Calculator Method Enter the augmented matrix of the system into the calculator. Use RREF to change it into reduced row echelon form, as in Figure 6.15. The answers may be read directly from the matrix: $x = 100$ vans, $y = 50$ small trucks, and $z = 50$ large trucks. ✓₁

EXAMPLE 2 **Finance** An investor plans to put a total of $100,000 in a money market account, a bond fund, an international stock fund, and a domestic stock fund. She wants 60% of her investment to be conservative (money market and bonds). She wants the amount in international stocks to be one-fourth of the amount in domestic stocks. Finally, she needs an annual return of $4000. Assuming she gets annual returns of 2.5% on the money market account, 3.5% on the bond fund, 5% on the international stock fund, and 6% on the domestic stock fund, how much should she put in each investment?

Solution Let x be the amount invested in the money market account, y the amount in the bond fund, z the amount in the international stock fund, and w the amount in the domestic stock fund. Then

$$x + y + z + w = \text{total amount invested} = 100{,}000. \tag{4}$$

Use her annual return to get a second equation:

$$\left(\begin{matrix}2.5\% \text{ return}\\ \text{on money}\\ \text{market}\end{matrix}\right) + \left(\begin{matrix}3.5\% \text{ return}\\ \text{on bond}\\ \text{fund}\end{matrix}\right) + \left(\begin{matrix}5\% \text{ return on}\\ \text{international}\\ \text{stock fund}\end{matrix}\right) + \left(\begin{matrix}6\% \text{ return}\\ \text{on domestic}\\ \text{stock fund}\end{matrix}\right) = 4000$$

$$.025x \quad + \quad .035y \quad + \quad .05z \quad + \quad .06w \quad = 4000. \tag{5}$$

Since she wants the amount in international stocks to be one-fourth of the amount in domestic stocks, we have

$$z = \frac{1}{4}w, \qquad \text{or equivalently,} \qquad z - .25w = 0. \tag{6}$$

Finally, the amount in conservative investments is $x + y$, and this quantity should be equal to 60% of $100,000—that is,

$$x + y = 60{,}000. \tag{7}$$

✓ **Checkpoint 2**

(a) Write the augmented matrix of the system given by equations (4)–(7) of Example 2.

(b) List a sequence of row operations that transforms this matrix into row echelon form.

(c) Display the final row echelon form matrix.

Now solve the system given by equations (4)–(7):

$$x + \quad y + \quad z + \quad w = 100{,}000$$
$$.025x + .035y + .05z + .06w = \quad 4{,}000$$
$$z - .25w = \quad 0$$
$$x + \quad y \qquad\qquad = \quad 60{,}000.$$

Manual Method Write the augmented matrix of the system and transform it into row echelon form, as in Checkpoint 2. ✓₂

The final matrix in Checkpoint 2 represented the system

$$
\begin{aligned}
x + y + z + w &= 100{,}000 \\
y + 2.5z + 3.5w &= 150{,}000 \\
z - .25w &= 0 \\
w &= 32{,}000.
\end{aligned}
$$

Back substitution shows that

$$
\begin{aligned}
w &= 32{,}000 \\
z &= .25w = .25(32{,}000) = 8000 \\
y &= -2.5z - 3.5w + 150{,}000 = -2.5(8000) - 3.5(32{,}000) + 150{,}000 = 18{,}000 \\
x &= -y - z - w + 100{,}000 = -18{,}000 - 8000 - 32{,}000 + 100{,}000 = 42{,}000.
\end{aligned}
$$

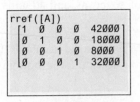

Figure 6.16

Therefore, she should put $42,000 in the money market account, $18,000 in the bond fund, $8000 in the international stock fund, and $32,000 in the domestic stock fund.

Calculator Method Enter the augmented matrix of the system into the calculator. Use RREF to change it into reduced row echelon form, as in Figure 6.16. The answers can easily be read from this matrix: $x = 42{,}000$, $y = 18{,}000$, $z = 8000$, and $w = 32{,}000$.

EXAMPLE 3   **Livestock Dietary Requirements** An animal feed is to be made from corn, soybeans, and cottonseed. Determine how many units of each ingredient are needed to make a feed that supplies 1800 units of fiber, 2800 units of fat, and 2200 units of protein, given that 1 unit of each ingredient provides the numbers of units shown in the table below. The table states, for example, that a unit of corn provides 10 units of fiber, 30 units of fat, and 20 units of protein.

	Corn	Soybeans	Cottonseed	Totals
Units of Fiber	10	20	30	1800
Units of Fat	30	20	40	2800
Units of Protein	20	40	25	2200

Solution Let x represent the required number of units of corn, y the number of units of soybeans, and z the number of units of cottonseed. Since the total amount of fiber is to be 1800, we have

$$10x + 20y + 30z = 1800.$$

The feed must supply 2800 units of fat, so

$$30x + 20y + 40z = 2800.$$

Finally, since 2200 units of protein are required, we have

$$20x + 40y + 25z = 2200.$$

Thus, we must solve this system of equations:

$$
\begin{aligned}
10x + 20y + 30z &= 1800 \\
30x + 20y + 40z &= 2800 \\
20x + 40y + 25z &= 2200.
\end{aligned}
\tag{8}
$$

Now solve the system, either manually or with technology.

✓ Checkpoint 3

(a) Write the augmented matrix of system (8) of Example 3.

(b) List a sequence of row operations that transforms this matrix into row echelon form.

(c) Display the final row echelon form matrix.

$$\begin{bmatrix} 10 & 20 & 30 & 1800 \\ 30 & 20 & 40 & 2800 \\ 20 & 40 & 25 & 2200 \end{bmatrix}$$
rref([A])
$$\begin{bmatrix} 1 & 0 & 0 & 30 \\ 0 & 1 & 0 & 15 \\ 0 & 0 & 1 & 40 \end{bmatrix}$$

Figure 6.17

Manual Method Write the augmented matrix and use row operations to transform it into row echelon form, as in Checkpoint 3. The resulting matrix represents the following system: ✓3

$$x + 2y + 3z = 180$$
$$y + \frac{5}{4}z = 65$$
$$z = 40. \tag{9}$$

Back substitution now shows that

$$z = 40, \quad y = 65 - \frac{5}{4}(40) = 15, \quad \text{and} \quad x = 180 - 2(15) - 3(40) = 30.$$

Thus, the feed should contain 30 units of corn, 15 units of soybeans, and 40 units of cottonseed.

Calculator Method Enter the augmented matrix of the system into the calculator (top of Figure 6.17). Use RREF to transform it into reduced row echelon form (bottom of Figure 6.17), which shows that $x = 30$, $y = 15$, and $z = 40$.

EXAMPLE 4   **Population Growth** The table shows Census Bureau projections for the population of the United States (in millions).

Year	2020	2040	2050
U.S. Population	334	380	400

(a) Use the given data to construct a quadratic function that gives the U.S. population (in millions) in year x.

Solution Let $x = 0$ correspond to the year 2000. Then the table represents the data points $(20,334)$, $(40,380)$, and $(50,400)$. We must find a function of the form

$$f(x) = ax^2 + bx + c$$

whose graph contains these three points. If $(20,334)$ is to be on the graph, we must have $f(20) = 334$; that is,

$$a(20)^2 + b(20) + c = 334$$
$$400a + 20b + c = 334.$$

The other two points lead to these equations:

$$f(40) = 380 \qquad\qquad f(50) = 400$$
$$a(40)^2 + b(40) + c = 380 \qquad a(50)^2 + b(50) + c = 400$$
$$1600a + 40b + c = 380 \qquad 2500a + 50b + c = 400.$$

Now work by hand or use technology to solve the following system:

$$400a + 20b + c = 334$$
$$1600a + 40b + c = 380$$
$$2500a + 50b + c = 400.$$

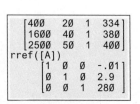

$$\begin{bmatrix} 400 & 20 & 1 & 334 \\ 1600 & 40 & 1 & 380 \\ 2500 & 50 & 1 & 400 \end{bmatrix}$$
rref([A])
$$\begin{bmatrix} 1 & 0 & 0 & -.01 \\ 0 & 1 & 0 & 2.9 \\ 0 & 0 & 1 & 280 \end{bmatrix}$$

Figure 6.18

The reduced row echelon form of the augmented matrix in Figure 6.18 shows that the solution is $a = -.01$, $b = 2.9$, and $c = 280$. So the function is

$$f(x) = -.01x^2 + 2.9x + 280.$$

(b) Use this model to estimate the U.S. population in the year 2030.

Solution The year 2030 corresponds to $x = 30$, so the U.S. population is projected to be

$$f(30) = -.01(30)^2 + 2.9(30) + 280 = 358 \text{ million}.$$

EXAMPLE 5   **Delivery Logistics** A specialty wholesaler sells espresso machines. The EZ model weighs 10 pounds and comes in a 10-cubic-foot box. The compact model weighs 20 pounds and comes in an 8-cubic-foot box. The commercial model weighs 60 pounds and comes in a 28-cubic-foot box. Each delivery van has 248 cubic feet of space and can hold a maximum of 440 pounds. In order for a van to be fully loaded, how many of each model should it carry?

Solution Let x be the number of EZ, y the number of compact, and z the number of commercial models carried by a van. Then we can summarize the given information in this table.

Model	Number	Weight	Volume
EZ	x	10	10
Compact	y	20	8
Commercial	z	60	28
Total for a load		440	248

Since a fully loaded van can carry 440 pounds and 248 cubic feet, we must solve this system of equations:

$$10x + 20y + 60z = 440 \qquad \text{Weight equation}$$
$$10x + 8y + 28z = 248. \qquad \text{Volume equation}$$

The augmented matrix of the system is

$$\begin{bmatrix} 10 & 20 & 60 & | & 440 \\ 10 & 8 & 28 & | & 248 \end{bmatrix}.$$

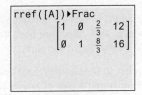

Figure 6.19

If you are working by hand, transform the matrix into row echelon form and use back substitution. If you are using a graphing calculator, use RREF to transform the matrix into reduced row echelon form, as in Figure 6.19 (which also uses FRAC in the MATH menu to eliminate long decimals).

The system corresponding to Figure 6.19 is

$$x + \frac{2}{3}z = 12$$
$$y + \frac{8}{3}z = 16,$$

which is easily solved: $x = 12 - \frac{2}{3}z$ and $y = 16 - \frac{8}{3}z$.

Hence, all solutions of the system are given by $\left(12 - \frac{2}{3}z, 16 - \frac{8}{3}z, z \right)$. The only solutions that apply in this situation, however, are those given by $z = 0, 3$, and 6, because all other values of z lead to fractions or negative numbers. (You can't deliver part of a box or a negative number of boxes). Hence, there are three ways to have a fully loaded van:

Solution	Van Load
(12, 16, 0)	12 EZ, 16 compact, 0 commercial
(10, 8, 3)	10 EZ, 8 compact, 3 commercial
(8, 0, 6)	8 EZ, 0 compact, 6 commercial.

6.3 Exercises

Use systems of equations to work these applied problems. (See Examples 1–5.)

1. **Lyft Rates** Lyft charges customers a flat fee per ride (which is higher during rush hour traffic) plus charges for each minute and each mile. (Data from: www.lyft.com.) Suppose that, in a certain metropolitan area during rush hour, the flat fee is $4, the cost per minute is $.20, and the cost per mile is $1.40. Let x be the number of minutes and y the number of miles. At the end of a ride, the driver said that you owed $14 and remarked that the number of minutes was three times the number of miles. Find the number of minutes and the number of miles for this trip.

2. **Uber Rates** Uber charges customers a flat fee per ride (which is higher during rush hour traffic) plus charges for each minute and each mile. (Data from: www.uber.com.) For a certain trip in Birmingham, Alabama, in 2016, Uber charged a $3.05 flat fee plus $0.18 per minute and $1.20 per mile. Let x be the number of minutes and y the number of miles. At the end of the ride, the driver said that the bill was $10.85 and remarked that the number of minutes was twice the number of miles. Find the number of minutes and the number of miles for this trip.

3. **Euro and Pounds** Abigail Henderson has $1000 budgeted for spending money on an upcoming trip to London and Paris. The European currency is trading at $1.20 per euro, and British pounds sterling are trading at $1.40 per pound. She plans to spend more time in France, so she wants to have three times as many euros as pounds. Set up and solve a system of equations to model this problem, and explain what your answer means in practical terms.

4. **Pesos and Canadian Dollars** Emma Henderson is planning a North American odyssey and has $1500 budgeted for spending money. It costs $.60 to buy a peso and $.75 to buy a Canadian dollar. She expects her adventures in Mexico to take more time, so she wants to have five times as many pesos as Canadian dollars. Set up and solve a system of equations to model this problem, and explain what your answer means in practical terms.

5. **Stock Portfolio** You invested a total of $25,000 in Coca-Cola Company and Intel Corporation. You purchased Coca-Cola at $45 per share and Intel at $32 per share. If Coca-Cola doubles in value and Intel stock goes up by 50%, your portfolio will be worth $42,000. How many shares of each stock do you own?

6. **Bond Portfolio** You purchased a total of $10,000 in bond funds. The JP Morgan Emerging Markets bond fund costs $100 per share, and the Fidelity Conservative Income bond fund costs $10 per share. If the emerging markets fund goes up in value by 5% and the conservative bond fund increases by 10%, your portfolio will be worth $10,600. How many shares of each fund do you own?

7. **Basketball Statistics** In a championship basketball game, the center scored a total of 31 *points*. Let x be the number of free-throw *baskets* made (worth 1 point each), y the number

of regular *baskets* made (2 points each), and z the number of *baskets* made from behind the three-point line (3 points each). The player made a total of 12 baskets from the floor (2-point and 3-point shots) and 4 baskets from the free-throw line.

(a) Set up a system of three equations to model this situation.

(b) Solve the system and interpret your answer.

8. **Baseball Statistics** In the final game of a baseball series that went to extra innings, the most valuable player (MVP) rounded a total of 12 *bases*. Let x be the number of singles, y the number of doubles, z the number of triples, and w the number of home runs hit. The MVP had 1 single and 4 hits for extra bases, including 1 home run.

(a) Set up a system of four equations to model this situation.

(b) Solve the system and interpret your answer.

9. **Truck Rentals** The rent-a-truck company of Example 1 learns that each van now costs $25,000, each small truck costs $30,000, and each large truck costs $40,000. The company still needs twice as many vans as small trucks and wants to spend $5 million, but it decides to buy only 175 vehicles. How many of each kind should it buy?

10. **Finance** Suppose that the investor in Example 2 finds that her annual return on the international stock fund will be only 4%. Now how much should she put in each investment?

11. **Livestock Dietary Requirements** To meet consumer demand, the animal feed in Example 3 now supplies only 2400 units of fat. How many units of corn, soybeans, and cottonseed are now needed?

12. **Delivery Logistics** Suppose that the wholesaler in Example 5 finds a way to pack the EZ model in an 8-cubic-foot box. Now how many of each model should a fully loaded van carry?

13. **Dry Cleaning** Mario took clothes to the cleaners three times last month. First, he brought 3 shirts and 1 pair of slacks and paid $10.96. Then he brought 7 shirts, 2 pairs of slacks, and a sport coat and paid $30.40. Finally, he brought 4 shirts and 1 sports coat and paid $14.45. How much was he charged for each shirt, each pair of slacks, and each sports coat?

14. **Online Retailing** Shipping charges at an online bookstore are $4 for one book, $6 for two books, and $7 for three to five books. Last week, there were 6400 orders of five or fewer books, and total shipping charges for these orders were $33,600. The number of shipments with $7 charges was 1000 less than the number with $6 charges. How many shipments were made in each category (one book, two books, three to five books)?

15. **Concert Tickets** Tickets to a band concert cost $5 for adults, $3 for teenagers, and $2 for preteens. There were 570 people at the concert, and total ticket receipts were $1950. Three-fourths as many teenagers as preteens attended. How many adults, teenagers, and preteens attended?

16. **Minor League Baseball** A minor league baseball team had 7000 fans in attendance, and total ticket receipts were $26,400.

Box seats cost $6, grandstand seats cost $4, and bleacher seats cost $2. If the number of tickets sold for box seats was one-third the number of bleacher seats sold, how many tickets of each type were sold?

17. **Finance** An investment firm recommends that a client invest in bonds rated AAA, A, and B. The average yield on AAA bonds is 6%, on A bonds 7%, and on B bonds 10%. The client wants to invest twice as much in AAA bonds as in B bonds. How much should be invested in each type of bond if the total investment is $25,000, and the investor wants an annual return of $1810 on the three investments?

18. **Finance** An investor wants to invest $30,000 in corporate bonds that are rated AAA, A, and B. The lower rated ones pay higher interest, but pose a higher risk as well. The average yield is 5% on AAA bonds, 6% on A bonds, and 10% on B bonds. Being conservative, the investor wants to have twice as much in AAA bonds as in B bonds. How much should she invest in each type of bond to have an interest income of $2000?

19. **Finance** An investor plans to invest $70,000 in a mutual fund, corporate bonds, and a fast-food franchise. She plans to put twice as much in bonds as in the mutual fund. On the basis of past performance, she expects the mutual fund to pay a 2% dividend, the bonds 6%, and the franchise 10%. She would like a dividend income of $4800. How much should she put in each of three investments?

20. **Finance** Makayla borrows $10,000. Some is from her friend at 8% annual interest, twice as much as that from her bank at 9%, and the remainder from her insurance company at 5%. She pays a total of $830 in interest for the first year. How much did she borrow from each source?

21. **Trail Mix Recipe** Pretzels cost $3 per pound, dried fruit $4 per pound, and nuts $8 per pound. How many pounds of each should be used to produce 140 pounds of trail mix costing $6 per pound in which there are twice as many pretzels (by weight) as dried fruit?

22. **Auto Manufacturing** An auto manufacturer sends cars from two plants, I and II, to dealerships A and B, located in a midwestern city. Plant I has a total of 28 cars to send, and plant II has 8. Dealer A needs 20 cars, and dealer B needs 16. Transportation costs based on the distance of each dealership from each plant are $220 from I to A, $300 from I to B, $400 from II to A, and $180 from II to B. The manufacturer wants to limit transportation costs to $10,640. How many cars should be sent from each plant to each of the two dealerships?

23. **Animal Feed** An animal breeder can buy four types of tiger food. Each case of Brand A contains 25 units of fiber, 30 units of protein, and 30 units of fat. Each case of Brand B contains 50 units of fiber, 30 units of protein, and 20 units of fat. Each case of Brand C contains 75 units of fiber, 30 units of protein, and 20 units of fat. Each case of Brand D contains 100 units of fiber, 60 units of protein, and 30 units of fat. How many cases of each brand should the breeder mix together to obtain a food that provides 1200 units of fiber, 600 units of protein, and 400 units of fat?

24. **Physical Science** The stopping distance for a car traveling 25 mph is 61.7 feet, and for a car traveling 35 mph it is 106 feet. The stopping distance in feet can be described by the equation

$y = ax^2 + bx$, where x is the speed in mph. (Data from: National Traffic Safety Institute.)

(a) Find the values of a and b.

(b) Use your answers from part (a) to find the stopping distance for a car traveling 55 mph.

Work these exercises. (See Example 4.)

25. **Female Population** The table shows Census Bureau projections for the female population of the United States (in millions).

Year	2015	2035	2040
Female Population	163	187	192

(a) Find a quadratic function $f(x) = ax^2 + bx + c$ that gives the female population (in millions) in year x, where $x = 0$ corresponds to the year 2000.

(b) Estimate the female population in the year 2020.

26. **Male Population** The table shows Census Bureau projections for the male population of the United States (in millions).

Year	2015	2035	2040
Male Population	158	183	188

(a) Find a quadratic function $f(x) = ax^2 + bx + c$ that gives the male population (in millions) in year x, where $x = 0$ corresponds to the year 2000.

(b) Estimate the male population in the year 2020.

27. **Fuel Consumption** At a pottery factory, fuel consumption for heating the kilns varies with the size of the order being fired. In the past, the company recorded the figures in the table.

$x = $ **Number of Platters**	$y = $ **Fuel Cost per Platter**
6	$2.80
8	2.48
10	2.24

(a) Find an equation of the form $y = ax^2 + bx + c$ whose graph contains the three points corresponding to the data in the table.

(b) How many platters should be fired at one time in order to minimize the fuel cost per platter? What is the minimum fuel cost per platter?

28. **Alzheimer's Disease** The number of Alzheimer's cases in people 85 and older was 2 million in 2004 and is projected to be 3 million in 2025 and 8 million in 2050. (Data from: Alzheimer's Association.)

(a) Let $x = 0$ correspond to 2000. Find a quadratic function that models the given data.

(b) How many people 85 or older will have Alzheimer's disease in 2020 and in 2034?

(c) In the year you turn 85, how many people your age or older are expected to have Alzheimer's disease?

📟 *A graphing calculator or other technology is recommended for the following exercises.*

29. Health The table shows the calories, sodium, and protein in one cup of various kinds of soup.

	Progresso™ Hearty Chicken Rotini	Healthy Choice™ Hearty Chicken	Campbell's™ Chunky Chicken Noodle
Calories	100	130	110
Sodium (mg)	960	480	890
Protein (g)	7	9	8

How many cups of each kind of soup should be mixed together to produce 10 servings of soup, each of which provides 171 calories, 1158 milligrams of sodium, and 12.1 grams of protein? What is the serving size (in cups)? (*Hint:* In 10 servings, there must be 1710 calories, 11,580 milligrams of sodium, and 121 grams of protein.)

30. Health The table shows the calories, sodium, and fat in 1 ounce of various snack foods (all produced by Planters™):

	Sweet N' Crunchy Peanuts	Dry Roasted Honey Peanuts	Kettle Roasted Honey BBQ Peanuts
Calories	140	160	180
Sodium (mg)	20	110	55
Fat (g)	8	13	15

How many ounces of each kind of snack should be combined to produce 10 servings, each of which provides 284 calories, 93 milligrams of sodium, and 20.6 grams of fat? What is the serving size?

✓**Checkpoint Answers**

1.
$$x + y + z = 150$$
$$4x + 5y + 7z = 400$$
$$x - 3y = 0$$

2. (a)
$$\begin{bmatrix} 1 & 1 & 1 & 1 & 100,000 \\ .025 & .035 & .05 & .06 & 4000 \\ 0 & 0 & 1 & -.25 & 0 \\ 1 & 1 & 0 & 0 & 60,000 \end{bmatrix}$$

(b) Many sequences are possible, including this one:

Replace R_2 by $-.025R_1 + R_2$;

replace R_4 by $-R_1 + R_4$;

replace R_2 by $\dfrac{1}{.01}R_2$;

replace R_4 by $R_3 + R_4$;

replace R_4 by $\dfrac{-1}{1.25}R_4$.

(c)
$$\begin{bmatrix} 1 & 1 & 1 & 1 & 100,000 \\ 0 & 1 & 2.5 & 3.5 & 150,000 \\ 0 & 0 & 1 & -.25 & 0 \\ 0 & 0 & 0 & 1 & 32,000 \end{bmatrix}$$

3. (a)
$$\begin{bmatrix} 10 & 20 & 30 & 1800 \\ 30 & 20 & 40 & 2800 \\ 20 & 40 & 25 & 2200 \end{bmatrix}$$

(b) Many sequences are possible, including this one:

Replace R_1 by $\dfrac{1}{10}R_1$;

replace R_2 by $\dfrac{1}{10}R_2$;

replace R_3 by $\dfrac{1}{5}R_3$;

replace R_2 by $-3R_1 + R_2$;
replace R_3 by $-4R_1 + R_3$;

replace R_2 by $-\dfrac{1}{4}R_2$;

replace R_3 by $-\dfrac{1}{7}R_3$.

(c)
$$\begin{bmatrix} 1 & 2 & 3 & 180 \\ 0 & 1 & \dfrac{5}{4} & 65 \\ 0 & 0 & 1 & 40 \end{bmatrix}$$

6.4 Basic Matrix Operations

Until now, we have used matrices only as a convenient shorthand to solve systems of equations. However, matrices are also important in the fields of management, natural science, engineering, and social science as a way to organize data, as Example 1 demonstrates.

EXAMPLE 1 **Furniture Manufacturing** The EZ Life Company manufactures sofas and armchairs in three models: *A*, *B*, and *C*. The company has regional warehouses in New York, Chicago, and San Francisco. In its August shipment, the company sends 10 model *A* sofas, 12 model *B* sofas, 5 model *C* sofas, 15 model *A* chairs, 20 model *B* chairs, and 8 model *C* chairs to each warehouse.

These data might be organized by first listing them as follows:

Sofas	10 model A	12 model B	5 model C;
Chairs	15 model A	20 model B	8 model C.

Alternatively, we might tabulate the data:

		MODEL		
		A	B	C
FURNITURE	Sofa	10	12	5
	Chair	15	20	8

With the understanding that the numbers in each row refer to the type of furniture (sofa or chair) and the numbers in each column refer to the model (A, B, or C), the same information can be given by a matrix as follows:

$$M = \begin{bmatrix} 10 & 12 & 5 \\ 15 & 20 & 8 \end{bmatrix}. \quad ✓_1$$

A matrix with m horizontal rows and n vertical columns has dimension, or size, $m \times n$. The number of rows is always given first.

✓ Checkpoint 1

Rewrite matrix M in Example 1 in a matrix with three rows and two columns.

EXAMPLE 2

(a) The matrix $\begin{bmatrix} 6 & 5 \\ 3 & 4 \\ 5 & -1 \end{bmatrix}$ is a 3×2 matrix.

(b) $\begin{bmatrix} 5 & 8 & 9 \\ 0 & 5 & -3 \\ -4 & 0 & 5 \end{bmatrix}$ is a 3×3 matrix.

(c) $[1 \quad 6 \quad 5 \quad -2 \quad 5]$ is a 1×5 matrix.

(d) A graphing calculator displays a 4×1 matrix like this:

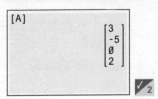

✓ Checkpoint 2

Give the size of each of the following matrices.

(a) $\begin{bmatrix} 2 & 1 & -5 & 6 \\ 3 & 0 & 7 & -4 \end{bmatrix}$

(b) $\begin{bmatrix} 1 & 2 & 3 \\ 4 & 5 & 6 \\ 9 & 8 & 7 \end{bmatrix}$

A matrix with only one row, as in Example 2(c), is called a **row matrix,** or **row vector.** A matrix with only one column, as in Example 2(d), is called a **column matrix,** or **column vector.** A matrix with the same number of rows as columns is called a **square matrix.** The matrix in Example 2(b) is a square matrix, as are

✓ Checkpoint 3

Use the numbers 2, 5, −8, and 4 to write

(a) a row matrix;

(b) a column matrix;

(c) a square matrix.

$$A = \begin{bmatrix} -5 & 6 \\ 8 & 3 \end{bmatrix} \quad \text{and} \quad B = \begin{bmatrix} 0 & 0 & 0 & 0 \\ -2 & 4 & 1 & 3 \\ 0 & 0 & 0 & 0 \\ -5 & -4 & 1 & 8 \end{bmatrix}. \quad ✓_3$$

When a matrix is denoted by a single letter, such as the matrix A above, the element in row i and column j is denoted a_{ij}. For example, $a_{21} = 8$ (the element in row 2, column 1). Similarly, in matrix B, $b_{42} = -4$ (the element in row 4, column 2).

Addition

The matrix given in Example 1,

$$M = \begin{bmatrix} 10 & 12 & 5 \\ 15 & 20 & 8 \end{bmatrix},$$

shows the August shipment from the EZ Life plant to each of its warehouses. If matrix N below gives the September shipment to the New York warehouse, what is the total shipment for each item of furniture to the New York warehouse for the two months?

$$N = \begin{bmatrix} 45 & 35 & 20 \\ 65 & 40 & 35 \end{bmatrix}.$$

If 10 model A sofas were shipped in August and 45 in September, then altogether 55 model A sofas were shipped in the two months. Adding the other corresponding entries gives a new matrix, Q, that represents the total shipment to the New York warehouse for the two months:

$$Q = \begin{bmatrix} 55 & 47 & 25 \\ 80 & 60 & 43 \end{bmatrix}.$$

It is convenient to refer to Q as the *sum* of M and N.

The way these two matrices were added illustrates the following definition of addition of matrices.

Matrix Addition

The **sum** of two $m \times n$ matrices X and Y is the $m \times n$ matrix $X + Y$ in which each element is the sum of the corresponding elements of X and Y.

It is important to remember that only matrices that are the same size can be added.

EXAMPLE 3 Find each sum if possible.

(a) $\begin{bmatrix} 5 & -6 \\ 8 & 9 \end{bmatrix} + \begin{bmatrix} -4 & 6 \\ 8 & -3 \end{bmatrix} = \begin{bmatrix} 5 + (-4) & -6 + 6 \\ 8 + 8 & 9 + (-3) \end{bmatrix} = \begin{bmatrix} 1 & 0 \\ 16 & 6 \end{bmatrix}.$

(b) The matrices

$$A = \begin{bmatrix} 5 & 8 \\ 6 & 2 \end{bmatrix} \quad \text{and} \quad B = \begin{bmatrix} 3 & 9 & 1 \\ 4 & 2 & 5 \end{bmatrix}$$

are of different sizes, so it is not possible to find the sum $A + B$. ☑ 4

✓ **Checkpoint 4**

Find each sum when possible.

(a) $\begin{bmatrix} 2 & 5 & 7 \\ 3 & -1 & 4 \end{bmatrix}$

$+ \begin{bmatrix} -1 & 2 & 0 \\ 10 & -4 & 5 \end{bmatrix}$

(b) $\begin{bmatrix} 1 \\ 2 \\ 3 \end{bmatrix} + \begin{bmatrix} 2 & -1 \\ 4 & 5 \\ 6 & 0 \end{bmatrix}$

(c) $[5 \quad 4 \quad -1] + [-5 \quad 2 \quad 3]$

▦ **TECHNOLOGY TIP** Graphing calculators can find matrix sums, as illustrated in Figure 6.20.

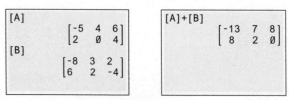

Figure 6.20

EXAMPLE 4 **Furniture Manufacturing** The September shipments of the three models of sofas and chairs from the EZ Life Company to the New York, San Francisco, and Chicago warehouses are given respectively in matrices N, S, and C as follows:

$$N = \begin{bmatrix} 45 & 35 & 20 \\ 65 & 40 & 35 \end{bmatrix}; \quad S = \begin{bmatrix} 30 & 32 & 28 \\ 43 & 47 & 30 \end{bmatrix}; \quad C = \begin{bmatrix} 22 & 25 & 38 \\ 31 & 34 & 35 \end{bmatrix}.$$

What was the total amount shipped to the three warehouses in September?

Solution The total of the September shipments is represented by the sum of the three matrices N, S, and C:

$$N + S + C = \begin{bmatrix} 45 & 35 & 20 \\ 65 & 40 & 35 \end{bmatrix} + \begin{bmatrix} 30 & 32 & 28 \\ 43 & 47 & 30 \end{bmatrix} + \begin{bmatrix} 22 & 25 & 38 \\ 31 & 34 & 35 \end{bmatrix}$$

$$= \begin{bmatrix} 97 & 92 & 86 \\ 139 & 121 & 100 \end{bmatrix}.$$

For example, from this sum, the total number of model C sofas shipped to the three warehouses in September was 86. ✔5

✔ **Checkpoint 5**

From the result of Example 4, find the total number of the following shipped to the three warehouses.

(a) Model A chairs

(b) Model B sofas

(c) Model C chairs

EXAMPLE 5 **Drug Testing** A drug company is testing 200 patients to see if Painoff (a new headache medicine) is effective. Half the patients receive Painoff and half receive a placebo. The data on the first 50 patients is summarized in this matrix:

Pain Relief Obtained

Yes No

Patient took Painoff $\begin{bmatrix} 22 & 3 \\ Patient\ took\ placebo & 8 & 17 \end{bmatrix}$

For example, row 2 shows that, of the people who took the placebo, 8 got relief, but 17 did not. The test was repeated on three more groups of 50 patients each, with the results summarized by these matrices:

$$\begin{bmatrix} 21 & 4 \\ 6 & 19 \end{bmatrix}; \quad \begin{bmatrix} 19 & 6 \\ 10 & 15 \end{bmatrix}; \quad \begin{bmatrix} 23 & 2 \\ 3 & 22 \end{bmatrix}.$$

The total results of the test can be obtained by adding these four matrices:

$$\begin{bmatrix} 22 & 3 \\ 8 & 17 \end{bmatrix} + \begin{bmatrix} 21 & 4 \\ 6 & 19 \end{bmatrix} + \begin{bmatrix} 19 & 6 \\ 10 & 15 \end{bmatrix} + \begin{bmatrix} 23 & 2 \\ 3 & 22 \end{bmatrix} = \begin{bmatrix} 85 & 15 \\ 27 & 73 \end{bmatrix}.$$

The final matrix shows that 85 of 100 patients got relief with Painoff, but only 27 of 100 did so with the placebo. ✔6

✔ **Checkpoint 6**

Later, it was discovered that the data in the last group of 50 patients in Example 5 was invalid. Use a matrix to represent the total test results after those data were eliminated.

Subtraction

Subtraction of matrices can be defined in a manner similar to matrix addition.

Matrix Subtraction

The **difference** of two $m \times n$ matrices X and Y is the $m \times n$ matrix $X - Y$ in which each element is the difference of the corresponding elements of X and Y.

EXAMPLE 6 Find the following.

(a) $\begin{bmatrix} 1 & 2 & 3 \\ 0 & -1 & 5 \end{bmatrix} - \begin{bmatrix} -2 & 3 & 0 \\ 1 & -7 & 2 \end{bmatrix}$

Solution

$$\begin{bmatrix} 1 & 2 & 3 \\ 0 & -1 & 5 \end{bmatrix} - \begin{bmatrix} -2 & 3 & 0 \\ 1 & -7 & 2 \end{bmatrix} = \begin{bmatrix} 1-(-2) & 2-3 & 3-0 \\ 0-1 & -1-(-7) & 5-2 \end{bmatrix}$$

$$= \begin{bmatrix} 3 & -1 & 3 \\ -1 & 6 & 3 \end{bmatrix}.$$

(b) $[8 \quad 6 \quad -4] - [3 \quad 5 \quad -8]$

Solution

$$[8 \quad 6 \quad -4] - [3 \quad 5 \quad -8] = [8-3 \quad 6-5 \quad -4-(-8)]$$

$$= [5 \quad 1 \quad 4].$$

(c) $\begin{bmatrix} -2 & 5 \\ 0 & 1 \end{bmatrix} - \begin{bmatrix} 3 \\ 5 \end{bmatrix}$

Solution The matrices are of different sizes and therefore cannot be subtracted.

✓ **Checkpoint 7**

Find each of the following differences when possible.

(a) $\begin{bmatrix} 2 & 5 \\ -1 & 0 \end{bmatrix} - \begin{bmatrix} 6 & 4 \\ 3 & -2 \end{bmatrix}$

(b) $\begin{bmatrix} 1 & 5 & 6 \\ 2 & 4 & 8 \end{bmatrix} - \begin{bmatrix} 2 & 1 \\ 10 & 3 \end{bmatrix}$

(c) $[5 \quad -4 \quad 1] - [6 \quad 0 \quad -3]$

EXAMPLE 7 **Furniture Manufacturing** During September, the Chicago warehouse of the EZ Life Company shipped out the following numbers of each model, where the entries in the matrix have the same meaning as given earlier:

$$K = \begin{bmatrix} 5 & 10 & 8 \\ 11 & 14 & 15 \end{bmatrix}.$$

What was the Chicago warehouse's inventory at the end of September, taking into account only the number of items received and sent out that month?

Solution The number of each kind of item received during September is given by matrix C from Example 4; the number of each model sent out during September is given by matrix K. The inventory at the end of September is thus represented by the matrix $C - K$:

$$C - K = \begin{bmatrix} 22 & 25 & 38 \\ 31 & 34 & 35 \end{bmatrix} - \begin{bmatrix} 5 & 10 & 8 \\ 11 & 14 & 15 \end{bmatrix} = \begin{bmatrix} 17 & 15 & 30 \\ 20 & 20 & 20 \end{bmatrix}.$$

Scalar Multiplication

Suppose one of the EZ Life Company warehouses receives the following order, written in matrix form, where the entries have the same meaning as given earlier:

$$\begin{bmatrix} 5 & 4 & 1 \\ 3 & 2 & 3 \end{bmatrix}.$$

Later, the store that sent the order asks the warehouse to send six more of the same order. The six new orders can be written as one matrix by multiplying each element in the matrix by 6, giving the product

$$6\begin{bmatrix} 5 & 4 & 1 \\ 3 & 2 & 3 \end{bmatrix} = \begin{bmatrix} 30 & 24 & 6 \\ 18 & 12 & 18 \end{bmatrix}.$$

In work with matrices, a real number, like the 6 in the preceding multiplication, is called a **scalar**.

🖩 **TECHNOLOGY TIP**

Graphing calculators can do matrix subtraction, as illustrated in Figure 6.21.

```
[B]
    [-4   5   Ø   2]
    [ 6   3  -7   1]
[C]
    [ 4  12   8  -5]
    [ 2   8   5   Ø]
```

```
[B]-[C]
    [-8  -7  -8   7]
    [ 4  -5 -12   1]
```

Figure 6.21

[A]
$$\begin{bmatrix} 3 & -4 \\ 5 & 2 \end{bmatrix}$$
2[A]
$$\begin{bmatrix} 6 & -8 \\ 1\emptyset & 4 \end{bmatrix}$$

Figure 6.22

✓ Checkpoint 8

Find each product.

(a) $-3\begin{bmatrix} 4 & -2 \\ 1 & 5 \end{bmatrix}$

(b) $4\begin{bmatrix} 2 & 4 & 7 \\ 8 & 2 & 1 \\ 5 & 7 & 3 \end{bmatrix}$

🖩 TECHNOLOGY TIP

To compute the negative of matrix A on a calculator, use the "negative" key (−), as shown in Figure 6.23.

[A]
$$\begin{bmatrix} 2 & 3 \\ 1 & 1.5 \end{bmatrix}$$
- [A]
$$\begin{bmatrix} -2 & -3 \\ -1 & -1.5 \end{bmatrix}$$

Figure 6.23

✓ Checkpoint 9

Let A and B be the matrices in Example 9. Find

(a) $A + (-B)$;

(b) $A - B$.

(c) What can you conclude from parts (a) and (b)?

Scalar Multiplication

The **product** of a scalar k and a matrix X is the matrix kX in which each element is k times the corresponding element of X.

EXAMPLE 8

(a) $(-3)\begin{bmatrix} 2 & -5 \\ 1 & 7 \\ 4 & -6 \end{bmatrix} = \begin{bmatrix} -6 & 15 \\ -3 & -21 \\ -12 & 18 \end{bmatrix}$.

(b) Graphing calculators can also do scalar multiplication (Figure 6.22). ✓8

 Recall that the *negative* of a real number a is the number $-a = (-1)a$. The negative of a matrix is defined similarly.

The **negative** (or *additive inverse*) of a matrix A is the matrix $(-1)A$ which is obtained by multiplying each element of A by -1. It is denoted $-A$.

EXAMPLE 9   Find $-A$ and $-B$ when

$$A = \begin{bmatrix} 1 & 2 & 3 \\ 0 & -1 & 5 \end{bmatrix} \quad \text{and} \quad B = \begin{bmatrix} -2 & 3 & 0 \\ 1 & -7 & 2 \end{bmatrix}.$$

Solution By the preceding definition,

$$-A = \begin{bmatrix} -1 & -2 & -3 \\ 0 & 1 & -5 \end{bmatrix} \quad \text{and} \quad -B = \begin{bmatrix} 2 & -3 & 0 \\ -1 & 7 & -2 \end{bmatrix}. \quad ✓9$$

A matrix consisting only of zeros is called a **zero matrix** (or *additive identity*) and is denoted O. There is an $m \times n$ zero matrix for each pair of values of m and n—for instance,

$$\begin{bmatrix} 0 & 0 \\ 0 & 0 \end{bmatrix}; \qquad \begin{bmatrix} 0 & 0 & 0 & 0 \\ 0 & 0 & 0 & 0 \end{bmatrix}.$$

2 × 2 zero matrix   2 × 4 zero matrix

Matrices satisfy a number of properties that are similar to properties of real numbers, as summarized below and illustrated in Checkpoint 9 and Exercises 26–29.

If A, B, and C are any $m \times n$ matrices and O is the $m \times n$ zero matrix, then the following properties are satisfied.

Commutative Property of Addition of Matrices	$A + B = B + A$
Associative Property of Addition of Matrices	$A + (B + C) = (A + B) + C$
Identity Property of Addition of Matrices	$A + O = A = O + A$
Inverse Properties of Addition of Matrices	$A + (-A) = O = (-A) + A$
	$A + (-B) = A - B$

6.4 Exercises

Find the size of each of the given matrices. Identify any square, column, or row matrices. Give the negative (additive inverse) of each matrix. (See Examples 2 and 9.)

1. $\begin{bmatrix} 7 & -8 & 4 \\ 0 & 13 & 9 \end{bmatrix}$

2. $\begin{bmatrix} -7 & 23 \\ 5 & -6 \end{bmatrix}$

3. $\begin{bmatrix} -3 & 0 & 11 \\ 1 & \frac{1}{4} & -7 \\ 5 & -3 & 9 \end{bmatrix}$

4. $[6 \quad -4 \quad \frac{2}{3} \quad 12 \quad 2]$

5. $\begin{bmatrix} 7 \\ 11 \end{bmatrix}$
6. $[-5]$

7. If A is a 5×3 matrix and $A + B = A$, what do you know about B?

8. If C is a 3×3 matrix and D is a 3×4 matrix, then $C + D$ is _____.

Perform the indicated operations where possible. (See Examples 3–7.)

9. $\begin{bmatrix} 1 & 2 & 7 & -1 \\ 8 & 0 & 2 & -4 \end{bmatrix} + \begin{bmatrix} -8 & 12 & -5 & 5 \\ -2 & -3 & 0 & 0 \end{bmatrix}$

10. $\begin{bmatrix} 1 & 7 \\ 2 & -3 \\ 3 & 7 \end{bmatrix} + \begin{bmatrix} 2 & 8 \\ 6 & 8 \\ -1 & 9 \end{bmatrix}$

11. $\begin{bmatrix} -1 & -5 & 9 \\ 2 & 2 & 3 \end{bmatrix} + \begin{bmatrix} 4 & 4 & -7 \\ 1 & -1 & 2 \end{bmatrix}$

12. $\begin{bmatrix} 2 & 4 \\ -8 & 2 \end{bmatrix} + \begin{bmatrix} 9 & -5 \\ 8 & 5 \end{bmatrix}$

13. $\begin{bmatrix} 0 & -2 \\ 1 & 9 \\ -5 & -9 \end{bmatrix} - \begin{bmatrix} 8 & 6 \\ 9 & 17 \\ 3 & -1 \end{bmatrix}$

14. $\begin{bmatrix} -3 & -2 & 5 \\ 3 & 9 & 0 \end{bmatrix} - \begin{bmatrix} 1 & 5 & -2 \\ -3 & 6 & 8 \end{bmatrix}$

15. $\begin{bmatrix} 9 & 1 \\ 0 & -3 \\ 4 & 10 \end{bmatrix} - \begin{bmatrix} 1 & 9 & -4 \\ -1 & 1 & 0 \end{bmatrix}$

16. $\begin{bmatrix} 3 & -8 & 0 \\ 1 & 5 & -7 \end{bmatrix} - \begin{bmatrix} 2 & 6 \\ 9 & -4 \end{bmatrix}$

Let $A = \begin{bmatrix} -2 & 0 \\ 5 & 3 \end{bmatrix}$ and $B = \begin{bmatrix} 0 & 2 \\ 4 & -6 \end{bmatrix}$. Find each of the following. (See Examples 8 and 9.)

17. $2A$
18. $-3B$
19. $-4B$

20. $5A$
21. $-4A + 5B$
22. $3A - 10B$

Let $A = \begin{bmatrix} 1 & -2 \\ 4 & 3 \end{bmatrix}$ and $B = \begin{bmatrix} 2 & -1 \\ 0 & 5 \end{bmatrix}$. Find a matrix X satisfying the given equation.

23. $2X = 2A + 3B$
24. $3X = A - 3B$

Using matrices

$$O = \begin{bmatrix} 0 & 0 \\ 0 & 0 \end{bmatrix}, P = \begin{bmatrix} m & n \\ p & q \end{bmatrix}, T = \begin{bmatrix} r & s \\ t & u \end{bmatrix}, \text{and } X = \begin{bmatrix} x & y \\ z & w \end{bmatrix},$$

verify that the statements in Exercises 25–29 are true.

25. $X + T$ is a 2×2 matrix.

26. $X + T = T + X$ (commutative property of addition of matrices).

27. $X + (T + P) = (X + T) + P$ (associative property of addition of matrices).

28. $X + (-X) = O$ (inverse property of addition of matrices).

29. $P + O = P$ (identity property of addition of matrices).

30. Which of the preceding properties are valid for matrices that are not square?

Work the following exercises. (See Example 1.)

31. **Concessions and Merchandise** When ticket holders fail to attend, major league sports teams lose the money these fans would have spent on refreshments and merchandise. The percentage of fans who don't show up is 16% in basketball and hockey, 20% in football, and 18% in baseball. The lost revenue per fan is $18.20 in basketball, $18.25 in hockey, $19 in football, and $15.40 in baseball. The total annual lost revenue is $22.7 million in basketball, $35.8 million in hockey, $51.9 million in football, and $96.3 million in baseball. Express this information in matrix form; specify what the rows and columns represent. (Data from: American Demographics.)

32. **Earnings for College Graduates** In 2015, median weekly earnings were $493 for workers without a high school diploma, $678 for high school graduates, and $1137 for those who had earned a bachelor's degree. The unemployment rate was 2.8% for college graduates, 5.4% for high school graduates, and 8.0% for those with less than a high school diploma. Write this information as a 3×2 matrix, labeling rows and columns. (Data from: Bureau of Labor Statistics.)

33. **Organ Transplants** The shortage of organs for transplants is a continuing problem in the United States. In 2016, there were 4154 candidates waiting for a heart transplant, 1470 for a lung transplant, 14,794 for a liver transplant, and 100,467 for a kidney transplant. Corresponding figures for 2010 were 2874, 1759, 15,394, and 83,919, respectively. Express this information as a matrix, labeling rows and columns. (Data from: Organ Procurement and Transplant Network.)

34. **Household Debt** In 2015, only 38% of American households had credit card debt, but for households who *carried credit card*

debt, the average amount of debt per household was $15,762. Averages for households with other types of debt included $27,141 in auto loans, $48,172 in student loan, and $168,614 in mortgage debt. The corresponding total figures for all consumers in the United States are $733 billion, $1060 billion, $1230 billion, and $8250 billion, respectively. Write this information in a 4 × 2 matrix, labeling rows and columns. (Data from: Federal Reserve Bank of New York.)

Work these exercises.

35. **Convenience Stores** There are three convenience stores in Gambier. This week, Store I sold 88 loaves of bread, 48 quarts of milk, 16 jars of peanut butter, and 112 pounds of cold cuts. Store II sold 105 loaves of bread, 72 quarts of milk, 21 jars of peanut butter, and 147 pounds of cold cuts. Store III sold 60 loaves of bread, 40 quarts of milk, no peanut butter, and 50 pounds of cold cuts.

    (a) Use a 3 × 4 matrix to express the sales information for the three stores.

    (b) During the following week, sales on these products at Store I increased by 25%, sales at Store II increased by one-third, and sales at Store III increased by 10%. Write the sales matrix for that week.

    (c) Write a matrix that represents total sales over the two-week period.

36. **Female and Male Graduates** The following table gives educational attainment as a percent of the U.S. population 25 years and older in various years. (Data from: U.S. Census Bureau.)

Year	MALE		FEMALE	
	Four Years of High School or More	Four Years of College or More	Four Years of High School or More	Four Years of College or More
1940	22.7%	5.5%	26.3%	3.8%
1970	55.0%	14.1%	55.4%	8.2%
2000	84.2%	27.8%	84.0%	23.6%
2014	87.7%	31.9%	88.9%	32.0%

(a) Write a 2 × 4 matrix for the educational attainment of males.

(b) Write a 2 × 4 matrix for the educational attainment of females.

(c) Use the matrices from parts (a) and (b) to write a matrix showing how much more (or less) males have reached educational attainment than females.

✓**Checkpoint Answers**

1. $\begin{bmatrix} 10 & 15 \\ 12 & 20 \\ 5 & 8 \end{bmatrix}$

2. (a) 2 × 4   (b) 3 × 3

3. (a) $[2 \quad 5 \quad -8 \quad 4]$

   (b) $\begin{bmatrix} 2 \\ 5 \\ -8 \\ 4 \end{bmatrix}$ (c) $\begin{bmatrix} 2 & 5 \\ -8 & 4 \end{bmatrix}$ or $\begin{bmatrix} 2 & -8 \\ 5 & 4 \end{bmatrix}$
   (Other answers are possible.)

4. (a) $\begin{bmatrix} 1 & 7 & 7 \\ 13 & -5 & 9 \end{bmatrix}$ (b) Not possible

   (c) $[0 \quad 6 \quad 2]$

5. (a) 139   (b) 92   (c) 100

6. $\begin{bmatrix} 62 & 13 \\ 24 & 51 \end{bmatrix}$

7. (a) $\begin{bmatrix} -4 & 1 \\ -4 & 2 \end{bmatrix}$ (b) Not possible

   (c) $[-1 \quad -4 \quad 4]$

8. (a) $\begin{bmatrix} -12 & 6 \\ -3 & -15 \end{bmatrix}$ (b) $\begin{bmatrix} 8 & 16 & 28 \\ 32 & 8 & 4 \\ 20 & 28 & 12 \end{bmatrix}$

9. (a) $\begin{bmatrix} 3 & -1 & 3 \\ -1 & 6 & 3 \end{bmatrix}$ (b) $\begin{bmatrix} 3 & -1 & 3 \\ -1 & 6 & 3 \end{bmatrix}$

   (c) $A + (-B) = A - B$.

6.5 Matrix Products and Inverses

To understand the reasoning behind the definition of matrix multiplication, look again at the EZ Life Company. Suppose sofas and chairs of the same model are often sold as sets, with matrix W showing the number of each model set in each warehouse:

$$\begin{array}{c} \\ \text{New York} \\ \text{Chicago} \\ \text{San Francisco} \end{array} \begin{array}{c} \begin{array}{ccc} A & B & C \end{array} \\ \begin{bmatrix} 10 & 7 & 3 \\ 5 & 9 & 6 \\ 4 & 8 & 2 \end{bmatrix} \end{array} = W.$$

If the selling price of a model A set is \$800, of a model B set is \$1000, and of a model C set is \$1200, find the total value of the sets in the New York warehouse as follows:

Type	Number of Sets		Price of Set		Total
A	10	$\times$	\$ 800	=	\$ 8,000
B	7	$\times$	1000	=	7,000
C	3	$\times$	1200	=	3,600
			Total for New York		\$18,600

✓ **Checkpoint 1**

In this example of the EZ Life Company, find the total value of the New York sets if model A sets sell for \$1200, model B for \$1600, and model C for \$1300.

The total value of the three kinds of sets in New York is \$18,600. ✓ 1
The work done in the preceding table is summarized as

$$10(\$800) + 7(\$1000) + 3(\$1200) = \$18,600.$$

In the same way, the Chicago sets have a total value of

$$5(\$800) + 9(\$1000) + 6(\$1200) = \$20,200,$$

and in San Francisco, the total value of the sets is

$$4(\$800) + 8(\$1000) + 2(\$1200) = \$13,600.$$

The selling prices can be written as a column matrix P and the total value in each location as a column matrix V:

$$P = \begin{bmatrix} 800 \\ 1000 \\ 1200 \end{bmatrix} \quad \text{and} \quad V = \begin{bmatrix} 18,600 \\ 20,200 \\ 13,600 \end{bmatrix}.$$

Consider how the first row of the matrix W and the single column P lead to the first entry of V:

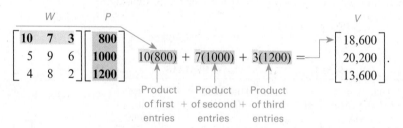

Similarly, adding the products of corresponding entries in the second row of W and the column P produces the second entry in V. The third entry in V is obtained in the same way by using the third row of W and column P. This suggests that it is reasonable to *define* the product WP to be V:

$$WP = \begin{bmatrix} 10 & 7 & 3 \\ 5 & 9 & 6 \\ 4 & 8 & 2 \end{bmatrix} \begin{bmatrix} 800 \\ 1000 \\ 1200 \end{bmatrix} = \begin{bmatrix} 18,600 \\ 20,200 \\ 13,600 \end{bmatrix} = V$$

Note the sizes of the matrices here: The product of a 3×3 matrix and a 3×1 matrix is a 3×1 matrix.

Multiplying Matrices

In order to define matrix multiplication in the general case, we first define the **product of a row of a matrix and a column of a matrix** (with the same number of entries in each) to be the *number* obtained by multiplying the corresponding entries (first by first, second by second, etc.) and adding the results. For instance,

$$[3 \quad -2 \quad 1] \cdot \begin{bmatrix} 4 \\ 5 \\ 0 \end{bmatrix} = 3 \cdot 4 + (-2) \cdot 5 + 1 \cdot 0 = 12 - 10 + 0 = 2.$$

Now **matrix multiplication** is defined as follows.

> ## Matrix Multiplication
>
> Let A be an $m \times n$ matrix and let B be an $n \times k$ matrix. The **product matrix** AB is the $m \times k$ matrix whose entry in the ith row and jth column is
>
> the product of the ith row of A and the jth column of B.

⬡ **CAUTION** Be careful when multiplying matrices. Remember that the number of *columns* of A must equal the number of *rows* of B in order to get the product matrix AB. The final product will have as many rows as A and as many columns as B.

EXAMPLE 1 Suppose matrix A is 2×2 and matrix B is 2×4. Can the product AB be calculated? If so, what is the size of the product?

Solution The following diagram helps decide the answers to these questions:

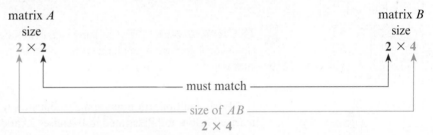

The product AB can be calculated because A has two columns and B has two rows. The product will be a 2×4 matrix. ✔2

EXAMPLE 2 Find the product CD when

$$C = \begin{bmatrix} -3 & 4 & 1 \\ 5 & 0 & 4 \end{bmatrix} \quad \text{and} \quad D = \begin{bmatrix} -6 & 4 \\ 2 & 3 \\ 3 & -2 \end{bmatrix}.$$

Solution Here, matrix C is 2×3 and matrix D is 3×2, so matrix CD can be found and will be 2×2.

Step 1 row 1, column 1

$$\begin{bmatrix} -3 & 4 & 1 \\ 5 & 0 & 4 \end{bmatrix} \begin{bmatrix} -6 & 4 \\ 2 & 3 \\ 3 & -2 \end{bmatrix} \quad (-3) \cdot (-6) + 4 \cdot 2 + 1 \cdot 3 = 29.$$

Hence, 29 is the entry in row 1, column 1, of CD, as shown in Step 5 on the following page.

Step 2 row 1, column 2

$$\begin{bmatrix} -3 & 4 & 1 \\ 5 & 0 & 4 \end{bmatrix} \begin{bmatrix} -6 & 4 \\ 2 & 3 \\ 3 & -2 \end{bmatrix} \quad (-3) \cdot 4 + 4 \cdot 3 + 1 \cdot (-2) = -2.$$

So -2 is the entry in row 1, column 2, of CD, as shown in Step 5. Continue in this manner to find the remaining entries of CD.

Step 3 row 2, column 1

$$\begin{bmatrix} -3 & 4 & 1 \\ 5 & 0 & 4 \end{bmatrix} \begin{bmatrix} -6 & 4 \\ 2 & 3 \\ 3 & -2 \end{bmatrix} \quad 5 \cdot (-6) + 0 \cdot 2 + 4 \cdot 3 = -18.$$

✔ **Checkpoint 2**

Matrix A is 4×6 and matrix B is 2×4.

(a) Can AB be found? If so, give its size.

(b) Can BA be found? If so, give its size.

✓**Checkpoint 3**

Find the product CD, given that

$$C = \begin{bmatrix} 1 & 3 & 5 \\ 2 & -4 & -1 \end{bmatrix}$$

and

$$D = \begin{bmatrix} 2 & -1 \\ 4 & 3 \\ 1 & -2 \end{bmatrix}.$$

✓**Checkpoint 4**

Give the size of each of the following products, if the product can be found.

(a) $\begin{bmatrix} 2 & 4 \\ 6 & 8 \end{bmatrix}\begin{bmatrix} 1 & 2 & 3 \\ 0 & -1 & 2 \end{bmatrix}$

(b) $\begin{bmatrix} 1 & 2 \\ 5 & 10 \\ 12 & 7 \end{bmatrix}\begin{bmatrix} 2 & 4 \\ 3 & 6 \\ 9 & 1 \end{bmatrix}$

(c) $\begin{bmatrix} 5 \\ 2 \\ 4 \end{bmatrix}\begin{bmatrix} 1 & 0 & 6 \end{bmatrix}$

Step 4 row 2, column 2

$$\begin{bmatrix} -3 & 4 & 1 \\ 5 & 0 & 4 \end{bmatrix}\begin{bmatrix} -6 & 4 \\ 2 & 3 \\ 3 & -2 \end{bmatrix} \qquad 5 \cdot 4 + 0 \cdot 3 + 4 \cdot (-2) = 12.$$

Step 5 The product is

$$CD = \begin{bmatrix} -3 & 4 & 1 \\ 5 & 0 & 4 \end{bmatrix}\begin{bmatrix} -6 & 4 \\ 2 & 3 \\ 3 & -2 \end{bmatrix} = \begin{bmatrix} 29 & -2 \\ -18 & 12 \end{bmatrix}.\ ✓_3$$

EXAMPLE 3 Find BA, given that

$$A = \begin{bmatrix} 1 & 7 \\ -3 & 2 \end{bmatrix} \quad \text{and} \quad B = \begin{bmatrix} 1 & 0 & -1 \\ 3 & 1 & 4 \end{bmatrix}.$$

Since B is a 2×3 matrix and A is a 2×2 matrix, the product BA is not defined. ✓_4

TECHNOLOGY TIP Graphing calculators can find matrix products (see Figure 6.24). However, if you use a graphing calculator to try to find the product in Example 3, the calculator will display an error message.

Matrix multiplication has some similarities to the multiplication of numbers, as summarized below and illustrated in Exercises 23 and 24.

> If A, B, and C are any matrices such that all the indicated sums and products exist, then the following properties are satisfied.
>
> Associative Property of Multiplication of Matrices $\qquad A(BC) = (AB)C$
> Distributive Properties $\qquad\qquad\qquad\qquad\qquad A(B + C) = AB + AC$
> $\qquad\qquad\qquad\qquad\qquad\qquad\qquad\qquad\qquad (B + C)A = BA + CA$

However, there are important differences between matrix multiplication and the multiplication of numbers (see Exercises 19–22). In particular, matrix multiplication is *not* commutative.

> If A and B are matrices such that the products AB and BA exist,
>
> $$AB \text{ may not equal } BA.$$

Figure 6.24 shows an example of this situation.

```
[A]                        [A][B]
        ⌈1  -4⌉                    ⌈-22  -8 ⌉
        ⌊3   2⌋                    ⌊ 4   18 ⌋
[B]                        [B][A]
        ⌈-2  4⌉                    ⌈10  16 ⌉
        ⌊ 5  3⌋                    ⌊14  -14⌋
```

Figure 6.24

EXAMPLE 4 **Construction Costs** A contractor builds three kinds of houses, models A, B, and C, with a choice of two styles, Spanish or contemporary. Matrix P shows the number of each kind of house planned for a new 100-home subdivision:

$$\begin{array}{c} \\ \text{Model } A \\ \text{Model } B \\ \text{Model } C \end{array} \begin{array}{c} \text{Spanish} \quad \text{Contemporary} \\ \left[\begin{array}{cc} 0 & 30 \\ 10 & 20 \\ 20 & 20 \end{array}\right] = P. \end{array}$$

The amounts for each of the exterior materials used depend primarily on the style of the house. These amounts are shown in matrix Q (concrete is measured in cubic yards, lumber in units of 1000 board feet, brick in thousands, and shingles in units of 100 square feet):

$$\begin{array}{c} \\ \text{Spanish} \\ \text{Contemporary} \end{array} \begin{array}{c} \text{Concrete} \quad \text{Lumber} \quad \text{Brick} \quad \text{Shingles} \\ \left[\begin{array}{cccc} 10 & 2 & 0 & 2 \\ 50 & 1 & 20 & 2 \end{array}\right] = Q. \end{array}$$

Matrix R gives the cost for each kind of material:

$$\begin{array}{c} \\ \text{Concrete} \\ \text{Lumber} \\ \text{Brick} \\ \text{Shingles} \end{array} \begin{array}{c} \text{Cost per Unit} \\ \left[\begin{array}{c} 20 \\ 180 \\ 60 \\ 25 \end{array}\right] = R. \end{array}$$

(a) What is the total cost for each model of house?

Solution First find the product PQ, which shows the amount of each material needed for each model of house:

$$PQ = \left[\begin{array}{cc} 0 & 30 \\ 10 & 20 \\ 20 & 20 \end{array}\right]\left[\begin{array}{cccc} 10 & 2 & 0 & 2 \\ 50 & 1 & 20 & 2 \end{array}\right]$$

$$PQ = \begin{array}{c} \text{Concrete} \quad \text{Lumber} \quad \text{Brick} \quad \text{Shingles} \\ \left[\begin{array}{cccc} 1500 & 30 & 600 & 60 \\ 1100 & 40 & 400 & 60 \\ 1200 & 60 & 400 & 80 \end{array}\right] \begin{array}{c} \text{Model } A \\ \text{Model } B. \\ \text{Model } C \end{array} \end{array}$$

Now multiply PQ and R, the cost matrix, to get the total cost for each model of house:

$$\left[\begin{array}{cccc} 1500 & 30 & 600 & 60 \\ 1100 & 40 & 400 & 60 \\ 1200 & 60 & 400 & 80 \end{array}\right]\left[\begin{array}{c} 20 \\ 180 \\ 60 \\ 25 \end{array}\right] = \begin{array}{c} \text{Cost} \\ \left[\begin{array}{c} 72{,}900 \\ 54{,}700 \\ 60{,}800 \end{array}\right] \begin{array}{c} \text{Model } A \\ \text{Model } B. \\ \text{Model } C \end{array} \end{array}$$

So materials costs are \$72,900 for a Model A home, \$54,700 for a Model B home, and \$60,800 for a Model C home.

(b) How much of each of the four kinds of material must be ordered?

Solution The totals of the columns of matrix PQ will give a matrix whose elements represent the total amounts of each material needed for the subdivision. Call this matrix T, and write it as a row matrix:

$$T = [3800 \quad 130 \quad 1400 \quad 200].$$

So the amount of material needed is 3800 units of concrete, 130 units of lumber, 1400 units of brick, and 200 units of shingles.

(c) What is the total cost for material?

Solution Find the total cost of all the materials by taking the product of matrix T, the matrix showing the total amounts of each material, and matrix R, the cost matrix. [To multiply these and get a 1×1 matrix representing total cost, we must multiply a 1×4 matrix by a 4×1 matrix. This is why T was written as a row matrix in (b).] So, we have

$$TR = [3800 \quad 130 \quad 1400 \quad 200]\begin{bmatrix} 20 \\ 180 \\ 60 \\ 25 \end{bmatrix} = [188,400].$$

So the total cost of all materials is $188,400.

(d) Suppose the contractor builds the same number of homes in five subdivisions. What is the total amount of each material needed for each model in this case?

Solution Determine the total amount of each material for each model for all five subdivisions. Multiply PQ by the scalar 5 as follows:

$$5\begin{bmatrix} 1500 & 30 & 600 & 60 \\ 1100 & 40 & 400 & 60 \\ 1200 & 60 & 400 & 80 \end{bmatrix} = \begin{bmatrix} 7500 & 150 & 3000 & 300 \\ 5500 & 200 & 2000 & 300 \\ 6000 & 300 & 2000 & 400 \end{bmatrix}.$$

So the amount of concrete needed for Model A homes in all five subdivisions is $7500 (see row 1, column 1), and so on.

We can introduce notation to help keep track of the quantities a matrix represents. For example, we can say that matrix P from Example 4 represents models/styles, matrix Q represents styles/materials, and matrix R represents materials/cost. In each case, the meaning of the rows is written first and the columns second. When we found the product PQ in Example 4, the rows of the matrix represented models and the columns represented materials. Therefore, we can say that the matrix product PQ represents models/materials. The common quantity, styles, in both P and Q was eliminated in the product PQ. Do you see that the product $(PQ)R$ represents models/cost?

In practical problems, this notation helps decide in what order to multiply two matrices so that the results are meaningful. In Example 4(c), we could have found either product RT or product TR. However, since T represents subdivisions/materials and R represents materials/cost, the product TR gives subdivisions/cost. ✔5

⬡ Identity Matrices

Recall from Section 1.1 that the real number 1 is the identity element for multiplication of real numbers: For any real number a, $a \cdot 1 = 1 \cdot a = a$. In this section, an **identity matrix I** is defined that has properties similar to those of the number 1.

If I is to be the identity matrix for multiplication of matrices, the products AI and IA must both equal A. The 2×2 identity matrix that satisfies these conditions is

$$I = \begin{bmatrix} 1 & 0 \\ 0 & 1 \end{bmatrix}. \quad ✔6$$

To check that I is really the 2×2 identity matrix, let

$$A = \begin{bmatrix} a & b \\ c & d \end{bmatrix}.$$

✓ **Checkpoint 5**

Let matrix A be

$$\begin{array}{c} & \text{Vitamin} \\ & \begin{array}{ccc} \text{C} & \text{E} & \text{K} \end{array} \\ \text{Brand} \begin{array}{c} X \\ Y \end{array} & \begin{bmatrix} 2 & 7 & 5 \\ 4 & 6 & 9 \end{bmatrix} \end{array}$$

and matrix B be

$$\begin{array}{c} & \text{Cost} \\ & \begin{array}{cc} X & Y \end{array} \\ \text{Vitamin} \begin{array}{c} C \\ E \\ K \end{array} & \begin{bmatrix} 12 & 14 \\ 18 & 15 \\ 9 & 10 \end{bmatrix}. \end{array}$$

(a) What quantities do matrices A and B represent?

(b) What quantities does the product AB represent?

(c) What quantities does the product BA represent?

✓ **Checkpoint 6**

Let $A = \begin{bmatrix} 3 & -2 \\ 4 & -1 \end{bmatrix}$ and

$I = \begin{bmatrix} 1 & 0 \\ 0 & 1 \end{bmatrix}$.

Find IA and AI.

Then AI and IA should both equal A:

$$AI = \begin{bmatrix} a & b \\ c & d \end{bmatrix}\begin{bmatrix} 1 & 0 \\ 0 & 1 \end{bmatrix} = \begin{bmatrix} a(1) + b(0) & a(0) + b(1) \\ c(1) + d(0) & c(0) + d(1) \end{bmatrix} = \begin{bmatrix} a & b \\ c & d \end{bmatrix} = A;$$

$$IA = \begin{bmatrix} 1 & 0 \\ 0 & 1 \end{bmatrix}\begin{bmatrix} a & b \\ c & d \end{bmatrix} = \begin{bmatrix} 1(a) + 0(c) & 1(b) + 0(d) \\ 0(a) + 1(c) & 0(b) + 1(d) \end{bmatrix} = \begin{bmatrix} a & b \\ c & d \end{bmatrix} = A.$$

This verifies that I has been defined correctly. (It can also be shown that I is the only 2×2 identity matrix.)

The identity matrices for 3×3 matrices and 4×4 matrices are, respectively,

$$I = \begin{bmatrix} 1 & 0 & 0 \\ 0 & 1 & 0 \\ 0 & 0 & 1 \end{bmatrix} \quad \text{and} \quad I = \begin{bmatrix} 1 & 0 & 0 & 0 \\ 0 & 1 & 0 & 0 \\ 0 & 0 & 1 & 0 \\ 0 & 0 & 0 & 1 \end{bmatrix}.$$

TECHNOLOGY TIP

An $n \times n$ identity matrix can be displayed on most graphing calculators by using IDENTITY n or IDENT n or IDENMAT(n). Look in the MATH or OPS submenu of the MATRIX menu.

By generalizing these findings, an identity matrix can be found for any n by n matrix. This identity matrix will have 1s on the main diagonal from upper left to lower right, with all other entries equal to 0.

Inverse Matrices

Recall that for every nonzero real number a, the equation $ax = 1$ has a solution, namely, $x = 1/a = a^{-1}$. Similarly, for a square matrix A, we consider the matrix equation $AX = I$. This equation does not always have a solution, but when it does, we use special terminology. If there is a matrix A^{-1} satisfying

$$AA^{-1} = I,$$

(that is, A^{-1} is a solution of $AX = I$), then A^{-1} is called the **inverse matrix** of A. In this case, it can be proved that $A^{-1}A = I$ and that A^{-1} is unique (that is, a square matrix has no more than one inverse). When a matrix has an inverse, it can be found by using the row operations given in Section 6.2, as we shall see later.

CAUTION Only square matrices have inverses, but not every square matrix has one. A matrix that does not have an inverse is called a **singular matrix**. Note that the symbol A^{-1} (read "A-inverse") does *not* mean $1/A$; the symbol A^{-1} is just the notation for the inverse of matrix A. There is no such thing as matrix division.

EXAMPLE 5 Given matrices A and B as follows, determine whether B is the inverse of A:

$$A = \begin{bmatrix} 1 & 2 \\ 4 & 6 \end{bmatrix}; \qquad B = \begin{bmatrix} -3 & 1 \\ 2 & -\frac{1}{2} \end{bmatrix}.$$

Solution B is the inverse of A if $AB = I$, so we find that product:

$$AB = \begin{bmatrix} 1 & 2 \\ 4 & 6 \end{bmatrix}\begin{bmatrix} -3 & 1 \\ 2 & -\frac{1}{2} \end{bmatrix} = \begin{bmatrix} 1 & 0 \\ 0 & 1 \end{bmatrix} = I.$$

Therefore, B is the inverse of A; that is, $A^{-1} = B$. (It is also true that A is the inverse of B, or $B^{-1} = A$.) ✔ 7

✔ **Checkpoint 7**

Given $A = \begin{bmatrix} 2 & 1 \\ 3 & 8 \end{bmatrix}$ and $B = \begin{bmatrix} -1 & 3 \\ 1 & -2 \end{bmatrix}$, determine whether they are inverses.

EXAMPLE 6 Find the multiplicative inverse of

$$A = \begin{bmatrix} 2 & 4 \\ 1 & -1 \end{bmatrix}.$$

Solution Let the unknown inverse matrix be

$$A^{-1} = \begin{bmatrix} x & y \\ z & w \end{bmatrix}.$$

By the definition of matrix inverse, $AA^{-1} = I$, or

$$AA^{-1} = \begin{bmatrix} 2 & 4 \\ 1 & -1 \end{bmatrix} \begin{bmatrix} x & y \\ z & w \end{bmatrix} = \begin{bmatrix} 1 & 0 \\ 0 & 1 \end{bmatrix}.$$

Use matrix multiplication to get

$$\begin{bmatrix} 2x + 4z & 2y + 4w \\ x - z & y - w \end{bmatrix} = \begin{bmatrix} 1 & 0 \\ 0 & 1 \end{bmatrix}.$$

Setting corresponding elements equal to each other gives the system of equations

$$2x + 4z = 1 \tag{1}$$
$$2y + 4w = 0 \tag{2}$$
$$x - z = 0 \tag{3}$$
$$y - w = 1. \tag{4}$$

Since equations (1) and (3) involve only x and z, while equations (2) and (4) involve only y and w, these four equations lead to two systems of equations:

$$\begin{array}{ccc} 2x + 4z = 1 & & 2y + 4w = 0 \\ x - z = 0 & \text{and} & y - w = 1. \end{array}$$

Writing the two systems as augmented matrices gives

$$\begin{bmatrix} 2 & 4 & | & 1 \\ 1 & -1 & | & 0 \end{bmatrix} \quad \text{and} \quad \begin{bmatrix} 2 & 4 & | & 0 \\ 1 & -1 & | & 1 \end{bmatrix}.$$

Note that the row operations needed to transform both matrices into reduced row echelon form are the same because the first two columns of both matrices are identical. Consequently, we can save time by combining these matrices into the single matrix

$$\begin{bmatrix} 2 & 4 & | & 1 & 0 \\ 1 & -1 & | & 0 & 1 \end{bmatrix}. \tag{5}$$

Columns 1–3 represent the first system and columns 1, 2, and 4 represent the second system. Now use row operations as follows:

$$\begin{bmatrix} 1 & -1 & | & 0 & 1 \\ 2 & 4 & | & 1 & 0 \end{bmatrix} \qquad \text{Interchange } R_1 \text{ and } R_2.$$

$$\begin{bmatrix} 1 & -1 & | & 0 & 1 \\ 0 & 6 & | & 1 & -2 \end{bmatrix} \qquad -2R_1 + R_2$$

$$\begin{bmatrix} 1 & -1 & | & 0 & 1 \\ 0 & 1 & | & \frac{1}{6} & -\frac{1}{3} \end{bmatrix} \qquad \frac{1}{6}R_2$$

$$\begin{bmatrix} 1 & 0 & | & \frac{1}{6} & \frac{2}{3} \\ 0 & 1 & | & \frac{1}{6} & -\frac{1}{3} \end{bmatrix}. \qquad R_2 + R_1 \tag{6}$$

The left half of the augmented matrix in reduced row echelon form in equation (6) is the identity matrix, so the Gauss–Jordan process is finished and the solutions can be read from the right half of the augmented matrix. The numbers in the first column to the right of the vertical bar give the values of x and z. The second column to the right of the bar gives the values of y and w. That is,

$$\begin{bmatrix} 1 & 0 & | & x & y \\ 0 & 1 & | & z & w \end{bmatrix} = \begin{bmatrix} 1 & 0 & | & \frac{1}{6} & \frac{2}{3} \\ 0 & 1 & | & \frac{1}{6} & -\frac{1}{3} \end{bmatrix},$$

so that

$$A^{-1} = \begin{bmatrix} x & y \\ z & w \end{bmatrix} = \begin{bmatrix} \frac{1}{6} & \frac{2}{3} \\ \frac{1}{6} & -\frac{1}{3} \end{bmatrix}.$$

Check by multiplying A and A^{-1}. The result should be I:

$$AA^{-1} = \begin{bmatrix} 2 & 4 \\ 1 & -1 \end{bmatrix}\begin{bmatrix} \frac{1}{6} & \frac{2}{3} \\ \frac{1}{6} & -\frac{1}{3} \end{bmatrix} = \begin{bmatrix} \frac{1}{3} + \frac{2}{3} & \frac{4}{3} - \frac{4}{3} \\ \frac{1}{6} - \frac{1}{6} & \frac{2}{3} + \frac{1}{3} \end{bmatrix} = \begin{bmatrix} 1 & 0 \\ 0 & 1 \end{bmatrix} = I.$$

Thus, the original augmented matrix in equation (5) has A as its left half and the identity matrix as its right half, while the final augmented matrix in equation (6), at the end of the Gauss–Jordan process, has the identity matrix as its left half and the inverse matrix A^{-1} as its right half:

$$[A \mid I] \rightarrow [I \mid A^{-1}]. \checkmark_8$$

The procedure in Example 6 can be generalized as follows.

✓ **Checkpoint 8** ▦

Carry out the process in Example 6 on a graphing calculator as follows: Enter matrix (5) in the example as matrix [B] on the calculator. Then find RREF [B]. What is the result?

Inverse Matrix

To obtain an **inverse matrix** A^{-1} for any $n \times n$ matrix A for which A^{-1} exists, follow these steps:

1. Form the augmented matrix $[A \mid I]$, where I is the $n \times n$ identity matrix.
2. Perform row operations on $[A \mid I]$ to get a matrix of the form $[I \mid B]$.
3. Matrix B is A^{-1}.

EXAMPLE 7 Find A^{-1} if

$$A = \begin{bmatrix} 1 & 0 & 1 \\ 2 & -2 & -1 \\ 5 & 0 & 0 \end{bmatrix}.$$

Solution First write the augmented matrix $[A \mid I]$:

$$[A \mid I] = \begin{bmatrix} 1 & 0 & 1 & 1 & 0 & 0 \\ 2 & -2 & -1 & 0 & 1 & 0 \\ 5 & 0 & 0 & 0 & 0 & 1 \end{bmatrix}.$$

Transform the left side of this matrix into the 3×3 identity matrix:

$$\begin{bmatrix} 1 & 0 & 1 & 1 & 0 & 0 \\ 0 & -2 & -3 & -2 & 1 & 0 \\ 0 & 0 & -5 & -5 & 0 & 1 \end{bmatrix} \quad \begin{array}{l} -2R_1 + R_2 \\ -5R_1 + R_3 \end{array}$$

$$\begin{bmatrix} 1 & 0 & 1 & 1 & 0 & 0 \\ 0 & 1 & \frac{3}{2} & 1 & -\frac{1}{2} & 0 \\ 0 & 0 & 1 & 1 & 0 & -\frac{1}{5} \end{bmatrix} \quad \begin{array}{l} -\frac{1}{2}R_2 \\ -\frac{1}{5}R_3 \end{array}$$

$$\begin{bmatrix} 1 & 0 & 0 & 0 & 0 & \frac{1}{5} \\ 0 & 1 & 0 & -\frac{1}{2} & -\frac{1}{2} & \frac{3}{10} \\ 0 & 0 & 1 & 1 & 0 & -\frac{1}{5} \end{bmatrix}. \quad \begin{array}{l} -1R_3 + R_1 \\ -\frac{3}{2}R_3 + R_2 \end{array}$$

Looking at the right half of the preceding matrix, we see that

$$A^{-1} = \begin{bmatrix} 0 & 0 & \frac{1}{5} \\ -\frac{1}{2} & -\frac{1}{2} & \frac{3}{10} \\ 1 & 0 & -\frac{1}{5} \end{bmatrix}.$$

Verify that AA^{-1} is I.

The best way to find matrix inverses on a graphing calculator is illustrated in the next example.

EXAMPLE 8 Use a graphing calculator to find the inverse of the following matrices (if they have inverses):

$$A = \begin{bmatrix} 1 & 0 & 1 \\ 2 & -2 & -1 \\ 5 & 0 & 0 \end{bmatrix} \quad \text{and} \quad B = \begin{bmatrix} 2 & 4 \\ 3 & 6 \end{bmatrix}.$$

Solution Enter matrix A into the calculator [Figure 6.25(a)]. Then use the x^{-1} key to find the inverse matrix, as in Figure 6.25(b). (Using $\wedge$ and -1 for the inverse results in an error message on most calculators.) Note that Figure 6.25(b) agrees with the answer we found working by hand in Example 7.

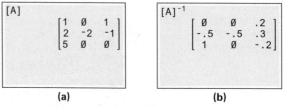

Figure 6.25

Now enter matrix B into the calculator and use the x^{-1} key. The result is an error message (Figure 6.26), which indicates that the matrix is singular; it does not have an inverse. (If you were working by hand, you would have found that the appropriate system of equations has no solution.) ✓9

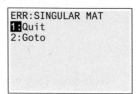

Figure 6.26

✓ Checkpoint 9 ▦

Use a graphing calculator to find the inverses of these matrices (if they exist).

(a) $A = \begin{bmatrix} 1 & 2 & 3 \\ 4 & -1 & 0 \\ 5 & 1 & 3 \end{bmatrix}$

(b) $B = \begin{bmatrix} 2 & 3 \\ -1 & 4 \end{bmatrix}$

▦ TECHNOLOGY TIP

Because of round-off error, a graphing calculator may sometimes display an "inverse" for a matrix that doesn't actually have one. So always verify your results by multiplying A and A^{-1}. If the product is not the identity matrix, then A does not have an inverse.

6.5 Exercises

In Exercises 1–6, the sizes of two matrices A and B are given. Find the sizes of the product AB and the product BA whenever these products exist. (See Example 1.)

1. A is 2×2 and B is 2×2.

2. A is 3×3 and B is 3×2.

3. A is 3×5 and B is 5×3.

4. A is 4×3 and B is 3×6.

5. A is 4×2 and B is 3×4.

6. A is 7×3 and B is 2×7.

7. To find the product matrix AB, the number of _____ of A must be the same as the number of _____ of B.

8. The product matrix AB has the same number of _____ as A and the same number of _____ as B.

Find each of the following matrix products, if they exist. (See Examples 2 and 3.)

9. $\begin{bmatrix} 1 & 2 \\ 3 & 4 \end{bmatrix}\begin{bmatrix} -1 \\ 3 \end{bmatrix}$

10. $\begin{bmatrix} -2 & 5 \\ 7 & 0 \end{bmatrix}\begin{bmatrix} 2 \\ 6 \end{bmatrix}$

11. $\begin{bmatrix} 2 & 2 & -1 \\ 5 & 0 & 1 \end{bmatrix}\begin{bmatrix} 0 & -2 \\ -1 & 5 \\ 0 & 2 \end{bmatrix}$

12. $\begin{bmatrix} -9 & 3 & 1 \\ 3 & 0 & 0 \end{bmatrix}\begin{bmatrix} 2 \\ -1 \\ 4 \end{bmatrix}$

13. $\begin{bmatrix} -4 & 1 \\ 2 & -3 \end{bmatrix}\begin{bmatrix} 1 & 0 \\ 0 & 1 \end{bmatrix}$

14. $\begin{bmatrix} 1 & 0 \\ 0 & 1 \end{bmatrix}\begin{bmatrix} 3 & -2 \\ 1 & -5 \end{bmatrix}$

15. $\begin{bmatrix} 1 & 0 & 0 \\ 0 & 1 & 0 \\ 0 & 0 & 1 \end{bmatrix}\begin{bmatrix} 3 & -5 & 7 \\ -2 & 1 & 6 \\ 0 & -3 & 4 \end{bmatrix}$

16. $\begin{bmatrix} -8 & 9 \\ 3 & -4 \\ -1 & 6 \end{bmatrix}\begin{bmatrix} 1 & 0 & 0 \\ 0 & 1 & 0 \end{bmatrix}$

17. $\begin{bmatrix} 1 & 2 & 3 \\ 4 & 0 & 6 \\ 7 & 8 & 9 \end{bmatrix} \begin{bmatrix} -1 & 4 \\ 7 & 0 \\ 1 & 2 \end{bmatrix}$

18. $\begin{bmatrix} -2 & 0 & 3 \\ 4 & -3 & -1 \end{bmatrix} \begin{bmatrix} 2 & 0 & -1 & 5 \\ 0 & 1 & 0 & -1 \\ 4 & 2 & 5 & -4 \end{bmatrix}$

In Exercises 19–21, use the matrices

$$A = \begin{bmatrix} -3 & -9 \\ 2 & 6 \end{bmatrix} \quad \text{and} \quad B = \begin{bmatrix} 4 & 6 \\ 2 & 3 \end{bmatrix}.$$

19. Show that $AB \neq BA$. Hence, matrix multiplication is not commutative.

20. Show that $(A + B)^2 \neq A^2 + 2AB + B^2$.

21. Show that $(A + B)(A - B) \neq A^2 - B^2$.

22. Show that $D^2 = D$, where

$$D = \begin{bmatrix} 1 & 0 & 0 \\ \frac{1}{2} & 0 & \frac{1}{2} \\ 0 & 0 & 1 \end{bmatrix}.$$

Given matrices

$$P = \begin{bmatrix} m & n \\ p & q \end{bmatrix}, \quad X = \begin{bmatrix} x & y \\ z & w \end{bmatrix}, \quad \text{and} \quad T = \begin{bmatrix} r & s \\ t & u \end{bmatrix},$$

verify that the statements in Exercises 23–26 are true.

23. $(PX)T = P(XT)$ (associative property of multiplication of matrices)

24. $P(X + T) = PX + PT$ (distributive property)

25. $k(X + T) = kX + kT$ for any real number k

26. $(k + h)P = kP + hP$ for any real numbers k and h

Determine whether the given matrices are inverses of each other by computing their product. (See Example 5.)

27. $\begin{bmatrix} 5 & 2 \\ 3 & -1 \end{bmatrix}$ and $\begin{bmatrix} -1 & 2 \\ 3 & -4 \end{bmatrix}$

28. $\begin{bmatrix} 3 & 2 \\ 7 & 5 \end{bmatrix}$ and $\begin{bmatrix} 5 & 0 \\ -7 & 3 \end{bmatrix}$

29. $\begin{bmatrix} 3 & -1 \\ -4 & 2 \end{bmatrix}$ and $\begin{bmatrix} 1 & \frac{1}{2} \\ 2 & \frac{3}{2} \end{bmatrix}$

30. $\begin{bmatrix} 3 & 5 \\ 7 & 9 \end{bmatrix}$ and $\begin{bmatrix} -\frac{9}{8} & \frac{5}{8} \\ \frac{7}{8} & -\frac{3}{8} \end{bmatrix}$

31. $\begin{bmatrix} 1 & 1 & 1 \\ 2 & 3 & 0 \\ 1 & 2 & 1 \end{bmatrix}$ and $\begin{bmatrix} 1.5 & .5 & -1.5 \\ -1 & 0 & 1 \\ .5 & -.5 & .5 \end{bmatrix}$

32. $\begin{bmatrix} 2 & 5 & 4 \\ 1 & 4 & 3 \\ 1 & 3 & 2 \end{bmatrix}$ and $\begin{bmatrix} 1 & 2 & 1 \\ -5 & 8 & 2 \\ 7 & -11 & -3 \end{bmatrix}$

Find the inverse, if it exists, for each of the given matrices. (See Examples 6 and 7.)

33. $\begin{bmatrix} 1 & 2 \\ 3 & 6 \end{bmatrix}$

34. $\begin{bmatrix} 2 & 1 \\ 1 & 1 \end{bmatrix}$

35. $\begin{bmatrix} -1 & 2 \\ 1 & -1 \end{bmatrix}$

36. $\begin{bmatrix} 1 & 2 \\ 3 & 4 \end{bmatrix}$

37. $\begin{bmatrix} 2 & -3 \\ -1 & 2 \end{bmatrix}$

38. $\begin{bmatrix} -3 & -5 \\ 6 & 10 \end{bmatrix}$

39. $\begin{bmatrix} 1 & -1 & 0 \\ -1 & 2 & 3 \\ 1 & 0 & 2 \end{bmatrix}$

40. $\begin{bmatrix} 0 & 1 & -1 \\ 2 & -2 & -1 \\ -1 & 1 & 1 \end{bmatrix}$

41. $\begin{bmatrix} 1 & -1 & 4 \\ 0 & 1 & 3 \\ 2 & -3 & 4 \end{bmatrix}$

42. $\begin{bmatrix} 1 & 4 & 3 \\ 1 & -3 & -2 \\ 2 & 5 & 4 \end{bmatrix}$

43. $\begin{bmatrix} 1 & 2 & 0 \\ 3 & -1 & 2 \\ -2 & 3 & -2 \end{bmatrix}$

44. $\begin{bmatrix} 1 & 0 & -5 \\ 4 & -7 & 3 \\ 3 & -7 & 8 \end{bmatrix}$

Use a graphing calculator to find the inverse of each matrix. (See Example 8.)

45. $\begin{bmatrix} 1 & 2 & 3 \\ 1 & 4 & 2 \\ 0 & 1 & -1 \end{bmatrix}$

46. $\begin{bmatrix} 2 & 2 & -4 \\ 2 & 6 & 0 \\ -3 & -3 & 5 \end{bmatrix}$

47. $\begin{bmatrix} 1 & 0 & -2 & 0 \\ -2 & 1 & 2 & 2 \\ 3 & -1 & -2 & -3 \\ 0 & 1 & 4 & 1 \end{bmatrix}$

48. $\begin{bmatrix} 1 & 1 & 0 & 2 \\ 2 & -1 & 1 & -1 \\ 3 & 3 & 2 & -2 \\ 1 & 2 & 1 & 0 \end{bmatrix}$

Work these exercises. (See Example 4.)

49. **Grocery Delivery** Several competing companies provide grocery home delivery services. You and your roommate agree to order together and are deciding between two local delivery companies, Timesaver and Wheels. Matrix A shows the number of frozen pizza and quarts of ice cream that you each want to order, and matrix B shows the prices for each company.

$$\begin{array}{c} \\ \text{You} \\ \text{Roommate} \end{array} \begin{array}{cc} \text{Pizza} & \text{Ice cream} \end{array} \\ \begin{bmatrix} 6 & 2 \\ 1 & 3 \end{bmatrix} = A$$

$$\begin{array}{c} \\ \text{Pizza} \\ \text{Ice cream} \end{array} \begin{array}{cc} \text{Timesaver} & \text{Wheels} \end{array} \\ \begin{bmatrix} \$8 & \$7 \\ \$4 & \$5 \end{bmatrix} = B$$

(a) Find the matrix product AB.

(b) What do the rows of AB represent?

(c) What do the columns of AB represent?

(d) What would be the total cost (for you and your roommate) if you order from Timesaver? From Wheels?

50. **Bulk Shipments** Boxed.com is a company that ships grocery items in bulk. As the purchasing manager for a start-up, you decide to compare Boxed.com prices with your current supplier for an order of Charmin toilet paper (30 count ultra soft) and Bounty paper towels (12 count double rolls). Matrix A shows the quantities that the production and sales departments need

to order, and matrix B shows the prices for each supplier. (Data from: www.boxed.com.)

$$\begin{array}{c} & \text{Charmin} \quad \text{Bounty} \\ \begin{array}{c} \text{Production} \\ \text{Sales} \end{array} \left[\begin{array}{cc} 15 & 25 \\ 60 & 10 \end{array} \right] = A \end{array}$$

$$\begin{array}{c} & \text{Boxed.com} \quad \text{Current} \\ \begin{array}{c} \text{Charmin} \\ \text{Bounty} \end{array} \left[\begin{array}{cc} \$23 & \$24 \\ \$20 & \$18 \end{array} \right] = B \end{array}$$

(a) Find the matrix product AB.

(b) What do the rows of AB represent?

(c) What do the columns of AB represent?

(d) What would the total cost be (for both departments) if you order from Boxed.com? From your current supplier?

51. Food Truck A restaurant has a main location and a traveling food truck. The first matrix A shows the number of managers and associates employed. The second matrix B shows the average annual cost of salary and benefits (in thousands of dollars).

$$\begin{array}{c} & \text{Managers} \quad \text{Associates} \\ \begin{array}{c} \text{Restaurant} \\ \text{Food Truck} \end{array} \left[\begin{array}{cc} 5 & 25 \\ 1 & 4 \end{array} \right] = A \end{array}$$

$$\begin{array}{c} & \text{Salary} \quad \text{Benefits} \\ \begin{array}{c} \text{Managers} \\ \text{Associates} \end{array} \left[\begin{array}{cc} 41 & 6 \\ 20 & 2 \end{array} \right] = B \end{array}$$

(a) Find the matrix product AB.

(b) Explain what AB represents.

(c) According to matrix AB, what is the total cost of salaries for all employees (managers and associates) at the restaurant? The food truck? What is the total cost of benefits for all employees at the restaurant? The food truck?

52. Teacher Salaries The average annual wage (in thousands of dollars) for teachers in the United States at various grade levels is shown in matrix W. (Data from: Bureau of Labor Statistics.) A small school district has both city schools and county schools; matrix N shows the number of teachers in each.

$$\begin{array}{c} & \text{Wage (thousands)} \\ \begin{array}{c} \text{Preschool and Kindergarten} \\ \text{Elementary and Middle School} \\ \text{Secondary School} \end{array} \left[\begin{array}{c} 34 \\ 61 \\ 51 \end{array} \right] = W \end{array}$$

$$\begin{array}{c} & \begin{array}{c} \text{Preschool and} \\ \text{Kindergarten} \end{array} \begin{array}{c} \text{Elementary and} \\ \text{Middle School} \end{array} \begin{array}{c} \text{Secondary} \\ \text{School} \end{array} \\ \begin{array}{c} \text{City} \\ \text{County} \end{array} \left[\begin{array}{ccc} 25 & 60 & 40 \\ 5 & 12 & 10 \end{array} \right] = N \end{array}$$

(a) Find the matrix product NW.

(b) Explain what NW represents.

(c) According to matrix NW, what is the total cost of teacher wages in city schools? In county schools? In the entire school district?

53. Gourmet Pops Steel City Pops sells fruity, creamy, and cookie pops at locations in Texas, Kentucky, and Alabama (www.steelcitypops.com). Suppose that, for a special event, sales (in hundreds of pops) and prices (in dollars per hundred) were as shown in the following matrices.

$$\begin{array}{c} & \text{Price} \\ \begin{array}{c} \text{Fruity} \\ \text{Creamy} \\ \text{Cookie} \end{array} \left[\begin{array}{c} 250 \\ 300 \\ 350 \end{array} \right] = P \end{array}$$

$$\begin{array}{c} & \text{Fruity} \quad \text{Creamy} \quad \text{Cookie} \\ \begin{array}{c} \text{Texas} \\ \text{Kentucky} \\ \text{Alabama} \end{array} \left[\begin{array}{ccc} 8 & 7 & 2 \\ 4 & 5 & 2 \\ 6 & 9 & 4 \end{array} \right] = S \end{array}$$

(a) Is it possible to find the matrix product PS? If so, find it and explain its meaning.

(b) Is it possible to find the matrix product SP? If so, find it and explain its meaning.

(c) What was the total revenue from all locations for this special event?

54. Franchise Costs According to Franchise Business Review, the minimum investments (in thousands) to launch certain franchises are listed in matrix M below. The numbers of franchises that your investment group would like to start this year in Pennsylvania and Indiana are shown in matrix N.

$$\begin{array}{c} & \text{Cost} \\ \begin{array}{c} \text{Deli Delicious} \\ \text{Taziki's Mediterranean Cafe} \\ \text{Your Pie (brick oven pizza)} \end{array} \left[\begin{array}{c} 150 \\ 313 \\ 400 \end{array} \right] = M \end{array}$$

$$\begin{array}{c} & \text{Deli} \quad \text{Taziki's} \quad \text{Pie} \\ \begin{array}{c} \text{Indiana} \\ \text{Pynnsylvania} \end{array} \left[\begin{array}{ccc} 3 & 0 & 5 \\ 2 & 7 & 1 \end{array} \right] = N \end{array}$$

(a) Is it possible to find the product MN? If so, find it and explain the meaning of each entry.

(b) Is it possible to find the product NM? If so, find it and explain the meaning of each entry.

(c) What is the total dollar amount that your group must invest for all locations?

📟 *Use a graphing calculator for Exercises 55 and 56.*

55. Birth and Death Rates Population estimates and projections (in millions) for the BRIC countries (Brazil, Russia, India, China) and birth and death rates per million people for those countries are shown in the tables. (Data from: United Nations Population Fund.)

	2010	2020
Brazil	195	210
Russia	143	141
India	1225	1387
China	1341	1388

	Births	Deaths
Brazil	.016	.006
Russia	.011	.014
India	.023	.008
China	.013	.007

(a) Write the data in the first table as a 2 × 4 matrix *A*.

(b) Write the data in the second table as a 4 × 2 matrix *B*.

(c) Find the matrix product *AB*.

(d) Explain what *AB* represents.

(e) According to matrix *AB*, what was the total number of deaths in the BRIC countries in 2010? What total number of births is projected for 2020?

56. Birth and Death Rates Population estimates and projections (in millions) for several regions of the world and the average birth and death rates per million people for those regions are shown in the tables below. (Data from: United Nations Population Fund.)

	Asia	Europe	Latin America	North America
2010	4164	738	590	345
2020	4566	744	652	374

	Births	Deaths
Asia	.019	.007
Europe	.011	.011
Latin America	.019	.006
North America	.024	.008

(a) Write the information in the first table as a 2 × 4 matrix *A*.

(b) Write the information in the second table as a 4 × 2 matrix *B*.

(c) Find the matrix product *AB*.

(d) Explain what *AB* represents.

(e) According to matrix *AB*, what was the total number of births in these four regions combined in 2010? What total number of deaths is projected for 2020?

6.6 Applications of Matrices

This section gives a variety of applications of matrices.

Solving Systems with Matrices

Consider this system of linear equations:

$$2x - 3y = 4$$
$$x + 5y = 2.$$

Let

$$A = \begin{bmatrix} 2 & -3 \\ 1 & 5 \end{bmatrix}, \quad X = \begin{bmatrix} x \\ y \end{bmatrix}, \quad \text{and} \quad B = \begin{bmatrix} 4 \\ 2 \end{bmatrix}.$$

Since

$$AX = \begin{bmatrix} 2 & -3 \\ 1 & 5 \end{bmatrix} \begin{bmatrix} x \\ y \end{bmatrix} = \begin{bmatrix} 2x - 3y \\ x + 5y \end{bmatrix} \quad \text{and} \quad B = \begin{bmatrix} 4 \\ 2 \end{bmatrix},$$

the original system is equivalent to the single matrix equation $AX = B$. Similarly, any system of linear equations can be written as a matrix equation $AX = B$. The matrix A is called the **coefficient matrix**, X is called the **matrix of variables**, and B is called the **matrix of constants**. ✓₁

A matrix equation $AX = B$ can be solved if A^{-1} exists. Assuming that A^{-1} exists and using the facts that $A^{-1}A = I$ and $IX = X$, along with the associative property of multiplication of matrices, gives

$$AX = B$$
$$A^{-1}(AX) = A^{-1}B \qquad \text{Multiply both sides by } A^{-1}.$$
$$(A^{-1}A)X = A^{-1}B \qquad \text{Associative property}$$
$$IX = A^{-1}B \qquad \text{Inverse property}$$
$$X = A^{-1}B. \qquad \text{Identity property}$$

When multiplying by matrices on both sides of a matrix equation, be careful to multiply in the same order on both sides, since multiplication of matrices is not commutative (unlike the multiplication of real numbers). This discussion is summarized below.

Suppose that a system of equations with the same number of equations as variables is written in matrix form as $AX = B$, where A is the square matrix of coefficients, X is the column matrix of variables, and B is the column matrix of constants. If A has an inverse, then the unique solution of the system is $X = A^{-1}B$.*

EXAMPLE 1   Consider this system of equations:

$$\begin{aligned} x + y + z &= 2 \\ 2x + 3y &= 5 \\ x + 2y + z &= -1. \end{aligned}$$

(a) Write the system as a matrix equation.

Solution We have these three matrices:

$$\underset{\text{Coefficient Matrix}}{A = \begin{bmatrix} 1 & 1 & 1 \\ 2 & 3 & 0 \\ 1 & 2 & 1 \end{bmatrix}}, \quad \underset{\text{Matrix of Variables}}{X = \begin{bmatrix} x \\ y \\ z \end{bmatrix}}, \quad \text{and} \quad \underset{\text{Matrix of Constants}}{B = \begin{bmatrix} 2 \\ 5 \\ -1 \end{bmatrix}}.$$

So the matrix equation is

$$AX = B$$

$$\begin{bmatrix} 1 & 1 & 1 \\ 2 & 3 & 0 \\ 1 & 2 & 1 \end{bmatrix} \begin{bmatrix} x \\ y \\ z \end{bmatrix} = \begin{bmatrix} 2 \\ 5 \\ -1 \end{bmatrix}.$$

---

*If A does not have an inverse, then the system either has no solution or has an infinite number of solutions. Use the methods of Sections 6.1 or 6.2.

✓ **Checkpoint 1**

Write the matrix of coefficients, the matrix of variables, and the matrix of constants for the system

$$\begin{aligned} 2x + 6y &= -14 \\ -x - 2y &= 3. \end{aligned}$$

(b) Find A^{-1} and solve the equation.

Solution Use Exercise 31 of Section 6.5, technology, or the method of Section 6.5 to find that

$$A^{-1} = \begin{bmatrix} 1.5 & .5 & -1.5 \\ -1 & 0 & 1 \\ .5 & -.5 & .5 \end{bmatrix}.$$

Hence,

$$X = A^{-1}B = \begin{bmatrix} 1.5 & .5 & -1.5 \\ -1 & 0 & 1 \\ .5 & -.5 & .5 \end{bmatrix} \begin{bmatrix} 2 \\ 5 \\ -1 \end{bmatrix} = \begin{bmatrix} 7 \\ -3 \\ -2 \end{bmatrix}.$$

Thus, the solution of the original system is $(7, -3, -2)$. ✓₂

✓ **Checkpoint 2**

Use the inverse matrix to solve the system in Example 1 if the constants for the three equations are 12, 0, and 8, respectively.

EXAMPLE 2 Use the inverse of the coefficient matrix to solve the system

$$x + 1.5y = 8$$
$$2x + 3y = 10.$$

Solution The coefficient matrix is $A = \begin{bmatrix} 1 & 1.5 \\ 2 & 3 \end{bmatrix}$. A graphing calculator will indicate that A^{-1} does not exist. If we try to carry out the row operations, we see why:

$$\begin{bmatrix} 1 & 1.5 & | & 1 & 0 \\ 2 & 3 & | & 0 & 1 \end{bmatrix}$$

$$\begin{bmatrix} 1 & 1.5 & | & 1 & 0 \\ 0 & 0 & | & -2 & 1 \end{bmatrix}. \quad -2R_1 + R_2$$

The next step cannot be performed because of the zero in the second row, second column. Verify that the original system has no solution. ✓₃

✓ **Checkpoint 3**

Solve the system in Example 2 if the constants are, respectively, 3 and 6.

Input–Output Analysis

An interesting application of matrix theory to economics was developed by Nobel Prize winner Wassily Leontief. His application of matrices to the interdependencies in an economy is called **input–output analysis.** In practice, input–output analysis is very complicated, with many variables. We shall discuss only simple examples with just a few variables.

Input–output models are concerned with the production and flow of goods and services. A typical economy is composed of a number of different sectors (such as manufacturing, energy, transportation, agriculture, etc.). Each sector requires input from other sectors (and possibly from itself) to produce its output. For instance, manufacturing output requires energy, transportation, and manufactured items (such as tools and machinery). If an economy has n sectors, then the inputs required by the various sectors from each other to produce their outputs can be described by an $n \times n$ matrix called the **input–output matrix** (or the **technological matrix**).

EXAMPLE 3 **Economics** Suppose a simplified economy involves just three sectors—agriculture, manufacturing, and transportation—all in appropriate units. The production of 1 unit of agriculture requires $\frac{1}{2}$ unit of manufacturing and $\frac{1}{4}$ unit of transportation. The production of 1 unit of manufacturing requires $\frac{1}{4}$ unit of agriculture and $\frac{1}{4}$ unit of transportation. The production of 1 unit of transportation requires $\frac{1}{3}$ unit of agriculture and $\frac{1}{4}$ unit of manufacturing. Write the input–output matrix of this economy.

Solution Since there are three sectors in the economy, the input–output matrix A is 3×3. Each row and each column is labeled by a sector of the economy, as shown below. The first column lists the units from each sector of the economy that are required to produce one unit of agriculture. The second column lists the units required from each sector to produce 1 unit of manufacturing, and the last column lists the units required from each sector to produce 1 unit of transportation.

$$
\begin{array}{c}
 \\
Input
\end{array}
\begin{array}{c}
\\
\text{Agriculture} \\
\text{Manufacturing} \\
\text{Transportation}
\end{array}
\overset{\displaystyle \begin{array}{ccc} \text{Agriculture} & \text{Manufacturing} & \text{Transportation} \end{array}}{\overset{\displaystyle Output}{\left[\begin{array}{ccc}
0 & \frac{1}{4} & \frac{1}{3} \\
\frac{1}{2} & 0 & \frac{1}{4} \\
\frac{1}{4} & \frac{1}{4} & 0
\end{array} \right]}} = A.
$$

Example 3 is a bit unrealistic in that no sector of the economy requires any input from itself. In an actual economy, most sectors require input from themselves as well as from other sectors to produce their output. Nevertheless, it is easier to learn the basic concepts from simplified examples, so we shall continue to use them.

The input–output matrix gives only a partial picture of an economy. We also need to know the amount produced by each sector and the amount of the economy's output that is used up by the sectors themselves in the production process. The remainder of the total output is available to satisfy the needs of consumers and others outside the production system.

✓ **Checkpoint 4**

Write a 2×2 input–output matrix in which 1 unit of electricity requires $\frac{1}{2}$ unit of water and $\frac{1}{3}$ unit of electricity, while 1 unit of water requires no water but $\frac{1}{4}$ unit of electricity.

EXAMPLE 4   **Economics** Consider the economy whose input–output matrix A was found in Example 3.

(a) Suppose this economy produces 60 units of agriculture, 52 units of manufacturing, and 48 units of transportation. Write this information as a column matrix.

Solution Listing the sectors in the same order as the rows of input–output matrix, we have

$$
X = \begin{bmatrix} 60 \\ 52 \\ 48 \end{bmatrix}.
$$

The matrix X is called the **production matrix.**

(b) How much from each sector is used up in the production process?

Solution Since $\frac{1}{4}$ unit of agriculture is used to produce each unit of manufacturing and there are 52 units of manufacturing output, the amount of agriculture used up by manufacturing is $\frac{1}{4} \times 52 = 13$ units. Similarly $\frac{1}{3}$ unit of agriculture is used to produce a unit of transportation, so $\frac{1}{3} \times 48 = 16$ units of agriculture are used up by transportation. Therefore $13 + 16 = 29$ units of agriculture are used up in the economy's production process.

A similar analysis shows that the economy's production process uses up

$$
\underset{\substack{\text{agricul-}\\\text{ture}}}{\tfrac{1}{2} \times 60} + \underset{\substack{\text{transpor-}\\\text{tation}}}{\tfrac{1}{4} \times 48} = 30 + 12 = 42 \text{ units of manufacturing}
$$

and

$$
\underset{\substack{\text{agricul-}\\\text{ture}}}{\tfrac{1}{4} \times 60} + \underset{\substack{\text{manufac-}\\\text{turing}}}{\tfrac{1}{4} \times 52} = 15 + 13 = 28 \text{ units of transportation.}
$$

(c) Describe the conclusions of part (b) in terms of the input–output matrix A and the production matrix X.

Solution The matrix product AX gives the amount from each sector that is used up in the production process, as shown here [with selected entries color coded as in part (b)]:

$$AX = \begin{bmatrix} 0 & \frac{1}{4} & \frac{1}{3} \\ \frac{1}{2} & 0 & \frac{1}{4} \\ \frac{1}{4} & \frac{1}{4} & 0 \end{bmatrix} \begin{bmatrix} 60 \\ 52 \\ 48 \end{bmatrix} = \begin{bmatrix} 0 \cdot 60 + \frac{1}{4} \cdot 52 + \frac{1}{3} \cdot 48 \\ \frac{1}{2} \cdot 60 + 0 \cdot 52 + \frac{1}{4} \cdot 48 \\ \frac{1}{4} \cdot 60 + \frac{1}{4} \cdot 52 + 0 \cdot 48 \end{bmatrix}$$

$$= \begin{bmatrix} \frac{1}{4} \cdot 52 + \frac{1}{3} \cdot 48 \\ \frac{1}{2} \cdot 60 + \frac{1}{4} \cdot 48 \\ \frac{1}{4} \cdot 60 + \frac{1}{4} \cdot 52 \end{bmatrix} = \begin{bmatrix} 29 \\ 42 \\ 28 \end{bmatrix}.$$

(d) Find the matrix $D = X - AX$ and explain what its entries represent.

Solution From parts (a) and (c), we have

$$D = X - AX = \begin{bmatrix} 60 \\ 52 \\ 48 \end{bmatrix} - \begin{bmatrix} 29 \\ 42 \\ 28 \end{bmatrix} = \begin{bmatrix} 31 \\ 10 \\ 20 \end{bmatrix}.$$

The matrix D lists the amount of each sector that is *not* used up in the production process and hence is available to groups outside the production process (such as consumers). For example, 60 units of agriculture are produced and 29 units are used up in the process, so the difference $60 - 29 = 31$ is the amount of agriculture that is available to groups outside the production process. Similar remarks apply to manufacturing and transportation. The matrix D is called the **demand matrix.** Matrices A and D show that the production of 60 units of agriculture, 52 units of manufacturing, and 48 units of transportation would satisfy an outside demand of 31 units of agriculture, 10 units of manufacturing, and 20 units of transportation. ✔5

✓**Checkpoint 5**

(a) Write a 2×1 matrix X to represent the gross production of 9000 units of electricity and 12,000 units of water.

(b) Find AX, using A from Checkpoint 4.

(c) Find D, using $D = X - AX$.

Example 4 illustrates the general situation. In an economy with n sectors, the input–output matrix A is $n \times n$. The production matrix X is a column matrix whose n entries are the outputs of each sector of the economy. The demand matrix D is also a column matrix with n entries. This matrix is defined by

$$D = X - AX.$$

D lists the amount from each sector that is available to meet the demands of consumers and other groups outside the production process.

In Example 4, we knew the input–output matrix A and the production matrix X and used them to find the demand matrix D. In practice, however, this process is reversed: The input–output matrix A and the demand matrix D are known, and we must find the production matrix X needed to satisfy the required demands. Matrix algebra can be used to solve the equation $D = X - AX$ for X:

$$D = X - AX$$
$$D = IX - AX \qquad \text{Identity property}$$
$$D = (I - A)X. \qquad \text{Distributive property}$$

If the matrix $I - A$ has an inverse, then

$$X = (I - A)^{-1}D.$$

▦ **EXAMPLE 5**   **Economics** Suppose, in the three-sector economy of Examples 3 and 4, there is a demand for 516 units of agriculture, 258 units of manufacturing, and 129 units of transportation. What should production be for each sector?

Solution The demand matrix is

$$D = \begin{bmatrix} 516 \\ 258 \\ 129 \end{bmatrix}.$$

Find the production matrix by first calculating $I - A$:

$$I - A = \begin{bmatrix} 1 & 0 & 0 \\ 0 & 1 & 0 \\ 0 & 0 & 1 \end{bmatrix} - \begin{bmatrix} 0 & \frac{1}{4} & \frac{1}{3} \\ \frac{1}{2} & 0 & \frac{1}{4} \\ \frac{1}{4} & \frac{1}{4} & 0 \end{bmatrix} = \begin{bmatrix} 1 & -\frac{1}{4} & -\frac{1}{3} \\ -\frac{1}{2} & 1 & -\frac{1}{4} \\ -\frac{1}{4} & -\frac{1}{4} & 1 \end{bmatrix}.$$

Using a calculator with matrix capability or row operations, find the inverse of $I - A$:

$$(I - A)^{-1} = \begin{bmatrix} 1.3953 & .4961 & .5891 \\ .8372 & 1.3643 & .6202 \\ .5581 & .4651 & 1.3023 \end{bmatrix}.$$

(The entries are rounded to four decimal places.[*]) Since $X = (I - A)^{-1}D$.

$$X = \begin{bmatrix} 1.3953 & .4961 & .5891 \\ .8372 & 1.3643 & .6202 \\ .5581 & .4651 & 1.3023 \end{bmatrix} \begin{bmatrix} 516 \\ 258 \\ 129 \end{bmatrix} = \begin{bmatrix} 924 \\ 864 \\ 576 \end{bmatrix}$$

(rounded to the nearest whole numbers).

From the last result, we see that the production of 924 units of agriculture, 864 units of manufacturing, and 576 units of transportation is required to satisfy demands of 516, 258, and 129 units, respectively.

⊞ **EXAMPLE 6** **Economics** An economy depends on two basic products: wheat and oil. To produce 1 metric ton of wheat requires .25 metric ton of wheat and .33 metric ton of oil. The production of 1 metric ton of oil consumes .08 metric ton of wheat and .11 metric ton of oil. Find the production that will satisfy a demand of 500 metric tons of wheat and 1000 metric tons of oil.

Solution The input–output matrix is

$$A = \begin{array}{cc} & \begin{array}{cc} \text{Wheat} & \text{Oil} \end{array} \\ \begin{array}{c} \end{array} & \begin{bmatrix} .25 & .08 \\ .33 & .11 \end{bmatrix} \begin{array}{c} \text{Wheat} \\ \text{Oil} \end{array} \end{array},$$

and we also have

$$I - A = \begin{bmatrix} 1 & 0 \\ 0 & 1 \end{bmatrix} - \begin{bmatrix} .25 & .08 \\ .33 & .11 \end{bmatrix} = \begin{bmatrix} .75 & -.08 \\ -.33 & .89 \end{bmatrix}.$$

Next, use technology or the methods of Section 6.5 to calculate $(I - A)^{-1}$:

$$(I - A)^{-1} = \begin{bmatrix} 1.3882 & .1248 \\ .5147 & 1.1699 \end{bmatrix} \quad \text{(rounded)}.$$

The demand matrix is

$$D = \begin{bmatrix} 500 \\ 1000 \end{bmatrix}.$$

---

[*]Although we show the matrix $(I - A)^{-1}$ with entries rounded to four decimal places, we did not round off in calculating $(I - A)^{-1}D$. If the rounded figures are used, the numbers in the product may vary slightly in the last digit.

Consequently, the production matrix is

$$X = (I - A)^{-1}D = \begin{bmatrix} 1.3882 & .1248 \\ .5147 & 1.1699 \end{bmatrix}\begin{bmatrix} 500 \\ 1000 \end{bmatrix} = \begin{bmatrix} 819 \\ 1427 \end{bmatrix},$$

where the production numbers have been rounded to the nearest whole numbers. The production of 819 metric tons of wheat and 1427 metric tons of oil is required to satisfy the indicated demand. ✓6

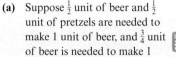

A simple economy depends on just two products: beer and pretzels.

(a) Suppose $\frac{1}{2}$ unit of beer and $\frac{1}{2}$ unit of pretzels are needed to make 1 unit of beer, and $\frac{3}{4}$ unit of beer is needed to make 1 unit of pretzels. Write the input–output matrix A for the economy.

(b) Find $I - A$.

(c) Find $(I - A)^{-1}$.

(d) Find the gross production X that will be needed to get a net production of

$$D = \begin{bmatrix} 100 \\ 1000 \end{bmatrix}.$$

TECHNOLOGY TIP If you are using a graphing calculator to determine X, you can calculate $(I - A)^{-1}D$ in one step without finding the intermediate matrices $I - A$ and $(I - A)^{-1}$.

Code Theory

Governments need sophisticated methods of coding and decoding messages. One example of such an advanced code uses matrix theory. This code takes the letters in the words and divides them into groups. (Each space between words is treated as a letter; punctuation is disregarded.) Then, numbers are assigned to the letters of the alphabet. For our purposes, let the letter a correspond to 1, b to 2, and so on. Let the number 27 correspond to a space. Then we use matrix methods to make the code more secure, as the next example illustrates.

EXAMPLE 7 **Encoding** Encode the message "mathematics is hip" with no punctuation.

Solution The message

$$mathematics\ is\ hip$$

can be divided into groups of three letters each:

$$mat\quad hem\quad ati\quad cs-\quad is-\quad hip$$

(We used − to represent a space.) We now write a column matrix for each group of three symbols, using the corresponding numbers instead of letters. For example, the first four groups can be written as

$$\overset{mat}{\begin{bmatrix} 13 \\ 1 \\ 20 \end{bmatrix}},\quad \overset{hem}{\begin{bmatrix} 8 \\ 5 \\ 13 \end{bmatrix}},\quad \overset{ati}{\begin{bmatrix} 1 \\ 20 \\ 9 \end{bmatrix}},\quad \overset{cs-}{\begin{bmatrix} 3 \\ 19 \\ 27 \end{bmatrix}}.$$

The entire message consists of six 3 × 1 column matrices:

$$\begin{bmatrix} 13 \\ 1 \\ 20 \end{bmatrix},\begin{bmatrix} 8 \\ 5 \\ 13 \end{bmatrix},\begin{bmatrix} 1 \\ 20 \\ 9 \end{bmatrix},\begin{bmatrix} 3 \\ 19 \\ 27 \end{bmatrix},\begin{bmatrix} 9 \\ 19 \\ 27 \end{bmatrix},\begin{bmatrix} 8 \\ 9 \\ 16 \end{bmatrix}.\ ✓7$$

✓ Checkpoint 7

Write the message "*when*" using 2 × 1 matrices.

Although you could transmit these matrices, a simple substitution code such as this is very easy to break.

To get a more reliable code, we choose a 3 × 3 matrix M that has an inverse. Suppose we choose

$$M = \begin{bmatrix} 1 & 3 & 3 \\ 1 & 4 & 3 \\ 1 & 3 & 4 \end{bmatrix}.$$

Then encode each message group by multiplying by M—that is,

$$\begin{bmatrix} 1 & 3 & 3 \\ 1 & 4 & 3 \\ 1 & 3 & 4 \end{bmatrix}\begin{bmatrix} 13 \\ 1 \\ 20 \end{bmatrix} = \begin{bmatrix} 76 \\ 77 \\ 96 \end{bmatrix}, \quad \begin{bmatrix} 1 & 3 & 3 \\ 1 & 4 & 3 \\ 1 & 3 & 4 \end{bmatrix}\begin{bmatrix} 8 \\ 5 \\ 13 \end{bmatrix} = \begin{bmatrix} 62 \\ 67 \\ 75 \end{bmatrix},$$

$$\begin{bmatrix} 1 & 3 & 3 \\ 1 & 4 & 3 \\ 1 & 3 & 4 \end{bmatrix}\begin{bmatrix} 1 \\ 20 \\ 9 \end{bmatrix} = \begin{bmatrix} 88 \\ 108 \\ 97 \end{bmatrix},$$

and so on. The coded message consists of the six 3×1 column matrices

$$\begin{bmatrix} 76 \\ 77 \\ 96 \end{bmatrix}, \quad \begin{bmatrix} 62 \\ 67 \\ 75 \end{bmatrix}, \quad \begin{bmatrix} 88 \\ 108 \\ 97 \end{bmatrix}, \dots, \begin{bmatrix} 83 \\ 92 \\ 99 \end{bmatrix}.$$

The message would be sent as a string of numbers:

$$76, 77, 96, 62, 67, 75, 88, 108, 97, \dots, 83, 92, 99.$$

Note that the same letter may be encoded by different numbers. For instance, the first a in "mathematics" is 77 and the second a is 88. This makes the code harder to break. ✔ 8

✓ **Checkpoint 8**

Use the following matrix to find the 2×1 matrices to be transmitted for the message in Checkpoint 7:

$$\begin{bmatrix} 2 & 1 \\ 5 & 0 \end{bmatrix}.$$

EXAMPLE 8 **Decoding** Decode the message sent in Example 7.

Solution The receiving agent rewrites the message as the six 3×1 column matrices shown in blue in Example 7. The agent then decodes the message by multiplying each of these column matrices by the matrix M^{-1}. Verify that

$$M^{-1} = \begin{bmatrix} 7 & -3 & -3 \\ -1 & 1 & 0 \\ -1 & 0 & 1 \end{bmatrix}.$$

So the first two matrices of the coded message are decoded as

$$\begin{bmatrix} 7 & -3 & -3 \\ -1 & 1 & 0 \\ -1 & 0 & 1 \end{bmatrix}\begin{bmatrix} 76 \\ 77 \\ 96 \end{bmatrix} = \begin{bmatrix} 13 \\ 1 \\ 20 \end{bmatrix}\begin{matrix} m \\ a \\ t \end{matrix}$$

and

$$\begin{bmatrix} 7 & -3 & -3 \\ -1 & 1 & 0 \\ -1 & 0 & 1 \end{bmatrix}\begin{bmatrix} 62 \\ 67 \\ 75 \end{bmatrix} = \begin{bmatrix} 8 \\ 5 \\ 13 \end{bmatrix}\begin{matrix} h \\ e \\ m \end{matrix}.$$

The other blocks are decoded similarly.

Routing

Many real-world applications involve networks such as connections on social media, airline routing maps, and networks of cell phone towers. The next example and the case study at the end of the chapter illustrate how matrices can be useful in these applications.

Figure 6.27

EXAMPLE 9 **Campus Pathways** The diagram in Figure 6.27 shows the pathways connecting five buildings on a college campus.

(a) Represent the pathways using a matrix A, where the entries represent the number of pathways connecting two buildings without passing through another building.

Solution From the diagram, we see that there are two pathways connecting building 1 (B1) to building 2 (B2) without passing through buildings 3, 4, or 5. This information is entered in row one, column two, and again in row two, column one of matrix A:

$$
\begin{array}{c}
 \quad \text{B1 B2 B3 B4 B5} \\
\begin{array}{c}
\text{B1} \\ \text{B2} \\ \text{B3} \\ \text{B4} \\ \text{B5}
\end{array}
\begin{bmatrix}
0 & 2 & 1 & 1 & 2 \\
2 & 0 & 1 & 0 & 1 \\
1 & 1 & 0 & 1 & 0 \\
1 & 0 & 1 & 0 & 1 \\
2 & 1 & 0 & 1 & 0
\end{bmatrix} = A.
\end{array}
$$

Note that there are no pathways connecting each building to itself, so the main diagonal of the matrix from upper left to lower right consists entirely of zeros.

(b) How many ways are there to go from building 1 to building 3 by going through exactly one other building?

Solution This can happen by going from building 1 to building 2 two ways, and then from building 2 to building 3. Another way is to go from building 1 to building 4 and then from building 4 to building 3. Thus, there are three total ways to go from building 1 to building 3 by going through exactly one other building.

(c) Given any pair of buildings, how many ways are there to go between them by going through exactly one other building?

Solution The matrix A^2 shows the number of ways to travel between any two buildings by going through exactly one other building.

$$
A^2 = \begin{bmatrix}
10 & 3 & 3 & 3 & 3 \\
3 & 6 & 2 & 4 & 4 \\
3 & 2 & 3 & 1 & 4 \\
3 & 4 & 1 & 3 & 2 \\
3 & 4 & 4 & 2 & 6
\end{bmatrix}.
$$

We ignore the values down the main diagonal for travel from one building into itself. We notice that the entry in row one, column three has a value of 3. This coincides with what we learned in part (b)—that there are three ways to go from building 1 to building 3 that go through exactly one other building.

In the previous example, we could have also found the number of ways to walk between any two buildings by passing through exactly two other buildings by calculating A^3. Additionally, $A + A^2$ gives the number of ways to walk between two buildings with at most one intermediate building. ✔9

The diagram can be given many other interpretations. For example, the lines could represent lines of mutual influence between people or nations, or they could represent communication lines, such as Ethernet cables.

✓ **Checkpoint 9** ▦

Use a graphing calculator to find the following.

(a) A^3

(b) $A + A^2$

6.6 Exercises

Solve the matrix equation $AX = B$ for X. (See Example 1.)

1. $A = \begin{bmatrix} 1 & -1 \\ 5 & 6 \end{bmatrix}, B = \begin{bmatrix} 4 \\ -2 \end{bmatrix}$

2. $A = \begin{bmatrix} 1 & 2 \\ 1 & 5 \end{bmatrix}, B = \begin{bmatrix} -3 \\ 5 \end{bmatrix}$

3. $A = \begin{bmatrix} 3 & 1 \\ 4 & 2 \end{bmatrix}, B = \begin{bmatrix} 3 & 4 \\ 5 & 6 \end{bmatrix}$

4. $A = \begin{bmatrix} 1 & 4 \\ 2 & 7 \end{bmatrix}, B = \begin{bmatrix} 0 & 8 \\ 4 & 1 \end{bmatrix}$

5. $A = \begin{bmatrix} 2 & 1 & 0 \\ -4 & -1 & 3 \\ 3 & 1 & -2 \end{bmatrix}$, $B = \begin{bmatrix} 1 \\ 4 \\ 0 \end{bmatrix}$

6. $A = \begin{bmatrix} 3 & -1 & 0 \\ 0 & 1 & 2 \\ 9 & 0 & 5 \end{bmatrix}$, $B = \begin{bmatrix} -3 \\ 6 \\ 12 \end{bmatrix}$

Use the inverse of the coefficient matrix to solve each system of equations. (See Example 1.)

7. $x + 2y + 3z = 10$
    $2x + 3y + 2z = 6$
    $-x - 2y - 4z = -1$

8. $-x + y - 3z = 3$
    $2x + 4y - 4z = 6$
    $-x + y + 4z = -1$

9. $x + 4y + 3z = -12$
    $x - 3y - 2z = 0$
    $2x + 5y + 4z = 7$

10. $5x \qquad + 2z = 3$
    $2x + 2y + z = 4$
    $-3x + y - z = 5$

11. $x + 2y + 3z = 4$
    $x + 4y + 2z = 8$
    $\quad\;\; y - z = -4$

12. $2x + 2y - 4z = 12$
    $2x + 6y \qquad = 16$
    $-3x - 3y + 5z = -20$

13. $x + y \qquad + 2w = 3$
    $2x - y + z - w = 3$
    $3x + 3y + 2z - 2w = 5$
    $x + 2y + z \qquad = 3$

14. $x \qquad - 2z \qquad = 4$
    $-2x + y + 2z + 2w = -8$
    $3x - y - 2z - 3w = 12$
    $\qquad y + 4z + w = -4$

Write a system of equations, and use the inverse of the coefficient matrix to solve each system of equations. (See Example 1.)

15. **E-Commerce** For online store sales, the payment processing service Square charges 2.9% of the total sales plus $.30 per transaction. For your online business, the cost of using Square last month was $291. A competitor to Square charges 3% of total sales plus $.25 per transaction, and your bill for the same month with that company would have been $295. Find the total sales amount and the number of transactions. (Data from: www.squareup.com.)

16. **Truck Rental** Budget Truck Rental charged $19.99 per day plus $.59 per mile to rent a 12-foot moving truck in Austin, Texas, in 2016. The total bill for your move was $86. The rates for a local moving company were $10 per day and $.95 per mile, and the bill for same trip (same number of miles in the same number of days) would have been $94.10. How many miles and how many days was this trip? (Data from: www.budgettruck.com.)

17. **Furniture Manufacturing** A Colorado-based company makes artisanal dining room furniture. A buffet requires 15 hours for cutting, 20 hours for assembly, and 5 hours for finishing. A chair requires 5 hours for cutting, 8 hours for assembly, and 5 hours for finishing. A table requires 10 hours for cutting, 6 hours for assembly, and 6 hours for finishing. The cutting department has 4900 hours of labor available each week, the assembly department has 6600 hours available, and the finishing department has 3900 hours available. How many pieces of each type of furniture should be produced each week if the factory is to run at full capacity?

18. **Nutrition** A hospital dietician is planning a special diet for a certain patient. The total amount per meal of food groups A, B,

and C must equal 400 grams. The diet should include one-third as much of group A as of group B, and the sum of the amounts of group A and group C should equal twice the amount of group B. How many grams of each food group should be included?

19. **Casual Clothing** A national chain of casual clothing stores recently sent shipments of jeans, jackets, sweaters, and shirts to its stores in various cities. The number of items shipped to each city and their total wholesale cost are shown in the table. Find the wholesale price of one pair of jeans, one jacket, one sweater, and one shirt.

City	Jeans	Jackets	Sweaters	Shirts	Total Cost
Cleveland	3000	3000	2200	4200	$507,650
St. Louis	2700	2500	2100	4300	459,075
Seattle	5000	2000	1400	7500	541,225
Phoenix	7000	1800	600	8000	571,500

20. **Materials Mixing** A company produces three combinations of mixed vegetables, which sell in 1-kilogram packages. Italian style combines .3 kilogram of zucchini, .3 of broccoli, and .4 of carrots. French style combines .6 kilogram of broccoli and .4 of carrots. California style combines .2 kilogram of zucchini, .5 of broccoli, and .3 of carrots. The company has a stock of 16,200 kilograms of zucchini, 41,400 kilograms of broccoli, and 29,400 kilograms of carrots. How many packages of each style should the company prepare in order to use up its supplies?

21. **Bacteria Colonies** Three species of bacteria are fed three foods: I, II, and III. A bacterium of the first species consumes 1.3 units each of foods I and II and 2.3 units of food III each day. A bacterium of the second species consumes 1.1 units of food I, 2.4 units of food II, and 3.7 units of food III each day. A bacterium of the third species consumes 8.1 units of I, 2.9 units of II, and 5.1 units of III each day. If 16,000 units of I, 28,000 units of II, and 44,000 units of III are supplied each day, how many of each species can be maintained in this environment?

22. **Assisted Living** A 100-bed assisted living facility provides two levels of long-term care: regular and maximum. Patients at each level have a choice of a private room or a less expensive semiprivate room. The tables (on this page and the following page) show the number of patients in each category at various times last year. The total daily costs for all patients were $24,040 in January, $23,926 in April, $23,760 in July, and $24,042 in October. Find the daily cost of each of the following: a private room (regular care), a private room (maximum care), a semiprivate room (regular care), and a semiprivate room (maximum care).

Month	REGULAR-CARE PATIENTS	
	Semiprivate	Private
January	22	8
April	26	8
July	24	14
October	20	10

MAXIMUM-CARE PATIENTS		
Month	Semiprivate	Private
January	60	10
April	54	12
July	56	6
October	62	8

Find the production matrix for the given input–output and demand matrices. (See Examples 3–6.).

23. $A = \begin{bmatrix} .1 & .03 \\ .07 & .6 \end{bmatrix}, D = \begin{bmatrix} 5 \\ 10 \end{bmatrix}$

24. $A = \begin{bmatrix} \frac{1}{2} & \frac{2}{5} \\ \frac{1}{4} & \frac{1}{5} \end{bmatrix}, D = \begin{bmatrix} 2 \\ 4 \end{bmatrix}$

Exercises 25 and 26 refer to Example 6.

25. **Input–Output Analysis** If the demand is changed to 690 metric tons of wheat and 920 metric tons of oil, how many units of each commodity should be produced?

26. **Input–Output Analysis** Change the input–output matrix so that the production of 1 metric ton of wheat requires $\frac{1}{5}$ metric ton of oil (and no wheat) and the production of 1 metric ton of oil requires $\frac{1}{3}$ metric ton of wheat (and no oil). To satisfy the same demand matrix, how many units of each commodity should be produced?

Work these problems. (See Examples 3–6.)

27. **Input–Output Analysis** A primitive economy depends on two basic goods: yams and pork. The production of 1 bushel of yams requires $\frac{1}{4}$ bushel of yams and $\frac{1}{2}$ of a pig. To produce 1 pig requires $\frac{1}{6}$ bushel of yams. Find the amount of each commodity that should be produced to get

    (a) 1 bushel of yams and 1 pig;

    (b) 100 bushels of yams and 70 pigs.

28. **Input–Output Analysis** A two-segment economy consists of manufacturing and agriculture. To produce 1 unit of manufacturing output requires .40 unit of its own output and .20 unit of agricultural output. To produce 1 unit of agricultural output requires .30 unit of its own output and .40 unit of manufacturing output. If there is a demand of 240 units of manufacturing and 90 units of agriculture, what should be the output of each segment?

29. **Input–Output Analysis** A simplified economy has only two industries: the electric company and the gas company. Each dollar's worth of the electric company's output requires $.40 of its own output and $.50 of the gas company's output. Each dollar's worth of the gas company's output requires $.25 of its own output and $.60 of the electric company's output. What should the production of electricity and gas be (in dollars) if there is a $12 million demand for gas and a $15 million demand for electricity?

30. **Input–Output Analysis** A simplified economy is based on agriculture, manufacturing, and transportation. Each unit of agricultural output requires .4 unit of its own output, .3 unit of manufacturing output, and .2 unit of transportation output.

One unit of manufacturing output requires .4 unit of its own output, .2 unit of agricultural output, and .3 unit of transportation output. One unit of transportation output requires .4 unit of its own output, .1 unit of agricultural output, and .2 unit of manufacturing output. There is demand for 35 units of agricultural, 90 units of manufacturing, and 20 units of transportation output. How many units should each segment of the economy produce?

31. **Input–Output Analysis** In his work *Input–Output Economics*, Leontief provides an example of a simplified economy with just three sectors: agriculture, manufacturing, and households (that is, the sector of the economy that produces labor).[*] It has the following input–output matrix:

	Agriculture	Manufacturing	Households
Agriculture	.25	.40	.133
Manufacturing	.14	.12	.100
Households	.80	3.60	.133

    (a) How many units from each sector does the manufacturing sector require to produce 1 unit?

    (b) What production levels are needed to meet a demand of 35 units of agriculture, 38 units of manufacturing, and 40 units of households?

    (c) How many units of agriculture are used up in the economy's production process?

32. **Input–Output Analysis** A much simplified version of Leontief's 42-sector analysis of the American economy has the following input–output matrix:[†]

	Agriculture	Manufacturing	Households
Agriculture	.245	.102	.051
Manufacturing	.099	.291	.279
Households	.433	.372	.011

    (a) What information about the needs of agricultural production is given by column 1 of the matrix?

    (b) Suppose the demand matrix (in billions of dollars) is

$$D = \begin{bmatrix} 2.88 \\ 31.45 \\ 30.91 \end{bmatrix}.$$

    Find the amount of each commodity that should be produced.

33. **Input–Output Analysis** An analysis of the Israeli economy is simplified here by grouping the economy into three sectors, with the following input–output matrix:[‡]

	Agriculture	Manufacturing	Energy
Agriculture	.293	0	0
Manufacturing	.014	.207	.017
Energy	.044	.010	.216

    (a) How many units from each sector does the energy sector require to produce one unit?

---

[*]Wassily Leontief, *Input–Output Economics*, 2d ed. (Oxford University Press, 1986), pp. 19–27.
[†]Ibid., pp. 6–9.
[‡]Ibid., pp. 174–175.

(b) If the economy's production (in thousands of Israeli pounds) is 175,000 of agriculture, 22,000 of manufacturing, and 12,000 of energy, how much is available from each sector to satisfy the demand from consumers and others outside the production process?

(c) The actual demand matrix is

$$D = \begin{bmatrix} 138,213 \\ 17,597 \\ 1786 \end{bmatrix}.$$

How much must each sector produce to meet this demand?

34. Input–Output Analysis The Chinese economy can be simplified to three sectors: agriculture, industry and construction, and transportation and commerce.[*] The input–output matrix is

	Agri.	Industry/Constr.	Trans./Commerce
Agri.	.158	.156	.009
Industry/Constr.	.136	.432	.071
Trans./Commerce	.013	.041	.011

The demand [in 100,000 *renminbi* (RMB), the unit of money in China] is

$$D = \begin{bmatrix} 106,674 \\ 144,739 \\ 26,725 \end{bmatrix}.$$

(a) Find the amount each sector should produce.

(b) Interpret the economic value of an increase in demand of 1 RMB in agriculture exports.

35. Input–Output Analysis The economy of the state of Washington has been simplified to four sectors: natural resource, manufacturing, trade and services, and personal consumption. The input–output matrix is as follows:[†]

	Natural Resources	Manufacturing	Trade and Services	Personal Consumption
Natural Resources	.1045	.0428	.0029	.0031
Manufacturing	.0826	.1087	.0584	.0321
Trade and Services	.0867	.1019	.2032	.3555
Personal Consumption	.6253	.3448	.6106	.0798

Suppose the demand (in millions of dollars) is

$$D = \begin{bmatrix} 450 \\ 300 \\ 125 \\ 100 \end{bmatrix}.$$

Find the amount each sector should produce.

36. Input–Output Analysis The economy of the state of Nebraska has been condensed to six sectors; livestock, crops, food products, mining and manufacturing, households, and other. The input–output matrix is as follows:[‡]

$$\begin{bmatrix} .178 & .018 & .411 & 0 & .005 & 0 \\ .143 & .018 & .088 & 0 & .001 & 0 \\ .089 & 0 & .035 & 0 & .060 & .003 \\ .001 & .010 & .012 & .063 & .007 & .014 \\ .141 & .252 & .088 & .089 & .402 & .124 \\ .188 & .156 & .103 & .255 & .008 & .474 \end{bmatrix}.$$

(a) Find the matrix $(I - A)^{-1}$ and interpret the value in row two, column one, of this matrix.

(b) Suppose the demand (in millions of dollars) is

$$D = \begin{bmatrix} 1980 \\ 650 \\ 1750 \\ 1000 \\ 2500 \\ 3750 \end{bmatrix}.$$

Find the dollar amount each sector should produce.

Work these coding exercises. (See Example 7 and 8.)

37. Encoding Encode the message

Anne is home.

Break the message into groups of two letters and use the matrix

$$M = \begin{bmatrix} 1 & 3 \\ 2 & 7 \end{bmatrix}.$$

38. Encoding Use the matrix of Exercise 37 to encode the message

Head for the hills!

39. Decoding Decode the following message, which was encoded using the matrix M of Exercise 37:

$$\begin{bmatrix} 57 \\ 129 \end{bmatrix}, \begin{bmatrix} 37 \\ 79 \end{bmatrix}.$$

40. Decoding Decode the following message, which was encoded by using the matrix M of Exercise 37:

$$\begin{bmatrix} 90 \\ 207 \end{bmatrix}, \begin{bmatrix} 39 \\ 87 \end{bmatrix}, \begin{bmatrix} 26 \\ 57 \end{bmatrix}, \begin{bmatrix} 66 \\ 145 \end{bmatrix}, \begin{bmatrix} 61 \\ 142 \end{bmatrix}, \begin{bmatrix} 89 \\ 205 \end{bmatrix}.$$

Work these routing problems. (See Example 9.)

41. Highway Networks The matrix

	City 1	City 2	City 3	City 4
City 1	0	1	2	2
City 2	1	0	1	0
City 3	2	1	0	1
City 4	2	0	1	0

$= A$

[*]*Input–Output Tables of China: 1981*, China Statistical Information and Consultancy Service Centre, 1987, pp. 17–19.

[†]Robert Chase, Philip Bourque, and Richard Conway, Jr., *The 1987 Washington State Input–Output Study*, Report to the Graduate School of Business Administration, University of Washington, September 1993.

[‡]F. Charles Lamphear and Theodore Roesler, "1970 Nebraska Input–Output Tables," *Nebraska Economic and Business Report No. 10*, Bureau of Business Research, University of Nebraska, Lincoln.

shows the number of roads between various cities. Find A^2, then use it to answer the following questions: How many ways are there to travel from

(a) city 1 to city 3 by passing through exactly one city?

(b) city 2 to city 4 by passing through exactly one city?

(c) city 1 to city 3 by passing through at most one city?

(d) city 2 to city 4 by passing through at most one city?

42. **Highway Networks** For matrix A in the previous exercise, find A^3, then use it to answer the following questions.

(a) How many ways are there to travel between cities 1 and 4 by passing through exactly two cities?

(b) How many ways are there to travel between cities 1 and 4 by passing through at most two cities?

43. **Food Webs** The figure shows a food web. The arrows indicate the food sources of each population. For example, cats feed on rats and on mice.

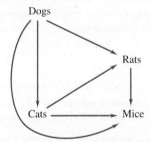

(a) Write a matrix C in which each row and corresponding column represent a population in the food chain. Enter a 1 when the population in a given row feeds on the population in the given column.

(b) Calculate and interpret C^2.

44. **Airline Routing** The figure shows four southern cities served by Supersouth Airlines.

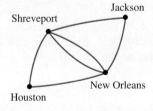

(a) Write a matrix to represent the number of nonstop routes between cities.

(b) Find the number of one-stop flights between Houston and Jackson.

(c) Find the number of flights between Houston and Shreveport that require at most one stop.

(d) Find the number of one-stop flights between New Orleans and Houston.

✓ **Checkpoint Answers**

1. $A = \begin{bmatrix} 2 & 6 \\ -1 & -2 \end{bmatrix}$, $X = \begin{bmatrix} x \\ y \end{bmatrix}$, and $B = \begin{bmatrix} -14 \\ 3 \end{bmatrix}$.

2. $(6, -4, 10)$

3. $(3 - 1.5y, y)$ for all real numbers y

4.
$$\begin{array}{c} \\ \text{Elec.} \\ \text{Water} \end{array} \begin{array}{cc} \text{Elec.} & \text{Water} \\ \left[\begin{array}{cc} \frac{1}{3} & \frac{1}{4} \\ \frac{1}{2} & 0 \end{array}\right. & \left.\right] \end{array}$$

5. (a) $\begin{bmatrix} 9000 \\ 12{,}000 \end{bmatrix}$ (b) $\begin{bmatrix} 6000 \\ 4500 \end{bmatrix}$ (c) $\begin{bmatrix} 3000 \\ 7500 \end{bmatrix}$

6. (a) $\begin{bmatrix} \frac{1}{2} & \frac{3}{4} \\ \frac{1}{2} & 0 \end{bmatrix}$ (b) $\begin{bmatrix} \frac{1}{2} & -\frac{3}{4} \\ -\frac{1}{2} & 1 \end{bmatrix}$

   (c) $\begin{bmatrix} 8 & 6 \\ 4 & 4 \end{bmatrix}$ (d) $\begin{bmatrix} 6800 \\ 4400 \end{bmatrix}$

7. $\begin{bmatrix} 23 \\ 8 \end{bmatrix}, \begin{bmatrix} 5 \\ 14 \end{bmatrix}$

8. $\begin{bmatrix} 54 \\ 115 \end{bmatrix}, \begin{bmatrix} 24 \\ 25 \end{bmatrix}$

9. (a)
```
[A]³
[18  26  16  16  26]
[26  12  13   9  16]
[16  13   6  10   9]
[16   9  10   6  13]
[26  16   9  13  12]
```
   (b)
```
[A]+[A]²
[10   5   4   4   5]
[ 5   6   3   4   5]
[ 4   3   3   2   4]
[ 4   4   2   3   3]
[ 5   5   4   3   6]
```

CHAPTER 6 Summary and Review

Key Terms and Symbols

6.1 linear equation
system of linear equations
solution of a system
substitution method
elimination method
independent system

dependent system
inconsistent system
6.2 equivalent systems
elementary operations
elimination method
row

column
matrix (matrices)
element (entry)
augmented matrix
row operations
row echelon form

Gauss–Jordan method
reduced row echelon form
independent system
inconsistent system
dependent system
parameter

6.3 applications of systems of linear equations

6.4 row matrix (row vector)

column matrix (column vector)

square matrix

additive inverse of a matrix

zero matrix

scalar

product of a scalar and a matrix

6.5 product matrix

identity matrix

inverse matrix

singular matrix

6.6 coefficient matrix

matrix of variables

matrix of constants

input–output analysis

input–output matrix

production matrix

demand matrix

code theory

routing theory

Chapter 6 Key Concepts

Solving Systems of Equations

The following **elementary operations** are used to transform a system of equations into a simpler equivalent system:

1. Interchange any two equations.

2. Multiply both sides of an equation by a nonzero constant.

3. Replace an equation by the sum of itself and a nonzero constant multiple of another equation in the system.

The **elimination method** is a systematic way of using elementary operations to transform a system into an equivalent one that can be solved by **back substitution.** See Section 6.2 for details.

The matrix version of the elimination method uses the following **matrix row operations,** which correspond to using elementary row operations with back substitution on a system of equations:

1. Interchange any two rows.

2. Multiply each element of a row by a nonzero constant.

3. Replace a row by the sum of itself and a nonzero constant multiple of another row in the matrix.

The **Gauss–Jordan method** is an extension of the elimination method for solving a system of linear equations. It uses row operations on the augmented matrix of the system. See Section 6.2 for details.

Operations on Matrices

The **sum** of two $m \times n$ matrices X and Y is the $m \times n$ matrix $X + Y$ in which each element is the sum of the corresponding elements of X and Y. The **difference** of two $m \times n$ matrices X and Y is the $m \times n$ matrix $X - Y$ in which each element is the difference of the corresponding elements of X and Y.

The **product** of a scalar k and a matrix X is the matrix kX, with each element being k times the corresponding element of X.

The **product matrix** AB of an $m \times n$ matrix A and an $n \times k$ matrix B is the $m \times k$ matrix whose entry in the ith row and jth column is the product of the ith row of A and the jth column of B.

The $n \times n$ **identity matrix** I has 1s on the main diagonal from upper left to lower right, with all other entries equal to 0.

If there is a matrix A^{-1} satisfying $AA^{-1} = I$, then A^{-1} is called the **inverse matrix** of A. The inverse matrix A^{-1} for any $n \times n$ matrix A for which A^{-1} exists is found as follows: Form the augmented matrix $[A|I]$, and perform row operations on $[A|I]$ to get the matrix $[I|A^{-1}]$.

Chapter 6 Review Exercises

Solve each of the following systems.

1. $-5x - 3y = -3$
    $2x + y = 4$

2. $3x - y = 8$
    $2x + 3y = 6$

3. $3x - 5y = 16$
    $2x + 3y = -2$

4. $\dfrac{1}{4}x - \dfrac{1}{3}y = -\dfrac{1}{4}$
    $\dfrac{1}{10}x + \dfrac{2}{5}y = \dfrac{2}{5}$

5. **Business** Juan Felipe plans to buy shares of two stocks. One costs $32 per share and pays dividends of $1.20 per share. The other costs $23 per share and pays dividends of $1.40 per share. He has $10,100 to spend and wants to earn dividends of $540. How many shares of each stock should he buy?

6. **Business** Maria has money in two investment funds. Last year, the first fund paid a dividend of 8% and the second a dividend of 2%, and she received a total of $780. This year, the first fund paid a 10% dividend and the second only 1%, and she received $810. How much does she have invested in each fund?

Solve each of the following systems.

7. $x - 2y = 1$
    $4x + 4y = 2$
    $10x + 8y = 4$

8. $4x - y - 2z = 4$
    $x - y - \dfrac{1}{2}z = 1$
    $2x - y - z = 8$

9. $3x + y - z = 3$
$x \qquad + 2z = 6$
$-3x - y + 2z = 9$

10. $x + y - 4z = 0$
$2x + y - 3z = 2$

Solve each of the following systems by using matrix methods.

11. $x + \quad z = -3$
$y - z = 6$
$2x + 3z = 5$

12. $2x + 3y + 4z = 8$
$-x + y - 2z = -9$
$2x + 2y + 6z = 16$

13. $x - 2y + 5z = 3$
$4x + 3y - 4z = 1$
$3x + 5y - 9z = 7$

14. $5x - 8y + z = 1$
$3x - 2y + 4z = 3$
$10x - 16y + 2z = 3$

15. $x - 2y + 3z = 4$
$2x + y - 4z = 3$
$-3x + 4y - z = -2$

16. $3x + 2y - 6z = 3$
$x + y + 2z = 2$
$2x + 2y + 5z = 0$

17. Finance You are given $144 in one-, five-, and ten-dollar bills. There are 35 bills. There are two more ten-dollar bills than five-dollar bills. How many bills of each type are there?

18. Social Services A social service agency provides counseling, meals, and shelter to clients referred by sources I, II, and III. Clients from source I require an average of $100 for food, $250 for shelter, and no counseling. Source II clients require an average of $100 for counseling, $200 for food, and nothing for shelter. Source III clients require an average of $100 for counseling, $150 for food, and $200 for shelter. The agency has funding of $25,000 for counseling, $50,000 for food, and $32,500 for shelter. How many clients from each source can be served?

19. Manufacturing A small business makes woven blankets, rugs, and skirts. Each blanket requires 24 hours for spinning the yarn, 4 hours for dying the yarn, and 15 hours for weaving. Rugs require 30, 5, and 18 hours, and skirts require 12, 3, and 9 hours, respectively. If there are 306, 59, and 201 hours available for spinning, dying, and weaving, respectively, how many of each item can be made? (*Hint:* Simplify the equations you write, if possible, before solving the system.)

20. Manufacturing Each week at a furniture factory, 2000 work hours are available in the construction department, 1400 work hours in the painting department, and 1300 work hours in the packing department. Producing a chair requires 2 hours of construction, 1 hour of painting, and 2 hours for packing. Producing a table requires 4 hours of construction, 3 hours of painting, and 3 hours for packing. Producing a chest requires 8 hours of construction, 6 hours of painting, and 4 hours for packing. If all available time is used in every department, how many of each item are produced each week?

For each of the following, find the dimensions of the matrix and identify any square, row, or column matrices.

21. $\begin{bmatrix} 2 & 3 \\ 5 & 9 \end{bmatrix}$

22. $\begin{bmatrix} 2 & -1 \\ 4 & 6 \\ 5 & 7 \end{bmatrix}$

23. $[12 \quad 4 \quad -8 \quad -1]$

24. $\begin{bmatrix} -7 & 5 & 6 & 4 \\ 3 & 2 & -1 & 2 \\ -1 & 12 & 8 & -1 \end{bmatrix}$

25. $\begin{bmatrix} 6 & 8 & 10 \\ 5 & 3 & -2 \end{bmatrix}$

26. $\begin{bmatrix} -9 \\ 15 \\ 4 \end{bmatrix}$

27. Stock Prices The opening stock price, high price, and closing price for the following companies were reported on April 12, 2016: Coca-Cola Bottling Company Consolidated, $162.45, $164.44, and $163.64; Intel Corporation: $31.88, $31.94, and $31.86; Verizon Communications Inc., $51.66, $52.19, and $51.95. Write these data as a 3×3 matrix. (Data from: www.morningstar.com.)

28. Animal Behavior The activities of a grazing animal can be classified roughly into three categories: grazing, moving, and resting. Suppose horses spend 8 hours grazing, 8 moving, and 8 resting; cattle spend 10 grazing, 5 moving, and 9 resting; sheep spend 7 grazing, 10 moving, and 7 resting; and goats spend 8 grazing, 9 moving, and 7 resting. Write this information as a 4×3 matrix.

Given the matrices

$$A = \begin{bmatrix} 4 & 6 \\ -2 & -2 \\ 5 & 9 \end{bmatrix}, \quad B = \begin{bmatrix} 1 & 2 & -3 \\ 2 & 3 & 0 \\ 0 & 1 & 4 \end{bmatrix}, \quad C = \begin{bmatrix} 5 & 0 \\ -1 & 3 \\ 4 & 7 \end{bmatrix},$$

$$D = \begin{bmatrix} 6 \\ 1 \\ 0 \end{bmatrix}, \quad E = [1 \quad 3 \quad -4], \quad F = \begin{bmatrix} -1 & 2 \\ 6 & 7 \end{bmatrix}, \quad and$$

$$G = \begin{bmatrix} 2 & 5 \\ 1 & 6 \end{bmatrix},$$

find each of the following (if possible).

29. $-B$ **30.** $-D$ **31.** $A + 2C$ **32.** $F + 3G$

33. $2B - 5C$ **34.** $D + E$ **35.** $3A - 2C$ **36.** $G - 2F$

37. Trading Data The change in price (in dollars) and volume of stocks traded (in millions of shares) for the following companies were reported on April 11, 2016: General Electric Company (GE), $-.08$ and 25.1M; Nike Inc. (NKE), -1.50 and 12.6M; United Health Group Incorporated (UNH), $.65 and 3.1M. On the following day, the changes in price and volume data for the same three companies were GE: $.10 and 27.2M, NKE: $.63 and 12.4M, UNH, $.47 and 2.9M. Write the daily data as 3×2 matrices, then use matrix addition to find a matrix representing the total change in price and volume of stock traded for these companies over the two days. (Data from: www.morningstar.com.)

38. Oil Distribution An oil refinery in Tulsa sent 110,000 gallons of oil to a Chicago distributor, 73,000 to a Dallas distributor, and 95,000 to an Atlanta distributor. Another refinery in New Orleans sent the following respective amounts to the same three distributors: 85,000, 108,000, and 69,000. The next month, the two refineries sent the same distributors new shipments of oil as follows: from Tulsa, 58,000 to Chicago, 33,000 to Dallas, and 80,000 to Atlanta; from New Orleans, 40,000, 52,000, and 30,000, respectively.

(a) Write the monthly shipments from the two distributors to the three refineries as 3×2 matrices.

(b) Use matrix addition to find the total amounts sent to the refineries from each distributor.

Use the matrices shown before Exercise 29 to find each of the following (if possible).

39. *AG*

40. *EB*

41. *GA*

42. *CA*

43. *AGF*

44. *EBD*

45. Hockey Injuries In a study, the numbers of head and neck injuries among hockey players wearing full face shields and half face shields were compared. Injury rates were recorded per 1000 athlete exposures for specific injuries that caused a player to miss one or more events.[*]

Players wearing a half shield had the following rates of injury:

Head and face injuries, excluding concussions: 3.54;
Concussions: 1.53;
Neck injuries: .34;
Other injuries: 7.53.

Players wearing a full shield had the following injury rates:

Head and face injuries, excluding concussions: 1.41;
Concussions: 1.57;
Neck injuries: .29;
Other injuries: 6.21.

If an equal number of players in a large league wear each type of shield and the total number of athlete exposures for the league in a season is 8000, use matrix operations to estimate the total number of injuries of each type.

46. Manufacturing An office supply manufacturer makes two kinds of paper clips: standard and extra large. To make a unit of standard paper clips requires $\frac{1}{4}$ hour on a cutting machine and $\frac{1}{2}$ hour on a machine that shapes the clips. A unit of extra-large paper clips requires $\frac{1}{3}$ hour on each machine.

(a) Write this information as a 2×2 matrix (size/machine).

(b) If 48 units of standard and 66 units of extra-large clips are to be produced, use matrix multiplication to find out how many hours each machine will operate. (*Hint:* Write the units as a 1×2 matrix.)

47. Price and Dividends The cost per share of stock was $60.94 for Starbucks Corporation, $113.22 for PepsiCo Inc., and $119.11 for Anheuser-Busch, as reported on May 8, 2017. The dividends per share were $1.00, $3.22, and $4.24, respectively. Your stock portfolio consists of 500 shares of Starbucks, 300 shares of PepsiCo, and 200 shares of Anheuser-Busch. (Data from: www.morningstar.com.)

(a) Write the cost per share and dividend per share data as a 3×2 matrix.

(b) Write the number of shares of each stock as a 1×3 matrix.

(c) Use matrix multiplication to find the total cost and total dividends for this portfolio.

48. Price and Dividends Stan Palla's stock portfolio consists of 1000 shares of Apple Inc., 500 shares of Texas Instruments Inc., and 700 shares of Microsoft Corporation. On May 8, 2017, the cost per share of stock was $153.01 for Apple, $79.46 for Texas Instruments, and $68.94 for Microsoft. The dividends per share

were $2.52, $2.00, and $1.56, respectively. (Data from: www.morningstar.com.)

(a) Write the cost per share and dividend per share data as a 3×2 matrix.

(b) Write the number of shares of each stock as a 1×3 matrix.

(c) Use matrix multiplication to find the total cost and total dividends for this portfolio.

49. If $A = \begin{bmatrix} 3 & 0 \\ 2 & 1 \end{bmatrix}$, find a matrix B such that both AB and BA are defined and $AB \neq BA$.

50. Is it possible to do Exercise 49 if $A = \begin{bmatrix} 4 & 0 \\ 0 & 4 \end{bmatrix}$? Explain.

Find the inverse of each of the following matrices, if an inverse exists for that matrix.

51. $\begin{bmatrix} -2 & 2 \\ 0 & 5 \end{bmatrix}$

52. $\begin{bmatrix} 3 & -1 \\ -5 & 2 \end{bmatrix}$

53. $\begin{bmatrix} 6 & 4 \\ 3 & 2 \end{bmatrix}$

54. $\begin{bmatrix} 15 & -12 \\ 5 & -4 \end{bmatrix}$

55. $\begin{bmatrix} 2 & 0 & 6 \\ 1 & -1 & 0 \\ 0 & 1 & -3 \end{bmatrix}$

56. $\begin{bmatrix} 2 & -1 & 0 \\ 1 & 0 & 2 \\ 1 & -4 & 0 \end{bmatrix}$

57. $\begin{bmatrix} 2 & 3 & 5 \\ -2 & -3 & -5 \\ 1 & 4 & 2 \end{bmatrix}$

58. $\begin{bmatrix} 1 & 3 & 6 \\ 4 & 0 & 9 \\ 5 & 15 & 30 \end{bmatrix}$

59. $\begin{bmatrix} 1 & 3 & -2 & -1 \\ 0 & 1 & 1 & 2 \\ -1 & -1 & 1 & -1 \\ 1 & -1 & -3 & -2 \end{bmatrix}$

60. $\begin{bmatrix} 3 & 2 & 0 & -1 \\ 2 & 0 & 1 & 2 \\ 1 & 2 & -1 & 0 \\ 2 & -1 & 1 & 1 \end{bmatrix}$

Refer again to the matrices shown before Exercise 29 to find each of the following (if possible).

61. F^{-1}

62. G^{-1}

63. $(G - F)^{-1}$

64. $(F + G)^{-1}$

65. B^{-1}

66. Explain why the matrix $\begin{bmatrix} a & 0 \\ c & 0 \end{bmatrix}$, where a and c are nonzero constants, cannot have an inverse.

Solve each of the following matrix equations $AX = B$ for X.

67. $A = \begin{bmatrix} -3 & 4 \\ -1 & 2 \end{bmatrix}$, $B = \begin{bmatrix} 3 \\ -1 \end{bmatrix}$

68. $A = \begin{bmatrix} 1 & 3 \\ -2 & 4 \end{bmatrix}$, $B = \begin{bmatrix} 9 \\ 6 \end{bmatrix}$

69. $A = \begin{bmatrix} 1 & 0 & 2 \\ -1 & 1 & 0 \\ 3 & 0 & 4 \end{bmatrix}$, $B = \begin{bmatrix} 8 \\ 4 \\ -6 \end{bmatrix}$

70. $A = \begin{bmatrix} 2 & 4 & 0 \\ 1 & -2 & 0 \\ 0 & 0 & 3 \end{bmatrix}$, $B = \begin{bmatrix} 72 \\ -24 \\ 48 \end{bmatrix}$

---

[*]Brian Benson, Nicholas Nohtadi, Sarah Rose, and Willem Meeuwisse, "Head and Neck Injuries among Ice Hockey Players Wearing Full Face Shields vs. Half Face Shields," *JAMA*, 282, no. 24, (December 22/29, 1999): 2328–2332.

Use the inverse of the coefficient matrix to solve each system of equations.

71. $x + y = -2$
$2x + 5y = 2$

72. $5x - 3y = -2$
$2x + 7y = -9$

73. $2x + y = 10$
$3x - 2y = 8$

74. $x - 2y = 7$
$3x + y = 7$

75. $x + y + z = 1$
$2x - y = -2$
$3y + z = 2$

76. $x = -3$
$y + z = 6$
$2x - 3z = -9$

77. $3x - 2y + 4z = 4$
$4x + y - 5z = 2$
$-6x + 4y - 8z = -2$

78. $-2x + 3y - z = 1$
$5x - 7y + 8z = 4$
$6x - 9y + 3z = 2$

Solve each of the following problems by any method.

79. Wine Blending A wine maker has two large casks of wine. One is 9% alcohol and the other is 14% alcohol. How many liters of each wine should be mixed to produce 40 liters of wine that is 12% alcohol?

80. Metal Blending A gold merchant has some 12-carat gold (12/24 pure gold) and some 22-carat gold (22/24 pure). How many grams of each could be mixed to get 25 grams of 15-carat gold?

81. Materials Mixing A chemist has a 40% acid solution and a 60% solution. How many liters of each should be used to get 40 liters of a 45% solution?

82. Materials Mixing How many pounds of tea worth $4.60 a pound should be mixed with tea worth $6.50 a pound to get 10 pounds of a mixture worth $5.74 a pound?

83. Manufacturing A machine in a pottery factory takes 3 minutes to form a bowl and 2 minutes to form a plate. The material for a bowl costs $.25, and the material for a plate costs $.20. If the machine runs for 8 hours and exactly $44 is spent for material, how many bowls and plates can be produced?

84. Physical Science A boat travels at a constant speed a distance of 57 kilometers downstream in 3 hours and then turns around and travels 55 kilometers upstream in 5 hours. What are the speeds of the boat and the current?

85. Finance Henry Miller invests $50,000 three ways—at 8%, $8\frac{1}{2}\%$, and 11%. In total, he receives $4436.25 per year in interest. The interest on the 11% investment is $80 more than the interest on the 8% investment. Find the amount he has invested at each rate.

86. Football Tickets Tickets to the homecoming football game cost $4 for students, $10 for alumni, $12 for other adults, and $6 for children. The total attendance was 3750, and the ticket receipts were $29,100. Six times more students than children attended. The number of alumni was $\frac{4}{5}$ of the number of students. How many students, alumni, other adults, and children were at the game?

87. Input–Output Analysis Given the input–output matrix
$A = \begin{bmatrix} 0 & \frac{1}{4} \\ \frac{1}{2} & 0 \end{bmatrix}$ and the demand matrix $D = \begin{bmatrix} 2100 \\ 1400 \end{bmatrix}$, find each of the following.

(a) $I - A$
(b) $(I - A)^{-1}$
(c) The production matrix $X = (I - A)^{-1}D$

88. Input–Output Analysis An economy depends on two commodities: goats and cheese. It takes $\frac{2}{3}$ unit of goats to produce 1 unit of cheese and $\frac{1}{2}$ unit of cheese to produce 1 unit of goats.

(a) Write the input–output matrix for this economy.

(b) Find the production required to satisfy a demand of 400 units of cheese and 800 units of goats.

Use technology to do Exercises 89–92.

89. Input–Output Analysis In a simple economic model, a country has two industries: agriculture and manufacturing. To produce $1 of agricultural output requires $.10 of agricultural output and $.40 of manufacturing output. To produce $1 of manufacturing output requires $.70 of agricultural output and $.20 of manufacturing output. If agricultural demand is $60,000 and manufacturing demand is $20,000, what must each industry produce? (Round answers to the nearest whole number.)

90. Input–Output Analysis Here is the input–output matrix for a small economy:

	Agriculture	Services	Mining	Manufacturing
Agriculture	.02	.9	0	.001
Services	0	.4	0	.06
Mining	.01	.02	.06	.07
Manufacturing	.25	.9	.9	.4

(a) How many units from each sector does the service sector require to produce 1 unit?

(b) What production levels are needed to meet a demand for 760 units of agriculture, 1600 units of services, 1000 units of mining, and 2000 units of manufacturing?

(c) How many units of manufacturing production are used up in the economy's production process?

91. Input–Output Analysis Use this input–output matrix to answer the questions below.

	Agriculture	Construction	Energy	Manufacturing	Transportation
Agriculture	.18	.017	.4	.005	0
Construction	.14	.018	.09	.001	0
Energy	.9	0	.4	.06	.002
Manufacturing	.19	.16	.1	.008	.5
Transportation	.14	.25	.9	.4	.12

(a) How many units from each sector does the energy sector require to produce 1 unit?

(b) If the economy produces 28,067 units of agriculture, 9383 units of construction, 51,372 units of energy, 61,364 units of manufacturing, and 90,403 units of transportation, how much is available from each sector to satisfy the demand from consumers and others outside the production system?

92. Input–Output Analysis A new demand matrix for the economy in the previous exercise is

$$\begin{bmatrix} 2400 \\ 850 \\ 1400 \\ 3200 \\ 1800 \end{bmatrix}.$$

How much must each sector produce to meet this demand?

93. Routing Matrix M represents the number of direct train lines between four stations in a metropolitan subway system.

$$\begin{array}{c} \\ A \\ B \\ C \\ D \end{array} \begin{array}{c} A \ B \ C \ D \\ \begin{bmatrix} 0 & 1 & 1 & 0 \\ 1 & 0 & 1 & 1 \\ 1 & 1 & 0 & 1 \\ 0 & 1 & 1 & 0 \end{bmatrix} = M \end{array}$$

(a) Find M^2.

(b) How many routes between station A and station D go through exactly one other station?

(c) Find M^3.

(d) How many routes between station A and station D go through exactly two other stations?

94. Routing The following matrix represents the number of direct flights between four cities:

$$\begin{array}{c} \\ A \\ B \\ C \\ D \end{array} \begin{array}{c} A \ B \ C \ D \\ \begin{bmatrix} 0 & 1 & 0 & 1 \\ 1 & 0 & 0 & 1 \\ 0 & 0 & 0 & 1 \\ 1 & 1 & 1 & 0 \end{bmatrix} . \end{array}$$

(a) Find the number of one-stop flights between cities A and C.

(b) Find the total number of flights between cities B and C that are either direct or one stop.

(c) Find the matrix that gives the number of two-stop flights between these cities.

95. Encoding Use the matrix $M = \begin{bmatrix} 2 & 6 \\ 1 & 4 \end{bmatrix}$ to encode the message "leave now."

96. Decoding What matrix should be used to decode the answer to the previous exercise?

Case Study 6 Airline Route Maps

Airline route maps are usually published on airline Web sites, as well as in in-flight magazines. The purpose of these maps is to show what cities are connected to each other by nonstop flights provided by the airline. We can think of these maps as another type of **graph**, and we can use matrix operations to answer questions of interest about the graph. To study these graphs, a bit of terminology will be helpful. In this context, a **graph** is a set of points called **vertices** or **nodes** and a set of lines called **edges** connecting some pairs of vertices. Two vertices connected by an edge are said to be **adjacent**. Suppose, for example, that Stampede Air is a short-haul airline serving the state of Texas whose route map is shown in Figure 1 below. The vertices are the cities to which Stampede Air flies, and two vertices are connected if there is a nonstop flight between them.

Some natural questions arise about graphs. It might be important to know if two vertices are connected by a sequence of two edges, even if they are not connected by a single edge. In the route map, Huston and Lubbock are connected by a two-edge sequence, meaning that a passenger would have to stop in Austin while flying between those cities on Stampede Air. It might be important to know if it is possible to get from a vertex to another vertex in a given number of flights. In the example, a passenger on Stampede Air can get from any city in the company's network to any other city, given enough flights. But how many flights are enough? This is another issue of interest: What is the minimum number of steps required to get from one vertex to another? What is the minimum number of steps required to get from any vertex

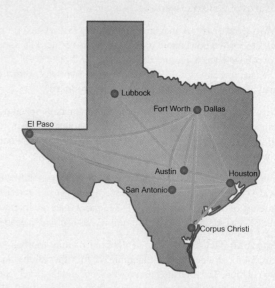

Figure 1 Stampede Air route map

on the graph to any other? While these questions are relatively easy to answer for a small graph, as the number of vertices and edges grows, it becomes harder to keep track of all the different ways the vertices are connected. Matrix notation and computation can help to answer these questions.

The **adjacency matrix** for a graph with n vertices is an $n \times n$ matrix whose (i, j) entry is 1 if the ith and jth vertices are connected and 0 if they are not. If the vertices in the Stampede Air graph respectively correspond to Austin (A), El Paso (E), Houston (H), Dallas-Fort Worth (D), San Antonio (S), Corpus Christi (C), and Lubbock (L), then the adjacency matrix for Stampede Air is as follows.

$$
A = \begin{array}{c} \\ \\ \\ \\ \\ \\ \\ \end{array}
\begin{array}{cccccccc}
\text{A} & \text{E} & \text{H} & \text{D} & \text{S} & \text{C} & \text{L} & \\
\left[\begin{array}{ccccccc} 0 & 1 & 1 & 1 & 0 & 0 & 1 \\ 1 & 0 & 1 & 1 & 0 & 0 & 0 \\ 1 & 1 & 0 & 1 & 1 & 1 & 0 \\ 1 & 1 & 1 & 0 & 1 & 1 & 0 \\ 0 & 0 & 1 & 1 & 0 & 0 & 0 \\ 0 & 0 & 1 & 1 & 0 & 0 & 0 \\ 1 & 0 & 0 & 0 & 0 & 0 & 0 \end{array}\right] & \begin{array}{c} \text{A} \\ \text{E} \\ \text{H} \\ \text{D} \\ \text{S} \\ \text{C} \\ \text{L} \end{array}
\end{array}
$$

Adjacency matrices can be used to address the questions about graphs raised earlier. Which vertices are connected by a two-edge sequence? How many different two-edge sequences connect each pair of vertices? Consider the matrix A^2, which is A multiplied by itself. For example, let the $(6, 2)$ entry in the matrix A^2 be named b_{62}. This entry in A^2 is the product of the 6th row of A and the 2nd column of A, or

$$
\begin{aligned}
b_{62} &= a_{61}a_{12} + a_{62}a_{22} + a_{63}a_{32} + a_{64}a_{42} + a_{65}a_{52} + a_{66}a_{62} + a_{67}a_{72} \\
&= 0 \cdot 1 + 0 \cdot 0 + 1 \cdot 1 + 1 \cdot 1 + 0 \cdot 0 + 0 \cdot 0 + 0 \cdot 0 \\
&= 2,
\end{aligned}
$$

which happens to be the number of two-flight sequences that connect city 6 (Corpus Christi) and city 2 (El Paso). A careful look at Figure 1 confirms this fact. This calculation works because, in order for a two-flight sequence to occur between Corpus Christi and El Paso, Corpus Christi and El Paso must each connect to an intermediate city. Since Corpus Christi connects to Houston and Houston connects to El Paso, $a_{63}a_{32} = 1 \cdot 1 = 1$. Thus, there is one two-flight sequence between Corpus Christi and El Paso that passes through Houston. Since Corpus Christi does not connect to Austin (city 1), but Austin does connect with El Paso, $a_{61}a_{12} = 0 \cdot 1 = 0$. Hence, there is no two-flight sequence from Corpus Christi to El Paso that passes through Austin. To find the total number of two-flight sequences between Corpus Christi and El Paso, simply sum over all intermediate points. Notice that this sum, which is

$$
a_{61}a_{12} + a_{62}a_{22} + a_{63}a_{32} + a_{64}a_{42} + a_{65}a_{52} + a_{66}a_{62} + a_{67}a_{72},
$$

is just b_{62}, the $(6, 2)$ entry in the matrix A^2. So we see that the number of two-step sequences between vertex i and vertex j in a graph with adjacency matrix A is the (i, j) entry in A^2. A more general result is the following:

> **The number of k-step sequences between vertex i and vertex j in a graph with adjacency matrix A is the (i, j) entry in A^k.**

If A is the adjacency matrix for Figure 1, then

$$
A^2 = \begin{array}{c} \\ \\ \\ \\ \\ \\ \\ \end{array}
\begin{array}{cccccccc}
\text{A} & \text{E} & \text{H} & \text{D} & \text{S} & \text{C} & \text{L} & \\
\left[\begin{array}{ccccccc} 4 & 2 & 2 & 2 & 2 & 2 & 0 \\ 2 & 3 & 2 & 2 & 2 & 2 & 1 \\ 2 & 2 & 5 & 4 & 1 & 1 & 1 \\ 2 & 2 & 4 & 5 & 1 & 1 & 1 \\ 2 & 2 & 1 & 1 & 2 & 2 & 0 \\ 2 & 2 & 1 & 1 & 2 & 2 & 0 \\ 0 & 1 & 1 & 1 & 0 & 0 & 1 \end{array}\right] & \begin{array}{c} \text{A} \\ \text{E} \\ \text{H} \\ \text{D} \\ \text{S} \\ \text{C} \\ \text{L} \end{array}
\end{array}
$$

and

$$
A^3 = \begin{array}{c} \\ \\ \\ \\ \\ \\ \\ \end{array}
\begin{array}{cccccccc}
\text{A} & \text{E} & \text{H} & \text{D} & \text{S} & \text{C} & \text{L} & \\
\left[\begin{array}{ccccccc} 6 & 8 & 12 & 12 & 4 & 4 & 4 \\ 8 & 6 & 11 & 11 & 4 & 4 & 2 \\ 12 & 11 & 10 & 11 & 9 & 9 & 2 \\ 12 & 11 & 11 & 10 & 9 & 9 & 2 \\ 4 & 4 & 9 & 9 & 2 & 2 & 2 \\ 4 & 4 & 9 & 9 & 2 & 2 & 2 \\ 4 & 2 & 2 & 2 & 2 & 2 & 0 \end{array}\right] & \begin{array}{c} \text{A} \\ \text{E} \\ \text{H} \\ \text{D} \\ \text{S} \\ \text{C} \\ \text{L} \end{array}
\end{array}
$$

Since the $(6, 3)$ entry in A^2 is 1, there is one two-step sequence from Corpus Christi to Houston. Likewise, there are four three-step sequences between El Paso and San Antonio, since the $(2, 5)$ entry in A^3 is 4.

In observing the figure, note that some two-step or three-step sequences may not be meaningful. On the Stampede Air route map, Dallas-Fort Worth is reachable in two steps from Austin (via El Paso or Houston), but in reality this does not matter, since there is a nonstop flight between the two cities. A better question to ask of a graph might be, "What is the least number of edges that must be traversed to go from vertex i to vertex j?"

To answer this question, consider the matrix $S_k = A + A^2 + \cdots + A^k$. The (i, j) entry in this matrix tallies the number of ways to get from vertex i to vertex j in k steps or less. If such a trip is impossible, this entry will be zero. Thus, to find the shortest number of steps between the vertices, continue to compute S_k as k increases; the first k for which the (i, j) entry in S_k is nonzero is the shortest number of steps between i and j. Note that although the shortest number of steps may be computed, the method does not determine what those steps are.

Exercises

1. Which Stampede Air cities may be reached by a two-flight sequence from San Antonio? Which may be reached by a three-flight sequence?

2. It was shown previously that there are four three-step sequences between El Paso and San Antonio. Describe these three-step sequences.

3. Which trips in the Stampede Air network take the greatest number of flights?

4. Suppose that Vacation Air is an airline serving Jacksonville, Orlando, West Palm Beach, Miami, and Tampa, Florida. The route map is given in Figure 2. Produce an adjacency matrix B for this map, listing the cities in the order given.

5. Find B^2, the square of the adjacency matrix found in Exercise 4. What does this matrix tell you?

6. Find $B + B^2$. What can you conclude from this calculation?

Figure 2 Vacation Air route map

Extended Project

Search the Internet for a small regional airline that serves your area and use their route map to produce an adjacency matrix. Determine which trips using this airline take the largest number of flights. How many flights did these trips take? Prepare a report that includes an overview of the airline, the route map, an explanation of the adjacency matrix, and your findings about which trips take the largest number of flights.

Linear Programming

7

CHAPTER

Linear programming is one of the most remarkable (and useful) mathematical techniques developed in the last 65 years. It is used to deal with a variety of issues faced by businesses, financial planners, medical personnel, sports leagues, and others. Typical applications include) maximizing company profits by adjusting production schedules, minimizing shipping costs by locating warehouses efficiently, and maximizing pension income by choosing the best mix of financial products.

Many real-world problems involve inequalities. For example, a factory may have no more than 200 workers on a shift and must manufacture at least 3000 units at a cost of no more than $35 each. How many workers should it have per shift in order to produce the required number of units at minimal cost? *Linear programming* is a method for finding the optimal (best possible) solution for such problems—if there is one.

In this chapter, we shall study two methods of solving linear programming problems: the graphical method and the simplex method. The graphical method requires a knowledge of **linear inequalities,** which are inequalities involving only first-degree polynomials in x and y. So we begin with a study of such inequalities.

7.1 Graphing Linear Inequalities in Two Variables

Examples of linear inequalities in two variables include

$$x + 2y < 4, \qquad 3x + 2y > 6, \qquad \text{and} \qquad 2x - 5y \geq 10.$$

A solution of a linear inequality is an ordered pair that satisfies the inequality. For example (4, 4) is a solution of

$$3x - 2y \leq 6.$$

(Check by substituting 4 for x and 4 for y.) A linear inequality has an infinite number of solutions, one for every choice of a value for x. The best way to show these solutions is to sketch the **graph of the inequality,** which consists of all points in the plane whose coordinates satisfy the inequality.

EXAMPLE 1 Graph the inequality $3x - 2y \leq 6$.

Solution First, solve the inequality for y:

$$3x - 2y \leq 6$$
$$-2y \leq -3x + 6$$
$$y \geq \frac{3}{2}x - 3 \qquad \text{Multiply by } -\frac{1}{2}; \text{ reverse the inequality.}$$
$$y \geq 1.5x - 3.$$

This inequality has the same solutions as the original one. To solve it, note that the points on the line $y = 1.5x - 3$ certainly satisfy $y \geq 1.5x - 3$. Plot some points, and graph this line, as in Figure 7.1.

The points on the line satisfy "y *equals* $1.5x - 3$." The points satisfying "y is *greater than* $1.5x - 3$" are the points *above* the line (because they have larger second coordinates than the points on the line; see Figure 7.2). Similarly, the points satisfying

$$y < 1.5x - 3$$

lie below the line (because they have smaller second coordinates), as shown in Figure 7.2. The line $y = 1.5x - 3$ is the **boundary line.**

Thus, the solutions of $y \geq 1.5x - 3$ are all points *on or above* the line $y = 1.5x - 3$. The line and the shaded region of Figure 7.3 make up the graph of the inequality $y \geq 1.5x - 3$. ✓ 1

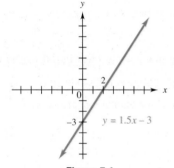

Figure 7.1

✓ **Checkpoint 1**

Graph the given inequalities.

(a) $2x + 5y \leq 10$

(b) $x - y \geq 4$

Answers to Checkpoint exercises are found at the end of the section.

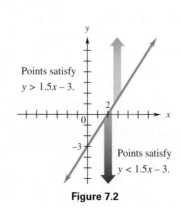

Points satisfy $y > 1.5x - 3$.

Points satisfy $y < 1.5x - 3$.

Figure 7.2

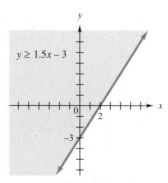

$y \geq 1.5x - 3$

Figure 7.3

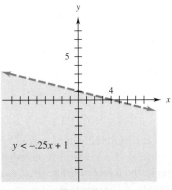

Figure 7.4

✓ Checkpoint 2

Graph the given inequalities.

(a) $2x + 3y > 12$

(b) $3x - 2y < 6$

EXAMPLE 2   Graph $x + 4y < 4$.

Solution First obtain an equivalent inequality by solving for y:

$$4y < -x + 4$$

$$y < -\frac{1}{4}x + 1$$

$$y < -.25x + 1.$$

The boundary line is $y = -.25x + 1$, but it is *not* part of the solution, since points *on* the line do not satisfy $y < -.25x + 1$. To indicate this, the line is drawn dashed in Figure 7.4. The points *below* the boundary line are the solutions of $y < -.25x + 1$, because they have smaller second coordinates than the points on the line $y = -.25x + 1$. The shaded region in Figure 7.4 (excluding the dashed line) is the graph of the inequality $y < -.25x + 1$. ✓₂

Examples 1 and 2 show that the solutions of a linear inequality form a **half-plane** consisting of all points on one side of the boundary line (and possibly the line itself). When an inequality is solved for y, the inequality symbol immediately tells whether the points above ($>$), on ($=$), or below ($<$) the boundary line satisfy the inequality, as summarized here.

Inequality	Solution Consists of All Points
$y \geq mx + b$	*on or above* the line $y = mx + b$
$y > mx + b$	*above* the line $y = mx + b$
$y \leq mx + b$	*on or below* the line $y = mx + b$
$y < mx + b$	*below* the line $y = mx + b$

When graphing by hand, draw the boundary line $y = mx + b$ solid when it is included in the solution ($\geq$ or $\leq$ inequalities) and dashed when it is not part of the solution ($>$ or $<$ inequalities).

⌨ TECHNOLOGY TIP To shade the area above or below the graph of Y_1 on TI-84+, go to the Y = menu and move the cursor to the left of Y_1. Press ENTER until the correct shading pattern appears (◥ for above the line and ◣ for below the line). Then press GRAPH. On TI-86/89, use the STYLE key in the Y = menu instead of the ENTER key. For other calculators, consult your instruction manual.

Figure 7.5

✓ Checkpoint 3 ⌨

Use a graphing calculator to graph $2x < y$.

EXAMPLE 3   Graph $5y - 2x \leq 10$.

Solution Solve the inequality for y:

$$5y \leq 2x + 10$$

$$y \leq \frac{2}{5}x + 2$$

$$y \leq .4x + 2.$$

The graph consists of all points on or below the boundary line $y \leq .4x + 2$, as shown in Figure 7.5. (See the Technology Tip.) ✓₃

❗ CAUTION You cannot tell from a calculator-produced graph whether the boundary line is included. It is included in Figure 7.5, but not in the answer to Checkpoint 3.

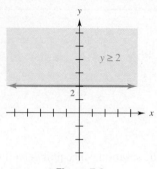

Figure 7.6

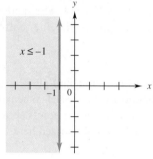

Figure 7.7

✓ Checkpoint 4

Graph each of the following.

(a) $x \geq 3$

(b) $y - 3 \leq 0$

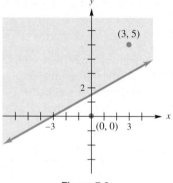

Figure 7.8

EXAMPLE 4 Graph each of the given inequalities.

(a) $y \geq 2$

Solution The boundary line is the horizontal line $y = 2$. The graph consists of all points on or above this line (Figure 7.6).

(b) $x \leq -1$

Solution This inequality does not fit the pattern discussed earlier, but it can be solved by a similar technique. Here, the boundary line is the vertical line $x = -1$, and it is included in the solution. The points satisfying $x < -1$ are all points to the left of this line (because they have x-coordinates smaller than -1). So the graph consists of the points that are *on or to the left of* the vertical line $x = -1$, as shown in Figure 7.7. ✓ 4

An alternative technique for solving inequalities that does not require solving for y is illustrated in the next example. Feel free to use it if you find it easier than the technique presented in Examples 1–4.

EXAMPLE 5 Graph $4y - 2x \geq 6$.

Solution The boundary line is $4y - 2x = 6$, which can be graphed by finding its intercepts:

$$x\text{-intercept: Let } y = 0. \qquad y\text{-intercept: Let } x = 0.$$
$$4(0) - 2x = 6 \qquad\qquad 4y - 2(0) = 6$$
$$x = -3. \qquad\qquad y = \frac{6}{4} = 1.5.$$

The graph contains the half-plane above or below this line. To determine which, choose a test point—any point not on the boundary line, say, $(0, 0)$. Letting $x = 0$ and $y = 0$ in the inequality produces

$$4(0) - 2(0) \geq 6, \qquad \text{a } \textit{false} \text{ statement.}$$

Therefore, $(0, 0)$ is not in the solution. So the solution is the half-plane that does *not* include $(0, 0)$, as shown in Figure 7.8. If a different test point is used, say, $(3, 5)$, then substituting $x = 3$ and $y = 5$ in the inequality produces

$$4(5) - 2(3) \geq 6, \qquad \text{a } \textit{true} \text{ statement.}$$

Therefore, the solution of the inequality is the half-plane containing $(3, 5)$, as shown in Figure 7.8.

📄 **NOTE** When using the method of Example 5, $(0, 0)$ is the best choice for the test point because it makes the calculation very easy. The only time that $(0, 0)$ cannot be used is for inequalities of the form $ax + by \geq 0$ (or $>$ or $<$ or $\leq$); in such cases, $(0, 0)$ is on the line $ax + by = 0$.

Systems of Inequalities

Real-world problems often involve many inequalities. For example, a manufacturing problem might produce inequalities resulting from production requirements, as well as inequalities about cost requirements. A set of at least two inequalities is called a **system of inequalities**. The **graph** of a system of inequalities is made up of all those points which satisfy *all* the inequalities of the system.

EXAMPLE 6 Graph the system

$$3x + y \leq 12$$
$$x \leq 2y.$$

Solution First, solve each inequality for y:

$$3x + y \leq 12 \qquad\qquad x \leq 2y$$
$$y \leq -3x + 12 \qquad y \geq \frac{x}{2}.$$

Then the original system is equivalent to this one:

$$y \leq -3x + 12$$
$$y \geq \frac{x}{2}.$$

✓ **Checkpoint 5**

Graph the system

$$x + y \leq 6$$
$$2x + y \geq 4.$$

The solutions of the first inequality are the points *on or below* the line $y = -3x + 12$ (Figure 7.9). The solutions of the second inequality are the points *on or above* the line $y = x/2$ (Figure 7.10). So the solutions of the *system* are the points that satisfy both of these conditions, as shown in Figure 7.11. ✓5

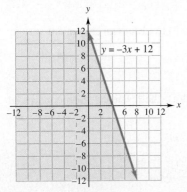

Figure 7.9

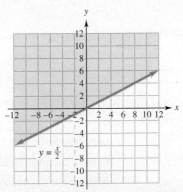

Figure 7.10

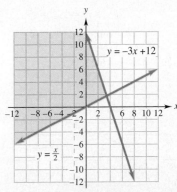

Figure 7.11

The shaded region in Figure 7.11 is sometimes called the **region of feasible solutions,** or just the **feasible region,** since it consists of all the points that satisfy (are feasible for) every inequality of the system.

EXAMPLE 7 Graph the feasible region for the system

$$2x - 5y \leq 10$$
$$x + 2y \leq 8$$
$$x \geq 0, y \geq 0.$$

Solution Begin by solving the first two inequalities for y:

$$2x - 5y \leq 10 \qquad\qquad x + 2y \leq 8$$
$$-5y \leq -2x + 10 \qquad 2y \leq -x + 8$$
$$y \geq .4x - 2 \qquad\qquad y \leq -.5x + 4.$$

Then the original system is equivalent to this one:

$$y \geq .4x - 2$$
$$y \leq -.5x + 4$$
$$x \geq 0, y \geq 0.$$

The inequalities $x \geq 0$ and $y \geq 0$ restrict the graph to the first quadrant. So the feasible region consists of all points in the first quadrant that are on or above the line $y = .4x - 2$ *and* on or below the line $y = -.5x + 4$. In the calculator-generated graph of Figure 7.12, the feasible region is the darkest region. This is confirmed by the hand-drawn graph of the feasible region in Figure 7.13.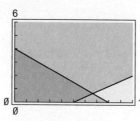

✓ **Checkpoint 6**

Graph the feasible region of the system

$$x + 4y \leq 8$$
$$x - y \geq 3$$
$$x \geq 0, y \geq 0.$$

Figure 7.12

Figure 7.13

EXAMPLE 8   Graph the feasible region for the system

$$5y - 2x \leq 10$$
$$x \geq 3, y \geq 2.$$

Solution Solve the first inequality for y (as in Example 3) to obtain an equivalent system:

$$y \leq .4x + 2$$
$$x \geq 3, y \geq 2.$$

As shown in Example 3, Checkpoint 4, and Example 4(a), the feasible region consists of all points that lie

on or below the line $y \leq .4x + 2$ *and*

on or to the right of the vertical line $x = 3$ *and*

on or above the horizontal line $y = 2$,

as shown in Figure 7.14.

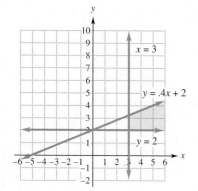

Figure 7.14

EXAMPLE 9   **Bond Investing** In 2017, the T. Rowe Price Tax-Free High-Yield Fund paid a 3-year return of 5%, and the BlackRock California Municipal Opportunities Fund paid a 3-year return of 4%. Sherri would like to invest up to $75,000 and obtain at least $2500 in interest over the 3-year period. Write a system of inequalities expressing these conditions, and graph the feasible region. (Data from: www.morningstar.com.)

Solution Let x represent the amount invested in the T. Rowe Price fund and y the amount invested in the BlackRock fund. Then make a chart that summarizes the given information.

Funds	Value	Return
T. Rowe Price	x	$5\% = .05$
BlackRock	y	$4\% = .04$
Maximum or minimum	$75,000	$2500

We must have $x \geq 0$ and $y \geq 0$ because Sherri cannot invest negative dollars into the two funds. In terms of investing up to $75,000, we have

$$x + y \leq 75,000$$
$$y \leq -x + 75,000. \quad \text{Add } -x \text{ to each side.}$$

Similarly, the interest requirement has that

$$.05x + .04y \geq 2500$$

$$5x + 4y \geq 250{,}000 \qquad \text{Multiply both sides by 100.}$$

$$4y \geq -5x + 250{,}000 \qquad \text{Add } -5x \text{ to each side.}$$

$$y \geq -\frac{5}{4}x + 62{,}500. \qquad \text{Divide both sides by 4.}$$

So we must solve the system

$$y \leq -x + 75{,}000,$$

$$y \geq -\frac{5}{4}x + 62{,}500,$$

$$x \geq 0, y \geq 0.$$

The feasible region is shown in Figure 7.15.

✓ **Checkpoint 7**

Mylene wishes to invest in two bond products. Bond A has an annual return of 4%, and Bond B has an annual return of 5.2%. She wants to earn at least $1800 annually. She can invest up to $40,000 in the two bonds. Write a system of inequalities expressing these conditions, and graph the feasible region.

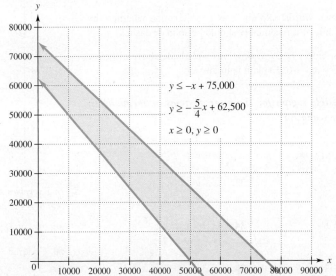

Figure 7.15

7.1 Exercises

Match the inequality with its graph, which is one of the ones shown.

1. $y \geq -x - 2$

2. $y \leq 2x - 2$

3. $y \leq x + 2$

4. $y \geq x + 1$

5. $6x + 4y \geq -12$

6. $3x - 2y \geq -4$

A.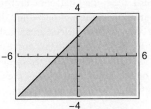

B.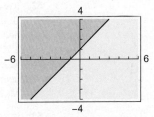

C.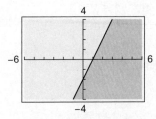

D.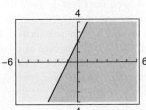

E.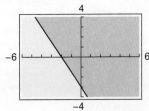

F.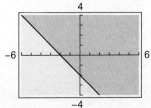

Graph each of the given linear inequalities. (See Examples 1–5.)

7. $y < 5 - 2x$ **8.** $y < x + 3$

9. $x - 2y \geq 3$ **10.** $2x + 5y \geq 10$

11. $2x - y \leq 4$ **12.** $4x - 3y \leq 24$

13. $y \leq -4$ **14.** $x \geq -2$

15. $3x - 2y \geq 18$ **16.** $3x + 2y \geq -4$

17. $3x + 4y \geq 12$ **18.** $4x - 3y > 9$

19. $2x - 4y \leq 3$ **20.** $4x - 3y < 12$

21. $x \leq 5y$ **22.** $2x \geq y$

23. $-3x \leq y$ **24.** $-x \geq 6y$

25. $y \leq x$ **26.** $y > -2x$

27. In your own words, explain how to determine whether the boundary line of an inequality should be solid or dashed.

28. When graphing $y \leq 3x - 6$, would you shade above or below the line $y = 3x - 6$? Explain your answer.

Graph the feasible region for the given systems of inequalities. (See Examples 6 and 7.)

29. $y \geq 3x - 6$
      $y \geq -x + 1$

30. $x + y \leq 4$
      $x - y \geq 2$

31. $2x + y \leq 5$
      $x + 2y \leq 5$

32. $x - y \geq 1$
      $x \leq 3$

33. $2x + y \geq 8$
      $4x - y \leq 3$

34. $4x + y \geq 9$
      $2x + 3y \leq 7$

35. $2x - y \leq 1$
      $3x + y \leq 6$

36. $x + 3y \leq 6$
      $2x + 4y \geq 7$

37. $-x - y \leq 5$
      $2x - y \leq 4$

38. $6x - 4y \geq 8$
      $3x + 2y \geq 4$

39. $3x + y \geq 6$
      $x + 2y \geq 7$
         $x \geq 0$
         $y \geq 0$

40. $2x + 3y \geq 12$
      $x + y \geq 4$
        $x \geq 0$
        $y \geq 0$

41. $-2 \leq x \leq 3$
      $-1 \leq y \leq 5$
      $2x + y \leq 6$

42. $-2 \leq x \leq 2$
         $y \geq 1$
      $x - y \geq 0$

43. $2y - x \geq -5$
        $y \leq 3 + x$
         $x \geq 0$
         $y \geq 0$

44. $2x + 3y \leq 12$
      $2x + 3y \geq -6$
      $3x + y \leq 4$
         $x \geq 0$
         $y \geq 0$

45. $3x + 4y \geq 12$
      $2x - 3y \leq 6$
     $0 \leq y \leq 2$
         $x \geq 0$

46. $0 \leq x \leq 9$
      $x - 2y \geq 4$
      $3x + 5y \leq 30$
         $y \geq 0$

Find a system of inequalities that has the given graph.

47.

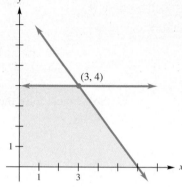

48.

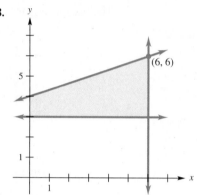

C *In Exercises 49 and 50, find a system of inequalities whose feasible region is the interior of the given polygon.*

49. Rectangle with vertices $(2, 3)$, $(2, -1)$, $(7, 3)$, and $(7, -1)$

50. Triangle with vertices $(2, 4)$, $(-4, 0)$, and $(2, -1)$

In each of the following, write a system of inequalities that describes all the given conditions, and graph the feasible region of the system. (See Example 9.)

51. Investing As of April 2017, the Nuveen Core Plus Bond C fund yielded an annual return of 3.6%, and the Nuveen Core Plus Bond R6 fund yielded an annual return of 5.0%. You would like to invest up to $50,000 and obtain at least $2100 in interest. Let x be the amount invested in the C fund, and let y be the amount invested in the R6 fund. (Data from: www.morningstar.com.)

52. Investing As of April 2017, the Guggenheim Total Return Bond A fund yielded an annual return of 6.0%, and the Quantified Managed Income Advisor fund yielded an annual return of 4.7%. You would like to invest up to $40,000 and earn at least $2250 in interest. Let x be the amount invested in the Guggenheim fund, and let y be the amount invested in the Quantified Managed fund. (Data from: www.morningstar.com.)

53. Investing As of April 2017, the Franklin Total Return Advisor fund yielded an annual return of 2.4%, and the Highland Fixed Income Y fund yielded an annual return of 4.2%. You would like to invest up to $60,000 and earn at least $2016 in interest. Let x be the amount invested in the Franklin fund, and let y be the amount invested in the Highland fund. (Data from: www.morningstar.com.)

54. Investing As of April 2017, the Brandes Core Plus Fixed Income fund yielded an annual return of 2.7%, and the Transamerica Institutional Asset Allocation Short Horizon fund yielded an annual return of 4.1%. You would like to invest up to $80,000 and earn at least $2214 in interest. Let x be the amount invested in the Brandes fund, and let y be the amount invested in the Transamerica fund. (Data from: www.morningstar.com.)

55. Business Joyce Blake is the marketing director for a new company selling a fashion collection for young women and she wishes to place ads in two magazines: *Vogue* and *Elle*. Joyce estimates that each one-page ad in *Vogue* will be read by 1.3 million people and each one-page ad in *Elle* will be read by 1.1 million people. Joyce wants to reach at least 8 million readers and to place at least 2 ads in each magazine. (Data from: www.nyjobsource.com.)

56. Business Corey Steinbruner is the marketing director for a new vitamin supplement and he wants to place ads in two magazines: *Sports Illustrated* and *Men's Health*. Corey estimates that each one-page ad in *Sports Illustrated* will be read by 3.2 million people and each one-page ad in *Men's Health* will be read by 1.9 million people. Corey wants to reach at least 20 million readers and to place at least 3 ads in each magazine. (Data from: www.nyjobsource.com.)

✓**Checkpoint Answers**

1. (a)

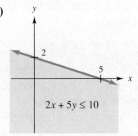

$2x + 5y \le 10$

(b)

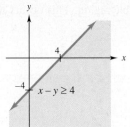

$x - y \ge 4$

2. (a)

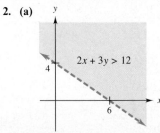

$2x + 3y > 12$

(b)

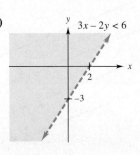

$3x - 2y < 6$

3.

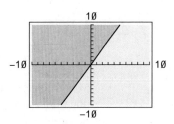

4. (a)

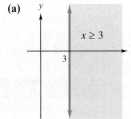

$x \ge 3$

(b)

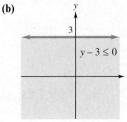

$y - 3 \le 0$

5.

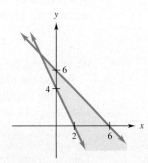

6.

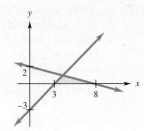

7.

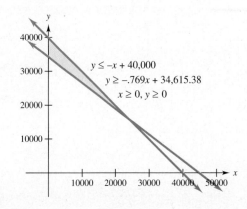

$y \le -x + 40{,}000$
$y \ge -.769x + 34{,}615.38$
$x \ge 0, y \ge 0$

7.2 Linear Programming: The Graphical Method

Many problems in business, science, and economics involve finding the optimal value of a function (for instance, the maximum value of the profit function or the minimum value of the cost function), subject to various **constraints** (such as transportation costs, environmental protection laws, availability of parts, and interest rates). **Linear programming** deals with such situations. In linear programming, the function to be optimized, called the **objective function,** is linear and the constraints are given by linear inequalities. Linear programming problems that involve only two variables can be solved by the graphical method, explained in Example 1.

EXAMPLE 1 Find the maximum and minimum values of the objective function $z = 2x + 5y$, subject to the following constraints:

$$3x + 2y \leq 6$$
$$-2x + 4y \leq 8$$
$$x + y \geq 1$$
$$x \geq 0, y \geq 0.$$

Solution First, graph the feasible region of the system of inequalities (Figure 7.16). The points in this region or on its boundaries are the only ones that satisfy all the constraints. However, each such point may produce a different value of the objective function. For instance, the points (.5, 1) and (1, 0) in the feasible region lead to the respective values

$$z = 2(.5) + 5(1) = 6 \quad \text{and} \quad z = 2(1) + 5(0) = 2.$$

We must find the points that produce the maximum and minimum values of z.

To find the maximum value, consider various possible values for z. For instance, when $z = 0$, the objective function is $0 = 2x + 5y$, whose graph is a straight line. Similarly, when z is 5, 10, and 15, the objective function becomes (in turn)

$$5 = 2x + 5y, \quad 10 = 2x + 5y, \quad \text{and} \quad 15 = 2x + 5y.$$

These four lines are graphed in Figure 7.17. (All the lines are parallel because they have the same slope.) The figure shows that z cannot take on the value 15, because the graph for $z = 15$ is entirely outside the feasible region. The maximum possible value of z will be obtained from a line parallel to the others and between the lines representing the objective function when $z = 10$ and $z = 15$. The value of z will be as large as possible, and all constraints will be satisfied, if this line just touches the feasible region. This occurs with the green line through point A.

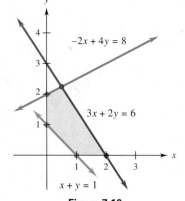

Figure 7.16

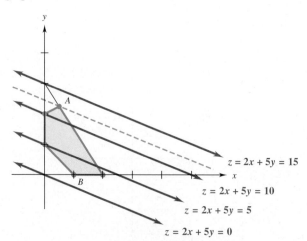

Figure 7.17

The point A is the intersection of the graphs of $3x + 2y = 6$ and $-2x + 4y = 8$. (See Figure 7.16.) Its coordinates can be found either algebraically or using a graphing calculator.

Algebraic Method	**Calculator Method**
Solve the system	Solve the two equations for y:

$$3x + 2y = 6$$
$$-2x + 4y = 8,$$

$$y = -1.5x + 3$$
$$y = .5x + 2.$$

as in Section 6.1, to get $x = \frac{1}{2}$ and $y = \frac{9}{4}$. Hence, A has coordinates $\left(\frac{1}{2}, \frac{9}{4}\right)$.

Graph both equations on the same screen and use the intersection finder to find that the coordinates of the intersection point A are (.5, 2.25).

The value of z at point A is

$$z = 2x + 5y$$

$$= 2\left(\frac{1}{2}\right) + 5\left(\frac{9}{4}\right) \qquad \text{Substitution}$$

$$= \frac{2}{2} + \frac{45}{4} \qquad \text{Multiplication}$$

$$= \frac{4}{4} + \frac{45}{4} \qquad \text{Addition of fractions}$$

$$= \frac{49}{4}.$$

Thus, the maximum possible value of z is 49/4. Similarly, the minimum value of z occurs at point B, which has coordinates (1, 0). The minimum value of z is $2(1) + 5(0) = 2$. ☑₁

Points such as A and B in Example 1 are called corner points. A **corner point** is a point in the feasible region where the boundary lines of two constraints cross. The feasible region in Figure 7.16 is **bounded** because the region is enclosed by boundary lines on all sides. Linear programming problems with bounded regions always have solutions. However, if Example 1 did not include the constraint $3x + 2y \leq 6$, the feasible region would be **unbounded,** and there would be no way to *maximize* the value of the objective function.

Some general conclusions can be drawn from the method of solution used in Example 1. Figure 7.18 shows various feasible regions and the lines that result from graphing an objective function with various values of z. (Figure 7.18 shows the situation in which the lines are in order from left to right as z increases.) In part (a) of Figure 7.18, the objective function takes on its minimum value at corner point Q and its maximum value at corner point P. The

✓ **Checkpoint 1**

Suppose the objective function in Example 1 is changed to $z = 5x + 2y$.

(a) Sketch the graphs of the objective function when $z = 0$, $z = 5$, and $z = 10$ on the region of feasible solutions given in Figure 7.16.

(b) From the graph, decide what values of x and y will maximize the objective function.

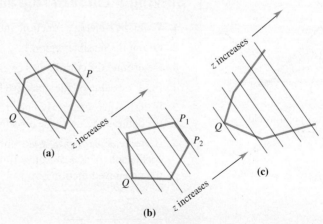

(a)

(b)

(c)

Figure 7.18

minimum is again at Q in part (b), but the maximum occurs at P_1 or P_2, or any point on the line segment connecting them. Finally, in part (c), the minimum value occurs at Q, but the objective function has no maximum value because the feasible region is unbounded.

The preceding discussion suggests the **corner point theorem.**

Corner Point Theorem

If the feasible region is bounded, then the objective function has both a maximum and a minimum value, and each occurs at one or more corner points.

If the feasible region is unbounded, the objective function may not have a maximum or minimum. But if a maximum or minimum value exists, it will occur at one or more corner points.

This theorem simplifies the job of finding an optimum value. First, graph the feasible region and find all corner points. Then test each point in the objective function. Finally, identify the corner point producing the optimum solution.

With the theorem, the problem in Example 1 could have been solved by identifying the five corner points of Figure 7.16: $(0, 1)$, $(0, 2)$, $(\frac{1}{2}, \frac{9}{4})$, $(2, 0)$, and $(1, 0)$. Then, substituting each of these points into the objective function $z = 2x + 5y$ would identify the corner points that produce the maximum and minimum values of z.

Corner Point	Value of $z = 2x + 5y$	
$(0, 1)$	$2(0) + 5(1) = 5$	
$(0, 2)$	$2(0) + 5(2) = 10$	
$(\frac{1}{2}, \frac{9}{4})$	$2(\frac{1}{2}) + 5(\frac{9}{4}) = \frac{49}{4}$	(maximum)
$(2, 0)$	$2(2) + 5(0) = 4$	
$(1, 0)$	$2(1) + 5(0) = 2$	(minimum)

From these results, the corner point $(\frac{1}{2}, \frac{9}{4})$ yields the maximum value of 49/4 and the corner point $(1, 0)$ gives the minimum value of 2. These are the same values found earlier. ✔2

A summary of the steps for solving a linear programming problem by the graphical method is given here.

✓ **Checkpoint 2**

(a) Identify the corner points in the given graph.

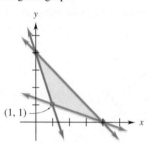

$(1, 1)$

(b) Which corner point would minimize $z = 2x + 3y$?

Solving a Linear Programming Problem Graphically

1. Write the objective function and all necessary constraints.
2. Graph the feasible region.
3. Determine the coordinates of each of the corner points.
4. Find the value of the objective function at each corner point.
5. If the feasible region is bounded, the solution is given by the corner point producing the optimum value of the objective function.
6. If the feasible region is an unbounded region in the first quadrant and both coefficients of the objective function are positive,* then the minimum value of the objective function occurs at a corner point and there is no maximum value.

---

*This is the only case of an unbounded region that occurs in the applications considered here.

EXAMPLE 2   Sketch the feasible region for the following set of constraints:

$$3y - 2x \geq 0$$
$$y + 8x \leq 52$$
$$y - 2x \leq 2$$
$$x \geq 3.$$

Then find the maximum and minimum values of the objective function $z = 5x + 2y$.

Solution   Graph the feasible region, as in Figure 7.19. To find the corner points, you must solve these four systems of equations:

A	B	C	D
$y - 2x = 2$	$3y - 2x = 0$	$3y - 2x = 0$	$y - 2x = 2$
$x = 3$	$x = 3$	$y + 8x = 52$	$y + 8x = 52$

The first two systems are easily solved by substitution, which shows that $A = (3, 8)$ and $B = (3, 2)$. The other two systems can be solved either with a calculator (as in Figure 7.20) or algebraically (see Checkpoint 3). Hence, $C = (6, 4)$ and $D = (5, 12)$. ✔ 3

✔ Checkpoint 3

Use the elimination method (see Section 6.1) to solve the last system and find the coordinates of D in Example 2.

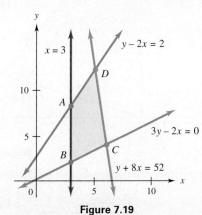

Figure 7.19

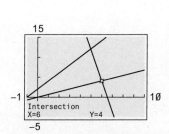

Figure 7.20

Use the corner points from the graph to find the maximum and minimum values of the objective function.

Corner Point	Value of $z = 5x + 2y$
(3, 8)	$5(3) + 2(8) = 31$
(3, 2)	$5(3) + 2(2) = 19$ (minimum)
(6, 4)	$5(6) + 2(4) = 38$
(5, 12)	$5(5) + 2(12) = 49$ (maximum)

The minimum value of $z = 5x + 2y$ is 19, at the corner point (3, 2). The maximum value is 49, at (5, 12). ✔ 4

✔ Checkpoint 4

Use the region of feasible solutions in the accompanying sketch to find the given values.

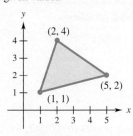

(a) The values of x and y that maximize $z = 2x - y$

(b) The maximum value of $z = 2x - y$

(c) The values of x and y that minimize $z = 4x + 3y$

(d) The minimum value of $z = 4x + 3y$

EXAMPLE 3   Solve the following linear programming problem:

$$\text{Minimize} \quad z = x + 3y$$
$$\text{subject to} \quad 2x + y \leq 10$$
$$5x + 2y \geq 20$$
$$-x + 2y \geq 0$$
$$x \geq 0, y \geq 0.$$

Solution First, we graph the feasible region, as in Figure 7.21. To find the corner points, you must solve these three systems of equations:

A	B	C
$2x + y = 10$	$2x + y = 10$	$5x + 2y = 20$
$5x + 2y = 20$	$-x + 2y = 0$	$-x + 2y = 0$

Solving the two equations to find point A by substitution yields $A = (0, 10)$. Similarly, solving the two equations to find point B by substitution yields $B = (4, 2)$. Finding the last corner point C can also be found by substitution, and we obtain $C = \left(\frac{10}{3}, \frac{5}{3}\right)$.

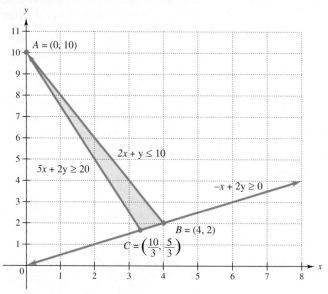

Figure 7.21

These corner points give the following values of z.

Corner Point	Value of $z = x + 3y$
$(0, 10)$	$0 + 3(10) = 30$
$(4, 2)$	$4 + 3(2) = 10$
$\left(\frac{10}{3}, \frac{5}{3}\right)$	$\left(\frac{10}{3}\right) + 3\left(\frac{5}{3}\right) = \frac{25}{3} \approx 8.33$ (minimum)

The minimum of z is $\frac{25}{3}$; it occurs at $\left(\frac{10}{3}, \frac{5}{3}\right)$. ✓ 5

✓ **Checkpoint 5**

The given sketch shows a feasible region. Let $z = x + 3y$. Use the sketch to find the values of x and y that

(a) minimize z;

(b) maximize z.

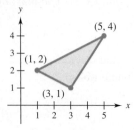

EXAMPLE 4 Solve the following linear programming problem:

$$\text{Minimize} \quad z = 2x + 4y$$
$$\text{subject to} \quad x + 2y \geq 10$$
$$3x + y \geq 10$$
$$x \geq 0, y \geq 0.$$

Solution Figure 7.22 shows the hand-drawn graph with corner points $(0, 10)$, $(2, 4)$, and $(10, 0)$, as well as the calculator graph with the corner point $(2, 4)$. Find the value of z for each point.

Corner Point	Value of $z = 2x + 4y$	
$(0, 10)$	$2(0) + 4(10) = 40$	
$(2, 4)$	$2(2) + 4(4) = 20$	(minimum)
$(10, 0)$	$2(10) + 4(0) = 20$	(minimum)

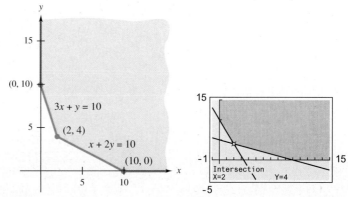

Figure 7.22

Checkpoint 6

The sketch shows a region of feasible solutions. From the sketch, decide what ordered pair would minimize $z = 2x + 4y$.

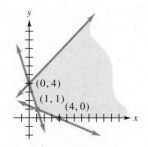

In this case, both (2, 4) and (10, 0), as well as all the points on the boundary line between them, give the same optimum value of z. So there is an infinite number of equally "good" values of x and y that give the same minimum value of the objective function $z = 2x + 4y$. The minimum value is 20. ✓ 6

7.2 Exercises

Exercises 1–6 show regions of feasible solutions. Use these regions to find maximum and minimum values of each given objective function. (See Examples 1 and 2.)

1. $z = 6x + y$

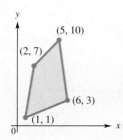

2. $z = 4x + y$

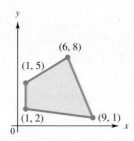

3. $z = .3x + .5y$

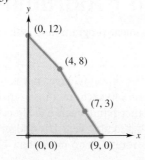

4. $z = .35x + 1.25y$

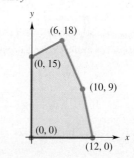

5. (a) $z = x + 5y$
  (b) $z = 2x + 3y$
  (c) $z = 2x + 4y$
  (d) $z = 4x + y$

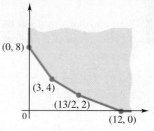

6. (a) $z = 5x + 2y$
  (b) $z = 5x + 6y$
  (c) $z = x + 2y$
  (d) $z = x + y$

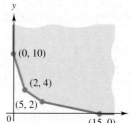

Use graphical methods to solve Exercises 7–12. (See Examples 2–4.)

7. Maximize $z = 4x + 3y$
   subject to $2x + 3y \le 6$
         $4x + y \le 6$
         $x \ge 0, y \ge 0.$

8. Minimize $z = x + 2y$
subject to $x + y \le 10$
$3x + 2y \ge 6$
$x \ge 0, y \ge 0.$

9. Minimize $z = 2x + y$
subject to $3x - y \ge 12$
$x + y \le 15$
$x \ge 2, y \ge 3.$

10. Maximize $z = x + 3y$
subject to $2x + 3y \le 100$
$5x + 4y \le 200$
$x \ge 10, y \ge 20.$

11. Maximize $z = 5x + y$
subject to $x - y \le 10$
$5x + 3y \le 75$
$x \ge 0, y \ge 0.$

12. Maximize $z = 4x + 5y$
subject to $10x - 5y \le 100$
$20x + 10y \ge 150$
$x \ge 0, y \ge 0.$

Find the minimum and maximum values of $z = 3x + 4y$ (if possible) for each of the given sets of constraints. (See Examples 2–4.)

13. $3x + 2y \ge 6$
$x + 2y \ge 4$
$x \ge 0, y \ge 0$

14. $2x + y \le 20$
$10x + y \ge 36$
$2x + 5y \ge 36$

15. $x + y \le 6$
$-x + y \le 2$
$2x - y \le 8$

16. $-x + 2y \le 6$
$3x + y \ge 3$
$x \ge 0, y \ge 0$

17. Find values of $x \ge 0$ and $y \ge 0$ that maximize $z = 10x + 12y$, subject to each of the following sets of constraints.

(a) $x + y \le 20$
$x + 3y \le 24$

(b) $3x + y \le 15$
$x + 2y \le 18$

(c) $x + 2y \ge 10$
$2x + y \ge 12$
$x - y \le 8$

18. Find values of $x \ge 0$ and $y \ge 0$ that minimize $z = 3x + 2y$, subject to each of the following sets of constraints.

(a) $10x + 7y \le 42$
$4x + 10y \ge 35$

(b) $6x + 5y \ge 25$
$2x + 6y \ge 15$

(c) $2x + 5y \ge 22$
$4x + 3y \le 28$
$2x + 2y \le 17$

19. Explain why it is impossible to maximize the function $z = 3x + 4y$ subject to the constraints

$x + y \ge 8$
$2x + y \le 10$
$x + 2y \le 8$
$x \ge 0, y \ge 0.$

20. Explain why it is impossible to maximize the function $z = 4x + 7y$ subject to the constraints

$8y + 5x \ge 40$
$4y + 9x \le 36$
$11y + 2x \le 22$
$x \ge 0, y \ge 0.$

✓ **Checkpoint Answers**

1. (a)

(b) $(2, 0)$

2. (a) $(0, 4), (1, 1), (4, 0)$ **(b)** $(1, 1)$

3. $D = (5, 12)$

4. (a) $(5, 2)$ **(b)** 8 **(c)** $(1, 1)$ **(d)** 7

5. (a) $(3, 1)$ **(b)** $(5, 4)$

6. $(1, 1)$

7.3 Applications of Linear Programming

In this section, we show several applications of linear programming with two variables.

EXAMPLE 1   **Office Purchases** An office manager needs to purchase new filing cabinets. He knows that Ace cabinets cost $40 each, require 6 square feet of floor space, and hold 8 cubic feet of files. On the other hand, each Excello cabinet costs $80, requires 8 square feet of floor space, and holds 12 cubic feet. His budget permits him to spend no more than $560, while the office has room for no more than 72 square feet of cabinets. The manager desires the greatest storage capacity within the limitations imposed by funds and space. How many of each type of cabinet should he buy?

Solution Let x represent the number of Ace cabinets to be bought, and let y represent the number of Excello cabinets. The information given in the problem can be summarized as follows.

	Number	Cost of Each	Space Required	Storage Capacity
Ace	x	$40	6 sq ft	8 cu ft
Excello	y	$80	8 sq ft	12 cu ft
Maximum Available		$560	72 sq ft	

The constraints imposed by cost and space are

$$40x + 80y \le 560 \quad \text{Cost}$$
$$6x + 8y \le 72. \quad \text{Floor space}$$

The number of cabinets cannot be negative, so $x \ge 0$ and $y \ge 0$. The objective function to be maximized gives the amount of storage capacity provided by some combination of Ace and Excello cabinets. From the information in the chart, the objective function is

$$z = \text{Storage capacity} = 8x + 12y.$$

In sum, the given problem has produced the following linear programming problem:

$$\text{Maximize} \quad z = 8x + 12y$$
$$\text{subject to} \quad 40x + 80y \le 560$$
$$6x + 8y \le 72$$
$$x \ge 0, y \ge 0.$$

A graph of the feasible region is shown in Figure 7.23. Three of the corner points can be identified from the graph as $(0, 0)$, $(0, 7)$, and $(12, 0)$. The fourth corner point, labeled Q in the figure, can be found algebraically or with a graphing calculator to be $(8, 3)$. ✔1

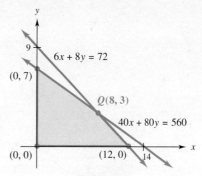

Figure 7.23

Use the corner point theorem to find the maximum value of z.

Corner Point	Value of $z = 8x + 12y$	
$(0, 0)$	0	
$(0, 7)$	84	
$(8, 3)$	100	(maximum)
$(12, 0)$	96	

The objective function, which represents storage space, is maximized when $x = 8$ and $y = 3$. The manager should buy 8 Ace cabinets and 3 Excello cabinets. ✔2

✓ **Checkpoint 1**

Find the corner point labeled P on the region of feasible solutions in the given graph.

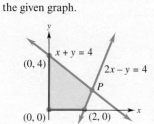

✓ **Checkpoint 2**

A popular cereal combines oats and corn. At least 27 tons of the cereal are to be made. For the best flavor, the amount of corn should be no more than twice the amount of oats. Oats cost $300 per ton, and corn costs $200 per ton. How much of each grain should be used to minimize the cost?

(a) Make a chart to organize the information given in the problem.

(b) Write an equation for the objective function.

(c) Write four inequalities for the constraints.

EXAMPLE 2 **Nutrition** Certain laboratory animals must have at least 30 grams of protein and at least 20 grams of fat per feeding period. These nutrients come from food *A*, which costs 18¢ per unit and supplies 2 grams of protein and 4 of fat, and food *B*, with 6 grams of protein and 2 of fat, costing 12¢ per unit. Food *B* is bought under a long-term contract requiring that at least 2 units of *B* be used per serving. How much of each food must be bought to produce the minimum cost per serving?

Solution Let *x* represent the amount of food *A* needed and *y* the amount of food *B*. Use the given information to produce the following table.

Food	Number of Units	Grams of Protein	Grams of Fat	Cost
A	*x*	2	4	18¢
B	*y*	6	2	12¢
Minimum Required		30	20	

Use the table to develop the linear programming problem. Since the animals must have *at least* 30 grams of protein and 20 grams of fat, use ≥ in the constraint inequalities for protein and fat. The long-term contract provides a constraint not shown in the table, namely, $y \geq 2$. So we have the following problem:

$$
\begin{aligned}
\text{Minimize} \quad & z = .18x + .12y && \text{Cost} \\
\text{subject to} \quad & 2x + 6y \geq 30 && \text{Protein} \\
& 4x + 2y \geq 20 && \text{Fat} \\
& y \geq 2 && \text{Contract} \\
& x \geq 0, y \geq 0.
\end{aligned}
$$

(The constraint $y \geq 0$ is redundant because of the constraint $y \geq 2$.) A graph of the feasible region with the corner points identified is shown in Figure 7.24.

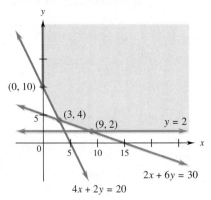

Figure 7.24

✓ Checkpoint 3

Use the information in Checkpoint 2 to do the following.

(a) Graph the feasible region and find the corner points.

(b) Determine the minimum value of the objective function and the point where it occurs.

(c) Is there a maximum cost?

Use the corner point theorem to find the minimum value of *z* as shown in the table.

Corner Points	$z = .18x + .12y$	
(0, 10)	.18(0) + .12(10) = 1.20	
(3, 4)	.18(3) + .12(4) = 1.02	(minimum)
(9, 2)	.18(9) + .12(2) = 1.86	

The minimum value of 1.02 occurs at (3, 4). Thus, 3 units of food *A* and 4 units of food *B* will produce a minimum cost of $1.02 per serving. ✓ 3

The feasible region in Figure 7.24 is an unbounded one: The region extends indefinitely to the upper right. With this region, it would not be possible to *maximize* the objective function, because the total cost of the food could always be increased by encouraging the animals to eat more.

One measure of the risk involved in investing in a stock or mutual fund is called the standard deviation. The standard deviation measures the volatility of investment returns relative to an historical average. If the return of an investment tool fluctuates a great deal from the historical average return, then there will be a higher standard deviation value for that stock. If an investment tool's value stays near the historical average, then it will have a small standard deviation value. Thus, a higher standard deviation for an investment tool can be one measure of higher risk. Investors often wish to obtain the highest profit while minimizing risk.

EXAMPLE 3   **Investing** Mariana wants to invest up to $5000 in stocks. The share price for CVS Health Corp. is $81 and the share price for Target Corp. is $57. Based on the average of 10-year returns, CVS would produce in a year a profit of $11 per share, and Target would produce a profit of $4 per share. Mariana would like to obtain at least $520 in profit. A measure of risk for each of these stocks is the beta value. The lower the beta value, the less risk associated with that stock. If the beta value for CVS is .62 and the beta value for Target is .24, how many shares of each stock should Mariana purchase to minimize the risk as measured by the beta value? (Data from: www.morningstar.com.)

Solution Let x represent the number of shares of CVS stock to be purchased, and let y be the number of shares of Target stock to be purchased. The information in the problem can be summarized in the following table.

	Number of Shares	Cost of Each	Profit	Risk
CVS	x	$81	$11	.62
Target	y	$57	$4	.24
Constraints		$5000	$520	

The constraints imposed by the cost of the shares and the profits are

$$81x + 57y \leq 5000$$
$$11x + 4y \geq 520.$$

The number of stocks to be purchased cannot be negative, so $x \geq 0$ and $y \geq 0$. The objective function to be minimized gives the amount of risk provided by some combination of shares in CVS and Target stocks. From the information in the chart, the objective function is

$$z = \text{beta} = .62x + .24y.$$

In sum, the given problem has produced the following linear programming problem:

$$\text{Minimize} \quad z = .62x + .24y$$
$$\text{subject to} \quad 81x + 57y \leq 5000$$
$$11x + 4y \geq 520$$
$$x \geq 0, y \geq 0.$$

A graph of the feasible region with the corner points identified is shown in Figure 7.25a. Using the calculator method from the previous section to find the intersection point of the two lines shows the TI-84 screenshot in Figure 7.25b.

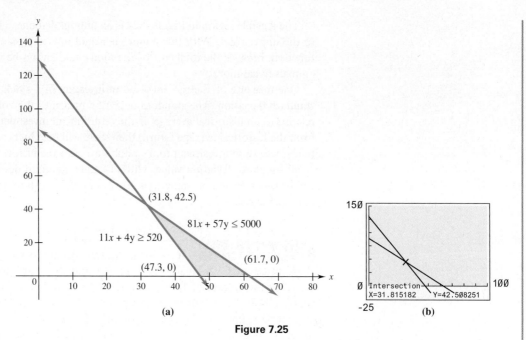

Figure 7.25

Use the corner point theorem to find the minimum z as shown in the table.

Corner Points	$z = .62x + .24y$
(47.3, 0)	29.326
(61.7, 0)	38.254
(31.8, 42.5)	29.916

The minimum value of 29.326 occurs at (47.3, 0). Mariana must buy a whole share of stock, so she would buy 47 shares of the CVS stock.

7.3 Exercises

Write the constraints in Exercises 1–4 as linear inequalities and identify all variables used. In some instances, not all of the information is needed to write the constraints. (See Examples 1–3.)

1. A canoe requires 8 hours of fabrication and a rowboat 5 hours. The fabrication department has at most 110 hours of labor available each week.

2. Jayson Gilbert needs at least 2800 milligrams of vitamin C per day. Each Supervite pill provides 250 milligrams, and each Vita-health pill provides 350 milligrams.

3. A candidate can afford to spend no more than $9500 on radio and TV advertising. Each radio spot costs $250, and each TV ad costs $750.

4. A hospital dietician has two meal choices: one for patients on solid food that costs $2.75 and one for patients on liquids that costs $3.75. There is a maximum of 600 patients in the hospital.

Solve these linear programming problems, which are somewhat simpler than the examples in the text.

5. **Business** A chain saw requires 4 hours of assembly and a wood chipper 6 hours. A maximum of 48 hours of assembly time is available. The profit is $150 on a chain saw and $220

on a chipper. How many of each should be assembled for maximum profit?

6. **Nutrition** Samar likes to snack frequently during the day, but he wants his snacks to provide at least 24 grams of protein per day. Each Snack-Pack provides 4 grams of protein, and each Minibite provides 1 gram. Snack-Packs cost 50 cents each and Minibites 12 cents. How many of each snack should he use to minimize his daily cost?

7. **Coffee Production** Deluxe coffee is to be mixed with regular coffee to make at least 50 pounds of a blended coffee. The mixture must contain at least 10 pounds of deluxe coffee. Deluxe coffee costs $6 per pound and regular coffee $5 per pound. How many pounds of each kind of coffee should be used to minimize costs?

8. **Outdoor Recreation** The company in Exercise 1 cannot sell more than 10 canoes each week and always sells at least 6 row-boats. The profit on a canoe is $400, and the profit on a rowboat is $225. Assuming the same situation as in Exercise 1, how many of each should be made per week to maximize profits?

9. **Business** Alejandro is an auto mechanic. He spends 3 hours when he replaces the shocks on a car and 2 hours when he

replaces the brakes. He works no more than 48 hours a week. He routinely completes at least 2 shocks replacements and 6 brake replacements a week. If he charges $500 for labor replacing shocks and $300 in labor for replacing brakes, how many jobs of each type should he complete a week to maximize his income?

10. **Vitamin Supplements** Jayson Gilbert of Exercise 2 pays 3 cents for each Supervite pill and 4 cents for each Vitahealth pill. Because of its other ingredients, he cannot take more than 7 Supervite pills per day. Assuming the same conditions as in Example 2, how many of each pill should he take to provide the desired level of vitamin C at minimum cost?

11. **Advertising** The candidate in Exercise 3 wants to have at least 8 radio spots and at least 3 TV ads. A radio spot reaches 600 people, and a TV ad reaches 2000 people. Assuming the monetary facts given in Exercise 3, how many of each kind should be used to reach the largest number of people?

12. **Nutrition** The hospital in Exercise 4 always has at least 100 patients on solid foods and at least 100 on liquids. Assuming the facts in Exercise 4, what number of each type of patient would minimize food costs?

Solve the following linear programming problems. (See Examples 1–3.)

13. **Nutrition** Mike May has been told that each day he needs at least 16 units of vitamin A, at least 5 units of vitamin B-1, and at least 20 units of vitamin C. Each Brand X pill contain 8 units of vitamin A, 1 of vitamin B-1, and 2 of vitamin C, while each Brand Z pill contains 2 units of vitamin A, 1 of vitamin B-1, and 7 of vitamin C. A Brand X pill costs 15 cents, and a Brand Z pill costs 30 cents. How many pills of each brand should he buy to minimize his daily cost? What is the minimum cost?

14. **Oil and Gas** The manufacturing process requires that oil refineries manufacture at least 2 gallons of gasoline for every gallon of fuel oil. To meet the winter demand for fuel oil, at least 3 million gallons a day must be produced. The demand for gasoline is no more than 12 million gallons per day. It takes .25 hour to ship each million gallons of gasoline and 1 hour to ship each million gallons of fuel oil out of the warehouse. No more than 6.6 hours are available for shipping. If the refinery sells gasoline for $1.25 per gallon and fuel oil for $1 per gallon, how much of each should be produced to maximize revenue? Find the maximum revenue.

15. **Bolt Manufacturing** A machine shop manufactures two types of bolts. The bolts require time on each of three groups of machines, but the time required on each group differs, as shown in the table:

		MACHINE GROUP	
		I	II	III
	Type 1	.1 min	.1 min	.1 min
Bolts	Type 2	.1 min	.4 min	.02 min

Production schedules are made up one day at a time. In a day, there are 240, 720, and 160 minutes available, respectively, on these machines. Type 1 bolts sell for 10¢ and type 2 bolts for 12¢. How many of each type of bolt should be manufactured per day to maximize revenue? What is the maximum revenue?

16. **Animal Nutrition** A zookeeper has a walrus with a nutritional deficiency. She wants to make sure the walrus consumes at least 2400 mg of iron, 2100 mg of vitamin B-1, and 1500 mg of vitamin B-2. One Maxivite pill contains 400 mg of iron, 100 mg of vitamin B-1, and 50 mg of vitamin B-2, and costs $.06. One Healthovite pill provides 100 mg of iron, 150 mg of vitamin B-1, and 150 mg of vitamin B-2, and costs $.08.

(a) What combination of Maxivite and Healthovite pills will meet the walrus's nutrition requirements at lowest cost? What is the lowest cost?

(b) In your solution to part (a), does the walrus receive more than the minimum amount needed of any vitamin? If so, which vitamin is it?

(c) Is there a way for the zookeeper to avoid having the walrus receive more than the minimum needed, still meet the other constraints, and minimize the cost? Explain.

17. **Greeting Cards** A greeting card manufacturer has 500 boxes of a particular card in warehouse I and 290 boxes of the same card in warehouse II. A greeting card shop in San Jose orders 350 boxes of the card, and another shop in Memphis orders 250 boxes. The shipping costs per box to these shops from the two warehouses are shown in the following table:

		DESTINATION	
		San Jose	Memphis
Warehouse	I	$.25	$.22
	II	$.23	$.21

How many boxes should be shipped to each city from each warehouse to minimize shipping costs? What is the minimum cost? (*Hint:* Use x, $350 - x$, y, and $250 - y$ as the variables.)

18. **Magazine Distribution** *Hotnews Magazine* publishes a U.S. and a Canadian edition each week. There are 30,000 subscribers in the United States and 20,000 in Canada. Other copies are sold at newsstands. Postage and shipping costs average $80 per thousand copies for the United States and $60 per thousand copies for Canada. Surveys show that no more than 120,000 copies of each issue can be sold (including subscriptions) and that the number of copies of the Canadian edition should not exceed twice the number of copies of the U.S. edition. The publisher can spend at most $8400 a month on postage and shipping. If the profit is $200 for each thousand copies of the U.S. edition and $150 for each thousand copies of the Canadian edition, how many copies of each version should be printed to earn as large a profit as possible? What is that profit?

19. **Investing** A pension fund manager decides to invest at most $50 million in U.S. Treasury Bonds paying 4% annual interest and in mutual funds paying 6% annual interest. He plans to invest at least $20 million in bonds and at least $6 million in mutual funds. Bonds have an initial fee of $300 per million dollars, while the fee for mutual funds is $100 per million. The fund manager is allowed to spend no more than $8400 on fees. How much should be invested in each to maximize annual interest? What is the maximum annual interest?

20. **Ecology** A certain predator requires at least 10 units of protein and 8 units of fat per day. One prey of Species I provides

5 units of protein and 2 units of fat; one prey of Species II provides 3 units of protein and 4 units of fat. Capturing and digesting each Species II prey requires 3 units of energy, and capturing and digesting each Species I prey requires 2 units of energy. How many of each prey would meet the predator's daily food requirements with the least expenditure of energy?

For Exercises 21–26, all prices are as of May 15, 2017.

21. **Investing** General Electric Co. stock sells for $29 a share and has a 3-year average annual return of $2 a share. The beta value is 1.10. Procter & Gamble Co. sells for $86 a share and has a 3-year average annual return of $4 a share. The beta value is .53. Tori wants to spend no more than $12,000 investing in these two stocks, but she wants to earn at least $700 in annual revenue. Tori also wants to minimize the risk. Determine the number of shares of each stock that Tori should buy. (Data from: www.morningstar.com.)

22. **Investing** Deere and Co. stock sells for $113 a share and has a 3-year average annual return of $10 a share. The beta value is .68. Boeing Co. sells for $185 a share and has a 3-year average annual return of $26 a share. The beta value is 1.19. Jermaine wants to spend no more than $10,000 investing in these two stocks, but he wants to earn at least $1050 in annual revenue. Jermaine also wants to minimize the risk. Determine the number of shares of each stock that Jermaine should buy. (Data from: www.morningstar.com.)

23. **Investing** Visa Inc. stock sells for $92 a share and has a 3-year average annual return of $20 a share. The beta value is 1.06. JP Morgan Chase and Co. sells for $87 a share and has a 3-year average annual return of $17 a share. The beta value is 1.21. Derek wants to spend no more than $15,000 investing in these two stocks, but he wants to earn at least $2500 in annual revenue. Derek also wants to minimize the risk. Determine the number of shares of each stock that Derek should buy. (Data from: www.morningstar.com.)

24. **Investing** Microsoft stock sells for $69 a share and has a 3-year average annual return of $16 a share. The beta value is 1.38. McDonald's Corp. sells for $144 a share and has a 3-year average annual return of $16 a share. The beta value is .77. Roselyn wants to spend no more than $12,000 investing in these two stocks, but she wants to earn at least $2250 in annual revenue. Roselyn also wants to minimize the risk. Determine the number of shares of each stock that Roselyn should buy. (Data from: www.morningstar.com.)

25. **Investing** The Vanguard Total Bond Market Index Inv (VBMFX) fund sells for $11 a share and has a 3-year average

annual return of $.25 a share. The beta value is 1.07. The Fidelity Freedom® 2020 (FFFDX) fund sells for $16 a share and has a 3-year average annual return of $.92. The beta value is .77. Jonquin wants to spend no more than $3000 investing in these two funds, but he wants to obtain at least $100 in annual revenue. Jonquin also wants to minimize his risk. Determine the number of shares of each stock Jonquin should buy. (Data from: www.morningstar.com.)

26. **Investing** The American Funds Growth Fund of America A (AGTHX) fund sells for $47 a share and has a 3-year average annual return of $5.30 a share. The beta value is .94. The Fidelity® Puritan® Fund (FPURX) sells for $22 a share and has a 3-year average annual return of $1.70. The beta value is .66. Brittany wants to spend no more than $7500 investing in these two funds, but she wants to obtain at least $700 in annual revenue. Brittany also wants to minimize her risk. Determine the number of shares of each stock Brittany should buy. (Data from: www.morningstar.com.)

✓**Checkpoint Answers**

1. $\left(\frac{8}{3}, \frac{4}{3}\right)$

2. (a)

	Number of Tons	Cost/ Ton
Oats	x	$300
Corn	y	$200
	27	

(b) $z = 300x + 200y$

(c) $x + y \geq 27$
$y \leq 2x$
$x \geq 0$
$y \geq 0$

3. (a)

Corner points: (27, 0), (9, 18)

(b) $6300 at (9, 18)

(c) No

7.4 The Simplex Method: Maximization

For linear programming problems with more than two variables or with two variables and many constraints, the graphical method is usually inefficient or impossible, so the **simplex method** is used. This method, which is introduced in this section, was developed for the U.S. Air Force by George B. Danzig in 1947. It is now used in industrial planning, factory design, product distribution networks, sports scheduling, truck routing, resource allocation, and a variety of other ways.

Because the simplex method is used for problems with many variables, it usually is not convenient to use letters such as x, y, z, or w as variable names. Instead, the symbols x_1 (read "x-sub-one"), x_2, x_3, and so on, up to x_n, are used. In the simplex method, all constraints must be expressed in the linear form

$$a_1 x_1 + a_2 x_2 + a_3 x_3 + \cdots + a_n x_n \leq b,$$

where $x_1, x_2, x_3, \ldots, x_n$ are variables, $a_1, a_2, a_3, \ldots, a_n$ are coefficients, and b is a constant.

We first discuss the simplex method for linear programming problems such as the following:

$$
\begin{aligned}
\text{Maximize} \quad & z = 2x_1 + 3x_2 + x_3 \\
\text{subject to} \quad & x_1 + x_2 + 4x_3 \leq 100 \\
& x_1 + 2x_2 + x_3 \leq 150 \\
& 3x_1 + 2x_2 + x_3 \leq 320, \\
\text{with} \quad & x_1 \geq 0, x_2 \geq 0, x_3 \geq 0.
\end{aligned}
$$

This example illustrates *standard maximum form*, which is defined as follows.

Standard Maximum Form

A linear programming problem is in **standard maximum form** if

1. the objective function is to be maximized;
2. all variables are nonnegative ($x_i \geq 0, i = 1, 2, 3, \ldots$);
3. all constraints involve $\leq$;
4. the constants on the right side in the constraints are all nonnegative ($b \geq 0$).

Problems that do not meet all of these conditions are considered in Sections 7.6 and 7.7.

The "mechanics" of the simplex method are demonstrated in Examples 1–5. Although the procedures to be followed will be made clear, as will the fact that they result in an optimal solution, the reasons these procedures are used may not be immediately apparent. Examples 6 and 7 will supply these reasons and explain the connection between the simplex method and the graphical method used in Section 7.3.

Setting up the Problem

The first step is to convert each constraint, a linear inequality, into a linear equation. This is done by adding a nonnegative variable, called a **slack variable,** to each constraint. For example, convert the inequality $x_1 + x_2 \leq 10$ into an equation by adding the slack variable s_1, to get

$$x_1 + x_2 + s_1 = 10, \qquad \text{where } s_1 \geq 0.$$

The inequality $x_1 + x_2 \leq 10$ says that the sum $x_1 + x_2$ is less than or equal to 10. The variable s_1 "takes up any slack" and represents the amount by which $x_1 + x_2$ fails to equal 10. For example, if $x_1 + x_2$ equals 8, then s_1 is 2. If $x_1 + x_2 = 10$, the value of s_1 is 0.

⚠ **CAUTION** A different slack variable must be used for each constraint.

EXAMPLE 1 Restate the following linear programming problem by introducing slack variables:

$$\text{Maximize} \quad z = 2x_1 + 3x_2 + x_3$$
$$\text{subject to} \quad x_1 + x_2 + 4x_3 \leq 100$$
$$x_1 + 2x_2 + x_3 \leq 150$$
$$3x_1 + 2x_2 + x_3 \leq 320,$$
$$\text{with} \quad x_1 \geq 0, x_2 \geq 0, x_3 \geq 0.$$

Solution Rewrite the three constraints as equations by introducing nonnegative slack variables s_1, s_2, and s_3, one for each constraint. Then the problem can be restated as

$$\text{Maximize} \quad z = 2x_1 + 3x_2 + x_3$$
$$\text{subject to} \quad x_1 + x_2 + 4x_3 + s_1 \qquad\qquad = 100 \qquad \text{Constraint 1}$$
$$x_1 + 2x_2 + x_3 \qquad + s_2 \qquad = 150 \qquad \text{Constraint 2}$$
$$3x_1 + 2x_2 + x_3 \qquad\qquad + s_3 = 320, \qquad \text{Constraint 3}$$

with $x_1 \geq 0, x_2 \geq 0, x_3 \geq 0, s_1 \geq 0, s_2 \geq 0, s_3 \geq 0.$ ✓₁

Adding slack variables to the constraints converts a linear programming problem into a system of linear equations. These equations should have all variables on the left of the equals sign and all constants on the right. All the equations of Example 1 satisfy this condition except for the objective function, $z = 2x_1 + 3x_2 + x_3$, which may be written with all variables on the left as

$$-2x_1 - 3x_2 - x_3 + z = 0. \qquad \text{Objective Function}$$

Now the equations of Example 1 (with the constraints listed first and the objective function last) can be written as the following augmented matrix.

$$
\begin{array}{ccccccc}
x_1 & x_2 & x_3 & s_1 & s_2 & s_3 & z \\
\end{array}
$$
$$
\left[
\begin{array}{ccccccc|c}
1 & 1 & 4 & 1 & 0 & 0 & 0 & 100 \\
1 & 2 & 1 & 0 & 1 & 0 & 0 & 150 \\
3 & 2 & 1 & 0 & 0 & 1 & 0 & 320 \\
\hline
-2 & -3 & -1 & 0 & 0 & 0 & 1 & 0 \\
\end{array}
\right]
\begin{array}{l}
\text{Constraint 1} \\
\text{Constraint 2} \\
\text{Constraint 3} \\
\text{Objective Function}
\end{array}
$$

Indicators

This matrix is the initial **simplex tableau.** Except for the last entries—the 1 and 0 on the right end—the numbers in the bottom row of a simplex tableau are called **indicators.** ✓₂

This simplex tableau represents a system of four linear equations in seven variables. Since there are more variables than equations, the system is dependent and has infinitely many solutions. Our goal is to find a solution in which all the variables are nonnegative and z is as large as possible. This will be done by using row operations to replace the given system by an equivalent one in which certain variables are eliminated from some of the equations. The process will be repeated until the optimum solution can be read from the matrix, as explained next.

Selecting the Pivot

Recall how row operations are used to eliminate variables in the Gauss–Jordan method: A particular nonzero entry in the matrix is chosen and changed to a 1; then all other entries in that column are changed to zeros. A similar process is used in the simplex method. The chosen entry is called the **pivot.** If we were interested only in solving the system, we could choose the various pivots in many different ways, as in Chapter 6. Here, however, it is not enough just to find a solution. We must find one that is nonnegative,

✓ **Checkpoint 1**

Rewrite the following set of constraints as equations by adding nonnegative slack variables:

$$x_1 + x_2 + x_3 \leq 12$$
$$2x_1 + 4x_2 \qquad \leq 15$$
$$x_2 + 3x_3 \leq 10.$$

✓ **Checkpoint 2**

Set up the initial simplex tableau for the following linear programming problem:

$$\text{Maximize} \quad z = 2x_1 + 3x_2$$
$$\text{subject to} \quad x_1 + 2x_2 \leq 85$$
$$2x_1 + x_2 \leq 92$$
$$x_1 + 4x_2 \leq 104,$$
$$\text{with} \quad x_1 \geq 0 \text{ and } x_2 \geq 0.$$

Locate and label the indicators.

satisfies all the constraints, *and* makes z as a large as possible. Consequently, the pivot must be chosen carefully, as explained in the next example. The reasons this procedure is used and why it works are discussed in Example 7.

EXAMPLE 2 Determine the pivot in the simplex tableau for the problem in Example 1.

Solution Look at the indicators (the last row of the tableau) and choose the most negative one:

$$
\begin{array}{ccccccc}
x_1 & x_2 & x_3 & s_1 & s_2 & s_3 & z \\
\end{array}
$$

$$
\left[
\begin{array}{ccccccc|c}
1 & 1 & 4 & 1 & 0 & 0 & 0 & 100 \\
1 & 2 & 1 & 0 & 1 & 0 & 0 & 150 \\
3 & 2 & 1 & 0 & 0 & 1 & 0 & 320 \\
\hline
-2 & -3 & -1 & 0 & 0 & 0 & 1 & 0 \\
\end{array}
\right].
$$

Most negative indicator

The most negative indicator identifies the variable that is to be eliminated from all but one of the equations (rows)—in this case, x_2. The column containing the most negative indicator is called the **pivot column.** Now, for each *positive* entry in the pivot column, divide the number in the far right column of the same row by the positive number in the pivot column:

$$
\begin{array}{ccccccc}
x_1 & x_2 & x_3 & s_1 & s_2 & s_3 & z \\
\end{array}
$$

$$
\left[
\begin{array}{ccccccc|c}
1 & 1 & 4 & 1 & 0 & 0 & 0 & 100 \\
1 & 2 & 1 & 0 & 1 & 0 & 0 & 150 \\
3 & 2 & 1 & 0 & 0 & 1 & 0 & 320 \\
\hline
-2 & -3 & -1 & 0 & 0 & 0 & 1 & 0 \\
\end{array}
\right].
$$

Quotients
$100/1 = 100$
$150/2 = 75 \leftarrow$ Smallest
$320/2 = 160$

The row with the smallest quotient (in this case, the second row) is called the **pivot row.** The entry in the pivot row and pivot column is the pivot:

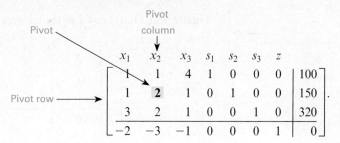

Pivot
Pivot column
Pivot row

$$
\begin{array}{ccccccc}
x_1 & x_2 & x_3 & s_1 & s_2 & s_3 & z \\
\end{array}
$$

$$
\left[
\begin{array}{ccccccc|c}
1 & 1 & 4 & 1 & 0 & 0 & 0 & 100 \\
1 & 2 & 1 & 0 & 1 & 0 & 0 & 150 \\
3 & 2 & 1 & 0 & 0 & 1 & 0 & 320 \\
\hline
-2 & -3 & -1 & 0 & 0 & 0 & 1 & 0 \\
\end{array}
\right].
$$

⬡ CAUTION In some simplex tableaus, the pivot column may contain zeros or negative entries. Only the positive entries in the pivot column should be used to form the quotients and determine the pivot row. If there are no positive entries in the pivot column (so that a pivot row cannot be chosen), then no maximum solution exists. ✓3

✓**Checkpoint 3**

Find the pivot for the following tableau:

$$
\begin{array}{cccccc}
x_1 & x_2 & s_1 & s_2 & s_3 & z \\
\end{array}
$$

$$
\left[
\begin{array}{cccccc|c}
0 & 1 & 1 & 0 & 0 & 0 & 50 \\
-2 & 3 & 0 & 1 & 0 & 0 & 78 \\
2 & 4 & 0 & 0 & 1 & 0 & 65 \\
\hline
-5 & -3 & 0 & 0 & 0 & 1 & 0 \\
\end{array}
\right].
$$

Pivoting

Once the pivot has been selected, row operations are used to replace the initial simplex tableau by another simplex tableau in which the pivot column variable is eliminated from all but one of the equations. Since this new tableau is obtained by row operations, it represents an equivalent system of equations (that is, a system with the same solutions as the original system). This process, which is called **pivoting,** is explained in the next example.

EXAMPLE 3 Use the indicated pivot, 2, to perform the pivoting on the simplex tableau of Example 2:

$$
\begin{array}{ccccccc}
x_1 & x_2 & x_3 & s_1 & s_2 & s_3 & z \\
\end{array}
$$

$$
\left[
\begin{array}{ccccccc|c}
1 & 1 & 4 & 1 & 0 & 0 & 0 & 100 \\
1 & 2 & 1 & 0 & 1 & 0 & 0 & 150 \\
3 & 2 & 1 & 0 & 0 & 1 & 0 & 320 \\
\hline
-2 & -3 & -1 & 0 & 0 & 0 & 1 & 0
\end{array}
\right].
$$

Solution Start by multiplying each entry of row 2 by $\frac{1}{2}$ in order to change the pivot to 1:

$$
\begin{array}{ccccccc}
x_1 & x_2 & x_3 & s_1 & s_2 & s_3 & z \\
\end{array}
$$

$$
\left[
\begin{array}{ccccccc|c}
1 & 1 & 4 & 1 & 0 & 0 & 0 & 100 \\
\frac{1}{2} & 1 & \frac{1}{2} & 0 & \frac{1}{2} & 0 & 0 & 75 \\
3 & 2 & 1 & 0 & 0 & 1 & 0 & 320 \\
\hline
-2 & -3 & -1 & 0 & 0 & 0 & 1 & 0
\end{array}
\right]. \quad \frac{1}{2}R_2
$$

Now use row operations to make the entry in row 1, column 2, a 0:

$$
\begin{array}{ccccccc}
x_1 & x_2 & x_3 & s_1 & s_2 & s_3 & z \\
\end{array}
$$

$$
\left[
\begin{array}{ccccccc|c}
\frac{1}{2} & 0 & \frac{7}{2} & 1 & -\frac{1}{2} & 0 & 0 & 25 \\
\frac{1}{2} & 1 & \frac{1}{2} & 0 & \frac{1}{2} & 0 & 0 & 75 \\
3 & 2 & 1 & 0 & 0 & 1 & 0 & 320 \\
\hline
-2 & -3 & -1 & 0 & 0 & 0 & 1 & 0
\end{array}
\right]. \quad -R_2 + R_1
$$

Change the 2 in row 3, column 2, to a 0 by a similar process:

$$
\begin{array}{ccccccc}
x_1 & x_2 & x_3 & s_1 & s_2 & s_3 & z \\
\end{array}
$$

$$
\left[
\begin{array}{ccccccc|c}
\frac{1}{2} & 0 & \frac{7}{2} & 1 & -\frac{1}{2} & 0 & 0 & 25 \\
\frac{1}{2} & 1 & \frac{1}{2} & 0 & \frac{1}{2} & 0 & 0 & 75 \\
2 & 0 & 0 & 0 & -1 & 1 & 0 & 170 \\
\hline
-2 & -3 & -1 & 0 & 0 & 0 & 1 & 0
\end{array}
\right]. \quad -2R_2 + R_3
$$

Finally, add 3 times row 2 to the last row in order to change the indicator -3 to 0:

$$
\begin{array}{ccccccc}
x_1 & x_2 & x_3 & s_1 & s_2 & s_3 & z \\
\end{array}
$$

$$
\left[
\begin{array}{ccccccc|c}
\frac{1}{2} & 0 & \frac{7}{2} & 1 & -\frac{1}{2} & 0 & 0 & 25 \\
\frac{1}{2} & 1 & \frac{1}{2} & 0 & \frac{1}{2} & 0 & 0 & 75 \\
2 & 0 & 0 & 0 & -1 & 1 & 0 & 170 \\
\hline
-\frac{1}{2} & 0 & \frac{1}{2} & 0 & \frac{3}{2} & 0 & 1 & 225
\end{array}
\right]. \quad 3R_2 + R_4
$$

The pivoting is now complete, because the pivot column variable x_2 has been eliminated from all equations except the one represented by the pivot row. The initial simplex tableau has been replaced by a new simplex tableau, which represents an equivalent system of equations.

✓ **Checkpoint 4**

For the given simplex tableau,

(a) find the pivot;

(b) perform the pivoting and write the new tableau.

$$
\begin{array}{cccccc}
x_1 & x_2 & x_3 & s_1 & s_2 & z \\
\end{array}
$$

$$
\left[
\begin{array}{cccccc|c}
1 & 2 & 6 & 1 & 0 & 0 & 16 \\
1 & 3 & 0 & 0 & 1 & 0 & 25 \\
\hline
-1 & -4 & -3 & 0 & 0 & 1 & 0
\end{array}
\right]
$$

⚠ **CAUTION** During pivoting, do not interchange rows of the matrix. Make the pivot entry 1 by multiplying the pivot row by an appropriate constant, as in Example 3. ✓4

When at least one of the indicators in the last row of a simplex tableau is negative (as is the case with the tableau obtained in Example 3), the simplex method requires that a new pivot be selected and the pivoting be performed again. This procedure is repeated until a simplex tableau with no negative indicators in the last row is obtained or a tableau is reached in which no pivot row can be chosen.

EXAMPLE 4   In the simplex tableau obtained in Example 3, select a new pivot and perform the pivoting.

Solution   First, locate the pivot column by finding the most negative indicator in the last row. Then locate the pivot row by computing the necessary quotients and finding the smallest one, as shown here:

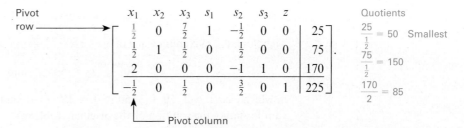

So the pivot is the number $\frac{1}{2}$ in row 1, column 1. Begin the pivoting by multiplying every entry in row 1 by 2. Then continue as indicated to obtain the following simplex tableau:

$$
\begin{array}{ccccccc}
x_1 & x_2 & x_3 & s_1 & s_2 & s_3 & z
\end{array}
$$
$$
\left[
\begin{array}{ccccccc|c}
1 & 0 & 7 & 2 & -1 & 0 & 0 & 50 \\
0 & 1 & -3 & -1 & 1 & 0 & 0 & 50 \\
0 & 0 & -14 & -4 & 1 & 1 & 0 & 70 \\
\hline
0 & 0 & 4 & 1 & 1 & 0 & 1 & 250
\end{array}
\right]
\qquad
\begin{array}{l}
2R_1 \\
-\frac{1}{2}R_1 + R_2 \\
-2R_1 + R_3 \\
\frac{1}{2}R_1 + R_4
\end{array}
$$

Since there are no negative indicators in the last row, no further pivoting is necessary, and we call this the **final simplex tableau.**

Reading the Solution

The next example shows how to read an optimal solution of the original linear programming problem from the final simplex tableau.

EXAMPLE 5   Solve the linear programming problem introduced in Example 1.

Solution   Look at the final simplex tableau for this problem, which was obtained in Example 4:

$$
\begin{array}{ccccccc}
x_1 & x_2 & x_3 & s_1 & s_2 & s_3 & z
\end{array}
$$
$$
\left[
\begin{array}{ccccccc|c}
1 & 0 & 7 & 2 & -1 & 0 & 0 & 50 \\
0 & 1 & -3 & -1 & 1 & 0 & 0 & 50 \\
0 & 0 & -14 & -4 & 1 & 1 & 0 & 70 \\
\hline
0 & 0 & 4 & 1 & 1 & 0 & 1 & 250
\end{array}
\right].
$$

The last row of this matrix represents the equation

$$
4x_3 + s_1 + s_2 + z = 250, \qquad \text{or equivalently,} \qquad z = 250 - 4x_3 - s_1 - s_2.
$$

If x_3, s_1, and s_2 are all 0, then the value of z is 250. If any one of x_3, s_1, or s_2 is positive, then z will have a smaller value than 250. (Why?) Consequently, since we want a solution for this system in which all the variables are nonnegative and z is as large as possible, we must have

$$
x_3 = 0, \qquad s_1 = 0, \qquad s_2 = 0.
$$

When these values are substituted into the first equation (represented by the first row of the final simplex tableau), the result is

$$x_1 + 7 \cdot 0 + 2 \cdot 0 - 1 \cdot 0 = 50; \quad \text{that is,} \quad x_1 = 50.$$

Similarly, substituting 0 for x_3, s_1, and s_2 in the last three equations represented by the final simplex tableau shows that

$$x_2 = 50, \quad s_3 = 70, \quad \text{and} \quad z = 250.$$

Therefore, the maximum value of $z = 2x_1 + 3x_2 + x_3$ occurs when

$$x_1 = 50, \quad x_2 = 50, \quad \text{and} \quad x_3 = 0,$$

in which case $z = 2 \cdot 50 + 3 \cdot 50 + 0 = 250$. (The values of the slack variables are irrelevant in stating the solution of the original problem.)

In any simplex tableau, some columns look like columns of an identity matrix (one entry is 1 and the rest are 0). The variables corresponding to these columns are called **basic variables** and the variables corresponding to the other columns are referred to as **nonbasic variables**. In the tableau of Example 5, for instance, the basic variables are x_1, x_2, s_3, and z (shown in blue), and the nonbasic variables are x_3, s_1, and s_2:

$$
\begin{array}{ccccccc}
x_1 & x_2 & x_3 & s_1 & s_2 & s_3 & z \\
\end{array}
$$

$$
\left[
\begin{array}{ccccccc|c}
1 & 0 & 7 & 2 & -1 & 0 & 0 & 50 \\
0 & 1 & -3 & -1 & 1 & 0 & 0 & 50 \\
0 & 0 & -14 & -4 & 1 & 1 & 0 & 70 \\
\hline
0 & 0 & 4 & 1 & 1 & 0 & 1 & 250 \\
\end{array}
\right].
$$

The optimal solution in Example 5 was obtained from the final simplex tableau by setting the nonbasic variables equal to 0 and solving for the basic variables. Furthermore, the values of the basic variables are easy to read from the matrix: Find the 1 in the column representing a basic variable; the last entry in that row is the value of that basic variable in the optimal solution. In particular, *the entry in the lower right-hand corner of the final simplex tableau is the maximum value of z.* ✓ 5

✓ Checkpoint 5

A linear programming problem with slack variables s_1 and s_2 has the following final simplex tableau:

$$
\begin{array}{cccccc}
x_1 & x_2 & x_3 & s_1 & s_2 & z \\
\end{array}
$$

$$
\left[
\begin{array}{cccccc|c}
0 & 3 & 1 & 5 & 2 & 0 & 9 \\
1 & -2 & 0 & 4 & 1 & 0 & 6 \\
\hline
0 & 5 & 0 & 1 & 0 & 1 & 21 \\
\end{array}
\right].
$$

What is the optimal solution?

! CAUTION If there are two identical columns in a tableau, each of which is a column in an identity matrix, only one of the variables corresponding to these columns can be a basic variable. The other is treated as a nonbasic variable. You may choose either one to be the basic variable, unless one of them is z, in which case z must be the basic variable.

The steps involved in solving a standard maximum linear programming problem by the simplex method have been illustrated in Examples 1–5 and are summarized here.

Simplex Method

1. Determine the objective function.
2. Write down all necessary constraints.
3. Convert each constraint into an equation by adding a slack variable.
4. Set up the initial simplex tableau.

5. Locate the most negative indicator. If there are two such indicators, choose one. This indicator determines the pivot column.

6. Use the positive entries in the pivot column to form the quotients necessary for determining the pivot. If there are no positive entries in the pivot column, no maximum solution exists. If two quotients are equally the smallest, let either determine the pivot.[*]

7. Multiply every entry in the pivot row by the reciprocal of the pivot to change the pivot to 1. Then use row operations to change all other entries in the pivot column to 0 by adding suitable multiples of the pivot row to the other rows.

8. If the indicators are all positive or 0, you have found the final tableau. If not, go back to Step 5 and repeat the process until a tableau with no negative indicators is obtained.[†]

9. In the final tableau, the *basic* variables correspond to the columns that have one entry of 1 and the rest 0. The *nonbasic* variables correspond to the other columns. Set each nonbasic variable equal to 0 and solve the system for the basic variables. The maximum value of the objective function is the number in the lower right-hand corner of the final tableau.

The solution found by the simplex method may not be unique, especially when choices are possible in steps 5, 6, or 9. There may be other solutions that produce the same maximum value of the objective function. (See Exercises 37 and 38 at the end of this section.) ✔ 6

✓ **Checkpoint 6**

A linear programming problem has the following initial tableau:

$$
\begin{array}{ccccc}
x_1 & x_2 & s_1 & s_2 & z \\
\end{array}
$$

$$
\left[
\begin{array}{ccccc|c}
1 & 1 & 1 & 0 & 0 & 40 \\
2 & 1 & 0 & 1 & 0 & 24 \\
\hline
-300 & -200 & 0 & 0 & 1 & 0 \\
\end{array}
\right].
$$

Use the simplex method to solve the problem.

The Simplex Method with Technology

Unless indicated otherwise, the simplex method is carried out by hand in the examples and exercises of this chapter, so you can see how and why it works. Once you are familiar with the method, however,

we strongly recommend that you use technology to apply the simplex method.

Doing so will eliminate errors that occur in manual computations. It will also give you a better idea of how the simplex method is used in the real world, where applications involve so many variables and constraints that the manual approach is impractical. Readily available technology includes the following:

Graphing Calculators A program called "SIMPLEX" is available for the TI-84. It pauses after each round of pivoting, so you can examine the intermediate simplex tableau. The program and details on its use can be found at https://goo.gl/3stavd.

Spreadsheets Most spreadsheets have a built-in simplex method program. Figures 7.26(a)–(c) show screenshots for using the Solver in Microsoft Excel (Microsoft Corporation Excel© 2016) to solve Example 1 in this section. Spreadsheets also provide a sensitivity analysis, which allows you to see how much the constraints can be varied without changing the maximal solution. For more information, see the *Spreadsheet Manual* available with this text.

Other Computer Programs A variety of simplex method programs, many of which are free, can be downloaded on the Internet. Google "simplex method program" for some possibilities.

---

[*]It may be that the first choice of a pivot does not produce a solution. In that case, try the other choice.

[†]Some special circumstances are noted at the end of Section 7.7.

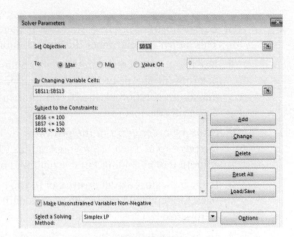

Figure 7.26(a)   Figure 7.26(b)   Figure 7.26(c)

Geometric Interpretation of the Simplex Method

Although it may not be immediately apparent, the simplex method is based on the same geometrical considerations as the graphical method. This can be seen by looking at a problem that can be readily solved by both methods.

EXAMPLE 6   In Example 1 of Section 7.3, the following problem was solved graphically (using x and y instead of x_1 and x_2, respectively):

$$\text{Maximize} \quad z = 8x_1 + 12x_2$$
$$\text{subject to} \quad 40x_1 + 80x_2 \leq 560$$
$$6x_1 + 8x_2 \leq 72$$
$$x_1 \geq 0, x_2 \geq 0.$$

Graphing the feasible region (Figure 7.27) and evaluating z at each corner point shows that the maximum value of z occurs at (8, 3).

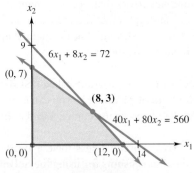

Corner Point	Value of $z = 8x_1 + 12x_2$
(0, 0)	0
(0, 7)	84
(8, 3)	100 (maximum)
(12, 0)	96

Figure 7.27

To solve the same problem by the simplex method, add a slack variable to each constraint:

$$40x_1 + 80x_2 + s_1 \qquad = 560$$
$$6x_1 + 8x_2 \qquad + s_2 = 72.$$

Then write the initial simplex tableau:

$$\begin{array}{ccccc} x_1 & x_2 & s_1 & s_2 & z \end{array}$$
$$\left[\begin{array}{ccccc|c} 40 & 80 & 1 & 0 & 0 & 560 \\ 6 & 8 & 0 & 1 & 0 & 72 \\ \hline -8 & -12 & 0 & 0 & 1 & 0 \end{array} \right].$$

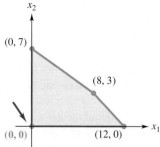

Figure 7.28

In this tableau, the basic variables are s_1, s_2, and z. (Why?) By setting the nonbasic variables (namely, x_1 and x_2) equal to 0 and solving for the basic variables, we obtain the following solution (which will be called a **basic feasible solution**):

$$x_1 = 0, \quad x_2 = 0, \quad s_1 = 560, \quad s_2 = 72, \quad \text{and} \quad z = 0.$$

Since $x_1 = 0$ and $x_2 = 0$, this solution corresponds to the corner point at the origin in the graphical solution (Figure 7.28).

The basic feasible solution $(0, 0)$ given by the initial simplex tableau has $z = 0$, which is obviously not maximal. Each round of pivoting in the simplex method will produce another corner point, with a larger value of z, until we reach a corner point that provides the maximum solution.

The most negative indicator in the initial tableau is -12, and it determines the pivot column. Then we form the necessary quotients and determine the pivot row:

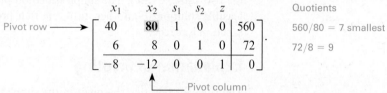

Thus, the pivot is 80 in row 1, column 2. Performing the pivoting leads to this tableau:

$$
\begin{bmatrix}
\frac{1}{2} & 1 & \frac{1}{80} & 0 & 0 & 7 \\
2 & 0 & -\frac{1}{10} & 1 & 0 & 16 \\
-2 & 0 & \frac{3}{20} & 0 & 1 & 84
\end{bmatrix}
\quad
\begin{matrix}
\frac{1}{80}R_1 \\
-8R_1 + R_2 \\
12R_1 + R_3
\end{matrix}
$$

The basic variables here are x_2, s_2, and z, and the basic feasible solution (found by setting the nonbasic variables equal to 0 and solving for the basic variables) is

$$x_1 = 0, \quad x_2 = 7, \quad s_1 = 0, \quad s_2 = 16, \quad \text{and} \quad z = 84,$$

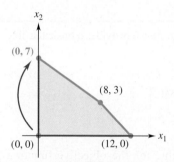

Figure 7.29

which corresponds to the corner point $(0, 7)$ in Figure 7.29. Note that the new value of the pivot variable x_2 is precisely the smallest quotient, 7, that was used to select the pivot row. Although this value of z is better, further improvement is possible.

Now the most negative indicator is -2. We form the necessary quotients and determine the pivot as usual:

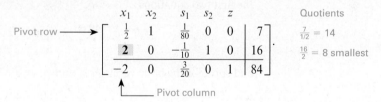

The pivot is 2 in row 2, column 1. Pivoting now produces the final tableau:

$$
\begin{bmatrix}
0 & 1 & \frac{3}{80} & -\frac{1}{4} & 0 & 3 \\
1 & 0 & -\frac{1}{20} & \frac{1}{2} & 0 & 8 \\
0 & 0 & \frac{1}{20} & 1 & 1 & 100
\end{bmatrix}
\quad
\begin{matrix}
-\frac{1}{2}R_2 + R_1 \\
\frac{1}{2}R_2 \\
2R_2 + R_3
\end{matrix}
$$

Here, the basic feasible solution is

$$x_1 = 8, \quad x_2 = 3, \quad s_1 = 0, \quad s_2 = 0, \quad \text{and} \quad z = 100,$$

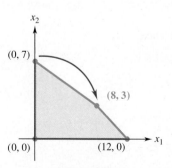

Figure 7.30

which corresponds to the corner point $(8, 3)$ in Figure 7.30. Once again, the new value of the pivot variable x_1 is the smallest quotient, 8, that was used to select the pivot. Because all the indicators in the last row of the final tableau are nonnegative, $(8, 3)$ is the maximum solution according to the simplex method. We know that this is the case, since this is the maximum solution found by the graphical method. An algebraic argument similar to the one in Example 5 could also be made.

As illustrated in Example 6, the basic feasible solution obtained from a simplex tableau corresponds to a corner point of the feasible region. Pivoting, which replaces one tableau with another, is a systematic way of moving from one corner point to another, each time improving the value of the objective function. The simplex method ends when a corner point that produces the maximum value of the objective function is reached (or when it becomes clear that the problem has no maximum solution).

When there are three or more variables in a linear programming problem, it may be difficult or impossible to draw a picture, but it can be proved that the optimal value of the objective function occurs at a basic feasible solution (corresponding to a corner point in the two-variable case). The simplex method provides a means of moving from one basic feasible solution to another until one that produces the optimal value of the objective function is reached.

Explanation of Pivoting

The rules for selecting the pivot in the simplex method can be understood by examining how the first pivot was chosen in Example 6.

EXAMPLE 7 The initial simplex tableau of Example 6 provides a basic feasible solution with $x_1 = 0$ and $x_2 = 0$:

$$
\begin{array}{ccccc}
x_1 & x_2 & s_1 & s_2 & z \\
\end{array}
$$
$$
\left[
\begin{array}{ccccc|c}
40 & 80 & 1 & 0 & 0 & 560 \\
6 & 8 & 0 & 1 & 0 & 72 \\
\hline
-8 & -12 & 0 & 0 & 1 & 0
\end{array}
\right].
$$

This solution certainly does not give a maximum value for the objective function $z = 8x_1 + 12x_2$. Since x_2 has the largest coefficient, z will be increased most if x_2 is increased. In other words, the most negative indicator in the tableau (which corresponds to the largest coefficient in the objective function) identifies the variable that will provide the greatest change in the value of z.

To determine how much x_2 can be increased without leaving the feasible region, look at the first two equations,

$$
\begin{aligned}
40x_1 + 80x_2 + s_1 \quad\quad &= 560 \\
6x_1 + 8x_2 \quad\quad + s_2 &= 72,
\end{aligned}
$$

and solve for the basic variables s_1 and s_2:

$$
\begin{aligned}
s_1 &= 560 - 40x_1 - 80x_2 \\
s_2 &= 72 - 6x_1 - 8x_2.
\end{aligned}
$$

Now x_2 is to be increased while x_1 is to keep the value 0. Hence,

$$
\begin{aligned}
s_1 &= 560 - 80x_2 \\
s_2 &= 72 - 8x_2.
\end{aligned}
$$

Since $s_1 \geq 0$ and $s_2 \geq 0$, we must have

$$
\begin{array}{ll}
0 \leq s_1 & 0 \leq s_2 \\
0 \leq 560 - 80x_2 \quad \text{and} & 0 \leq 72 - 8x_2 \\
80x_2 \leq 560 & 8x_2 \leq 72 \\
x_2 \leq \dfrac{560}{80} = 7 & x_2 \leq \dfrac{72}{8} = 9.
\end{array}
$$

The right sides of these last inequalities are the quotients used to select the pivot row. Since x_2 must satisfy both inequalities, x_2 can be at most 7. In other words, the smallest quotient formed from positive entries in the pivot column identifies the value of x_2 that produces the largest change in z while remaining in the feasible region. By pivoting with the pivot determined in this way, we obtain the second tableau and a basic feasible solution in which $x_2 = 7$, as was shown in Example 6.

An analysis similar to that in Example 7 applies to each occurrence of pivoting in the simplex method. The idea is to improve the value of the objective function by adjusting one variable at a time. The most negative indicator identifies the variable that will account for the largest increase in z. The smallest quotient determines the largest value of that variable which will produce a feasible solution. Pivoting leads to a solution in which the selected variable has this largest value.

7.4 Exercises

In Exercises 1–4, (a) determine the number of slack variables needed; (b) name them; (c) use the slack variables to convert each constraint into a linear equation. (See Example 1.)

1. Maximize $z = 32x_1 + 9x_2$
   subject to $4x_1 + 2x_2 \le 20$
   $5x_1 + x_2 \le 50$
   $2x_1 + 3x_2 \le 25$
   $x_1 \ge 0, x_2 \ge 0.$

2. Maximize $z = 3.7x_1 + 4.3x_2$
   subject to $2.4x_1 + 1.5x_2 \le 10$
   $1.7x_1 + 1.9x_2 \le 15$
   $x_1 \ge 0, x_2 \ge 0.$

3. Maximize $z = 8x_1 + 3x_2 + x_3$
   subject to $3x_1 - x_2 + 4x_3 \le 95$
   $7x_1 + 6x_2 + 8x_3 \le 118$
   $4x_1 + 5x_2 + 10x_3 \le 220$
   $x_1 \ge 0, x_2 \ge 0, x_3 \ge 0.$

4. Maximize $z = 12x_1 + 15x_2 + 10x_3$
   subject to $2x_1 + 2x_2 + x_3 \le 8$
   $x_1 + 4x_2 + 3x_3 \le 12$
   $x_1 \ge 0, x_2 \ge 0, x_3 \ge 0.$

Introduce slack variables as necessary and then write the initial simplex tableau for each of these linear programming problems.

5. Maximize $z = 5x_1 + x_2$
   subject to $2x_1 + 5x_2 \le 6$
   $4x_1 + x_2 \le 6$
   $5x_1 + 3x_2 \le 15$
   $x_1 \ge 0, x_2 \ge 0.$

6. Maximize $z = 5x_1 + 3x_2 + 7x_3$
   subject to $4x_1 + 3x_2 + 2x_3 \le 60$
   $3x_1 + 4x_2 + x_3 \le 24$
   $x_1 \ge 0, x_2 \ge 0, x_3 \ge 0.$

7. Maximize $z = x_1 + 5x_2 + 10x_3$
   subject to $x_1 + 2x_2 + 3x_3 \le 10$
   $2x_1 + x_2 + x_3 \le 8$
   $3x_1 + 4x_3 \le 6$
   $x_1 \ge 0, x_2 \ge 0, x_3 \ge 0.$

8. Maximize $z = 5x_1 - x_2 + 3x_3$
   subject to $3x_1 + 2x_2 + x_3 \le 36$
   $x_1 + 6x_2 + x_3 \le 24$
   $x_1 - x_2 - x_3 \le 32$
   $x_1 \ge 0, x_2 \ge 0, x_3 \ge 0.$

Find the pivot in each of the given simplex tableaus. (See Example 2.)

9.

x_1	x_2	x_3	s_1	s_2	z	
2	2	0	3	1	0	15
3	4	1	6	0	0	20
-2	-3	0	1	0	1	10

10.

x_1	x_2	x_3	s_1	s_2	z	
0	2	1	1	3	0	5
1	-5	0	1	2	0	8
0	-2	0	-3	1	1	10

11.

x_1	x_2	x_3	s_1	s_2	s_3	z	
6	2	1	3	0	0	0	8
0	2	0	1	0	1	0	7
6	1	0	3	1	0	0	6
-3	-2	0	2	0	0	1	12

12.

x_1	x_2	x_3	s_1	s_2	s_3	z	
0	2	0	1	2	2	0	3
0	3	1	0	1	2	0	4
1	4	0	0	3	5	0	5
0	-4	0	0	4	3	1	20

In Exercises 13–16, use the indicated entry as the pivot and perform the pivoting. (See Examples 3 and 4.)

13.

$$\begin{array}{cccccc} x_1 & x_2 & x_3 & s_1 & s_2 & z \end{array}$$
$$\begin{bmatrix} 1 & 2 & 4 & 1 & 0 & 0 & 56 \\ 2 & 2 & 1 & 0 & 1 & 0 & 40 \\ \hline -1 & -3 & -2 & 0 & 0 & 1 & 0 \end{bmatrix}$$

14.

$$\begin{array}{ccccccc} x_1 & x_2 & x_3 & s_1 & s_2 & s_3 & z \end{array}$$
$$\begin{bmatrix} 2 & 2 & 1 & 1 & 0 & 0 & 0 & 12 \\ 1 & 2 & 3 & 0 & 1 & 0 & 0 & 45 \\ 3 & 1 & 1 & 0 & 0 & 1 & 0 & 20 \\ \hline -2 & -1 & -3 & 0 & 0 & 0 & 1 & 0 \end{bmatrix}$$

15.

$$\begin{array}{ccccccc} x_1 & x_2 & x_3 & s_1 & s_2 & s_3 & z \end{array}$$
$$\begin{bmatrix} 1 & 1 & 1 & 1 & 0 & 0 & 0 & 60 \\ 3 & 1 & 2 & 0 & 1 & 0 & 0 & 100 \\ 1 & 2 & 3 & 0 & 0 & 1 & 0 & 200 \\ \hline -1 & -1 & -2 & 0 & 0 & 0 & 1 & 0 \end{bmatrix}$$

16.

$$\begin{array}{ccccccc} x_1 & x_2 & x_3 & s_1 & s_2 & s_3 & z \end{array}$$
$$\begin{bmatrix} 4 & 2 & 3 & 1 & 0 & 0 & 0 & 22 \\ 2 & 2 & 5 & 0 & 1 & 0 & 0 & 28 \\ 1 & 3 & 2 & 0 & 0 & 1 & 0 & 45 \\ \hline -3 & -2 & -4 & 0 & 0 & 0 & 1 & 0 \end{bmatrix}$$

For each simplex tableau in Exercises 17–20, (a) list the basic and the nonbasic variables, (b) find the basic feasible solution determined by setting the nonbasic variables equal to 0, and (c) decide whether this is a maximum solution. (See Examples 5 and 6.)

17.

$$\begin{array}{ccccccc} x_1 & x_2 & x_3 & s_1 & s_2 & z \end{array}$$
$$\begin{bmatrix} 3 & 2 & 0 & -3 & 1 & 0 & 29 \\ 4 & 0 & 1 & -2 & 0 & 0 & 16 \\ \hline -5 & 0 & 0 & -1 & 0 & 1 & 11 \end{bmatrix}$$

18.

$$\begin{array}{ccccccc} x_1 & x_2 & x_3 & s_1 & s_2 & s_3 & z \end{array}$$
$$\begin{bmatrix} -3 & 0 & \frac{1}{2} & 1 & -2 & 0 & 0 & 22 \\ 2 & 0 & -3 & 0 & 1 & 1 & 0 & 10 \\ 4 & 1 & 4 & 0 & \frac{3}{4} & 0 & 0 & 17 \\ \hline -1 & 0 & 0 & 0 & 1 & 0 & 1 & 120 \end{bmatrix}$$

19.

$$\begin{array}{ccccccc} x_1 & x_2 & x_3 & s_1 & s_2 & s_3 & z \end{array}$$
$$\begin{bmatrix} 1 & 0 & 2 & \frac{1}{2} & 0 & \frac{1}{3} & 0 & 6 \\ 0 & 1 & -1 & 5 & 0 & -1 & 0 & 13 \\ 0 & 0 & 1 & \frac{3}{2} & 1 & -\frac{1}{3} & 0 & 21 \\ \hline 0 & 0 & 2 & \frac{1}{2} & 0 & 3 & 1 & 18 \end{bmatrix}$$

20.

$$\begin{array}{cccccccc} x_1 & x_2 & x_3 & x_4 & s_1 & s_2 & s_3 & z \end{array}$$
$$\begin{bmatrix} -1 & 0 & 0 & 1 & 0 & 3 & -2 & 0 & 47 \\ 2 & 0 & 1 & 0 & 0 & 2 & -\frac{1}{2} & 0 & 37 \\ 3 & 5 & 0 & 0 & 1 & -1 & 6 & 0 & 43 \\ \hline 4 & 1 & 0 & 0 & 0 & 6 & 0 & 1 & 86 \end{bmatrix}$$

Use the simplex method to solve Exercises 21–36.

21. Maximize $z = x_1 + 3x_2$

subject to $x_1 + x_2 \le 10$

$5x_1 + 2x_2 \le 20$

$x_1 + 2x_2 \le 36$

$x_1 \ge 0, x_2 \ge 0.$

22. Maximize $z = 5x_1 + x_2$

subject to $2x_1 + 3x_2 \le 8$

$4x_1 + 8x_2 \le 12$

$5x_1 + 2x_2 \le 30$

$x_1 \ge 0, x_2 \ge 0.$

23. Maximize $z = 2x_1 + x_2$

subject to $x_1 + 3x_2 \le 12$

$2x_1 + x_2 \le 10$

$x_1 + x_2 \le 4$

$x_1 \ge 0, x_2 \ge 0.$

24. Maximize $z = 4x_1 + 2x_2$

subject to $-x_1 - x_2 \le 12$

$3x_1 - x_2 \le 15$

$x_1 \ge 0, x_2 \ge 0.$

25. Maximize $z = 5x_1 + 4x_2 + x_3$

subject to $-2x_1 + x_2 + 2x_3 \le 3$

$x_1 - x_2 + x_3 \le 1$

$x_1 \ge 0, x_2 \ge 0, x_3 \ge 0.$

26. Maximize $z = 3x_1 + 2x_2 + x_3$

subject to $2x_1 + 2x_2 + x_3 \le 10$

$x_1 + 2x_2 + 3x_3 \le 15$

$x_1 \ge 0, x_2 \ge 0, x_3 \ge 0.$

27. Maximize $z = 2x_1 + x_2 + x_3$

subject to $x_1 - 3x_2 + x_3 \le 3$

$x_1 - 2x_2 + 2x_3 \le 12$

$x_1 \ge 0, x_2 \ge 0, x_3 \ge 0.$

28. Maximize $z = 4x_1 + 5x_2 + x_3$

subject to $x_1 + 2x_2 + 4x_3 \le 10$

$2x_1 + 2x_2 + x_3 \le 10$

$x_1 \ge 0, x_2 \ge 0, x_3 \ge 0.$

29. Maximize $z = 2x_1 + 2x_2 - 4x_3$

subject to $3x_1 + 3x_2 - 6x_3 \le 51$

$5x_1 + 5x_2 + 10x_3 \le 99$

$x_1 \ge 0, x_2 \ge 0, x_3 \ge 0.$

30. Maximize $z = 4x_1 + x_2 + 3x_3$

subject to $x_1 + 3x_3 \le 6$

$6x_1 + 3x_2 + 12x_3 \le 40$

$x_1 \ge 0, x_2 \ge 0, x_3 \ge 0.$

31. Maximize $z = 300x_1 + 200x_2 + 100x_3$
subject to $x_1 + x_2 + x_3 \le 100$
$$2x_1 + 3x_2 + 4x_3 \le 320$$
$$2x_1 + x_2 + x_3 \le 160$$
$$x_1 \ge 0, x_2 \ge 0, x_3 \ge 0.$$

32. Maximize $z = x_1 + 5x_2 - 10x_3$
subject to $8x_1 + 4x_2 + 12x_3 \le 18$
$$x_1 + 6x_2 + 2x_3 \le 45$$
$$5x_1 + 7x_2 + 3x_3 \le 60$$
$$x_1 \ge 0, x_2 \ge 0, x_3 \ge 0.$$

33. Maximize $z = 4x_1 - 3x_2 + 2x_3$
subject to $2x_1 - x_2 + 8x_3 \le 40$
$$4x_1 - 5x_2 + 6x_3 \le 60$$
$$2x_1 - 2x_2 + 6x_3 \le 24$$
$$x_1 \ge 0, x_2 \ge 0, x_3 \ge 0.$$

34. Maximize $z = 3x_1 + 2x_2 - 4x_3$
subject to $x_1 - x_2 + x_3 \le 10$
$$2x_1 - x_2 + 2x_3 \le 30$$
$$-3x_1 + x_2 + 3x_3 \le 40$$
$$x_1 \ge 0, x_2 \ge 0, x_3 \ge 0.$$

35. Maximize $z = x_1 + 2x_2 + x_3 + 5x_4$
subject to $x_1 + 2x_2 + x_3 + x_4 \le 50$
$$3x_1 + x_2 + 2x_3 + x_4 \le 100$$
$$x_1 \ge 0, x_2 \ge 0, x_3 \ge 0, x_4 \ge 0.$$

36. Maximize $z = x_1 + x_2 + 4x_3 + 5x_4$
subject to $x_1 + 2x_2 + 3x_3 + x_4 \le 115$
$$2x_1 + x_2 + 8x_3 + 5x_4 \le 200$$
$$x_1 + x_3 \le 50$$
$$x_1 \ge 0, x_2 \ge 0, x_3 \ge 0, x_4 \ge 0.$$

37. The initial simplex tableau of a linear programming problem is

$$\begin{array}{ccccccc} x_1 & x_2 & x_3 & s_1 & s_2 & z \\ \left[\begin{array}{cccccc|c} 1 & 1 & 1 & 1 & 0 & 0 & 12 \\ 2 & 1 & 2 & 0 & 1 & 0 & 30 \\ \hline -2 & -2 & -1 & 0 & 0 & 1 & 0 \end{array}\right] \end{array}$$

(a) Use the simplex method to solve the problem with column 1 as the first pivot column.

(b) Now use the simplex method to solve the problem with column 2 as the first pivot column.

(c) Does this problem have a unique maximum solution? Why?

38. The final simplex tableau of a linear programming problem is

$$\begin{array}{ccccc} x_1 & x_2 & s_1 & s_2 & z \\ \left[\begin{array}{cccc|c} 1 & 1 & 2 & 0 & 0 & 24 \\ 2 & 0 & 2 & 1 & 0 & 8 \\ \hline 4 & 0 & 0 & 0 & 1 & 40 \end{array}\right] \end{array}.$$

(a) What is the solution given by this tableau?

(b) Even though all the indicators are nonnegative, perform one more round of pivoting on this tableau, using column 3 as the pivot column and choosing the pivot row by forming quotients in the usual way.

(c) Show that there is more than one solution to the linear programming problem by comparing your answer in part (a) with the basic feasible solution given by the tableau found in part (b). Does it give the same value of z as the solution in part (a)?

✓**Checkpoint Answers**

1. $x_1 + x_2 + x_3 + s_1 = 12$
$$2x_1 + 4x_2 + s_2 = 15$$
$$x_2 + 3x_3 + s_3 = 10$$

2.
$$\begin{array}{cccccc} x_1 & x_2 & s_1 & s_2 & s_3 & z \\ \left[\begin{array}{cccccc|c} 1 & 2 & 1 & 0 & 0 & 0 & 85 \\ 2 & 1 & 0 & 1 & 0 & 0 & 92 \\ 1 & 4 & 0 & 0 & 1 & 0 & 104 \\ \hline -2 & -3 & 0 & 0 & 0 & 1 & 0 \end{array}\right] \end{array}$$
$$\underbrace{\qquad\qquad\qquad\qquad}_{\text{Indicators}}$$

3. 2 (in first column)

4. **(a)** 2

(b)
$$\begin{array}{cccccc} x_1 & x_2 & x_3 & s_1 & s_2 & z \\ \left[\begin{array}{cccccc|c} \frac{1}{2} & 1 & 3 & \frac{1}{2} & 0 & 0 & 8 \\ -\frac{1}{2} & 0 & -9 & -\frac{3}{2} & 1 & 0 & 1 \\ \hline 1 & 0 & 9 & 2 & 0 & 1 & 32 \end{array}\right] \end{array}$$

5. $z = 21$ when $x_1 = 6, x_2 = 0$, and $x_3 = 9$.

6. $x_1 = 0, x_2 = 24, s_1 = 16, s_2 = 0, z = 4800$

7.5 Maximization Applications

Applications of the simplex method are considered in this section. First, however, we make a slight change in notation. You have noticed that the column representing the variable z in a simplex tableau never changes during pivoting. (Since all the entries except the last one in this column are 0, performing row operations has no effect on these entries—they remain 0.) Consequently, this column is unnecessary and can be omitted without causing any difficulty.

Hereafter in this text, the column corresponding to the variable z (representing the objective function) will be omitted from all simplex tableaus.

EXAMPLE 1   **Farming** A farmer has 110 acres of available land he wishes to plant with a mixture of potatoes, corn, and cabbage. It costs him $400 to produce an acre of potatoes, $160 to produce an acre of corn, and $280 to produce an acre of cabbage. He has a maximum of $20,000 to spend. He makes a profit of $120 per acre of potatoes, $40 per acre of corn, and $60 per acre of cabbage.

(a) How many acres of each crop should he plant to maximize his profit?

Solution Let the number of acres alloted to each of potatoes, corn, and cabbage be x_1, x_2, and x_3, respectively. Then summarize the given information as follows:

Crop	Number of Acres	Cost per Acre	Profit per Acre
Potatoes	x_1	$400	$120
Corn	x_2	$160	$ 40
Cabbage	x_3	$280	$ 60
Maximum Available	110	$20,000	

The constraints can be expressed as

$$x_1 + x_2 + x_3 \leq 110 \qquad \text{Number of acres}$$
$$400x_1 + 160x_2 + 280x_3 \leq 20{,}000, \qquad \text{Production costs}$$

where x_1, x_2, and x_3 are all nonnegative. The first of these constraints says that $x_1 + x_2 + x_3$ is less than or perhaps equal to 110. Use s_1 as the slack variable, giving the equation

$$x_1 + x_2 + x_3 + s_1 = 110.$$

Here, s_1 represents the amount of the farmer's 110 acres that will not be used. (s_1 may be 0 or any value up to 110.)

In the same way, the constraint $400x_1 + 160x_2 + 280x_3 \leq 20{,}000$ can be converted into an equation by adding a slack variable s_2:

$$400x_1 + 160x_2 + 280x_3 + s_2 = 20{,}000.$$

The slack variable s_2 represents any unused portion of the farmer's $20,000 capital. (Again, s_2 may have any value from 0 to 20,000.)

The farmer's profit on potatoes is the product of the profit per acre ($120) and the number x_1 of acres, that is, $120x_1$. His profits on corn and cabbage are computed similarly. Hence, his total profit is given by

$$z = \text{profit on potatoes} + \text{profit on corn} + \text{profit on cabbage}$$
$$z = 120x_1 + 40x_2 + 60x_3.$$

The linear programming problem can now be stated as follows:

$$\text{Maximize} \quad z = 120x_1 + 40x_2 + 60x_3$$
$$\text{subject to} \quad x_1 + x_2 + x_3 + s_1 \qquad = 110$$
$$400x_1 + 160x_2 + 280x_3 \qquad + s_2 = 20{,}000,$$
$$\text{with} \quad x_1 \geq 0, x_2 \geq 0, x_3 \geq 0, s_1 \geq 0, s_2 \geq 0.$$

The initial simplex tableau (without the z column) is

$$
\begin{array}{c}
\begin{array}{ccccc} x_1 & x_2 & x_3 & s_1 & s_2 \end{array} \\
\left[\begin{array}{ccccc|c}
1 & 1 & 1 & 1 & 0 & 110 \\
400 & 160 & 280 & 0 & 1 & 20{,}000 \\
\hline
-120 & -40 & -60 & 0 & 0 & 0
\end{array}\right].
\end{array}
$$

The most negative indicator is -120; column 1 is the pivot column. The quotients needed to determine the pivot row are $110/1 = 110$ and $20{,}000/400 = 50$. So the pivot is 400 in row 2, column 1. Multiplying row 2 by $1/400$ and completing the pivoting leads to the final simplex tableau:

$$
\begin{array}{c}
\begin{array}{ccccc} x_1 & x_2 & x_3 & s_1 & \quad s_2 \end{array} \\
\left[\begin{array}{ccccc|c}
0 & .6 & .3 & 1 & -.0025 & 60 \\
1 & .4 & .7 & 0 & .0025 & 50 \\
\hline
0 & 8 & 24 & 0 & .3 & 6000
\end{array}\right].
\end{array}
\quad
\begin{array}{l}
-1R_2 + R_1 \\[4pt]
\frac{1}{400}R_2 \\[4pt]
120R_2 + R_3
\end{array}
$$

Setting the nonbasic variables x_2, x_3, and s_2 equal to 0, solving for the basic variables x_1 and s_1, and remembering that the value of z is in the lower right-hand corner leads to this maximum solution:

$$x_1 = 50, \qquad x_2 = 0, \qquad x_3 = 0, \qquad s_1 = 60, \qquad s_2 = 0, \qquad \text{and} \qquad z = 6000.$$

Therefore, the farmer will make a maximum profit of $6000 by planting 50 acres of potatoes and no corn or cabbage.

(b) If the farmer maximizes his profit, how much land will remain unplanted? What is the explanation for this?

Solution Since 50 of 110 acres are planted, 60 acres will remain unplanted. Alternatively, note that the unplanted acres of land are represented by s_1, the slack variable in the "number of acres" constraint. In the maximal solution found in part (a), $s_1 = 60$, which means that 60 acres are left unplanted.

The amount of unused cash is represented by s_2, the slack variable in the "production costs" constraint. Since $s_2 = 0$, all the available money has been used. By using the maximal solution in part (a), the farmer has used his $20,000 most effectively. If he had more cash, he would plant more crops and make a larger profit.

EXAMPLE 2   **Advertising** Ana Pott, who is a candidate for the state legislature, has $96,000 to buy TV advertising time. Ads cost $400 per minute on a local cable channel, $4000 per minute on a regional independent channel, and $12,000 per minute on a national network channel. Because of existing contracts, the TV stations can provide at most a total of 30 minutes of advertising time, with a maximum of 6 minutes on the national network channel. At any given time during the evening, approximately 100,000 people watch the cable channel, 200,000 the independent channel, and 600,000 the network channel. To get maximum exposure, how much time should Ana buy from each station?

(a) Set up the initial simplex tableau for this problem.

Solution Let x_1 be the number of minutes of ads on the cable channel, x_2 the number of minutes on the independent channel, and x_3 the number of minutes on the network channel. Exposure is measured in viewer-minutes. For instance, 100,000 people watching x_1 minutes of ads on the cable channel produces $100{,}000x_1$ viewer-minutes. The amount of exposure is given by the total number of viewer-minutes for all three channels, namely,

$$100{,}000x_1 + 200{,}000x_2 + 600{,}000x_3.$$

Since 30 minutes are available,

$$x_1 + x_2 + x_3 \leq 30.$$

The fact that only 6 minutes can be used on the network channel means that

$$x_3 \leq 6.$$

Expenditures are limited to $96,000, so

$$\text{Cable cost} + \text{independent cost} + \text{network cost} \leq 96{,}000$$
$$400x_1 + 4000x_2 + 12{,}000x_3 \leq 96{,}000.$$

Therefore, Ana must solve the following linear programming problem:

$$\text{Maximize} \quad z = 100{,}000x_1 + 200{,}000x_2 + 600{,}000x_3$$
$$\text{subject to} \quad x_1 + x_2 + x_3 \leq 30$$
$$x_3 \leq 6$$
$$400x_1 + 4000x_2 + 12{,}000x_3 \leq 96{,}000,$$
$$\text{with} \quad x_1 \geq 0, x_2 \geq 0, x_3 \geq 0.$$

Introducing slack variables $s_1, s_2,$ and s_3 (one for each constraint), rewriting the constraints as equations, and expressing the objective function as

$$-100{,}000x_1 - 200{,}000x_2 - 600{,}000x_3 + z = 0$$

leads to the initial simplex tableau:

$$
\begin{array}{cccccc|c}
x_1 & x_2 & x_3 & s_1 & s_2 & s_3 & \\
1 & 1 & 1 & 1 & 0 & 0 & 30 \\
0 & 0 & 1 & 0 & 1 & 0 & 6 \\
400 & 4000 & 12{,}000 & 0 & 0 & 1 & 96{,}000 \\
\hline
-100{,}000 & -200{,}000 & -600{,}000 & 0 & 0 & 0 & 0
\end{array}
$$

(b) Use the simplex method to find the final simplex tableau.

Solution Work by hand, or use a graphing calculator's simplex program or a spreadsheet, to obtain this final tableau:

$$
\begin{array}{cccccc|c}
x_1 & x_2 & x_3 & s_1 & s_2 & s_3 & \\
1 & 0 & 0 & \frac{10}{9} & \frac{20}{9} & -\frac{25}{90{,}000} & 20 \\
0 & 0 & 1 & 0 & 1 & 0 & 6 \\
0 & 1 & 0 & -\frac{1}{9} & -\frac{29}{9} & \frac{25}{90{,}000} & 4 \\
\hline
0 & 0 & 0 & \frac{800{,}000}{9} & \frac{1{,}600{,}000}{9} & \frac{250}{9} & 6{,}400{,}000
\end{array}
$$

Therefore, the optimal solution is

$$x_1 = 20, \quad x_2 = 4, \quad x_3 = 6, \quad s_1 = 0, \quad s_2 = 0, \quad \text{and} \quad s_3 = 0.$$

Ana should buy 20 minutes of time on the cable channel, 4 minutes on the independent channel, and 6 minutes on the network channel.

(c) What do the values of the slack variables in the optimal solution tell you?

Solution All three slack variables are 0. This means that all the available minutes have been used ($s_1 = 0$ in the first constraint), the maximum possible 6 minutes on the national network have been used ($s_2 = 0$ in the second constraint), and all of the $96,000 has been spent ($s_3 = 0$ in the third constraint). ✓₁

✓ **Checkpoint 1**

In Example 2, what is the number of viewer-minutes in the optimal solution?

EXAMPLE 3   **Manufacturing** A chemical plant makes three products—glaze, solvent, and clay—each of which brings in different revenue per truckload. Production is limited, first by the number of air pollution units the plant is allowed to produce each day and second by the time available in the evaporation tank. The plant manager wants to maximize the daily revenue. Using information not given here, he sets up an initial simplex tableau and uses the simplex method to produce the following final simplex tableau:

$$
\begin{array}{ccccc}
x_1 & x_2 & x_3 & s_1 & s_2 \\
\end{array}
$$
$$
\begin{bmatrix}
-10 & -25 & 0 & 1 & -1 & 60 \\
3 & 4 & 1 & 0 & .1 & 24 \\
7 & 13 & 0 & 0 & .4 & 96
\end{bmatrix}.
$$

The three variables represent the number of truckloads of glaze, solvent, and clay, respectively. The first slack variable comes from the air pollution constraint and the second slack variable from the time constraint on the evaporation tank. The revenue function is given in hundreds of dollars.

(a) What is the optimal solution?

Solution

$$x_1 = 0, \qquad x_2 = 0, \qquad x_3 = 24, \qquad s_1 = 60, \qquad s_2 = 0, \qquad \text{and} \qquad z = 96.$$

(b) Interpret this solution. What do the variables represent, and what does the solution mean?

Solution The variable x_1 is the number of truckloads of glaze, x_2 the number of truckloads of solvent, x_3 the number of truckloads of clay to be produced, and z the revenue produced (in hundreds of dollars). The plant should produce 24 truckloads of clay and no glaze or solvent, for a maximum revenue of $9600. The first slack variable, s_1, represents the number of air pollution units below the maximum number allowed. Since $s_1 = 60$, the number of air pollution units will be 60 less than the allowable maximum. The second slack variable, s_2, represents the unused time in the evaporation tank. Since $s_2 = 0$, the evaporation tank is fully used.

7.5 Exercises

Set up the initial simplex tableau for each of the given problems. You will be asked to solve these problems in Exercises 19–22.

1. **Juice Business** A juice blender has the following amounts of three kinds of juice: 90 gallons of pineapple juice, 80 gallons of orange juice, and 50 gallons of mango juice. To create its signature juice WildSplash, the juice blender uses 2 gallons of pineapple juice, 1 gallon of orange juice, and 1 gallon of mango juice. To create the new mixture TastyTreat, the juice blender uses 1, 2, and 1 gallon, respectively, of each type of juice. If the juice blender sells the mixtures wholesale at the price of $12 for a batch of WildSplash and $10 for a batch of TastyTreat, how many batches of each product should be created to obtain maximum gross income? What is the maximum gross income?

2. **Window Manufacturing** A window manufacturing company can make 3 windows in small, medium, and large. Each window is worked on by three technicians: Todd, Donna, and Jackson. Todd can work up to 25 hours per week, Donna can work up to 45 hours a week, and Jackson can work 40 hours a week. Todd works 2 hours on the small windows, 1 hour on the medium windows, and 1 hour on the large windows. Donna works 2 hours on the small windows, 3 hours on the medium windows, and 2 hours on the large windows. Jackson works 1 hour on the small windows, 2 hours on the medium windows, and 4 hours on the large windows. The manager is worried about getting overextended, so no more than 4 large windows can be

done in a week. The small windows sell for $180, the medium windows for $200, and the large windows for $220. Find the number of each size of windows that should be made weekly in order to maximize profit. Find the maximum possible profit.

3. **Nutrition** A biologist has 500 kilograms of nutrient A, 600 kilograms of nutrient B, and 300 kilograms of nutrient C. These nutrients will be used to make 4 types of food—P, Q, R, and S—whose contents (in percent of nutrient per kilogram of food) and whose "growth values" are as shown in the following table:

	P	Q	R	S
A	0	0	37.5	62.5
B	0	75	50	37.5
C	100	25	12.5	0
Growth Value	90	70	60	50

How many kilograms of each food should be produced in order to maximize total growth value? Find the maximum growth value.

4. **Fish Nutrition** A lake is stocked each spring with three species of fish: A, B, and C. The average weights of the fish are 1.62, 2.12, and 3.01 kilograms for the three species, respectively. Three foods—I, II, and III—are available in the lake. Each fish of species A requires 1.32 units of food I, 2.9 units of food II, and

1.75 units of food III, on the average, each day. Species *B* fish require 2.1 units of food I, .95 units of food II, and .6 units of food III daily. Species *C* fish require .86, 1.52, and 2.01 units of I, II, and III per day, respectively. If 490 units of food I, 897 units of food II, and 653 units of food III are available daily, how should the lake be stocked to maximize the weight of the fish it supports?

In each of the given exercises, (a) use the simplex method to solve the problem and (b) explain what the values of the slack variables in the optimal solution mean in the context of the problem. (See Examples 1–3).

5. **Business** A manufacturer of bicycles builds 1-, 3-, and 10-speed models. The bicycles are made of both aluminum and steel. The company has available 91,800 units of steel and 42,000 units of aluminum. The 1-, 3-, and 10-speed models need, respectively, 20, 30, and 40 units of steel and 12, 21, and 16 units of aluminum. How many of each type of bicycle should be made in order to maximize profit if the company makes $80 per 1-speed bike, $120 per 3-speed bike, and $240 per 10-speed bike? What is the maximum possible profit?

6. **Fund-Raising** Liz is working to raise money for breast cancer research by sending informational letters to local neighborhood organizations and church groups. She discovered that each church group requires 2 hours of letter writing and 1 hour of follow-up, while each neighborhood group needs 2 hours of letter writing and 3 hours of follow-up. Liz can raise $1000 from each church group and $2000 from each neighborhood organization, and she has a maximum of 16 hours of letter-writing time and a maximum of 12 hours of follow-up time available per month. Determine the most profitable mixture of groups she should contact and the most money she can raise in a month.

7. **Local News** A local news channel plans a 27-minute Saturday morning news show. The show will be divided into three segments involving sports, news, and weather. Market research has shown that the sports segment should be twice as long as the weather segment. The total time taken by the sports and weather segments should be twice the time taken by the news segment. On the basis of the market research, it is believed that 40, 60, and 50 (in thousands) viewers will watch the program for each minute the sports, news, and weather segments, respectively, are on the air. Find the time that should be allotted to each segment in order to get the maximum number of viewers. Find the number of viewers.

8. **Food Distribution** A food wholesaler has three kinds of individual bags of potato chips: regular, barbeque, and salt and vinegar. She wants to sells the bags of chips in bulk packages. The bronze package consists of 20 bags of regular and 10 bags of barbeque. The silver package contains 20 bags of regular, 10 bags of barbeque, and 10 bags of salt and vinegar. The gold package consists of 30 bags of regular, 10 bags of barbeque, and 10 bags of salt and vinegar. The profit is $30 on each bronze package, $40 on each silver package, and $60 on each gold package. The food wholesaler has a total of 8000 bags of regular chips, 4000 bags of barbeque, and 2000 bags of salt and vinegar. Assuming all the packages will be sold, how many gold, silver, and bronze packages should be made up in order to maximize profit? What is the maximum profit?

9. **Tree Nursery** Mario Cekada owns a tree nursery business. Mario can sell a weeping Japanese maple of a certain size for $350 in profit. He can sell tricolor beech trees of a certain size for $500 profit. The travel time to acquire the Japanese maple trees is 5 hours while the travel time for the beech trees is 7 hours. The digging process takes 1 hour for the Japanese maple trees and 2 hours for the beech trees. Both kinds of trees require 4 hours of time to deliver to the client. In a particular season, Mario has available 3600 work hours for travel time to acquire the trees, 900 work-hours for digging, and 2600 work-hours for delivery to the clients. How many trees of each kind should he acquire to make a maximum profit? What is the maximum profit?

10. **Business** The Texas Poker Company assembles three different poker sets. Each Royal Flush poker set contains 1000 poker chips, 4 decks of cards, 10 dice, and 2 dealer buttons. Each Deluxe Diamond poker set contains 600 poker chips, 2 decks of cards, 5 dice, and 1 dealer button. The Full House poker set contains 300 poker chips, 2 decks of cards, 5 dice, and 1 dealer button. The Texas Poker Company has 2,800,000 poker chips, 10,000 decks of cards, 25,000 dice, and 6000 dealer buttons in stock. It earns a profit of $38 for each Royal Flush poker set, $22 for each Deluxe Diamond poker set, and $12 for each Full House poker set. How many of each type of poker set should it assemble to maximize profit? What is the maximum profit?

Use the simplex method to solve the given problems. (See Examples 1–3.)

11. **Advertising** The Fancy Fashions Store has $8000 available each month for advertising. Newspaper ads cost $400 each, and no more than 20 can be run per month. Radio ads cost $200 each, and no more than 30 can run per month. TV ads cost $1200 each, with a maximum of 6 available each month. Approximately 2000 women will see each newspaper ad, 1200 will hear each radio commercial, and 10,000 will see each TV ad. How much of each type of advertising should be used if the store wants to maximize its ad exposure?

12. **Candy Manufacturing** Caroline's Quality Candy Confectionery is famous for fudge, chocolate cremes, and pralines. Its candy-making equipment is set up to make 100-pound batches at a time. Currently there is a chocolate shortage, and the company can get only 120 pounds of chocolate in the next shipment. On a week's run, the confectionery's cooking and processing equipment is available for a total of 42 machine hours. During the same period, the employees have a total of 56 work hours available for packaging. A batch of fudge requires 20 pounds of chocolate, while a batch of cremes uses 25 pounds of chocolate. The cooking and processing take 120 minutes for fudge, 150 minutes for chocolate cremes, and 200 minutes for pralines. The packaging times, measured in minutes per 1-pound box, are 1, 2, and 3, respectively, for fudge, cremes, and pralines. Determine how many batches of each type of candy the confectionery

should make, assuming that the profit per 1-pound box is 50¢ on fudge, 40¢ on chocolate cremes, and 45¢ on pralines. Also, find the maximum profit for the week.

13. **Fund-Raising** A political party is planning its fund-raising activities for a coming election. It plans to raise money through large fund-raising parties, letters requesting funds, and dinner parties where people can meet the candidate personally. Each large fund-raising party costs $3000, each mailing costs $1000, and each dinner party costs $12,000. The party can spend up to $102,000 for these activities. From experience, it is known that each large party will raise $200,000, each letter campaign will raise $100,000, and each dinner party will raise $600,000. The party is able to carry out as many as 25 of these activities.

(a) How many of each should the party plan to raise the maximum amount of money? What is the maximum amount?

(b) Dinner parties are more expensive than letter campaigns, yet the optimum solution found in part (a) includes dinner parties, but no letter campaigns. Explain how this is possible.

14. **Bread Baking** A baker has 60 units of flour, 132 units of sugar, and 102 units of raisins. A loaf of raisin bread requires 1 unit of flour, 1 unit of sugar, and 2 units of raisins, while a raisin cake needs 2, 4, and 1 units, respectively. If raisin bread sells for $3 a loaf and a raisin cake for $4, how many of each should be baked so that the gross income is maximized? What is the maximum gross income?

15. **Exercise** Ramona, a fitness trainer, has an exercise regimen that includes running, biking, and walking. She has no more than 15 hours per week to devote to exercise, including at most 3 hours for running. She wants to walk at least twice as many hours as she bikes. A 130-pound person like Ramona will burn on average 531 calories per hour running, 472 calories per hour biking, and 354 calories per hour walking. How many hours per week should Ramona spend on each exercise to maximize the number of calories she burns? What is the maximum number of calories she will burn? (*Hint:* Write the constraint involving walking and biking in the form ≤0.)

16. **Exercise** Liam's exercise regimen includes light calisthenics, swimming, and playing the drums. He has at most 10 hours per week to devote to these activities. He wants the total time he does calisthenics and plays the drums to be at least twice as long as he swims. His neighbors, however, will tolerate no more than 4 hours per week on the drums. A 190-pound person like Liam will burn an average of 388 calories per hour doing calisthenics, 518 calories per hour swimming, and 345 calories per hour playing the drums. How many hours per week should Liam spend on each exercise to maximize the number of calories he burns? What is the maximum number of calories he will burn?

Use a graphing calculator or a computer program for the simplex method to solve the given linear programming problems.

17. Solve the problem in Exercise 1.

18. Exercise 2. Your final answer should consist of whole numbers (because you can't sell half a window).

19. Exercise 3

20. Exercise 4

✓**Checkpoint Answer**

1. $z = 6,400,000$

7.6 The Simplex Method: Duality and Minimization

📄 **NOTE** Sections 7.6 and 7.7 are independent of each other and may be read in either order.

Here, we present a method of solving *minimization* problems in which all constraints involve ≥ and all coefficients of the objective function are positive. When it applies, this method may be more efficient than the method discussed in Section 7.7. However, the method in Section 7.7 applies to a wider variety of problems—both minimization and maximization problems, even those that involve *mixed constraints* (≤, =, or ≥)—and it has no restrictions on the objective function.

We begin with a necessary tool from matrix algebra: if A is a matrix, then the **transpose** of A is the matrix obtained by interchanging the rows and columns of A.

 **EXAMPLE 1** Find the transpose of each matrix.

(a) $A = \begin{bmatrix} 2 & -1 & 5 \\ 6 & 8 & 0 \\ -3 & 7 & -1 \end{bmatrix}$.

Solution Write the rows of matrix A as the columns of the transpose:

$$\text{Transpose of } A = \begin{bmatrix} 2 & 6 & -3 \\ -1 & 8 & 7 \\ 5 & 0 & -1 \end{bmatrix}.$$

(b) $A = \begin{bmatrix} 1 & 2 & 4 & 0 \\ 2 & 1 & 7 & 6 \end{bmatrix}.$

Solution The transpose of $\begin{bmatrix} 1 & 2 & 4 & 0 \\ 2 & 1 & 7 & 6 \end{bmatrix}$ is $\begin{bmatrix} 1 & 2 \\ 2 & 1 \\ 4 & 7 \\ 0 & 6 \end{bmatrix}.$ ✓ 1

✓ **Checkpoint 1**

Give the transpose of each matrix.

(a) $\begin{bmatrix} 2 & 4 \\ 6 & 3 \\ 1 & 5 \end{bmatrix}$

(b) $\begin{bmatrix} 4 & 7 & 10 \\ 3 & 2 & 6 \\ 5 & 8 & 12 \end{bmatrix}$

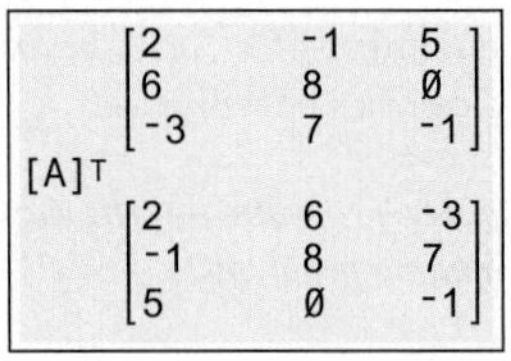

$[A]^T$

Figure 7.31

▦ **TECHNOLOGY TIP** Most graphing calculators can find the transpose of a matrix. Look for this feature in the MATRIX MATH menu (TI). The transpose of matrix A from Example 1(a) is shown in Figure 7.31.

We now consider linear programming problems satisfying the following conditions:

1. The objective function is to be minimized.

2. All the coefficients of the objective function are nonnegative.

3. All constraints involve $\geq$.

4. All variables are nonnegative.

The method of solving minimization problems presented here is based on an interesting connection between maximization and minimization problems: Any solution of a maximizing problem produces the solution of an associated minimizing problem, and vice versa. Each of the associated problems is called the **dual** of the other. Thus, duals enable us to solve minimization problems of the type just described by the simplex method introduced in Section 7.4.

When dealing with minimization problems, we use y_1, y_2, y_3, etc., as variables and denote the objective function by w. The next two examples show how to construct the dual problem. Later examples will show how to solve both the dual problem and the original one.

EXAMPLE 2 Construct the dual of this problem:

$$\text{Minimize} \quad w = 8y_1 + 16y_2$$
$$\text{subject to} \quad y_1 + 5y_2 \geq 9$$
$$2y_1 + 2y_2 \geq 10$$
$$y_1 \geq 0, y_2 \geq 0.$$

Solution Write the augmented matrix of the system of inequalities *and* include the coefficients of the objective function (not their negatives) as the last row of the matrix:

Constants

$$\begin{bmatrix} 1 & 5 & | & 9 \\ 2 & 2 & | & 10 \\ \hline 8 & 16 & | & 0 \end{bmatrix}.$$

Objective function

Now form the transpose of the preceding matrix:

$$\begin{bmatrix} 1 & 2 & | & 8 \\ 5 & 2 & | & 16 \\ \hline 9 & 10 & | & 0 \end{bmatrix}.$$

In this last matrix, think of the first two rows as constraints and the last row as the objective function. Then the dual maximization problem is as follows:

$$\text{Maximize} \quad z = 9x_1 + 10x_2$$
$$\text{subject to} \quad x_1 + 2x_2 \le 8$$
$$5x_1 + 2x_2 \le 16$$
$$x_1 \ge 0, x_2 \ge 0.$$

EXAMPLE 3 Write the duals of the given minimization linear programming problems.

(a) Minimize $\quad w = 10y_1 + 8y_2$
subject to $\quad y_1 + 2y_2 \ge 2$
$$y_1 + y_2 \ge 5$$
$$y_1 \ge 0, y_2 \ge 0.$$

Solution Begin by writing the augmented matrix for the given problem:

$$\left[\begin{array}{cc|c} 1 & 2 & 2 \\ 1 & 1 & 5 \\ \hline 10 & 8 & 0 \end{array}\right].$$

Form the transpose of this matrix to get

$$\left[\begin{array}{cc|c} 1 & 1 & 10 \\ 2 & 1 & 8 \\ \hline 2 & 5 & 0 \end{array}\right].$$

The dual problem is stated from this second matrix as follows (using x instead of y):

$$\text{Maximize} \quad z = 2x_1 + 5x_2$$
$$\text{subject to} \quad x_1 + x_2 \le 10$$
$$2x_1 + x_2 \le 8$$
$$x_1 \ge 0, x_2 \ge 0.$$

(b) Minimize $\quad w = 7y_1 + 5y_2 + 8y_3$
subject to $\quad 3y_1 + 2y_2 + y_3 \ge 10$
$$y_1 + y_2 + y_3 \ge 8$$
$$4y_1 + 5y_2 \qquad \ge 25$$
$$y_1 \ge 0, y_2 \ge 0, y_3 \ge 0.$$

Solution Find the augmented matrix for this problem, as in part (a). Form the dual matrix, which represents the following problem:

$$\text{Maximize} \quad z = 10x_1 + 8x_2 + 25x_3$$
$$\text{subject to} \quad 3x_1 + x_2 + 4x_3 \le 7$$
$$2x_1 + x_2 + 5x_3 \le 5$$
$$x_1 + x_2 \qquad \le 8$$
$$x_1 \ge 0, x_2 \ge 0, x_3 \ge 0. \quad ✓_2$$

✓ **Checkpoint 2**

Write the dual of the following linear programming problem:

Minimize $\quad w = 2y_1 + 5y_2 + 6y_3$
subject to $\quad 2y_1 + 3y_2 + y_3 \ge 15$
$$y_1 + y_2 + 2y_3 \ge 12$$
$$5y_1 + 3y_2 \qquad \ge 10$$
$$y_1 \ge 0, y_2 \ge 0, y_3 \ge 0.$$

In Example 3, all the constraints of the minimization problems were $\ge$ inequalities, while all those in the dual maximization problems were $\le$ inequalities. This is generally the case; inequalities are reversed when the dual problem is stated.

The following table shows the close connection between a problem and its dual.

Given Problem	Dual Problem
m variables	n variables
n constraints	m constraints (m slack variables)
Coefficients from objective function	Constraint constants
Constraint constants	Coefficients from objective function

Now that you know how to construct the dual problem, we examine how it is related to the original problem and how both may be solved.

EXAMPLE 4   Solve this problem and its dual:

$$\text{Minimize} \quad w = 8y_1 + 16y_2$$
$$\text{subject to} \quad y_1 + 5y_2 \geq 9$$
$$2y_1 + 2y_2 \geq 10$$
$$y_1 \geq 0, y_2 \geq 0.$$

Solution In Example 2, we saw that the dual problem is

$$\text{Maximize} \quad z = 9x_1 + 10x_2$$
$$\text{subject to} \quad x_1 + 2x_2 \leq 8$$
$$5x_1 + 2x_2 \leq 16$$
$$x_1 \geq 0, x_2 \geq 0.$$

In this case, both the original problem and the dual may be solved geometrically, as in Section 7.2. Figure 7.32(a) shows the region of feasible solutions for the original minimization problem, and Checkpoint 3 shows that

the minimum value of w is 48 at the vertex $(4, 1)$. ✓3

Figure 7.32(b) shows the region of feasible solutions for the dual maximization problem, and Checkpoint 4 shows that

the maximum value of z is 48 at the vertex $(2, 3)$. ✓4

Even though the regions and the corner points are different, the minimization problem and its dual have the same solution, 48.

✓ **Checkpoint 3**

Use the corner points in Figure 7.32(a) to find the minimum value of $w = 8y_1 + 16y_2$ and where it occurs.

✓ **Checkpoint 4**

Use Figure 7.32(b) to find the maximum value of $z = 9x_1 + 10x_2$ and where it occurs.

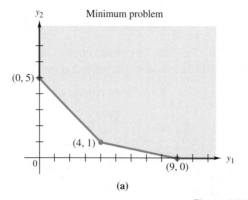

(a)

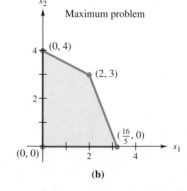

(b)

Figure 7.32

The next theorem, whose proof requires advanced methods, guarantees that what happened in Example 4 happens in the general case as well.

Theorem of Duality

The objective function w of a minimizing linear programming problem takes on a minimum value if, and only if, the objective function z of the corresponding dual maximizing problem takes on a maximum value. The maximum value of z equals the minimum value of w.

Geometric solution methods were used in Example 4, but the simplex method can also be used. In fact, the final simplex tableau shows the solutions for both the original minimization problem and the dual maximization problem, as illustrated in the next example.

EXAMPLE 5   Use the simplex method to solve the minimization problem in Example 4.

Solution First, set up the dual problem, as in Example 4:

$$\text{Maximize} \quad z = 9x_1 + 10x_2$$
$$\text{subject to} \quad x_1 + 2x_2 \leq 8$$
$$5x_1 + 2x_2 \leq 16$$
$$x_1 \geq 0, x_2 \geq 0.$$

This is a maximization problem in standard form, so it can be solved by the simplex method. Use slack variables to write the inequalities and the objective function as equations:

$$\begin{aligned} x_1 + 2x_2 + s_1 &= 8 \\ 5x_1 + 2x_2 + s_2 &= 16 \\ -9x_1 - 10x_2 + z &= 0, \end{aligned}$$

with $x_1 \geq 0$, $x_2 \geq 0$, $s_1 \geq 0$, and $s_2 \geq 0$. So the initial tableau is

$$\begin{array}{cccc|c} x_1 & x_2 & s_1 & s_2 & \\ 1 & \mathbf{2} & 1 & 0 & 8 \\ 5 & 2 & 0 & 1 & 16 \\ \hline -9 & -10 & 0 & 0 & 0 \end{array}$$

Quotients
$8/2 = 4$
$16/2 = 8$

The quotients show that the first pivot is the 2 shaded in blue. Pivoting is done as follows:

$$\begin{array}{cccc|c} x_1 & x_2 & s_1 & s_2 & \\ \frac{1}{2} & 1 & \frac{1}{2} & 0 & 4 \\ 4 & 0 & -1 & 1 & 8 \\ \hline -4 & 0 & 5 & 0 & 40 \end{array}$$

$\frac{1}{2}R_1$
$-2R_1 + R_2$
$10R_1 + R_3$

✓ Checkpoint 5

In the second tableau, find the next pivot.

Checkpoint 5 shows that the new pivot is the 4 in row 2, column 1. **✓5** Pivoting leads to the final simplex tableau:

$$\begin{array}{cccc|c} x_1 & x_2 & s_1 & s_2 & \\ 0 & 1 & \frac{5}{8} & -\frac{1}{8} & 3 \\ 1 & 0 & -\frac{1}{4} & \frac{1}{4} & 2 \\ \hline 0 & 0 & 4 & 1 & 48 \end{array}$$

$-\frac{1}{2}R_2 + R_1$
$\frac{1}{4}R_2$
$4R_2 + R_3$

The final simplex tableau shows that the maximum value of 48 occurs when $x_1 = 2$ and $x_2 = 3$. In Example 4, we saw that the minimum value of 48 occurs when $y_1 = 4$ and $y_2 = 1$. Note that this information appears in the last row (shown in blue). The minimum value of 48 is in the lower right-hand corner, and the values where this occurs ($y_1 = 4$ and $y_2 = 1$) are in the last row at the bottom of the slack-variable columns.

A minimization problem that meets the conditions listed after Example 1 can be solved by the method used in Example 5 and summarized here.

Solving Minimization Problems with Duals

1. Find the dual standard maximization problem.[*]

2. Use the simplex method to solve the dual maximization problem.

3. Read the optimal solution of the original minimization problem from the final simplex tableau:

y_1 is the last entry in the column corresponding to the first slack variable;

y_2 is the last entry in the column corresponding to the second slack variable; and so on.

These values of y_1, y_2, y_3, etc., produce the minimum value of w, which is the entry in the lower right-hand corner of the tableau.

EXAMPLE 6   Minimize $w = 3y_1 + 2y_2$

subject to $y_1 + 3y_2 \geq 6$

$2y_1 + y_2 \geq 3$

$y_1 \geq 0, y_2 \geq 0.$

Solution Use the given information to write the matrix:

$$\left[\begin{array}{cc|c} 1 & 3 & 6 \\ 2 & 1 & 3 \\ \hline 3 & 2 & 0 \end{array}\right].$$

Transpose to get the following matrix for the dual problem:

$$\left[\begin{array}{cc|c} 1 & 2 & 3 \\ 3 & 1 & 2 \\ \hline 6 & 3 & 0 \end{array}\right].$$

Write the dual problem from this matrix as follows:

Maximize $z = 6x_1 + 3x_2$

subject to $x_1 + 2x_2 \leq 3$

$3x_1 + x_2 \leq 2$

$x_1 \geq 0, x_2 \geq 0.$

Solve this standard maximization problem by the simplex method. Start by introducing slack variables, giving the system

$$x_1 + 2x_2 + s_1 \qquad\qquad = 3$$
$$3x_1 + x_2 \qquad + s_2 \qquad = 2$$
$$-6x_1 - 3x_2 - 0s_1 - 0s_2 + z = 0,$$

with $x_1 \geq 0, x_2 \geq 0, s_1 \geq 0$, and $s_2 \geq 0.$

---

[*]The coefficients of the objective function in the minimization problem are the constants on the right side of the constraints in the dual maximization problem. So when all these coefficients are nonnegative (condition 2), the dual problem is in standard maximum form.

The initial tableau for this system is

$$
\begin{array}{cccc}
x_1 & x_2 & s_1 & s_2 \\
\end{array}
$$

$$
\left[
\begin{array}{cccc|c}
1 & 2 & 1 & 0 & 3 \\
\mathbf{3} & 1 & 0 & 1 & 2 \\
\hline
-6 & -3 & 0 & 0 & 0 \\
\end{array}
\right],
\quad
\begin{array}{l}
\text{Quotients} \\
3/1 = 3 \\
2/3 \\
\end{array}
$$

with the pivot as indicated. Two rounds of pivoting produce the following final tableau:

$$
\begin{array}{cccc}
x_1 & x_2 & s_1 & s_2 \\
\end{array}
$$

$$
\left[
\begin{array}{cccc|c}
0 & 1 & \frac{3}{5} & -\frac{1}{5} & \frac{7}{5} \\
1 & 0 & -\frac{1}{5} & \frac{2}{5} & \frac{1}{5} \\
\hline
0 & 0 & \frac{3}{5} & \frac{9}{5} & \frac{27}{5} \\
\end{array}
\right].
$$

$$
\begin{array}{ccc}
\quad\; y_1 & y_2 & w \\
\end{array}
$$

As indicated in blue below the final tableau, the last entries in the columns corresponding to the slack variables (s_1 and s_2) give the values of the original variables y_1 and y_2 that produce the minimal value of w. This minimal value of w appears in the lower right-hand corner (and is the same as the maximal value of z in the dual problem). So the solution of the given minimization problem is as follows:

The minimum value of $w = 3y_1 + 2y_2$, subject to the given constraints, is $\frac{27}{5}$ and occurs when $y_1 = \frac{3}{5}$ and $y_2 = \frac{9}{5}$. ✓6

✓ **Checkpoint 6**

Minimize $\quad w = 10y_1 + 8y_2$

subject to $\quad y_1 + 2y_2 \geq 2$

$\qquad\qquad\; y_1 + \; y_2 \geq 5$

$\qquad\qquad\; y_1 \geq 0, y_2 \geq 0.$

EXAMPLE 7 A minimization problem in three variables was solved by the use of duals. The final simplex tableau for the dual maximization problem is shown here:

$$
\begin{array}{cccccc}
x_1 & x_2 & x_3 & s_1 & s_2 & s_3 \\
\end{array}
$$

$$
\left[
\begin{array}{cccccc|c}
3 & 1 & 1 & 0 & 9 & 0 & 1 \\
13 & -1 & 0 & 1 & -2 & 0 & 10 \\
9 & 10 & 0 & 0 & 7 & 1 & 7 \\
\hline
5 & 1 & 0 & 4 & 1 & 7 & 28 \\
\end{array}
\right].
$$

(a) What is the optimal solution of the dual minimization problem?

Solution Looking at the bottom of the columns corresponding to the slack variables s_1, s_2, and s_3, we see that the solution of the minimization problem is

$$y_1 = 4, \qquad y_2 = 1, \qquad \text{and} \qquad y_3 = 7, \text{ with a minimal value of } w = 28.$$

(b) What is the optimal solution of the dual maximization problem?

Solution Since s_1, s_2, and s_3 are slack variables by part (a), the variables in the dual problem are x_1, x_2, and x_3. Read the solution from the final tableau, as in Sections 7.4 and 7.5:

$$x_1 = 0, \qquad x_2 = 0, \qquad \text{and} \qquad x_3 = 1, \qquad \text{with a maximal value of } z = 28.$$

Further Uses of the Dual

The dual is useful not only in solving minimization problems, but also in seeing how small changes in one variable will affect the value of the objective function. For example, suppose an animal breeder needs at least 6 units per day of nutrient A and at least 3 units of nutrient B and that the breeder can choose between two different feeds: feed 1 and feed 2. Find the minimum cost for the breeder if each bag of feed 1 costs \$3 and provides 1 unit of nutrient A and 2 units of B, while each bag of feed 2 costs \$2 and provides 3 units of nutrient A and 1 of B.

If y_1 represents the number of bags of feed 1 and y_2 represents the number of bags of feed 2, the given information leads to the following minimization problem:

$$\text{Minimize} \quad w = 3y_1 + 2y_2$$
$$\text{subject to} \quad y_1 + 3y_2 \geq 6$$
$$2y_1 + y_2 \geq 3$$
$$y_1 \geq 0, y_2 \geq 0.$$

This minimization linear programming problem is the one we solved in Example 6 of this section. In that example, we formed the dual and reached the following final tableau:

$$
\begin{array}{cccc}
x_1 & x_2 & s_1 & s_2
\end{array}
$$
$$
\left[
\begin{array}{cccc|c}
0 & 1 & \frac{3}{5} & -\frac{1}{5} & \frac{7}{5} \\
1 & 0 & -\frac{1}{5} & \frac{2}{5} & \frac{1}{5} \\
0 & 0 & \frac{3}{5} & \frac{9}{5} & \frac{27}{5}
\end{array}
\right].
$$

This final tableau shows that the breeder will obtain minimum feed costs by using $\frac{3}{5}$ bag of feed 1 and $\frac{9}{5}$ bags of feed 2 per day, for a daily cost of $\frac{27}{5} = 5.40$ dollars.

Now look at the data from the feed problem shown in the following table:

	Units of Nutrient (per Bag)		Cost per Bag
	A	*B*	
Feed 1	1	2	$3
Feed 2	3	1	$2
Minimum Nutrient Needed	6	3	

If x_1 and x_2 are the cost *per unit* of nutrients A and B, the constraints of the dual problem can be stated as follows (see pages 378–379):

$$\text{Cost of feed 1:} \quad x_1 + 2x_2 \leq 3$$
$$\text{Cost of feed 2:} \quad 3x_1 + x_2 \leq 2.$$

The solution of the dual problem, which maximizes nutrients, also can be read from the final tableau above:

$$x_1 = \frac{1}{5} = .20 \quad \text{and} \quad x_2 = \frac{7}{5} = 1.40.$$

This means that a unit of nutrient A costs $\frac{1}{5}$ of a dollar $= \$.20$, while a unit of nutrient B costs $\frac{7}{5}$ dollars $= \$1.40$. The minimum daily cost, 5.40, is found as follows.

$$(\$.20 \text{ per unit of } A) \times (6 \text{ units of } A) = \$1.20$$
$$+ (\$1.40 \text{ per unit of } B) \times (3 \text{ units of } B) = \$4.20$$
$$\text{Minimum daily cost} = \$5.40.$$

The numbers .20 and 1.40 are called the **shadow costs** of the nutrients. These two numbers from the dual, $.20 and $1.40, also allow the breeder to estimate feed costs for "small" changes in nutrient requirements. For example, an increase of 1 unit in the requirement for each nutrient would produce a total cost as follows:

$5.40	6 units of A, 3 of B
.20	1 extra unit of A
1.40	1 extra unit of B
$7.00·	Total cost per day

7.6 Exercises

Find the transpose of each matrix. (See Example 1.)

1. $\begin{bmatrix} 3 & -4 & 5 \\ 1 & 10 & 7 \\ 0 & 3 & 6 \end{bmatrix}$ 2. $\begin{bmatrix} 3 & -5 & 9 & 4 \\ 1 & 6 & -7 & 0 \\ 4 & 18 & 11 & 9 \end{bmatrix}$

3. $\begin{bmatrix} 3 & 0 & 14 & -5 & 3 \\ 4 & 17 & 8 & -6 & 1 \end{bmatrix}$ 4. $\begin{bmatrix} 15 & -6 & -2 \\ 13 & -1 & 11 \\ 10 & 12 & -3 \\ 24 & 1 & 0 \end{bmatrix}$

State the dual problem for each of the given problems, but do not solve it. (See Examples 2 and 3.)

5. Minimize $w = 3y_1 + 5y_2$
   subject to $\quad 3y_1 + y_2 \geq 4$
   $\qquad\qquad -y_1 + 2y_2 \geq 6$
   $\qquad\qquad y_1 \geq 0, y_2 \geq 0.$

6. Minimize $w = 4y_1 + 7y_2$
   subject to $\quad y_1 + y_2 \geq 17$
   $\qquad\qquad 3y_1 + 6y_2 \geq 21$
   $\qquad\qquad 2y_1 + 4y_2 \geq 19$
   $\qquad\qquad y_1 \geq 0, y_2 \geq 0.$

7. Minimize $w = 2y_1 + 8y_2$
   subject to $\quad y_1 + 7y_2 \geq 18$
   $\qquad\qquad 4y_1 + y_2 \geq 15$
   $\qquad\qquad 5y_1 + 3y_2 \geq 20$
   $\qquad\qquad y_1 \geq 0, y_2 \geq 0.$

8. Minimize $w = y_1 + 2y_2 + 6y_3$
   subject to $\quad 3y_1 + 4y_2 + 6y_3 \geq 8$
   $\qquad\qquad y_1 + 5y_2 + 2y_3 \geq 12$
   $\qquad\qquad y_1 \geq 0, y_2 \geq 0, y_3 \geq 0.$

9. Minimize $w = 5y_1 + y_2 + 3y_3$
   subject to $\quad 7y_1 + 6y_2 + 8y_3 \geq 18$
   $\qquad\qquad 4y_1 + 5y_2 + 10y_3 \geq 20$
   $\qquad\qquad y_1 \geq 0, y_2 \geq 0, y_3 \geq 0.$

10. Minimize $w = 4y_1 + 3y_2 + y_3$
    subject to $\quad y_1 + 2y_2 + 3y_3 \geq 115$
    $\qquad\qquad 2y_1 + y_2 + 8y_3 \geq 200$
    $\qquad\qquad y_1 \qquad\quad - y_3 \geq 50$
    $\qquad\qquad y_1 \geq 0, y_2 \geq 0, y_3 \geq 0.$

11. Minimize $w = 8y_1 + 9y_2 + 3y_3$
    subject to $\quad y_1 + y_2 + y_3 \geq 5$
    $\qquad\qquad y_1 + y_2 \qquad\quad \geq 4$
    $\qquad\qquad 2y_1 + y_2 + 3y_3 \geq 15$
    $\qquad\qquad y_1 \geq 0, y_2 \geq 0, y_3 \geq 0.$

12. Minimize $w = y_1 + 2y_2 + y_3 + 5y_4$
    subject to $\quad y_1 + y_2 + y_3 + y_4 \geq 50$
    $\qquad\qquad 3y_1 + y_2 + 2y_3 + y_4 \geq 100$
    $\qquad\qquad y_1 \geq 0, y_2 \geq 0, y_3 \geq 0, y_4 \geq 0.$

Use duality to solve the problem that was set up in the given exercise.

13. Exercise 9   14. Exercise 8

15. Exercise 11   16. Exercise 12

Use duality to solve the given problems. (See Examples 5 and 6.)

17. Minimize $w = 2y_1 + y_2 + 3y_3$
    subject to $\quad y_1 + y_2 + y_3 \geq 100$
    $\qquad\qquad 2y_1 + y_2 \qquad\quad \geq 50$
    $\qquad\qquad y_1 \geq 0, y_2 \geq 0, y_3 \geq 0.$

18. Minimize $w = 2y_1 + 4y_2$
    subject to $\quad 4y_1 + 2y_2 \geq 10$
    $\qquad\qquad 4y_1 + y_2 \geq 8$
    $\qquad\qquad 2y_1 + y_2 \geq 12$
    $\qquad\qquad y_1 \geq 0, y_2 \geq 0.$

19. Minimize $w = 3y_1 + y_2 + 4y_3$
    subject to $\quad 2y_1 + y_2 + y_3 \geq 6$
    $\qquad\qquad y_1 + 2y_2 + y_3 \geq 8$
    $\qquad\qquad 2y_1 + y_2 + 2y_3 \geq 12$
    $\qquad\qquad y_1 \geq 0, y_2 \geq 0, y_3 \geq 0.$

20. Minimize $w = y_1 + y_2 + 3y_3$
    subject to $\quad 2y_1 + 6y_2 + y_3 \geq 8$
    $\qquad\qquad y_1 + 2y_2 + 4y_3 \geq 12$
    $\qquad\qquad y_1 \geq 0, y_2 \geq 0, y_3 \geq 0.$

21. Minimize $w = 6y_1 + 4y_2 + 2y_3$
    subject to $\quad 2y_1 + 2y_2 + y_3 \geq 2$
    $\qquad\qquad y_1 + 3y_2 + 2y_3 \geq 3$
    $\qquad\qquad y_1 + y_2 + 2y_3 \geq 4$
    $\qquad\qquad y_1 \geq 0, y_2 \geq 0, y_3 \geq 0.$

22. Minimize $w = 12y_1 + 10y_2 + 7y_3$
    subject to $\quad 2y_1 + y_2 + y_3 \geq 7$
    $\qquad\qquad y_1 + 2y_2 + y_3 \geq 4$
    $\qquad\qquad y_1 \geq 0, y_2 \geq 0, y_3 \geq 0.$

23. Minimize $w = 20y_1 + 12y_2 + 40y_3$
    subject to $\quad y_1 + y_2 + 5y_3 \geq 20$
    $\qquad\qquad 2y_1 + y_2 + y_3 \geq 30$
    $\qquad\qquad y_1 \geq 0, y_2 \geq 0, y_3 \geq 0.$

24. Minimize $w = 4y_1 + 5y_2$
    subject to $\quad 10y_1 + 5y_2 \geq 100$
    $\qquad\qquad 20y_1 + 10y_2 \geq 150$
    $\qquad\qquad y_1 \geq 0, y_2 \geq 0.$

25. Minimize $w = 4y_1 + 2y_2 + y_3$
    subject to $\quad y_1 + y_2 + y_3 \geq 4$
    $\qquad\qquad 3y_1 + y_2 + 3y_3 \geq 6$
    $\qquad\qquad y_1 + y_2 + 3y_3 \geq 5$
    $\qquad\qquad y_1 \geq 0, y_2 \geq 0, y_3 \geq 0.$

26. Minimize $w = 3y_1 + 2y_2$

subject to $2y_1 + 3y_2 \geq 60$

$y_1 + 4y_2 \geq 40$

$y_1 \geq 0, y_2 \geq 0.$

27. Nutrition Glenn Russell, who is dieting, requires two food supplements: I and II. He can get these supplements from two different products—A and B—as shown in the following table:

Product	Supplement (Grams per Serving)	
	I	II
A	4	2
B	2	5

Glenn's physician has recommended that he include at least 20 grams of supplement I and 18 grams of supplement II in his diet. If product A costs 24¢ per serving and product B costs 40¢ per serving, how can he satisfy these requirements most economically?

28. Animal Food An animal food must provide at least 54 units of vitamins and 60 calories per serving. One gram of soybean meal provides at least 2.5 units of vitamins and 5 calories. One gram of meat by-products provides at least 4.5 units of vitamins and 3 calories. One gram of grain provides at least 5 units of vitamins and 10 calories. If a gram of soybean meal costs 8¢, a gram of meat by-products 9¢, and a gram of grain 10¢, what mixture of these three ingredients will provide the required vitamins and calories at minimum cost?

29. Beer Production A brewery produces regular beer and a lower-carbohydrate "light" beer. Steady customers of the brewery buy 12 units of regular beer and 10 units of light beer monthly. While setting up the brewery to produce the beers, the management decides to produce extra beer, beyond the need to satisfy the steady customers. The cost per unit of regular beer is $36,000, and the cost per unit of light beer is $48,000. Every unit of regular beer brings in $100,000 in revenue, while every unit of light beer brings in $300,000. The brewery wants at least $7,000,000 in revenue. At least 20 additional units of beer can be sold. How much of each type of beer should be made so as to minimize total production costs?

30. Business Joan has a part-time job conducting public-opinion interviews. She has found that a political interview takes 45 minutes and a market interview takes 55 minutes. To allow more time for her full-time job, she needs to minimize the time she spends doing interviews. Unfortunately, to keep her part-time job, she must complete at least 8 interviews each week. Also, she must earn at least $60 per week at this job, at which she earns $8 for each political interview and $10 for each market interview. Finally, to stay in good standing with her supervisor, she must earn at least 40 bonus points per week; she receives 6 bonus points for each political interview and 5 points for each market interview. How many of each interview should she do each week to minimize the time spent?

31. Feed Production Refer to the end of this section in the text, on minimizing the daily cost of feeds.

(a) Find a combination of feeds that will cost $7.00 and give 7 units of A and 4 units of B.

(b) Use the dual variables to predict the daily cost of feed if the requirements change to 5 units of A and 4 units of B. Find a combination of feeds to meet these requirements at the predicted price.

32. Refer to Example 1 in Section 7.5.

(a) Give the dual problem.

(b) Use the shadow values to estimate the farmer's profit if land is cut to 90 acres, but capital increases to $21,000.

(c) Suppose the farmer has 110 acres, but only $19,000. Find the optimal profit and the planting strategy that will produce this profit.

Toy Production *For Exercises 33 and 34, use the following information. A small toy-manufacturing firm has 200 squares of felt, 600 ounces of stuffing, and 90 feet of trim available to make two types of toys: a small bear and a monkey. The bear requires 1 square of felt and 4 ounces of stuffing. The monkey requires 2 squares of felt, 3 ounces of stuffing, and 1 foot of trim. The firm makes $1 profit on each bear and $1.50 profit on each monkey. The linear programming problem to maximize profit is*

Maximize $z = x_1 + 1.5x_2$

subject to $x_1 + 2x_2 \leq 200$

$4x_1 + 3x_2 \leq 600$

$x_2 \leq 90$

$x_1 \geq 0, x_2 \geq 0.$

The final simplex tableau is

$$\begin{bmatrix} 1 & 0 & -.6 & .4 & 0 & | & 120 \\ 0 & 0 & -.8 & .2 & 1 & | & 50 \\ 0 & 1 & .8 & -.2 & 0 & | & 40 \\ \hline 0 & 0 & .6 & .1 & 0 & | & 180 \end{bmatrix}.$$

33. (a) What is the corresponding dual problem?

(b) What is the optimal solution to the dual problem?

(c) Use the shadow values to estimate the profit the firm will make if its supply of felt increases to 210 squares.

34. How much profit will the firm make if its supply of stuffing is cut to 590 ounces and its supply of trim is cut to 80 feet?

✓ **Checkpoint Answers**

1. (a) $\begin{bmatrix} 2 & 6 & 1 \\ 4 & 3 & 5 \end{bmatrix}$ **(b)** $\begin{bmatrix} 4 & 3 & 5 \\ 7 & 2 & 8 \\ 10 & 6 & 12 \end{bmatrix}$

2. Maximize $z = 15x_1 + 12x_2 + 10x_3$

subject to $2x_1 + x_2 + 5x_3 \leq 2$

$3x_1 + x_2 + 3x_3 \leq 5$

$x_1 + 2x_2 \leq 6$

$x_1 \geq 0, x_2 \geq 0, x_3 \geq 0.$

3. 48 when $y_1 = 4$ and $y_2 = 1$ **4.** 48 when $x_1 = 2$ and $x_2 = 3$

5. The 4 in row 2, column 1

6. $y_1 = 0$ and $y_2 = 5$, for a minimum of 40

7.7 The Simplex Method: Nonstandard Problems

📄 **NOTE** Section 7.7 is independent of Section 7.6 and may be read first, if desired.

So far, the simplex method has been used to solve problems in which the variables are nonnegative and all the other constraints are of one type (either all $\leq$ or all $\geq$). Now we extend the simplex method to linear programming problems with nonnegative variables and mixed constraints ($\leq$, $=$, and $\geq$).

The solution method to be used here requires that all inequality constraints be written so that the constant on the right side is nonnegative. For instance, the inequality

$$4x_1 + 5x_2 - 12x_3 \leq -30$$

can be replaced by the equivalent one obtained by multiplying both sides by -1 and reversing the direction of the inequality sign:

$$-4x_1 - 5x_2 + 12x_3 \geq 30.$$

Maximization with $\leq$ and $\geq$ Constraints

As is always the case when the simplex method is involved, each inequality constraint must be written as an equation. Constraints involving $\leq$ are converted to equations by adding a nonnegative slack variable, as in Section 7.4. Similarly, constraints involving $\geq$ are converted to equations by *subtracting* a nonnegative **surplus variable**. For example, the inequality $2x_1 - x_2 + 5x_3 \geq 12$ is written as

$$2x_1 - x_2 + 5x_3 - s_1 = 12,$$

where $s_1 \geq 0$. The surplus variable s_1 represents the amount by which $2x_1 - x_2 + 5x_3$ exceeds 12.

EXAMPLE 1   Restate the following problem in terms of equations, and write its initial simplex tableau:

$$\text{Maximize} \quad z = 4x_1 + 10x_2 + 6x_3$$
$$\text{subject to} \quad x_1 + 4x_2 + 4x_3 \geq 8$$
$$x_1 + 3x_2 + 2x_3 \leq 6$$
$$3x_1 + 4x_2 + 8x_3 \leq 22$$
$$x_1 \geq 0, x_2 \geq 0, x_3 \geq 0.$$

Solution In order to write the constraints as equations, subtract a surplus variable from the $\geq$ constraint and add a slack variable to each $\leq$ constraint. So the problem becomes

$$\text{Maximize} \quad z = 4x_1 + 10x_2 + 6x_3$$
$$\text{subject to} \quad x_1 + 4x_2 + 4x_3 - s_1 \qquad\qquad = 8$$
$$x_1 + 3x_2 + 2x_3 \qquad + s_2 \qquad = 6$$
$$3x_1 + 4x_2 + 8x_3 \qquad\qquad + s_3 = 22$$
$$x_1 \geq 0, x_2 \geq 0, x_3 \geq 0, s_1 \geq 0, s_2 \geq 0, s_3 \geq 0.$$

Write the objective function as $-4x_1 - 10x_2 - 6x_3 + z = 0$ and use the coefficients of the four equations to write the initial simplex tableau (omitting the z column):

$$
\begin{array}{cccccc}
x_1 & x_2 & x_3 & s_1 & s_2 & s_3 \\
\end{array}
$$

$$
\left[
\begin{array}{cccccc|c}
1 & 4 & 4 & -1 & 0 & 0 & 8 \\
1 & 3 & 2 & 0 & 1 & 0 & 6 \\
3 & 4 & 8 & 0 & 0 & 1 & 22 \\
\hline
-4 & -10 & -6 & 0 & 0 & 0 & 0 \\
\end{array}
\right]. \quad \boxed{\diagdown}_1
$$

✓ **Checkpoint 1**

(a) Restate this problem in terms of equations:

$$\text{Maximize} \quad z = 3x_1 - 2x_2$$
$$\text{subject to} \quad 2x_1 + 3x_2 \leq 8$$
$$6x_1 - 2x_2 \geq 3$$
$$x_1 + 4x_2 \geq 1$$
$$x_1 \geq 0, x_2 \geq 0.$$

(b) Write the initial simplex tableau.

The tableau in Example 1 resembles those which have appeared previously, and similar terminology is used. The variables whose columns have one entry that is ± 1 and the rest that are 0 will be called **basic variables**; the other variables are nonbasic. A solution obtained by setting the nonbasic variables equal to 0 and solving for the basic variables (by looking at the constants in the right-hand column) will be called a **basic solution**. A basic solution that is feasible is called a **basic feasible solution**. In the tableau of Example 1, for instance, the basic variables are s_1, s_2, and s_3, and the basic solution is

$$x_1 = 0, \quad x_2 = 0, \quad x_3 = 0, \quad s_1 = -8, \quad s_2 = 6, \quad \text{and} \quad s_3 = 22.$$

However, because one variable is negative, this solution is not feasible. ✔️2

The solution method for problems such as the one in Example 1 consists of two stages. **Stage I** consists of finding a basic *feasible* solution that can be used as the starting point for the simplex method. (This stage is unnecessary in a standard maximization problem, because the solution given by the initial tableau is always feasible.) There are many systematic ways of finding a feasible solution, all of which depend on the fact that row operations produce a tableau that represents a system with the same solutions as the original one. One such technique is explained in the next example. Since the immediate goal is to find a feasible solution, not necessarily an optimal one, the procedures for choosing pivots differ from those in the ordinary simplex method.

✓ **Checkpoint 2**

State the basic solution given by each tableau. Is it feasible?

(a)
$$\begin{array}{ccccc} x_1 & x_2 & s_1 & s_2 & s_3 \\ \left[\begin{array}{ccccc|c} 3 & -5 & 1 & 0 & 0 & 12 \\ 4 & 7 & 0 & 1 & 0 & 6 \\ 1 & 3 & 0 & 0 & -1 & 5 \\ \hline -7 & 4 & 0 & 0 & 0 & 0 \end{array}\right] \end{array}$$

(b)
$$\begin{array}{ccccc} x_1 & x_2 & x_3 & s_1 & s_2 \\ \left[\begin{array}{ccccc|c} 9 & 8 & -1 & 1 & 0 & 12 \\ -5 & 3 & 0 & 0 & 1 & 7 \\ \hline 4 & 2 & 3 & 0 & 0 & 0 \end{array}\right] \end{array}$$

EXAMPLE 2 Find a basic feasible solution for the problem in Example 1, whose initial tableau is

$$\begin{array}{cccccc} x_1 & x_2 & x_3 & s_1 & s_2 & s_3 \\ \left[\begin{array}{cccccc|c} 1 & 4 & 4 & -1 & 0 & 0 & 8 \\ 1 & 3 & 2 & 0 & 1 & 0 & 6 \\ 3 & 4 & 8 & 0 & 0 & 1 & 22 \\ \hline -4 & -10 & -6 & 0 & 0 & 0 & 0 \end{array}\right] \end{array}.$$

Solution In the basic solution given by this tableau, s_1 has a negative value. The only nonzero entry in its column is the -1 in row 1. Choose any *positive* entry in row 1 except the entry on the far right. The column that the chosen entry is in will be the pivot column. We choose the first positive entry in row 1: the 1 in column 1. The pivot row is determined in the usual way by considering quotients (the constant at the right end of the row, divided by the positive entry in the pivot column) in each row except the objective row:

$$\frac{8}{1} = 8, \qquad \frac{6}{1} = 6, \qquad \frac{22}{3} = 7\frac{1}{3}.$$

The smallest quotient is 6, so the pivot is the 1 in row 2, column 1. Pivoting in the usual way leads to the tableau

$$\begin{array}{cccccc} x_1 & x_2 & x_3 & s_1 & s_2 & s_3 \\ \left[\begin{array}{cccccc|c} 0 & 1 & 2 & -1 & -1 & 0 & 2 \\ 1 & 3 & 2 & 0 & 1 & 0 & 6 \\ 0 & -5 & 2 & 0 & -3 & 1 & 4 \\ \hline 0 & 2 & 2 & 0 & 4 & 0 & 24 \end{array}\right] \end{array} \quad \begin{array}{l} -R_2 + R_1 \\ \\ -3R_2 + R_3 \\ 4R_2 + R_4 \end{array}$$

and the basic solution

$$x_1 = 6, \qquad x_2 = 0, \qquad x_3 = 0, \qquad s_1 = -2, \qquad s_2 = 0, \qquad \text{and} \qquad s_3 = 4.$$

Since the basic variable s_1 is negative, this solution is not feasible. So we repeat the pivoting process. The s_1 column has a -1 in row 1, so we choose a positive entry in that row, namely, the 1 in row 1, column 2. This choice makes column 2 the pivot column. The pivot row is determined by the quotients $\frac{2}{1} = 2$ and $\frac{6}{3} = 2$. (Negative entries in the pivot column and the entry in the objective row are not used.) Since there is a tie, we can choose either row 1 or row 2. We choose row 1 and use the 1 in row 1, column 2, as the pivot. Pivoting produces the tableau

$$
\begin{array}{c}
\begin{array}{cccccc} x_1 & x_2 & x_3 & s_1 & s_2 & s_3 \end{array} \\
\left[\begin{array}{cccccc|c}
0 & 1 & 2 & -1 & -1 & 0 & 2 \\
1 & 0 & -4 & 3 & 4 & 0 & 0 \\
0 & 0 & 12 & -5 & -8 & 1 & 14 \\
0 & 0 & -2 & 2 & 6 & 0 & 20
\end{array}\right]
\begin{array}{l}
\\
-3R_1 + R_2 \\
5R_1 + R_3 \\
-2R_1 + R_4
\end{array}
\end{array}
$$

and the basic *feasible* solution

$$x_1 = 0, \quad x_2 = 2, \quad x_3 = 0, \quad s_1 = 0, \quad s_2 = 0, \quad \text{and} \quad s_3 = 14.$$

Once a basic feasible solution has been found, Stage I is ended. The procedures used in Stage I are summarized here.[*]

Finding a Basic Feasible Solution

1. If any basic variable has a negative value, locate the -1 in that variable's column and note the row it is in.
2. In the row determined in Step 1, choose a positive entry (other than the one at the far right) and note the column it is in. This is the pivot column.
3. Use the positive entries in the pivot column (except in the objective row) to form quotients and select the pivot.
4. Pivot as usual, which results in the pivot column having one entry that is 1 and the rest that are 0's.
5. Repeat Steps 1–4 until every basic variable is nonnegative, so that the basic solution given by the tableau is feasible. If it ever becomes impossible to continue, then the problem has no feasible solution.

One way to make the required choices systematically is to choose the first possibility in each case (going from the top for rows and from the left for columns). However, any choice meeting the required conditions may be used. For maximum efficiency, it is usually best to choose the pivot column in Step 2, so that the pivot is in the same row chosen in Step 1, if this is possible. ✓₃

In **Stage II**, the simplex method is applied as usual to the tableau that produced the basic feasible solution in Stage I. Just as in Section 7.4, each round of pivoting replaces the basic feasible solution of one tableau with the basic feasible solution of a new tableau in such a way that the value of the objective function is increased, until an optimal value is obtained (or it becomes clear that no optimal solution exists).

✓ **Checkpoint 3**

The initial tableau of a maximization problem is shown. Use column 1 as the pivot column for carrying out Stage I, and state the basic feasible solution that results.

$$
\begin{array}{c}
\begin{array}{cccc} x_1 & x_2 & s_1 & s_2 \end{array} \\
\left[\begin{array}{cccc|c}
1 & 3 & 1 & 0 & 70 \\
2 & 4 & 0 & -1 & 50 \\
-8 & -10 & 0 & 0 & 0
\end{array}\right]
\end{array}
$$

EXAMPLE 3 Solve the linear programming problem in Example 1 of this section.

Solution A basic feasible solution for this problem was found in Example 2 by using the tableau shown below. However, this solution is not maximal, because there is a negative indicator in the objective row. So we use the simplex method. The most negative indicator

---

[*]Except in rare cases that do not occur in this book, this method either eventually produces a basic feasible solution or shows that one does not exist. The *two-phase method* using artificial variables, which is discussed in more advanced texts, works in all cases and often is more efficient.

determines the pivot column, and the usual quotients determine that the number 2 in row 1, column 3, is the pivot:

$$
\begin{array}{c}
\begin{array}{cccccc} x_1 & x_2 & x_3 & s_1 & s_2 & s_3 \end{array} \\
\left[\begin{array}{cccccc|c}
0 & 1 & \mathbf{2} & -1 & -1 & 0 & 2 \\
1 & 0 & -4 & 3 & 4 & 0 & 0 \\
0 & 0 & 12 & -5 & -8 & 1 & 14 \\
\hline
0 & 0 & -2 & 2 & 6 & 0 & 20
\end{array}\right]
\end{array}
\qquad
\begin{array}{l}
\text{Quotients} \\
2/2 \leftarrow \text{Smallest} \\
\\
14/12 \\
\end{array}
$$

Most negative indicator

Pivoting leads to the final tableau:

$$
\begin{array}{c}
\begin{array}{cccccc} x_1 & x_2 & x_3 & s_1 & s_2 & s_3 \end{array} \\
\left[\begin{array}{cccccc|c}
0 & \frac{1}{2} & 1 & -\frac{1}{2} & -\frac{1}{2} & 0 & 1 \\
1 & 0 & -4 & 3 & 4 & 0 & 0 \\
0 & 0 & 12 & -5 & -8 & 1 & 14 \\
\hline
0 & 0 & -2 & 2 & 6 & 0 & 20
\end{array}\right]
\end{array}
\qquad \frac{1}{2}R_1
$$

$$
\begin{array}{c}
\begin{array}{cccccc} x_1 & x_2 & x_3 & s_1 & s_2 & s_3 \end{array} \\
\left[\begin{array}{cccccc|c}
0 & \frac{1}{2} & 1 & -\frac{1}{2} & -\frac{1}{2} & 0 & 1 \\
1 & 2 & 0 & 1 & 2 & 0 & 4 \\
0 & -6 & 0 & 1 & -2 & 1 & 2 \\
\hline
0 & 1 & 0 & 1 & 5 & 0 & 22
\end{array}\right]
\end{array}
\qquad
\begin{array}{l}
4R_1 + R_2 \\
-12R_1 + R_3 \\
2R_1 + R_4
\end{array}
$$

Therefore, the maximum value of z occurs when $x_1 = 4$, $x_2 = 0$, and $x_3 = 1$, in which case $z = 22$. ✔ 4

✓ **Checkpoint 4**

Complete Stage II and find an optimal solution for Checkpoint 3 on page 385. What is the optimal value of the objective function z?

Minimization Problems

When dealing with minimization problems, we use y_1, y_2, y_3, etc., as variables and denote the objective function by w. The two-stage method for maximization problems illustrated in Examples 1–3 also provides a means of solving minimization problems. To see why, consider this simple fact: When a number t gets smaller, $-t$ gets larger, and vice versa. For instance, if t goes from 6 down to -8, then $-t$ goes from -6 up to 8. Thus, if w is the objective function of a linear programming problem, the feasible solution that produces the minimum value of w also produces the maximum value of $-w$, and vice versa. Therefore, to solve a minimization problem with objective function w, we need only solve the maximization problem with the same constraints and objective function $z = -w$.

EXAMPLE 4   Minimize $w = 2y_1 + y_2 - y_3$

subject to $-y_1 - y_2 + y_3 \le -4$

$y_1 + 3y_2 + 3y_3 \ge 6$

$y_1 \ge 0, y_2 \ge 0, y_3 \ge 0.$

Solution Make the constant in the first constraint positive by multiplying both sides by -1. Then solve this maximization problem:

Maximize $z = -w = -2y_1 - y_2 + y_3$

subject to $y_1 + y_2 - y_3 \ge 4$

$y_1 + 3y_2 + 3y_3 \ge 6$

$y_1 \ge 0, y_2 \ge 0, y_3 \ge 0.$

Convert the constraints to equations by subtracting surplus variables, and set up the first tableau:

$$
\begin{array}{c}
\begin{array}{ccccc} y_1 & y_2 & y_3 & s_1 & s_2 \end{array} \\
\left[\begin{array}{ccccc|c}
1 & 1 & -1 & -1 & 0 & 4 \\
\hline
1 & 3 & 3 & 0 & -1 & 6 \\
\hline
2 & 1 & -1 & 0 & 0 & 0
\end{array}\right].
\end{array}
$$

The basic solution given by this tableau, namely, $y_1 = 0$, $y_2 = 0$, $y_3 = 0$, $s_1 = -4$, and $s_2 = -6$, is not feasible, so the procedures of Stage I must be used to find a basic feasible solution. In the column of the negative basic variable s_1, there is a -1 in row 1; we choose the first positive entry in that row, so that column 1 will be the pivot column. The quotients $\frac{4}{1} = 4$ and $\frac{6}{1} = 6$ show that the pivot is the 1 in row 1, column 1. Pivoting produces this tableau:

$$
\begin{array}{c}
\begin{array}{ccccc} y_1 & y_2 & y_3 & s_1 & s_2 \end{array} \\
\left[\begin{array}{ccccc|c}
1 & 1 & -1 & -1 & 0 & 4 \\
\hline
0 & 2 & 4 & 1 & -1 & 2 \\
\hline
0 & -1 & 1 & 2 & 0 & -8
\end{array}\right].
\end{array}
\qquad
\begin{array}{l}
 \\
-R_1 + R_2 \\
-2R_1 + R_3
\end{array}
$$

The basic solution $y_1 = 4$, $y_2 = 0$, $y_3 = 0$, $s_1 = 0$, and $s_2 = -2$ is not feasible because s_2 is negative, so we repeat the process. We choose the first positive entry in row 2 (the row containing the -1 in the s_2 column), which is in column 2, so that column 2 is the pivot column. The relevant quotients are $\frac{4}{1} = 4$ and $\frac{2}{2} = 1$, so the pivot is the 2 in row 2, column 2. Pivoting produces a new tableau:

$$
\begin{array}{c}
\begin{array}{ccccc} y_1 & y_2 & y_3 & s_1 & s_2 \end{array} \\
\left[\begin{array}{ccccc|c}
1 & 1 & -1 & -1 & 0 & 4 \\
\hline
0 & 1 & 2 & \frac{1}{2} & -\frac{1}{2} & 1 \\
\hline
0 & -1 & 1 & 2 & 0 & -8
\end{array}\right]
\end{array}
\qquad
\begin{array}{l}
 \\
\frac{1}{2}R_2 \\
 \\
\end{array}
$$

$$
\begin{array}{c}
\begin{array}{ccccc} y_1 & y_2 & y_3 & s_1 & s_2 \end{array} \\
\left[\begin{array}{ccccc|c}
1 & 0 & -3 & -\frac{3}{2} & \frac{1}{2} & 3 \\
\hline
0 & 1 & 2 & \frac{1}{2} & -\frac{1}{2} & 1 \\
\hline
0 & 0 & -3 & \frac{5}{2} & -\frac{1}{2} & -7
\end{array}\right].
\end{array}
\qquad
\begin{array}{l}
-R_2 + R_1 \\
 \\
R_2 + R_3
\end{array}
$$

The basic solution $y_1 = 3$, $y_2 = 1$, $y_3 = 0$, $s_1 = 0$, and $s_2 = 0$ is feasible, so Stage I is complete. However, this solution is not optimal, because the objective row contains the negative indicator $-\frac{1}{2}$ in column 5. According to the simplex method, column 5 is the next pivot column. The only positive ratio, $3/\frac{1}{2} = 6$, is in row 1, so the pivot is $\frac{1}{2}$ in row 1, column 5. Pivoting produces the final tableau:

$$
\begin{array}{c}
\begin{array}{ccccc} y_1 & y_2 & y_3 & s_1 & s_2 \end{array} \\
\left[\begin{array}{ccccc|c}
2 & 0 & -6 & -3 & 1 & 6 \\
\hline
0 & 1 & 2 & \frac{1}{2} & -\frac{1}{2} & 1 \\
\hline
0 & 0 & 3 & \frac{5}{2} & -\frac{1}{2} & -7
\end{array}\right]
\end{array}
\qquad
\begin{array}{l}
2R_1 \\
 \\
 \\
\end{array}
$$

$$
\begin{array}{c}
\begin{array}{ccccc} y_1 & y_2 & y_3 & s_1 & s_2 \end{array} \\
\left[\begin{array}{ccccc|c}
2 & 0 & -6 & -3 & 1 & 6 \\
\hline
1 & 1 & -1 & -1 & 0 & 4 \\
\hline
1 & 0 & 0 & 1 & 0 & -4
\end{array}\right].
\end{array}
\qquad
\begin{array}{l}
 \\
\frac{1}{2}R_1 + R_2 \\
\frac{1}{2}R_1 + R_3
\end{array}
$$

Since there are no negative indicators, the solution given by this tableau ($y_1 = 0$, $y_2 = 4$, $y_3 = 0$, $s_1 = 0$, and $s_2 = 6$) is optimal. The maximum value of $z = -w$ is -4. Therefore, the minimum value of the original objective function w is $-(-4) = 4$, which occurs when $y_1 = 0$, $y_2 = 4$, and $y_3 = 0$. ✓5

✓ **Checkpoint 5**

Minimize $w = 2y_1 + 3y_2$

subject to $y_1 + y_2 \geq 10$

$2y_1 + y_2 \geq 16$

$y_1 \geq 0, y_2 \geq 0$.

Equation Constraints

Recall that, for any real numbers a and b,

$$a = b \qquad \text{exactly when } a \geq b \text{ and simultaneously } a \leq b.$$

Thus, an equation such as $y_1 + 3y_2 + 3y_3 = 6$ is equivalent to this pair of inequalities:

$$y_1 + 3y_2 + 3y_3 \geq 6$$
$$y_1 + 3y_2 + 3y_3 \leq 6.$$

In a linear programming problem, each equation constraint should be replaced in this way by a pair of inequality constraints. Then the problem can be solved by the two-stage method.

EXAMPLE 5

Minimize $w = 2y_1 + y_2 - y_3$

subject to $\quad -y_1 - y_2 + y_3 \leq -4$

$\quad\quad\quad\quad y_1 + 3y_2 + 3y_3 = 6$

$\quad\quad\quad\quad y_1 \geq 0, y_2 \geq 0, y_3 \geq 0.$

Solution Multiply the first inequality by -1 and replace the equation by an equivalent pair of inequalities, as just explained, to obtain this problem:

Maximize $z = -w = -2y_1 - y_2 + y_3$

subject to $\quad y_1 + y_2 - y_3 \geq 4$

$\quad\quad\quad\quad y_1 + 3y_2 + 3y_3 \geq 6$

$\quad\quad\quad\quad y_1 + 3y_2 + 3y_3 \leq 6$

$\quad\quad\quad\quad y_1 \geq 0, y_2 \geq 0, y_3 \geq 0.$

Convert the constraints to equations by subtracting surplus variables s_1 and s_2 from the first two inequalities and adding a slack variable s_3 to the third. Then the first tableau is

$$\begin{array}{cccccc} y_1 & y_2 & y_3 & s_1 & s_2 & s_3 \\ \left[\begin{array}{cccccc|c} 1 & 1 & -1 & -1 & 0 & 0 & 4 \\ 1 & 3 & 3 & 0 & -1 & 0 & 6 \\ \hline 1 & 3 & 3 & 0 & 0 & 1 & 6 \\ \hline 2 & 1 & -1 & 0 & 0 & 0 & 0 \end{array}\right] \end{array}.$$

The basic solution given by this tableau is $y_1 = 0, y_2 = 0, y_3 = 0, s_1 = -4, s_2 = -6$, and $s_3 = 6$, which is not feasible. So we begin Stage I. The basic variable s_1 has a negative value because of the -1 in row 1. We choose the first positive entry in that row. So column 1 will be the pivot column. The quotients are $\frac{4}{1} = 4$ in row 1 and $\frac{6}{1} = 6$ in rows 2 and 3, which means that the pivot is the 1 in row 1, column 1. Pivoting produces this tableau:

$$\begin{array}{cccccc} y_1 & y_2 & y_3 & s_1 & s_2 & s_3 \\ \left[\begin{array}{cccccc|c} 1 & 1 & -1 & -1 & 0 & 0 & 4 \\ 0 & 2 & 4 & 1 & -1 & 0 & 2 \\ \hline 0 & 2 & 4 & 1 & 0 & 1 & 2 \\ \hline 0 & -1 & 1 & 2 & 0 & 0 & -8 \end{array}\right] \end{array}.$$

Now the basic variable s_2 has a negative value because of the -1 in row 2. We choose the first nonzero entry in this row and form the quotients $\frac{4}{1} = 4$ in row 1 and $\frac{2}{2} = 1$ in rows 2 and 3. Thus, there are two choices for the pivot, and we take the 2 in row 2, column 2. Pivoting produces the tableau

$$
\begin{array}{ccccccc}
y_1 & y_2 & y_3 & s_1 & s_2 & s_3 & \\
\left[\begin{array}{cccccc|c}
1 & 0 & -3 & -1.5 & .5 & 0 & 3 \\
0 & 1 & 2 & .5 & -.5 & 0 & 1 \\
0 & 0 & 0 & 0 & 1 & 1 & 0 \\
\hline
0 & 0 & 3 & 2.5 & -.5 & 0 & -7
\end{array}\right].
\end{array}
$$

This tableau gives the basic feasible solution $y_1 = 3$, $y_2 = 1$, $y_3 = 0$, $s_1 = 0$, $s_2 = 0$, and $s_3 = 0$, so Stage I is complete. Now apply the simplex method. One round of pivoting produces the final tableau:

$$
\begin{array}{ccccccc}
y_1 & y_2 & y_3 & s_1 & s_2 & s_3 & \\
\left[\begin{array}{cccccc|c}
1 & 0 & -3 & -1.5 & 0 & -.5 & 3 \\
0 & 1 & 2 & .5 & 0 & .5 & 1 \\
0 & 0 & 0 & 0 & 1 & 1 & 0 \\
\hline
0 & 0 & 3 & 2.5 & 0 & .5 & -7
\end{array}\right].
\end{array}
$$

Therefore, the minimum value of $w = -z$ is $w = -(-7) = 7$, which occurs when $y_1 = 3$, $y_2 = 1$, and $y_3 = 0$.

You may have noticed that Example 5 is just Example 4 with the last inequality constraint replaced by an equation constraint. Note, however, that the optimal solutions are different in the two examples. The minimal value of w found in Example 4 is smaller than the one found in Example 5, but does not satisfy the equation constraint in Example 5.

The two-stage method used in Examples 1–5 is summarized here.

Solving Nonstandard Problems

1. Replace each equation constraint by an equivalent pair of inequality constraints.
2. If necessary, write each constraint with a positive constant.
3. Convert a minimization problem to a maximization problem by letting $z = -w$.
4. Add slack variables and subtract surplus variables as needed to convert the constraints into equations.
5. Write the initial simplex tableau.
6. Find a basic feasible solution for the problem if such a solution exists (Stage I).
7. When a basic feasible solution is found, use the simplex method to solve the problem (Stage II).

NOTE It may happen that the tableau which gives the basic feasible solution in Stage I has no negative indicators in its last row. In this case, the solution found is already optimal and Stage II is not necessary.

Applications

Many real-world applications of linear programming involve mixed constraints. Since they typically include a large number of variables and constraints, technology is normally required to solve such problems.

EXAMPLE 6 **Supply Management** A college textbook publisher has received orders from two colleges: C_1 and C_2. C_1 needs 500 books, and C_2 needs 1000. The publisher can supply the books from either of two warehouses. Warehouse W_1 has 900 books available, and warehouse W_2 has 700. The costs to ship a book from each warehouse to each college are as follows:

		To	
		C_1	C_2
From	W_1	$1.20	$1.80
	W_2	$2.10	$1.50

How many books should be sent from each warehouse to each college to minimize the shipping costs?

Solution To begin, let

$$y_1 = \text{the number of books shipped from } W_1 \text{ to } C_1;$$
$$y_2 = \text{the number of books shipped from } W_2 \text{ to } C_1;$$
$$y_3 = \text{the number of books shipped from } W_1 \text{ to } C_2;$$
$$y_4 = \text{the number of books shipped from } W_2 \text{ to } C_2.$$

C_1 needs 500 books, so $y_1 + y_2 = 500$, which is equivalent to this pair of inequalities:

$$y_1 + y_2 \geq 500$$
$$y_1 + y_2 \leq 500.$$

Similarly, $y_3 + y_4 = 1000$, which is equivalent to

$$y_3 + y_4 \geq 1000$$
$$y_3 + y_4 \leq 1000.$$

Since W_1 has 900 books available and W_2 has 700 available,

$$y_1 + y_3 \leq 900 \quad \text{and} \quad y_2 + y_4 \leq 700.$$

The company wants to minimize shipping costs, so the objective function is

$$w = 1.20y_1 + 2.10y_2 + 1.80y_3 + 1.50y_4.$$

Now write the problem as a system of linear equations, adding slack or surplus variables as needed, and let $z = -w$:

$$
\begin{aligned}
y_1 + y_2 & - s_1 &&&&&&= 500 \\
y_1 + y_2 & + s_2 &&&&&&= 500 \\
&y_3 + y_4 - s_3 &&&&&&= 1000 \\
&y_3 + y_4 + s_4 &&&&&&= 1000 \\
y_1 + y_3 &+ s_5 &&&&&&= 900 \\
y_2 + y_4 &+ s_6 &&&&&&= 700 \\
1.20y_1 + 2.10y_2 + 1.80y_3 + 1.50y_4 &+ z &&&&&&= 0.
\end{aligned}
$$

Set up the initial simplex tableau:

y_1	y_2	y_3	y_4	s_1	s_2	s_3	s_4	s_5	s_6	
1	1	0	0	-1	0	0	0	0	0	500
1	1	0	0	0	1	0	0	0	0	500
0	0	1	1	0	0	-1	0	0	0	1000
0	0	1	1	0	0	0	1	0	0	1000
1	0	1	0	0	0	0	0	1	0	900
0	1	0	1	0	0	0	0	0	1	700
1.20	2.10	1.80	1.50	0	0	0	0	0	0	0

The basic solution here is not feasible, because $s_1 = -500$ and $s_3 = -1000$. Stages I and II could be done by hand here, but because of the large size of the matrix, it is more efficient to use technology, such as the program in the Graphing Calculator Appendix. Stage I takes four rounds of pivoting and produces the feasible solution in Figure 7.33.

$$
\begin{array}{cccccc}
y_1 & y_2 & y_3\ y_4 & s_1 & s_2 \\
\end{array}
$$

$$
\begin{bmatrix}
1 & 1 & 0 & 0 & -1 & 0 \\
0 & 0 & 0 & 0 & 1 & 1 \\
0 & 1 & 0 & 1 & 0 & 0 \\
0 & 0 & 0 & 0 & 0 & 0 \\
0 & -1 & 1 & 0 & 0 & 0 \\
0 & 0 & 0 & 0 & 1 & 0 \\
0 & 1.2 & 0 & 0 & 1.2 & 0 \\
\end{bmatrix}
$$

$$
\begin{array}{cccc}
s_3 & s_4\ s_5 & s_6 \\
\end{array}
$$

$$
\begin{bmatrix}
0 & 0 & 0 & 500 \\
0 & 0 & 0 & 0 \\
0 & 0 & 1 & 700 \\
1 & 1 & 0 & 0 \\
-1 & 0 & 0 & -1 & 300 \\
1 & 0 & 1 & 1 & 100 \\
1.8 & 0 & 0 & .3 & -2190 \\
\end{bmatrix}
$$

Figure 7.34

$$
\begin{array}{cccccc}
y_1 & y_2 & y_3 & y_4 & s_1 & s_2 \\
\end{array}
$$

$$
\begin{bmatrix}
1 & 1 & 0 & 0 & -1 & 0 \\
0 & 0 & 0 & 0 & 1 & 1 \\
-1 & 0 & 0 & 1 & 0 & 0 \\
0 & 0 & 0 & 0 & 0 & 0 \\
1 & 0 & 1 & 0 & 0 & 0 \\
0 & 0 & 0 & 1 & 0 & 0 \\
-1.2 & 0 & 0 & 0 & 2.1 & 0 \\
\end{bmatrix}
$$

$$
\begin{array}{cccc}
s_3 & s_4\ s_5 & s_6 \\
\end{array}
$$

$$
\begin{bmatrix}
0 & 0 & 0 & 0 & 500 \\
0 & 0 & 0 & 1 & 0 \\
-1 & 0 & -1 & 0 & 100 \\
1 & 1 & 0 & 0 & 0 \\
0 & 0 & 1 & 0 & 900 \\
1 & 0 & 1 & 1 & 100 \\
1.5 & 0 & -.3 & 0 & -2820 \\
\end{bmatrix}
$$

Figure 7.33

Because of the small size of a calculator screen, you must scroll to the right to see the entire matrix. Now Stage II begins. Two rounds of pivoting produce the final tableau (Figure 7.34).

The optimal solution is $y_1 = 500$, $y_2 = 0$, $y_3 = 300$, and $y_4 = 700$, which results in a minimum shipping cost of \$2190. (Remember that the optimal value for the original minimization problem is the negative of the optimal value of the associated maximization problem.)

7.7 Exercises

In Exercises 1–4, (a) restate the problem in terms of equations by introducing slack and surplus variables and (b) write the initial simplex tableau. (See Example 1.)

1. Maximize $z = -5x_1 + 4x_2 - 2x_3$

    subject to $-2x_2 + 5x_3 \geq 8$

                 $4x_1 - x_2 + 3x_3 \leq 12$

                 $x_1 \geq 0, x_2 \geq 0, x_3 \geq 0.$

2. Maximize $z = -x_1 + 4x_2 - 2x_3$

    subject to $2x_1 + 2x_2 + 6x_3 \leq 10$

                 $-x_1 + 2x_2 + 4x_3 \geq 7$

                 $x_1 \geq 0, x_2 \geq 0, x_3 \geq 0.$

3. Maximize $z = 2x_1 - 3x_2 + 4x_3$

    subject to $x_1 + x_2 + x_3 \leq 100$

                 $x_1 + x_2 + x_3 \geq 75$

                 $x_1 + x_2 \qquad \geq 27$

                 $x_1 \geq 0, x_2 \geq 0, x_3 \geq 0.$

4. Maximize $z = -x_1 + 5x_2 + x_3$

    subject to $2x_1 \qquad + x_3 \leq 40$

                 $x_1 + x_2 \qquad \geq 18$

                 $x_1 \qquad + x_3 = 20$

                 $x_1 \geq 0, x_2 \geq 0, x_3 \geq 0.$

Convert Exercises 5–8 into maximization problems with positive constants on the right side of each constraint, and write the initial simplex tableau. (See Examples 4 and 5.)

5. Minimize $w = 2y_1 + 5y_2 - 3y_3$

    subject to $y_1 + 2y_2 + 3y_3 \geq 115$

                 $2y_1 + y_2 + y_3 \leq 200$

                 $y_1 \qquad + y_3 \geq 50$

                 $y_1 \geq 0, y_2 \geq 0, y_3 \geq 0.$

6. Minimize $w = 7y_1 + 6y_2 + y_3$

    subject to $y_1 + y_2 + y_3 \geq 5$

                 $-y_1 + y_2 \qquad \leq -4$

                 $2y_1 + y_2 + 3y_3 \geq 15$

                 $y_1 \geq 0, y_2 \geq 0, y_3 \geq 0.$

7. Minimize $w = 10y_1 + 8y_2 + 15y_3$

    subject to $y_1 + y_2 + y_3 \geq 12$

                 $5y_1 + 4y_2 + 9y_3 \geq 48$

                 $y_1 \geq 0, y_2 \geq 0, y_3 \geq 0.$

8. Minimize $w = y_1 + 2y_2 + y_3 + 5y_4$

    subject to $-y_1 - y_2 + y_3 - y_4 \leq -50$

                 $3y_1 + y_2 + 2y_3 + y_4 = 100$

                 $y_1 \geq 0, y_2 \geq 0, y_3 \geq 0, y_4 \geq 0.$

Use the two-stage method to solve Exercises 9–20. (See Examples 1–5.)

9. Maximize $z = 12x_1 + 10x_2$
subject to $x_1 + 2x_2 \geq 24$
$x_1 + x_2 \leq 40$
$x_1 \geq 0, x_2 \geq 0.$

10. Find $x_1 \geq 0, x_2 \geq 0,$ and $x_3 \geq 0$ such that
$x_1 + x_2 + x_3 \leq 150$
$x_1 + x_2 + x_3 \geq 100$
and $z = 2x_1 + 5x_2 + 3x_3$ is maximized.

11. Find $x_1 \geq 0, x_2 \geq 0,$ and $x_3 \geq 0$ such that
$x_1 + x_2 + 2x_3 \leq 38$
$2x_1 + x_2 + x_3 \geq 24$
and $z = 3x_1 + 2x_2 + 2x_3$ is maximized.

12. Maximize $z = 6x_1 + 8x_2$
subject to $3x_1 + 12x_2 \geq 48$
$2x_1 + 4x_2 \leq 60$
$x_1 \geq 0, x_2 \geq 0.$

13. Find $x_1 \geq 0$ and $x_2 \geq 0$ such that
$x_1 + 2x_2 \leq 18$
$x_1 + 3x_2 \geq 12$
$2x_1 + 2x_2 \leq 30$
and $z = 5x_1 + 10x_2$ is maximized.

14. Find $y_1 \geq 0, y_2 \geq 0$ such that
$10y_1 + 5y_2 \geq 100$
$20y_1 + 10y_2 \geq 160$
and $w = 4y_1 + 5y_2$ is minimized.

15. Minimize $w = 3y_1 + 2y_2$
subject to $2y_1 + 3y_2 \geq 60$
$y_1 + 4y_2 \geq 40$
$y_1 \geq 0, y_2 \geq 0.$

16. Minimize $w = 3y_1 + 4y_2$
subject to $y_1 + 2y_2 \geq 10$
$y_1 + y_2 \geq 8$
$2y_1 + y_2 \leq 22$
$y_1 \geq 0, y_2 \geq 0.$

17. Maximize $z = 3x_1 + 2x_2$
subject to $x_1 + x_2 = 50$
$4x_1 + 2x_2 \geq 120$
$5x_1 + 2x_2 \leq 200$
with $x_1 \geq 0, x_2 \geq 0.$

18. Maximize $z = 10x_1 + 9x_2$
subject to $x_1 + x_2 = 30$
$x_1 + x_2 \geq 25$
$2x_1 + x_2 \leq 40$
with $x_1 \geq 0, x_2 \geq 0.$

19. Minimize $w = 32y_1 + 40y_2$
subject to $20y_1 + 10y_2 = 200$
$25y_1 + 40y_2 \leq 500$
$18y_1 + 24y_2 \geq 300$
with $y_1 \geq 0, y_2 \geq 0.$

20. Minimize $w = 15y_1 + 12y_2$
subject to $y_1 + 2y_2 \leq 12$
$3y_1 + y_2 \geq 18$
$y_1 + y_2 = 10$
with $y_1 \geq 0, y_2 \geq 0.$

Use the two-stage method to solve the problem that was set up in the given exercise.

21. Exercise 1   **22.** Exercise 2

23. Exercise 3   **24.** Exercise 4

25. Exercise 5   **26.** Exercise 6

27. Exercise 7   **28.** Exercise 8

In Exercises 29–32, set up the initial simplex tableau, but do not solve the problem. (See Example 6.)

29. Gasoline Additive A company is developing a new additive for gasoline. The additive is a mixture of three liquid ingredients: I, II, and III. For proper performance, the total amount of additive must be at least 10 ounces per barrel of gasoline. However, for safety reasons, the amount of additive should not exceed 15 ounces per barrel of gasoline. At least $\frac{1}{4}$ ounce of ingredient I must be used for every ounce of ingredient II, and at least 1 ounce of ingredient III must be used for every ounce of ingredient I. If the costs of I, II, and III are \$.30, \$.09, and \$.27 per ounce, respectively, find the mixture of the three ingredients that produces the minimum cost of the additive. What is the minimum cost?

30. Soft Drink Production A popular soft drink called Sugarlo, which is advertised as having a sugar content of no more than 10%, is blended from five ingredients, each of which has some sugar content. Water may also be added to dilute the mixture. The sugar content of the ingredients and their costs per gallon are given in the table:

	Ingredient					
	1	**2**	**3**	**4**	**5**	**Water**
Sugar content (%)	.28	.19	.43	.57	.22	0
Cost (\$/gal.)	.48	.32	.53	.28	.43	.04

At least .01 of the content of Sugarlo must come from ingredient 3 or 4, .01 must come from ingredient 2 or 5, and .01 from ingredient 1 or 4. How much of each ingredient should be used in preparing at least 15,000 gallons of Sugarlo to minimize the cost? What is the minimum cost?

31. Supply Management The manufacturer of a popular personal computer has orders from two dealers. Dealer D_1 wants 32 computers, and dealer D_2 wants 20 computers. The manufacturer can fill the orders from either of two warehouses, W_1 or W_2. W_1 has 25 of the computers on hand, and W_2 has 30. The costs (in dollars) to ship one computer to each dealer from each warehouse are as follows:

		To	
		D_1	D_2
From	W_1	$14	$22
	W_2	$12	$10

How should the orders be filled to minimize shipping costs? What is the minimum cost?

32. **Nutrition** Mark, who is ill, takes vitamin pills. Each day, he must have at least 16 units of vitamin A, 5 units of vitamin B_1, and 20 units of vitamin C. He can choose between pill 1, which costs 10¢ and contains 8 units of vitamin A, 1 unit of vitamin B_1, and 2 units of vitamin C, and pill 2, which costs 20¢ and contains 2 units of vitamin A, 1 unit of vitamin B_1, and 7 units of vitamin C. How many of each pill should he buy in order to minimize his cost?

Use the two-stage method to solve Exercises 33–40. (See Examples 5 and 6.)

33. **Supply Management** Southwestern Oil supplies two distributors in the Northwest from two outlets: S_1 and S_2. Distributor D_1 needs at least 3000 barrels of oil, and distributor D_2 needs at least 5000 barrels. The two outlets can each furnish up to 5000 barrels of oil. The costs per barrel to ship the oil are given in the table:

		Distributors	
		D_1	D_2
Outlets	S_1	$30	$20
	S_2	$25	$22

There is also a shipping tax per barrel as given in the table below.

	D_1	D_2
S_1	$2	$6
S_2	$5	$4

Southwestern Oil is determined to spend no more than $40,000 on shipping tax. How should the oil be supplied to minimize shipping costs?

34. **Supply Management** Change Exercise 33 so that the two outlets each furnish exactly 5000 barrels of oil, with everything else the same. Solve the problem as in Example 5.

35. **Seed Production** Topgrade Turf lawn seed mixture contains three types of seeds: bluegrass, rye, and Bermuda. The costs per pound of the three types of seed are 12¢, 15¢, and 5¢, respectively. In each batch, there must be at least 20% bluegrass seed, and the amount of Bermuda seed must be no more than two-thirds the amount of rye seed. To fill current orders, the company must make at least 5000 pounds of the mixture. How much of each kind of seed should be used to minimize costs?

36. **Seed Production** Change Exercise 35 so that the company must make exactly 5000 pounds of the mixture. Solve the problem as in Example 5.

37. **Loan Allocation** A bank has set aside a maximum of $25 million for commercial and home loans. Every million dollars in commercial loans requires 2 lengthy application forms, while every million dollars in home loans requires 3 lengthy application forms. The bank cannot process more than 72 application forms at this time. The bank's policy is to loan at least four times as much for home loans as for commercial loans. Because of prior commitments, at least $10 million will be used for these two types of loans. The bank earns 12% on home loans and 10% on commercial loans. What amount of money should be allotted for each type of loan to maximize the interest income?

38. **Investing** Virginia has decided to invest a $100,000 inheritance in government securities that earn 7% per year, municipal bonds that earn 6% per year, and mutual funds that earn an average of 10% per year. She will spend at least $40,000 on government securities, and she wants at least half the inheritance to go to bonds and mutual funds. Government securities have an initial fee of 2%, municipal bonds an initial fee of 1%, and mutual funds an initial fee of 3%. Virginia has $2400 available to pay initial fees. How much money should go into each type of investment to maximize the interest while meeting the constraints? What is the maximum interest she can earn?

39. **Beer Production** A brewery produces regular beer and a lower-carbohydrate "light" beer. Steady customers of the brewery buy 12 units of regular beer and 10 units of light beer. While setting up the brewery to produce the beers, the management decides to produce extra beer, beyond that needed to satisfy the steady customers. The cost per unit of regular beer is $36,000, and the cost per unit of light beer is $48,000. The number of units of light beer should not exceed twice the number of units of regular beer. At least 20 additional units of beer can be sold. How much of each type of beer should be made so as to minimize total production costs?

40. **Supply Management** The chemistry department at a local college decides to stock at least 800 small test tubes and 500 large test tubes. It wants to buy at least 1500 test tubes to take advantage of a special price. Since the small tubes are broken twice as often as the large, the department will order at least twice as many small tubes as large. If the small test tubes cost 15¢ each and the large ones, made of a cheaper glass, cost 12¢ each, how many of each size should the department order to minimize cost?

Use technology to solve the following exercises, whose initial tableaus were set up in Exercises 29–32.

41. Exercise 29   42. Exercise 30
43. Exercise 31   44. Exercise 32

✓ Checkpoint Answers

1. (a) Maximize $z = 3x_1 - 2x_2$
   subject to $2x_1 + 3x_2 + s_1 = 8$
   $6x_1 - 2x_2 - s_2 = 3$
   $x_1 + 4x_2 - s_3 = 1$
   $x_1 \geq 0, x_2 \geq 0, s_1 \geq 0, s_2 \geq 0, s_3 \geq 0.$

$$
\begin{array}{c}
\quad\; x_1 \quad x_2 \quad s_1 \quad s_2 \quad s_3 \\
\textbf{(b)}\;\left[\begin{array}{ccccc|c}
2 & 3 & 1 & 0 & 0 & 8 \\
6 & -2 & 0 & -1 & 0 & 3 \\
1 & 4 & 0 & 0 & -1 & 1 \\
\hline
-3 & 2 & 0 & 0 & 0 & 0
\end{array}\right]
\end{array}
$$

2. (a) $x_1 = 0,\ x_2 = 0,\ s_1 = 12,\ s_2 = 6,\ s_3 = -5$; no

(b) $x_1 = 0,\ x_2 = 0,\ x_3 = 0,\ s_1 = 12,\ s_2 = 7$; yes

$$
\begin{array}{c}
\quad\; x_1 \quad x_2 \quad s_1 \quad s_2 \\
\textbf{3.}\;\left[\begin{array}{cccc|c}
0 & 1 & 1 & \frac{1}{2} & 45 \\
1 & 2 & 0 & -\frac{1}{2} & 25 \\
\hline
0 & 6 & 0 & -4 & 200
\end{array}\right]
\end{array}
$$

$x_1 = 25,\ x_2 = 0,\ s_1 = 45$, and $s_2 = 0$.

4. The optimal value $z = 560$ occurs when $x_1 = 70,\ x_2 = 0,\ s_1 = 0$, and $s_2 = 90$.

5. $y_1 = 10$ and $y_2 = 0$; $w = 20$

CHAPTER 7 Summary and Review

Key Terms and Symbols

7.1 linear inequality
graphs of linear
inequalities
boundary
half-plane
system of inequalities
region of feasible solutions
(feasible region)

7.2 linear programming
objective function

constraints
corner point
bounded feasible region
unbounded feasible region
corner point theorem

7.3 applications of linear
programming

7.4 standard maximum form
slack variable
simplex tableau

indicator
pivot and pivoting
final simplex tableau
basic variables
nonbasic variables
basic feasible solution

7.6 transpose of a matrix
dual
theorem of duality
shadow costs

7.7 surplus variable
basic variables
basic solution
basic feasible solution
Stage I
Stage II

Chapter 7 Key Concepts

Graphing a Linear Inequality

Graph the boundary line as a solid line if the inequality includes "or equal to," and as a dashed line otherwise. Shade the half-plane for which the inequality is true. The graph of a system of inequalities, called the **region of feasible solutions**, includes all points that satisfy all the inequalities of the system at the same time.

Solving Linear Programming Problems

Graphical Method: Determine the objective function and all necessary constraints. Graph the region of feasible solutions. The maximum or minimum value will occur at one or more of the corner points of this region.

Simplex Method: Determine the objective function and all necessary constraints. Convert each constraint into an equation by adding slack variables. Set up the initial simplex tableau. Locate the most negative indicator. Form the quotients to determine the pivot. Use row operations to change the pivot to 1 and all other numbers in that column to 0. If the indicators are all positive or 0, this is the final tableau. If not, choose a new pivot and repeat the process until no indicators are negative. Read the solution from the final tableau. The optimum value of the objective function is the number in the lower right corner of the final tableau. For problems with **mixed constraints**, replace each equation constraint by a pair of inequality constraints. Then add slack variables and subtract surplus variables as needed to convert each constraint into an equation. In Stage I, use row operations to transform the matrix until the solution is feasible. In Stage II, use the simplex method as just described. For **minimization** problems, let the objective function be w and set $-w = z$. Then proceed as with mixed constraints.

Solving Minimization Problems with Duals

Find the dual maximization problem. Solve the dual problem with the simplex method. The minimum value of the objective function w is the maximum value of the dual objective function z. The optimal solution is found in the entries in the bottom row of the columns corresponding to the slack variables.

Chapter 7 Review Exercises

Graph each of the given linear inequalities.

1. $y \leq 3x + 2$ **2.** $2x - y \geq 6$

3. $3x + 4y \geq 12$ **4.** $y \leq 4$

Graph the solution of each of the given systems of inequalities.

5. $x + y \leq 6$
  $2x - y \geq 3$

6. $4x + y \geq 8$
  $2x - 3y \leq 6$

7. $2 \leq x \leq 5$
  $1 \leq y \leq 6$
  $x - y \leq 3$

8. $x + 2y \leq 4$
  $2x - 3y \leq 6$
  $x \geq 0, y \geq 0$

Set up a system of inequalities for each of the given problems, and then graph the region of feasible solutions.

9. Consulting Linda Quinn is a statistical consultant who creates two kind of reports: summary reports and inference reports. Each summary report requires 2 hours of analysis time and 1.5 hours to write the report. Each inference report needs 4 hours of analysis time and 2 hours of writing time. Suppose Linda has only 15 hours available for analysis and 9 hours for writing.

10. Pizza Production A company makes two kinds of pizza: basic (cheese and tomatoes) and margherita (cheese, tomatoes, and basil). The company sells at least 60 units a day of the pizza margherita and 40 units a day of the basic. The cost of the tomatoes and basil is $2 per unit for the pizza margherita and the cost of tomatoes is $1 per unit for the basic. No more than $320 a day can be spent on tomatoes and basil together. The cheese used for the pizza margherita is $5 per unit, and the cheese for the basic is $4 per unit. The company spends no more than $1000 a day on cheese.

Use the given regions to find the maximum and minimum values of the objective function $z = 3x + 4y$.

11.

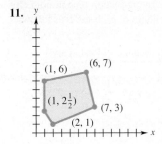

12.

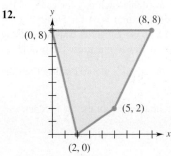

Use the graphical method to solve Exercises 13–16.

13. Maximize $z = 6x + 2y$
subject to $2x + 7y \leq 14$
  $2x + 3y \leq 10$
  $x \geq 0, y \geq 0.$

14. Find $x \geq 0$ and $y \geq 0$ such that

$$8x + 9y \geq 72$$
$$6x + 8y \geq 72$$

and $w = 2x + 10y$ is minimized.

15. Find $x \geq 0$ and $y \geq 0$ such that

$$x + y \leq 50$$
$$2x + y \geq 20$$
$$x + 2y \geq 30$$

and $w = 5x + 2y$ is minimized.

16. Maximize $z = 5x - 2y$
subject to $3x + 2y \leq 12$
  $5x + y \geq 5$
  $x \geq 0, y \geq 0.$

17. Consulting How many of each kind of statistical report (summary or inference) should Linda Quinn create in Exercise 9 in order to maximize profit if the summary reports earn a profit of $500 and the inference reports produce a profit of $750? (Assume it is possible to complete a fractional part of a project.)

18. Pizza Production How many units of each pizza in Exercise 10 should the company make in order to maximize revenue if the basic pizza sells for $15 per unit and the pizza margherita sells for $20 per unit.

19. Finance The BlackRock Equity Dividend Fund sells at $24 a share and has a 3-year average annual return of $3 per share. The risk measure of beta is .95. The Columbia Dividend Income Fund sells for $20 a share and has a 3-year average annual return of $2 a share. The risk measure of beta is .89. Donald wants to spend no more than $10,000 investing in these two funds, but he wants to obtain at least $1100 in annual revenue. Donald also wants to minimize his risk. Determine how many shares of each fund Donald should buy. (Data from: www.morningstar.com.)

20. Finance The Alger Spectra Z Fund sells at $15 a share and has a 3-year average annual return of $2 per share. The risk measure of beta is .96. The JPMorgan U.S. Dynamic Plus Fund sells for $22 a share and has a 3-year average annual return of $3.50 a share. The risk measure of beta is 1.14. Ming-Na wants to spend no more than $5016 inventing in these two funds, but she wants to obtain at least $700 in revenue. Ming-Na also wants to minimize risk. Determine how many shares of each fund Ming-Na should buy. (Data from: www.morningstar.com.)

For each of the following problems, (a) add slack variables and (b) set up the initial simplex tableau.

21. Maximize $z = 5x_1 + 6x_2 + 3x_3$

   subject to $x_1 + x_2 + x_3 \le 100$

$$2x_1 + 3x_2 \quad\quad \le 500$$

$$x_1 \quad\quad + 2x_3 \le 350$$

$$x_1 \ge 0, x_2 \ge 0, x_3 \ge 0.$$

22. Maximize $z = 2x_1 + 9x_2$

   subject to $3x_1 + 5x_2 \le 47$

$$x_1 + x_2 \le 25$$

$$5x_1 + 2x_2 \le 35$$

$$2x_1 + x_2 \le 30$$

$$x_1 \ge 0, x_2 \ge 0.$$

23. Maximize $z = x_1 + 8x_2 + 2x_3$

   subject to $x_1 + x_2 + x_3 \le 90$

$$2x_1 + 5x_2 + x_3 \le 120$$

$$x_1 + 3x_2 \quad\quad \le 80$$

$$x_1 \ge 0, x_2 \ge 0, x_3 \ge 0.$$

24. Maximize $z = 15x_1 + 12x_2$

   subject to $2x_1 + 5x_2 \le 50$

$$x_1 + 3x_2 \le 25$$

$$4x_1 + x_2 \le 18$$

$$x_1 + x_2 \le 12$$

$$x_1 \ge 0, x_2 \ge 0.$$

For each of the following, use the simplex method to solve the maximization linear programming problems with initial tableaus as given.

25.
$$\begin{array}{ccccc}
x_1 & x_2 & x_3 & s_1 & s_2 \\
\end{array}$$
$$\left[\begin{array}{ccccc|c}
1 & 2 & 3 & 1 & 0 & 28 \\
2 & 4 & 8 & 0 & 1 & 32 \\
\hline
-5 & -2 & -3 & 0 & 0 & 0
\end{array}\right]$$

26.
$$\begin{array}{cccc}
x_1 & x_2 & s_1 & s_2 \\
\end{array}$$
$$\left[\begin{array}{cccc|c}
2 & 1 & 1 & 0 & 10 \\
9 & 3 & 0 & 1 & 15 \\
\hline
-2 & -3 & 0 & 0 & 0
\end{array}\right]$$

27.
$$\begin{array}{cccccc}
x_1 & x_2 & x_3 & s_1 & s_2 & s_3 \\
\end{array}$$
$$\left[\begin{array}{cccccc|c}
1 & 2 & 2 & 1 & 0 & 0 & 50 \\
4 & 24 & 0 & 0 & 1 & 0 & 20 \\
1 & 0 & 2 & 0 & 0 & 1 & 15 \\
\hline
-5 & -3 & -2 & 0 & 0 & 0 & 0
\end{array}\right]$$

28.
$$\begin{array}{cccccc}
x_1 & x_2 & s_1 & s_2 & s_3 \\
\end{array}$$
$$\left[\begin{array}{ccccc|c}
1 & -2 & 1 & 0 & 0 & 38 \\
1 & -1 & 0 & 1 & 0 & 12 \\
2 & 1 & 0 & 0 & 1 & 30 \\
\hline
-1 & -2 & 0 & 0 & 0 & 0
\end{array}\right]$$

Use the simplex method to solve the problem that was set up in the given exercise.

29. Exercise 21   **30.** Exercise 22

31. Exercise 23   **32.** Exercise 24

For Exercises 33–36, (a) select appropriate variables, (b) write the objective function, and (c) write the constraints as inequalities.

33. Sales Roberta Hernandez sells three items—A, B, and C—in her gift shop. Each unit of A costs her \$2 to buy, \$1 to sell, and \$2 to deliver. For each unit of B, the costs are \$3, \$2, and \$2, respectively, and for each unit of C, the costs are \$6, \$2, and \$4, respectively. The profit on A is \$4, on B it is \$3, and on C it is \$3. How many of each item should she order to maximize her profit if she can spend \$1200 to buy, \$800 to sell, and \$500 to deliver?

34. Investing An investor is considering three types of investment: a high-risk venture into oil leases with a potential return of 15%, a medium-risk investment in bonds with a 9% return, and a relatively safe stock investment with a 5% return. He has \$50,000 to invest. Because of the risk, he will limit his investment in oil leases and bonds to 30% and his investment in oil leases and stock to 50%. How much should he invest in each to maximize his return, assuming investment returns are as expected?

35. Wine Production The Aged Wood Winery makes two white wines—Fruity and Crystal—from two kinds of grapes and sugar. The wines require the amounts of each ingredient per gallon and produce a profit per gallon as shown in the following table:

	Grape A (bushels)	Grape B (bushels)	Sugar (pounds)	Profit (dollars)
Fruity	2	2	2	12
Crystal	1	3	1	15

The winery has available 110 bushels of grape A, 125 bushels of grape B, and 90 pounds of sugar. How much of each wine should be made to maximize profit?

36. Packaging A company makes three sizes of plastic bags: 5 gallon, 10 gallon, and 20 gallon. The production time in hours for cutting, sealing, and packaging a unit of each size is as follows:

Size	Cutting	Sealing	Packaging
5 gallon	1	1	2
10 gallon	1.1	1.2	3
20 gallon	1.5	1.3	4

There are at most 8 hours available each day for each of the three operations. If the profit per unit is \$1 for 5-gallon bags, \$.90 for 10-gallon bags, and \$.95 for 20-gallon bags, how many of each size should be made per day to maximize the profit?

37. When is it necessary to use the simplex method rather than the graphical method?

38. What types of problems can be solved with the use of slack variables and surplus variables?

39. What kind of problem can be solved with the method of duals?

40. In solving a linear programming problem, you are given the following initial tableau:

$$\begin{bmatrix} 4 & 2 & 3 & 1 & 0 & 9 \\ 5 & 4 & 1 & 0 & 1 & 10 \\ \hline -6 & -7 & -5 & 0 & 0 & 0 \end{bmatrix}.$$

(a) What is the problem being solved?

(b) If the 1 in row 1, column 4, were a -1 rather than a 1, how would it change your answer to part (a)?

(c) After several steps of the simplex algorithm, the following tableau results:

$$\begin{bmatrix} .6 & 0 & 1 & .4 & -.2 & 1.6 \\ 1.1 & 1 & 0 & -.1 & .3 & 2.1 \\ \hline 4.7 & 0 & 0 & 1.3 & 1.1 & 22.7 \end{bmatrix}.$$

What is the solution? (List only the values of the original variables and the objective function. Do not include slack or surplus variables.)

(d) What is the dual of the problem you found in part (a)?

(e) What is the solution of the dual you found in part (d)? [Do not perform any steps of the simplex algorithm; just examine the tableau given in part (c).]

The tableaus in Exercises 41–43 are the final tableaus of minimization problems solved by the method of duals. State the solution and the minimum value of the objective function for each problem.

41. $$\begin{bmatrix} 1 & 0 & 0 & 3 & 1 & 2 & 12 \\ 0 & 0 & 1 & 4 & 5 & 3 & 5 \\ 0 & 1 & 0 & -2 & 7 & -6 & 8 \\ \hline 0 & 0 & 0 & 5 & 7 & 3 & 172 \end{bmatrix}$$

42. $$\begin{bmatrix} 0 & 0 & 1 & 6 & 3 & 1 & 2 \\ 1 & 0 & 0 & 4 & -2 & 2 & 8 \\ 0 & 1 & 0 & 10 & 7 & 0 & 12 \\ \hline 0 & 0 & 0 & 9 & 5 & 8 & 62 \end{bmatrix}$$

43. $$\begin{bmatrix} 1 & 0 & 7 & -1 & 100 \\ 0 & 1 & 1 & 3 & 27 \\ \hline 0 & 0 & 7 & 2 & 640 \end{bmatrix}$$

Use the method of duals to solve these minimization problems.

44. Minimize $w = 5y_1 + 2y_2$

subject to $2y_1 + 3y_2 \geq 6$

$\qquad\quad 2y_1 + \; y_2 \geq 7$

$\qquad\quad y_1 \geq 0, y_2 \geq 0.$

45. Minimize $w = 18y_1 + 10y_2$

subject to $y_1 + \; y_2 \geq 17$

$\qquad\quad 5y_1 + 8y_2 \geq 42$

$\qquad\quad y_1 \geq 0, y_2 \geq 0.$

46. Minimize $w = 4y_1 + 5y_2$

subject to $10y_1 + \; 5y_2 \geq 100$

$\qquad\quad 20y_1 + 10y_2 \geq 150$

$\qquad\quad y_1 \geq 0, y_2 \geq 0.$

Write the initial simplex tableau for each of these mixed-constraint problems.

47. Maximize $z = 20x_1 + 30x_2$

subject to $5x_1 + 10x_2 \leq 120$

$\qquad\quad 10x_1 + 15x_2 \geq 200$

$\qquad\quad x_1 \geq 0, x_2 \geq 0.$

48. Minimize $w = 4y_1 + 2y_2$

subject to $y_1 + 3y_2 \geq 6$

$\qquad\quad 2y_1 + 8y_2 \leq 21$

$\qquad\quad y_1 \geq 0, y_2 \geq 0.$

49. Minimize $w = 12y_1 + 20y_2 - 8y_3$

subject to $y_1 + y_2 + 2y_3 \geq 48$

$\qquad\quad y_1 + y_2 \qquad\quad \leq 12$

$\qquad\qquad\qquad\quad y_3 \geq 10$

$\qquad\quad 3y_1 \qquad + \; y_3 \geq 30$

$\qquad\quad y_1 \geq 0, y_2 \geq 0, y_3 \geq 0.$

50. Maximize $w = 6x_1 - 3x_2 + 4x_3$

subject to $2x_1 + x_2 + x_3 \leq 112$

$\qquad\quad x_1 + x_2 + x_3 \geq 80$

$\qquad\quad x_1 + x_2 \qquad\quad \leq 45$

$\qquad\quad x_1 \geq 0, x_2 \geq 0, x_3 \geq 0.$

The given tableaus are the final tableaus of minimization problems solved by letting $w = -z$. Give the solution and the minimum value of the objective function for each problem.

51. $$\begin{bmatrix} 0 & 1 & 0 & 2 & 5 & 0 & 17 \\ 0 & 0 & 1 & 3 & 1 & 1 & 25 \\ 1 & 0 & 0 & 4 & 2 & \frac{1}{2} & 8 \\ \hline 0 & 0 & 0 & 2 & 5 & 0 & -427 \end{bmatrix}$$

52. $$\begin{bmatrix} 0 & 0 & 2 & 1 & 0 & 6 & 6 & 92 \\ 1 & 0 & 3 & 0 & 0 & 0 & 2 & 47 \\ 0 & 1 & 0 & 0 & 0 & 1 & 0 & 68 \\ 0 & 0 & 4 & 0 & 1 & 0 & 3 & 35 \\ \hline 0 & 0 & 5 & 0 & 0 & 2 & 9 & -1957 \end{bmatrix}$$

Use the two-stage method to solve these mixed-constraint problems.

53. Exercise 47   **54.** Exercise 48

55. Minimize $w = 4y_1 - 8y_2$

subject to $y_1 + \; y_2 \leq 50$

$\qquad\quad 2y_1 - 4y_2 \geq 20$

$\qquad\quad y_1 - \; y_2 \leq 22$

$\qquad\quad y_1 \geq 0, y_2 \geq 0.$

56. Maximize $z = 2x_1 + 4x_2$
subject to $3x_1 + 2x_2 \leq 12$
$5x_1 + x_2 \geq 5$
$x_1 \geq 0, x_2 \geq 0.$

Business *Solve the following maximization problems, which were begun in Exercises 33–36.*

57. Exercise 33 **58.** Exercise 34

59. Exercise 35 **60.** Exercise 36

Solve the following minimization problems.

61. Vegetable Production Cauchy Canners produces canned corn, beans, and carrots. Demand for vegetables requires the company to produce at least 1000 cases per month. Based on past sales, it should produce at least twice as many cases of corn as of beans and at least 340 cases of carrots. It costs $10 to produce a case of corn, $15 to produce a case of beans, and $25 to produce a case of carrots. How many cases of each vegetable should be produced to minimize costs? What is the minimum cost?

62. Construction A contractor builds boathouses in two basic models: the Atlantic and the Pacific. Each Atlantic model requires 1000 feet of framing lumber, 3000 cubic feet of concrete, and $2000 for advertising. Each Pacific model requires 2000 feet of framing lumber, 3000 cubic feet of concrete, and $3000 for advertising. Contracts call for using at least 8000 feet of framing lumber, 18,000 cubic feet of concrete, and $15,000 worth of advertising. If the total spent on each Atlantic model is $3000 and the total spent on each Pacific

model is $4000, how many of each model should be built to minimize costs?

Solve these mixed-constraint problems.

63. Food Production Brand X Cannery produces canned whole tomatoes and tomato sauce. This season, the company has available 3,000,000 kilograms of tomatoes for these two products. To meet the demands of regular customers, it must produce at least 80,000 kilograms of sauce and 800,000 kilograms of whole tomatoes. The cost per kilogram is $4 to produce canned whole tomatoes and $3.25 to produce tomato sauce. Labor agreements require that at least 110,000 person-hours be used. Each 1-kilogram can of sauce requires 3 minutes for one worker to produce, and each 1-kilogram can of whole tomatoes requires 6 minutes for one worker. How many kilograms of tomatoes should Brand X use for each product to minimize cost? (For simplicity, assume that the production of y_1 kilograms of canned whole tomatoes and y_2 kilograms of tomato sauce requires $y_1 + y_2$ kilograms of tomatoes.)

64. Steel Production A steel company produces two types of alloys. A run of type I requires 3000 pounds of molybdenum and 2000 tons of iron ore pellets, as well as $2000 in advertising. A run of type II requires 3000 pounds of molybdenum and 1000 tons of iron ore pellets, as well as $3000 in advertising. Total costs are $15,000 on a run of type I and $6000 on a run of type II. The company has on hand 18,000 pounds of molybdenum and 7000 tons of iron ore pellets and wants to use all of it. It plans to spend at least $14,000 on advertising. How much of each type should be produced to minimize cost? What is the minimum cost?

Case Study 7 Cooking with Linear Programming

Constructing a nutritious recipe can be a difficult task. The recipe must produce food that tastes good, and it must also balance the nutrients that each ingredient brings to the dish. This balancing of nutrients is very important in several diet plans that are currently popular. Many of these plans restrict the intake of certain nutrients (usually fat) while allowing for large amounts of other nutrients (protein and carbohydrates are popular choices). The number of calories in the dish is also often minimized. Linear programming can be used to help create recipes that balance nutrients.

In order to develop solutions to this type of problem, we will need to have nutritional data for the ingredients in our recipes. This data can be found in the U.S. Department of Agriculture's USDA Nutrient Database for Standard Reference, available at www.ndb.nal.usda.gov. This database contains the nutrient levels for hundreds of basic foods. The nutrient levels are given per 100 grams of food. Unfortunately, grams are not often used in recipes; instead, kitchen measures like cups, tablespoons, and fractions of vegetables are used. Table 1 shows the conversion factors from grams to more familiar kitchen units and gives serving sizes for various food.

Table 1   Serving Sizes of Various Food

Food	Serving Size
Beef	6 oz $= 170$ g
Egg	1 egg $= 61$ g
Feta Cheese	$\frac{1}{4}$ cup $= 38$ g
Lettuce	$\frac{1}{2}$ cup $= 28$ g
Milk	1 cup $= 244$ g
Oil	1 Tbsp $= 13.5$ g
Onion	1 onion $= 110$ g
Salad Dressing	1 cup $= 250$ g
Soy Sauce	1 Tbsp $= 18$ g
Spinach	1 cup $= 180$ g
Tomato	1 tomato $= 123$ g

Consider creating a recipe for a spinach omelet from eggs, milk, vegetable oil, and spinach. The nutrients of interest will be protein, fat, and carbohydrates. Calories will also be monitored. The amounts of the nutrients and calories for these ingredients are given in Table 2.

Table 2 Nutritional Values per 100 g of Food

Nutrient (units)	Eggs	Milk	Oil	Spinach
Calories (kcal)	152	61.44	884	23
Protein (g)	10.33	3.29	0	2.9
Fat (g)	11.44	3.34	100	.26
Carbohydrates (g)	1.04	4.66	0	3.75

Let x_1 be the number of 100-gram units of eggs to use in the recipe, x_2 be the number of 100-gram units of milk, x_3 be the number of 100-gram units of oil, and x_4 be the number of 100-gram units of spinach. We will want to minimize the number of calories in the dish while providing at least 15 grams of protein, 4 grams of carbohydrates, and 20 grams of fat. The cooking technique specifies that at least $\frac{1}{8}$ of a cup of milk (30.5 grams) must be used in the recipe. We should thus minimize the objective function (using 100-gram units of food)

$$z = 152x_1 + 61.44x_2 + 884x_3 + 23x_4$$

subject to

$$10.33x_1 + 3.29x_2 + \ 0x_3 + 2.90x_4 \geq 15$$
$$11.44x_1 + 3.34x_2 + 100x_3 + \ .26x_4 \geq 20$$
$$1.04x_1 + 4.66x_2 + \ 0x_3 + 3.75x_4 \geq 4$$
$$0x_1 + \ 1x_2 + \ 0x_3 + \ 0x_4 \geq .305.$$

Of course, all variables are subject to nonnegativity constraints:

$$x_1 \geq 0, \quad x_2 \geq 0, \quad x_3 \geq 0, \quad \text{and} \quad x_4 \geq 0.$$

Using a graphing calculator or a computer with linear programming software, we get the following solution:

$$x_1 = 1.2600, \quad x_2 = .3050, \quad x_3 = .0448, \quad \text{and} \quad x_4 = .338.$$

This recipe produces an omelet with 257.63 calories. The amounts of each ingredient are 126 grams of eggs, 30.5 grams of milk, 4.48 grams of oil, and 33.8 grams of spinach. Converting to kitchen units using Table 1, we find the recipe to be approximately 2 eggs, $\frac{1}{8}$ cup milk, 1 teaspoon oil, and $\frac{1}{4}$ cup spinach.

Exercises

1. Consider preparing a high-carbohydrate Greek salad using feta cheese, lettuce, salad dressing, and tomato. The amount of

carbohydrates in the salad should be maximized. In addition, the salad should have less than 260 calories, over 210 milligrams of calcium, and over 6 grams of protein. The salad should also weigh less than 400 grams and be dressed with at least 2 tablespoons ($\frac{1}{8}$ cup) of salad dressing. The amounts of the nutrients and calories for these ingredients are given in Table 3.

Table 3 Nutritional Values per 100 g of Food

Nutrient (units)	Feta Cheese	Lettuce	Salad Dressing	Tomato
Calories (kcal)	263	14	448.8	21
Calcium (mg)	492.5	36	0	5
Protein (g)	10.33	1.62	0	.85
Carbohydrates (g)	4.09	2.37	2.5	4.64

Use linear programming to find the number of 100-gram units of each ingredient in such a Greek salad, and convert to kitchen units by using Table 1. (*Hint:* Since the ingredients are measured in 100-gram units, the constant in the weight contraint is 4, not 400.)

2. Consider preparing a stir-fry using beef, oil, onion, and soy sauce. A low-calorie stir-fry is desired, which contains less than 10 grams of carbohydrates, more than 50 grams of protein, and more than 3.5 grams of vitamin C. In order for the wok to function correctly, at least one teaspoon (or 4.5 grams) of oil must be used in the recipe. The amounts of the nutrients and calories for these ingredients are given in Table 4.

Table 4 Nutritional Values per 100 g of Food

Nutrient (units)	Beef	Oil	Onion	Soy Sauce
Calories (kcal)	215	884	38	60
Protein (g)	26	0	1.16	10.51
Carbohydrates (g)	0	1	8.63	5.57
Vitamin C (g)	0	0	6.4	0

Use linear programming to find the number of 100-gram units of each ingredient to be used in the stir-fry, and convert to kitchen units by using Table 1.

Extended Project

Use the USDA Nutrient Database for Standard Reference to determine the calorie content and the amount of protein, fat, and carbohydrates for a recipe of your choice. You can set your own constraints on the total amounts of calories, protein, fat, and car-

bohydrates desired. Determine the amount of each of the ingredients that make up the recipe that satisfies your constraints. Do you think the meal would taste good if you actually made the recipe?

Sets and Probability

8

CHAPTER

CHAPTER OUTLINE

We often use the relative frequency of an event from a study or survey to estimate unknown probabilities. For example, we can estimate the probability that a U.S. worker is highly satisfied with his or her job. Other applications of probability occur in business, marketing, health, and the social sciences. Examples include estimating the percentage of subscribers to different cable companies, whether a stock pays a dividend yield, the effectiveness of a medication, and marital happiness, just to name a few.

Federal officials cannot predict exactly how the number of traffic deaths is affected by the trend toward fewer drunken drivers and the increased use of seat belts. Economists cannot tell exactly how stricter federal regulations on bank loans affect the U.S. economy. The number of traffic deaths and the growth of the economy are subject to many factors that cannot be predicted precisely.

Probability theory enables us to deal with uncertainty. The basic concepts of probability are discussed in this chapter, and applications of probability are discussed in the next chapter. Sets and set operations are the basic tools for the study of probability, so we begin with them.

8.1 Sets

Think of a **set** as a well-defined collection of objects. A set of coins might include one of each type of coin now put out by the U.S. government. Another set might be made up of all the students in your English class. By contrast, a collection of young adults does not constitute a set unless the designation "young adult" is clearly defined. For example, this set might be defined as those aged 18–29.

In mathematics, sets are often made up of numbers. The set consisting of the numbers 3, 4, and 5 is written as

$$\{3, 4, 5\},$$

where **set braces**, { }, are used to enclose the numbers belonging to the set. The numbers 3, 4, and 5 are called the **elements,** or **members,** of this set. To show that 4 is an element of the set {3, 4, 5}, we use the symbol $\in$ and write

$$4 \in \{3, 4, 5\},$$

read, "4 is an element of the set containing 3, 4, and 5."

Also, $5 \in \{3, 4, 5\}$. Place a slash through the symbol $\in$ to show that 8 is *not* an element of this set:

$$8 \notin \{3, 4, 5\}.$$

This statement is read, "8 is not an element of the set {3, 4, 5}."

Sets are often named with capital letters, so that if

$$B = \{5, 6, 7\},$$

then, for example, $6 \in B$ and $10 \notin B$.

Sometimes a set has no elements. Some examples are the set of female presidents of the United States in the period 1789–2016, the set of natural numbers less than 1, and the set of men more than 10 feet tall. A set with no elements is called the **empty set.** The symbol $\varnothing$ is used to represent the empty set.

> **! CAUTION** Be careful to distinguish between the symbols 0, $\varnothing$, and {0}. The symbol 0 represents a *number;* $\varnothing$ represents a *set* with no elements; and {0} represents a *set* with one element, the number 0. Do not confuse the empty set symbol $\varnothing$ with the zero symbol 0 on a computer screen or printout.

Two sets are **equal** if they contain exactly the same elements. The sets {5, 6, 7}, {7, 6, 5}, and {6, 5, 7} all contain exactly the same elements and are equal. In symbols,

$$\{5, 6, 7\} = \{7, 6, 5\} = \{6, 5, 7\}.$$

This means that the ordering of the elements in a set is unimportant. Sets that do not contain exactly the same elements are *not equal.* For example, the sets {5, 6, 7} and {5, 6, 7, 8} do not contain exactly the same elements and are not equal. We show this by writing

$$\{5, 6, 7\} \neq \{5, 6, 7, 8\}.$$

Sometimes we describe a set by a common property of its elements rather than by a list of its elements. This common property can be expressed with **set-builder notation;** for example,

$$\{x | x \text{ has property } P\}$$

(read, "the set of all elements x such that x has some property P") represents the set of all elements x having some property P.

Checkpoint 1

Indicate whether each statement is *true* or *false*.

(a) $9 \in \{8, 4, -3, -9, 6\}$.

(b) $4 \notin \{3, 9, 7\}$.

(c) If $M = \{0, 1, 2, 3, 4\}$, then $0 \in M$.

Answers to Checkpoint exercises are found at the end of the section.

Checkpoint 2

List the elements in the given sets.

(a) $\{x | x$ is a natural number more than 5 and less than 8}

(b) $\{x | x$ is an integer and $-3 < x \leq 1\}$

EXAMPLE 1   List the elements belonging to each of the given sets.

(a) $\{x | x$ is a natural number less than 5}

Solution The natural numbers less than 5 make up the set {1, 2, 3, 4}.

(b) $\{x | x$ is a state that borders Florida}

Solution The states that border Florida make up the set {Alabama, Georgia}.

The **universal set** in a particular discussion is a set that contains all of the objects being discussed. In primary school arithmetic, for example, the set of whole numbers might be the universal set, whereas in a college calculus class the universal set might be the set of all real numbers. When it is necessary to consider the universal set being used, it will be clearly specified or easily understood from the context of the problem.

Sometimes, every element of one set also belongs to another set. For example, if

$$A = \{3, 4, 5, 6\}$$

and

$$B = \{2, 3, 4, 5, 6, 7, 8\},$$

then every element of A is also an element of B. This is an example of the following definition.

A set A is a **subset** of a set B (written $A \subseteq B$) provided that every element of A is also an element of B. A set A is said to be a **proper subset** of a set B (written $A \subset B$) if every element of A is an element of B, but B contains at least one element that is not a member of A. When A contains an element that is not in B, then A is not a subset of B, and we write $A \nsubseteq B$

Thus, for the sets A and B above, we note that there are elements of B not in A (namely $\{2, 7, 8\}$), so A is a proper subset of B, which is written $A \subset B$.

EXAMPLE 2   **Stock Performance** For each case, decide if $M \subseteq N$, $M \subset N$, or $M \nsubseteq N$.

(a) M is the set of all stocks that at least doubled in value over the course of a year. N is the set of all stocks that at least tripled in value over the course of a year.

Solution Some stocks that at least doubled in value did not achieve the threshold of also tripling in value, so there are elements in M that are not in N. Thus, M is a proper subset of N, written $M \subset N$.

(b) M is the set of all stocks that increased in value over the course of the year. N is the set of all stocks that increased in value by a penny or more as listed on a stock exchange over the course of the year.

Solution Because stock prices are listed to the nearest penny on stock exchanges, for a stock to increase in value it must increase by a penny or more. Thus, M is a subset of N. Because all the stocks that increased in value by a penny or more are stocks that increased in value, there are no stocks in N that are not also in M. For this case, we can write $M \subseteq N$.

(c) M is the set of all stocks priced $50.00 or higher on January 3, 2017. N is the set of stocks priced $50.00 or higher on January 2, 2018.

Solution Invariably, some stocks that were at $50.00 or higher will fall below $50.00. So there are elements in M that are not in N. Thus, M is not a subset of N, and we write $M \nsubseteq N$.

Every set A is a subset of itself, because the statement "every element of A is also an element of A" is always true. It is also true that the empty set is a subset of every set.[*]

---

[*]This fact is not intuitively obvious to most people. If you wish, you can think of it as a convention that we agree to adopt in order to simplify the statements of several results later.

For any set A,

$$\emptyset \subset A \quad \text{and} \quad A \subseteq A.$$

EXAMPLE 3   Decide whether $E \subset F$.

(a) $E = \{2, 4, 6, 8\}$ and $F = \{1, 2, 3, 4, 5, 6, 7, 8, 9, 10\}$.

Solution Since each element of E is an element of F and F contains several elements not in E, $E \subset F$.

(b) E is the set of registered voters in Texas. F is the set of adults aged 18 years or older.

Solution To register to vote, one must be at least 18 years old. Not all adults at least 18 years old, however, are registered. Thus, every element of E is contained in F and F contains elements not in E. Therefore, $E \subset F$.

(c) E is the set of diet soda drinks. F is the set of diet soda drinks sweetened with Nutrasweet®.

Solution Some diet soda drinks are sweetened with the sugar substitute Splenda®. In this case, E is not a subset of F ($E \not\subseteq F$). ✔3

Checkpoint 3

Indicate whether each statement is *true* or *false*.

(a) $\{10, 20, 30\} \subseteq \{20, 30, 40, 50\}$.

(b) $\{x | x \text{ is a minivan}\} \subseteq \{x | x \text{ is a motor vehicle}\}$.

(c) $\{a, e, i, o, u\} \subset \{a, e, i, o, u, y\}$.

(d) $\{x | x \text{ is a U.S. state that begins with the letter "A"}\} \subset$ {Alabama, Alaska, Arizona, Arkansas}.

EXAMPLE 4   List all possible subsets for each of the given sets.

(a) $\{7, 8\}$

Solution A good way to find the subsets of $\{7, 8\}$ is to use a **tree diagram**—a systematic way of listing all the subsets of a given set. The tree diagram in Figure 8.1(a) shows there are four subsets of $\{7, 8\}$:

$$\emptyset, \quad \{7\}, \quad \{8\}, \quad \text{and} \quad \{7, 8\}.$$

(b) $\{a, b, c\}$

Solution The tree diagram in Figure 8.1(b) shows that there are 8 subsets of $\{a, b, c\}$:

$$\emptyset, \quad \{a\}, \quad \{b\}, \quad \{c\}, \quad \{a, b\}, \quad \{a, c\}, \quad \{b, c\}, \quad \text{and} \quad \{a, b, c\}. ✔4$$

Checkpoint 4

List all subsets of $\{w, x, y, z\}$.

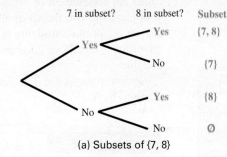

(a) Subsets of {7, 8}

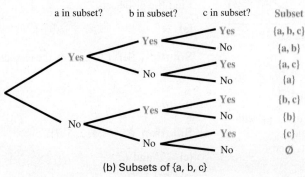

(b) Subsets of {a, b, c}

Figure 8.1

By using the fact that there are two possibilities for each element (either it is in the subset or it is not), we have found that a set with 2 elements has 4 ($=2^2$) subsets and a set with 3 elements has 8 ($=2^3$) subsets. Similar arguments work for any finite set and lead to the following conclusion.

✓ **Checkpoint 5**

Find the number of subsets for each of the given sets.

(a) $\{x \mid x$ is a season of the year$\}$

(b) $\{-6, -5, -4, -3, -2, -1, 0\}$

(c) $\{6\}$

> A set of n distinct elements has 2^n subsets.

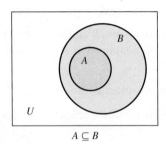

$A \subseteq B$

Figure 8.2

EXAMPLE 5 Find the number of subsets for each of the given sets.

(a) {blue, brown, hazel, green}

Solution Since this set has 4 elements, it has $2^4 = 16$ subsets.

(b) $\{x \mid x$ is a month of the year$\}$

Solution This set has 12 elements and therefore has $2^{12} = 4096$ subsets.

(c) $\varnothing$

Solution Since the empty set has 0 elements, it has $2^0 = 1$ subset, $\varnothing$ itself. ✓ 5

✓ **Checkpoint 6**

Refer to sets A, B, C, and U in the diagram.

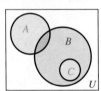

(a) Is $A \subseteq B$?

(b) Is $C \subset B$?

(c) Is $C \subset U$?

(d) Is $\varnothing \subset A$?

Venn diagrams are sometimes used to illustrate relationships among sets. The Venn diagram in Figure 8.2 shows a set A that is a proper subset of a set B, because A is entirely in B. (The areas of the regions are not meant to be proportional to the sizes of the corresponding sets.) The rectangle represents the universal set U.

Some sets have infinitely many elements. We often use the notation "..." to indicate such sets. One example of an infinite set is the set of natural numbers, $\{1, 2, 3, 4, \ldots\}$. Another infinite set is the set of integers, $\{\ldots, -3, -2, -1, 0, 1, 2, 3, \ldots\}$.

Operations on Sets

Given a set A and a universal set U, the set of all elements of U that do *not* belong to A is called the **complement** of set A. For example, if A is the set of all the female students in your class and U is the set of all students in the class, then the complement of A would be the set of all male students in the class. The complement of set A is written A' (read "A-prime"). The Venn diagram in Figure 8.3 shows a set B. Its complement, B', is shown in color.

Some textbooks use $\overline{A}$ to denote the complement of A. This notation conveys the same meaning as A'.

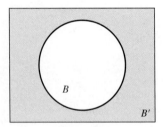

B'

Figure 8.3

✓ **Checkpoint 7**

Let $U = \{a, b, c, d, e, f, g\}$, with $K = \{c, d, f, g\}$ and $R = \{a, c, d, e, g\}$. Find

(a) K';

(b) R'.

EXAMPLE 6 Let $U = \{1, 2, 3, 4, 5, 6, 7\}$, $A = \{1, 3, 5, 7\}$, and $B = \{3, 4, 6\}$. Find the given sets.

(a) A'

Solution Set A' contains the elements of U that are not in A:

$$A' = \{2, 4, 6\}.$$

(b) B'

Solution $B' = \{1, 2, 5, 7\}$.

(c) $\varnothing'$ and U'

Solution $\varnothing' = U$ and $U' = \varnothing$. ✓ 7

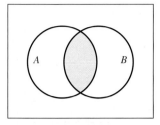

$A \cap B$

Figure 8.4

Given two sets A and B, the set of all elements belonging to *both* set A and set B is called the **intersection** of the two sets, written $A \cap B$. For example, the elements that belong to both $A = \{1, 2, 4, 5, 7\}$ and $B = \{2, 4, 5, 7, 9, 11\}$ are 2, 4, 5, and 7, so

$$A \text{ and } B = A \cap B$$
$$= \{1, 2, 4, 5, 7\} \cap \{2, 4, 5, 7, 9, 11\}$$
$$= \{2, 4, 5, 7\}.$$

The Venn diagram in Figure 8.4 shows two sets A and B, with their intersection, $A \cap B$, shown in color.

EXAMPLE 7   Find the given sets.

(a) $\{9, 15, 25, 36\} \cap \{15, 20, 25, 30, 35\}$

Solution $\{15, 25\}$. The elements 15 and 25 are the only ones belonging to both sets.

(b) $\{x \mid x \text{ is a teenager}\} \cap \{x \mid x \text{ is a senior citizen}\}$

Solution $\varnothing$ since no teenager is a senior citizen. ✓ 8

✓ **Checkpoint 8**

Find $\{1, 2, 3, 4\} \cap \{3, 5, 7, 9\}$.

Two sets that have no elements in common are called **disjoint sets.** For example, there are no elements common to both $\{50, 51, 54\}$ and $\{52, 53, 55, 56\}$, so these two sets are disjoint, and

$$\{50, 51, 54\} \cap \{52, 53, 55, 56\} = \varnothing.$$

The result of this example can be generalized as follows.

For any sets A and B,

if A and B are disjoint sets, then $A \cap B = \varnothing$.

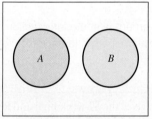

A and B are disjoint sets.

Figure 8.5

Figure 8.5 is a Venn diagram of disjoint sets.

The set of all elements belonging to set A or to set B, or to both sets, is called the **union** of the two sets, written $A \cup B$. For example, for sets $A = \{1, 3, 5\}$ and $B = \{3, 5, 7, 9\}$,

$$A \text{ or } B = A \cup B$$
$$= \{1, 3, 5\} \cup \{3, 5, 7, 9\}$$
$$= \{1, 3, 5, 7, 9\}.$$

The Venn diagram in Figure 8.6 shows two sets A and B, with their union, $A \cup B$, shown in color.

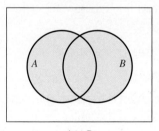

$A \cup B$

Figure 8.6

EXAMPLE 8   Find the given sets.

(a) $\{1, 2, 5, 9, 14\} \cup \{1, 3, 4, 8\}$.

Solution Begin by listing the elements of the first set, $\{1, 2, 5, 9, 14\}$. Then include any elements from the second set *that are not already listed.* Doing this gives

$$\{1, 2, 5, 9, 14\} \cup \{1, 3, 4, 8\} = \{1, 2, 3, 4, 5, 8, 9, 14\}.$$

(b) {terriers, spaniels, chows, dalmatians} $\cup$ {spaniels, collies, bulldogs}

Solution {terriers, spaniels, chows, dalmatians, collies, bulldogs}. ✓ 9

✓ **Checkpoint 9**

Find $\{a, b, c\} \cup \{a, c, e\}$.

Finding the complement of a set, the intersection of two sets, or the union of two sets is an example of a *set operation*.

Operations on Sets

Let A and B be any sets, with U signifying the universal set. Then

the **complement** of A, written A', is

$$A' = \{x \mid x \notin A \text{ and } x \in U\};$$

the **intersection** of A and B is

$$A \cap B = \{x \mid x \in A \textbf{ and } x \in B\};$$

the **union** of A and B is

$$A \cup B = \{x \mid x \in A \textbf{ or } x \in B \textbf{ or both}\}.$$

⊘ CAUTION As shown in the preceding definitions, an element is in the intersection of sets A and B if it is in *both* A and B at the same time, but an element is in the union of sets A and B if it is in *either* set A or set B, *or* in both sets A and B.

EXAMPLE 9 **Stock Prices** The following table gives the 52-week high and low prices, the last price, and the change in price from the day before for six stocks on a recent day.

Stock	High	Low	Last	Change
First Solar	51.33	25.56	36.38	−.24
Ford	14.04	10.90	10.92	−.09
IBM	182.79	142.50	150.37	−.28
Netflix	161.10	84.50	160.81	+2.27
Procter & Gamble	92.00	79.41	86.19	+.02
Verizon	56.95	45.76	45.84	−.18

Let the universal set U consist of the six stocks listed in the table. Let A contain all stocks with a high price greater than \$50, B all stocks with a last price between \$25 and \$80, and C all stocks with a negative value for the change. Find the results of the given set operations. (Data from: www.morningstar.com.)

(a) B'

Solution Set B consists of First Solar and Verizon. Set B' contains all the listed stocks that are not in set B, so

$$B' = \{\text{Ford, IBM, Netflix, Procter \& Gamble}\}.$$

(b) $A \cap C$

Solution Set A consists of First Solar, IBM, Netflix, Procter & Gamble, and Verizon, and set C consists of First Solar, Ford, IBM, and Verizon. Hence the stocks in both A and C are

$$A \cap C = \{\text{First Solar, IBM, Verizon}\}.$$

(c) $A \cup B$

Solution $A \cup B = \{\text{First Solar, IBM, Netflix, Procter & Gamble, and Verizon}\}$ ✓10

✓**Checkpoint 10**

In Example 9, find the given set of stocks.

(a) $B \cap C$

(b) $B \cup C$

8.1 Exercises

Write true or false for each statement. (See Example 1.)

1. $3 \in \{2, 5, 7, 9, 10\}$

2. $6 \in \{-2, 6, 9, 5\}$

3. $9 \notin \{2, 1, 5, 8\}$

4. $3 \notin \{7, 6, 5, 4\}$

5. $\{2, 5, 8, 9\} = \{2, 5, 9, 8\}$

6. $\{3, 7, 12, 14\} = \{3, 7, 12, 14, 0\}$

7. $\{\text{all whole numbers greater than 7 and less than 10}\} = \{8, 9\}$

8. $\{\text{all natural numbers not greater than 3}\} = \{0, 1, 2\}$

9. $\{x | x \text{ is an odd integer, } 6 \leq x \leq 18\} = \{7, 9, 11, 15, 17\}$

10. $\{x | x \text{ is a vowel}\} = \{a, e, i, o, u\}$

11. The elements of a set may be sets themselves, as in $\{1, \{1, 3\}, \{2\}, 4\}$. Explain why the set $\{\varnothing\}$ is not the same set as $\{0\}$.

12. What is set-builder notation? Give an example.

Let $A = \{-3, 0, 3\}$, $B = \{-2, -1, 0, 1, 2\}$, $C = \{-3, -1\}$, $D = \{0\}$, $E = \{-2\}$, and $U = \{-3, -2, -1, 0, 1, 2, 3\}$. Insert $\subseteq$ or $\not\subseteq$ to make the given statements true. (See Example 2.)

13. A____U

14. E____A

15. A____E

16. B____C

17. $\varnothing$____A

18. $\{0, 2\}$____D

19. D____B

20. A____C

Find the number of subsets of the given set. (See Example 5.)

21. $\{A, B, C\}$

22. $\{\text{red, yellow, blue, black, white}\}$

23. $\{x | x \text{ is an integer strictly between 0 and 8}\}$

24. $\{x | x \text{ is a whole number less than 4}\}$

Find the complement of each set. (See Example 6.)

25. The set in Exercise 23 if U is the set of all integers.

26. The set in Exercise 24 if U is the set of all whole numbers.

27. Describe the intersection and union of sets. How do they differ?

Insert $\cap$ or $\cup$ to make each statement true. (See Examples 7 and 8.)

28. $\{5, 7, 9, 19\}$____$\{7, 9, 11, 15\} = \{7, 9\}$

29. $\{8, 11, 15\}$____$\{8, 11, 19, 20\} = \{8, 11\}$

30. $\{2, 1, 7\}$____$\{1, 5, 9\} = \{1\}$

31. $\{6, 12, 14, 16\}$____$\{6, 14, 19\} = \{6, 14\}$

32. $\{3, 5, 9, 10\}$____$\varnothing = \varnothing$

33. $\{3, 5, 9, 10\}$____$\varnothing = \{3, 5, 9, 10\}$

34. $\{1, 2, 4\}$____$\{1, 2, 4\} = \{1, 2, 4\}$

35. $\{1, 2, 4\}$____$\{1, 2\} = \{1, 2, 4\}$

36. Is it possible for two nonempty sets to have the same intersection and union? If so, give an example.

Let $U = \{a, b, c, d, e, f, 1, 2, 3, 4, 5, 6\}$, $X = \{a, b, c, 1, 2, 3\}$, $Y = \{b, d, f, 1, 3, 5\}$, and $Z = \{b, d, 2, 3, 5\}$.

List the members of each of the given sets, using set braces. (See Examples 6–8.)

37. $X \cap Y$

38. $X \cup Y$

39. X'

40. Y'

41. $X' \cap Y'$

42. $X' \cap Z$

43. $X \cup (Y \cap Z)$

44. $Y \cap (X \cup Z)$

Let $U = \{\text{all students in this school}\}$, $M = \{\text{all students taking this course}\}$, $N = \{\text{all students taking accounting}\}$, and $P = \{\text{all students taking philosophy}\}$.

Describe each of the following sets in words.

45. M'

46. $M \cup N$

47. $N \cap P$

48. $N' \cap P'$

49. Refer to the sets listed in the directions for Exercises 13–20. Which pairs of sets are disjoint?

50. Refer to the sets listed in the directions for Exercises 37–44. Which pairs of sets are disjoint?

Refer to Example 9 in the text. Describe each of the sets in Exercises 51–54 in words; then list the elements of each set.

51. A'

52. $B \cup C'$

53. $A' \cap B'$

54. $B' \cup C$

Business *A stock broker classifies her clients by sex, marital status, and employment status. Let the universal set be the set of all clients, M be the set of male clients, S be the set of single clients, and E be the set of employed clients. Describe the following sets in words.*

55. $M \cap E$

56. $M' \cap S$

57. $M' \cup S'$

58. $E' \cup S'$

Food and Drink Sales *Sales (in millions of dollars) in the United States at eating and drinking establishments in the years 2014 and 2015 are shown in the following table. (Data from: National Restaurant Association.)*

Type of Place or Service	2014	2015
Table-service restaurants (T)	213,498	219,690
Quick-service and fast-casual restaurants (Q)	192,836	201,128
Cafeterias, grill-buffets, and buffets (C)	8,429	8,336
Snack and nonalcoholic beverage bars (S)	31,221	32,845
Bars and taverns (B)	19,943	20,562
Other (O)	8,637	9,076

List the elements of each set.

59. The set of eating and drinking establishments that generated more than $200,000 million in revenue in both 2014 and 2015.

60. The set of eating and drinking establishments that generated less than $20,000 million in revenue in 2014 or 2015.

61. The set of eating and drinking establishments that had revenues fall from 2014 to 2015.

62. The set of eating and drinking establishments that had revenues rise from 2014 to 2015 and had less than $35,000 million in revenue.

Cable Subscribers *The top six basic cable television providers in the year 2016 are listed in the following table. Use this information for Exercises 63–72.*

Rank	Cable Provider	Subscribers (millions)
1	Direct TV	25.3
2	Comcast	22.5
3	Charter	17.5
4	Dish	13.9
5	Verizon FIOS	4.7
6	Altice	4.6

List the elements of the following sets. (Data from: www.ncta.com.)

63. F, the set of cable providers with more than 5 million subscribers.

64. G, the set of cable providers with between 10 million and 20 million subscribers.

65. H, the set of cable providers with less than 20 million subscribers.

66. I, the set of cable providers with more than 15 million subscribers.

67. $F \cup G$ **68.** $H \cap F$ **69.** I' **70.** $I' \cap H$

71. J, the set of cable providers with more than 30 million subscribers

72. $(F \cup H)'$

Farm Products *The following table gives the amount (in millions of metric tons) of several farm products exported from the United States in 2011 and 2016. (Data from: U.S. Department of Agriculture.)*

Product	2011	2016
Wheat	28.6	28.2
Corn	39.4	56.5
Rice	3.2	3.6
Barley	.2	.1

List the elements of the following sets.

73. The set of farm products where exports increased from 2011 to 2016

74. The set of farm products that had more than 40 million metric tons of exports in both 2011 and 2016

75. The set of farm products that had more than 50 million metric tons of exports in both 2011 and 2016

76. The set of farm products that had less than 40 million metric tons of exports in either 2011 or 2016

77. The set of farm products that had more than 3.5 million metric tons of exports in either 2011 or 2016

78. The set of farm products that decreased in exports from 2011 to 2016 and had more than 30 million metric tons of exports in 2016

✓ Checkpoint Answers

1. (a) False   **(b)** True   **(c)** True

2. (a) $\{6, 7\}$ **(b)** $\{-2, -1, 0, 1\}$

3. (a) False   **(b)** True   **(c)** True   **(d)** False

4. $\varnothing$, $\{w\}$, $\{x\}$, $\{y\}$, $\{z\}$, $\{w, x\}$, $\{w, y\}$, $\{w, z\}$, $\{x, y\}$, $\{x, z\}$, $\{y, z\}$, $\{w, x, y\}$, $\{w, x, z\}$, $\{w, y, z\}$, $\{x, y, z\}$, $\{w, x, y, z\}$

5. (a) 16   **(b)** 128   **(c)** 2

6. (a) No   **(b)** Yes   **(c)** Yes   **(d)** Yes

7. (a) $\{a, b, e\}$ **(b)** $\{b, f\}$

8. $\{3\}$

9. $\{a, b, c, e\}$

10. (a) {First Solar, Verizon}

(b) {First Solar, Ford, IBM, Verizon}

8.2 Applications of Venn Diagrams and Contingency Tables

We used Venn diagrams in the previous section to illustrate set union and intersection. The rectangular region in a Venn diagram represents the universal set U. Including only a single set A inside the universal set, as in Figure 8.7, divides U into two nonoverlapping regions. Region 1 represents A', those elements outside set A, while region 2 represents those elements belonging to set A. (The numbering of these regions is arbitrary.)

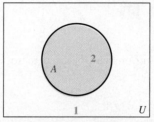

One set leads to 2 regions.
(Numbering is arbitrary.)

Figure 8.7

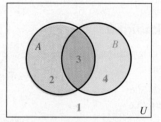

Two sets lead to 4 regions.
(Numbering is arbitrary.)

Figure 8.8

The Venn diagram of Figure 8.8 shows two sets inside U. These two sets divide the universal set into four nonoverlapping regions. As labeled in Figure 8.8, region 1 includes those elements outside both set A and set B. Region 2 includes those elements belonging to A and not to B. Region 3 includes those elements belonging to both A and B. Which elements belong to region 4? (Again, the numbering is arbitrary.)

EXAMPLE 1   Draw a Venn diagram similar to Figure 8.8, and shade the regions representing the given sets.

(a) $A' \cap B$

Solution Set A' contains all the elements outside set A. As labeled in Figure 8.8, A' is represented by regions 1 and 4. Set B is represented by the elements in regions 3 and 4. The intersection of sets A' and B, the set $A' \cap B$, is given by the region common to regions 1 and 4 and regions 3 and 4. The result, region 4, is shaded in Figure 8.9.

(b) $A' \cup B'$

Solution Again, set A' is represented by regions 1 and 4 and set B' by regions 1 and 2. To find $A' \cup B'$, identify the region that represents the set of all elements in A', B', or both. The result, which is shaded in Figure 8.10, includes regions 1, 2, and 4. ✓ **1**

✓ **Checkpoint 1**

Draw Venn diagrams for the given set operations.

(a) $A \cup B'$

(b) $A' \cap B'$

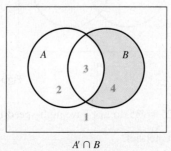

$A' \cap B$

Figure 8.9

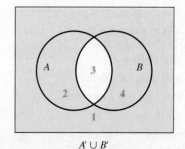

$A' \cup B'$

Figure 8.10

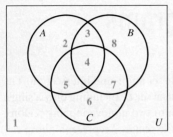

Three sets lead to 8 regions.

Figure 8.11

✓**Checkpoint 2**

Draw Venn diagrams for the given set operations.

(a) $(B' \cap A) \cup C$

(b) $(A \cup B)' \cap C$

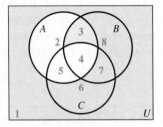

$A' \cup (B \cap C')$

Figure 8.12

Venn diagrams also can be drawn with three sets inside U. These three sets divide the universal set into eight nonoverlapping regions that can be numbered (arbitrarily) as in Figure 8.11.

EXAMPLE 2   Shade $A' \cup (B \cap C')$ in a Venn diagram.

Solution   First find $B \cap C'$. Set B is represented by regions 3, 4, 7, and 8, and set C' by regions 1, 2, 3, and 8. The overlap of these regions, regions 3 and 8, represents the set $B \cap C'$. Set A' is represented by regions 1, 6, 7, and 8. The union of regions 3 and 8 and regions 1, 6, 7, and 8 contains regions 1, 3, 6, 7, and 8, which are shaded in Figure 8.12. ✓₂

Venn diagrams can be used to solve problems that result from surveying groups of people.

EXAMPLE 3   **Consumer Profiling** A market researcher collecting data on 100 household finds that

81 have cable television (CT);

65 have high-speed Internet (HSI);

56 have both.

The researcher wants to answer the following questions:

(a) How many households do not have high-speed Internet?

(b) How many households have neither cable television nor high-speed Internet?

(c) How many have cable television, but not high-speed Internet?

Solution   A Venn diagram like the one in Figure 8.13 will help sort out the information. In Figure 8.13(a), we place the number 56 in the region common to both cable television and high-speed Internet, because 56 households have both. Of the 81 with cable television, $81 - 56 = 25$ do not have high-speed Internet, so in Figure 8.13(b) we place 25 in the region for cable television, but not high-speed Internet. Similarly, $65 - 56 = 9$ households have high-speed Internet, but not cable television, so we place 9 in that region. Finally, the diagram shows that $100 - 9 - 56 - 25 = 10$ have neither high-speed Internet nor cable television. Now we can answer the questions.

✓**Checkpoint 3**

(a) Place numbers in the regions on a Venn diagram if the data on the 100 households in Example 3 showed

78 with cable television;

52 with high-speed Internet;

48 with both.

(b) How many have high-speed Internet, but not cable television?

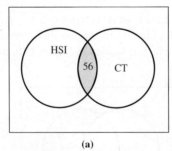

(a)

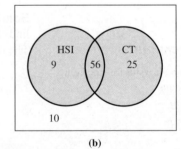

(b)

Figure 8.13

(a) $10 + 25 = 35$ do not have high-speed Internet.

(b) 10 have neither.

(c) 25 have cable television, but not high-speed Internet.

EXAMPLE 4 **Media** A researcher surveyed 67 business majors at a university to determine which business-focused media websites the students regularly visited. The results are summarized below.

21 of the students visited *Business Week*;

20 visited the *Wall Street Journal*;

46 visited *Fortune*;

12 visited *Business Week* and the *Journal*;

11 visited the *Journal* and *Fortune*;

13 visited *Business Week* and *Fortune*;

9 visited all three websites.

Use the preceding data to answer the following questions:

(a) How many students visited none of the websites?

(b) How many visited only *Fortune*?

(c) How many visited *Business Week* and the *Journal*, but not *Fortune*?

Solution Once again, use a Venn diagram to represent the data. Since 9 students visited all three websites, begin by placing 9 in the area in Figure 8.14(a) that belongs to all three regions.

Of the 13 students who visited *Business Week* and *Fortune*, 9 also visited the *Journal*. Therefore, only $13 - 9 = 4$ students visited just *Business Week* and *Fortune*. So place a 4 in the region common only to *Business Week* and *Fortune* readers, as in Figure 8.14(b).

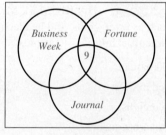

Figure 8.14(a)

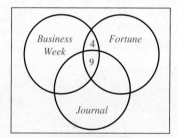

Figure 8.14(b)

In the same way, place a 3 in the region of Figure 8.14(c) common only to *Business Week* and the *Journal* visitors, and 2 in the region common only to *Fortune* and the *Journal* visitors.

The data shows that 21 students visited *Business Week*. However, $4 + 9 + 3 = 16$ visitors have already been placed in the *Business Week* region. The balance of this region in Figure 8.14(d) will contain only $21 - 16 = 5$ students. These 5 students visited *Business Week* only—not *Fortune* and not the *Journal*.

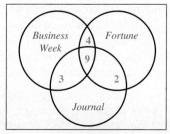

Figure 8.14(c)

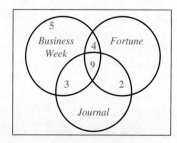

Figure 8.14(d)

In the same way, 6 students visited only the *Journal* and 31 visited only *Fortune*, as shown in Figure 8.14(e).

A total of $5 + 4 + 31 + 3 + 9 + 2 + 6 = 60$ students are placed in the various regions of Figure 8.14(e). Since 67 students were surveyed, $67 - 60 = 7$ students visited none of the three websites, and so 7 is placed outside the other regions in Figure 8.14(f).

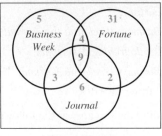

Figure 8.14(e)

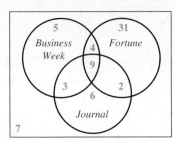

Figure 8.14(f)

Figure 8.14(f) can now be used to answer the questions asked at the beginning of this example:

(a) There are 7 students who did not visit any of the 3 websites.

(b) There are 31 students who visited only *Fortune*.

(c) The overlap of the regions representing *Business Week* and the *Journal* shows that 3 students regularly visited the sites for *Business Week* and the *Journal*, but not *Fortune*. ✓₄

✓ Checkpoint 4

In Example 4, how many students visited exactly,

(a) 1 of the websites?

(b) 2 of the websites?

EXAMPLE 5 **Health** Mark McCloney, M.D., saw 100 patients exhibiting flu symptoms such as fever, chills, and headache. Dr. McCloney reported the following information on patients exhibiting symptoms:

Of the 100 patients,

74 reported a fever;

72 reported chills;

67 reported a headache;

55 reported both a fever and chills;

47 reported both a fever and a headache;

49 reported both chills and a headache;

35 reported all three;

3 thought they had the flu, but did not report fever, chills, or headache.

Create a Venn diagram to represent this data. It should show the number of people in each region.

Solution Begin with the 35 patients who reported all three symptoms. This leaves $55 - 35 = 20$ who reported fever and chills, but not headache; $47 - 35 = 12$ who reported fever and headache, but not chills; and $49 - 35 = 14$ who reported chills and headache, but not fever. With this information, we have $74 - (35 + 20 + 12) = 7$ who reported fever alone; $72 - (35 + 20 + 14) = 3$ with chills alone; and $67 - (35 + 12 + 14) = 6$ with headache alone. The remaining 3 patients who thought they had the flu, but did not report fever, chills, or headache are denoted outside the 3 circles. See Figure 8.15. ✓₅

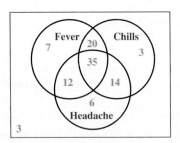

Figure 8.15

✓ Checkpoint 5

In Example 5, suppose 75 patients reported a fever and only 2 thought they had the flu, but did not report fever, chills, or headache. Then how many

(a) reported only a fever?

(b) reported a fever or chills?

(c) reported a fever, chills, or headache?

📝 **NOTE** In all the preceding examples, we started in the innermost region with the intersection of the categories. This is usually the best way to begin solving problems of this type.

We use the symbol $n(A)$ to denote the *number* of elements in A. For instance, if $A = \{w, x, y, z\}$, then $n(A) = 4$. Next, we prove the following useful fact.

> ## Addition Rule for Counting
> $$n(A \cup B) = n(A) + n(B) - n(A \cap B).$$

For example, if $A = \{r, s, t, u, v\}$ and $B = \{r, t, w\}$, then $A \cap B = \{r, t\}$, so that $n(A) = 5$, $n(B) = 3$, and $n(A \cap B) = 2$. By the formula in the box, $n(A \cup B) = 5 + 3 - 2 = 6$, which is certainly true, since $A \cup B = \{r, s, t, u, v, w\}$.

Here is a proof of the statement in the box: Let x be the number of elements in A that are not in B, y be the number of elements in $A \cap B$, and z be the number of elements in B that are not in A, as indicated in Figure 8.16. That diagram shows that $n(A \cup B) = x + y + z$. It also shows that $n(A) = x + y$ and $n(B) = y + z$, so that

$$n(A) + n(B) - n(A \cap B) = (x + y) + (z + y) - y$$
$$= x + y + z$$
$$= n(A \cup B).$$

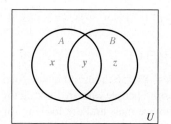

Figure 8.16

EXAMPLE 6   **College Majors** A group of 10 students meets to plan a school function. All are majoring in accounting or economics or both. Five of the students are economics majors, and 7 are majors in accounting. How many major in both subjects?

Solution Let A represent the set of economics majors and B represent the set of accounting majors. Use the union rule, with $n(A) = 5$, $n(B) = 7$, and $n(A \cup B) = 10$. We must find $n(A \cap B)$:

$$n(A \cup B) = n(A) + n(B) - n(A \cap B)$$
$$10 = 5 + 7 - n(A \cap B).$$

So,

$$n(A \cap B) = 5 + 7 - 10 = 2. \checkmark_6$$

✓ **Checkpoint 6**

If $n(A) = 10$, $n(B) = 7$, and $n(A \cap B) = 3$, find $n(A \cup B)$.

In addition to Venn diagrams, we can use a *contingency table*, which is sometimes called a *cross tabulation*, to summarize counts from several different groups. A contingency table is a table in a matrix format that gives the frequency distribution of several variables. With such tables, we can compute the number of elements in sets of interest, the intersection of two sets, and the union of two sets.

EXAMPLE 7   **Employment Status** The following contingency table gives the number (in thousands) of males (denoted with M) and females (denoted with F) who were working full-time (denoted A), working part-time (denoted B), or looking for work (denoted C) in the year 2016. (Data from: U.S. Census Bureau.)

	A Worked Full-Time	*B* Worked Part-Time	*C* Looked for Work
M Males	70,567	10,002	4187
F Females	53,194	17,674	3565

Find the number of people in the given sets.

(a) *A*

Solution The set *A* consists of males and females who worked full-time. From the table, we see that there were 70,567 thousand males and 53,194 thousand females, which yields 123,761 thousand people who worked full-time in the year 2016.

(b) $F \cap A$

Solution The set $F \cap A$ consists of all the people who are female *and* who worked full-time. We see that that there were 53,194 thousand females who worked full-time.

(c) $M \cup C$

Solution The set $M \cup C$ consists of people who are male or people who were looking for work. Using the addition rule for counting, we have

$$n(M \cup C) = n(M) + n(C) - n(M \cap C).$$

From the table, we have $n(M) = 70{,}567 + 10{,}002 + 4187 = 84{,}756$ thousand males. Also from the table we have $n(C) = 4187 + 3565 = 7752$ thousand people who are looking for work. We also have $n(M \cap C)$, the number of people who are male and are looking for work, is 4187 thousand. Thus,

$$
\begin{aligned}
n(M \cup C) &= n(M) + n(C) - n(M \cap C) \\
&= 84{,}756 + 7752 - 4187 \\
&= 88{,}321 \text{ thousand people}.
\end{aligned}
$$

✓Checkpoint 7

Refer to Example 7 and find the number of people in the following sets.

(a) $A \cup B$

(b) $F \cup B$

(d) $(B \cup C) \cap F$

Solution Begin with the set $B \cup C$, which contains all those who worked part-time or were looking for work. Of this set, take only those who were females, for a total of 17,674 + 3565 = 21,239 thousand. ✓₇

8.2 Exercises

Sketch a Venn diagram like the one shown, and use shading to show each of the given sets. (See Example 1.)

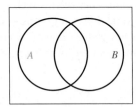

1. $A \cap B'$

2. $A \cup B'$

3. $B' \cup A'$

4. $A' \cap B'$

5. $B' \cup (A \cap B')$

6. $(A \cap B) \cup A'$

7. U'

8. $\varnothing'$

9. Three sets divide the universal set into at most _____ regions.

10. What does the notation $n(A)$ represent?

Sketch a Venn diagram like the one shown, and use shading to show each of the given sets. (See Example 2.)

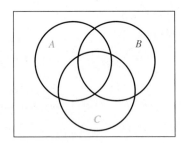

11. $(A \cap C') \cup B$

12. $A \cap (B \cup C')$

13. $A' \cap (B \cap C)$

14. $(A' \cap B') \cap C$

15. $(A \cap B') \cup C$

16. $(A \cap B') \cap C$

Use Venn diagrams to answer the given questions. (See Examples 3, 4, and 5.)

17. **Movies** In 2016, among the top 100 grossing movies in the United States, 15 were rated PG-13 and earned over $100 million. The number of movies that were rated PG-13 that earned less than $100 million was 38. The number of movies that were

not rated PG-13 that earned less than $100 million was 34. What was the number of movies that earned over $100 million and that were not rated PG-13? (Data from: www.the-numbers.com.)

18. Movies In 2016, among the top 100 grossing movies in the United States, 20 were classified as adventure and earned more than $75 million. The number of movies that earned over $75 million was 40. The number of movies that were not adventure movies and earned less than $75 million was 55. What was the number of movies that were adventure movies that earned less than $75 million? (Data from: www.the-numbers.com.)

19. Job Applicants The human resources director for a commercial real estate company received the following numbers of applications from people with the given information:

66 with sales experience;

40 with a college degree;

23 with a real estate license;

26 with sales experience and a college degree;

16 with sales experience and a real estate license;

15 with a college degree and a real estate license;

11 with sales experience, a college degree, and a real estate license;

22 with neither sales experience, a college degree, nor a real estate license.

(a) How many applicants were there?

(b) How many applicants did not have sales experience?

(c) How many had sales experience and a college degree, but not a real estate license?

(d) How many had only a real estate license?

20. Product Sales A pet store keeps track of the purchases of customers over a four-hour period. The store manager classifies purchases as containing a dog product, a cat product, a fish product, or a product for a different kind of pet. She found that:

83 customers purchased a dog product;

101 customers purchased a cat product;

22 customers purchased a fish product;

31 customers purchased a dog and a cat product;

8 customers purchased a dog and a fish product;

10 customers purchased a cat and a fish product;

6 customers purchased a dog, a cat, and a fish product;

34 customers purchased a product for a pet other than a dog, cat, or fish.

(a) How many purchases were for a dog product only?

(b) How many purchases were for a cat product only?

(c) How many purchases were for a dog or fish product?

(d) How many purchases were there in total?

21. Recreational Fishing A marine biologist surveys people who fish on Lake Erie and caught at least one fish to determine whether they had caught a walleye, a smallmouth bass, or a yellow perch in the last year. He finds:

124 caught at least one walleye;

133 caught at least one smallmouth bass;

146 caught at least one yellow perch;

75 caught at least one walleye and at least one smallmouth bass;

67 caught at least one walleye and at least one yellow perch;

79 caught at least one smallmouth bass and at least one yellow perch;

45 caught all three.

(a) Find the total number of people surveyed.

(b) How many caught at least one walleye or at least one smallmouth bass?

(c) How many caught only walleye?

22. Blood Type Human blood can contain either no antigens, the A antigen, the B antigen, or both the A and B antigens. A third antigen, called the Rh antigen, is important in human reproduction and, like the A and B antigens, may or may not be present in an individual. Blood is called type A positive if the individual has the A and Rh antigens, but not the B antigen. A person having only the A and B antigens is said to have type AB-negative blood. A person having only the Rh antigen has type O-positive blood. Other blood types are defined in a similar manner. Identify the blood type of the individuals in regions (a)–(g) of the Venn diagram.

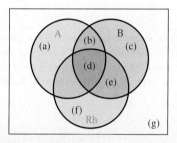

23. Blood Type Use the diagram from Exercise 22. In a certain hospital, the following data was recorded:

25 patients had the A antigen;

17 had the A and B antigens;

27 had the B antigen;

22 had the B and Rh antigens;

30 had the Rh antigen;

12 had none of the antigens;

16 had the A and Rh antigens;

15 had all three antigens.

How many patients

(a) were represented?

(b) had exactly one antigen?

(c) had exactly two antigens?

(d) had O-positive blood?

(e) had AB-positive blood?

(f) had B-negative blood?

(g) had O-negative blood?

(h) had A-positive blood?

24. **Portfolio Management** In reviewing the portfolios of 365 of its clients, a mutual funds company categorized whether the clients were invested in international stock funds, domestic stock funds, or bond funds. It found that,

125 were invested in domestic stocks, international stocks, and bond funds;

145 were invested in domestic stocks and bond funds;

300 were invested in domestic stocks;

200 were invested in international and domestic stocks;

18 were invested in international stocks and bond funds, but not domestic stocks;

35 were invested in bonds, but not in international or domestic stocks;

87 were invested in international stocks, but not in bond funds.

(a) How many were invested in international stocks?

(b) How many were invested in bonds, but not international stocks?

(c) How many were not invested in bonds?

(d) How many were invested in international or domestic stocks?

For Exercises 25–30, use the given contingency tables. (See Example 7.)

Partnerships *In Exercises 25 and 26, the given contingency table lists the cross tabulation of the number of business partnerships by industry and whether the partnership reported net income or net loss for the year 2014. (Data from: U.S. Internal Revenue Service.)*

Industry	Number of Partnerships	
	With Net Income (A)	With Net Loss (B)
Construction (C)	75,071	67,561
Manufacturing (M)	24,389	42,386
Wholesale Trade (W)	37,023	45,369
Retail Trade (R)	86,749	81,878

Using the letter given in the table, find the number of partnerships in each set.

25. (a) $A \cup C$ (b) $M \cup W$
    (c) $R \cap B$ (d) C'

26. (a) $B \cup R$ (b) $W \cup R$
    (c) $M \cap A$ (d) W'

Life Classification *In Exercises 27 and 28, the given contingency table lists the cross tabulation of whether adults (for whom answering the question was appropriate) feel their life is exciting, routine, or dull for males and females among the 1873 appropriate adults in the 2016 General Social Survey. (Data from: gss.norc.org.)*

Life Classification	Males (M)	Females (F)
Exciting (E)	435	510
Routine (R)	357	478
Dull (D)	38	55

Using the letters in the table, find the number of respondents in each set.

27. (a) E (b) $D \cap F$
    (c) $E \cup M$ (d) E'

28. (a) D (b) $R \cap M$
    (c) $D \cup F$ (d) R'

Job Satisfaction *In Exercises 29–32, the given contingency table lists the cross tabulation of how satisfied or dissatisfied adults (for whom answering the question was appropriate) are with their job for males and females among the 933 appropriate adults in the 2016 General Social Survey. (Data from: gss.norc.org.)*

Job Satisfaction	Males (M)	Females (F)
Completely Satisfied (CS)	86	78
Very Satisfied (VS)	176	167
Fairly Satisfied (FS)	149	144
Neither Satisfied nor Dissatisfied (N)	26	28
Fairly Dissatisfied (FD)	20	32
Very Dissatisfied (VD)	9	11
Completely Dissatisfied (CD)	3	4

Using the letters given in the table, find the number of respondents in each set.

29. (a) $CS \cup VS$ (b) $(CS \cup VS) \cap F$
    (c) Describe your answer to part (b) in words.

30. (a) $VD \cup CD$ (b) $(VD \cup CD) \cap M$
    (c) Describe your answer to part (b) in words.

31. (a) $CS \cup VS \cup FS$ (b) $(CS \cup VS \cup FS) \cap M$
    (c) Describe your answer to part (b) in words.

32. (a) $FD \cup VD \cup CD$ (b) $(FD \cup VD \cup CD) \cap F$
    (c) Describe your answer to part (b) in words.

33. Restate the union rule in words.

Use Venn diagrams to answer the given questions. (See Example 6.)

34. If $n(A) = 5, n(B) = 8$, and $n(A \cap B) = 4$, what is $n(A \cup B)$?

35. If $n(A) = 12, n(B) = 27$, and $n(A \cup B) = 30$, what is $n(A \cap B)$?

36. Suppose $n(B) = 7, n(A \cap B) = 3$, and $n(A \cup B) = 20$. What is $n(A)$?

37. Suppose $n(A \cap B) = 5, n(A \cup B) = 35$, and $n(A) = 13$. What is $n(B)$?

Draw a Venn diagram and use the given information to fill in the number of elements for each region.

38. $n(U) = 48, n(A) = 26, n(A \cap B) = 12, n(B') = 30$

39. $n(A) = 28, n(B) = 12, n(A \cup B) = 30, n(A') = 19$

40. $n(A \cup B) = 17, n(A \cap B) = 3, n(A) = 8, n(A' \cup B') = 21$

41. $n(A') = 28, n(B) = 25, n(A' \cup B') = 45, n(A \cap B) = 12$

42. $n(A) = 28, n(B) = 34, n(C) = 25, n(A \cap B) = 14, n(B \cap C) = 15, n(A \cap C) = 11, n(A \cap B \cap C) = 9, n(U) = 59$

43. $n(A) = 54, n(A \cap B) = 22, n(A \cup B) = 85, n(A \cap B \cap C) = 4, n(A \cap C) = 15, n(B \cap C) = 16, n(C) = 44, n(B') = 63$

In Exercises 44–47, show that the statements are true by drawing Venn diagrams and shading the regions representing the sets on each side of the equals signs.*

44. $(A \cup B)' = A' \cap B'$

45. $(A \cap B)' = A' \cup B'$

46. $A \cap (B \cup C) = (A \cap B) \cup (A \cap C)$

47. $A \cup (B \cap C) = (A \cup B) \cap (A \cup C)$

48. Explain in words the statement about sets in question 44.

49. Explain in words the statement about sets in question 45.

50. Explain in words the statement about sets in question 46.

51. Explain in words the statement about sets in question 47.

---

*The statements in Exercises 44 and 45 are known as De Morgan's laws. They are named for the English mathematician Augustus De Morgan (1806–1871).

✓**Checkpoint Answers**

1. (a)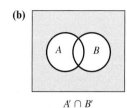

(b)

$A \cup B'$ $A' \cap B'$

2. (a)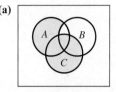

(b)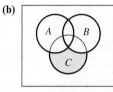

$(B' \cap A) \cup C$ $(A \cup B)' \cap C$

3. (a)

HSI
4   48   30

CT

18

(b) 4

4. (a) 42   **(b)** 9

5. (a) 8   **(b)** 92   **(c)** 98

6. 14

7. (a) 151,437 thousand

(b) 84,435 thousand

8.3 Introduction to Probability

If you go to a pizzeria and order two large pizzas at $14.99 each, you can easily find the *exact* price of your purchase: $29.98. For the manager at the pizzeria, however, it is impossible to predict the *exact* number of pizzas to be purchased daily. The number of pizzas purchased during a day is *random*: The quantity cannot be predicted exactly. A great many problems that come up in applications of mathematics involve random phenomena—phenomena for which exact prediction is impossible. The best that we can do is determine the *probability* of the possible outcomes.

Random Experiments and Sample Spaces

A **random experiment** (sometimes called a random phenomenon) has outcomes that we cannot predict, but that nonetheless have a regular distribution in a large number of repetitions. We call a repetition from a random experiment a **trial.** The possible results of each trial are called **outcomes.** For instance, when we flip a coin, the outcomes are heads and tails. We do not know whether a particular flip will yield heads or tails, but we do know that if we flip the coin a large a number of times, about half the flips will be heads and half will be tails. Each flip of the coin is a trial. The **sample space** (denoted by S) for a random experiment is the set of all possible outcomes. For the coin flipping, the sample space is

$$S = \{\text{heads, tails}\}.$$

EXAMPLE 1   Give the sample space for each random experiment.

(a) Use the spinner in Figure 8.17.

Figure 8.17

Solution The 7 outcomes are 1, 2, 3, . . . 7, so the sample space is

$$\{1, 2, 3, 4, 5, 6, 7\}.$$

(b) For the purposes of a public opinion poll, respondents are classified as young, middle aged, or senior and as male or female.

Solution A sample space for this poll could be written as a set of ordered pairs:

{(young, male), (young, female), (middle aged, male),

(middle aged, female), (senior, male), (senior, female)}.

(c) An experiment consists of studying the numbers of boys and girls in families with exactly 3 children. Let *b* represent *boy* and *g* represent *girl*.

Solution For this experiment, drawing a tree diagram can be helpful. First, we draw two branches starting on the left to indicate that the first child can be either a boy or a girl. From each of those outcomes, we draw two branches to indicate that the second child can be either a boy or girl. Last, we draw two branches from each of those outcomes to indicate that after the second child, the third child can be either a boy or a girl. The result is the tree in Figure 8.18.

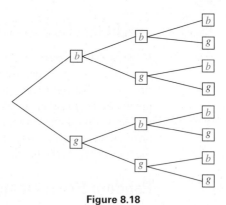

Figure 8.18

We can now easily list the members of the sample space *S*. We follow the eight paths of the branches to yield

$$S = \{bbb, bbg, bgb, bgg, gbb, gbg, ggb, ggg\}.$$

✓ Checkpoint 1

Draw a tree diagram for the random experiment of flipping a coin two times, and determine the sample space.

Events

An **event** is an outcome, or a set of outcomes, of a random experiment. Thus, an event is a subset of the sample space. For example, if the sample space for tossing a coin is $S = \{h, t\}$, then one event is $E = \{h\}$, which represents the outcome "heads."

An ordinary die is a cube whose six different faces show the following numbers of dots: 1, 2, 3, 4, 5, and 6. If the die is fair (not "loaded" to favor certain faces over others), then any one of the faces is equally likely to come up when the die is rolled. The sample space for the experiment of rolling a single fair die is $S = \{1, 2, 3, 4, 5, 6\}$. Some possible events are as follows:

The die shows an even number: $E_1 = \{2, 4, 6\}$.

The die shows a 1: $E_2 = \{1\}$.

The die shows a number less than 5: $E_3 = \{1, 2, 3, 4\}$.

The die shows a multiple of 3: $E_4 = \{3, 6\}$.

EXAMPLE 2   For the sample space S in Example 1(c) on the previous page, write the given events in set notation.

(a) Event H: The family has exactly two girls.

Solution Families with three children can have exactly two girls with either *bgg*, *gbg*, or *ggb*, so that event H is

$$H = \{bgg, gbg, ggb\}.$$

(b) Event K: The three children are the same sex.

Solution Two outcomes satisfy this condition: all boys and all girls, or

$$K = \{bbb, ggg\}.$$

(c) Event J: The family has three girls.

Solution Only *ggg* satisfies this condition, so

$$J = \{ggg\}. \quad \boxed{2}$$

If an event E equals the sample space S, then E is a **certain event.** If event $E = \varnothing$, then E is an **impossible event.**

EXAMPLE 3   Suppose a fair die is rolled. Then the sample space is $\{1, 2, 3, 4, 5, 6\}$. Find the requested events.

(a) The event "the die shows a 4."

Solution $\{4\}$.

(b) The event "the number showing is less than 10."

Solution The event is the entire sample space $\{1, 2, 3, 4, 5, 6\}$. This event is a certain event; if a die is rolled, the number showing (either 1, 2, 3, 4, 5, or 6) must be less than 10.

(c) The event "the die shows a 7."

Solution The empty set, $\varnothing$; this is an impossible event. $\boxed{3}$

Since events are sets, we can use set operations to find unions, intersections, and complements of events. Here is a summary of the set operations for events.

✓**Checkpoint 2**

Suppose a die is tossed. Write the given events in set notation.

(a) The number showing is less than 3.

(b) The number showing is 5.

(c) The number showing is 8.

✓**Checkpoint 3**

Which of the events listed in Checkpoint 2 is

(a) certain?

(b) impossible?

Set Operations for Events

Let E and F be events for a sample space S. Then

$E \cap F$ occurs when both E **and** F occur;

$E \cup F$ occurs when E **or** F **or both** occur;

E' occurs when E does **not** occur.

EXAMPLE 4   A study of college students grouped the students into various categories that can be interpreted as events when a student is selected at random. Consider the following events:

E: The student is under 20 years old;

F: The student is male;

G: The student is a business major.

Describe each of the following events in words.

(a) E'

Solution E' is the event that the student is 20 years old or older.

(b) $F' \cap G$

Solution $F' \cap G$ is the event that the student is not male and the student is a business major—that is, the student is a female business major.

(c) $E' \cup G$

Solution $E' \cup G$ is the event that the student is 20 or over or is a business major. Note that this event includes all students 20 or over, regardless of major. ✓4

Two events that cannot both occur at the same time, such as getting both a head and a tail on the same toss of a coin, are called **disjoint events.** (Disjoint events are sometimes referred to as *mutually exclusive events.*)

✓ Checkpoint 4

Write the set notation for the given events for the experiment of rolling a fair die if $E = \{1, 3\}$ and $F = \{2, 3, 4, 5\}$.

(a) $E \cap F$

(b) $E \cup F$

(c) E'

✓ Checkpoint 5

In Example 5, let $F = \{2, 4, 6\}$, $K = \{1, 3, 5\}$, and G remain the same. Are the given events disjoint?

(a) F and K

(b) F and G

Disjoint Events

Events E and F are disjoint events if $E \cap F = \emptyset$.

For any event E, E and E' are disjoint events.

EXAMPLE 5   Let $S = \{1, 2, 3, 4, 5, 6\}$, the sample space for tossing a die. Let $E = \{4, 5, 6\}$, and let $G = \{1, 2\}$. Are E and G disjoint events?

Solution Yes, because they have no outcomes in common; $E \cap G = \emptyset$. See Figure 8.19. ✓5

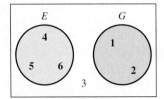

Figure 8.19

Probability

For sample spaces with *equally likely* outcomes, the probability of an event is defined as follows.

Basic Probability Principle

Let S be a sample space of equally likely outcomes, and let event E be a subset of S. Then the **probability that event E occurs** is

$$P(E) = \frac{n(E)}{n(S)}.$$

By this definition, the **probability of an event** is a number that indicates the relative likelihood of the event.

> **CAUTION** The basic probability principle applies only when the outcomes are equally likely.

EXAMPLE 6   Suppose a single fair die is rolled, with the sample space $S = \{1, 2, 3, 4, 5, 6\}$. Give the probability of each of the following events.

(a) E: The die shows an even number.

Solution Here, $E = \{2, 4, 6\}$, a set with three elements. Because S contains six elements,

$$P(E) = \frac{3}{6} = \frac{1}{2}.$$

(b) F: The die shows a number less than 10.

Solution Event F is a certain event, with

$$F = \{1, 2, 3, 4, 5, 6\},$$

so that

$$P(F) = \frac{6}{6} = 1.$$

(c) G: The die shows an 8.

Solution This event is impossible, so

$$P(G) = \frac{0}{6} = 0. \quad \boxed{6}$$

✓ **Checkpoint 6**

A fair die is rolled. Find the probability of rolling

(a) an odd number;

(b) 2, 4, 5, or 6;

(c) a number greater than 5;

(d) the number 7.

A standard deck of 52 cards has four suits—hearts (♥), clubs (♣), diamonds (♦), and spades (♠)—with 13 cards in each suit. The hearts and diamonds are red, and the spades and clubs are black. Each suit has an ace (A), a king (K), a queen (Q), a jack (J), and cards numbered from 2 to 10. The jack, queen, and king are called face cards and for many purposes can be thought of as having values 11, 12, and 13, respectively. The ace can be thought of as the low card (value 1) or the high card (value 14). See Figure 8.20. We will refer to this standard deck of cards often in our discussion of probability.

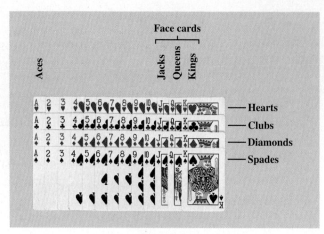

Figure 8.20

EXAMPLE 7 If a single card is drawn at random from a standard, well-shuffled, 52-card deck, find the probability of each of the given events.

(a) Drawing an ace

Solution There are 4 aces in the deck. The event "drawing an ace" is

{heart ace, diamond ace, club ace, spade ace}.

Therefore,

$$P(\text{ace}) = \frac{4}{52} = \frac{1}{13}.$$

(b) Drawing a face card

Solution Since there are 12 face cards,

$$P(\text{face card}) = \frac{12}{52} = \frac{3}{13}.$$

(c) Drawing a spade

Solution The deck contains 13 spades, so

$$P(\text{spade}) = \frac{13}{52} = \frac{1}{4}.$$

(d) Drawing a spade or a heart

Solution Besides the 13 spades, the deck contains 13 hearts, so

$$P(\text{spade or heart}) = \frac{26}{52} = \frac{1}{2}.$$

✓ **Checkpoint 7**

A single playing card is drawn at random from an ordinary 52-card deck. Find the probability of drawing

(a) a queen;

(b) a diamond;

(c) a red card.

In the preceding examples, the probability of each event was a number between 0 and 1, inclusive. The same thing is true in general. Any event E is a subset of the sample space S, so $0 \le n(E) \le n(S)$. Since $P(E) = n(E)/n(S)$, it follows that $0 \le P(E) \le 1$.

For any event E,

$$0 \le P(E) \le 1.$$

In many real-life problems, the events in the sample space are not all equally likely. When this is the case, we can estimate probabilities by determining the long-run proportion that an outcome of interest will occur given many repetitions under identical, and independent, circumstances. We do this often when we perform a statistical study. The long-run proportion is called the **relative frequency probability.** Estimates based on relative frequency probability are sometimes called *empirical probabilities.* Independence in this context refers to the idea that what occurs in one run has no effect on the outcome of a subsequent run.

For example, imagine that we want to determine the probability that a newly manufactured porcelain sink contains a defect. We examine the first sink before it is shipped and see that it has no defects. So we estimate the probability of a defect as 0 because the relative frequency of defects is 0 out of 1 trial. The second sink, however, has a defect. Now our number of defects is 1 out of two trials, so our relative frequency is 1 out of 2, which is .5. The table shows the results for these two sinks and eight additional sinks.

Sink Number	1	2	3	4	5	6	7	8	9	10
Defect Y/N	N	Y	N	N	N	Y	Y	N	N	N
Relative Frequency	$0/1 =$ 0	$1/2 =$ .5	$1/3 \approx$ .333	$1/4 =$ .25	$1/5 =$ .2	$2/6 \approx$ .333	$3/7 \approx$ .429	$3/8 =$ .375	$3/9 \approx$ .333	$3/10 =$ .3

Notice that the relative frequency fluctuates a great deal. To get a better estimate of the probability of a defect, we need to examine the *long-run* proportion of defects. When the same conditions are repeated a large number of times, the relative frequency will stabilize to the long-run frequency. Figure 8.21 shows how the relative frequency stabilizes to some degree after 50 trials and Figure 8.22 shows the relative frequency after 1000 trials. We can see in Figure 8.22 that the long-run proportion of defects is close to .25.

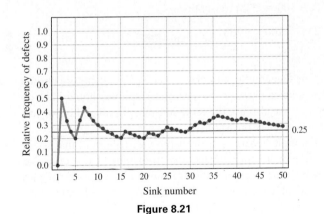

Figure 8.21

Figure 8.22

📄 **NOTE** Independence here pertains to the assumption that one sink having, or not having, a defect has no influence on whether another sink has a defect or not. We will investigate independence in greater depth in Section 8.5.

Thus, when we examine many trials of a given phenomenon of interest we can obtain a long-run estimate of the probability of interest. The next example shows one approach to finding such relative frequency probabilities.

EXAMPLE 8 **Standard of Living** The General Social Survey picks U.S. residents at random and asks them a great many questions. One of the questions they asked respondents in 2016 was "Do you feel your standard of living is much better, somewhat better, about the same, somewhat worse or much worse than your parents?" The table categorizes the responses.

Response	Frequency
Much Better	573
Somewhat Better	555
About the Same	466
Somewhat Worse	238
Much Worse	100

(a) Estimate the probability a U.S. resident feels his or her standard of living is much better than that of his or her parents.

Solution Let us define the set A to be the event that a resident feels his or her standard of living is much better. To find the relative frequency probability, we first need to find the total number of respondents. This is $573 + 555 + 466 + 238 + 100 = 1932$. We then divide the frequency in the much better category (in this case 573) by 1932 to obtain

$$P(A) = \frac{573}{1932} \approx .2966.$$

We use this relative frequency as our estimate of the long-run frequency and say that the estimated probability that a U.S. resident feels his or her standard of living is much better is approximately .2966.

(b) Estimate the probability of the event B that a U.S. resident feels that his or her standard of living is somewhat worse or much worse.

Solution Here, we add together the number of respondents in the categories of somewhat worse and much worse ($238 + 100 = 338$) to obtain

$$P(B) = \frac{238 + 100}{1932} = \frac{338}{1932} \approx .1749. \;\boxed{8}$$

✓ **Checkpoint 8**

From the data given in Example 8, estimate the probability that a U.S. resident believes his or her standard of living is much better, somewhat better, or about the same as that of his or her parents.

After conducting a study such as the General Social Survey where respondents are chosen at random, we can assume independence because one person's response should have no relation to another person's response. We also use the term "estimated probability" or just "probability" rather than "relative frequency probability."

A table of frequencies, as in Example 8, sets up a probability distribution; that is, for each possible outcome of an experiment, a number, called the probability of that outcome, is assigned. This assignment may be done in any reasonable way (on a relative frequency basis, as in Example 8, or by theoretical reasoning, as in Example 6), provided that it satisfies the following conditions.

Properties of Probability

Let S be a sample space consisting of n distinct outcomes $s_1, s_2, \ldots, s_n$. An acceptable probability assignment consists of assigning to each outcome s_i a number p_i (the probability of s_i) according to the following rules:

1. The probability of each outcome is a number between 0 and 1:

$$0 \le p_1 \le 1, \quad 0 \le p_2 \le 1, \ldots, \quad 0 \le p_n \le 1.$$

2. The sum of the probabilities of all possible outcomes is 1:

$$p_1 + p_2 + p_3 + \cdots + p_n = 1.$$

8.3 Exercises

1. What is meant by a "fair" coin or die?

2. What is the sample space for a random experiment?

Write sample spaces for the random experiments in Exercises 3–9. (See Example 1.)

3. A month of the year is chosen for a wedding.

4. A day in April is selected for a bicycle race.

5. A student is asked how many points she earned on a recent 80-point test.

6. A person is asked the number of hours (to the nearest hour) he watched television yesterday.

7. The management of an oil company must decide whether to go ahead with a new oil shale plant or to cancel it.

8. A coin is tossed and a die is rolled.

9. The quarter of the year in which a company's profits were highest.

10. Define an event.

11. Define disjoint events in your own words.

Decide whether the events are disjoint. (See Example 5.)

12. Owning an SUV and owning a Jeep

13. Wearing a hat and wearing glasses

14. Being married and being under 30 years old

15. Being a doctor and being under 5 years old

16. Being male and being a nurse

17. Being female and being a pilot

For the random experiments in Exercises 18–20, write out an equally likely sample space, and then write the indicated events in set notation. (See Examples 2 and 3.)

18. A marble is drawn at random from a bowl containing 3 yellow, 4 white, and 8 blue marbles.
    (a) A yellow marble is drawn.
    (b) A blue marble is drawn.
    (c) A white marble is drawn.
    (d) A black marble is drawn.

19. Six people live in a dorm suite. Two are to be selected to go to the campus café to pick up a pizza. Of course, no one wants to go, so the six names (Connie, Kate, Lindsey, Jackie, Taisa, and Nicole) are placed in a hat. After the hat is shaken, two names are selected.
    (a) Taisa is selected.
    (b) The two names selected have the same number of letters.

20. An unprepared student takes a three-question true-or-false quiz in which he flips a coin to guess the answers. If the coin is heads, he guesses true, and if the coin is tails, he guesses false.

(a) The student guesses true twice and guesses false once.

(b) The student guesses all false.

(c) The student guesses true once and guesses false twice.

In Exercises 21–24, write out the sample space and assume each outcome is equally likely. Then give the probability of the requested outcomes. (See Examples 7 and 8.)

21. In deciding what color and style to paint a room, Greg has narrowed his choices to three colors—forest sage, evergreen whisper, and opaque emerald—and two styles—rag painting and colorwash.
    (a) Greg picks a combination with colorwash.
    (b) Greg picks a combination with opaque emerald or rag painting.

22. Tami goes shopping and sees three kinds of shoes: flats, 2″ heels, and 3″ heels. They come in two shades of beige (light and dark) and black.
    (a) The shoe selected has a heel and is black.
    (b) The shoe selected has no heel and is beige.
    (c) The shoe selected has a heel and is beige.

23. Doug is shopping for a new patio umbrella. There is a 10-foot and a 12-foot model, and each is available in beige, forest green, and rust.
    (a) Doug buys a 12-foot forest green umbrella.
    (b) Doug buys a 10-foot umbrella.
    (c) Doug buys a rust-colored umbrella.

24. Sharonda is deciding between four brands of cell phones: Samsung, Motorola, Nokia, and Sony. Each of these phones comes with the option of insurance or no insurance.
    (a) Sharonda picks a phone with insurance.
    (b) Sharonda picks a phone that begins with the letter "S" and no insurance.

A single fair die is rolled. Find the probabilities of the given events. (See Example 6.)

25. Getting a number less than 4

26. Getting a number greater than 4

27. Getting a 2 or a 5

28. Getting a multiple of 3

Alejandro wants to adopt a puppy from an animal shelter. At the shelter, he finds eight puppies that he likes: a male and female puppy from each of the four breeds of beagle, boxer, collie, and Labrador. The puppies are each so cute that Alejandro cannot make up his mind, so he decides to pick the dog randomly.

29. Write the sample space for the outcomes, assuming each outcome is equally likely.

Find the probability that Alejandro chooses the given puppy.

30. A male boxer

31. A male puppy

32. A collie

33. A female Labrador

34. A beagle or a boxer

35. Anything except a Labrador

36. Anything except a male beagle or male boxer.

Fatal Work Injuries *The following table gives the number of fatal work injuries categorized by cause from 2015. (Data from: U.S. Bureau of Labor Statistics.)*

Cause	Number of Fatalities
Transportation accidents	2054
Assaults and violent acts	703
Contacts with objects and equipment	722
Falls	800
Exposure to harmful substances or environments	424
Fires and explosions	121
Other	12

Find the probability that a randomly chosen work fatality had the given cause. (See Example 8.)

37. A fall

38. Fires and explosions

39. **Religious Services** Respondents to the 2016 General Social Survey (GSS) indicated the following categorizations pertaining to attendance at religious services: (Data from: gss.norc.org.)

Attendance	Number of Respondents
Never	723
Less than once a year	173
Once a year	385
Several times a year	312
Once a month	190
2–3 times a month	250
Nearly every week	126
Every week	481
More than once a week	210
Don't know/no answer	17
Total	**2867**

Find the probability that a randomly chosen person in the United States attends religious services

(a) several times a year;

(b) 2–3 times a month;

(c) nearly every week or more frequently.

40. **Gender Roles** Respondents to the 2016 General Social Survey (GSS) indicated whether they strongly agreed, agreed, disagreed, or strongly disagreed with the question, "It is better for

men to work and for women to tend to the home." The responses are categorized in the following table. (Data from: gss.norc.org.)

Response	Number of Respondents
Strongly Agree	110
Agree	405
Disagree	853
Strongly Disagree	504
Don't Know/No Answer	995

Find the probability that a person in the United States

(a) strongly agreed or agreed with the statement;

(b) strongly disagreed with the statement;

(c) did not strongly agree with the statement.

S&P 500 *As of March 7, 2017, the 505 stocks in the S&P 500 were classified in 11 different sectors. The following table gives the numbers of stocks in each sector. (Data from: www.morningstar.com.)*

Sector	Number of Stocks in Sector
Consumer Discretionary	86
Consumer Staples	37
Energy	35
Financials	65
Health Care	60
Industrials	66
Information Technology	68
Materials	25
Real Estate	30
Telecommunications Services	5
Utilities	28
Total	**505**

Find the probability that a stock in the S&P 500 is in the requested sector.

41. Energy

42. Financials

43. Real Estate or Utilities

44. Health Care or Information Technology

S&P 500 *The following table gives the counts for three price categories of the 505 stocks in the S&P 500. (Data from: www.morningstar.com.)*

Price	Count
$0–$49.99	159
$50–$99.99	207
$100 or higher	139

Find the probability that a stock in the S&P 500 is in the requested price range.

45. Below $50   **46.** Below $100

An experiment is conducted for which the sample space is $S = \{s_1, s_2, s_3, s_4, s_5\}$. Which of the probability assignments in Exercises 47–52 is possible for this experiment? If an assignment is not possible, tell why.

47.

Outcomes	s_1	s_2	s_3	s_4	s_5
Probabilities	.09	.32	.21	.25	.13

48.

Outcomes	s_1	s_2	s_3	s_4	s_5
Probabilities	.92	.03	0	.02	.03

49.

Outcomes	s_1	s_2	s_3	s_4	s_5
Probabilities	1/3	1/4	1/6	1/8	1/10

50.

Outcomes	s_1	s_2	s_3	s_4	s_5
Probabilities	1/5	1/3	1/4	1/5	1/10

51.

Outcomes	s_1	s_2	s_3	s_4	s_5
Probabilities	.64	−.08	.30	.12	.02

52.

Outcomes	s_1	s_2	s_3	s_4	s_5
Probabilities	.05	.35	.5	.2	−.3

✓ Checkpoint Answers

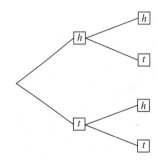

1. $\{hh, ht, th, tt\}$

2. **(a)** $\{1, 2\}$ **(b)** $\{5\}$ **(c)** $\varnothing$

3. **(a)** None   **(b)** Part (c)

4. **(a)** $\{3\}$ **(b)** $\{1, 2, 3, 4, 5\}$ **(c)** $\{2, 4, 5, 6\}$

5. **(a)** Yes   **(b)** No

6. **(a)** 1/2   **(b)** 2/3   **(c)** 1/6   **(d)** 0

7. **(a)** 1/13   **(b)** 1/4   **(c)** 1/2

8. About .8251

8.4 Basic Concepts of Probability

We determine the probability of more complex events in this section.

To find the probability of the union of two sets E and F in a sample space S, we use the union rule for counting given in Section 8.2:

$$n(E \cup F) = n(E) + n(F) - n(E \cap F).$$

Dividing both sides by $n(S)$ yields

$$\frac{n(E \cup F)}{n(S)} = \frac{n(E)}{n(S)} + \frac{n(F)}{n(S)} - \frac{n(E \cap F)}{n(S)}$$

$$P(E \cup F) = P(E) + P(F) - P(E \cap F).$$

This discussion is summarized in the next rule.

Addition Rule for Probability

For any events E and F from a sample space S,

$$P(E \cup F) = P(E) + P(F) - P(E \cap F).$$

In words, we have

$$P(E \text{ or } F) = P(E) + P(F) - P(E \text{ and } F).$$

(Although the addition rule applies to any events E and F from any sample space, the derivation we have given is valid only for sample spaces with equally likely simple events.)

EXAMPLE 1 When playing American roulette, the croupier (attendant) spins a marble that lands in one of the 38 slots in a revolving turntable. The slots are numbered 1 to 36, with two additional slots labeled 0 and 00 that are painted green. Half the remaining slots are colored red, and half are black. (See Figure 8.23.)

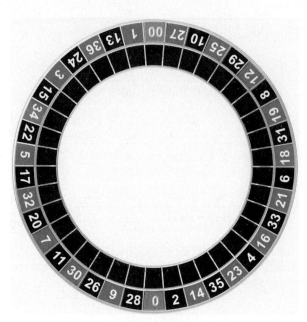

Figure 8.23

In roulette, the slots labeled 0 or 00 are not considered as even or odd. Find the probability that the marble will land in a red or even number.

Solution Let R represent the event of the marble landing in a red slot and E the event of the marble landing in an even-numbered slot. There are 18 slots that are colored red, so $P(R) = 18/38$. There are also 18 even numbers between 1 and 36, so $P(E) = 18/38$. In order to use the addition rule, we also need to know the number of slots that are red and even numbered. Looking at Figure 8.23, we can see there are 8 such slots, which implies that $P(R \cap E) = 8/38$. Using the addition rule, we find the probability that the marble will land in a slot that is red or even numbered is

$$P(R \cup E) = P(R) + P(E) - P(R \cap E)$$
$$= \frac{18}{38} + \frac{18}{38} - \frac{8}{38} = \frac{28}{38} = \frac{14}{19}.$$

✓ Checkpoint 1

If an American roulette wheel is spun, find the probability of the marble landing in a black slot or a slot (excluding 0 and 00) whose number is divisible by 3.

EXAMPLE 2 Suppose two fair dice (plural of *die*) are rolled. Find each of the given probabilities.

(a) The first die shows a 2 or the sum of the results is 6 or 7.

Solution The sample space for the throw of two dice is shown in Figure 8.24, on the following page, where 1-1 represents the event "the first die shows a 1 and the second die shows a 1," 1-2 represents the event "the first die shows a 1 and the second die shows a 2," and so on. Let A represent the event "the first die shows a 2" and B represent the event "the sum of the results is 6 or 7." These events are indicated in color in Figure 8.24.

From the diagram, event A has 6 elements, B has 11 elements, and the sample space has 36 elements. Thus,

$$P(A) = \frac{6}{36}, \quad P(B) = \frac{11}{36}, \quad \text{and} \quad P(A \cap B) = \frac{2}{36}.$$

By the addition rule,

$$P(A \cup B) = P(A) + P(B) - P(A \cap B),$$

$$P(A \cup B) = \frac{6}{36} + \frac{11}{36} - \frac{2}{36} = \frac{15}{36} = \frac{5}{12}.$$

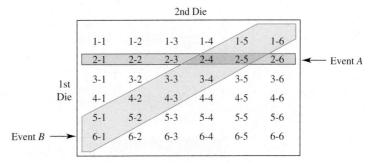

Figure 8.24

(b) The sum is 11 or the second die shows a 5.

Solution $P(\text{sum is } 11) = 2/36$, $P(\text{second die shows } 5) = 6/36$, and $P(\text{sum is } 11$ and second die shows $5) = 1/36$, so

$$P(\text{sum is 11 or second die shows 5}) = \frac{2}{36} + \frac{6}{36} - \frac{1}{36} = \frac{7}{36}.$$

If events E and F are disjoint, then $E \cap F = \varnothing$ by definition; hence, $P(E \cap F) = 0$. Applying the addition rule yields the useful fact that follows.

Addition Rule for Disjoint Events

For disjoint events E and F,

$$P(E \cup F) = P(E) + P(F).$$

EXAMPLE 3   Assume that the probability of a couple having a baby boy is the same as the probability of the couple having a baby girl. If the couple has 3 children, find the probability that at least 2 of them are girls.

Solution The event of having at least 2 girls is the union of the disjoint events $E =$ "the family has exactly 2 girls" and $F =$ "the family has exactly 3 girls." Using the equally likely sample space

$$\{ggg, ggb, gbg, bgg, gbb, bgb, bbg, bbb\},$$

where b represents a boy and g represents a girl, we see that $P(2 \text{ girls}) = 3/8$ and $P(3 \text{ girls}) = 1/8$. Therefore,

$$P(\text{at least 2 girls}) = P(2 \text{ girls}) + P(3 \text{ girls})$$

$$= \frac{3}{8} + \frac{1}{8} = \frac{1}{2}.$$

✓ **Checkpoint 2**

In the random experiment of Example 2, find the given probabilities.

(a) The sum is 5 or the second die shows a 3.

(b) Both dice show the same number, or the sum is at least 11.

✓ **Checkpoint 3**

In Example 3, find the probability of having no more than 2 girls.

By definition of E', for any event E from a sample space S,

$$E \cup E' = S \quad \text{and} \quad E \cap E' = \varnothing.$$

Because $E \cap E' = \varnothing$, events E and E' are disjoint, so that

$$P(E \cup E') = P(E) + P(E').$$

However, $E \cup E' = S$, the sample space, and $P(S) = 1$. Thus,

$$P(E \cup E') = P(E) + P(E') = 1.$$

Rearranging these terms gives the following useful rule.

Complement Rule

For any event E,

$$P(E') = 1 - P(E) \quad \text{and} \quad P(E) = 1 - P(E').$$

EXAMPLE 4
If a fair die is rolled, what is the probability that any number but 5 will come up?

Solution If E is the event that 5 comes up, then E' is the event that any number but 5 comes up. $P(E) = 1/6$, so we have $P(E') = 1 - 1/6 = 5/6$. ✔ 4

✓ **Checkpoint 4**

(a) Let $P(K) = 2/3$. Find $P(K')$.

(b) If $P(X') = 3/4$, find $P(X)$.

EXAMPLE 5
If two fair dice are rolled, find the probability that the sum of the numbers showing is greater than 3.

Solution To calculate this probability directly, we must find each of the probabilities that the sum is 4, 5, 6, 7, 8, 9, 10, 11, and 12 and then add them. It is much simpler to first find the probability of the complement, the event that the sum is less than or equal to 3:

$$P(\text{sum} \leq 3) = P(\text{sum is } 2) + P(\text{sum is } 3)$$

$$= \frac{1}{36} + \frac{2}{36} = \frac{3}{36} = \frac{1}{12}.$$

Now use the fact that $P(E) = 1 - P(E')$ to get

$$P(\text{sum} > 3) = 1 - P(\text{sum} \leq 3) = 1 - \frac{1}{12} = \frac{11}{12}. \quad ✔ 5$$

✓ **Checkpoint 5**

In Example 5, find the probability that the sum of the numbers rolled is at least 5.

Odds

Sometimes probability statements are given in terms of **odds**: a comparison of $P(E)$ with $P(E')$. For example, suppose $P(E) = \frac{4}{5}$. Then $P(E') = 1 - \frac{4}{5} = \frac{1}{5}$. These probabilities predict that E will occur 4 out of 5 times and E' will occur 1 out of 5 times. Then we say that the **odds in favor** of E are 4 to 1, or 4:1.

Odds

The **odds in favor** of an event E are defined as the ratio of $P(E)$ to $P(E')$, or

$$\frac{P(E)}{P(E')}, \quad P(E') \neq 0.$$

EXAMPLE 6   Suppose the weather forecaster says that the probability of rain tomorrow is 1/3. Find the odds in favor of rain tomorrow.

Solution   Let E be the event "rain tomorrow." Then E' is the event "no rain tomorrow." Since $P(E) = 1/3$, $P(E') = 2/3$. By the definition of odds, the odds in favor of rain are

$$\frac{1/3}{2/3} = \frac{1}{2}, \qquad \text{written 1 to 2 or 1:2.}$$

On the other hand, the odds that it will *not* rain, or the odds *against* rain, are

$$\frac{2/3}{1/3} = \frac{2}{1}, \qquad \text{written 2 to 1.}$$

If the odds in favor of an event are, say, 3 to 5, then the probability of the event is $3/8$, while the probability of the complement of the event is $5/8$. (Odds of 3 to 5 indicate 3 outcomes in favor of the event out of a total of 8 outcomes.) The above example suggests the following generalization:

> If the odds favoring event E are m to n, then
>
> $$P(E) = \frac{m}{m+n} \qquad \text{and} \qquad P(E') = \frac{n}{m+n}.$$

EXAMPLE 7   Often, weather forecasters give probability in terms of percentage. Suppose the weather forecaster says that there is a 40% chance that it will snow tomorrow. Find the odds of snow tomorrow.

Solution   In this case, we can let E be the event "snow tomorrow." Then E' is the event "no snow tomorrow." Now, we have $P(E) = .4 = 4/10$ and $P(E') = .6 = 6/10$. By the definition of odds in favor, the odds in favor of snow are

$$\frac{4/10}{6/10} = \frac{4}{6} = \frac{2}{3}, \qquad \text{written 2 to 3 or 2:3.}$$

It is important to put the final fraction into lowest terms in order to communicate the odds. ✓ 6

EXAMPLE 8   The odds that a particular bid will be the low bid are 4 to 5.

(a)   Find the probability that the bid will be the low bid.

Solution   Odds of 4 to 5 show 4 favorable chances out of $4 + 5 = 9$ chances altogether, so

$$P(\text{bid will be low bid}) = \frac{4}{4+5} = \frac{4}{9}.$$

(b)   Find the odds against that bid being the low bid.

Solution   There is a 5/9 chance that the bid will not be the low bid, so the odds against a low bid are

$$\frac{P(\text{bid will not be low})}{P(\text{bid will be low})} = \frac{5/9}{4/9} = \frac{5}{4},$$

or 5:4. ✓ 7

✓**Checkpoint 6**

In Example 7, suppose $P(E) = 9/10$. Find the odds

(a)   in favor of E;

(b)   against E.

Suppose the chance of snow is 80%. Find the odds

(c)   in favor of snow;

(d)   against snow.

✓**Checkpoint 7**

If the odds in favor of event E are 1 to 5, find

(a) $P(E)$;

(b) $P(E')$.

Applications

EXAMPLE 9   **Fatal Vehicle Accidents** Let A represent that the driver in a fatal vehicle accident was age 24 or younger and let B represent that the driver had a blood alcohol content (BAC) of .08% or higher. From data on fatal crashes in the year 2015 from the U.S. National Highway Traffic Safety Administration, we have the following probabilities.

$$P(A) = .1882, \quad P(B) = .1648, \quad P(A \cap B) = .0330.$$

(a) Find the probability that the driver was 25 or older and had a BAC level below .08%.

Solution Place the given information on a Venn diagram, starting with .0330 in the intersection of the regions A and B. (See Figure 8.25.) As stated earlier, event A has probability .1882. Since .0330 has already been placed inside the intersection of A and B,

$$.1882 - .0330 = .1552$$

goes inside region A, but outside the intersection of A and B. In the same way,

$$.1648 - .0330 = .1318$$

goes inside region B, but outside the overlap.

Since being age 25 or older is the complement of event A and having a BAC below .08% is the complement of event B, we need to find $P(A' \cap B')$. From the Venn diagram in Figure 8.25, the labeled regions have a total probability of

$$.1552 + .0330 + .1318 = .3200.$$

Since the entire region of the Venn diagram must have probability 1, the region outside A and B, namely $A' \cap B'$, has the probability

$$1 - .3200 = .6800.$$

The probability that a driver of a fatal crash is 25 or older and has a BAC below .08% is .6800.

(b) Find the probability that a driver in a fatal crash is 25 or older or has a BAC below .08%.

Solution The corresponding region for $A' \cup B'$ from Figure 8.25 has probability

$$.6800 + .1318 + .1552 = .9670. \quad \boxed{8}$$

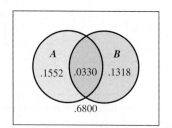

Figure 8.25

✓**Checkpoint 8**

Using the data from Example 9, find the probability that a driver in a fatal crash

(a) has a BAC below .08%;

(b) is age 25 or older and has a BAC of .08% or higher.

EXAMPLE 10   **S&P 500 Stocks** Data from March 7, 2017, for all 505 stocks in the S&P 500 regarding the price of the stock and the dividend yield are cross classified in the table below. The price of the stock is categorized into three groups: $0–$49.99, $50.00–$99.99, and $100.00 or higher. The dividend yield is categorized as to whether there was no dividend, a dividend yield between 0.01% and 1.99%, and a dividend yield of 2.00% or higher. (Data from: www.okfn.org.)

Price of Stock	Dividend Yield			Total
	No Dividend	.01%–1.99%	2.00% or higher	
$0–$49.99	21	69	68	158
$50.00–$99.99	18	87	102	207
$100.00 or higher	26	69	45	140
Total	65	225	215	505

(a) Let E be the event that the dividend yield is 2.00% or higher. Find $P(E)$.

Solution We need to find the total number of stocks in the S&P 500 that paid a dividend yield of 2.0% or higher (see shaded column below.)

Price of Stock	Dividend Yield			Total
	No Dividend	.01%–1.99%	2.00% or higher	
$0–$49.99	21	69	68	158
$50.00–$99.99	18	87	102	207
$100.00 or Higher	26	69	45	140
Total	65	225	215	505

Thus, we have $68 + 102 + 45 = 215$ is the total number of stocks with a dividend yield of 2.00% or higher, and

$$P(E) = \frac{215}{505} \approx .4257.$$

(b) Let F be the event that the price of the stock is between $50.00 and $99.99. Find $P(E \cap F)$.

Solution We want to find the number that satisfies both conditions of E and F. We see that there are 102 stocks in the column for a dividend yield of 2.00% or higher and the row for stock price of $50.00–99.99 (see the area shaded in green in the table below).

Price of Stock	Dividend Yield			Total
	No Dividend	.01%–1.99%	2.00% or higher	
$0–$49.99	21	69	68	158
$50.00–$99.99	18	87	102	207
$100.00 or Higher	26	69	45	140
Total	65	225	215	505

Thus,

$$P(E \cap F) = \frac{102}{505} \approx .2020.$$

(c) Find $P(E \cup F)$.

Solution We can use the Additive Rule for Probability to find $P(E \cup F)$. We know from part (a) that $P(E) \approx .4257$. In a similar manner to part (a), we can find $P(F)$. There are $18 + 87 + 102 = 207$ stocks that have a price between $50.00 and $99.99, so

$$P(F) = \frac{207}{505} \approx .4099.$$

With the answer to (b), using the Additive Rule yields

$$P(E \cup F) = P(E) + P(F) - P(E \cap F) \approx .4257 + .4099 - .2020 = .6336. \quad ✓_9$$

✓ **Checkpoint 9**

Let G be the event that the price of the stock is under $50 and H be the event that the dividend yield is between .01% and 1.99%. Find the following.

(a) $P(G)$

(b) $P(G \cap H)$

(c) $P(G \cup H)$

8.4 Exercises

Assume a single spin of the roulette wheel is made. Find the probability for the given events. (See Example 1.)

1. The marble lands in a green or black slot.

2. The marble lands in a green or even slot.

3. The marble lands in an odd or black slot.

Also with a single spin of the roulette wheel, find the probability of winning with the given bets.

4. The marble will land in a slot numbered 13–18.

5. The marble will land in slots 0, 00, 1, 2, or 3.

6. The marble will land in a slot that is a positive multiple of 3.

7. The marble will land in a slot numbered 25–36.

Two dice are rolled. Find the probabilities of rolling the given sums. (See Examples 2, 4, and 5.)

8. (a) 2 **(b)** 4 **(c)** 5 **(d)** 6

9. (a) 8 **(b)** 9 **(c)** 10 **(d)** 13

10. (a) 9 or more

(b) Less than 7

(c) Between 5 and 8 (exclusive)

11. (a) Not more than 5

(b) Not less than 8

(c) Between 3 and 7 (exclusive)

Tami goes shopping and sees three kinds of shoes: flats, 2″ heels, and 3″ heels. The shoes come in two shades of beige (light and dark) and black. If each option has an equal chance of being selected, find the probabilities of the given events.

12. The shoes Tami buys have a heel.

13. The shoes Tami buys are black.

14. The shoes Tami buys have a 2″ heel and are beige.

Ms. Elliott invites 10 relatives to a party: her mother, 3 aunts, 2 uncles, 2 sisters, 1 male cousin, and 1 female cousin. If the chances of any one guest arriving first are equally likely, find the probabilities that the given guests will arrive first.

15. (a) A sister or an aunt

(b) A sister or a cousin

(c) A sister or her mother

16. (a) An aunt or a cousin

(b) A male or an uncle

(c) A female or a cousin

Use Venn diagrams to work Exercises 17–22. (See Example 9.)

17. Suppose $P(E) = .30$, $P(F) = .51$, and $P(E \cap F) = .19$. Find each of the given probabilities.

(a) $P(E \cup F)$ **(b)** $P(E' \cap F)$

(c) $P(E \cap F')$ **(d)** $P(E' \cup F')$

18. Let $P(Z) = .40$, $P(Y) = .30$, and $P(Z \cup Y) = .58$. Find each of the given probabilities.

(a) $P(Z' \cap Y')$ **(b)** $P(Z' \cup Y')$

(c) $P(Z' \cup Y)$ **(d)** $P(Z \cap Y')$

19. Fatal Vehicle Accidents According to 2015 data from the National Highway Traffic Safety Administration, the probability that the driver in a fatal vehicle accident is female (event F) is .2584. The probability that the driver is 24 years old or less (event A) is .1914. The probability that the driver is female and is 24 years old or less is .0536. Find the probability of the given events.

(a) $F \cup A$ **(b)** $F' \cup A$

(c) $F \cap A'$ **(d)** $F' \cup A'$

20. Fatal Vehicle Accidents According to 2015 data from the National Highway Traffic Safety Administration, the probability that an airbag was deployed (event A) during an accident is .4364.

The probability that the driver was impaired (event I) is .1403. The probability that the airbag was deployed and the driver was impaired is .0778. Find the probability of the given events.

(a) $A \cup I$ **(b)** $A \cup I'$

(c) $A \cap I'$ **(d)** $A' \cap I'$

21. Marriage and Happiness According to the 2016 General Social Survey, the probability that a U.S. resident is currently married is .4227 and the probability that a U.S. resident describes himself or herself as very happy is .2820. The probability of being married and being very happy is .1627. Find the probability of the described events. (Data from: gss.norc.org.)

(a) Not being very happy and being married

(b) Not being married and not being very happy

(c) Not being married or not being very happy

22. Education and Earnings According to the 2016 General Social Survey, the probability that a U.S. resident has attended junior college or higher is .3845, and the probability that annual household income is $50,000 or higher is .5040. The probability of having attended junior college or higher and being in a household with income $50,000 or higher is .2700. Find the probability of the described events. (Data from: gss.norc.org.)

(a) Not having attended junior college or higher and being in a household earning $50,000 or more.

(b) Having attended junior college or higher or not being in a household earning $50,000 or more.

(c) Not having attended junior college and not being in a household earning $50,000 or more.

A single fair die is rolled. Find the odds in favor of getting the results in Exercises 23–26. (See Examples 6 and 7.)

23. 2 **24.** 2, 3, 4

25. 2, 3, 5, or 6 **26.** Some number greater than 5

27. A marble is drawn from a box containing 3 yellow, 4 white, and 8 blue marbles. Find the odds in favor of drawing the given marbles.

(a) A yellow marble

(b) A blue marble

(c) A white marble

28. Find the odds of *not* drawing a white marble in Exercise 27.

29. Two dice are rolled. Find the odds of rolling a 7 or an 11.

30. Define what is meant by odds.

Education *For Exercises 31–34, find the odds of the event occurring from the given probability that a bachelor's degree recipient in 2014 majored in the given discipline. (Data from: Digest of Educational Statistics.)*

31. Business; probability 19/100.

32. Natural sciences and mathematics; probability 2/25.

33. Education; probability 1/20.

34. Social and behavioral sciences; 3/20.

Government Revenue *For Exercises 35–38, convert the given odds to the probability of the event. The events involve revenue for the U.S. government from taxes. (Data from: U.S. Census Bureau.)*

35. The odds that tax revenue comes from property taxes are 31:69.

36. The odds that tax revenue comes from sales and gross receipts are 7:13.

37. The odds that tax revenue comes from individual income tax are 23:77.

38. The odds that tax revenue comes from corporate income tax are 37:963.

S&P 500 Stocks *For Exercises 39–46, use the following table, which cross classifies the price of the 505 stocks in the S&P 500 on March 7, 2017, with whether the earnings per share ratio was positive or not.*

	Earnings per Share	
Price of Stock	**Negative or 0**	**Positive**
$0–$49.99	30	128
$50.00–$99.99	18	189
$100.00 or higher	5	135

Find the probability of the described events. (Data from: www.okfn.org.)

39. The earnings per share ratio is positive.

40. The earnings per share ratio is not positive.

41. The price of the stock is less than $50 and the earnings per share ratio is negative or 0.

42. The price of the stock is $100 or higher and the earning per share ratio is positive.

43. The price of the stock is $50.00–$99.99 or the earning per share ratio is positive.

44. The price of the stock is $100 or higher or the earnings per share ratio is positive.

45. The price of the stock is not below $50.00 and the earnings per share ratio is positive.

46. The price of the stock is not $100 or higher and the earnings per share ratio is negative or 0.

Hours Worked *Data from the 2016 General Social Survey can allow us to estimate how many hours people who classified their work status as part-time or full-time actually worked in the last week. Use the following table to find the probabilities of the events in Exercises 47–54. (Data from: gss.norc.org.)*

Labor Force Status	**Hours Worked in the Last Week**					
	0–19	**20–29**	**30–39**	**40–49**	**50 or More**	**Total**
Full-Time Worker	26	43	137	691	377	1274
Part-Time Worker	74	82	87	41	24	308
Total	100	125	224	732	401	1582

47. Full-time worker

48. Part-time worker

49. Worked 50 hours or more

50. Worked 0–19 hours

51. Part-time worker and someone who worked 40 hours or more

52. Full-time worker and someone who worked less than 40 hours

53. Full-time worker or someone who worked 40 hours or more

54. Part-time worker or someone who worked 29 hours or less

One way to solve a probability problem is to repeat the experiment many times, keeping track of the results. Then the probability can be approximated by using the basic definition of the probability of an event E, which is $P(E) = n(E)/n(S)$, where E occurs $n(E)$ times out of $n(S)$ trials of an experiment. This is called the Monte Carlo method of finding probabilities. If physically repeating the experiment is too tedious, it may be simulated with the use of a random-number generator, available on most computers and scientific or graphing calculators. To simulate a coin toss or the roll of a die on a graphing calculator, change the setting to fixed decimal mode with 0 digits displayed. To simulate multiple tosses of a coin, press RAND (or RANDOM or RND#) in the PROB submenu of the MATH (or OPTN) menu, and then press ENTER repeatedly. Interpret 0 as a head and 1 as a tail. To simulate multiple rolls of a die, press RAND × 6 + .5, and then press ENTER repeatedly.

55. Suppose two dice are rolled. Use the Monte Carlo method with at least 50 repetitions to approximate the given probabilities. Compare them with the results of Exercise 10.
  (a) P(the sum is 9 or more)
  (b) P(the sum is less than 7)

56. Suppose two dice are rolled. Use the Monte Carlo method with at least 50 repetitions to approximate the given probabilities. Compare them with the results of Exercise 11.
  (a) P(the sum is not more than 5)
  (b) P(the sum is not less than 8)

57. Suppose three dice are rolled. Use the Monte Carlo method with at least 100 repetitions to approximate the given probabilities.
  (a) P(the sum is 5 or less)
  (b) P(neither a 1 nor a 6 is rolled)

58. Suppose a coin is tossed 5 times. Use the Monte Carlo method with at least 50 repetitions to approximate the given probabilities.
  (a) P(exactly 4 heads)
  (b) P(2 heads and 3 tails)

Genetics *Gregor Mendel, an Austrian monk, was the first to use probability in the study of genetics. In an effort to understand the mechanism of characteristic transmittal from one generation to the next in plants, he counted the number of occurrences of various characteristics. Mendel found that the flower color in certain pea plants obeyed this scheme:*

Pure red crossed with pure white produces red.

From its parents, the red offspring received genes for both red (R) and white (W), but in this case red is dominant and white recessive, so the offspring exhibits the color red. However, the offspring still carries both

genes, and when two such offspring are crossed, several things can happen in the third generation. The following table, called a Punnett square, shows the equally likely outcomes:

		Second Parent	
		R	W
First Parent	R	RR	RW
	W	WR	WW

Use the fact that red is dominant over white to find each of the given probabilities. Assume that there are an equal number of red and white genes in the population.

59. P(a flower is red)

60. P(a flower is white)

8.5 Conditional Probability and Independent Events

In Section 8.4, we examined the data from March 7, 2017, for all 505 stocks in the S&P 500 regarding the price of the stock and the dividend yield. (Data from: www.okfn.org.)

Price of Stock	Dividend Yield			Total
	No Dividend	.01%–1.99%	2.00% or Higher	
$0–$49.99	21	69	68	158
$50.00–$99.99	18	87	102	207
$100.00 or Higher	26	69	45	140
Total	65	225	215	505

Suppose we want to examine that data further to learn if higher-priced stocks are more likely to pay a dividend yield of 2.00% or higher. To find this out, we will need to calculate what are called *conditional probabilities*.

First, let us review. Let A be the event that the dividend yield is 2.00% or higher, and let B be the event that the stock price is $100 or higher. Because 215 of the stocks out of the total 505 have a dividend yield of 2.00% or higher, and 140 of the 505 stocks are priced $100.00 or higher, we have

$$P(A) = \frac{215}{505} \approx .4257 \text{ and } P(B) = \frac{140}{505} \approx 2772. \text{ ✓}_1$$

Now let's calculate the probability that a stock priced $100.00 or more has a dividend yield of 2.00% or higher. From the highlighted row in the table below, we see there are a total of 140 stocks priced $100.00 or higher, of which 45 have a dividend yield of 2.0% or higher (notice that the number 45 is circled).

✓**Checkpoint 1**

Use the data in the table to find

(a) $P(A \cap B)$;

(b) $P(A')$;

(c) $P(B')$.

Price of Stock	Dividend Yield			Total
	No Dividend	.01%–1.99%	2.00% or higher	
$0–$49.99	21	69	68	158
$50.00–$99.99	18	87	102	207
$100.00 or Higher	26	69	㊺	140
Total	65	225	215	505

Thus,

$$P(\text{a stock priced \$100.00 or more has a dividend yield 2.00\% or higher}) = \frac{45}{140} \approx 3214.$$

This probability (.3214) is a different value from the probability, in general, of a stock with a dividend yield of 2.00% or higher (.4257) because *we have additional information* (that the stock is priced $100 or higher) *that has reduced the sample size.* In other words, we found the probability the stock has a dividend yield of 2.00% or higher, *A*, given the additional information that the stock sells for $100.00 or higher, *B*. This is called the *conditional probability* of event *A*, given that event *B* has occurred, which is written $P(A|B)$ and read as "the probability of *A* given *B*."

In the preceding discussion,

$$P(A|B) = \frac{45}{140} \approx .3214.$$

If we divide the numerator and denominator by 505 (the number of stocks in the S&P 500), this can be written as

$$P(A|B) = \frac{\dfrac{45}{505}}{\dfrac{140}{505}} = \frac{P(A \cap B)}{P(B)},$$

where $P(A \cap B)$ represents, as usual, the probability that both *A* and *B* occur.

To generalize this result, assume that *E* and *F* are two events for a particular experiment. Assume also that the sample space *S* for the experiment has *n* possible equally likely outcomes. Suppose event *F* has *m* elements and $E \cap F$ has *k* elements $(k \le m)$. Then, using the fundamental principle of probability yields

$$P(F) = \frac{m}{n} \quad \text{and} \quad P(E \cap F) = \frac{k}{n}.$$

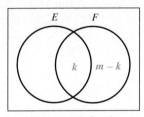

Event *F* has a total of *m* elements.

Figure 8.26

We now want to find $P(E|F)$: the probability that *E* occurs, given that *F* has occurred. Since we assume that *F* has occurred, we reduce the sample space to *F*; that is, we look only at the *m* elements inside *F*. (See Figure 8.26.) Of these *m* elements, there are *k* elements for which *E* also occurs, because $E \cap F$ has *k* elements. This yields

$$P(E|F) = \frac{k}{m}.$$

Divide numerator and denominator by *n* to get

$$P(E|F) = \frac{k/n}{m/n} = \frac{P(E \cap F)}{P(F)}.$$

The last result motivates the following definition of conditional probability. The **conditional probability** of an event *E*, given event *F*, written $P(E|F)$, is

$$P(E|F) = \frac{P(E \cap F)}{P(F)} = \frac{P(E \text{ and } F)}{P(F)}, \quad P(F) \neq 0.$$

This definition tells us that, for equally likely outcomes, conditional probability is found by *reducing the sample space to event F* and then finding the number of outcomes in *F* that are also in event *E*. Thus,

$$P(E|F) = \frac{n(E \cap F)}{n(F)}.$$

Although the definition of conditional probability was motivated by an example with equally likely outcomes, it is valid in all cases. For an intuitive explanation, think of the formula as giving the probability that both *E* and *F* occur, compared with the entire probability of *F* occuring.

EXAMPLE 1 **S&P 500 Stocks** Use the information at the beginning of this section from the table on the price of S&P 500 stocks and dividend yields to find the following probabilities.

(a) Let event C be that the price of the stock was $0 to $49.99. Find $P(A|C)$.

Solution In words, $P(A|C)$ is the probability that a stock has a dividend yield of 2.00% or higher given that the price of the stock is between $0 and $49.99.

$$P(A|C) = \frac{n(A \cap C)}{n(C)} = \frac{68}{158} \approx .4304$$

(b) Let event D be that the price of the stock was $50.00 to $99.99. Find $P(A|D)$.

Solution In words, $P(A|D)$ is the probability that a stock has a dividend yield of 2.00% or higher given that the price of the stock is between $50.00 and $99.99.

$$P(A|D) = \frac{n(A \cap D)}{n(D)} = \frac{102}{207} \approx .4928$$

Notice that each stock price has a very different probability of having a dividend yield of 2.00% or higher. If the price of the stock is low ($0–$49.99), we have $P(A|C) \approx .4304$. If the price of the stock is in the middle category ($50.00–$99.99), we have $P(A|D) \approx .4928$. Finally, if the price of the stock is high ($100.00 or higher), $P(A|B) \approx .3214$. The additional information of knowing the price of the stock changes our estimate of the probability of having a dividend yield of 2.00% or higher.

(c) Find $P(B|A)$.

Solution It is important to note that $P(B|A)$ does not usually equal $P(A|B)$. In this case, $P(B|A)$ in words is the probability that a stock has a price of $100.00 or more given that the dividend yield is 2.00% or higher. The count for the numerator of the number of stocks with a price of $100 or more and a dividend yield of 2.00% or higher is still 45, but now our denominator is the number of stocks with a dividend yield of 2.00% or higher (215). Thus, we have

$$P(B|A) = \frac{n(B \cap A)}{n(A)} = \frac{n(A \cap B)}{n(A)} = \frac{45}{215} \approx .2093.$$

(d) Find $P(A'|B')$

Solution In words, we are looking for the probability that the dividend yield is less than 2.00% given that the price of the stock is less than $100. To find $n(A' \cap B')$ and $n(B')$, we examine the table below. The numbers colored in pink indicate those where the stock price is less than $100. The circled numbers indicate, of those, which do not have dividend yield of 2.00% or higher.

Price of Stock	Dividend Yield			
	No Dividend	**.01%–1.99%**	**2.00% or Higher**	**Total**
$0–$49.99	21	69	68	158
$50.00–$99.99	18	87	102	207
$100.00 or Higher	26	69	45	140
Total	65	225	215	505

Thus, we have

$$P(A'|B') = \frac{n(A' \cap B')}{n(B')} = \frac{21 + 69 + 18 + 87}{158 + 207} = \frac{195}{365} \approx .5342. \quad ✓_2$$

✓ **Checkpoint 2**

The table shows the results of the 2016 General Social Survey regarding happiness for married and never married respondents.

	Very Happy	Pretty Happy or Not Too Happy	Total
Married	465	743	1208
Never Married	156	647	803
Total	621	1390	2011

Let M represent married respondents and V represent very happy respondents. Find each of the given probabilities.

(a) $P(V|M)$

(b) $P(V|M')$

(c) $P(M|V)$

(d) $P(M'|V')$

(e) State the probability of part (d) in words.

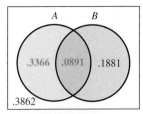

Figure 8.27

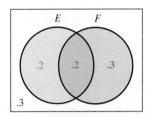

Figure 8.28

✓**Checkpoint 3**

Find $P(F|E)$ if $P(E) = .3$, $P(F)=.4$, and $P(E \cup F) = .6$.

✓**Checkpoint 4**

In Example 3, find the probability that exactly one coin showed a head, given that at least one was a head.

Venn diagrams can be used to illustrate problems in conditional probability. A Venn diagram for Example 1, in which the probabilities are used to indicate the number in the set defined by each region, is shown in Figure 8.27. In the diagram, $P(B|A)$ is found by *reducing the sample space to just set A*. Then $P(B|A)$ is the ratio of the number in that part of set B which is also in A to the number in set A, or $.0891/(.0891 + .3366) = .0891/.4257 \approx .2093$. This agrees with part (c) of Example 1.

EXAMPLE 2   Given $P(E) = .4$, $P(F) = .5$, and $P(E \cup F) = .7$, find $P(E|F)$.

Solution Find $P(E \cap F)$ first. Then use a Venn diagram to find $P(E|F)$. By the addition rule,

$$P(E \cup F) = P(E) + P(F) - P(E \cap F)$$
$$.7 = .4 + .5 - P(E \cap F)$$
$$P(E \cap F) = .2.$$

Now use the probabilities to indicate the number in each region of the Venn diagram in Figure 8.28. $P(E|F)$ is the ratio of the probability of that part of E which is in F to the probability of F, or

$$P(E|F) = \frac{P(E \cap F)}{P(F)} = \frac{.2}{.5} = \frac{2}{5} = .4. \;\; ✓_3$$

EXAMPLE 3   Two fair coins were tossed, and it is known that at least one was a head. Find the probability that both were heads.

Solution The sample space has four equally likely outcomes: $S = \{hh, ht, th, tt.\}$ Define two events

$$E_1 = \text{at least 1 head} = \{hh, ht, th\}$$

and

$$E_2 = 2 \text{ heads} = \{hh\}.$$

Because there are four equally likely outcomes, $P(E_1) = 3/4$. Also, $P(E_1 \cap E_2) = 1/4$. We want the probability that both were heads, given that at least one was a head; that is, we want to find $P(E_2|E_1)$. Because of the condition that at least one coin was a head, the reduced sample space is

$$\{hh, ht, th\}.$$

Since only one outcome in this reduced sample space is two heads,

$$P(E_2|E_1) = \frac{1}{3}.$$

Alternatively, use the definition given earlier:

$$P(E_2|E_1) = \frac{P(E_2 \cap E_1)}{P(E_1)} = \frac{1/4}{3/4} = \frac{1}{3}. \;\; ✓_4$$

It is important not to confuse $P(A|B)$ with $P(B|A)$. For example, in a criminal trial, a prosecutor may point out to the jury that the probability of the defendant's DNA profile matching that of a sample taken at the scene of the crime, given that the defendant is innocent, is very small. What the jury must decide, however, is the probability that the defendant is innocent, given that the defendant's DNA profile matches the sample. Confusing the two is an error sometimes called "the prosecutor's fallacy," and the 1990 conviction of a rape suspect in England was overturned by a panel of judges who ordered a retrial, because the fallacy made the original trial unfair.[*] This mistake is often called "confusion of the inverse."

In the next section, we will see how to compute $P(A|B)$ when we know $P(B|A)$.

Product Rule

If $P(E) \neq 0$ and $P(F) \neq 0$, then the definition of conditional probability shows that

$$P(E|F) = \frac{P(E \cap F)}{P(F)} \quad \text{and} \quad P(F|E) = \frac{P(F \cap E)}{P(E)}.$$

Using the fact that $P(E \cap F) = P(F \cap E)$, and solving each of these equations for $P(E \cap F)$, we obtain the following rule.

Product Rule of Probability

If E and F are events, then $P(E \cap F)$ may be found by either of these formulas:

$$P(E \cap F) = P(F) \cdot P(E|F) \quad \text{or} \quad P(E \cap F) = P(E) \cdot P(F|E).$$

The **product rule** gives a method for finding the probability that events E and F both occur. Here is a simple way to remember the ordering of E and F in the probability rule:

$$P(E \cap F) = P(F) \cdot P(E|F) \quad \text{or} \quad P(E \cap F) = P(E) \cdot P(F|E).$$

EXAMPLE 4 **Business** According to data from the U.S. Census Bureau, we can estimate the probability that a business is female owned as .3595. The probability that the business has one to four employees given that it is female owned is .5709. What is the probability that a business is female owned *and* has one to four employees?

Solution Let F represent the event of "having a female-owned business" and E represent the event of "having one to four employees." We want to find $P(F \cap E)$. By the product rule,

$$P(F \cap E) = P(F)P(E|F).$$

From the given information, $P(F) = .3595$ and the probability that a business has one to four employees given that it is a female-owned business is $P(E|F) = .5709$. Thus,

$$P(F \cap E) = (.3595)(.5709) \approx .2052 \; \boxed{✓}_5$$

✓ **Checkpoint 5**

In a litter of puppies, 3 were female and 4 were male. Half the males were black. Find the probability that a puppy chosen at random from the litter would be a black male.

In Section 8.1, we used a tree diagram to find the number of subsets of a given set. By including the probabilities for each branch of a tree diagram, we convert it to a **probability tree**. The following examples show how a probability tree is used with the product rule to find the probability of a sequences of events.

EXAMPLE 5 **Hiring Decision** A company needs to hire a new director of advertising. It has decided to try to hire either person A or person B, both of whom are assistant advertising directors for its major competitor. To decide between A and B, the company does research on the campaigns managed by A or B (none are managed by both) and finds that A is in charge of twice as many advertising campaigns as B. Also, A's campaigns have yielded satisfactory results three out of four times, while B's campaigns have

---

[*]David Pringle, "Who's the DNA Fingerprinting Pointing At?," *New Scientist*, January 29, 1994, pp. 51–52.

yielded satisfactory results only two out of five times. Suppose one of the competitor's advertising campaigns (managed by A or B) is selected randomly.

We can represent this situation schematically as follows: Let A denote the event "Person A does the job" and B the event "Person B does the job." Let S be the event "satisfactory results" and U the event "unsatisfactory results." Then the given information can be summarized in the probability tree in Figure 8.29. Since A does twice as many jobs as B, $P(A) = 2/3$ and $P(B) = 1/3$, as noted on the first-stage branches of the tree. When A does a job, the probability of satisfactory results is 3/4 and of unsatisfactory results 1/4, as noted on the second-stage branches. Similarly, the probabilities when B does the job are noted on the remaining second-stage branches. The composite branches labeled 1–4 represent the four disjoint possibilities for the running and outcome of the campaign.

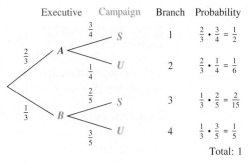

Figure 8.29

(a) Find the probability that A is in charge of a campaign and it produces satisfactory results.

Solution   We are asked to find $P(A \cap S)$. We know that when A does the job, the probability of success is 3/4; that is, $P(S|A) = 3/4$. Hence, by the product rule,

$$P(A \cap S) = P(A) \cdot P(S|A) = \frac{2}{3} \cdot \frac{3}{4} = \frac{1}{2}.$$

The event $(A \cap S)$ is represented by branch 1 of the tree, and as we have just seen, its probability is the product of the probabilities that make up that branch.

(b) Find the probability that B runs a campaign and it produces satisfactory results.

Solution   We must find $P(B \cap S)$. This event is represented by branch 3 of the tree, and as before, its probability is the product of the probabilities of the pieces of that branch:

$$P(B \cap S) = P(B) \cdot P(S|B) = \frac{1}{3} \cdot \frac{2}{5} = \frac{2}{15}.$$

(c) What is the probability that the selected campaign is satisfactory?

Solution   The event S is the union of the disjoint events $A \cap S$ and $B \cap S$, which are represented by branches 1 and 3 of the tree diagram. By the addition rule,

$$P(S) = P(A \cap S) + P(B \cap S) = \frac{1}{2} + \frac{2}{15} = \frac{19}{30}.$$

Thus, the probability of an event that appears at the end of several branches is the sum of the probabilities of each of these branches.

(d) What is the probability that the selected campaign is unsatisfactory?

Solution $P(U)$ can be read from branches 2 and 4 of the tree:

$$P(U) = \frac{1}{6} + \frac{1}{5} = \frac{11}{30}.$$

Alternatively, because U is the complement of S,

$$P(U) = 1 - P(S) = 1 - \frac{19}{30} = \frac{11}{30}.$$

(e) Find the probability that either A runs the campaign or the results are satisfactory (or possibly both).

Solution Event A combines branches 1 and 2, while event S combines branches 1 and 3, so use branches 1, 2, and 3:

$$P(A \cup S) = \frac{1}{2} + \frac{1}{6} + \frac{2}{15} = \frac{4}{5}. \quad \boxed{✓}_{6}$$

✓ **Checkpoint 6**

Find each of the given probabilities for the scenario in Example 5.

(a) $P(U|A)$

(b) $P(U|B)$

EXAMPLE 6 **Jury Selection** Suppose 6 potential jurors remain in a jury pool and 2 are to be selected to sit on the jury for the trial. The races of the 6 potential jurors are 1 Hispanic, 3 Caucasian, and 2 African-American. If we select one juror at a time, find the probability that one Caucasian and one African-American are drawn.

Solution A probability tree showing the various possible outcomes is given in Figure 8.30 on the next page. In this diagram, C represents the event "selecting a Caucasian juror" and A represents "selecting an African-American juror." On the first draw, $P(C$ on the 1st$) = 3/6 = 1/2$ because three of the six jurors are Caucasian. On the second draw, $P(A$ on the 2nd$|C$ on the 1st$) = 2/5$. One Caucasian juror has been removed, leaving 5, of which 2 are African-American.

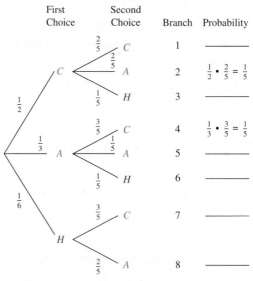

Figure 8.30

We want to find the probability of selecting exactly one Caucasian and exactly one African-American. Two events satisfy this condition: selecting a Caucasian first and then selecting an African-American (branch 2 of the tree) and drawing an African-American juror first and then selecting a Caucasian juror (branch 4). For branch 2,

✓ **Checkpoint 7**

In Example 6, find the probability of selecting an African-American juror and then a Caucasian juror.

$$P(C \text{ on 1st}) \cdot P(A \text{ on 2nd}|C \text{ on 1st}) = \frac{1}{2} \cdot \frac{2}{5} = \frac{1}{5}. \quad \boxed{✓}_{7}$$

For branch 4, on which the African-American juror is selected first,

$$P(A \text{ first}) \cdot P(C \text{ second}|A \text{ first}) = \frac{1}{3} \cdot \frac{3}{5} = \frac{1}{5}.$$

Since these two events are disjoint, the final probability is the sum of the two probabilities.

$$P(\text{one } C \text{ and one } A) = P(C \text{ on 1st}) \cdot P(A \text{ on 2nd} | C \text{ on 1st})$$

$$+ P(A \text{ on 1st}) \cdot P(C \text{ on 2nd} | A \text{ on 1st}) = \frac{2}{5}. \quad \boxed{8}$$

The product rule is often used in dealing with *stochastic processes*, which are mathematical models that evolve over time in a probabilistic manner. For example, selecting different jurors is such a process, because the probabilities change with each successive selection. (Particular stochastic processes are studied further in Section 9.5.)

EXAMPLE 7   Two cards are drawn without replacement from an ordinary deck (52 cards). Find the probability that the first card is a heart and the second card is red.

Solution Start with the probability tree of Figure 8.31. (You may wish to refer to the deck of cards shown on page 422.) On the first draw, since there are 13 hearts in the 52 cards, the probability of drawing a heart first is $13/52 = 1/4$. On the second draw, since a (red) heart has been drawn already, there are 25 red cards in the remaining 51 cards. Thus, the probability of drawing a red card on the second draw, given that the first is a heart, is $25/51$. By the product rule of probability,

$$P(\text{heart on 1st and red on 2nd})$$

$$= P(\text{heart on 1st}) \cdot P(\text{red on 2nd} | \text{heart on 1st})$$

$$= \frac{1}{4} \cdot \frac{25}{51} = \frac{25}{204} \approx .1225. \quad \boxed{9}$$

EXAMPLE 8   Three cards are drawn, without replacement, from an ordinary deck. Find the probability that exactly 2 of the cards are red.

Solution Here, we need a probability tree with three stages, as shown in Figure 8.32. The three branches indicated with arrows produce exactly 2 red cards from the draws. Multiply the probabilities along each of these branches and then add:

$$P(\text{exactly 2 red cards}) = \frac{26}{52} \cdot \frac{25}{51} \cdot \frac{26}{50} + \frac{26}{52} \cdot \frac{26}{51} \cdot \frac{25}{50} + \frac{26}{52} \cdot \frac{26}{51} \cdot \frac{25}{50}$$

$$= \frac{50,700}{132,600} = \frac{13}{34} \approx .3824. \quad \boxed{10}$$

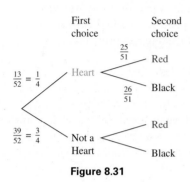

✓ **Checkpoint 8**

In Example 6, find the probability of selecting a Caucasian juror and then a Hispanic juror.

First choice / **Second choice**

$\frac{13}{52} = \frac{1}{4}$ Heart $\frac{25}{51}$ Red $\frac{26}{51}$ Black

$\frac{39}{52} = \frac{3}{4}$ Not a Heart Red Black

Figure 8.31

✓ **Checkpoint 9**

Find the probability of drawing a heart on the first draw and a black card on the second if two cards are drawn without replacement.

✓ **Checkpoint 10**

Use the tree in Example 8 to find the probability that exactly one of the cards is red.

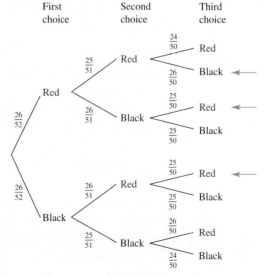

Figure 8.32

Independent Events

Suppose in Example 7 that we draw the two cards *with* replacement rather than without replacement. (That is, we put the first card back, shuffle the deck, and then draw the second card.) If the first card is a heart, then the probability of drawing a red card on the second draw is 26/52, rather than 25/51, because there are still 52 cards in the deck, 26 of them red. In this case, P(red second|heart first) is the same as P(red second). The value of the second card is not affected by the value of the first card. We say that the event that the second card is red is *independent* of the event that the first card is a heart, since knowledge of the first card does not influence what happens to the second card. On the other hand, when we draw *without* replacement, the events that the first card is a heart and the second is red are *dependent* events. The fact that the first card is a heart means that there is one less red card in the deck, influencing the probability that the second card is red.

As another example, consider tossing a fair coin twice. If the first toss shows heads, the probability that the next toss is heads is still 1/2. Coin tosses are independent events, since the outcome of one toss does not influence the outcome of the next toss. Similarly, rolls of a fair die are independent events. On the other hand, the events "the milk is old" and "the milk is sour" are dependent events: If the milk is old, there is an increased chance that it is sour. Also, in the example at the beginning of this section, the events A (the stock's dividend yield is 2.00% or higher) and B (the stock price is $100 or higher) are dependent events, because $P(A|B)$ is different from $P(A)$.

If events E and F are independent, then the knowledge that E has occurred gives no (probability) information about the occurrence or nonoccurrence of event F. That is, $P(F)$ is exactly the same as $P(F|E)$, or

$$P(F|E) = P(F).$$

This, in fact, is the formal definition of independent events.

> E and F are **independent events** if
> $$P(F|E) = P(F) \quad \text{or} \quad P(E|F) = P(E).$$

If the events are not independent, they are **dependent events.**

When E and F are independent events, $P(F|E) = P(F)$, and the product rule becomes

$$P(E \cap F) = P(E) \cdot P(F|E) = P(E) \cdot P(F).$$

Conversely, if this equation holds, it follows that $P(F) = P(F|E)$. Consequently, we have the useful rule that follows.

> ## Product Rule for Independent Events
> E and F are independent events if and only if
> $$P(E \cap F) = P(E) \cdot P(F).$$

EXAMPLE 9 **Manufacturing** A USB flash drive requires a two-step manufacturing process. The first step involves the circuit board, and let us assume that 98% of the circuit board assemblies are satisfactory. The second step involves the plastic casing. Assume that 96.5% of the casings are satisfactory. Assume the manufacturing process of the circuit boards has no relationship with the manufacturing of the plastic casings, and find the probability that both components are satisfactory.

Solution Since the manufacturing processes of the two components have no relationship, we can assume that event A (the circuit board is satisfactory) and event B (the plastic

✓ **Checkpoint 11**

Find the probability of getting 4 successive heads on 4 tosses of a fair coin.

casing is satisfactory) are two independent events. Thus, the probability that both components are satisfactory is

$$P(A \cap B) = P(A) \cdot P(B) = (.98)(.965) = .9457.$$ ✓₁₁

⚠ **CAUTION** It is common to confuse the ideas of *disjoint* events and *independent* events. Events E and F are disjoint if $E \cap F = \varnothing$. For example, if a family has exactly one child, the only possible outcomes are $B = \{boy\}$ and $G = \{girl\}$. The events B and G are disjoint and $P(B) = P(G) = .5$. However, the events are *not* independent, since $P(G|B) = 0$ (if a family with only one child has a boy, the probability that it has a girl is then 0). Since $P(G|B) \neq P(G)$, the events are not independent. Of all the families with exactly *two* children, the events $G_1 = \{first\ child\ is\ a\ girl\}$ and $G_2 = \{second\ child\ is\ a\ girl\}$ are independent, because $P(G_2|G_1)$ equals $P(G_2)$. However, G_1 and G_2 are not disjoint, since $G_1 \cap G_2 = \{both\ children\ are\ girls\} \neq \varnothing$.

To show that two events E and F are independent, we can show that $P(F|E) = P(F)$, that $P(E|F) = P(E)$, or that $P(E \cap F) = P(E) \cdot P(F)$. Another way is to observe that knowledge of one outcome does not influence the probability of the other outcome, as we did for coin tosses.

📄 **NOTE** In some cases, it may not be apparent from the physical description of the problem whether two events are independent or not. For example, it is not obvious whether the event that a baseball player gets a hit tomorrow is independent of the event that he got a hit today. In such cases, it is necessary to use the definition and calculate whether $P(F|E) = P(F)$, or, equivalently, whether $P(E \cap F) = P(E) \cdot P(F)$.

EXAMPLE 10   **Types of Businesses** Among partnerships and corporations that file tax returns in the United States, the probability that the enterprise consists of a firm dedicated to real estate, rentals, or leasing is .2498. The probability that the firm is a partnership is .3672. The probability that a firm is dedicated to real estate, rentals, or leasing or is a partnership is .4367. Are the events of a firm being dedicated to real estate, rentals, or leasing and a firm being a partnership independent?

Solution Let E represent the event the firm is dedicated to real estate, rentals, or leasing and let F represent the event the firm is a partnership. We must determine whether

$$P(E|F) = P(E) \quad \text{or} \quad P(F|E) = P(F).$$

We know that $P(E) = .2498$, $P(F) = .3672$, and $P(E \cup F) = .4367$. By the addition rule, we know that

$$P(E) + P(F) - P(E \cap F) = P(E \cup F)$$
$$.2498 + .3672 - P(E \cap F) = .4367$$
$$.6170 - P(E \cap F) = .4367$$
$$P(E \cap F) = .1803.$$

Therefore,

$$P(E|F) = \frac{P(E \cap F)}{P(F)} = \frac{.1803}{.3672} \approx .4910 \neq .2498 = P(E).$$

✓ **Checkpoint 12**

The probability of living in Texas is .08, the probability of speaking English at home is .83, and the probability of living in Texas or speaking English at home is .854. Are the events "speaking English at home" and "living in Texas" independent?

Also,

$$P(F|E) = \frac{P(E \cap F)}{P(E)} = \frac{.1803}{.2498} \approx .7218 \neq .3672 = P(F).$$

Since $P(E|F) \neq P(E)$ and $P(F|E) \neq P(F)$ these two events are not independent. ✓₁₂

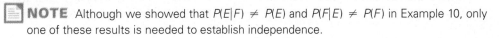 📄 **NOTE** Although we showed that $P(E|F) \neq P(E)$ and $P(F|E) \neq P(F)$ in Example 10, only one of these results is needed to establish independence.

8.5 Exercises

If a single fair die is rolled, find the probability of rolling the given events. (See Examples 1 and 2.)

1. 3, given that the number rolled was odd

2. 5, given that the number rolled was even

3. An odd number, given that the number rolled was 3

If two fair dice are rolled (recall the 36-outcome sample space), find the probability of rolling the given events.

4. A sum of 8, given that the sum was greater than 7

5. A sum of 6, given that the roll was a "double" (two identical numbers)

6. A double, given that the sum was 9

If two cards are drawn without replacement from an ordinary deck, find the probabilities of the given event. (See Example 7.)

7. The second is a heart, given that the first is a heart

8. The second is black, given that the first is a spade

9. A jack and a 10 are drawn.

10. An ace and a 4 are drawn.

11. In your own words, explain how to find the conditional probability $P(E|F)$.

12. Your friend asks you to explain how the product rule for independent events differs from the product rule for dependent events. How would you respond?

13. Another friend asks you to explain how to tell whether two events are dependent or independent. How would you reply? (Use your own words.)

14. A student reasons that the probability in Example 3 of both coins being heads is just the probability that the other coin is a head—that is, $1/2$. Explain why this reasoning is wrong.

Decide whether the two events listed are independent.

15. S is the event that it is snowing and L is the event that the instructor is late for class.

16. R is the event that four semesters of theology are required to graduate from a certain college and A is the event that the college is a religiously affiliated school.

17. R is the event that it rains in the Amazon jungle and H is the event that an instructor in New York City writes a difficult exam.

18. T is the event that Tom Cruise's next movie grosses over $200 million and R is the event that the Republicans have a majority in Congress in the year 2024.

19. In a two-child family, if we assume that the probabilities of a child being male and a child being female are each .5, are the events "the children are the same sex" and "at most one child is male" independent? Are they independent for a three-child family?

20. Let A and B be independent events with $P(A) = \dfrac{1}{4}$ and $P(B) = \dfrac{1}{5}$. Find $P(A \cap B)$ and $P(A \cup B)$.

Firm Size *According to data from the U.S. Small Business Administration, the probability that a firm has one to four employees given that it is a construction firm is .6930. The probability that a firm has one to four employees given that it is a manufacturing firm is .4196. The probability the firm is a construction firm is .1135, and the probability that a firm is a manufacturing firm is .0449. The probability that a firm has one to four employees is .6214. Use this information to find the probabilities for Exercises 21–24. (See Example 4.)*

21. A firm has one to four employees and is a construction firm.

22. A firm has one to four employees and is a manufacturing firm.

23. A firm is a construction firm given that it has one to four employees.

24. A firm is a manufacturing firm given that it has one to four employees.

Hours Worked *Data from the 2016 General Social Survey can allow us to estimate how many hours people who classified their work status as part-time or full-time actually worked in the last week. Use the following table to find the probabilities of the events in Exercises 25–30. (Data from: gss.norc.org.)*

Labor Force Status	Hours Worked in the Last Week					
	0–19	20–29	30–39	40–49	50 or More	Total
Full-Time Worker	26	43	137	691	377	1274
Part-Time Worker	74	82	87	41	24	308
Total	100	125	224	732	401	1582

25. The worker was full-time given that he or she worked 50 or more hours.

26. The worker was part-time given that she or he worked 50 or more hours.

27. The hours worked were 0–19 given that the worker was full-time.

28. The hours worked were 20–29 given that the worker was part-time.

29. The hours worked were 20 or less given that the worker was part-time.

30. The hours worked were 40 or more given that the worker was full-time.

Consumer Confidence *The 2016 General Social Survey asked respondents about their confidence in banks and financial institutions as well as their confidence in major corporations. Each question allowed three responses: "a great deal, only some, or hardly any." The table below summarizes the responses.*

Confidence in Banks and Financial Institutions	Confidence in Major Corporations			
	A Great Deal	Only Some	Hardly Any	Total
A Great Deal	125	119	25	269
Only Some	172	759	113	1044
Hardly Any	43	362	201	606
Total	340	1240	339	1919

Find the probability of the described events. (Data from: gss.norc.org.)

31. A respondent has a great deal of confidence in banks and financial institutions given that he or she has hardly any confidence in major corporations.

32. A respondent has hardly any confidence in banks and financial institutions given that he or she has hardly any confidence in major corporations.

33. A respondent has a great deal of confidence in major corporations given that he or she has only some confidence in banks and financial institutions.

34. A respondent has hardly any confidence in major corporations given that he or she has hardly any confidence in banks and financial institutions.

Color Blindness *The following table shows frequencies for red–green color blindness, where M represents that a person is male and C represents that a person is color blind.*

	M	M'	Totals
C	.042	.007	.049
C'	.485	.466	.951
Totals	.527	.473	1.000

Use the table to find the given probabilities.

35. $P(M)$

36. $P(C)$

37. $P(M \cap C)$

38. $P(M \cup C)$

39. $P(M|C)$

40. $P(M'|C)$

41. Are the events C and M dependent? Recall that two events E and F are dependent if $P(E|F) \neq P(E)$. (See Example 10.)

42. Are the events M' and C dependent?

Color Blindness and Deafness *A scientist wishes to determine whether there is any dependence between color blindness (C) and deafness (D). Use the probabilities in the table to answer Exercises 43 and 44.*

	D	D'	Total
C	.0004	.0796	.0800
C'	.0046	.9154	.9200
Total	.0050	.9950	1.0000

43. Find $P(C)$, $P(D)$, $P(C \cap D)$, and $P(C|D)$. Does $P(C) = P(C|D)$? What does that result imply regarding independence of C and D?

44. Find $P(D')$, $P(C)$, $P(D' \cap C)$, and $P(D'|C)$. Does $P(D') = P(D'|C)$? What does that result imply regarding independence of D' and C?

Driver's Test *The Motor Vehicle Department in a certain state has found that the probability of a person passing the test for a driver's license on the first try is .75. The probability that an individual who fails on the first test will pass on the second try is .80, and the probability that an individual who fails the first and second tests will pass the third time is .70. Find the probability of the given event.*

45. A person fails both the first and second tests

46. A person will fail three times in a row

Airlines *The number of on-time flights and the number of delayed or cancelled flights for four major airlines for March 2017 is given in the table below.*

Airline	On Time	Delayed or Cancelled
American	62,642	15,472
Delta	68,721	10,349
Southwest	91,664	23,480
United	39,262	9235

Find the probability of the described event. (Data from: www.transportation.gov.)

47. An on-time flight given the airline was American.

48. An on-time flight given the airline was Delta.

49. An on-time flight given the airline was Southwest.

50. An on-time flight given the airline was United.

51. The airline was United given that the flight was on time.

52. The airline was American given that the flight was on time.

53. The airline was not Delta given that the flight was delayed or cancelled.

54. The airline was not American given that the flight was delayed or cancelled.

55. Which airline had the highest probability of an on-time flight?

56. Which airline had the lowest probability of an on-time flight?

Suppose the probability that the first record by a singing group will be a hit is .32. If the first record is a hit, so are all the group's subsequent records. If the first record is not a hit, the probability of the group's second record and all subsequent ones being hits is .16. If the first two records are not hits, the probability that the third is a hit is .08. The probability that a record is a hit continues to decrease by half with each successive nonhit record. Find the probability of the given event.

57. The group will have at least one hit in its first four records.

58. The group will have a hit in its first six records if the first three are not hits.

Work the given problems on independent events. (See Examples 9 and 10.)

59. Computer Failure Corporations such as banks, where a computer is essential to day-to-day operations, often have a second, backup computer in case of failure by the main computer. Suppose that there is a .003 chance that the main computer will fail in a given period and a .005 chance that the backup computer will fail while the main computer is being repaired. Suppose these failures represent independent events, and find the fraction of the time the corporation can assume that it will have computer service.

60. Multiple Flights According to data from the U.S. Department of Transportation, Delta Airlines was on time approximately 87% of the time in March 2017. Use this information, and assume that the event that a given flight takes place on time is independent of the event that another flight is on time to answer the following questions

(a) Elisabeta plans to visit her company's branch offices; her journey requires 3 separate flights on Delta Airlines. What is the probability that all of these flights will be on time?

(b) How reasonable do you believe it is to suppose the independence of being on time from flight to flight?

61. Rocket Failure The probability that a key component of a space rocket will fail is .03.

(a) How many such components must be used as backups to ensure that the probability that at least one of the components will work is .9999999?

(b) Is it reasonable to assume independence here?

62. Medication Effectiveness A medical experiment showed that the probability that a new medicine is effective is .75, the probability that a patient will have a certain side effect is .4, and the probability that both events will occur is .3. Decide whether these events are dependent or independent.

Consumer Confidence *Refer to the table included for Exercises 31–34.*

63. Are the events of having hardly any confidence in banks and financial institutions and having hardly any confidence in major corporations independent?

64. Are the events of having a great deal of confidence in banks and financial institutions and having a great deal of confidence in major corporations independent?

✓**Checkpoint Answers**

1. (a) About .0891 **(b)** About .5743 **(c)** About .7228

2. (a) About .3849 **(b)** About .1943 **(c)** About .7488

(d) About .4655

(e) The probability of never being married, given that the person is partially or not happy

3. 1/3   **4.** 2/3   **5.** 2/7

6. (a) 1/4   **(b)** 3/5

7. 1/5   **8.** 1/10   **9.** 13/102 ≈ .1275

10. 13/34 ≈ .3824   **11.** 1/16   **12.** No

8.6 Bayes' Formula

Suppose the probability that a person gets lung cancer, given that the person smokes a pack or more of cigarettes daily, is known. For a research project, it might be necessary to know the probability that a person smokes a pack or more of cigarettes daily, given that the person has lung cancer. More generally, if $P(E|F)$ is known for two events E and F, can $P(F|E)$ be found? The answer is yes, we can find $P(F|E)$ by using the formula to be developed in this section. To develop this formula, we can use a probability tree to find $P(F|E)$. Since $P(E|F)$ is known, the first outcome is either F or F'. Then, for each of these outcomes, either E or E' occurs, as shown in Figure 8.33.

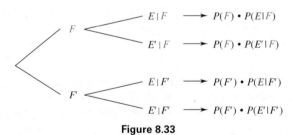

Figure 8.33

The four cases have the probabilities shown on the right. By the definition of conditional probability and the product rule,

$$P(E) = P(F \cap E) + P(F' \cap E),$$

$$P(F \cap E) = P(F) \cdot P(E|F), \quad \text{and} \quad P(F' \cap E) = P(F') \cdot P(E|F').$$

By substitution,

$$P(E) = P(F) \cdot P(E|F) + P(F') \cdot P(E|F') \tag{1}$$

and

$$P(F|E) = \frac{P(F \cap E)}{P(E)} = \frac{P(F) \cdot P(E|F)}{P(F) \cdot P(E|F) + P(F') \cdot P(E|F')}.$$

We have proved a special case of Bayes' formula, which is generalized later in this section. ✔₁

✓ Checkpoint 1

Use the special case of Bayes' formula to find $P(F|E)$ if $P(F) = .2$, $P(E|F) = .1$, and $P(E|F') = .3$. [*Hint:* $P(F') = 1 - P(F)$.]

Bayes' Formula (Special Case)

$$P(F|E) = \frac{P(F) \cdot P(E|F)}{P(F) \cdot P(E|F) + P(F') \cdot P(E|F')}.$$

EXAMPLE 1 **Start-Up Success** Data reported by the *Harvard Business Review* states that start-ups with at least one female founder are more likely to succeed and that 18% of start-ups now have at least one female founder. Additionally, the probability that a start-up succeeds given that there is at least one female founder is .15. The probability that a start-up succeeds given there are only male founders is .09.

(a) Find the probability that a start-up succeeds.

Solution Let E represent the event that the start-up succeeds, and let F represent the event of having at least one female founder. From the given information, we have

$$P(F) = .18, \ P(E|F) = .15, P(E|F') = .09.$$

We can use the complement rule so that we have

$$P(F') = 1 - .18 = .82, P(E'|F) = 1 - .15 = .85, P(E'|F') = 1 - .09 = .91.$$

These probabilities are shown on the probability tree in Figure 8.34.

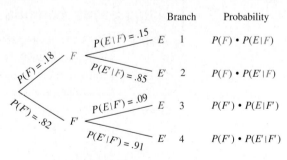

Figure 8.34

From Equation **(1)** earlier, we have

$$P(E) = P(F) \cdot P(E|F) + P(F') \cdot P(E|F').$$

Substituting values from our tree diagram in Figure 8.34, we have

$$P(E) = (.18)(.15) + (.82)(.09) = .1008.$$

Thus, the probability that a start-up is successful is about .1.

(b) Find the probability that there is at least one female founder given that the start-up succeeds.

Solution Using our notation, we want to find $P(F|E)$. Applying Bayes' formula, we find that

$$P(F|E) = \frac{P(F) \cdot P(E|F)}{P(F) \cdot P(E|F) + P(F') \cdot P(E|F')}$$

$$= \frac{(.18)(.15)}{(.18)(.15) + (.82)(.09)} \approx .2679.$$

The probability that a start-up has at least one female founder given that the start-up succeeds is about .2679. ✔ 2

If we rewrite the special case of Bayes' formula, replacing F by F_1 and F' by F_2, then it says:

$$P(F_1|E) = \frac{P(F_1) \cdot P(E|F_1)}{P(F_1) \cdot P(E|F_1) + P(F_2) \cdot P(E|F_2)}.$$

Since $F_1 = F$ and $F_2 = F'$ we see that F_1 and F_2 are disjoint and that their union is the entire sample space. The generalization of Bayes' formula to more than two possibilities follows this same pattern.

✓ **Checkpoint 2**

In Example 1, find $P(F'|E)$.

Bayes' Formula

Suppose $F_1, F_2, \ldots, F_n$ are pairwise disjoint events (meaning that any two of them are disjoint) whose union is the sample space. Then for an event E and for each i with $1 \le i \le n$,

$$P(F_i|E) = \frac{P(F_i) \cdot P(E|F_i)}{P(F_1) \cdot P(E|F_1) + \cdots + P(F_n) \cdot P(E|F_n)}.$$

This result is known as Bayes' formula, after the Reverend Thomas Bayes (1702–61), whose paper on probability was published in 1764 after his death.

The statement of Bayes' formula can be daunting. It may be easier to remember the formula by thinking of the probability tree that produced it. Go through the following steps.

Using Bayes' Formula

Step 1 Start a probability tree with branches representing events $F_1, F_2, \ldots, F_n$. Label each branch with its corresponding probability.

Step 2 From the end of each of these branches, draw a branch for event E. Label this branch with the probability of getting to it, or $P(E|F_i)$.

Step 3 There are now n different paths that result in event E. Next to each path, put its probability: the product of the probabilities that the first branch occurs, $P(F_i)$, and that the second branch occurs, $P(E|F_i)$: that is, $P(F_i) \cdot P(E|F_i)$.

Step 4 $P(F_i|E)$ is found by dividing the probability of the branch for F_i by the sum of the probabilities of all the branches producing event E.

Example 2 illustrates this process.

EXAMPLE 2   **Marital Status and Number of Children** The 2016 General Social Survey of women who are age 18 or older indicated that 86% of married women have one or more children, 42% of never married women have one or more children, and 90% of women who are divorced, separated, or widowed have one or more children. The survey also indicated that 41% of women age 18 or older were currently married, 25% had never been married, and 34% were divorced, separated, or widowed (labeled "other"). Find the probability that a woman who has one or more children is married.

Solution Let E represent the event "having one or more children," with F_1 representing "married women," F_2 representing "never married women," and F_3 "other." Then

$$P(F_1) = .41; \quad P(E|F_1) = .86;$$
$$P(F_2) = .25; \quad P(E|F_2) = .42;$$
$$P(F_3) = .34; \quad P(E|F_3) = .90.$$

We need to find $P(F_1|E)$, the probability that a woman is married, given that she has one or more children. First, draw a probability tree using the given information, as in Figure 8.35. The steps leading to event E are shown.

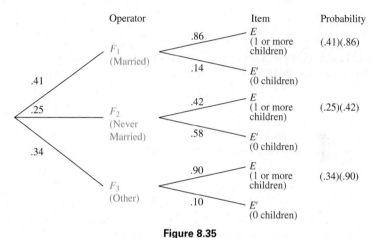

Figure 8.35

Find $P(F_1|E)$, using the top branch of the tree shown in Figure 8.35, by dividing the probability of this branch by the sum of the probabilities of all the branches leading to E:

$$P(F_1|E) = \frac{(.41)(.86)}{(.41)(.86) + (.25)(.42) + (.34)(.90)} \approx .4618.$$

✓**Checkpoint 3**

In Example 2, find

(a) $P(F_2|E)$;

(b) $P(F_3|E)$.

EXAMPLE 3 **Business** A manufacturer buys items from six different suppliers. The fraction of the total number of items obtained from each supplier, along with the probability that an item purchased from that supplier is defective, is shown in the following table:

Supplier	Fraction of Total Supplied	Probability of Being Defective
1	.05	.04
2	.12	.02
3	.16	.07
4	.23	.01
5	.35	.03
6	.09	.05

Find the probability that a defective item came from supplier 5.

Solution Let F_1 be the event that an item came from supplier 1, with $F_2, F_3, F_4, F_5,$ and F_6 defined in a similar manner. Let E be the event that an item is defective. We want to find $P(F_5|E)$. By Bayes' formula,

$$P(F_5|E) = \frac{(.35)(.03)}{(.05)(.04) + (.12)(.02) + (.16)(.07) + (.23)(.01) + (.35)(.03) + (.09)(.05)}$$

$$= \frac{.0105}{.0329} \approx .3191.$$

There is about a 32% chance that a defective item came from supplier 5. Even though supplier 5 has only 3% defectives, his probability of being "guilty" is relatively high, about 32%, because of the large fraction of items he supplies. ✓ 4

✓ **Checkpoint 4**

In Example 3, find the probability that the defective item came from

(a) supplier 3;

(b) supplier 6.

8.6 Exercises

For two events M and N, P(M) = .4, P(N|M) = .3, and P(N|M') = .4. Find each of the given probabilities. (See Example 1.)

1. $P(M|N)$

2. $P(M'|N)$

For disjoint events $R_1, R_2,$ and $R_3, P(R_1) = .05, P(R_2) = .6,$ and $P(R_3) = .35.$ In addition, $P(Q|R_1) = .40, P(Q|R_2) = .30,$ and $P(Q|R_3) = .60.$ Find each of the given probabilities. (See Examples 2 and 3.)

3. $P(R_1|Q)$

4. $P(R_2|Q)$

5. $P(R_3|Q)$

6. $P(R_1'|Q)$

Suppose three jars have the following contents: 2 black balls and 1 white ball in the first; 1 black ball and 2 white balls in the second; 1 black ball and 1 white ball in the third. If the probability of selecting one of the three jars is 1/2, 1/3, and 1/6, respectively, find the probability that if a white ball is drawn, it came from the given jar.

7. The second jar

8. The third jar

S&P 500 Stocks *On March 7, 2017, the probability that a stock was in the Information Technology sector was .1347. The probability that the stock had a dividend yield of 2.00% or higher given that it was in the Information Technology sector was .2941. The probability that a stock had a dividend yield of 2.00% or higher given that it was not in the Information Technology sector was .4462. Find the following probabilities. (Data from: www.okfn.org.)*

9. The stock was in the Information Technology sector given that it had a dividend yield of 2.00% or higher.

10. The stock was not in the Information Technology sector given that it did not have a dividend yield of 2.00% or higher.

Educational Attainment and Earnings *Data from the U.S. Census Bureau indicates that for the year 2015, 33.3% of the labor force age 25 and over had a high school diploma or fewer years of education, 27.9% had some college or an associate's degree, and 38.9% had a bachelor's degree or more education. Of those with a high school diploma or fewer years of education, 7.4% earned $75,000 or more. Of those with some college or an associate's degree, 13.5% earned $75,000 or more, and of those with a bachelor's degree or more education, 37.0% earned $75,000 or more. Find the following probabilities that a randomly chosen labor force participant has the given characteristics. (Data from: www.census.gov.)*

11. Some college or an associate's degree, given that he or she earns $75,000 or more annually

12. A bachelor's degree or more education, given that he or she earns $75,000 or more annually

13. A high school diploma or less education, given that he or she earns *less* than $75,000 annually

14. A bachelor's degree or more education, given that he or she earns *less* than $75,000 annually

Airlines *In March 2017, the probability that a flight was an American Airlines flight was .1599, the probability that it was a Delta flight was .1618, and the probability that it was another airline was .6783. The probability that the flight was on time given that it was an American Airlines flight was .8019, given that it was a Delta flight was .8691, and given that it was another airline was .7820. Find the probability of the following events. (Data from: www.transportation.gov.)*

15. American Airlines given that it was on time.

16. Delta given that it was on time.

17. An airline not American or Delta given that it was not on time.

18. Delta given that it was not on time.

Health *In a test for toxemia, a disease that affects pregnant women, a woman lies on her left side and then rolls over on her back. The test is considered positive if there is a 20-mm rise in her blood pressure within 1 minute. The results have produced the following probabilities, where T represents having toxemia at some time during the pregnancy and N represents a negative test:*

$$P(T'|N) = .90 \quad and \quad P(T|N') = .75.$$

Assume that $P(N') = .11,$ and find each of the given probabilities.

19. $P(N|T)$

20. $P(N'|T)$

Business *In April 2017, the total sales from General Motors, Ford, or Chrysler was 630,596 cars or light trucks. The probability that the vehicle sold was made by General Motors was .3873, by Ford .3385, and by Chrysler .2742. Additionally, the probability that a General Motors* *vehicle sold was a car was .2765, a Ford vehicle sold was a car was .2336, and a Chrysler vehicle sold was a car was .1253.*

21. Given the vehicle sold was a car, find the probability it was made by General Motors.

22. Given the vehicle sold was a car, find the probability it was made by Chrysler.

23. Given the vehicle sold was not a car, find the probability it was made by Ford.

24. Given the vehicle sold was not a car, find the probability it was made by General Motors.

Business *At the close of the markets on May 19, 2017, there were 6100 companies listed with the New York Stock Exchange (NYSE) and NASDAQ. For the day May 19, 2017, based on data from the Wall Street Journal, the probabilities of advancing, declining, and unchanged company stocks appears below for the two exchanges.*

Exchange	Percentage of Companies Listed	Probability of Stock Advancing	Probability of Stock Declining	Probability of Stock Remaining Unchanged
NYSE	.5085	.7240	.2453	.0306
NASDAQ	.4915	.6024	.3436	.0540

Find the probability of each of the following events.

25. The stock was listed on the NYSE given that it advanced.

26. The stock was listed on NASDAQ given that it declined.

27. The stock was listed on the NYSE given that it remained unchanged.

28. The stock was listed on NASDAQ given that it remained unchanged.

29. The stock was listed on NASDAQ given that it advanced.

30. The stock was listed on the NYSE given that it declined.

Happiness *The proportion of adult males is .492 and the proportion of adult females is .508. Data from the 2016 General Social Survey has the following probabilities of happiness for each sex. (Data from: U.S. Census Bureau and gss.norc.org.)*

Sex	Proportion	Very Happy	Pretty Happy	Not Too Happy
Male	.492	.287	.564	.149
Female	.508	.277	.558	.165

Find the probability of the described events.

31. Given that the person is very happy, the respondent is male.

32. Given that the person is pretty happy, the respondent is female.

33. Given that the person is not too happy, the respondent is female.

34. Given that the person is very happy, the respondent is female.

35. Given that the person is pretty happy, the respondent is male.

36. Given that the person is not too happy, the respondent is male.

✓ **Checkpoint Answers**

1. $1/13 \approx .077$

2. About .7321

3. (a) About .1375   (b) About .4007

4. (a) About .3404   (b) About .1368

CHAPTER 8 Summary and Review

Key Terms and Symbols

{ }	set braces
∈	is an element of
∉	is not an element of
∅	empty set
⊆	is a subset of
⊈	is not a subset of
⊂	is a proper subset of
A'	complement of set A
∩	set intersection
∪	set union
$P(E)$	probability of event E
$P(F \mid E)$	probability of event F, given that event E has occurred

8.1 set
element (member)
empty set
set-builder notation
universal set
subset
set operations
tree diagram
Venn diagram
complement
intersection
disjoint sets
union

8.2 addition rule for counting

8.3 experiment
trial
outcome
sample space
event
certain event
impossible event
disjoint events
basic probability principle
relative frequency probability
probability distribution

8.4 addition rule for probability
complement rule
odds

8.5 conditional probability
product rule of probability
probability tree
independent events
dependent events

8.6 Bayes' formula

Chapter 8 Key Concepts

Sets

Set A is a **subset** of set B if every element of A is also an element of B. A set of n elements has 2^n subsets.

Let A and B be any sets with universal set U.

The **complement** of A is $A' = \{x \mid x \notin A \text{ and } x \in U\}$.
The **intersection** of A and B is $A \cap B = \{x \mid x \in A \text{ and } x \in B\}$.
The **union** of A and B is $A \cup B = \{x \mid x \in A \text{ or } x \in B \text{ or both}\}$.
$n(A \cup B) = n(A) + n(B) - n(A \cap B)$, where $n(X)$ is the number of elements in set X.

Basic Probability Principle

Let S be a sample space of equally likely outcomes, and let event E be a subset of S. Then the probability that event E occurs is

$$P(E) = \frac{n(E)}{n(S)}.$$

Addition Rule

For any events E and F from a sample space S,

$$P(E \cup F) = P(E) + P(F) - P(E \cap F).$$

For disjoint events E and F,

$$P(E \cup F) = P(E) + P(F).$$

Complement Rule

$P(E) = 1 - P(E')$ and $P(E') = 1 - P(E)$.

Odds

The odds in favor of event E are $\dfrac{P(E)}{P(E')}$, $P(E') \neq 0$.

Properties of Probability

1. For any event E in sample space S, $0 \leq P(E) \leq 1$.
2. The sum of the probabilities of all possible distinct outcomes is 1.

Conditional Probability

The conditional probability of event E, given that event F has occurred, is

$$P(E|F) = \frac{P(E \cap F)}{P(F)}, \quad \text{where} \quad P(F) \neq 0.$$

For equally likely outcomes, conditional probability is found by reducing the sample space to event F; then

$$P(E|F) = \frac{n(E \cap F)}{n(F)}.$$

Product Rule of Probability

If E and F are events, then $P(E \cap F)$ may be found by either of these formulas:

$$P(E \cap F) = P(F) \cdot P(E|F) \quad \text{or} \quad P(E \cap F) = P(E) \cdot P(F|E).$$

Product Rule for Independent Events

E and F are independent events if and only if

$$P(E \cap F) = P(E)P(F).$$

Bayes' Formula

$$P(F_i|E) = \frac{P(F_i) \cdot P(E|F_i)}{P(F_1) \cdot P(E|F_1) + P(F_2) \cdot P(E|F_2) + \cdots + P(F_n) \cdot P(E|F_n)}.$$

Chapter 8 Review Exercises

Write true or false for each of the given statements.

1. $9 \in \{8, 4, -3, -9, 6\}$

2. $4 \in \{3, 9, 7\}$

3. $2 \notin \{0, 1, 2, 3, 4\}$

4. $0 \notin \{0, 1, 2, 3, 4\}$

5. $\{3, 4, 5\} \subset \{2, 3, 4, 5, 6\}$

6. $\{1, 2, 5, 8\} \subseteq \{1, 2, 5, 10, 11\}$

7. $\{1, 5, 9\} \subset \{1, 5, 6, 9, 10\}$

8. $0 \subseteq \varnothing$

List the elements in the given sets.

9. $\{x \mid x \text{ is a national holiday}\}$

10. $\{x \mid x \text{ is an integer}, -3 \leq x < 1\}$

11. $\{\text{all natural numbers less than 5}\}$

12. $\{x \mid x \text{ is a leap year between 1999 and 2018}\}$

Let $U = \{Vitamins\ A,\ B_1,\ B_2,\ B_3,\ B_6,\ B_{12},\ C,\ D,\ E\}$, $M = \{Vitamins\ A,\ C,\ D,\ E\}$, and $N = \{Vitamins\ A,\ B_1,\ B_2,\ C,\ E\}$. Find the given sets.

13. M'

14. N'

15. $M \cap N$

16. $M \cup N$

17. $M \cup N'$

18. $M' \cap N$

Consider these sets:

$U = \{Students\ taking\ Intermediate\ Accounting\}$;
$A = \{Females\}$;
$B = \{Finance\ majors\}$;
$C = \{Students\ older\ than\ 22\}$;
$D = \{Students\ with\ a\ GPA > 3.5\}$.

Describe each of the following in words.

19. $A \cap C$

20. $B \cap D$

21. $A \cup D$

22. $A' \cap D$

23. $B' \cap C'$

Draw a Venn diagram and shade the given set in it.

24. $B \cup A'$

25. $A' \cap B$

26. $A' \cap (B' \cap C)$

Box Office Receipts *As of May 2017, the 50 highest grossing movies as measured by domestic box office receipts could be categorized as follows. (Data from: www.the-movie-times.com.)*

28 were action movies;
30 were rated PG-13;
36 were made in 2005 or more recently;
24 were action and PG-13;
23 were action and made in 2005 or more recently;
23 were PG-13 and made in 2005 or more recently;
20 were action, PG-13, and made in 2005 or more recently.

27. How many movies were action movies and not rated PG-13?

28. How many movies were action or rated PG-13?

29. How many movies were made in the year 2005 or more recently or rated PG-13?

30. How many movies were not action, not rated PG-13, and made prior to the year 2005?

Write sample spaces for the given scenarios.

31. A die is rolled and the number of dots showing is noted.

32. A color is selected from the set {red, blue, green}, and then a number is chosen from the set {10, 20, 30}.

Music Selection *A student purchases a digital music player and installs 10 songs on the device to see how it works. The genres of the songs are rock (3 songs), pop (4 songs), and alternative (3 songs). She listens to the first two songs on shuffle mode.*

33. Write the sample space for the genre if shuffle mode picks songs at random. (*Note:* Shuffle mode is allowed to play the same song twice in a row.)

34. Are the outcomes in the sample space for Exercise 33 equally likely?

Business *A customer wants to purchase a computer and printer. She has narrowed her selection among 2 Dell models, 1 Apple model, and 2 HP models for the computer and 2 Epson models and 3 HP models for the printer.*

35. Write the sample space for the brands among which she can choose for the computer and printer.

36. Are the outcomes in the sample space for Exercise 35 equally likely?

Business *A company sells computers and copiers. Let E be the event "a customer buys a computer," and let F be the event "a customer buys a copier." In Exercises 37 and 38, write each of the given scenarios, using $\cap$, $\cup$, or ' as necessary.*

37. A customer buys neither a computer nor a copier.

38. A customer buys at least one computer or copier.

39. A student gives the answer to a probability problem as 6/5. Explain why this answer must be incorrect.

40. Describe what is meant by disjoint sets, and give an example.

41. Describe what is meant by mutually exclusive events, and give an example.

42. How are disjoint sets and mutually exclusive events related?

Stock Fund Allocation *The American Century Global Small Cap Fund had the following sector allocations as of May 2017. (Data from: personal.vanguard.com.)*

Sector	Percentage
Consumer Discretionary	17.46
Consumer Staples	4.11
Energy	4.54
Financials	16.16
Health Care	9.89
Industrials	19.06
Information Technology	15.94
Materials	9.89
Real Estate	1.76
Telecommunication Services	0
Utilities	0

Find the probability that an investment selected at random from this fund is in the given sector.

43. Consumer Discretionary or Consumer Staples

44. Information Technology or Industrials

45. Not Financials

46. Not Information Technology

47. Neither Consumer Discretionary nor Consumer Staples

48. Neither Materials nor Industrials

Confidence in Major Corporations *The following table cross classifies years of education and confidence in major companies from adult participants in the 2016 General Social Survey. (Data from: gss.norc.org.)*

	Confidence in Major Corporations			
Years of Education	A Great Deal	Only Some	Hardly Any	Total
Less than 12	42	149	57	248
12 Years	95	389	82	566
13–16 Years	153	502	153	808
More Than 16 Years	49	197	45	291
Total	339	1237	337	1913

If a participant is chosen at random, find the probability that he or she has the following characteristics.

49. 12 years or less of school

50. A great deal of confidence or only some confidence in major corporations

51. More than 16 years of education and only some confidence in major corporations

52. Less than 12 years of education and hardly any confidence in major corporations

53. 13–16 years of education or only some confidence in major corporations

54. More than 16 years of education or hardly any confidence in major corporations

Confidence in Major Corporations *Use the table above to determine if the probability changes substantially that someone has a great deal of confidence in major corporations depending on her or his years of education. Do this by finding the following.*

55. The probability a person has a great deal of confidence in major corporations given that the person has less than 12 years of school

56. The probability a person has a great deal of confidence in major corporations given that the person has 12 years of school

57. The probability a person has a great deal of confidence in major corporations given that the person has 13–16 years of school

58. The probability a person has a great deal of confidence in major corporations given that the person has more than 16 years of school

59. Which of your answers in 55–58 is the lowest? Which is the highest?

60. Do you think years of education seems to influence whether someone has a great deal of confidence in major corporations? Explain.

Health *The partial table shows the four possible (equally likely) combinations when both parents are carriers of the sickle-cell anemia trait. Each carrier parent has normal cells (N) and trait cells (T).*

		2nd Parent	
		N_2	T_2
1st Parent	N_1		$N_1 T_2$
	T_1		

61. Complete the table.

62. If the disease occurs only when two trait cells combine, find the probability that a child born to these parents will have sickle-cell anemia.

63. Find the probability that the child will carry the trait, but not have the disease, if a normal cell combines with a trait cell.

64. Find the probability that the child neither is a carrier nor has the disease.

Find the probabilities for the given sums when two fair dice are rolled.

65. 8

66. No more than 4

67. At least 9

68. Odd and greater than 8

69. 2, given that the sum is less than 4

70. 7, given that at least one die shows a 4

Suppose $P(E) = .62$, $P(F) = .45$, and $P(E \cap F) = .28$. Find each of the given probabilities.

71. $P(E \cup F)$

72. $P(E \cap F')$

73. $P(E' \cup F)$

74. $P(E' \cap F')$

For the events E and F, $P(E) = .2$, $P(E|F) = .3$, and $P(F) = .4$. Find each of the given probabilities.

75. $P(E'|F)$

76. $P(E|F')$

77. Define independent events, and give an example of one.

78. Are independent events always disjoint? Are they ever disjoint? Give examples.

79. **Stock Market Performance** On May 12, 2017, the odds of a stock on the NYSE declining in value were 419:353. What is the probability that a stock declined in value?

80. **Stock Market Performance** On May 12, 2017, the odds of a stock on the NASDAQ declining in value were 1635:1348. What is the probability that a stock declined in value?

81. **Stock Market Performance** On May 18, 2017, the probability that a stock on the NASDAQ advanced in value was .5452. What are the odds that a stock increased in value?

82. **Stock Market Performance** On May 18, 2017, the probability that a stock on the NYSE advanced in value was .4879. What are the odds that a stock advanced in value?

Business *Of the appliance repair shops listed in the phone book, 80% are competent and 20% are not. A competent shop can repair an appliance correctly 95% of the time; an incompetent shop can repair an appliance correctly 60% of the time. Suppose an appliance was repaired correctly. Find the probability that it was repaired by the given type of shop.*

83. A competent shop

84. An incompetent shop

Suppose an appliance was repaired incorrectly. Find the probability that it was repaired by the given type of shop.

85. A competent shop

86. An incompetent shop

Manufacturing *A manufacturer buys items from four different suppliers. The fraction of the total number of items that is obtained*

from each supplier, along with the probability that an item purchased from that supplier is defective, is shown in the following table:

Supplier	Fraction of Total Supplied	Probability of Defective
1	.17	.04
2	.39	.02
3	.35	.07
4	.09	.03

87. Find the probability that a defective item came from supplier 4.

88. Find the probability that a defective item came from supplier 2.

Titanic Survival *The following tables list the number of passengers who were on the Titanic and the number of passengers who survived, according to class of ticket:*[*]

	CHILDREN		WOMEN	
	On	Survived	On	Survived
First Class	6	6	144	140
Second Class	24	24	165	76
Third Class	79	27	93	80
Total	109	57	402	296

---

[*] Sandra L. Takis, "Titanic: A Statistical Exploration," *Mathematics Teacher* 92, no. 8, (November 1999): pp. 660–664. Reprinted with permission. ©1999 by the National Council of Teachers of Mathematics. All rights reserved.

	MEN		TOTALS	
	On	Survived	On	Survived
First Class	175	57	325	203
Second Class	168	14	357	114
Third Class	462	75	634	182
Total	805	146	1316	499

Use this information to determine the given probabilities. (Round answers to three decimal places.)

89. What is the probability that a randomly selected passenger was in second class?

90. What is the overall probability of surviving?

91. What is the probability of a first-class passenger surviving?

92. What is the probability of a child who was in third class surviving?

93. Given that a survivor is from first class, what is the probability that she was a woman?

94. Given that a male has survived, what is the probability that he was in third class?

95. Are the events "third-class survival" and "male survival" independent events? What does this imply?

96. Are the events "first-class survival" and "child survival" independent events? What does this imply?

Case Study 8 Medical Diagnosis

When patients undergo medical testing, a positive test result for a disease or condition can be emotionally devastating. In many cases, however, testing positive does not necessarily imply that the patient actually has the disease. Bayes' formula can be very helpful in determining the probability of actually having the disease when a patient tests positive.

Let us label the event of having the disease as D and not having the disease as D'. We will denote testing positive for the disease as T and testing negative as T'. Suppose a medical test is calibrated on patients so that we know that among patients with the disease, the test is positive 99.95% of the time. (This quantity is often called the **sensitivity** of the test.) Among patients known not to have the disease, 99.90% of the time the test gave a negative result. (This quantity is often called the **specificity** of the test.) In summary, we have

Sensitivity $= P(T|D) = .9995$ and Specificity $= P(T'|D') = .9990$.

Using the complement rule, we find that the probability the test will give a negative result when a patient has the disease is

$$P(T'|D) = 1 - P(T|D) = 1 - .9995 = .0005.$$

Similarly, for those patients without the disease, the probability of testing positive is .0010, calculated by

$$P(T|D') = 1 - P(T'|D') = 1 - .9990 = .0010.$$

These results do not yet answer the question of interest: If a patient tests positive for the disease, what is the probability the patient actually has the disease? Using our notation, we want to know $P(D|T)$. There are two steps to finding this probability. The first is that we need an estimate of the prevalence of the disease in the general population. Let us assume that one person in a thousand has the disease. We can then calculate that

$$P(D) = \frac{1}{1000} = .001 \quad \text{and} \quad P(D') = 1 - .001 = .999.$$

With this information, and the previous results from testing, we can now use Bayes' formula to find $P(D|T)$:

$$P(D|T) = \frac{P(D)P(T|D)}{P(D)P(T|D) + P(D')P(T|D')}.$$

Using $P(D) = .001$, $P(D') = .999$, the sensitivity $P(T|D) = .9995$ and the complement to the specificity $P(T|D') = .0010$, we have

$$P(D|T) = \frac{(.001)(.9995)}{(.001)(.9995) + (.999)(.0010)} = \frac{.0009995}{.0019985} \approx .5001.$$

Hence, the probability the patient actually has the disease after testing positive for the disease is only about .5. This is approximately the same probability as guessing "heads" when flipping a coin. It seems paradoxical that a test which has such high sensitivity (in this case, .9995) and specificity (in this case, .999) could lead to a probability of merely .5 that a person who tests positive for the disease actually has the disease. This is why it is imperative to have confirmatory tests run after testing positive.

The other factor in the calculation is the prevalence of the disease among the general population. In our example, we used $P(D) = .001$. Often, it is very difficult to know how prevalent a disease is among the general population. If the disease is more prevalent, such as 1 in 100, or $P(D) = .01$, we find the probability of a patient's having the disease, given that the patient tests positive, as

$$P(D|T) = \frac{(.01)(.9995)}{(.01)(.9995) + (.99)(.0010)} = \frac{.009995}{.010985} \approx .9099.$$

So when the disease has higher prevalence, then the probability of having the disease after testing positive is also higher. If the disease has a lower prevalence (as in the case of our first example), then the probability of having the disease after testing positive could be much lower than one might otherwise think.

Exercises

1. Suppose the specificity of a test is 0.999. Find $P(T|D')$.

2. If the sensitivity of a test for a disease is .99 and the prevalence of the disease is .005, use your answer to Exercise 1 to find the probability of a patient's having the disease, given that the patient tested positive, or $P(D|T)$.

3. Recalculate your answer to Exercise 2 using a prevalence of disease of .0005.

4. According to some experts, the sensitivity of a mammogram in detecting breast cancer can range from .75 to .85, while the specificity might range from .85 to .95. If the prevalence of actually having breast cancer is .0145, find the probability of having the disease, given that the sensitivity is .8, the specificity is .9, and the patient tests positive. (Data from: www.cancer.gov and breast-cancer.ca/est-incidn.)

5. Use the data from Exercise 4, but assume that the sensitivity is .90 and the specificity is .95.

Extended Project

For a disease or ailment of interest, find in the medical literature the prevalence of the disease or ailment among the general population, the specificity of a particular test to detect the disease or ailment, and the sensitivity of the same test. Then find the probability that a patient actually has the disease, given that the patient tested positive with the particular test. Did your results surprise you? Why or why not?

Counting, Probability Distributions, and Further Topics in Probability

9

CHAPTER OUTLINE

Probability has applications to quality control in manufacturing and to decision making in business. It plays a role in testing new medications, in evaluating DNA evidence in criminal trials, and in a host of other situations. Sophisticated counting techniques are often necessary for determining the probabilities used in these applications.

Probability distributions enable us to compute the "average value" or "expected outcome" when an experiment or process is repeated a number of times. These distributions are introduced in Section 9.1 and used in Sections 9.4 and 9.6. The other focus of this chapter is the development of effective ways to count the possible outcomes of an experiment without actually listing them all (which can be *very* tedious when large numbers are involved). These counting techniques are introduced in Section 9.2 and are used to find probabilities throughout the rest of the chapter.

9.1 Probability Distributions and Expected Value

Probability distributions were introduced briefly in Section 8.3. Now we take a more complete look at them. In this section, we shall see that the *expected value* of a probability distribution is a type of average. A probability distribution depends on the idea of a *random variable*, so we begin with that.

Random Variables

One of the questions asked in the 2016 General Social Survey (GSS) had to do with respondents' daily hours of TV viewing. For those who watched 6 hours or less, respondent's were given the choice of choosing the values 0 through 6. We will label their response as x. Since the value of x is random, x is called a random variable.

> ### Random Variable
>
> A **random variable** is a function that assigns a real number to each outcome of an experiment.

The following table gives each possible outcome of the study question on TV and video use together with the probability $P(x)$ of each outcome x. (Data from: gss.norc.org.)

x	0	1	2	3	4	5	6
$P(x)$	.10	.21	.28	.17	.13	.06	.05

A table that lists all the outcomes with the corresponding probabilities is called a **probability distribution.** The sum of the probabilities in a probability distribution must always equal 1. (The sum in some distributions may vary slightly from 1 because of rounding.)

Instead of writing the probability distribution as a table, we could write the same information as a set of ordered pairs:

$$\{(0, .10), (1, .21), (2, .28), (3, .17), (4, .13), (5, .06), (6, .05)\}.$$

There is just one probability for each value of the random variable.

The information in a probability distribution is often displayed graphically as a special kind of bar graph called a **histogram.** The bars of a histogram all have the same width, usually 1 unit. The heights of the bars are determined by the probabilities. A histogram for the data in the probability distribution presented above is given in Figure 9.1. A histogram shows important characteristics of a distribution that may not be readily apparent in tabular form, such as the relative sizes of the probabilities and any symmetry in the distribution.

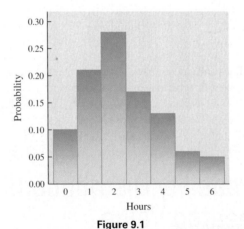

Figure 9.1

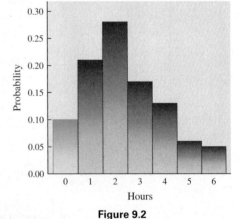

Figure 9.2

The area of the bar above $x = 0$ in Figure 9.1 is the product of 1 and .10, or $1 \cdot .10 = .10$. Since each bar has a width of 1, its area is equal to the probability that corresponds to its x-value. The probability that a particular value will occur is thus given by the area of the appropriate bar of the graph. For example, the probability that one or more hours are spent watching TV is the sum of the areas for $x = 1, x = 2, x = 3, x = 4, x = 5$, and $x = 6$. This area, shown in red in Figure 9.2, corresponds to .90 of the total area, since

$$
\begin{aligned}
P(x \geq 1) = P(x = 1) &+ P(x = 2) + P(x = 3) + P(x = 4) \\
&+ P(x = 5) + P(x = 6) \\
= .10 + .21 &+ .28 + .17 + .13 + .06 + .05 \\
= .90
\end{aligned}
$$

EXAMPLE 1

(a) Give the probability distribution for the number of heads showing when two coins are tossed.

Solution Let x represent the random variable "number of heads." Then x can take on the value 0, 1, or 2. Now find the probability of each outcome. When two coins are tossed, the sample space is {TT, TH, HT, HH}. So the probability of getting one head is $2/4 = 1/2$. Similar analysis of the other cases produces this table.

x	0	1	2
$P(x)$	1/4	1/2	1/4

(b) Draw a histogram for the distribution in the table. Find the probability that at least one coin comes up heads.

Solution The histogram is shown in Figure 9.3. The portion in red represents

$$
P(x \geq 1) = P(x = 1) + P(x = 2)
$$
$$
= \frac{3}{4}.
$$

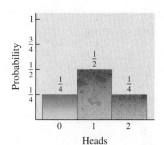

Figure 9.3

✓ **Checkpoint 1**

(a) Give the probability distribution for the number of heads showing when three coins are tossed.

(b) Draw a histogram for the distribution in part (a). Find the probability that no more than one coin comes up heads.

Answers to Checkpoint exercises are found at the end of the section.

▦ **TECHNOLOGY TIP** Virtually all graphing calculators can produce histograms. The procedures differ on various calculators, but you usually are required to enter the outcomes in one list and the corresponding frequencies in a second list. For specific details, check your instruction manual under "statistics graphs" or "statistical plotting." To get the histogram in Figure 9.3 with a TI-84+ calculator, we entered the outcomes 0, 1, and 2 in the first list and entered the probabilities .25, .5, and .25 in a second list. Two versions of the histogram are shown in Figure 9.4. They differ slightly because different viewing windows were used. With some calculators, the probabilities must be entered as integers, so make the entries in the second list 1, 2, and 1 (corresponding to 1/4, 2/4, and 1/4, respectively), and use a window with $0 \leq y \leq 4$.

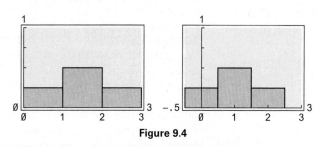

Figure 9.4

Expected Value

In working with probability distributions, it is useful to have a concept of the typical or average value that the random variable takes on. In Example 1, for instance, it seems reasonable that, on the average, one head shows when two coins are tossed. This does not tell what will happen the next time we toss two coins; we may get two heads, or we may get none. If we tossed two coins many times, however, we would expect that, in the long run, we would average about one head for each toss of two coins.

A way to solve such problems in general is to imagine flipping two coins 4 times. Based on the probability distribution in Example 1, we would expect that 1 of the 4 times we would get 0 heads, 2 of the 4 times we would get 1 head, and 1 of the 4 times we would get 2 heads. The total number of heads we would get, then, is

$$0 \cdot 1 + 1 \cdot 2 + 2 \cdot 1 = 4.$$

The expected number of heads per toss is found by dividing the total number of heads by the total number of tosses:

$$\frac{0 \cdot 1 + 1 \cdot 2 + 2 \cdot 1}{4} = 0 \cdot \frac{1}{4} + 1 \cdot \frac{1}{2} + 2 \cdot \frac{1}{4} = 1.$$

Notice that the expected number of heads turns out to be the sum of the three values of the random variable x, multiplied by their corresponding probabilities. We can use this idea to define the *expected value* of a random variable as follows.

Expected Value

Suppose that the random variable x can take on the n values $x_1, x_2, x_3, \ldots, x_n$. Suppose also that the probabilities that these values occur are, respectively, $p_1, p_2, p_3, \ldots, p_n$. Then the **expected value** of the random variable is

$$E(x) = x_1 p_1 + x_2 p_2 + x_3 p_3 + \cdots + x_n p_n.$$

EXAMPLE 2 **TV Viewing** In the case of TV viewing at the beginning of this section, find the expected number of hours per day of viewing.

Solution Multiply each outcome in the table on page 460 by its probability, and sum the products:

$$E(x) = (0)(.10) + (1)(.21) + (2)(.28) + (3)(.17) + (4)(.13) + (5)(.06) + (6)(.05)$$
$$= 2.4.$$

On the average, a respondent of the survey will indicate 2.4 hours of TV viewing.

✓ Checkpoint 2

Find the expected value of the number of heads showing when four coins are tossed.

Physically, the expected value of a probability distribution represents a balance point. If we think of the histogram in Figure 9.1 as a series of weights with magnitudes represented by the heights of the bars, then the system would balance if supported at the point corresponding to the expected value.

EXAMPLE 3 **Fund-Raising** Suppose a local symphony decides to raise money by raffling off concert box seats worth $400, a dinner for two worth $100, and two CD recordings worth $20 each. A total of 2000 tickets are sold at $1 each. Find the expected value of winning for a person who buys one ticket in the raffle.

Solution Here, the random variable represents the possible amounts of net winnings, where net winnings = amount won − cost of ticket. For example, the net winnings of the person winning the box seats are $400 (amount won) − $1 (cost of ticket) = $399, and the net winnings for each losing ticket are $0 − $1 = −$1.

The net winnings of the various prizes, as well as their respective probabilities, are shown in the table on the next page. The probability of winning $19 is 2/2000, because there are 2 prizes worth $20. (We have not reduced the fractions in order to keep all the denominators equal.) Because there are 4 winning tickets, there are 1996 losing tickets, so the probability of winning $-\$1$ is $1996/2000$.

x	\$399	\$99	\$19	$-\$1$
$P(x)$	1/2000	1/2000	2/2000	1996/2000

The expected winnings for a person buying one ticket are

$$399\left(\frac{1}{2000}\right) + 99\left(\frac{1}{2000}\right) + 19\left(\frac{2}{2000}\right) + (-1)\left(\frac{1996}{2000}\right) = -\frac{1460}{2000}$$
$$= -.73.$$

On average, a person buying one ticket in the raffle will lose \$.73, or 73¢.

It is not possible to lose 73¢ in this raffle: Either you lose \$1, or you win a prize worth \$400, \$100, or \$20, minus the \$1 you paid to play. But if you bought tickets in many such raffles over a long time, you would lose 73¢ per ticket, on average. It is important to note that the expected value of a random variable may be a number that can never occur in any one trial of the experiment. ☑ 3

✓ **Checkpoint 3**

Suppose you buy 1 of 10,000 tickets at \$1 each in a lottery where the prize is \$5,000. What are your expected net winnings? What does this answer mean?

📄 **NOTE** An alternative way to compute expected value in this and other such examples is to calculate the expected amount won and then subtract the cost of the ticket afterward. The amount won is either \$400 (with probability 1/2000), \$100 (with probability 1/2000), \$20 (with probability 2/2000), or \$0 (with probability 1996/2000). The expected winnings for a person buying one ticket are then

$$400\left(\frac{1}{2000}\right) + 100\left(\frac{1}{2000}\right) + 20\left(\frac{2}{2000}\right) + 0\left(\frac{1996}{2000}\right) - 1 = -\frac{1460}{2000}$$
$$= -.73.$$

EXAMPLE 4   Each day, Lynette and Tanisha toss a coin to see who buys coffee (at \$1.75 a cup). One tosses, while the other calls the outcome. If the person who calls the outcome is correct, the other buys the coffee; otherwise the caller pays. Find Lynette's expected winnings.

Solution Assume that an honest coin is used, that Tanisha tosses the coin, and that Lynette calls the outcome. The possible results and corresponding probabilities are shown in the following table:

	Possible Results			
Result of Toss	Heads	Heads	Tails	Tails
Call	Heads	Tails	Heads	Tails
Caller Wins?	Yes	No	No	Yes
Probability	1/4	1/4	1/4	1/4

Lynette wins a \$1.75 cup of coffee whenever the results and calls match, and she loses \$1.75 when there is no match. Her expected winnings are

$$1.75\left(\frac{1}{4}\right) + (-1.75)\left(\frac{1}{4}\right) + (-1.75)\left(\frac{1}{4}\right) + 1.75\left(\frac{1}{4}\right) = 0.$$

On the average, over the long run, Lynette breaks even. ☑ 4

✓ **Checkpoint 4**

Find Tanisha's expected winnings.

A game with an expected value of 0 (such as the one in Example 4) is called a **fair game.** Casinos do not offer fair games. If they did, they would win (on the average) \$0 and have a hard time paying their employees! Casino games have expected winnings for the house that vary from 1.5 cents per dollar to 60 cents per dollar. The next example examines the popular game of roulette.

EXAMPLE 5 **Roulette** As we saw in Chapter 8, an American roulette wheel has 38 slots. Two of the slots are marked 0 and 00 and are colored green. The remaining slots are numbered 1–36 and are colored red and black (18 slots are red and 18 slots are black). One simple wager is to bet $1 on the color red. If the marble lands in a red slot, the player gets his or her dollar back, plus $1 of winnings. Find the expected winnings for a $1 bet on red.

Solution For this bet, there are only two possible outcomes: winning or losing. The random variable has outcomes $+1$ if the marble lands in a red slot and -1 if it does not. We need to find the probability for these two outcomes. Since there are 38 total slots, 18 of which are colored red, the probability of winning a dollar is $18/38$. The player will lose if the marble lands in any of the remaining 20 slots, so the probability of losing the dollar is $20/38$. Thus, the probability distribution is

x	-1	$+1$
$P(x)$	$\dfrac{20}{38}$	$\dfrac{18}{38}$

The expected winnings are

$$E(x) = -1\left(\frac{20}{38}\right) + 1\left(\frac{18}{38}\right) = -\frac{2}{38} \approx -.053.$$

The winnings on a dollar bet for red average out to losing about a nickel on every spin of the roulette wheel. In other words, a casino earns, on average, 5.3 cents on every dollar bet on red. ☑5

Exercises 17–20 at the end of the section ask you to find the expected winnings for other bets on games of chance. The idea of expected value can be very useful in decision making, as shown by the next example.

EXAMPLE 6 **Real Estate Development** A real estate developer has two potential projects. With Project 1, there is a 30% chance of earning $1.25 million in profits and a 70% chance of earning $1.10 million in profits. With Project 2, there is a 25% chance of earning $1.45 million in profits and a 75% chance of earning $1.05 million in profits. Based strictly on expected value, which job should be developed?

Solution The probability distribution for Project 1 is shown below.

Profits x	1.10	1.25
$P(x)$	.7	.3

The expected value for Project 1 is then

$$E(x) = 1.10(.7) + 1.25(.3)$$
$$= 1.145.$$

The probability distribution for Project 2 is shown below.

Profits x	1.05	1.45
$P(x)$	.75	.25

The expected value for Project 2 is

$$E(x) = 1.05(.75) + 1.45(.25)$$
$$= 1.15.$$

The expected value for Project 2 of $1.15 million is slightly higher than Project 1. The developer should develop Project 2 based only on expected value. ☑6

✓**Checkpoint 5**

A gambling game requires a $5 bet. If the player wins, she gets $1000, but if she loses, she loses her $5. The probability of winning is .001. What are the expected winnings of this game?

✓**Checkpoint 6**

After college, a person is offered two jobs. With job A, after five years, there is a 50% chance of making $60,000 per year and a 50% chance of making $45,000. With job B, after five years, there is a 30% chance of making $80,000 per year and a 70% chance of making $35,000. Based strictly on expected value, which job should be taken?

EXAMPLE 7   **Social Science** The table gives the probability distribution for the number of children of respondents to the 2016 General Social Survey, for those with 7 or fewer children. (Data from: gss.norc.org.)

x	0	1	2	3	4	5	6	7
P(x)	.281	.162	.258	.165	.075	.032	.018	.009

Find the expected value for the number of children.

Solution Using the formula for the expected value, we have

$$E(x) = 0(.281) + 1(.162) + 2(.258) + 3(.165) + 4(.075) + 5(.032) + 6(.018) + 7(.009)$$
$$= 1.804.$$

For those respondents with 7 or fewer children, the number of children, on average, is 1.804.

9.1 Exercises

For each of the experiments described, let x determine a random variable and use your knowledge of probability to prepare a probability distribution. (Hint: Use a tree diagram.)

1. Four children are born, and the number of boys is noted. (Assume an equal chance of a boy or a girl for each birth.)

2. Two dice are rolled, and the total number of dots is recorded.

3. Three cards are drawn from a deck. The number of Queens are counted.

4. Two names are drawn from a hat, signifying who should go pick up pizza. Three of the names are on the swim team and two are not. The number of swimmers selected is counted.

Draw a histogram for each of the given exercises, and shade the region that gives the indicated probability. (See Example 1.)

5. Exercise 1; $P(x \le 2)$

6. Exercise 2; $P(x \ge 11)$

7. Exercise 3; P(at least one queen)

8. Exercise 4; P(fewer than two swimmers)

Find the expected value for each random variable. (See Example 2.)

9.

x	1	3	5	7
P(x)	.1	.5	.2	.2

10.

y	0	15	30	40
P(y)	.15	.20	.40	.25

11.

z	0	2	4	8	16
P(z)	.21	.24	.21	.17	.17

12.

x	5	10	15	20	25
P(x)	.40	.30	.15	.10	.05

Find the expected values for the random variables x whose probability functions are graphed.

13.
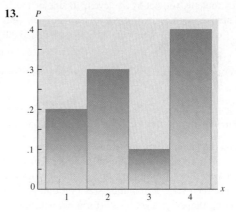

14.
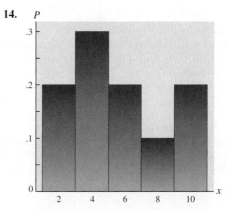

15.

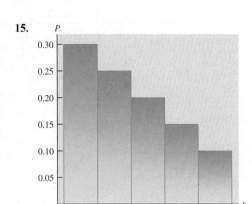

16.

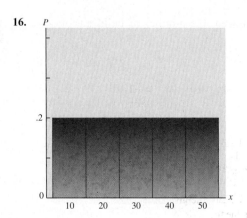

Games of Chance *Find the expected winnings for the games of chance described in Exercises 17–20. (See Example 5.)*

17. In one form of roulette, you bet $1 on "even." If one of the 18 positive even numbers comes up, you get your dollar back, plus another one. If one of the 20 other numbers (18 odd, 0, and 00) comes up, you lose your dollar.

18. Repeat Exercise 17 if there are only 19 noneven numbers (no 00).

19. *Numbers* is a game in which you bet $1 on any three-digit number from 000 to 999. If your number comes up, you get $500.

20. In one form of the game Keno, the house has a pot containing 80 balls, each marked with a different number from 1 to 80. You buy a ticket for $1 and mark one of the 80 numbers on it. The house then selects 20 numbers at random. If your number is among the 20, you get $3.20 (for a net winning of $2.20).

21. Online Gambling An online gambling site offers a first prize of $50,000 and two second prizes of $10,000 each for registered users when they place a bet. A random bet will be selected over a 24-hour period. Two million bets are received in the contest. Find the expected winnings if you can place one registered bet of $1 in the given period.

22. Online Gambling An online gambling site offers a lottery with a first prize of $25,000, a second prize of $10,000, and a third prize of $5000. It takes $5 to enter and a winner is selected at random. What are the expected winnings if 20,000 people enter the lottery?

Prize Promotion *A contest at a fast-food restaurant offered the following cash prizes and probabilities of winning when one buys a large order of French fries for $1.89:*

Prize	Probability
$100,000	1/8,504,860
$50,000	1/302,500
$10,000	1/282,735
$1000	1/153,560
$100	1/104,560
$25	1/9,540

23. Find the expected winnings if the player buys one large order of French fries.

24. Find the expected winnings if the player buys 25 large orders of French fries in multiple visits.

25. Unbanked Households According to the Federal Deposit Insurance Corporation (FDIC), 28.3% of all U.S. households conduct some or all of their financial transactions outside the mainstream banking system. Such people are sometimes called the "unbanked." If we select four households at random, and let x denote the number of unbanked households, the probability distribution of x is:

x	0	1	2	3	4
$P(x)$	.2643	.4173	.2470	.0650	.0064

Find the expected value for the number of unbanked households.

26. Retirement Funds Approximately 54% of U.S. residents own individual stocks, stock mutual funds, or stocks in a retirement fund. If we select five U.S. residents at random and let x denote the number who own stocks, the probability distribution of x is:

x	0	1	2	3	4	5
$P(x)$	.0206	.1209	.2838	.3332	.1956	.0459

Find the expected value for the number of U.S. residents who own stocks. (Data from: www.gallup.com.)

For Exercises 27–30, determine whether the probability distributions are valid or not. If not, explain why.

27.

x	5	10	15	20	25	30	35
$P(x)$	.01	.09	.25	.45	.05	.20	−.05

28.

x	−2	−1	0	1	2	3	4
$P(x)$	.05	.10	.75	.02	.03	.04	.01

29.

x	1	3	5	7	9	11
$P(x)$	.01	.02	.03	.04	.05	.85

30.

x	−10	−5	0	5	10
P(x)	.50	.10	−.20	.30	.30

For Exercises 31–35, fill in the missing value(s) to make a valid probability distribution.

31.

x	5	10	15	20	25	30
P(x)	.01	.09	.25	.45	.05	

32.

x	−3	−2	−1	0	1	2	3
P(x)		.15	.15	.15	.15	.15	.15

33.

x	10	20	30	40
P(x)	.20		.25	.30

34.

x	−50	−40	−30	−20	−10	0	10
P(x)	.05	.25	.10	.10	.05		

35.

x	1	2	3	6	12	24	48
P(x)	.10	.10	.20	.25	.05		

36. Retail Returns During the month of July, a home improvement store sold a great many air-conditioning units, but some were returned. The following table shows the probability distribution for the daily number of returns of air-conditioning units sold in July:

x	0	1	2	3	4	5
P(x)	.55	.31	.08	.04	.01	.01

Find the expected number of returns per day.

37. Life Insurance An insurance company has written 100 policies of $15,000, 250 of $10,000, and 500 of $5000 for people age 20. If experience shows that the probability that a person will die in the next year at age 20 is .0007, how much can the company expect to pay out during the year after the policies were written?

38. Business A market researcher came upon a recent survey by the Pew Research Center that found that 73% of U.S. teenagers owned a smartphone. If six teens are selected at random, the probability distribution for x, the number of teens with smartphones, is as follows.

x	0	1	2	3	4	5	6
P(x)	.0004	.0063	.0425	.1531	.3105	.3358	.1513

Find the expected number of teens with smartphones.

39. Auto Sales In April 2017, General Motors captured 17.1% of the light vehicles (cars and light trucks) sales market. If three new

vehicle sales are selected at random, the probability distribution for x, the number of sales that are General Motors vehicles, is given in the following table.

x	0	1	2	3
P(x)	.5697	.3526	.0727	.0050

Find the expected number of sales among the three sales selected that are General Motors vehicles. (Data from: www.wsj.com.)

40. Gaming Device According to a recent report of the Entertainment Software Association, 80% of U.S. households own a gaming device of some sort. If five households are selected at random, the probability distribution for x, the number of households that own a gaming device, is given in the following table.

x	0	1	2	3	4	5
P(x)	.0003	.0064	.0512	.2048	.4096	.3277

Find the expected number of households among the 5 selected that own a gaming device. (Data from: venturebeat.com.)

41. Drug Cost Effectiveness Two antibiotics are used to treat common infections. Researchers wish to compare the two antibiotics—Drug A and Drug B—for their cost effectiveness. Drug A is inexpensive, safe, and effective. Drug B is also safe. However, it is considerably more expensive and is generally more effective. Use the given tree diagram (in which costs are estimated as the total cost of the medication, a visit to a medical office, and hours of lost work) to complete the following tasks:

(a) Find the expected cost of using each antibiotic to treat a middle-ear infection.

(b) To minimize the total expected cost, which antibiotic should be chosen?

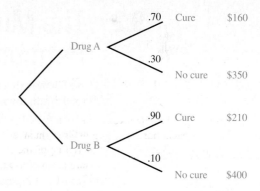

42. Hurricane Seeding One of the few methods that can be used in an attempt to cut the severity of a hurricane is to *seed* the storm. In this process, silver iodide crystals are dropped into the storm in order to decrease the wind speed. Unfortunately, silver iodide crystals sometimes cause the storm to *increase* its speed. Wind speeds may also increase or decrease even with no seeding. Use the given tree diagram to complete the following tasks.

(a) Find the expected amount of damage under each of the options, "seed" and "do not seed."

(b) To minimize total expected damage, which option should be chosen?

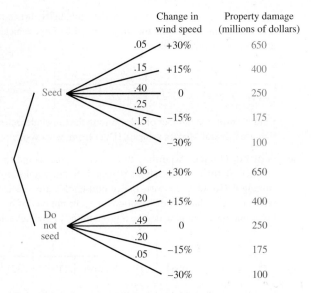

	Change in wind speed	Property damage (millions of dollars)

43. In the 2016 Gentlemen's Singles Wimbledon Championships, Andy Murray and Milos Raonic played in the finals. The prize money for the winner was £2,000,000 (British pounds sterling), and the prize money for the runner-up was £1,000,000. Find the expected winnings for Andy Murray if we

   (a) assume both players had an equal chance of winning;

   (b) use the players' prior head-to-head match record whereby Murray had a .64 probability of winning.

44. In the 2016 Ladies' Singles Wimbledon Championships, Serena Williams and Angelique Kerber played in the finals. The prize money for the winner was £2,000,000 (British pounds sterling), and the prize money for the runner-up was £1,000,000. Find the

expected winnings for Serena Williams if we use the players' prior head-to-head match record whereby Williams had a .71 probability of winning.

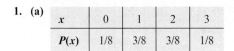

Checkpoint Answers

1. (a)

x	0	1	2	3
$P(x)$	1/8	3/8	3/8	1/8

   (b) 1/2

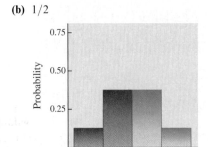

2. 2

3. $-\$.50$. On the average, you lose $.50 per ticket purchased.

4. 0

5. $-\$4$

6. Job A has an expected salary of $52,500, and job B has an expected salary of $48,500. Take job A.

9.2 The Multiplication Principle, Permutations, and Combinations

We begin with a simple example. If there are three roads from town A to town B and two roads from town B to town C, in how many ways can someone travel from A to C by way of B? We can solve this simple problem with the help of Figure 9.5, which lists all the possible ways to go from A to C.

The possible ways to go from A through B to C are a1, a2, b1, b2, c1, and c2. So there are 6 possible ways. Note that 6 is the product of $3 \cdot 2$, where 3 is the number of ways to go from A to B, and 2 is the number of ways to go from B to C.

Another way to solve this problem is to use a tree diagram, as shown in Figure 9.6. This diagram shows that, for each of the 3 roads from A, there are 2 different routes leading from B to C, making $3 \cdot 2 = 6$ different ways.

This example is an illustration of the *multiplication principle*.

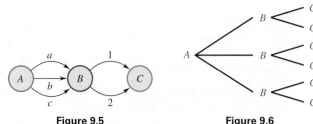

Figure 9.5   Figure 9.6

Multiplication Principle

Suppose n choices must be made, with

$$m_1 \text{ ways to make choice 1,}$$
$$m_2 \text{ ways to make choice 2,}$$
$$\vdots$$
$$m_n \text{ ways to make choice } n.$$

Then there are

$$m_1 \cdot m_2 \cdots \cdot m_n$$

different ways to make the entire sequence of choices.

EXAMPLE 1 Suppose Angela has 9 skirts, 8 blouses, and 13 different pairs of shoes. If she is willing to wear any combination, how many different skirt–blouse–shoe choices does she have?

Solution By the multiplication principle, there are 9 skirt choices, 8 blouse choices, and 13 shoe choices, for a total of $9 \cdot 8 \cdot 13 = 936$ skirt–blouse–shoe outfits.

EXAMPLE 2 **Kitchen Appliances** In May 2017, there were 111 built-in dishwashers, 224 gas stoves, and 82 side-by-side refrigerators available for purchase at the Home Depot website. How many different ways can a homeowner outfit his or her kitchen with these three appliances?

Solution A tree (or other diagram) would be far too complicated to use here, but the multiplication principle answers the question easily. There are

$$111 \cdot 224 \cdot 82 = 2,038,848$$

ways. ✓ₗ

EXAMPLE 3 A combination lock can be set to open to any 3-letter sequence.

(a) How many sequences are possible?

Solution Since there are 26 letters of the alphabet, there are 26 choices for each of the 3 letters, and, by the multiplication principle, $26 \cdot 26 \cdot 26 = 17,576$ different sequences.

(b) How many sequences are possible if no letter is repeated?

Solution There are 26 choices for the first letter. It cannot be used again, so there are 25 choices for the second letter and then 24 choices for the third letter. Consequently, the number of such sequences is $26 \cdot 25 \cdot 24 = 15,600$. ✓₂

Factorial Notation

The use of the multiplication principle often leads to products such as $5 \cdot 4 \cdot 3 \cdot 2 \cdot 1$, the product of all the natural numbers from 5 down to 1. If n is a natural number, the symbol $n!$ (read "n factorial") denotes the product of all the natural numbers from n down to 1. The factorial is an algebraic shorthand. For example, instead of writing $5 \cdot 4 \cdot 3 \cdot 2 \cdot 1$, we simply write $5!$. If $n = 1$, this formula is understood to give $1! = 1$.

✓ **Checkpoint 1**

On the same day described in Example 2, there were 6497 bathroom vanities with tops, 417 medicine cabinets, and 718 bathroom sink faucets available for purchase at the Home Depot website. How many different ways can a homeowner outfit her or his bathroom with these items?

✓ **Checkpoint 2**

(a) In how many ways can 6 business tycoons line up their golf carts at the country club?

(b) How many ways can 4 pupils be seated in a row with 4 seats?

n-Factorial

For any natural number *n*,

$$n! = n(n - 1)(n - 2) \cdots (3)(2)(1).$$

By definition, $0! = 1$.

Note that $6! = 6 \cdot 5 \cdot 4 \cdot 3 \cdot 2 \cdot 1 = 6 \cdot (5 \cdot 4 \cdot 3 \cdot 2 \cdot 1) = 6 \cdot 5!$. Similarly, the definition of $n!$ shows that

$$n! = n \cdot (n - 1)!.$$

One reason that $0!$ is defined to be 1 is to make the preceding formula valid when $n = 1$, for when $n = 1$, we have $1! = 1$ and $1 \cdot (1 - 1)! = 1 \cdot 0! = 1 \cdot 1 = 1$, so $n! = n \cdot (n - 1)!$. ✓3

Almost all calculators have an $n!$ key. A calculator with a 10-digit display and scientific-notation capability will usually give the exact value of $n!$ for $n \leq 13$ and approximate values of $n!$ for $14 \leq n \leq 69$. The value of $70!$ is approximately $1.198 \cdot 10^{100}$, which is too large for most calculators. So how would you simplify $\dfrac{100!}{98!}$? Depending on the type of calculator, there may be an overflow problem. The next two examples show how to avoid this problem.

Checkpoint 3

Evaluate:

(a) 4!

(b) 6!

(c) 1!

(d) 6!/4!

📖 **TECHNOLOGY TIP**

The factorial key on a graphing calculator is usually located in the PRB or PROB submenu of the MATH or OPTN menu.

EXAMPLE 4   Evaluate

$$\frac{100!}{98!}.$$

Solution We use the fact that $n! = n \cdot (n - 1)!$ several times:

$$\frac{100!}{98!} = \frac{100 \cdot 99!}{98!} = \frac{100 \cdot 99 \cdot 98!}{98!} = 100 \cdot 99 = 9900.$$

EXAMPLE 5   Evaluate

$$\frac{5!}{2! \, 3!}.$$

Solution $\dfrac{5!}{2! \, 3!} = \dfrac{5 \cdot 4!}{2! \, 3!} = \dfrac{5 \cdot 4 \cdot 3!}{2! \, 3!} = \dfrac{5 \cdot 4}{2 \cdot 1} = 10.$

EXAMPLE 6   **Morse Code** Morse code uses a sequence of dots and dashes to represent letters and words. How many sequences are possible with at most 3 symbols?

Solution "At most 3" means "1 or 2 or 3." Each symbol may be either a dot or a dash. Thus, the following numbers of sequences are possible in each case:

Number of Symbols	Number of Sequences
1	2
2	$2 \cdot 2 = 4$
3	$2 \cdot 2 \cdot 2 = 8$

Checkpoint 4

How many Morse code sequences are possible with at most 4 symbols?

Altogether, $2 + 4 + 8 = 14$ different sequences of at most 3 symbols are possible. Because there are 26 letters in the alphabet, some letters must be represented by sequences of 4 symbols in Morse code. ✓4

Permutations

A **permutation** of a set of elements is an ordering of the elements. For instance, there are six permutations (orderings) of the letters A, B, and C, namely,

$$ABC, ACB, BAC, BCA, CAB, \text{ and } CBA,$$

as you can easily verify. As this listing shows, order counts when determining the number of permutations of a set of elements. By saying "order counts," we mean that the event ABC is indeed distinct from CBA or any other ordering of the three letters. We can use the multiplication principle to determine the number of possible permutations of any set.

EXAMPLE 7   **Batting Order** How many batting orders are possible for a 9-person baseball team?

Solution There are 9 possible choices for the first batter, 8 possible choices for the second batter, 7 for the third batter, and so on, down to the eighth batter (2 possible choices) and the ninth batter (1 possibility). So the total number of batting orders is

$$9 \cdot 8 \cdot 7 \cdot 6 \cdot 5 \cdot 4 \cdot 3 \cdot 2 \cdot 1 = 362,880.$$

In other words, the number of permutations of a 9-person set is 9!.

The argument in Example 7 applies to any set, leading to the conclusion that follows.

> The number of permutations of an n element set is $n!$.

Sometimes we want to order only some of the elements in a set, rather than all of them.

EXAMPLE 8   A teacher has 5 books and wants to display 3 of them side by side on her desk. How many arrangements of 3 books are possible?

Solution The teacher has 5 ways to fill the first space, 4 ways to fill the second space, and 3 ways to fill the third space. Because she wants only 3 books on the desk, there are only 3 spaces to fill, giving $5 \cdot 4 \cdot 3 = 60$ possible arrangements. ☑5

In Example 8, we say that the possible arrangements are *the permutations of 5 things taken 3 at a time*, and we denote the number of such permutations by $_5P_3$. In other words, $_5P_3 = 60$. More generally, an ordering of r elements from a set of n elements is called a **permutation of n things taken r at a time**, and the number of such permutations is denoted $_nP_r$.[*] To see how to compute this number, look at the answer in Example 8, which can be expressed like this:

$$_5P_3 = 5 \cdot 4 \cdot 3 = 5 \cdot 4 \cdot 3 \cdot \frac{2 \cdot 1}{2 \cdot 1} = \frac{5 \cdot 4 \cdot 3 \cdot 2 \cdot 1}{2 \cdot 1} = \frac{5!}{2!} = \frac{5!}{(5-3)!}.$$

A similar analysis in the general case leads to this useful fact:

> ## Permutations
>
> If $_nP_r$ (where $r \le n$) is the number of permutations of n elements taken r at a time, then
>
> $$_nP_r = \frac{n!}{(n-r)!}.$$

✓**Checkpoint 5**

How many ways can a merchant with limited space display 4 fabric samples side by side from her collection of 8?

---
[*] Another notation that is sometimes used is $P(n, r)$.

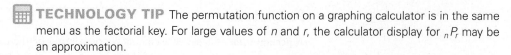

TECHNOLOGY TIP The permutation function on a graphing calculator is in the same menu as the factorial key. For large values of n and r, the calculator display for $_nP_r$ may be an approximation.

To find $_nP_r$, we can either use the preceding rule or apply the multiplication principle directly, as the next example shows.

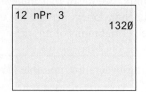

Figure 9.7

EXAMPLE 9 Early in 2016, 12 candidates sought the Republican nomination for president at the Iowa caucus. In a poll, how many ways could voters rank their first, second, and third choices?

Solution This is the same as finding the number of permutations of 12 elements taken 3 at a time. Since there are 3 choices to be made, the multiplication principle gives $_{12}P_3 = 12 \cdot 11 \cdot 10 = 1320$. Alternatively, by the formula for $_nP_r$,

$$_{12}P_3 = \frac{12!}{(12-3)!} = \frac{12 \cdot 11 \cdot 10 \cdot 9 \cdot 8 \cdot 7 \cdot 6 \cdot 5 \cdot 4 \cdot 3 \cdot 2 \cdot 1}{9 \cdot 8 \cdot 7 \cdot 6 \cdot 5 \cdot 4 \cdot 3 \cdot 2 \cdot 1} = 12 \cdot 11 \cdot 10 = 1320.$$

Figure 9.7 shows this result on a TI-84+ graphing calculator. ✓ 6

✓ **Checkpoint 6**

Find the number of permutations of

(a) 5 things taken 2 at a time;

(b) 9 things taken 3 at a time.

Find each of the following:

(c) $_3P_1$;

(d) $_7P_3$;

(e) $_{12}P_2$.

EXAMPLE 10 In a college admissions forum, 5 female and 4 male sophomore panelists discuss their college experiences with high school seniors.

(a) In how many ways can the panelists be seated in a row of 9 chairs?

Solution Find $_9P_9$, the total number of ways to seat 9 panelists in 9 chairs:

$$_9P_9 = \frac{9!}{(9-9)!} = \frac{9!}{0!} = \frac{9!}{1} = 9 \cdot 8 \cdot 7 \cdot 6 \cdot 5 \cdot 4 \cdot 3 \cdot 2 \cdot 1 = 362{,}880.$$

So, there are 362,880 ways to seat the 9 panelists.

(b) In how many ways can the panelists be seated if the males and females are to be alternated?

Solution Use the multiplication principle. In order to alternate males and females, a female must be seated in the first chair (since there are 5 females and only 4 males), any of the males next, and so on. Thus, there are 5 ways to fill the first seat, 4 ways to fill the second seat, 4 ways to fill the third seat (with any of the 4 remaining females), and so on, or

$$5 \cdot 4 \cdot 4 \cdot 3 \cdot 3 \cdot 2 \cdot 2 \cdot 1 \cdot 1 = 2880.$$

So, there are 2880 ways to seat the panelists.

(c) In how many ways can the panelists be seated if the males must sit together and the females sit together?

Solution Use the multiplication principle. We first must decide how to arrange the two groups (males and females). There are 2! ways of doing this. Next, there are 5! ways of arranging the females and 4! ways of arranging the men, for a total of

$$2!\,5!\,4! = 2 \cdot 120 \cdot 24 = 5760$$

✓ **Checkpoint 7**

A collection of 3 paintings by one artist and 2 by another is to be displayed. In how many ways can the paintings be shown

(a) in a row?

(b) if the works of the artists are to be alternated?

(c) if one painting by each artist is displayed?

ways. ✓ 7

Combinations

In Example 8, we found that there are 60 ways a teacher can arrange 3 of 5 different books on a desk. That is, there are 60 permutations of 5 things taken 3 at a time. Suppose now that the teacher does not wish to arrange the books on her desk, but rather wishes to choose, at random, any 3 of the 5 books to give to a book sale to raise money for her school. In how many ways can she do this?

At first glance, we might say 60 again, but that is incorrect. The number 60 counts all possible *arrangements* of 3 books chosen from 5. However, the following arrangements, for example, would all lead to the same set of 3 books being given to the book sale:

mystery–biography–textbook	biography–textbook–mystery
mystery–textbook–biography	textbook–biography–mystery
biography–mystery–textbook	textbook–mystery–biography

The foregoing list shows 6 different *arrangements* of 3 books, but only one *subset* of 3 books. A subset of items selected *without regard to order* is called a **combination**. The number of combinations of 5 things taken 3 at a time is written $_5C_3$. Since they are subsets, combinations are *not ordered.*

To evaluate $_5C_3$, start with the $5 \cdot 4 \cdot 3$ *permutations* of 5 things taken 3 at a time. Combinations are unordered; therefore, we find the number of combinations by dividing the number of permutations by the number of ways each group of 3 can be ordered—that is, by 3!:

$$_5C_3 = \frac{5 \cdot 4 \cdot 3}{3!} = \frac{5 \cdot 4 \cdot 3}{3 \cdot 2 \cdot 1} = 10.$$

There are 10 ways that the teacher can choose 3 books at random for the book sale.

Generalizing this discussion gives the formula for the number of combinations of n elements taken r at a time, written $_nC_r$.[*] In general, a set of r elements can be ordered in $r!$ ways, so we divide $_nP_r$ by $r!$ to get $_nC_r$:

$$_nC_r = \frac{_nP_r}{r!}$$

$$= {_nP_r}\frac{1}{r!}$$

$$= \frac{n!}{(n-r)!} \cdot \frac{1}{r!} \qquad \text{Definition of } _nP_r$$

$$= \frac{n!}{(n-r)!\, r!}.$$

✓ **Checkpoint 8**

Evaluate $\dfrac{_nP_r}{r!}$ for the given values.

(a) $n = 6, r = 2$

(b) $n = 8, r = 4$

(c) $n = 7, r = 0$

This last form is the most useful for setting up the calculation. ▸8

Combinations

The number of combinations of n elements taken r at a time, where $r \leq n$, is

$$_nC_r = \frac{n!}{(n-r)!\,r!}.$$

EXAMPLE 11 **Committee Selection** From a group of 10 students, a committee is to be chosen to meet with the dean. How many different 3-person committees are possible?

Solution A committee is not ordered, so we compute

$$_{10}C_3 = \frac{10!}{(10-3)!\,3!} = \frac{10!}{7!\,3!} = \frac{10 \cdot 9 \cdot 8 \cdot 7!}{7!\,3!} = \frac{10 \cdot 9 \cdot 8}{3 \cdot 2 \cdot 1} = 120.$$

⊞ **TECHNOLOGY TIP** The key for obtaining $_nC_r$ on a graphing calculator is located in the same menu as the key for obtaining $_nP_r$.

---

[*] Another notation that is sometimes used in place of $_nC_r$ is $\binom{n}{r}$.

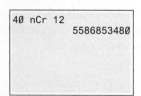

Figure 9.8

✓ **Checkpoint 9**

Use $\dfrac{n!}{(n-r)!\,r!}$ to evaluate $_nC_r$.

(a) $_6C_2$

(b) $_8C_4$

(c) $_7C_0$

Compare your answers with the answers to Checkpoint 8.

EXAMPLE 12 **Jury Selection** In how many ways can a 12-person jury be chosen from a pool of 40 people?

Solution Since the order in which the jurors are chosen does not matter, we use combinations. The number of combinations of 40 things taken 12 at a time is

$$_{40}C_{12} = \frac{40!}{(40-12)!\,12!} = \frac{40!}{28!\,12!}.$$

Using a calculator to compute this number (Figure 9.8), we see that there are 5,586,853,480 possible ways to choose a jury. ✓ 9

EXAMPLE 13 **Selection of Managers** Three managers are to be selected from a group of 30 to work on a special project.

(a) In how many different ways can the managers be selected?

Solution Here, we wish to know the number of 3-element combinations that can be formed from a set of 30 elements. (We want combinations, not permutations, since order within the group of 3 does not matter.) So, we calculate

$$_{30}C_3 = \frac{30!}{27!\,3!} = 4060.$$

There are 4060 ways to select the project group.

(b) In how many ways can the group of 3 be selected if a certain manager must work on the project?

Solution Since 1 manager has already been selected for the project, the problem is reduced to selecting 2 more from the remaining 29 managers:

$$_{29}C_2 = \frac{29!}{27!\,2!} = 406.$$

In this case, the project group can be selected in 406 ways.

(c) In how many ways can a nonempty group of at most 3 managers be selected from these 30 managers?

Solution The group is to be nonempty; therefore, "at most 3" means "1 or 2 or 3." Find the number of ways for each case:

Case	Number of Ways
1	$_{30}C_1 = \dfrac{30!}{29!\,1!} = \dfrac{30 \cdot 29!}{29!\,1!} = 30$
2	$_{30}C_2 = \dfrac{30!}{28!\,2!} = \dfrac{30 \cdot 29 \cdot 28!}{28! \cdot 2 \cdot 1} = 435$
3	$_{30}C_3 = \dfrac{30!}{27!\,3!} = \dfrac{30 \cdot 29 \cdot 28 \cdot 27!}{27! \cdot 3 \cdot 2 \cdot 1} = 4060$

✓ **Checkpoint 10**

Five orchids from a collection of 20 are to be selected for a flower show.

(a) In how many ways can this be done?

(b) In how many different ways can the group of 5 be selected if 2 particular orchids must be included?

(c) In how many ways can at least 1 and at most 5 orchids be selected? (*Hint:* Use a calculator.)

The total number of ways to select at most 3 managers will be the sum

$$30 + 435 + 4060 = 4525. \quad ✓ 10$$

Choosing a Method

The formulas for permutations and combinations given in this section will be very useful in solving probability problems in later sections. Any difficulty in using these formulas usually comes from being unable to differentiate between them. Both permutations and combinations give the number of ways to choose r objects from a set of n objects. The differences between permutations and combinations are outlined in the following summary.

Permutations

Different orderings or arrangements of the r objects are different permutations.

$$_nP_r = \frac{n!}{(n-r)!}$$

Clue words: arrangement, schedule, order
Order matters!

Combinations

Each choice or subset of r objects gives 1 combination. Order within the r objects does not matter.

$$_nC_r = \frac{n!}{(n-r)!\,r!}$$

Clue words: group, committee, set, sample
Order does not matter!

In the examples that follow, concentrate on recognizing which of the formulas should be applied.

EXAMPLE 14 For each of the given problems, tell whether permutations or combinations should be used to solve the problem.

(a) How many 4-digit numbers are possible if no digits are repeated?

Solution Since changing the order of the 4 digits results in a different number, we use permutations.

(b) A sample of 3 lightbulbs is randomly selected from a batch of 15 bulbs. How many different samples are possible?

Solution The order in which the 3 lightbulbs are selected is not important. The sample is unchanged if the bulbs are rearranged, so combinations should be used.

(c) In a basketball conference with 8 teams, how many games must be played so that each team plays every other team exactly once?

Solution The selection of 2 teams for a game is an *unordered* subset of 2 from the set of 8 teams. Use combinations again.

(d) In how many ways can 4 patients be assigned to 6 hospital rooms so that each patient has a private room?

Solution The room assignments are an *ordered* selection of 4 rooms from the 6 rooms. Exchanging the rooms of any 2 patients within a selection of 4 rooms gives a different assignment, so permutations should be used. ✓11

✓ **Checkpoint 11**

Solve the problems in Example 14.

EXAMPLE 15 A manager must select 4 employees for promotion. Twelve employees are eligible.

(a) In how many ways can the 4 employees be chosen?

Solution Because there is no reason to consider the order in which the 4 are selected, we use combinations:

$$_{12}C_4 = \frac{12!}{4!\,8!} = 495.$$

(b) In how many ways can 4 employees be chosen (from 12) to be placed in 4 different jobs?

Solution In this case, once a group of 4 is selected, its members can be assigned in many different ways (or arrangements) to the 4 jobs. Therefore, this problem requires permutations:

$$_{12}P_4 = \frac{12!}{8!} = 11,880. \quad ☑_{12}$$

✓ **Checkpoint 12**

A postal worker has special-delivery mail for 7 customers.

(a) In how many ways can he arrange his schedule to deliver to all 7?

(b) In how many ways can he schedule deliveries if he can deliver to only 4 of the 7?

```
(69 ncr 5)*26
          292201338
69!*26
64!5!
          292201338
```

Figure 9.9

EXAMPLE 16 **Powerball** Powerball is a lottery game played in 44 states (plus the District of Columbia, Puerto Rico, and the U.S. Virgin Islands). For a $2 ticket, a player selects five different numbers from 1 to 69 and one powerball number from 1 to 26 (which may be the same as one of the first five chosen). A match of all six numbers wins the jackpot. How many different selections are possible?

Solution The order in which the first five numbers are chosen does not matter. So we use combinations to find the number of combinations of 69 things taken 5 at a time—that is, $_{69}C_5$. There are 26 ways to choose one powerball number from 1 to 26. By the multiplication principle, the number of different selections is

$$(_{69}C_5)(26) = \frac{69!}{(69-5)!5!}(26) = \frac{69!(26)}{64!5!} = 292,201,338,$$

as shown in two ways on a graphing calculator in Figure 9.9. ☑_{13}

✓ **Checkpoint 13**

Under earlier Powerball rules, you had to choose five different numbers from 1 to 59 and then choose one powerball number from 1 to 35. Under those rules, how many different selections were possible?

EXAMPLE 17 A male student going on spring break at Daytona Beach has 8 tank tops and 12 pairs of shorts. He decides he will need 5 tank tops and 6 pairs of shorts for the trip. How many ways can he choose the tank tops and the shorts?

Solution We can break this problem into two parts: finding the number of ways to choose the tank tops, and finding the number of ways to choose the shorts. For the tank tops, the order is not important, so we use combinations to obtain

$$_8C_5 = \frac{8!}{3!\,5!} = 56.$$

Likewise, order is not important for the shorts, so we use combinations to obtain

$$_{12}C_6 = \frac{12!}{6!\,6!} = 924.$$

We now know there are 56 ways to choose the tank tops and 924 ways to choose the shorts. The total number of ways to choose the tank tops and shorts can be found using the multiplication principle to obtain $56 \cdot 924 = 51,744.$ ☑_{14}

✓ **Checkpoint 14**

Lacy wants to pack 4 of her 10 blouses and 2 of her 4 pairs of jeans for her trip to Europe. How many ways can she choose the blouses and jeans?

As Examples 16 and 17 show, often both combinations and the multiplication principle must be used in the same problem.

EXAMPLE 18 To illustrate the differences between permutations and combinations in another way, suppose 2 cans of soup are to be selected from 4 cans on a shelf: noodle (*N*), bean (*B*), mushroom (*M*), and tomato (*T*). As shown in Figure 9.10(a), there are 12 ways to select 2 cans from the 4 cans if the order matters (if noodle first and bean second is considered different from bean and then noodle, for example). However, if order is unimportant, then there are 6 ways to choose 2 cans of soup from the 4, as illustrated in Figure 9.10(b).

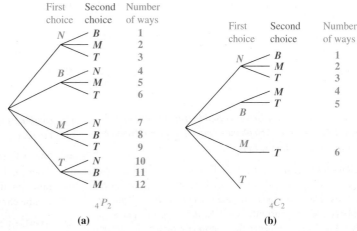

Figure 9.10

⚠ **CAUTION** It should be stressed that not all counting problems lend themselves to either permutations or combinations.

9.2 Exercises

Evaluate the given factorials, permutations, and combinations.

1. $_4P_2$ **2.** $3!$ **3.** $_8C_5$

4. $7!$ **5.** $_8P_1$ **6.** $_7C_2$

7. $4!$ **8.** $_4P_4$ **9.** $_9C_6$

10. $_8C_2$ **11.** $_{13}P_3$ **12.** $_9P_5$

Use a calculator to find values for Exercises 13–20.

13. $_{25}P_5$ **14.** $_{40}P_5$ **15.** $_{14}P_5$

16. $_{17}P_8$ **17.** $_{18}C_5$ **18.** $_{32}C_9$

19. $_{28}C_{14}$ **20.** $_{35}C_{30}$

21. Some students find it puzzling that $0! = 1$, and they think that $0!$ should equal 0. If this were true, what would be the value of $_4P_4$ according to the permutations formula?

22. If you already knew the value of $8!$, how could you find the value of $9!$ quickly?

Use the multiplication principle to solve the given problems. (See Examples 1–6.)

23. Tile Project As of May 2017, the Lowe's home improvement store website listed 264 different kinds of glass wall tile, 31 colors of acrylic grout, 12 kinds of mortar trowels, and 17 different kinds of mortar. How many ways are there to buy the products needed for a tile project?

24. Heating and Cooling How many different heating–cooling units are possible if a home owner has 3 choices for the efficiency rating of the furnace, 3 options for the fan speed, and 6 options for the air condenser?

25. Auto Manufacturing An auto manufacturer produces 6 models, each available in 8 different colors, with 4 different upholstery fabrics and 3 interior colors. How many varieties of the auto are available?

26. Radio Call Letters How many different 4-letter radio station call letters can be made

  (a) if the first letter must be K or W and no letter may be repeated?

  (b) if repeats are allowed (but the first letter still must be K or W)?

  (c) How many of the 4-letter call letters (starting with K or W) with no repeats end in R?

27. Social Security Numbers A Social Security number has 9 digits. How many Social Security numbers are possible? The U.S. population in 2016 was approximately 323 million. Was it possible for every U.S. resident to have a unique Social Security number? (Assume no restrictions.)

28. Zip Codes The United States Postal Service currently uses 5-digit zip codes in most areas. How many zip codes are possible if there are no restrictions on the digits used? How many would be possible if the first number could not be 0?

29. Zip Codes The Postal Service is encouraging the use of 9-digit zip codes in some areas, adding 4 digits after the usual 5-digit code. How many such zip codes are possible with no restrictions?

30. License Plates For many years, the state of California used 3 letters followed by 3 digits on its automobile license plates.

  (a) How many different license plates are possible with this arrangement?

  (b) When the state ran out of new numbers, the order was reversed to 3 digits followed by 3 letters. How many new license plate numbers were then possible?

  (c) Several years ago, the numbers described in part (b) were also used up. The state then issued plates with 1 letter followed by 3 digits and then 3 letters. How many new license plate numbers will this arrangement provide?

31. Hair Products A recent trip to the drug store revealed 12 different kinds of Pantene® shampoo and 10 different kinds of Pantene® conditioner. How many ways can Sherri buy 1 Pantene® shampoo and 1 Pantene® conditioner?

32. Sales Route A pharmaceutical salesperson has 6 doctors' offices to call on.

(a) In how many ways can she arrange her schedule if she calls on all 6 offices?

(b) In how many ways can she arrange her schedule if she decides to call on 4 of the 6 offices?

Phone Numbers *The United States is rapidly running out of telephone numbers. In large cities, telephone companies have introduced new area codes as numbers are used up.*

33. (a) Until recently, all area codes had a 0 or a 1 as the middle digit and the first digit could not be 0 or 1. How many area codes were possible with this arrangement? How many telephone numbers does the current 7-digit sequence permit per area code? (The 3-digit sequence that follows the area code cannot start with 0 or 1. Assume that there are no other restrictions.)

(b) The actual number of area codes under the previous system was 152. Explain the discrepancy between this number and your answer to part (a).

34. The shortage of area codes under the previous system was avoided by removing the restriction on the second digit. How many area codes are available under this new system?

35. A problem with the plan in Exercise 34 was that the second digit in the area code had been used to tell phone company equipment that a long-distance call was being made. To avoid changing all equipment, an alternative plan proposed a 4-digit area code and restricted the first and second digits as before. How many area codes would this plan have provided?

36. Still another solution to the area-code problem is to increase the local dialing sequence to 8 digits instead of 7. How many additional numbers would this plan create? (Assume the same restrictions.)

37. Define permutation in your own words.

Use permutations to solve each of the given problems. (See Examples 7–10.)

38. A baseball team has 15 players. How many 9-player batting orders are possible?

39. Tim is a huge fan of the latest album by country-music singer Kenny Chesney. If Tim has time to listen to only 5 of the 12 songs on the album, how many ways can he listen to the 5 songs?

40. From a cooler with 8 cans of different kinds of soda, 3 are selected for 3 people. In how many ways can this be done?

41. The Greek alphabet has 24 letters. How many ways can one name a fraternity using 3 Greek letters (with no repeats)?

42. A customer speaks to his financial advisor about investment products. The advisor has 9 products available, but knows he will only have time to speak about 4. How many different ways can he give the details on 4 different investment products to the customer?

43. The student activity club at the college has 32 members. In how many different ways can the club select a president, a vice president, a treasurer, and a secretary?

44. A student can take only one class a semester, and she needs to take 4 more electives in any order. If there are 20 courses from which she can choose, how many ways can she take her 4 electives?

45. In a club with 17 members, how many ways can the club elect a president and a treasurer?

Use combinations to solve each of the given problems. (See Examples 11–13.)

46. Manufacturing Four items are to be randomly selected from the first 25 items on an assembly line in order to determine the defect rate. How many different samples of 4 items can be chosen?

47. Student Selection A group of 4 students is to be selected from a group of 10 students to take part in a class in cell biology.

(a) In how many ways can this be done?

(b) In HOW many ways can the group that will *not* take part be chosen?

48. Study Design From a group of 15 smokers and 21 nonsmokers, a researcher wants to randomly select 7 smokers and 6 nonsmokers for a study. In how many ways can the study group be selected?

49. Co-Captain Selection The college football team has 11 seniors. The team needs to elect a group of 4 senior co-captains. How many different 4-person groups of co-captains are possible?

50. Drama Role Selection The drama department holds auditions for a play with a cast of 5 roles. If 33 students audition, how many casts of 5 people are possible?

51. Explain the difference between a permutation and a combination.

52. Padlocks with digit dials are often referred to as "combination locks." According to the mathematical definition of combination, is this an accurate description? Explain.

Exercises 53–70 are mixed problems that may require permutations, combinations, or the multiplication principle. (See Examples 14–18.)

53. Use a tree diagram to find the number of ways 2 letters can be chosen from the set $\{P, Q, R\}$ if order is important and

(a) if repetition is allowed;

(b) if no repeats are allowed.

(c) Find the number of combinations of 3 elements taken 2 at a time. Does this answer differ from that in part (a) or (b)?

54. Repeat Exercise 53, using the set $\{P, Q, R, S\}$ and 4 in place of 3 in part (c).

55. Delegation Selection The U.S. Senate Foreign Relations Committee in 2017 had 10 Democrats and 11 Republicans. A delegation of 5 people is to be selected to visit Iraq.

(a) How many delegations are possible?

(b) How many delegations would have all Republicans?

(c) How many delegations would have 3 Democrats and 2 Republicans?

(d) How many delegations would have at least one Democrat?

56. Plant Research In an experiment on plant hardiness, a researcher gathers 6 wheat plants, 5 barley plants, and 3 rye plants. She wishes to select 4 plants at random.

(a) In how many ways can this be done?

(b) In how many ways can this be done if exactly 2 wheat plants must be included?

57. Ice Cream According to the Baskin-Robbins® Web site, there are 25 "classic flavors" of ice cream.

(a) How many different double-scoop cones (of 2 different flavors) can be made if order does not matter (for example, putting chocolate on top of vanilla is equivalent to putting vanilla on top of chocolate)?

(b) How many different triple-scoop (of 3 different flavors) cones can be made if order does matter?

58. Finance A financial advisor offers 8 mutual funds in the high-risk category, 7 in the moderate-risk category, and 10 in the low-risk category. An investor decides to invest in 3 high-risk funds, 4 moderate-risk funds, and 3 low-risk funds. How many ways can the investor do this?

59. Lottery Game A lottery game requires that you pick 6 different numbers from 1 to 99. If you pick all 6 winning numbers, you win $4 million.

(a) How many ways are there to choose 6 numbers if order is not important?

(b) How many ways are there to choose 6 numbers if order matters?

60. Lottery Game In Exercise 59, if you pick 5 of the 6 numbers correctly, you win $5,000. In how many ways can you pick exactly 5 of the 6 winning numbers without regard to order?

61. The game of Set* consists of a special deck of cards. Each card has on it either one, two, or three shapes. The shapes on each card are all the same color, either green, purple, or red. The shapes on each card are the same style, either solid, shaded, or outline. There are three possible shapes—squiggle, diamond, and oval—and only one type of shape appears on a card. The deck consists of all possible combinations of shape, color, style, and number. How many cards are in a deck?

62. Patient Care Over the course of the previous nursing shift, 16 new patients were admitted onto a hospital ward. If a nurse begins her shift by caring for 6 of the new patients, how many possible ways could the 6 patients be selected from the 16 new arrivals?

63. Insect Classification A biologist is attempting to classify 52,000 species of insects by assigning 3 initials to each species. Is it possible to classify all the species in this way? If not, how many initials should be used?

64. Lottery Game One play in a state lottery consists of choosing 6 numbers from 1 to 44. If your 6 numbers are drawn (in any order), you win the jackpot.

(a) How many possible ways are there to draw the 6 numbers?

(b) If you get 2 plays for a dollar, how much would it cost to guarantee that one of your choices would be drawn?

(c) Assuming that you work alone and can fill out a betting ticket (for 2 plays) every second, and assuming that the lotto drawing will take place 3 days from now, can you place enough bets to guarantee that 1 of your choices will be drawn?

65. A cooler contains 5 cans of Pepsi®, 1 can of Diet Coke®, and 3 cans of 7UP®; you pick 3 cans at random. How many samples are possible in which the soda cans picked are

(a) only Pepsi;

(b) only Diet Coke;

(c) only 7UP;

(d) 2 Pepsi, 1 Diet Coke;

(e) 2 Pepsi, 1 7UP;

(f) 2 7UP, 1 Pepsi;

(g) 2 Diet Coke, 1 7UP?

66. A class has 9 male students and 8 female students. How many ways can the class select a committee of four people to petition the teacher not to make the final exam cumulative if the committee has to have 2 males and 2 females?

67. Study Enrollment A hospital wants to test the viability of a new medication for attention deficit disorder. It has 35 adults volunteer for the study, but can only enroll 20 in the study. How many ways can it choose the 20 volunteers to enroll in the study?

68. Pizza Making Suppose a pizza shop offers 4 choices of cheese and 9 toppings. If the order of the cheeses and toppings does not matter, how many different pizza selections are possible when choosing 2 cheeses and 2 toppings?

If the n objects in a permutations problem are not all distinguishable— that is, if there are n_1 of type 1, n_2 of type 2, and so on, for r different types—then the number of distinguishable permutations is

$$\frac{n!}{n_1!\, n_2! \cdots n_r!}.$$

Example *In how many ways can you arrange the letters in the word Mississippi?*

This word contains 1 m, 4 i's, 4 s's, and 2 p's. To use the formula, let n = 11, $n_1 = 1$, $n_2 = 4$, $n_3 = 4$, and $n_4 = 2$ to get

$$\frac{11!}{1!\, 4!\, 4!\, 2!} = 34{,}650$$

arrangements. The letters in a word with 11 different letters can be arranged in 11! = 39,916,800 ways.

69. Find the number of distinguishable permutations of the letters in each of the given words.

(a) martini

(b) nunnery

(c) grinding

---

70. A printer has 5 W's, 4 X's, 3 Y's, and 2 Z's. How many different "words" are possible that use all these letters? (A "word" does not have to have any meaning here.)

71. Shirley is a shelf stocker at the local grocery store. She has 4 varieties of Stouffer's® frozen dinners, 3 varieties of Lean Cuisine® frozen dinners, and 5 varieties of Weight Watchers® frozen dinners. In how many distinguishable ways can she stock the shelves if

(a) the dinners can be arranged in any order?

(b) dinners from the same company are considered alike and have to be shelved together?

(c) dinners from the same company are considered alike, but do not have to be shelved together?

72. A child has a set of different-shaped plastic objects. There are 2 pyramids, 5 cubes, and 6 spheres. In how many ways can she arrange them in a row

(a) if they are all different colors?

(b) if the same shapes must be grouped, but each object is a different color?

(c) In how many distinguishable ways can they be arranged in a row if objects of the same shape are also the same color, but need not be grouped?

✓**Checkpoint Answers**

1. 1,945,240,782

2. (a) 720   (b) 24

3. (a) 24   (b) 720   (c) 1   (d) 30

4. 30

5. $8 \cdot 7 \cdot 6 \cdot 5 = 1680$

6. (a) 20   (b) 504   (c) 3   (d) 210   (e) 132

7. (a) 120   (b) 12   (c) 6

8. (a) 15   (b) 70   (c) 1

9. (a) 15   (b) 70   (c) 1

10. (a) 15,504   (b) 816   (c) 21,699

11. (a) 5040   (b) 455   (c) 28   (d) 360

12. (a) 5040   (b) 840

13. 175,223,510

14. 1260

9.3 Applications of Counting

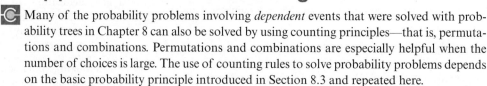 Many of the probability problems involving *dependent* events that were solved with probability trees in Chapter 8 can also be solved by using counting principles—that is, permutations and combinations. Permutations and combinations are especially helpful when the number of choices is large. The use of counting rules to solve probability problems depends on the basic probability principle introduced in Section 8.3 and repeated here.

If event E is a subset of sample space S with equally likely outcomes then the probability that event E occurs, written $P(E)$, is

$$P(E) = \frac{n(E)}{n(S)}.$$

EXAMPLE 1 **Jury Selection** From a potential jury pool with 1 Hispanic, 3 Caucasian, and 2 African-American members, 2 jurors are selected one at a time without replacement. Find the probability that 1 Caucasian and 1 African-American are selected.

Solution In Example 6 of Section 8.5, it was necessary to consider the order in which the jurors were selected. With combinations, it is not necessary: Simply count the number of ways in which 1 Caucasian and 1 African-American juror can be selected. The Caucasian can be selected in $_3C_1$ ways, and the African-American juror can be selected in $_2C_1$ ways. By the multiplication principle, both results can occur in

$$_3C_1 \cdot {_2C_1} = 3 \cdot 2 = 6 \text{ ways},$$

giving the numerator of the probability fraction. For the denominator, 2 jurors are selected from a total of 6 candidates. This can occur in $_6C_2 = 15$ ways. The required probability is

$$P(\text{1 Caucasian and 1 African American}) = \frac{_3C_1 \cdot {_2C_1}}{_6C_2} = \frac{3 \cdot 2}{15} = \frac{6}{15} = \frac{2}{5} = .40.$$

This result agrees with the answer found earlier.

EXAMPLE 2   **Grievance Committee** From a baseball team of 15 players, 4 are to be selected to present a list of grievances to the coach.

(a) In how many ways can this be done?

Solution Four players from a group of 15 can be selected in $_{15}C_4$ ways. (Use combinations, since the order in which the group of 4 is selected is unimportant.) So,

$$_{15}C_4 = \frac{15!}{4! \ 11!} = \frac{15(14)(13)(12)}{4(3)(2)(1)} = 1365.$$

There are 1365 ways to choose 4 players from 15.

(b) One of the players is Michael Branson. Find the probability that Branson will be among the 4 selected.

Solution The probability that Branson will be selected is the number of ways the chosen group includes him, divided by the total number of ways the group of 4 can be chosen. If Branson must be one of the 4 selected, the problem reduces to finding the number of ways the additional 3 players can be chosen. There are 3 chosen from 14 players; this can be done in

$$_{14}C_3 = \frac{14!}{3! \ 11!} = 364$$

ways. The number of ways 4 players can be selected from 15 is

$$n = {}_{15}C_4 = 1365.$$

The probability that Branson will be one of the 4 chosen is

$$P(\text{Branson is chosen}) = \frac{364}{1365} \approx .2667.$$

(c) Find the probability that Branson will not be selected.

Solution The probability that he will not be chosen is $1 - .2667 = .7333.$

✓**Checkpoint 1**

The ski club has 8 women and 7 men. What is the probability that if the club elects 3 officers at random, all 3 of them will be women?

EXAMPLE 3   **Quality Control** A manufacturing company performs a quality-control analysis on the ceramic tile it produces. It produces the tile in batches of 24 pieces. In the quality-control analysis, the company tests 3 pieces of tile per batch. Suppose a batch of 24 tiles has 4 defective tiles.

(a) What is the probability that exactly 1 of the 3 tested tiles is defective?

Solution Let $P(1 \text{ defective})$ represent the probability of there being exactly 1 defective tile among the 3 tested tiles. To find this probability, we need to know how many ways we can select 3 tiles for testing. Since order does not matter, there are $_{24}C_3$ ways to choose 3 tiles:

$$_{24}C_3 = \frac{24!}{21! \ 3!} = \frac{24 \cdot 23 \cdot 22}{3 \cdot 2 \cdot 1} = 2024.$$

There are $_4C_1$ ways of choosing 1 defective tile from the 4 in the batch. If we choose 1 defective tile, we must then choose 2 good tiles among the 20 good tiles in the batch. We can do this in $_{20}C_2$ ways. By the multiplication principle, there are

$$_4C_1 \cdot {}_{20}C_2 = \frac{4!}{3! 1!} \cdot \frac{20!}{18! \ 2!} = 4 \cdot 190 = 760$$

ways to choose exactly 1 defective tile.
   Thus,

$$P(1 \text{ defective}) = \frac{760}{2024} \approx .3755.$$

(b) If at least one of the tiles in a batch is defective, the company will not ship the batch. What is the probability that the batch is not shipped?

Solution The batch will not be shipped if 1, 2, or 3 of the tiles sampled are defective. We already found the probability of there being exactly 1 defective tile in part (a). We now need to find $P(2 \text{ defective})$ and $P(3 \text{ defective})$. To find $P(2 \text{ defective})$, we need to count the number of ways to choose 2 from the 4 defective tiles in the batch and choose 1 from the 20 good tiles in the batch:

$$_4C_2 \cdot {}_{20}C_1 = \frac{4!}{2!\,2!} \cdot \frac{20!}{19!\,1!} = 6 \cdot 20 = 120.$$

To find $P(3 \text{ defective})$, we need to count the number of ways to choose 3 from the 4 defective tiles in the batch and choose 0 from the 20 good tiles in the batch:

$$_4C_3 \cdot {}_{20}C_0 = \frac{4!}{1!\,3!} \cdot \frac{20!}{20!\,0!} = 4 \cdot 1 = 4.$$

We now have

$$P(2 \text{ defective}) = \frac{120}{2024} \approx .0593 \text{ and } P(3 \text{ defective}) = \frac{4}{2024} \approx .0020.$$

Thus, the probability of rejecting the batch because 1, 2, or 3 tiles are defective is

$$P(1 \text{ defective}) + P(2 \text{ defective}) + P(3 \text{ defective}) \approx .3755 + .0593 + .0020$$
$$= .4368.$$

(c) Use the complement rule to find the probability the batch will be rejected.

Solution We reject the batch if at least 1 of the sampled tiles is defective. The opposite of at least 1 tile being defective is that none are defective. We can find the probability that none of the 3 sampled tiles is defective by choosing 0 from the 4 defective tiles and choosing 3 from the 20 good tiles:

$$_4C_0 \cdot {}_{20}C_3 = 1 \cdot 1140 = 1140.$$

Therefore, the probability that none of the sampled tiles is defective is

$$P(0 \text{ defective}) = \frac{1140}{2024} \approx .5632.$$

Using the complement rule, we have

$$P(\text{at least 1 defective}) \approx 1 - .5632 = .4368,$$

the same answer as in part (b). Using the complement rule can often save time when multiple probabilities need to be calculated for problems involving "at least 1." ✔ 2

✓ **Checkpoint 2**

A batch of 15 granite slabs is mined, and 4 have defects. If the manager spot-checks 3 slabs at random, what is the probability that at least 1 slab is defective?

EXAMPLE 4 **Poker** In a common form of 5-card draw poker, a hand of 5 cards is dealt to each player from a deck of 52 cards. (For a review of a standard deck, see Figure 8.20 in Section 8.3.) There is a total of

$$_{52}C_5 = \frac{52!}{5!\,47!} = 2{,}598{,}960$$

such hands possible. Find the probability of being dealt each of the given hands.

(a) Heart-flush hand (5 hearts)

Solution There are 13 hearts in a deck; there are

$$_{13}C_5 = \frac{13!}{5!\,8!} = \frac{13(12)(11)(10)(9)}{5(4)(3)(2)(1)} = 1287$$

different hands containing only hearts. The probability of a heart flush is

$$P(\text{heart flush}) = \frac{1287}{2,598,960} \approx .000495.$$

(b) A flush of any suit (5 cards, all from 1 suit)

Solution There are 4 suits to a deck, so

$$P(\text{flush}) = 4 \cdot P(\text{heart flush}) = 4(.000495) \approx .00198.$$

(c) A full house of aces and eights (3 aces and 2 eights)

Solution There are $_4C_3$ ways to choose 3 aces from among the 4 in the deck and $_4C_2$ ways to choose 2 eights, so

$$P(\text{3 aces, 2 eights}) = \frac{_4C_3 \cdot _4C_2}{_{52}C_5} \approx .00000923.$$

(d) Any full house (3 cards of one value, 2 of another)

Solution There are 13 values in a deck, so there are 13 choices for the first value mentioned, leaving 12 choices for the second value. (Order *is* important here, since a full house of aces and eights, for example, is not the same as a full house of eights and aces.)

$$P(\text{full house}) = \frac{13 \cdot _4C_3 \cdot 12 \cdot _4C_2}{_{52}C_5} \approx .00144. \quad ✔_3$$

✔**Checkpoint 3**

Find the probability of being dealt a poker hand (5 cards) with 4 kings.

EXAMPLE 5   **Soda Selection** A cooler contains 8 different kinds of soda, among which 3 cans are Pepsi®, Coke®, and Sprite®. What is the probability, when picking at random, of selecting the 3 cans in the particular order listed in the previous sentence?

Solution Use permutations to find the number of arrangements in the sample, because order matters:

$$n = {_8P_3} = 8(7)(6) = 336.$$

Since each can is different, there is only 1 way to choose Pepsi, Coke, and Sprite in that order, so the probability is

$$\frac{1}{336} \approx .0030. \quad ✔_4$$

✔**Checkpoint 4**

Martha, Leonard, Calvin, and Sheila will be handling the officer duties of president, vice president, treasurer, and secretary.

(a) If the offices are assigned randomly, what is the probability that Calvin is the president?

(b) If the offices are assigned randomly, what is the probability that Sheila is president, Martha is vice president, Calvin is treasurer, and Leonard is secretary?

EXAMPLE 6   **Birthday Matches** Suppose a group of 5 people is in a room. Find the probability that at least 2 of the people have the same birthday.

Solution "Same birthday" refers to the month and the day, not necessarily the same year. Also, ignore leap years, and assume that each day in the year is equally likely as a birthday. First find the probability that *no 2 people* among 5 people have the same birthday. There are 365 different birthdays possible for the first of the 5 people, 364 for the second (so that the people have different birthdays), 363 for the third, and so on. The number of ways the 5 people can have different birthdays is thus the number of permutations of 365 things (days) taken 5 at a time, or

$$_{365}P_5 = 365 \cdot 364 \cdot 363 \cdot 362 \cdot 361.$$

The number of ways that the 5 people can have the same or different birthdays is

$$365 \cdot 365 \cdot 365 \cdot 365 \cdot 365 = (365)^5.$$

Finally, the *probability* that none of the 5 people have the same birthday is

$$\frac{_{365}P_5}{(365)^5} = \frac{365 \cdot 364 \cdot 363 \cdot 362 \cdot 361}{365 \cdot 365 \cdot 365 \cdot 365 \cdot 365} \approx .973.$$

The probability that at least 2 of the 5 people *do* have the same birthday is $1 - .973 = .027$.

Example 6 can be extended for more than 5 people. In general, the probability that no 2 people among n people have the same birthday is

$$\frac{_{365}P_n}{(365)^n}.$$

The probability that at least 2 of the n people *do* have the same birthday is

$$1 - \frac{_{365}P_n}{(365)^n}. \quad \boxed{\checkmark\ 5}$$

Checkpoint 5

Evaluate $1 - \dfrac{_{365}P_n}{(365)^n}$ for

(a) $n = 3$;

(b) $n = 6$.

The following table shows this probability for various values of n:

Number of People, n	Probability That at Least 2 Have the Same Birthday
5	.027
10	.117
15	.253
20	.411
22	.476
23	**.507**
25	.569
30	.706
35	.814
40	.891
50	.970

The probability that 2 people among 23 have the same birthday is .507, a little more than half. Many people are surprised at this result; somehow it seems that a larger number of people should be required. $\boxed{\checkmark\ 6}$

Checkpoint 6

Set up (but do not calculate) the probability that at least 2 of the 9 members of the Supreme Court have the same birthday.

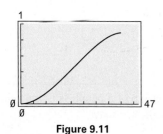

Figure 9.11

Using a graphing calculator, we can graph the probability formula in the previous example as a function of n, but the graphing calculator must be set to evaluate the function at integer points. Figure 9.11 was produced on a TI-84+ by letting $Y_1 = 1 - (365\,\text{nPr}\,X)/365^X$ on the interval $0 \le x \le 47$. (This domain ensures integer values for x.) Notice that the graph does not extend past $x = 39$. This is because $P(365, n)$ and 365^n are too large for the calculator when $n \ge 40$.

9.3 Exercises

Business *Suppose in Example 3 that the number of defective tiles in a batch of 24 is 5 rather than 4. If 3 tiles are sampled at random, the management would like to know the probability that*

1. exactly 1 of the sampled tiles is defective;

2. the batch is rejected (that is, at least 1 sampled tile is defective).

Computer Defects *A shipment of 8 computers contains 3 with defects. Find the probability that a sample of the given size, drawn from the 8, will not contain a defective computer. (See Example 3.)*

3. 1

4. 2

5. 3

6. 5

Radio Promotion *A radio station runs a promotion at an auto show with a money box with 10 $100 tickets, 12 $50 tickets, and 20 $25 tickets. The box contains an additional 200 "dummy" tickets with no value. Three tickets are randomly drawn. Find the given probabilities. (See Examples 1 and 2.)*

7. All $100 tickets

8. All $50 tickets

9. Exactly two $25 tickets and no other money winners

10. One ticket of each money amount

11. No tickets with money

12. At least one money ticket

Two cards are drawn at random from an ordinary deck of 52 cards. (See Example 4.)

13. How many 2-card hands are possible?

Find the probability that the 2-card hand in Exercise 13 contains the given cards.

14. 2 kings

15. No deuces (2's)

16. 2 face cards

17. Different suits

18. At least 1 black card

19. No more than 1 diamond

20. Discuss the relative merits of using probability trees versus combinations to solve probability problems. When would each approach be most appropriate?

21. Several examples in this section used the rule $P(E') = 1 - P(E)$. Explain the advantage (especially in Example 6) of using this rule.

Geology *A shipment contains 8 igneous, 7 sedimentary, and 7 metamorphic rocks. If we select 5 rocks at random, find the probability that*

22. all 5 are igneous;

23. exactly 3 are sedimentary;

24. only 1 is metamorphic.

25. **Lottery Game** In Exercise 59 in Section 9.2, we found the number of ways to pick 6 different numbers from 1 to 99 in a state lottery.

    (a) Assuming that order is unimportant, what is the probability of picking all 6 numbers correctly to win the big prize?

    (b) What is the probability if order matters?

26. **Lottery Game** In Exercise 25 (a), what is the probability of picking exactly 5 of the 6 numbers correctly?

27. Example 16 in Section 9.2 shows that the probability of winning the Powerball lottery is 1/292,201,338. If Juanita and Michelle each play Powerball on one particular evening, what is the probability that both will select the winning numbers if they make their selections independently of each other?

28. **Cell Phone Manufacturing** A cell phone manufacturer randomly selects 5 phones from every batch of 50 produced. If at least one of the phones is found to be defective, then each phone in the batch is tested individually. Find the probability that the entire batch will need testing if the batch contains

    (a) 8 defective phones;

    (b) 2 defective phones.

29. In 2017, the House of Representatives Financial Services committee had 60 members, of which 9 were women. If 8 members are selected at random, find the probability that the group of 8 would be composed as follows:

    (a) 7 men and 1 woman;

    (b) 5 men and 3 women;

    (c) at least one woman.

30. **Car Sales** A car dealership has 8 red, 9 silver, and 5 black cars on the lot. Ten cars are randomly chosen to be displayed in front of the dealership. Find the probability that

    (a) 4 are red and the rest are silver;

    (b) 5 are red and 5 are black;

    (c) exactly 8 are red.

31. **Study Volunteers** Twenty subjects volunteer for a study of a new cold medicine. Ten of the volunteers are ages 20–39, 8 are ages 40–59, and 2 are age 60 or older. If we select 7 volunteers at random, find the probability that

    (a) all the volunteers selected are ages 20–39;

    (b) 5 of the volunteers are ages 20–39 and 2 are age 60 or older;

    (c) exactly 3 of the volunteers are ages 40–59.

Birthday Problem *For Exercises 32–34, refer to Example 6 in this section.*

32. Set up the probability that at least 2 of the 44 men who have served as president of the United States have had the same birthday.[*]

33. Set up the probability that at least 2 of the 100 U.S. senators have the same birthday.

34. Set up the probability that at least 2 of the 50 U.S. governors have the same birthday.

Lottery Game *One version of the New York State lottery game Quick Draw has players selecting 4 numbers at random from the numbers 1–80.*

---

[*]In fact, James Polk and Warren Harding were both born on November 2. Although Donald Trump is the 45th president, the 22nd and 24th presidents were the same man: Grover Cleveland.

The state picks 20 winning numbers. If the player's 4 numbers are selected by the state, the player wins $55. (Data from: www.nylottery.org.)

35. What is the probability of winning?

36. If the state picks 3 of the player's numbers, the player wins $5. What is the probability of winning $5?

37. What is the probability of having none of your 4 numbers selected by the state?

38. What is the probability of having the state pick 2 of the player's numbers?

✓ **Checkpoint Answers**

1. About .1231

2. About .6374

3. .00001847

4. (a) 1/4   **(b)** 1/24

5. (a) .008   **(b)** .040

6. $1 - {}_{365}P_9/365^9$

9.4   Binomial Probability

In Section 9.1, we learned about probability distributions where we listed each outcome and its associated probability. After learning in Sections 9.2 and 9.3 how to count the number of possible outcomes, we are now ready to understand a special probability distribution known as the *binomial distribution*. This distribution occurs when the same experiment is repeated many times and each repetition is independent of previous ones. One outcome is designated a success and any other outcome is considered a failure. For example, you might want to find the probability of rolling 8 twos in 12 rolls of a die (rolling two is a success; rolling anything else is a failure). The individual trials (rolling the die once) are called **Bernoulli trials**, or **Bernoulli processes**, after the Swiss mathematician Jakob Bernoulli (1654–1705).

If the probability of a success in a single trial is p, then the probability of failure is $1 - p$ (by the complement rule). When Bernoulli trials are repeated a fixed number of times, and each trial's outcome is independent of the other trials' outcomes, the resulting distribution of outcomes is called a **binomial distribution**, or a **binomial experiment**. A binomial experiment must satisfy the following conditions.

> ## Binomial Experiment
>
> **1.** The same experiment is repeated a fixed number of times.
>
> **2.** There are only two possible outcomes: success and failure.
>
> **3.** The probability of success for each trial is constant.
>
> **4.** The repeated trials are independent.

The basic characteristics of binomial experiments are illustrated by a recent look at the New York Stock Exchange (NYSE). On May 23, 2017, the probability that a stock advanced (rather than declined or stayed at the same value) was .57. (Data from: www.wsj.com.) We use Y to denote the event that a stock on the exchange advanced, and N to denote the event that the stock did not advance. If we sample 5 stocks at random and use .57 as the probability for Y, we will generate a binomial experiment because all of the requirements are satisfied:

The sampling is repeated a fixed number of times (5);

there are only two outcomes of interest (Y or N);

the probability of success is constant ($p = .57$);

"at random" guarantees that the trials are independent.

To calculate the probability that all 5 randomly chosen stocks advance, we use the product rule for independent events (Section 8.5) and $P(Y) = .57$ to obtain

$$P(YYYYY) = P(Y) \cdot P(Y) \cdot P(Y) \cdot P(Y) \cdot P(Y) = (.57)^5 \approx .0602.$$

Determining the probability that 4 out of 5 stocks advanced is slightly more complicated. The stock that did not advance could be the first, second, third, fourth, or fifth stock chosen. So we have the following possible outcomes:

$$N\,Y\,Y\,Y\,Y$$
$$Y\,N\,Y\,Y\,Y$$
$$Y\,Y\,N\,Y\,Y$$
$$Y\,Y\,Y\,N\,Y$$
$$Y\,Y\,Y\,Y\,N$$

So the total number of ways in which 4 successes (and 1 failure) can occur is 5, which is the number $_5C_4$. The probability of each of these 5 outcomes is

$$P(Y) \cdot P(Y) \cdot P(Y) \cdot P(Y) \cdot P(N) = (.57)^4(1 - .57)^1 = (.57)^4(.43).$$

Since the 5 outcomes where there are 4 Ys and one N represent disjoint events, we multiply by $_5C_4 = 5$:

$$P(4\ Y\text{s out of 5 trials}) = {}_5C_4(.57)^4(.43)^{5-4} = 5(.57)^4(.43)^1 \approx .2270.$$

The probability of obtaining exactly 3 Ys and 2 Ns can be computed in a similar way. The probability of any one way of achieving 3 Ys and 2 Ns will be

$$(.57)^3(.43)^2.$$

Again, the desired outcome can occur in more than one way. Using combinations, we find that the number of ways in which 3 Ys and 2 Ns can occur is $_5C_3 = 10$. So we have

$$P(3\ Y\text{s out of 5 trials}) = {}_5C_3(.57)^3(.43)^{5-3} = 10(.57)^3(.43)^2 \approx .3424.\ \boxed{\checkmark\,_1}$$

With the probabilities just generated and the answers to Checkpoint 1, we can write the probability distribution for the number of stocks that advance when 5 stocks are selected at random:

x	0	1	2	3	4	5
$P(x)$	.0147	.0974	.2583	.3424	.2270	.0602

When the outcomes and their associated probabilities are written in this form, it is very easy to calculate answers to questions such as, What is the probability that 3 or more stocks advance? We see from the table that

$$P(3\text{ or more }Y\text{s}) = .3424 + .2270 + .0602 = .6296.$$

Similarly, the probability of one or fewer stocks advancing is

$$P(1\text{ or fewer }Y\text{s}) = .0974 + .0147 = .1121.$$

The example illustrates the following fact.

Binomial Probability

If p is the probability of success in a single trial of a binomial experiment, the probability of x successes and $n - x$ failures in n independent repeated trials of the experiment is

$$_nC_x p^x(1 - p)^{n-x}.$$

EXAMPLE 1   **Business** According to a 2016 study conducted by the Society for Human Resource Management, 38% of U.S. employees reported being very satisfied with their job. Suppose a random sample of 6 employees is chosen. Find the probability of the given scenarios. (Data from: www.shrm.org.)

✓ Checkpoint 1

Find the probability of obtaining

(a) exactly 2 stocks advance;

(b) exactly 1 stock advances;

(c) exactly no stocks advance.

▦ TECHNOLOGY TIP

On the TI-84+ calculator, use "binompdf(n,p,x)" in the DISTR menu to compute the probability of exactly x successes in n trials (where p is the probability of success in a single trial). Use "binomcdf(n,p,x)" to compute the probability of at most x successes in n trials. Figure 9.12 shows the probability of exactly 3 successes in 5 trials and the probability of at most 3 successes in 5 trials, with the probability of success set at .57 for each case.

```
binompdf(5,.57,3)
            .342421857
binomcdf(5,.57,3)
            .7128767728
```

Figure 9.12

(a) Exactly 4 of the 6 employees were very satisfied.

Solution We can think of the 6 employees as independent trials, and a success occurs if a worker is very satisfied with their job. This is a binomial experiment with $p = .38, n = 6$, and $x = 4$. By the binomial probability rule,

$$P(\text{exactly } 4) = {}_6C_4(.38)^4(1 - .38)^{6-4}$$
$$= 15(.38)^4(.62)^2$$
$$\approx .1202.$$

(b) All 6 employees are very satisfied with their job.

Solution Let $x = 6$. Then we have

$$P(\text{exactly } 6) = {}_6C_6(.38)^6(1 - .38)^{6-6}$$
$$= 1(.38)^6(.62)^0$$
$$\approx .0030.$$

✓ **Checkpoint 2**

According to the study in Example 1, 65% of employees indicated that respectful treatment of all employees was a very important contributor to their job satisfaction. If 4 employees are selected at random, find the probability that exactly the given number feel respectful treatment of all employees contributes to their job satisfaction.

(a) 1 of the 4;

(b) 3 of the 4.

EXAMPLE 2   Suppose a family has 3 children.

(a) Find the probability distribution for the number of girls.

Solution Let $x =$ the number of girls in three births. According to the binomial probability rule, the probability of exactly one girl being born is

$$P(x = 1) = {}_3C_1\left(\frac{1}{2}\right)^1\left(\frac{1}{2}\right)^2 = 3\left(\frac{1}{2}\right)^3 = \frac{3}{8}.$$

The other probabilities in this distribution are found similarly, as shown in the following table:

x	0	1	2	3
$P(x)$	${}_3C_0\left(\frac{1}{2}\right)^0\left(\frac{1}{2}\right)^3 = \frac{1}{8}$	${}_3C_1\left(\frac{1}{2}\right)^1\left(\frac{1}{2}\right)^2 = \frac{3}{8}$	${}_3C_2\left(\frac{1}{2}\right)^2\left(\frac{1}{2}\right)^1 = \frac{3}{8}$	${}_3C_3\left(\frac{1}{2}\right)^3\left(\frac{1}{2}\right)^0 = \frac{1}{8}$

(b) Find the expected number of girls in a 3-child family.

Solution For a binomial distribution, we can use the following method (which is presented here with a "plausibility argument," but not a full proof): Because 50% of births are girls, it is reasonable to expect that 50% of a sample of children will be girls. Since 50% of 3 is $3(.50) = 1.5$, we conclude that the expected number of girls is 1.5.

✓ **Checkpoint 3**

Find the probability of getting 2 fours in 8 tosses of a die.

The expected value in Example 2(b) was the product of the number of births and the probability of a single birth being a girl—that is, the product of the number of trials and the probability of success in a single trial. The same conclusion holds in the general case.

Expected Value for a Binomial Distribution

When an experiment meets the four conditions of a binomial experiment with n fixed trials and constant probability of success p, the expected value is

$$E(x) = np.$$

EXAMPLE 3 **Student Loan Defaults** In September 2016, the U.S. Department of Education released the latest data on student loan defaults. The national student loan default rate for the 2013 national cohort of students was 11.3%. If we select 10 students at random, find the probability of the following events.

(a) Exactly 3 of the students defaulted on their student loan.

Solution The experiment is repeated 10 times with defaulting on the loan considered a success. The probability of success is .113. Because the selection is done at random, the trials are considered independent. Thus, we have a binomial experiment, and

$$P(x = 3) = {}_{10}C_3(.113)^3(1 - .113)^7 = 120(.113)^3(.887)^7 \approx .0748.$$

(b) At most, 3 students defaulted on their student loan.

Solution "At most 3" means 0, 1, 2, or 3 successes. We have already calculated the probability of exactly 3 successes in part (a), so we need to calculate the probability of 0, 1, or 2 successes. We can then add those probabilities because they are disjoint events. We have

$$P(x = 0) = {}_{10}C_0(.113)^0(.887)^{10} \approx .3015;$$
$$P(x = 1) = {}_{10}C_1(.113)^1(.887)^9 \approx .3841;$$
$$P(x = 2) = {}_{10}C_2(.113)^2(.887)^8 \approx .2202.$$

Thus, using our answer to part (a) and the above, we have

$$P(\text{at most } 3) \approx .3015 + .3841 + .2202 + .0748 = .9806.$$

(c) The expected number of students who defaulted.

Solution Because this is a binomial experiment, we can use the formula $E(x) = np = 10(.113) = 1.13$. In repeated samples of 10 student borrowers, the average number of those who defaulted is 1.13. ✓4

✓ **Checkpoint 4**

The study described in Example 3 reported a default rate of 18.5% for public schools whose longest degree is 2 or 3 years in duration. If 8 students are selected at random, find the probability that

(a) exactly 2 default;

(b) at most 2 default.

(c) What is the expected number of students who default?

EXAMPLE 4 **New Business Failures** According to the U.S. Bureau of Labor Statistics, about 33% (.33) of new businesses fail within two years.

(a) If 7 new businesses are selected at random, find the probability that at least 3 will fail within two years.

Solution Because the businesses are selected at random, we can treat this problem as a binomial experiment, with $n = 7, p = .33$, and x representing the number of new businesses that fail. "At least 3 of 7" means 3 or 4 or 5, and so on, up to all 7. Thus, we have

$$P(\text{at least } 3) = {}_7C_3(.33)^3(.67)^4 + {}_7C_4(.33)^4(.67)^3 + {}_7C_5(.33)^5(.67)^2 +$$
$$\qquad {}_7C_6(.33)^6(.67)^1 + {}_7C_7(.33)^7(.67)^0$$
$$\approx .2535 + .1248 + .0369 + .0061 + .0004$$
$$= .4217.$$

(b) If 7 new businesses are selected at random, what is the probability that at least one of the new businesses will fail?

Solution As above, "At least 1 of 7" means 1 or 2 or 3, and so on, up to all 7. However, with problems that ask for "at least 1," we can employ the complement rule to save time. It is simpler to find the probability that *none* of the 7 selected start-ups succeed [namely: $P(x = 0)$] and then find the probability that at least one of the new businesses fails by using the complement rule and calculating $1 - P(x = 0)$. The calculations are:

$$P(x = 0) = {}_7C_0(.33)^0(.67)^7 \approx .0606;$$
$$P(x \geq 1) = 1 - P(x = 0) \approx 1 - .0606 = .9394. ✓5$$

✓ **Checkpoint 5**

In Example 4, find the probability that

(a) at least 2 businesses fail within 2 years;

(b) at most 4 of the businesses fail within 2 years.

(c) What is the expected value?

TECHNOLOGY TIP Some spreadsheets provide binomial probabilities. In Microsoft Excel, for example, the command "=BINOM.DIST (0,7,.33,false)" gives .060607, which is the probability $x = 0$ in Example 4(b). Alternatively, the command "=BINOM.DIST (2,7,.33,true)" gives .578326 as the probability of 2 or fewer businesses failing. Subtract .578326 from 1 to get .421674 as the probability that at least 3 businesses will fail. This agrees with the answer we found for Example 4(a).

9.4 Exercises

In Exercises 1–39, see Examples 1–4.

Student Loan Debt *According to a report from the American Association of University Women released in 2017, 44% of female undergraduates take on debt. Find the probability that the given number of female undergraduates have taken on debt if 10 female undergraduates were selected at random. (Data from: www.aauw.org.)*

1. Exactly 6
2. Exactly 5
3. None
4. All
5. At least 1
6. At most 4

Student Loan Debt *The report cited for Exercises 1–6 reported that 39% of male undergraduates take on debt. Find the probability that the given number of male undergraduates have taken on debt if 8 male undergraduates were selected at random. (Data from: www.aauw.org.)*

7. Exactly 2
8. Exactly 1
9. None
10. All
11. At least 1
12. At most 2

A coin is tossed 5 times. Find the probability of getting

13. all heads;
14. exactly 3 heads;
15. no more than 3 heads;
16. at least 3 heads.

17. How do you identify a probability problem that involves a binomial experiment?

18. Why do combinations occur in the binomial probability formula?

Labor Force *According to data from the U.S. Bureau of Labor Statistics, 27.2% of the civilian labor force had some college or an associate degree. For Exercises 19–21, find the probability for the described number of workers that had some college or an associate degree if 15 workers are selected at random.*

19. exactly 3;
20. none;
21. at most 2.
22. If 200 workers are selected at random, what would be the expected number of workers with some college or an associate degree?

Stock Ownership *In a poll conducted by Gallup in early 2017, 31% of 18- to 29-year-olds had money invested in the stock market (either in an individual stock, a stock market fund, or a self-directed 401(k) or IRA). If 9 adults age 18–29 are selected at random, find the probability that the given number of them have money invested in the stock market. (Data from: www.gallup.com.)*

23. none of them;
24. at most 3;
25. 5 or more;
26. at least 1.

27. If 500 adults aged 18–29 are selected at random, what is the expected number who will have money invested in the stock market?

28. In Exercise 27, what would the expected number be if 1250 adults in that age range are selected at random?

Twins *The probability that a birth will result in twins is .027. Assuming independence (perhaps not a valid assumption), what are the probabilities that, out of 100 births in a hospital, there will be the given numbers of sets of twins? (Data from: The World Almanac and Book of Facts, 2001.)*

29. Exactly 2 sets of twins
30. At most 2 sets of twins

Lefties *According to the Web site Answers.com, 10–13% of Americans are left handed. Assume that the percentage is 11%. If we select 9 people at random, find the probability that the number who are left handed is*

31. exactly 2;
32. at least 2;
33. none;
34. at most 3.

35. In a class of 35 students, how many left-handed students should the instructor expect?

Confidence in Banks *The 2016 General Social Survey indicated that 14% of U.S. residents have a great deal of confidence in banks and financial institutions. If 16 U.S. residents were chosen at random, find the probability that the number given had a great deal of confidence in banks and financial institutions. (Data from: gss.norc.org.)*

36. exactly 2;
37. at most 3;
38. at least 4.

39. If we select 300 U.S. residents at random, what is the expected number who have a great deal of confidence in banks and financial institutions?

40. If we select 800 U.S. residents at random, what is the expected number who have a great deal of confidence in banks and financial institutions?

Vehicle Sales *In April 2017, 44% of car and light truck sales were for vehicles made by the "Big 3" American auto makers: General Motors, Ford, and Chrysler. If 100 vehicle sales are selected at random, find the probability of each of the following. (Data from: www.wsj.com.)*

41. What is the probability that at most 35 of the selected vehicles were made by one of the Big 3 automakers?

42. What is the probability that 50 or more of the selected vehicles were made by one of the Big 3 automakers?

43. What is the expected number of vehicles that were made by one of the Big 3 automakers?

44. If 2000 vehicles sales are selected at random, what is the expected number made by one of the Big 3 automakers?

✓**Checkpoint Answers**

1. **(a)** About .2583   **(b)** About .0974   **(c)** About .0147

2. **(a)** About .1115   **(b)** About .3845

3. About .2605

4. **(a)** About .2808   **(b)** About .8290   **(c)** 1.48

5. **(a)** About .7304   **(b)** About .9567   **(c)** 2.31

9.5 Markov Chains

In Section 8.5, we touched on **stochastic processes**—mathematical models that evolve over time in a probabilistic manner. In the current section, we study a special kind of stochastic process called a **Markov chain**, in which the outcome of an experiment depends only on the outcome of the previous experiment. In other words, the next state of the system depends only on the present state, not on preceding states. Such experiments are common enough in applications to make their study worthwhile. Markov chains are named after the Russian mathematician A. A. Markov (1856–1922), who started the theory of stochastic processes. To see how Markov chains work, we look at an example.

EXAMPLE 1   **Wireless Customer Churn** The wireless industry uses the term *churn* to describe customers who switch wireless firms. The discount wireless firm BUU hopes to lure customers from another discount wireless company (UMobile). BUU hires a market research firm to determine if an extensive marketing campaign will be effective. After the campaign, the market research firm finds that there is a probability of .8 that a BUU customer will stay with BUU and a .35 chance that a UMobile customer will switch to BUU. Assume that, because there is a .8 probability that a BUU customer will stay with BUU, there is a $1 - .8 = .2$ probability that the customer will switch to UMobile. In a similar way, let us assume there is a $1 - .35 = .65$ chance that a UMobile customer will stay with UMobile. If a customer currently with BUU is said to be in state 1 and a customer currently with UMobile is said to be in state 2, then these probabilities of change from one company to the other are as shown in the following table:

	State	Carrier after Campaign	
		1	2
Current Carrier	1	.8	.2
	2	.35	.65

The information from the table can be written in other forms. Figure 9.13 is a **transition diagram** that shows the two states and the probabilities of going from one to another.

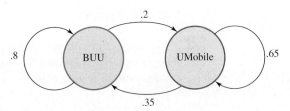

Figure 9.13

In a **transition matrix**, the states are indicated at the side and top, as follows:

$$
\begin{array}{cc}
 & \textit{Carrier after Campaign} \\
 & \begin{array}{cc} \text{BUU} & \text{UMobile} \end{array} \\
\textit{Current Carrier} \begin{array}{c} \text{BUU} \\ \text{UMobile} \end{array} & \begin{bmatrix} .8 & .2 \\ .35 & .65 \end{bmatrix}.
\end{array}
$$

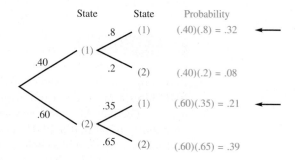

✓1

✓ **Checkpoint 1**

You are given the transition matrix

$$
\begin{array}{cc}
 & \text{State} \\
 & \begin{array}{cc} 1 & 2 \end{array} \\
\text{State} \begin{array}{c} 1 \\ 2 \end{array} & \begin{bmatrix} .3 & .7 \\ .1 & .9 \end{bmatrix}.
\end{array}
$$

(a) What is the probability of changing from state 1 to state 2?

(b) What does the number .1 represent?

(c) Draw a transition diagram for this information.

A transition matrix has the following features:

1. It is square, since all possible states must be used both as rows and as columns.

2. All entries are between 0 and 1, inclusive, because all entries represent probabilities.

3. The sum of the entries in any row must be 1, because the numbers in the row give the probability of changing from the state at the left to one of the states indicated across the top.

EXAMPLE 2   **Wireless Customer Churn** Suppose that when the new promotional campaign began, BUU had 40% of the market and UMobile had 60%. Use the probability tree in Figure 9.14 to find how these proportions would change after the marketing campaign.

State 1: using carrier BUU
State 2: using carrier UMobile

Figure 9.14

Solution Add the numbers indicated with arrows to find the proportion of customers using BUU after the marketing campaign:

$$.32 + .21 = .53$$

Similarly, the proportion using UMobile after the marketing campaign is

$$.08 + .39 = .47.$$

The initial distribution of 40% and 60% becomes 53% and 47%, respectively, after the marketing campaign.

These distributions can be written as the *probability vectors*

$$[.40 \quad .60] \quad \text{and} \quad [.53 \quad .47].$$

A **probability vector** is a one-row matrix, with nonnegative entries, in which the sum of the entries is equal to 1.

The results from the probability tree of Figure 9.14 are exactly the same as the result of multiplying the initial probability vector by the transition matrix (multiplication of matrices was discussed in Section 6.5):

$$[.4 \quad .6]\begin{bmatrix} .8 & .2 \\ .35 & .65 \end{bmatrix} = [.53 \quad .47].$$

Checkpoint 2

Find the product

$$[.53 \ .47]\begin{bmatrix} .8 & .2 \\ .35 & .65 \end{bmatrix}.$$

Checkpoint 3

Find each company's market share after 3 marketing campaigns.

If v denotes the original probability vector [.4 .6] and P denotes the transition matrix, then the market share vector after one week is $vP = [.53 \ .47]$. To find the market share vector after two weeks, multiply the vector $vP = [.53 \ .47]$ by P; this amounts to finding vP^2. ☑₂

Checkpoint 2 shows that after 2 weeks, the market share vector is $vP^2 = [.59 \ .41]$. To get the market share vector after three weeks, multiply this vector by P; that is, find vP^3. Do not use the rounded answer from Checkpoint 2. ☑₃

Continuing this process gives each company's share of the market after additional weeks:

Number of Marketing Campaigns	BUU	UMobile	
0	.4	.6	v
1	.53	.47	vP^1
2	.59	.41	vP^2
3	.61	.39	vP^3
4	.63	.37	vP^4
12	.64	.36	vP^{12}
13	.64	.36	vP^{13}

The results seem to approach the probability vector [.64 .36].

What happens if the initial probability vector is different from [.4 .6]? Suppose [.75 .25] is used; then the same powers of the transition matrix as before give the following results:

Number of Marketing Campaigns	BUU	UMobile	
0	.75	.25	v
1	.69	.31	vP^1
2	.66	.34	vP^2
3	.65	.35	vP^3
4	.64	.36	vP^4
5	.64	.36	vP^5
6	.64	.36	vP^6

The results again seem to be approaching the numbers in the probability vector [.64 .36], the same numbers approached with the initial probability vector [.4 .6]. In either case, the long-range trend is for a market share of about 64% for BUU and 36% for UMobile. The example suggests that this long-range trend does not depend on the initial distribution of market shares. This means that if the initial market share for BUU was less than 64%, the advertising campaign has paid off in terms of a greater long-range market share. If the initial share was more than 64%, the campaign did not pay off.

Regular Transition Matrices

One of the many applications of Markov chains is in finding long-range predictions. It is not possible to make long-range predictions with all transition matrices, but for a large set of transition matrices, long-range predictions *are* possible. Such predictions are always possible with **regular transition matrices.** A transition matrix is **regular** if some power of the matrix contains all positive entries. A Markov chain is a **regular Markov chain** if its transition matrix is regular.

EXAMPLE 3   Decide whether the given transition matrices are regular.

(a) $A = \begin{bmatrix} .3 & .1 & .6 \\ 0 & .2 & .8 \\ .3 & .7 & 0 \end{bmatrix}$.

Solution   Square A:

$$A^2 = \begin{bmatrix} .27 & .47 & .26 \\ .24 & .60 & .16 \\ .09 & .17 & .74 \end{bmatrix}.$$

Since all entries in A^2 are positive, matrix A is regular.

(b) $B = \begin{bmatrix} .3 & 0 & .7 \\ 0 & 1 & 0 \\ 0 & 0 & 1 \end{bmatrix}$.

Solution   Find various powers of B:

$$B^2 = \begin{bmatrix} .09 & 0 & .91 \\ 0 & 1 & 0 \\ 0 & 0 & 1 \end{bmatrix}; \quad B^3 = \begin{bmatrix} .027 & 0 & .973 \\ 0 & 1 & 0 \\ 0 & 0 & 1 \end{bmatrix}; \quad B^4 = \begin{bmatrix} .0081 & 0 & .9919 \\ 0 & 1 & 0 \\ 0 & 0 & 1 \end{bmatrix}.$$

Notice that all of the powers of B shown here have zeros in the same locations. Thus, further powers of B will still give the same zero entries, so that no power of matrix B contains all positive entries. For this reason, B is not regular. ✔ 4

✔ **Checkpoint 4**

Decide whether the given transition matrices are regular.

(a) $\begin{bmatrix} 0 & 1 \\ 1 & 0 \end{bmatrix}$

(b) $\begin{bmatrix} .45 & .55 \\ 1 & 0 \end{bmatrix}$

📄 **NOTE**   If a transition matrix P has some zero entries, and P^2 does as well, you may wonder how far you must compute P^n to be certain that the matrix is not regular. The answer is that if zeros occur in identical places in both P^n and P^{n+1} for any n, then they will appear in those places for all higher powers of P, so P is not regular.

Suppose that v is any probability vector. It can be shown that, for a regular Markov chain with a transition matrix P, there exists a single vector V that does not depend on v, such that $v \cdot P^n$ gets closer and closer to V as n gets larger and larger.

Equilibrium Vector of a Markov Chain

If a Markov chain with transition matrix P is regular, then there is a unique vector V such that, for any probability vector v and for large values of n,

$$v \cdot P^n \approx V.$$

Vector V is called the **equilibrium vector**, or the **fixed vector**, of the Markov chain.

In the example with BUU, the equilibrium vector V is approximately $[.64 \quad .36]$. Vector V can be determined by finding P^n for larger and larger values of n and then looking for a vector that the product $v \cdot P^n$ approaches. Such a strategy can be very tedious, however, and is prone to error. To find a better way, start with the fact that, for a large value of n,

$$v \cdot P^n \approx V,$$

as mentioned in the preceding box. We can multiply both sides of this result by P, $v \cdot P^n \cdot P \approx V \cdot P$, so that

$$v \cdot P^n \cdot P = v \cdot P^{n+1} \approx VP.$$

Since $v \cdot P^n \approx V$ for large values of n, it is also true that $v \cdot P^{n+1} \approx V$ for large values of n. (The product $v \cdot P^n$ approaches V, so that $v \cdot P^{n+1}$ must also approach V.) Thus, $v \cdot P^{n+1} \approx V$ and $v \cdot P^{n+1} \approx VP$, which suggests that

$$VP = V.$$

> If a Markov chain with transition matrix P is regular, then the equilibrium vector V satisfies
>
> $$VP = V.$$

The equilibrium vector V can be found by solving a system of linear equations, as shown in the remaining examples.

EXAMPLE 4   Find the long-range trend for the Markov chain in Examples 1 and 2, with transition matrix

$$P = \begin{bmatrix} .8 & .2 \\ .35 & .65 \end{bmatrix}.$$

Solution This matrix is regular, since all entries are positive. Let P represent this transition matrix and let V be the probability vector $[v_1 \quad v_2]$. We want to find V such that

$$VP = V,$$

or

$$[v_1 \quad v_2] \begin{bmatrix} .8 & .2 \\ .35 & .65 \end{bmatrix} = [v_1 \quad v_2].$$

Multiply on the left to get

$$[.8v_1 + .35v_2 \quad .2v_1 + .65v_2] = [v_1 \quad v_2].$$

Set corresponding entries from the two matrices equal to obtain

$$.8v_1 + .35v_2 = v_1; \quad .2v_1 + .65v_2 = v_2.$$

Simplify each of these equations:

$$-.2v_1 + .35v_2 = 0; \quad .2v_1 - .35v_2 = 0.$$

These last two equations are really the same. (The equations in the system obtained from $VP = V$ are always dependent.) To find the values of v_1 and v_2, recall that $V = [v_1 \quad v_2]$ is a probability vector, so that

$$v_1 + v_2 = 1.$$

Find v_1 and v_2 by solving the system

$$-.2v_1 + .35v_2 = 0$$
$$v_1 + v_2 = 1.$$

We can rewrite the second equation as $v_1 = 1 - v_2$. Now substitute for v_1 in the first equation:

$$-.2(1 - v_2) + .35v_2 = 0.$$

Solving for v_2 yields

$$-.2 + .2v_2 + .35v_2 = 0$$
$$-.2 + .55v_2 = 0$$
$$.55v_2 = .2$$
$$v_2 \approx .364.$$

Since $v_2 \approx .364$ and $v_1 = 1 - v_2$, it follows that $v_1 \approx 1 - .364 = .636$, and the equilibrium vector is $[.636 \quad .364] \approx [.64 \quad .36]$.

EXAMPLE 5   **Manufacturing** The probability that a complex assembly line works correctly depends on whether the line worked correctly the last time it was used. The various probabilities are as given in the following transition matrix:

$$
\begin{array}{c}
 \\
\text{Worked Properly Before} \\
\text{Did Not}
\end{array}
\begin{array}{c}
\text{Works} \\
\begin{array}{cc}
\text{Properly Now} & \text{Does Not}
\end{array} \\
\begin{bmatrix}
.79 & .21 \\
.68 & .32
\end{bmatrix}
\end{array}
$$

Find the long-range probability that the assembly line will work properly.

Solution Begin by finding the equilibrium vector $[v_1 \quad v_2]$, where

$$
[v_1 \quad v_2]\begin{bmatrix} .79 & .21 \\ .68 & .32 \end{bmatrix} = [v_1 \quad v_2].
$$

Multiplying on the left and setting corresponding entries equal gives the equations

$$
.79v_1 + .68v_2 = v_1 \quad \text{and} \quad .21v_1 + .32v_2 = v_2,
$$

or

$$
-.21v_1 + .68v_2 = 0 \quad \text{and} \quad .21v_1 - .68v_2 = 0.
$$

Substitute $v_1 = 1 - v_2$ into the first of these equations to get

$$
\begin{aligned}
-.21(1 - v_2) + .68v_2 &= 0 \\
-.21 + .21v_2 + .68v_2 &= 0 \\
-.21 + .89v_2 &= 0 \\
.89v_2 &= .21 \\
v_2 &= \frac{.21}{.89} = \frac{21}{89},
\end{aligned}
$$

and $v_1 = 1 - \dfrac{21}{89} = \dfrac{68}{89}$. The equilibrium vector is $[68/89 \quad 21/89]$. In the long run, the company can expect the assembly line to run properly $\dfrac{68}{89} \approx 76\%$ of the time. ✓₅

✓ **Checkpoint 5**

In Example 5, suppose the company modifies the line so that the transition matrix becomes

$$
\begin{bmatrix} .85 & .15 \\ .75 & .25 \end{bmatrix}.
$$

Find the long-range probability that the assembly line will work properly.

EXAMPLE 6   **Vehicle Sales** Data from the *Wall Street Journal* web site showed that the probability that a vehicle purchased in the United States in April 2017 was from General Motors (*GM*) was .171, from Ford (*F*) was .150, from Chrysler (*C*) was .121, and from other car manufacturers (*O*) was .558. The following transition matrix indicates market share changes from year to year.

$$
\begin{array}{c}
\begin{array}{cccc}
GM & F & C & O
\end{array} \\
\begin{array}{c}
GM \\ F \\ C \\ O
\end{array}
\begin{bmatrix}
.85 & .04 & .05 & .06 \\
.02 & .91 & .03 & .04 \\
.01 & .01 & .95 & .03 \\
.03 & .02 & .06 & .89
\end{bmatrix}
\end{array}
$$

(a) Find the probability that a vehicle purchased in the next year was from Ford.

Solution To find the market share for each company in the next year we multiply the current market share vector [.171 .150 .121 .558] by the transition matrix. We obtain

$$
[.171 \quad .150 \quad .121 \quad .558]\begin{bmatrix}
.85 & .04 & .05 & .06 \\
.02 & .91 & .03 & .04 \\
.01 & .01 & .95 & .03 \\
.03 & .02 & .06 & .89
\end{bmatrix}
$$

$$
= [.16630 \quad .15571 \quad .16148 \quad .51651].
$$

The second entry (.15571) indicates that the probability that the vehicle purchased in the next year was from Ford is .15571.

(b) If the trend continues, find the long-term probability that when a new vehicle is purchased, it is from Ford.

Solution We need to find the equilibrium vector $[v_1 \quad v_2 \quad v_3 \quad v_4]$ where

$$[v_1 \quad v_2 \quad v_3 \quad v_4] \begin{bmatrix} .85 & .04 & .05 & .06 \\ .02 & .91 & .03 & .04 \\ .01 & .01 & .95 & .03 \\ .03 & .02 & .06 & .89 \end{bmatrix} = [v_1 \quad v_2 \quad v_3 \quad v_4].$$

Multiplying on the left side and setting corresponding entries equal gives the equations:

$$.85v_1 + .02v_2 + .01v_3 + .03v_4 = v_1$$
$$.04v_1 + .91v_2 + .01v_3 + .02v_4 = v_2$$
$$.05v_1 + .03v_2 + .95v_3 + .06v_4 = v_3$$
$$.06v_1 + .04v_2 + .03v_3 + .89v_4 = v_4.$$

Simplifying these equations yields:

$$-.15v_1 + .02v_2 + .01v_3 + .03v_4 = 0$$
$$.04v_1 - .09v_2 + .01v_3 + .02v_4 = 0$$
$$.05v_1 + .03v_2 - .05v_3 + .06v_4 = 0$$
$$.06v_1 + .04v_2 + .03v_3 - .11v_4 = 0.$$

Since V is a probability vector, the entries have to add up to 1, so we have

$$v_1 + v_2 + v_3 + v_4 = 1.$$

This gives five equations and four unknown values:

$$v_1 + v_2 + v_3 + v_4 = 1$$
$$-.15v_1 + .02v_2 + .01v_3 + .03v_4 = 0$$
$$.04v_1 - .09v_2 + .01v_3 + .02v_4 = 0$$
$$.05v_1 + .03v_2 - .05v_3 + .06v_4 = 0$$
$$.06v_1 + .04v_2 + .03v_3 - .11v_4 = 0.$$

The system can be solved with the Gauss-Jordan method set forth in Section 6.2. Start with the augmented matrix

$$\begin{bmatrix} 1 & 1 & 1 & 1 & 1 \\ -.15 & .02 & .01 & .03 & 0 \\ .04 & -.09 & .01 & .02 & 0 \\ .05 & .03 & -.05 & .06 & 0 \\ .06 & .04 & .03 & -.11 & 0 \end{bmatrix}.$$

The solution of this system is

$$V \approx [.103 \quad .156 \quad .494 \quad .248].$$

The second entry (.156) tells us the long-term probability that a purchased vehicle is from Ford. ✓ 6

In Example 4, we found that [.64 .36] was the equilibrium vector for the regular transition matrix

$$P = \begin{bmatrix} .8 & .2 \\ .35 & .65 \end{bmatrix}.$$

✓ **Checkpoint 6**

Find the equilibrium vector for the transition matrix

$$P = \begin{bmatrix} .3 & .7 \\ .5 & .5 \end{bmatrix}.$$

Observe what happens when you take powers of the matrix P (the displayed entries have been rounded for easy reading, but the full decimals were used in the calculations):

$$P^2 = \begin{bmatrix} .71 & .29 \\ .51 & .49 \end{bmatrix}; \quad P^3 = \begin{bmatrix} .67 & .33 \\ .58 & .42 \end{bmatrix}; \quad P^4 = \begin{bmatrix} .65 & .35 \\ .61 & .39 \end{bmatrix};$$

$$P^5 = \begin{bmatrix} .64 & .36 \\ .62 & .38 \end{bmatrix}; \quad P^6 = \begin{bmatrix} .64 & .36 \\ .63 & .37 \end{bmatrix}; \quad P^{10} = \begin{bmatrix} .64 & .36 \\ .64 & .36 \end{bmatrix}.$$

As these results suggest, higher and higher powers of the transition matrix P approach a matrix having all identical rows—rows that have as entries the entries of the equilibrium vector V.

If you have the technology to compute matrix powers easily (such as a graphing calculator), you can approximate the equilibrium vector by taking higher and higher powers of the transition matrix until all its rows are identical. Figure 9.15 shows part of this process for the transition matrix

$$B = \begin{bmatrix} .79 & .21 \\ .68 & .32 \end{bmatrix}$$

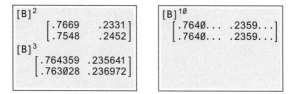

Figure 9.15

Figure 9.15 indicates that the equilibrium vector is approximately [.764 .236], which is what was found algebraically in Example 5.

The results of this section can be summarized as follows.

Properties of Regular Markov Chains

Suppose a regular Markov chain has a transition matrix P.

1. As n gets larger and larger, the product $v \cdot P^n$ approaches a unique vector V for any initial probability vector v. Vector V is called the *equilibrium vector*, or *fixed vector*.

2. Vector V has the property that $VP = V$.

3. To find V, solve a system of equations obtained from the matrix equation $VP = V$ and from the fact that the sum of the entries of V is 1.

4. The powers P^n come closer and closer to a matrix whose rows are made up of the entries of the equilibrium vector V.

9.5 Exercises

Decide which of the given vectors could be a probability vector.

1. $\begin{bmatrix} \frac{1}{4} & \frac{3}{4} \end{bmatrix}$

2. $\begin{bmatrix} \frac{11}{16} & \frac{5}{16} \end{bmatrix}$

3. $[0 \quad 1]$

4. $[.3 \quad .3 \quad .3]$

5. $[.3 \quad -.1 \quad .6]$

6. $\begin{bmatrix} \frac{2}{5} & \frac{3}{10} & .3 \end{bmatrix}$

Decide which of the given matrices could be a transition matrix. Sketch a transition diagram for any transition matrices.

7. $\begin{bmatrix} .7 & .2 \\ .5 & .5 \end{bmatrix}$

8. $\begin{bmatrix} \frac{1}{4} & \frac{3}{4} \\ 0 & 1 \end{bmatrix}$

9. $\begin{bmatrix} \frac{4}{9} & \frac{1}{3} \\ \frac{1}{5} & \frac{7}{10} \end{bmatrix}$

10. $\begin{bmatrix} 0 & 1 & 0 \\ .3 & .3 & .3 \\ 1 & 0 & 0 \end{bmatrix}$

11. $\begin{bmatrix} \frac{1}{2} & \frac{1}{4} & 1 \\ \frac{2}{3} & 0 & \frac{1}{3} \\ \frac{1}{3} & 1 & 0 \end{bmatrix}$

12. $\begin{bmatrix} .2 & .3 & .5 \\ 0 & 0 & 1 \\ .1 & .9 & 0 \end{bmatrix}$

In Exercises 13–15, write any transition diagrams as transition matrices.

13.
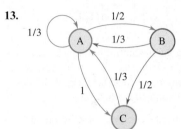

14.
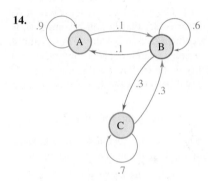

15.
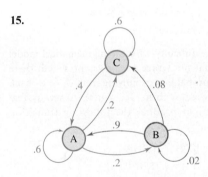

Decide whether the given transition matrices are regular. (See Example 3.)

16. $\begin{bmatrix} 1 & 0 \\ .25 & .75 \end{bmatrix}$

17. $\begin{bmatrix} .2 & .8 \\ .9 & .1 \end{bmatrix}$

18. $\begin{bmatrix} .3 & .5 & .2 \\ 1 & 0 & 0 \\ .5 & .1 & .4 \end{bmatrix}$

19. $\begin{bmatrix} .25 & .40 & .30 & .05 \\ .18 & .23 & .59 & 0 \\ 0 & .15 & .36 & .49 \\ .28 & .32 & .24 & .16 \end{bmatrix}$

20. $\begin{bmatrix} .23 & .41 & 0 & .36 \\ 0 & .27 & .21 & .52 \\ 0 & 0 & 1 & 0 \\ .48 & 0 & .39 & .13 \end{bmatrix}$

Find the equilibrium vector for each of the given transition matrices. (See Examples 4 and 5.)

21. $\begin{bmatrix} .3 & .7 \\ .4 & .6 \end{bmatrix}$

22. $\begin{bmatrix} .55 & .45 \\ .19 & .81 \end{bmatrix}$

23. $\begin{bmatrix} \frac{5}{8} & \frac{3}{8} \\ \frac{7}{9} & \frac{2}{9} \end{bmatrix}$

24. $\begin{bmatrix} \frac{2}{3} & \frac{1}{3} \\ \frac{1}{8} & \frac{7}{8} \end{bmatrix}$

25. $\begin{bmatrix} .25 & .35 & .4 \\ .1 & .3 & .6 \\ .55 & .4 & .05 \end{bmatrix}$

26. $\begin{bmatrix} .16 & .28 & .56 \\ .43 & .12 & .45 \\ .86 & .05 & .09 \end{bmatrix}$

27. $\begin{bmatrix} .15 & .15 & .70 \\ .42 & .38 & .20 \\ .16 & .28 & .56 \end{bmatrix}$

28. $\begin{bmatrix} .44 & .31 & .25 \\ .80 & .11 & .09 \\ .26 & .31 & .43 \end{bmatrix}$

For each of the given transition matrices, use a graphing calculator or computer to find the first five powers of the matrix. Then find the probability that state 2 changes to state 4 after 5 repetitions of the experiment.

29. $\begin{bmatrix} .3 & .2 & .3 & .1 & .1 \\ .4 & .2 & .1 & .2 & .1 \\ .1 & .3 & .2 & .2 & .2 \\ .2 & .1 & .3 & .2 & .2 \\ .1 & .1 & .4 & .2 & .2 \end{bmatrix}$

30. $\begin{bmatrix} .1 & .2 & .2 & .3 & .2 \\ .2 & .1 & .1 & .2 & .4 \\ .2 & .1 & .4 & .2 & .1 \\ .3 & .1 & .1 & .2 & .3 \\ .1 & .3 & .1 & .1 & .4 \end{bmatrix}$

31. Flu Shots In a recent year, the percentage of patients at a doctor's office who received a flu shot was 26%. A campaign by the doctors and nurses was designed to increase the percentage of patients who obtain a flu shot. The doctors and nurses believed there was an 85% chance that someone who received a shot in year 1 would obtain a shot in year 2. They also believed that there was a 40% chance that a person who did not receive a shot in year 1 would receive a shot in year 2.

(a) Give the transition matrix for this situation.

(b) Find the percentages of patients in year 2 who received, and did not receive, a flu shot.

(c) Find the long-range trend for the Markov chain representing receipt of flu shots at the doctor's office.

32. Media Impact Six months prior to an election, a poll found that only 35% of state voters planned to vote for a casino gambling initiative. After a media blitz emphasizing the new jobs that are created as a result of casinos, a new poll found that among those who did not favor it previously, 30% now favored the initiative. Among those who favored the initiative initially, 90% still favored it.

(a) Give the transition matrix for this situation.

(b) Find the percentage who favored the gambling initiative after the media blitz.

(c) Find the long-term percentage who favor the initiative if the trends and media blitz continue.

33. Manufacturing The probability that a complex assembly line works correctly depends on whether the line worked correctly the last time it was used. There is a .91 chance that the line will work correctly if it worked correctly the time before and a .68 chance that it will work correctly if it did *not* work correctly the time before. Set up a transition matrix with this information, and find the long-run probability that the line will work correctly. (See Example 5.)

34. Manufacturing Suppose something unplanned occurred to the assembly line of Exercise 33, so that the transition matrix becomes

$$\begin{array}{c} \\ \text{Works} \\ \text{Doesn't Work} \end{array} \begin{array}{cc} \text{Works} & \text{Doesn't Work} \\ \begin{bmatrix} .81 & .19 \\ .77 & .23 \end{bmatrix} \end{array}.$$

Find the new long-run probability that the line will work properly.

35. Genetics In Exercises 59 and 60 of Section 8.4 (pp. 435–436), we discussed the effect on flower color of cross-pollinating pea plants. As shown there, since the gene for red is dominant and the gene for white is recessive, 75% of the pea plants have red flowers and 25% have white flowers, because plants with 1 red and 1 white gene appear red. If a red-flowered plant is crossed with a red-flowered plant known to have 1 red and 1 white gene, then 75% of the offspring will be red and 25% will be white. Crossing a red-flowered plant that has 1 red and 1 white gene with a white-flowered plant produces 50% red-flowered offspring and 50% white-flowered offspring.

(a) Write a transition matrix using this information.

(b) Write a probability vector for the initial distribution of colors.

(c) Find the distribution of colors after 4 generations.

(d) Find the long-range distribution of colors.

36. Genetics Snapdragons with 1 red gene and 1 white gene produce pink-flowered offspring. If a red snapdragon is crossed with a pink snapdragon, the probabilities that the offspring will be red, pink, or white are 1/2, 1/2, and 0, respectively. If 2 pink snapdragons are crossed, the probabilities of red, pink, or white offspring are 1/4, 1/2, and 1/4, respectively. For a cross between a white and a pink snapdragon, the corresponding probabilities are 0, 1/2, and 1/2. Set up a transition matrix and find the long-range prediction for the fraction of red, pink, and white snapdragons.

37. Housing Approximately 58.2% of the residents of San Francisco are homeowners (O), 41.0% are renters (R), and .8% are homeless. Assume the following transition probabilities for residents of San Francisco per year. (Data from: www.nmhc.org and sfist.com.)

$$\begin{array}{c} \\ O \\ R \\ H \end{array} \begin{array}{ccc} O & R & H \\ \begin{bmatrix} .90 & .10 & 0 \\ .09 & .909 & .001 \\ 0 & .34 & .66 \end{bmatrix} \end{array}$$

(a) Find the probability that residents own, rent, and are homeless after one year.

(b) Find the long-range probabilities for the three categories.

38. Earthquakes Markov chains can be utilized in research into earthquakes. Researchers in Italy give the following example of a transition table in which the rows are magnitudes of an earthquake and the columns are magnitudes of the next earthquake in the sequence.[*]

	2.5	2.6	2.7	2.8
2.5	3/7	1/7	2/7	1/7
2.6	1/2	0	1/4	1/4
2.7	1/3	1/3	0	1/3
2.8	1/4	1/2	0	1/4

Thus, the probability of a 2.5-magnitude earthquake being followed by a 2.8-magnitude earthquake is 1/7. If these trends were to persist, find the long-range trend for the probabilities of each magnitude for the subsequent earthquake.

39. Car Insurance An insurance company classifies its drivers into three groups: G_0 (no accidents), G_1 (one accident), and G_2 (more than one accident). The probability that a driver in G_0 will stay in G_0 after 1 year is .85, that he will become a G_1 is .10, and that he will become a G_2 is .05. A driver in G_1 cannot move to G_0. (This insurance company has a long memory!) There is a .80 probability that a G_1 driver will stay in G_1 and a .20 probability that he will become a G_2. A driver in G_2 must stay in G_2.

(a) Write a transition matrix using this information.

Suppose that the company accepts 50,000 new policy-holders, all of whom are in group G_0. Find the number in each group

(b) after 1 year;   (c) after 2 years;

(d) after 3 years;   (e) after 4 years.

(f) Find the equilibrium vector here.

40. Car Insurance The difficulty with the mathematical model of Exercise 39 is that no "grace period" is provided; there should be a certain probability of moving from G_1 or G_2 back to G_0 (say, after 4 years with no accidents). A new system with this feature might produce the following transition matrix:

$$\begin{bmatrix} .85 & .10 & .05 \\ .15 & .75 & .10 \\ .10 & .30 & .60 \end{bmatrix}.$$

Suppose that when this new policy is adopted, the company has 50,000 policyholders in group G_0. Find the number in each group

(a) after 1 year;

(b) after 2 years;

(c) after 3 years.

(d) Find the equilibrium vector here.

41. Wireless Carriers In the final quarter of 2016, Verizon had 35.1% of the wireless market, while AT&T had 32.4%, and all other carriers had 32.5% of the market. Assume the following transition matrix for customers switching carriers in a year. (Data from: www.statistica.com.)

$$\begin{array}{c} \\ \\ \text{Now has} \\ \\ \end{array} \begin{array}{c} \\ \text{Verizon} \\ \text{AT\&T} \\ \text{Other} \end{array} \begin{array}{ccc} \text{Verizon} & \text{AT\&T} & \text{Other} \\ \begin{bmatrix} .91 & .07 & .02 \\ .03 & .87 & .10 \\ .14 & .04 & .82 \end{bmatrix} \end{array}.$$

Will Switch to

---

[*]Michele Lovallo, Vincenzo Lapenna, and Luciano Telesca, "Transition matrix analysis of earthquake magnitude sequences," *Chaos, Solitons, and Fractals* 24 (2005): 33–43.

Find the share of the market held by each company after

(a) 1 year; **(b)** 2 years; **(c)** 3 years.

(d) What is the long-range prediction?

42. **Changing Majors** At one liberal arts college, students are classified as humanities majors, science majors, or undecided. There is a 23% chance that a humanities major will change to a science major from one year to the next and a 40% chance that a humanities major will change to undecided. A science major will change to humanities with probability .12 and to undecided with probability .38. An undecided student will switch to humanities or science with probabilities of .45 and .28, respectively. Find the long-range prediction for the fraction of students in each of these three majors.

43. **Vehicle Sales** Using data similar to those in Example 6, the probability that a new vehicle purchased was from Toyota (T) was .142, American Honda Motor Corporation (H) was .097, and all other manufacturers (O) was .761. Use the transition matrix below for market share changes from year to year.

$$\begin{array}{c} \\ T \\ H \\ O \end{array} \begin{array}{ccc} T & H & O \\ \begin{bmatrix} .95 & .02 & .03 \\ .04 & .92 & .04 \\ .09 & .07 & .84 \end{bmatrix} \end{array}$$

(a) Find the probability that a new vehicle purchased in the next year was from Toyota.

(b) Find the long-term probability that a new vehicle purchase was from Honda.

44. **Queuing** In a queuing chain, we assume that people are queuing up to be served by, say, a bank teller. For simplicity, let us assume that once two people are in line, no one else can enter the line. Let us further assume that one person is served every minute, as long as someone is in line. Assume further that, in any minute, there is a probability of .4 that no one enters the line, a probability of .3 that exactly one person enters the line, and a probability of .3 that exactly two people enter the line, assuming that there is room. If there is not enough room for two people, then the probability that one person enters the line is .5. Let the state be given by the number of people in line.

(a) Give the transition matrix for the number of people in line:

$$\begin{array}{c} \\ 0 \\ 1 \\ 2 \end{array} \begin{array}{ccc} 0 & 1 & 2 \\ \begin{bmatrix} ? & ? & ? \\ ? & ? & ? \\ ? & ? & ? \end{bmatrix} \end{array}.$$

(b) Find the transition matrix for a 2-minute period.

(c) Use your result from part (b) to find the probability that a queue with no one in line has two people in line 2 minutes later.

Use a graphing calculator or computer for Exercises 45 and 46.

45. **Corporate Training** A company with a new training program classified each employee in one of four states: s_1, never in the program; s_2, currently in the program; s_3, discharged; s_4, completed the program. The transition matrix for this company is as follows.

$$\begin{array}{c} \\ s_1 \\ s_2 \\ s_3 \\ s_4 \end{array} \begin{array}{cccc} s_1 & s_2 & s_3 & s_4 \\ \begin{bmatrix} .4 & .2 & .05 & .35 \\ 0 & .45 & .05 & .5 \\ 0 & 0 & 1 & 0 \\ 0 & 0 & 0 & 1 \end{bmatrix} \end{array}.$$

(a) What percentage of employees who had never been in the program (state s_1) completed the program (state s_4) after the program had been offered five times?

(b) If the initial percentage of employees in each state was [.5 .5 0 0], find the corresponding percentages after the program had been offered four times.

46. **Corporate Training** Find the long-range prediction for the percentage of employees in each state for the company training program in Exercise 45.

✓**Checkpoint Answers**

1. **(a)** .7

   (b) The probability of changing from state 2 to state 1

   (c)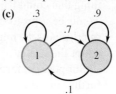

2. [.59 .41] (rounded)

3. [.61 .39] (rounded)

4. **(a)** No **(b)** Yes

5. $5/6 \approx 83\%$

6. [5/12 7/12]

9.6 Decision Making

John F. Kennedy once remarked that he had assumed that, as president, it would be difficult to choose between distinct, opposite alternatives when a decision needed to be made. Actually, he found that such decisions were easy to make; the hard decisions came when he was faced with choices that were not as clear cut. Most decisions fall into this last category—decisions that must be made under conditions of uncertainty. In Section 9.1, we saw how to use expected values to help make a decision. Those ideas are extended in this section, where we consider decision making in the face of uncertainty. We begin with an example.

EXAMPLE 1 **Fruit Production** Freezing temperatures are endangering the orange crop in central California. A farmer can protect his crop by burning smudge pots; the heat from the pots keeps the oranges from freezing. However, burning the pots is expensive, costing $20,000. The farmer knows that if he burns smudge pots, he will be able to sell his crop for a net profit (after the costs of the pots are deducted) of $50,000, provided that the freeze does develop and wipes out other orange crops in California. If he does nothing, he will either lose the $10,000 he has already invested in the crop if it does freeze or make a profit of $46,000 if it does not freeze. (If it does not freeze, there will be a large supply of oranges, and thus his profit will be lower than if there were a small supply.) What should the farmer do?

Solution He should begin by carefully defining the problem. First, he must decide on the **states of nature**—the possible alternatives over which he has no control. Here, there are two: freezing temperatures and no freezing temperatures. Next, the farmer should list the things he can control—his actions or **strategies.** He has two possible strategies: to use smudge pots or not. The consequences of each action under each state of nature, called **payoffs,** are summarized in a **payoff matrix,** as follows, where the payoffs in this case are the profits for each possible combination of events:

$$
\begin{array}{c}
\text{Strategies of Farmer} \\
\end{array}
\begin{array}{c}
 \\
\text{Use Smudge Pots} \\
\text{Do Not Use Pots} \\
\end{array}
\begin{array}{c}
\text{States of Nature} \\
\begin{array}{cc}
\text{Freeze} & \text{No Freeze} \\
\end{array} \\
\left[\begin{array}{cc}
\$50,000 & \$26,000 \\
-\$10,000 & \$46,000 \\
\end{array} \right]
\end{array}
$$

To get the $26,000 entry in the payoff matrix, use the profit if there is no freeze, namely, $46,000, and subtract the $20,000 cost of using the pots.

Once the farmer makes the payoff matrix, what then? The farmer might be an optimist (some might call him a gambler); in this case, he might assume that the best will happen and go for the biggest number of the matrix ($50,000). For that profit, he must adopt the strategy "use smudge pots."

On the other hand, if the farmer is a pessimist, he would want to minimize the worst thing that could happen. If he uses smudge pots, the worst thing that could happen to him would be a profit of $26,000, which will result if there is no freeze. If he does not use smudge pots, he might face a loss of $10,000. To minimize the worst, he once again should adopt the strategy "use smudge pots."

Suppose the farmer decides that he is neither an optimist nor a pessimist, but would like further information before choosing a strategy. For example, he might call the weather forecaster and ask for the probability of a freeze. Suppose the forecaster says that this probability is only .2. What should the farmer do? He should recall the discussion of expected value and work out the expected profit for each of his two possible strategies. If the probability of a freeze is .2, then the probability that there is no freeze is .8. This information leads to the following expected values:

If smudge pots are used: $50,000(.2) + 26,000(.8) = 30,800;$

If no smudge pots are used: $-10,000(.2) + 46,000(.8) = 34,800.$

Here, the maximum expected profit, $34,800, is obtained if smudge pots are not used. ☑²

As the example shows, the farmer's beliefs about the probabilities of a freeze affect his choice of strategies.

EXAMPLE 2 **Merchandising** An owner of several greeting-card stores must decide in July about the type of displays to emphasize for Sweetest Day in October. She has three possible choices: emphasize chocolates, emphasize collectible gifts, or emphasize gifts that can be engraved. Her success is dependent on the state of the economy in October. If the economy is strong, she will do well with the collectible gifts, while in a weak economy, the chocolates do very well. In a mixed economy, the gifts that can be engraved

---

✓ Checkpoint 1

Explain how each of the given payoffs in the matrix were obtained.

(a) −$10,000

(b) $50,000

✓ Checkpoint 2

What should the farmer do if the probability of a freeze is .6? What is his expected profit?

will do well. She first prepares a payoff matrix for all three possibilities, where the numbers in the matrix represent her profits in thousands of dollars:

$$
\begin{array}{c}
 & \textit{States of Nature} \\
 & \begin{array}{ccc} \text{Weak} & \text{Mixed} & \text{Strong} \\ \text{Economy} & & \text{Economy} \end{array}
\end{array}
$$

$$
\textit{Strategies} \quad
\begin{array}{r}
\text{Chocolates} \\
\text{Collectibles} \\
\text{Engraved}
\end{array}
\left[
\begin{array}{ccc}
85 & 30 & 75 \\
45 & 45 & 110 \\
60 & 95 & 85
\end{array}
\right].
$$

(a) What would an optimist do?

Solution If the owner is an optimist, she should aim for the biggest number on the matrix, 110 (representing $110,000 in profit). Her strategy in this case would be to display collectibles.

(b) How would a pessimist react?

Solution A pessimist wants to find the best of the worst things that can happen. If she displays collectibles, the worst that can happen is a profit of $45,000. For displaying engravable items, the worst is a profit of $60,000, and for displaying chocolates, the worst is a profit of $30,000. Her strategy here is to use the engravable items.

(c) Suppose the owner reads in a business magazine that leading experts believe that there is a 50% chance of a weak economy in October, a 20% chance of a mixed economy, and a 30% chance of a strong economy. How might she use this information?

Solution The owner can now find her expected profit for each possible strategy.

$$
\begin{array}{ll}
\text{Chocolates} & 85(.5) + 30(.20) + 75(.30) = 71; \\
\text{Collectibles} & 45(.5) + 45(.20) + 110(.30) = 64.5; \\
\text{Engraved} & 60(.5) + 95(.20) + 85(.30) = 74.5.
\end{array}
$$

Here, the best strategy is to display gifts that can be engraved; the expected profit is 74.5, or $74,500. ✔️₃

✔️ **Checkpoint 3**

Suppose the owner reads another article, which gives the following predictions: a 35% chance of a weak economy, a 25% chance of an in-between economy, and a 40% chance of a strong economy. What is the best strategy now? What is the expected profit?

9.6 Exercises

1. **Real Estate Development** A developer has $100,000 to invest in land. He has a choice of two parcels (at the same price): one on the highway and one on the coast. With both parcels, his ultimate profit depends on whether he faces light opposition from environmental groups or heavy opposition. He estimates that the payoff matrix is as follows (the numbers represent his profit):

$$
\begin{array}{c}
\textit{Opposition} \\
\begin{array}{cc} \text{Light} & \text{Heavy} \end{array}
\end{array}
$$

$$
\begin{array}{r}
\text{Highway} \\
\text{Coast}
\end{array}
\left[
\begin{array}{cc}
\$70,000 & \$30,000 \\
\$150,000 & -\$40,000
\end{array}
\right].
$$

What should the developer do if he is

(a) an optimist?

(b) a pessimist?

(c) Suppose the probability of heavy opposition is .8. What is his best strategy? What is the expected profit?

(d) What is the best strategy if the probability of heavy opposition is only .4?

2. **Concert Seating** Mount Union College has sold out all tickets for a jazz concert to be held in the stadium. If it rains, the show will have to be moved to the gym, which has a much smaller seating capacity. The dean must decide in advance whether to set up the seats and the stage in the gym, in the stadium, or in both, just in case. The following payoff matrix shows the net profit in each case:

$$
\begin{array}{c}
\textit{States of Nature} \\
\begin{array}{cc} \text{Rain} & \text{No Rain} \end{array}
\end{array}
$$

$$
\textit{Strategies} \quad
\begin{array}{r}
\text{Set up in Stadium} \\
\text{Set up in Gym} \\
\text{Set up in Both}
\end{array}
\left[
\begin{array}{cc}
-\$1550 & \$1500 \\
\$1000 & \$1000 \\
\$750 & \$1400
\end{array}
\right].
$$

What strategy should the dean choose if she is

(a) an optimist?

(b) a pessimist?

(c) If the weather forecaster predicts rain with a probability of .6, what strategy should she choose to maximize the expected profit? What is the maximum expected profit?

3. **Tire Manufacturing** An analyst must decide what fraction of the automobile tires produced at a particular manufacturing plant are defective. She has already decided that there are three possibilities for the fraction of defective items: .02, .09, and .16. She may recommend two courses of action: upgrade the equipment at the plant or make no upgrades. The following payoff matrix represents the *costs* to the company in each case, in hundreds of dollars:

$$\begin{array}{c} \textit{Strategies} \end{array} \begin{array}{cc} & \textit{Defectives} \\ & \begin{array}{ccc} .02 & .09 & .16 \end{array} \\ \begin{array}{c} \text{Upgrade} \\ \text{No Upgrade} \end{array} & \left[\begin{array}{ccc} 130 & 130 & 130 \\ 28 & 180 & 450 \end{array} \right]. \end{array}$$

What strategy should the analyst recommend if she is

(a) an optimist?

(b) a pessimist?

(c) Suppose the analyst is able to estimate probabilities for the three states of nature as follows:

Fraction of Defectives	Probability
.02	.70
.09	.20
.16	.10

Which strategy should she recommend? Find the expected cost to the company if that strategy is chosen.

4. **Marketing** The research department of the Allied Manufacturing Company has developed a new process that it believes will result in an improved product. Management must decide whether to go ahead and market the new product or not. The new product may be better than the old one, or it may not be better. If the new product is better and the company decides to market it, sales should increase by $50,000. If it is not better and the old product is replaced with the new product on the market, the company will lose $25,000 to competitors. If management decides not to market the new product, the company will lose $40,000 if it is better and will lose research costs of $10,000 if it is not.

(a) Prepare a payoff matrix.

(b) If management believes that the probability that the new product is better is .4, find the expected profits under each strategy and determine the best action.

5. **Machine Productivity** A businessman is planning to ship a used machine to his plant in Nigeria. He would like to use it there for the next 4 years. He must decide whether to overhaul the machine before sending it. The cost of overhaul is $2600. If the machine fails when it is in operation in Nigeria, it will cost him $6000 in lost production and repairs. He estimates that the probability that it will fail is .3 if he does not overhaul it and .1 if he does overhaul it. Neglect the possibility that the machine might fail more than once in the 4 years.

(a) Prepare a payoff matrix.

(b) What should the businessman do to minimize his expected costs?

6. **Business** A contractor prepares to bid on a job. If all goes well, his bid should be $25,000, which will cover his costs plus his usual profit margin of $4000. However, if a threatened labor strike actually occurs, his bid should be $35,000 to give him the same profit. If there is a strike and he bids $25,000, he will lose $5500. If his bid is too high, he may lose the job entirely, while if it is too low, he may lose money.

(a) Prepare a payoff matrix.

(b) If the contractor believes that the probability of a strike is .6, how much should he bid?

7. **Shelter Costs** An artist travels to craft fairs all summer long. She must book her booth at a June craft show six months in advance and decide if she wishes to rent a tent for an extra $500 in case it rains on the day of the show. If it does not rain, she believes she will earn $3000 at the show. If it rains, she believes she will earn only $2000, provided she has a tent. If she does not have a tent and it does rain, she will have to pack up and go home and will thus earn $0. Weather records over the last 10 years indicate that there is a .4 probability of rain in June.

(a) Prepare a profit matrix.

(b) What should the artist do to maximize her expected revenue?

8. **Investing** An investor has $50,000 to invest in stocks. She has two possible strategies: buy conservative blue-chip stocks or buy highly speculative stocks. There are two states of nature: the market goes up and the market goes down. The following payoff matrix shows the net amounts she will have under the various circumstances.

$$\begin{array}{c} \begin{array}{c} \text{Buy Blue Chip} \\ \text{Buy Speculative} \end{array} \begin{array}{cc} \text{Market Up} & \text{Market Down} \\ \left[\begin{array}{cc} \$60,000 & \$46,000 \\ \$80,000 & \$32,000 \end{array} \right]. \end{array} \end{array}$$

What should the investor do if she is

(a) an optimist?

(b) a pessimist?

(c) Suppose there is a .6 probability of the market going up. What is the best strategy? What is the expected profit?

(d) What is the best strategy if the probability of a market rise is .2?

Sometimes the numbers (or payoffs) in a payoff matrix do not represent money (profits or costs, for example). Instead, they may represent utility. A utility is a number that measures the satisfaction (or lack of it) that results from a certain action. Utility numbers must be assigned by each individual, depending on how he or she feels about a situation. For example, one person might assign a utility of +20 for a week's vacation in San Francisco, with −6 being assigned if the vacation were moved to Sacramento. Work the problems that follow in the same way as the preceding ones.

9. **Campaign Strategizing** A politician must plan her reelection strategy. She can emphasize jobs or she can emphasize the environment. The voters can be concerned about jobs or about the environment. Following is a payoff matrix showing the utility of each possible outcome.

$$\begin{array}{c} \textit{Candidate} \end{array} \begin{array}{cc} & \textit{Voters} \\ & \begin{array}{cc} \text{Jobs} & \text{Environment} \end{array} \\ \begin{array}{c} \text{Jobs} \\ \text{Environment} \end{array} & \left[\begin{array}{cc} +40 & -10 \\ -12 & +30 \end{array} \right] \end{array}$$

The political analysts feel that there is a .35 chance that the voters will emphasize jobs. What strategy should the candidate adopt? What is its expected utility?

10. Learning Aides In an accounting class, the instructor permits the students to bring a calculator or a reference book (but not both) to an examination. The examination itself can emphasize either numbers or definitions. In trying to decide which aid to take to an examination, a student first decides on the utilities shown in the following payoff matrix:

$$
\begin{array}{c}
 & & \textit{Exam Emphasizes} \\
 & & \text{Numbers} \quad \text{Definition} \\
\textit{Student Chooses} & \begin{array}{c} \text{Calculator} \\ \text{Book} \end{array} & \begin{bmatrix} +50 & 0 \\ +15 & +35 \end{bmatrix}
\end{array}
$$

(a) What strategy should the student choose if the probability that the examination will emphasize numbers is .6? What is the expected utility in this case?

(b) Suppose the probability that the examination emphasizes numbers is .4. What strategy should the student choose?

✓ Checkpoint Answers

1. (a) If the crop freezes and smudge pots are not used, the farmer's profit is $-\$10,000$ for labor costs.

   (b) If the crop freezes and smudge pots are used, the farmer makes a profit of $50,000.

2. Use smudge pots; $40,400

3. Engravable; $78,750

CHAPTER 9 Summary and Review

Key Terms and Symbols

9.1 random variable
probability distribution
histogram
expected value
fair game

9.2 *n!* (*n* factorial)
multiplication principle

permutations
combinations

9.4 Bernoulli trials (processes)
binomial experiment
binomial probability

9.5 stochastic processes
Markov chain

state
transition diagram
transition matrix
probability vector
regular transition
   matrix
regular Markov chain

equilibrium vector
   (fixed vector)

9.6 states of nature
strategies
payoffs
payoff matrix

Chapter 9 Key Concepts

Expected Value

For a random variable x with values $x_1, x_2, \ldots, x_n$ and probabilities $p_1, p_2, \ldots, p_n$, the expected value is

$$E(x) = x_1 p_1 + x_2 p_2 + \cdots + x_n p_n.$$

Multiplication Principle

If there are m_1 ways to make a first choice, m_2 ways to make a second choice, and so on, then there are $m_1 m_2 \cdots m_n$ different ways to make the entire sequence of choices.

Permutations

The number of **permutations** of n elements taken r at a time is $_n P_r = \dfrac{n!}{(n-r)!}$.

Combinations

The number of **combinations** of n elements taken r at a time is

$$_n C_r = \dfrac{n!}{(n-r)!\, r!}.$$

Binomial Experiments

Binomial Experiments have the following characteristics: (1) The same experiment is repeated a finite number of times; (2) there are only *two* outcomes, labeled success and failure; (3) the probability of success is the same for each trial; and (4) the trials are independent. If the probability of success in a single trial is p, the probability of x successes in n trials is

$$_n C_x p^x (1 - p)^{n - x}.$$

Markov Chains

A **transition matrix** must be square, with all entries between 0 and 1 inclusive, and the sum of the entries in any row must be 1. A Markov chain is *regular* if some power of its transition matrix P contains all positive entries. The long-range probabilities for a regular Markov chain are given by the **equilibrium,** or **fixed, vector** V, where, for any initial probability vector v, the products vP^n approach V as n gets larger and $VP = V$. To find V, solve the system of equations formed by $VP = V$ and the fact that the sum of the entries of V is 1.

Decision Making

A **payoff matrix,** which includes all available strategies and states of nature, is used in decision making to define the problem and the possible solutions. The expected value of each strategy can help to determine the best course of action.

Chapter 9 Review Exercises

In Exercises 1–4, (a) sketch the histogram of the given probability distribution, and (b) find the expected value.

1.

x	0	1	2	3
P(x)	.22	.54	.16	.08

2.

x	−3	−2	−1	0	1	2	3
P(x)	.15	.20	.25	.18	.12	.06	.04

3.

x	−10	0	10
P(x)	$\frac{1}{3}$	$\frac{1}{3}$	$\frac{1}{3}$

4.

x	0	2	4	6
P(x)	.35	.15	.2	.3

5. Vehicles per Household The probability distribution for the number of vehicles in a single household is given below. Find the expected number of vehicles in a household. (Data from: www.census.gov.)

x	0	1	2	3	4	5	6
P(x)	.079	.321	.381	.151	.049	.013	.006

6. House Size The probability distribution of the number of bedrooms in a given household is presented below. Find the expected number of bedrooms. (Data from: www.census.gov.)

x	0	1	2	3	4	5	6	7	8
P(x)	.0216	.0939	.2510	.4136	.1740	.0390	.0042	.0026	.0001

In Exercises 7 and 8, (a) give the probability distribution, and (b) find the expected value.

7. A grocery store has 10 bouquets of flowers for sale, 3 of which are rose displays. Two bouquets are selected at random, and the number of rose bouquets is noted.

8. In a class of 10 students, 3 did not do their homework. The professor selects 3 members of the class to present solutions to homework problems on the board and records how many of those selected did not do their homework.

Games of Chance *Solve the given problems.*

9. Suppose someone offers to pay you $100 if you draw 3 cards from a standard deck of 52 cards and all the cards are hearts. What should you pay for the chance to win if it is a fair game?

10. You pay $2 to play a game of "Over/Under," in which you will roll two dice and note the sum of the results. You can bet that the sum will be less than 7 (under), exactly 7, or greater than 7 (over). If you bet "under" and you win, you get your $2 back, plus $2 more. If you bet 7 and you win, you get your $2 back, plus $4, and if you

bet "over" and win, you get your $2 back, plus $2 more. What are the expected winnings for each type of bet?

11. Business In 2016, Apple's estimated share of the personal computer market was 7.1%. If we select 3 new computer purchases at random and define x to be the number of computers that are Apple, the probability distribution is as given below. Find the expected value for x. (Data from: www.businessinsider.com.)

x	0	1	2	3
P(x)	.8018	.1838	.0140	.0004

12. Smartwatches In the third quarter of 2016, Apple had 41.4% of the worldwide smartwatch market. If we select 5 smartwatch sales at random and define x to be the number that are Apple watches, the probability distribution for x appears below. Find the expected value for x. (Data from: www.icd.com.)

x	0	1	2	3	4	5
P(x)	.069	.244	.345	.244	.086	.012

13. In how many ways can 8 different taxis line up at the airport?

14. How many variations are possible for gold, silver, and bronze medalists in the 50-meter swimming race if there are 8 finalists?

15. In how many ways can a sample of 3 computer monitors be taken from a batch of 12 identical monitors?

16. If 4 of the 12 monitors in Exercise 15 are broken, in how many ways can the sample of 3 include the following?

  (a) 1 broken monitor;

  (b) no broken monitors;

  (c) at least 1 broken monitor.

17. In how many ways can 6 students from a class of 30 be arranged in the first row of seats? (There are 6 seats in the first row.)

18. In how many ways can the six students in Exercise 17 be arranged in a row if a certain student must be first?

19. In how many ways can the 6 students in Exercise 17 be arranged if half the students are science majors and the other half are business majors and if

  (a) like majors must be together?

  (b) science and business majors are alternated?

20. Explain under what circumstances a permutation should be used in a probability problem and under what circumstances a combination should be used.

21. Discuss under what circumstances the binomial probability formula should be used in a probability problem.

Suppose 2 cards are drawn without replacement from an ordinary deck of 52 cards. Find the probabilities of the given results.

22. Both cards are black.

23. Both cards are hearts.

24. Exactly 1 is a face card.

25. At most 1 is an ace.

An ice cream stand contains 4 custard flavors, 6 ice cream flavors, and 2 frozen yogurt selections. Three customers come to the window. If each customer's selection is random, find the probability that the selections include

26. all ice cream;

27. all custard;

28. at least one frozen yogurt;

29. one custard, one ice cream, and one frozen yogurt;

30. at most one ice cream.

Ⓒ *In Exercises 31 and 32, we study the connection between sets (from Chapter 8) and combinations.*

31. Given a set with n elements,

    (a) what is the number of subsets of size 0? of size 1? of size 2? of size n?

    (b) Using your answer from part (a) give an expression for the total number of subsets of a set with n elements.

32. Using your answers from Exercise 31 and the results from Chapter 8

    (a) explain why the following equation must be true:

    $$_nC_0 + {}_nC_1 + {}_nC_2 + \cdots + {}_nC_n = 2^n.$$

    (b) verify the equation in part (a) for $n = 4$ and $n = 5$.

33. **Stock Performance** On May 25, 2016, the percentage of stocks on the NASDAQ stock exchange that advanced was 50.7%. (Data from. www.wsj.com.) Suppose 7 stocks from the NASDAQ are selected at random, find

    (a) the probability distribution for x, the number of advancing stocks among the 7 selected;

    (b) the expected value of x.

34. **Stock Performance** On May 25, 2016, the percentage of stocks on the NYSE stock exchange that declined was 45.7%. (Data from: www.wsj.com.) Suppose 6 stocks from the NYSE are selected at random, find

    (a) the probability distribution for x, the number of declining stocks among the 6 selected;

    (b) the expected value of x.

35. **Internet Access** In 2016, the percentage of the U.S. population with access to the Internet at home was 88.5%. If 4 people are selected at random, find the probability that the given numbers have Internet access at home. (Data from: www.internetlivestats.com.)

    (a) None

    (b) At least 2 people

    (c) All 4 people

36. **Smartphone Ownership** The percentage of the U.S. population who own a smartphone in 2016 was 77%. If 5 people are

selected at random, find the probability that the given numbers own a smartphone. (Data from: www.pewinternet.org.)

    (a) Exactly 4

    (b) At most 4

    (c) At least 1

37. **Bank Earnings** From 2015 to 2016, 65.2% of banks insured by the Federal Deposit Insurance Corporation (FDIC) reported higher earnings. Suppose we select 5 banks insured by the FDIC at random. (Data from: www.fdic.gov.)

    (a) Give the probability distribution for x, the number of banks that reported higher earnings.

    (b) Find the probability that at least 2 banks reported higher earnings.

    (c) Give the expected value for the number of banks reporting higher earnings.

    ◩(d) Explain what your answer in part (c) means.

38. **Household Earnings** Approximately 22.5% of American households earn $100,000 or more a year. Suppose we select 5 households at random. (Data from: statisticalatlas.com.)

    (a) Give the probability distribution for x, the number of households earning $100,000 or more.

    (b) Find the probability that at most 3 households earned $100,000 or more.

    (c) Give the expected value for the number of households earning $100,000 or more.

    ◩(d) Explain what your answer in part (c) means.

Decide whether each matrix is a regular transition matrix.

39. $\begin{bmatrix} 0 & 1 \\ .77 & .23 \end{bmatrix}$ 40. $\begin{bmatrix} -.2 & .4 \\ .3 & .7 \end{bmatrix}$

41. $\begin{bmatrix} .21 & .15 & .64 \\ .50 & .12 & .38 \\ 1 & 0 & 0 \end{bmatrix}$ 42. $\begin{bmatrix} .22 & 0 & .78 \\ .40 & .33 & .27 \\ 0 & .61 & .39 \end{bmatrix}$

43. **E-Mail Usage** Using e-mail for professional correspondence has become a major component of a worker's day. A study classified e-mail use into 3 categories for an office day: no use, light use (1–60 minutes), and heavy use (more than 60 minutes). Researchers observed a pool of 100 office workers over a month and developed the following transition matrix of probabilities from day to day:

		Current Day		
		No Use	Light Use	Heavy Use
	No use	.35	.15	.50
Previous Day	Light Use	.30	.35	.35
	Heavy Use	.15	.30	.55

Suppose the initial distribution for the three categories is [.2, .4, .4]. Find the distribution after

    (a) 1 day;

    (b) 2 days.

    (c) What is the long-range prediction for the distribution of e-mail use?

44. **International Investing** An analyst at a major brokerage firm that invests in Europe, North America, and Asia has

examined the investment records for a particular international stock mutual fund over several years. The analyst constructed the following transition matrix for the probability of switching the location of an equity from year to year:

$$
\begin{array}{c}
& & \textit{Current Year} \\
& & \text{North} \\
& & \begin{array}{ccc} \text{Europe} & \text{America} & \text{Asia} \end{array} \\
\textit{Previous} \;\; \begin{array}{c} \text{Europe} \\ \text{North America} \\ \text{Year} \;\; \text{Asia} \end{array} &
\begin{bmatrix}
.80 & .14 & .06 \\
.04 & .85 & .11 \\
.03 & .13 & .84
\end{bmatrix}.
\end{array}
$$

If the initial investment vector is 15% in Europe, 60% in North America, and 25% in Asia,

(a) find the percentages in Europe, North America, and Asia after 1 year;

(b) find the percentages in Europe, North America, and Asia after 3 years;

(c) find the long-range percentages in Europe, North America, and Asia.

45. **Campaign Strategy** A candidate for city council can come out in favor of a new factory, be opposed to it, or waffle on the issue. The change in votes for the candidate depends on what her opponent does, with payoffs as shown in the following matrix:

$$
\begin{array}{c}
& & \textit{Opponent} \\
& & \begin{array}{ccc} \text{Favors} & \text{Waffles} & \text{Opposes} \end{array} \\
\textit{Candidate} \;\; \begin{array}{c} \text{Favors} \\ \text{Waffles} \\ \text{Opposes} \end{array} &
\begin{bmatrix}
0 & -1000 & -4000 \\
1000 & 0 & -500 \\
5000 & 2000 & 0
\end{bmatrix}.
\end{array}
$$

(a) What should the candidate do if she is an optimist?

(b) What should she do if she is a pessimist?

(c) Suppose the candidate's campaign manager feels that there is a 40% chance that the opponent will favor the plant and a

35% chance that he will waffle. What strategy should the candidate adopt? What is the expected change in the number of votes?

(d) The opponent conducts a new poll that shows strong opposition to the new factory. This changes the probability that he will favor the factory to 0 and the probability that he will waffle to .7. What strategy should our candidate adopt? What is the expected change in the number of votes now?

46. **Teaching Strategies** When teaching, an instructor can adopt a strategy using either active learning or lecturing to help students learn best. A class often reacts very differently to these two strategies. A class can prefer lecturing or active learning. A department chair constructs the following payoff matrix of the average point gain (out of 500 possible points) on the final exam after studying many classes that use active learning and many that use lecturing and polling students as to their preference:

$$
\begin{array}{c}
& & \textit{Students in class prefer} \\
& & \begin{array}{cc} \text{Lecture} & \text{Active Learning} \end{array} \\
\textit{Instructor} \;\; \begin{array}{c} \text{Lecture} \\ \textit{uses} \;\; \text{Active Learning} \end{array} &
\begin{bmatrix}
50 & -80 \\
-30 & 100
\end{bmatrix}.
\end{array}
$$

(a) If the department chair uses the preceding information to decide how to teach her own classes, what should she do if she is an optimist?

(b) What about if she is a pessimist?

(c) If the polling data shows that there is a 75% chance that a class will prefer the lecture format, what strategy should she adopt? What is the expected payoff?

(d) If the chair finds out that her next class has had more experience with active learning, so that there is now a 60% chance that the class will prefer active learning, what strategy should she adopt? What is the expected payoff?

Case Study 9 Quick Draw® from the New York State Lottery

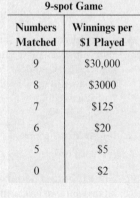

At bars and restaurants in the state of New York, patrons can play an electronic lottery game called Quick Draw.[*] A similar game is available in many other states. There are 10 ways for a patron to play this game. Prior to the draw, a person may bet $1 on games called 10-spot, 9-spot, 8-spot, 7-spot, 6-spot, 5-spot, 4-spot, 3-spot, 2-spot, and 1-spot. Depending on the game, the player will choose numbers from 1 to 80. For the 10-spot game, the player chooses 10 numbers; for a 9-spot game, the player chooses 9 numbers; etc. Every four minutes, the State of New York chooses 20 numbers at random from the numbers 1 to 80. For example, if a player chose the 6-spot game, he or she will have picked 6 numbers. If 3, 4, 5, or 6 of the numbers the player picked are also numbers the state picked randomly, then the player will win money. Each game has different ways to win, with differing payoff amounts. Notice with the 10-spot, 9-spot, 8-spot, and 7-spot, a player can win by matching 0 numbers correctly. The accompanying tables show the

payoffs for the different games. Notice that a player does not have to match all the numbers he or she picked in order to win.

10-spot Game	
Numbers Matched	Winnings per $1 Played
10	$100,000
9	$5000
8	$300
7	$45
6	$10
5	$2
0	$5

9-spot Game	
Numbers Matched	Winnings per $1 Played
9	$30,000
8	$3000
7	$125
6	$20
5	$5
0	$2

[*]More information on Quick Draw can be found at www.nylottery.org; click on "Daily Games."

8-spot Game	
Numbers Matched	Winnings per $1 Played
8	$10,000
7	$550
6	$75
5	$6
0	$2

7-spot Game	
Numbers Matched	Winnings per $1 Played
7	$5000
6	$100
5	$20
4	$2
0	$1

6-spot Game	
Numbers Matched	Winnings per $1 Played
6	$1000
5	$55
4	$6
3	$1

5-spot Game	
Numbers Matched	Winnings per $1 Played
5	$300
4	$20
3	$2

4-spot Game	
Numbers Matched	Winnings per $1 Played
4	$55
3	$5
2	$1

3-spot Game	
Numbers Matched	Winnings per $1 Played
3	$23
2	$2

2-spot Game	
Numbers Matched	Winnings per $1 Played
2	$10

1-spot Game	
Numbers Matched	Winnings per $1 Played
1	$2

With our knowledge of counting, it is possible for us to calculate the probability of winning for these different games.

EXAMPLE 1 Find the probability distribution for the number of matches for the 6-spot game.

Solution Let us define x to be the number of matches when playing 6-spot. The outcomes of x are then 0, 1, 2, . . . , 6. To find the probabilities of these matches, we need to do a little thinking. First, we need to know how many ways a player can pick 6 numbers from the selection of 1 to 80. Since the order in which the player picks the numbers does not matter, the number of ways to pick 6 numbers is

$$_{80}C_6 = \frac{80!}{74!\,6!} = 300{,}500{,}200.$$

To find the probability of the outcomes of 0 to 6, we can think of the 80 choices broken into groups: 20 winning numbers the state picked and 60 losing numbers the state did not pick. If x, the number of matches, is 0, then the player picked 0 numbers from the 20 winning

numbers and 6 from the 60 losing numbers. Using the multiplication principle, we find that this quantity is

$$_{20}C_0 \cdot {}_{60}C_6 = \left(\frac{20!}{20!\,0!}\right)\left(\frac{60!}{54!\,6!}\right) = (1)(50{,}063{,}860) = 50{,}063{,}860.$$

Therefore,

$$P(x = 0) = \frac{{}_{20}C_0 \cdot {}_{60}C_6}{{}_{80}C_6} = \frac{50{,}063{,}860}{300{,}500{,}200} \approx .16660.$$

Similarly for $x = 1, 2, \ldots, 6$, and completing the probability distribution table, we have

x	$P(x)$
0	$\dfrac{{}_{20}C_0 \cdot {}_{60}C_6}{{}_{80}C_6} \approx .16660$
1	$\dfrac{{}_{20}C_1 \cdot {}_{60}C_5}{{}_{80}C_6} \approx .36349$
2	$\dfrac{{}_{20}C_2 \cdot {}_{60}C_4}{{}_{80}C_6} \approx .30832$
3	$\dfrac{{}_{20}C_3 \cdot {}_{60}C_3}{{}_{80}C_6} \approx .12982$
4	$\dfrac{{}_{20}C_4 \cdot {}_{60}C_2}{{}_{80}C_6} \approx .02854$
5	$\dfrac{{}_{20}C_5 \cdot {}_{60}C_1}{{}_{80}C_6} \approx .00310$
6	$\dfrac{{}_{20}C_6 \cdot {}_{60}C_0}{{}_{80}C_6} \approx .00013$

EXAMPLE 2 Find the expected winnings for a $1 bet on the 6-spot game.

Solution To find the expected winnings, we take the winnings for each number of matches and subtract our $1 initial payment fee. Thus, we have the following:

x	Net Winnings	$P(x)$
0	−$1	.16660
1	−$1	.36349
2	−$1	.30832
3	$0	.12982
4	$5	.02854
5	$54	.00310
6	$999	.00013

The expected winnings are

$$\begin{aligned}
E(\text{winnings}) = {}&(-1)(.16660) + (-1)(.36349) \\
&+ (-1)(.30832) + 0(.12982) + 5(.02854) \\
&+ 54(.00310) + 999(.00013) \\
= {}&-.39844.
\end{aligned}$$

Thus, for every $1 bet on the 6-spot game, a player would lose about 40 cents. Put another way, the state gets about 40 cents, on average, from every $1 bet on 6-spot.

Exercises

1. If New York State initiates a promotion where players earn "double payoffs" for the 6-spot game (that is, if a player matched 3 numbers, she would win $2; if she matched 4 numbers, she would win $12; etc.), find the expected winnings.

2. Would it be in the state's interest to offer such a promotion? Why or why not?

3. Find the probability distribution for the 4-spot game.

4. Find the expected winnings for the 4-spot game.

5. If the state offers double payoffs for the 4-spot game, what are the expected winnings?

Extended Projects

1. In Ohio, a game similar to Quick Draw is KENO. Investigate the rules, bets, and payoffs of this game. Determine in what ways it is similar to Quick Draw and in what way it differs. Examine the expected value if any of the games were to offer double payoffs. Are any of the expected values with double payoffs positive?

2. Investigate whether your own state or a state near you has a game similar to Quick Draw. Determine if the rules, bets, and payoffs of this game are similar to Quick Draw. Determine in what ways the game in your state is similar to Quick Draw and in what way it differs. Examine the consequences on the expected value if the games were to offer double payoffs.

Introduction to Statistics

CHAPTER OUTLINE

Statistics has applications to almost every aspect of modern life. The digital age is creating a wealth of data that needs to be summarized, visualized, and analyzed, from the payroll of major-league baseball teams to movies' box-office receipts and the revenue of the home improvement industry.

Statistics is the science that deals with the collection and summarization of data. Methods of statistical analysis make it possible to draw conclusions about a population on the basis of data from a sample of the population. Statistical models have become increasingly useful in manufacturing, government, agriculture, medicine, and the social sciences and in all types of research. An Indianapolis race-car team, for example, is using statistics to improve its performance by gathering data on each run around the track. The team samples data 300 times a second and uses computers to process the data. In this chapter, we give a brief introduction to some of the key topics from statistical methodology.

10.1 Frequency Distributions

Researchers often wish to learn characteristics or traits of a specific **population** of individuals, objects, or units. The traits of interest are called **variables,** and it is these that we measure or label. Often, however, a population of interest is very large or constantly changing, so measuring each unit is impossible. Thus, researchers are forced to collect data on a subset of the population of interest, called a **sample.**

Sampling is a complex topic, but the universal aim of all sampling methods is to obtain a sample that "represents" the population of interest. One common way of obtaining a representative sample is to perform simple random sampling, in which every unit of the population has an equal chance to be selected to be in the sample. Suppose we wanted to study the height of students enrolled in a class. To obtain a random sample, we could place slips of paper containing the names of everyone in class in a hat, mix the papers, and draw 10 names blindly. We would then record the height (the variable of interest) for each student selected.

A simple random sample can be difficult to obtain in real life. For example, suppose you want to take a random sample of voters in your congressional district to see which candidate they prefer in the next election. If you do a telephone survey, you have a representative sample of people who are at home to answer the telephone, but those who are rarely home to answer the phone, those who have an unlisted number or only a cell phone, those who cannot afford a telephone, and those who refuse to answer telephone surveys are underrepresented. Such people may have an opinion different from those of the people you interview.

A famous example of an inaccurate poll was made by the *Literary Digest* in 1936. Its survey indicated that Alfred Landon would win the presidential election; in fact, Franklin Roosevelt won with 62% of the popular vote. The *Digest*'s major error was mailing its surveys to a sample of those listed in telephone directories. During the Depression, many poor people did not have telephones, and the poor voted overwhelmingly for Roosevelt. Modern pollsters use sophisticated techniques in an attempt to make their sample as representative as possible.

Once a sample has been collected and all data of interest are recorded, the data must be organized so that conclusions may be more easily drawn. With numeric responses, one method of organization is to group the data into intervals, usually of equal size.

EXAMPLE 1 **Tuition Costs** The following list consists of the 2017–2018 annual tuition (in thousands of dollars) for a random sample of 30 community colleges. (Data from: www.collegetuitioncompare.com.)

1.1	1.4	1.4	3.7	3.6	2.4	4.2	2.8	3.5	3.8
5.9	2.8	3.5	3.7	3.7	4.4	3.8	2.7	4.7	1.2
2.8	4.2	4.9	5.3	4.4	2.9	3.1	1.4	3.9	5.1

Identify the population and the variable, group the data into intervals, and find the frequency of each interval.

Solution The population is all community colleges. The variable of interest is the tuition (in thousands of dollars). The lowest value is 1.1 (corresponding to $1100) and the highest value is 5.9 (corresponding to $5900). One convenient way to group the intervals is in intervals of size .5, starting with 1.0–1.4 and ending with 5.0–5.4. This grouping provides an interval for each value in the list and results in 9 equally sized intervals of a convenient size. Too many intervals of smaller size would not simplify the data enough, while too few intervals of larger size would conceal information that the data might provide.

The first step in summarizing the data is to tally the number of schools in each interval. Then total the tallies in each interval, as in the following table:

Tuition Amount (in thousands of dollars)	Tally	Frequency
1.0–1.4	ⅣⅠ	5
1.5–1.9		0
2.0–2.4	Ⅰ	1
2.5–2.9	ⅣⅠ	5
3.0–3.4	Ⅰ	1
3.5–3.9	ⅣⅠ ⅠⅠⅠⅠ	9
4.0–4.4	ⅠⅠⅠⅠ	4
4.5–4.9	ⅠⅠ	2
5.0–5.4	ⅠⅠ	2
5.5–5.9	Ⅰ	1

✓ Checkpoint 1

A restaurant trade group commissioned a study to find how often families eat out. The following data gives the number of times a family ate a meal at a restaurant in the last month for 24 families surveyed at random.

8	12	0	6	10	8	0	14
8	12	14	16	4	14	7	11
9	12	7	15	11	21	22	19

Prepare a grouped frequency distribution for these data. Use intervals 0–4, 5–9, and so on.

Answers to Checkpoint exercises are found at the end of the section.

This table is an example of a **grouped frequency distribution.**

The frequency distribution in Example 1 shows information about the data that might not have been noticed before. For example, the interval with the largest number of colleges is 3.5–3.9. However, some information has been lost; for example, we no longer know exactly how many colleges charged 3.9 ($3900) for tuition.

Picturing Data

The information in a grouped frequency distribution can be displayed graphically with a **histogram,** which is similar to a bar graph. In a histogram, the number of observations in each interval determines the height of each bar, and the size of each interval determines the width of each bar. If equally sized intervals are used, all the bars have the same width.

A **frequency polygon** is another form of graph that illustrates a grouped frequency distribution. The polygon is formed by joining consecutive midpoints of the tops of the histogram bars with straight-line segments. Sometimes the midpoints of the first and last bars are joined to endpoints on the horizontal axis where the next midpoint would appear. (See Figure 10.1.)

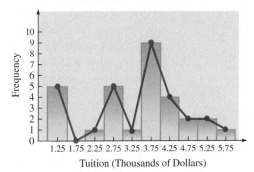

Tuition (Thousands of Dollars)

Figure 10.1

EXAMPLE 2 **Tuition Costs** A grouped frequency distribution of community college tuition was found in Example 1. Draw a histogram and a frequency polygon for this distribution.

Solution First, draw a histogram, shown in blue in Figure 10.1. To get a frequency polygon, connect consecutive midpoints of the tops of the bars. The frequency polygon is shown in red. ✓₂

✓ **Checkpoint 2**

Make a histogram and a frequency polygon for the distribution found in Checkpoint 1.

TECHNOLOGY TIP As noted in Section 9.1, most graphing calculators can display histograms. Many will also display frequency polygons (which are usually labeled LINE or xyLINE in calculator menus). When dealing with grouped frequency distributions, however, certain adjustments must be made on a calculator:

1. *The histogram's bar width affects the shape of the graph.* If you use a bar width of .4 in Example 1, the calculator may produce a histogram with gaps in it. To avoid this, use the interval $1.0 \leq x < 1.5$ in place of $1.1 \leq x \leq 1.4$, and similarly for other intervals, and make .5 the bar width.

2. *A calculator list of outcomes must consist of single numbers, not intervals.* The table in Example 1, for instance, cannot be entered as shown. To convert the first column of the table for calculator use, choose the midpoint number in each interval—1.25 for the first interval, 1.75 for the second interval, 2.25 for the third interval, etc. Then use 1.25, 1.75, 2.25, . . . , 5.75 as the list of outcomes to be entered into the calculator. The frequency list (the last column of the table) remains the same.

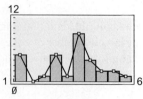

Figure 10.2

Following this procedure, we obtain the calculator-generated histogram and frequency polygon in Figure 10.2 for the data from Example 1. Note that the width of each histogram bar is .5. Some calculators cannot display both the histogram and the frequency polygon on the same screen, as is done here.

Stem-and-leaf plots allow us to organize the data into a distribution without the disadvantage of losing the original information. In a **stem-and-leaf plot,** we separate the digits in each data point into two parts consisting of the first one or two digits (the stem) and the remaining digit (the leaf). We also provide a key to show the reader the units of the data that were recorded.

EXAMPLE 3 **Tuition Costs** Construct a stem-and-leaf plot for the data in Example 1.

Solution Since the data is made up of two-digit numbers, we use the first digit for the stems: 1, 2, 3, 4, and 5 (which in this case represents thousands of dollars). The second digit provides the leaves (which here represents hundreds of dollars). For example, if we look at the seventh row of the stem-and-leaf plot, we have a stem value of 4 and leaf values of 2, 2, 4, and 4. These values correspond to entries of 4.2, 4.2, 4.4, and 4.4, meaning that one college had tuition of $4200, another had tuition of $4200, a third college had tuition of $4400, and a fourth college also had tuition of $4400. In this example, each row corresponds to an interval in the frequency table. The stems and leaves are separated by a vertical line.

Stem	Leaves
1	12444
1	
2	4
2	78889
3	1
3	556777889
4	2244
4	79
5	13
5	9

Units: 5|9 = $5900

If we turn the page on its side, the distribution looks like a histogram, but a stem-and-leaf plot still retains each of the original data values. We used each stem digit twice because, as with a histogram, using too few intervals conceals useful information about the shape of the distribution. ✓₃

✓ **Checkpoint 3**

Make a stem-and-leaf plot for the data in Example 1, using one stem each for 1, 2, 3, 4, and 5.

NOTE In this example we have split the leaves into two groups: those with leaf values of 0–4 and 5–9. We leave a row empty if there are no data values. For example, since there were no schools in Example 1 with tuition between $1500 and $1900, that row is left blank. We do not put a zero, however, as that would imply that a school had tuition of $1000. Not all stem-and-leaf plots split the leaves with two rows per stem, but it can help the reader discern the values more easily. Often there is only a single row per stem, but another common approach is to use five rows per stem where leaves of 0–1, 2–3, 4–5, 6–7, and 8–9 are grouped together.

EXAMPLE 4 **Resting Pulse** List the original data for the following stem-and-leaf plot of resting pulses taken on the first day of class for 36 students:

Stem	Leaves
4	8
5	278
6	034455688888
7	02222478
8	2269
9	00002289

Units: 9|0 = 90 beats per minute

The first stem and its leaf correspond to the data point 48 beats per minute. Similarly, the rest of the data are 52, 57, 58, 60, 63, 64, 64, 65, 65, 66, 68, 68, 68, 68, 68, 70, 72, 72, 72, 72, 74, 77, 78, 82, 82, 86, 89, 90, 90, 90, 90, 92, 92, 98, and 99 beats per minute. ✓₄

✓ **Checkpoint 4**

List the original data for the following heights (inches) of students:

Stem	Leaves
5	9
6	00012233334444
6	55567777799
7	0111134
7	558

Units: 7|5 = 75 inches

Assessing the Shape of a Distribution

Histograms and stem-and-leaf plots are very useful in assessing what is called the **shape** of the distribution. One common shape of data is seen in Figure 10.3(a). When all the bars of a histogram are approximately the same height, we say the data has a **uniform** shape. In Figure 10.3(b), we see a histogram that is said to be bell shaped, or **normal.** We use the "normal" label when the frequency peaks in the middle and tapers off equally on each side. When the data do not taper off equally on each side, we say the data are **skewed.** If the data taper off further to the left, we say the data are **left skewed** (Figure 10.3(c)). When the data taper off further to the right, we say the data are **right skewed** (Figure 10.3(d)). (Notice that with skewed data, we say "left skewed" or "right skewed" to refer to the tail, and not the peak of the data.)

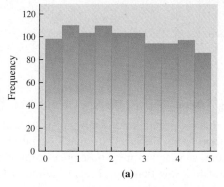

(a)

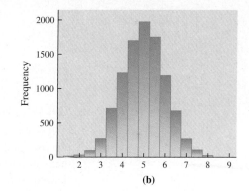

(b)

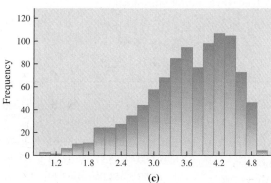

(c)

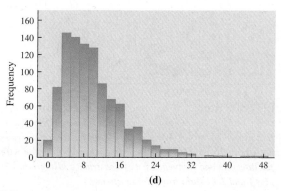

(d)

Figure 10.3

EXAMPLE 5   **Height and BMI** Characterize the shapes of the given distributions for 1000 adult males. (Data from: www.cdc.gov/nchs/nhanes.htm.)

(a) Height (inches); see Figure 10.4.

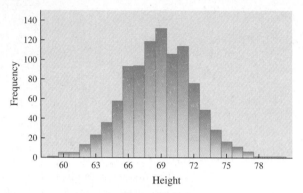

Figure 10.4

Checkpoint 5

Characterize the shape of the distribution from the following stem-and-leaf plot of ages (years):

Stem	Leaves
1	88888
2	22334
2	5579
3	333344
3	59
4	3344
4	667779
5	34
5	5679
6	1
6	
7	13
7	678

Units: 7|6 = 76 years

Solution The shape is **normal** because the shape peaks in the middle and tapers equally on each side.

(b) Body mass index (kg/m^2); see Figure 10.5.

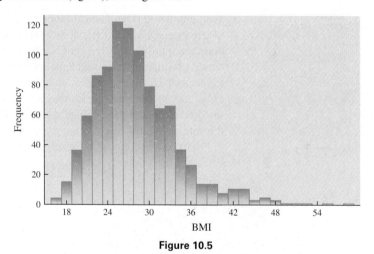

Figure 10.5

Solution The shape is **right skewed** because the tail is to the right. ✓5

It is important to note that most data are *not* normal, as we will see in the upcoming exercises and the next section. Using the label "normal" is a bit of a misnomer, because many important distributions, such as income, house prices, and infant birth weights, are skewed. It is also important to know that not all distributions have an easy-to-classify shape. This is especially true when samples are small.

10.1 Exercises

S&P 500 Stocks *The data for Exercises 1–4 consist of a random sample of 20 companies that were part of the S&P 500 and information was current as of March 7, 2017. For each variable, (a) group the data as indicated; (b) prepare a frequency distribution with columns for intervals and frequencies; (c) construct a histogram. (See Examples 1 and 2.) (Data from: data.okfn.org.)*

1. The variable is the price of the stock. Use 6 intervals starting with 0–39.99, 40–79.99, and so on.

71.95	239.60	182.02	78.94	73.72
167.00	75.97	27.20	71.36	62.55
82.77	58.56	67.39	143.60	37.52
126.26	41.47	26.66	30.81	23.50

2. The variable is the dividend yield (percent). Use 10 intervals with 0–.49, .5–.99, and so on.

2.38	1.14	3.12	2.26	2.12
1.06	1.50	1.62	1.37	3.23
1.64	3.26	0	2.14	3.98
2.11	2.66	0.94	4.75	2.95

3. The variable is the price/earnings ratio (calculated as the market price per share divided by the earnings per share). Use 6 intervals starting with 0–14.99, 15–29.99, and so on.

11.21	6.28	23.92	21.99	27.10
30.81	16.55	13.90	16.74	5.36
28.60	19.78	12.50	25.92	6.25
20.36	33.7	79.11	15.48	23.02

4. The variable is market capitalization (in billions of dollars).

28.88	80.32	111.48	11.15	65.08
73.30	20.48	15.92	19.71	22.96
30.33	18.56	40.82	9.30	56.20
96.11	38.93	9.59	9.37	7.60

The data for Exercises 5–10 consist of random samples of 20 households from the 2015 American Community Survey. Construct a frequency distribution and a histogram for each data set.

5. Annual household income (in thousands of dollars).

146	100	15	9	129	49	3	75	62	141
65	25	117	157	98	45	29	221	180	74

6. Amount (in dollars) of a monthly mortgage payment for homeowners.

930	280	600	2100	670	700	1300	640	1600	3200
390	1100	500	1200	2600	310	980	900	370	350

7. Amount (in dollars) of monthly electric bill.

180	110	180	350	200	360	60	230	100	100
60	460	270	30	150	90	170	100	140	140

8. Amount (in dollars) of annual water and sewage bill.

1200	380	940	750	300	1200	150	600	600	600
600	1000	500	960	700	1100	360	750	2000	1100

9. Amount (in dollars) of a monthly rent payment for nonhomeowners.

900	860	780	1900	1500	1400	600	800	600	1200
720	1300	800	1500	1500	480	800	300	1200	910

10. Amount (in dollars) of a monthly gas payment.

180	40	170	140	130	90	200	200	50	40
90	50	90	100	20	80	30	90	40	130

For Exercises 11–20, construct a stem-and-leaf plot for the data in the indicated exercise. (See Examples 3 and 4.)

11. **Stocks** Stock prices (rounded to the nearest dollar).

25	64	32	99	74	27	66	72	83	54
59	67	72	66	54	52	33	75	43	39

12. **Stocks** Stock prices (rounded to the nearest dollar).

85	110	76	98	120	105	84	77	94	100
102	125	84	93	98	132	84	139	112	77

13. **Height** Heights (in inches) of adult females.

62	64	64	69	66	62	65	68	61	64
60	61	64	66	54	61	62	62	63	65

14. **Height** Heights (in inches) of adult males

69	70	68	72	72	70	73	71	72	69
68	72	73	69	75	71	71	66	68	73

15. **Stocks** The price/earnings ratio for the 20 stocks in Exercise 3 (rounded to the nearest integer).

11	6	24	22	27	31	17	14	17	5
29	20	13	26	6	20	34	79	15	23

16. **Stocks** The dividend yield for the 20 stocks in Exercise 2 (rounded to the nearest tenth).

2.4	1.1	3.1	2.3	2.1	1.1	1.5	1.6	1.4	3.2
1.6	3.3	0	2.1	4.0	2.1	2.7	.9	4.8	3.0

17. **High School Graduates** The percentage of residents in the first 25 states alphabetically with a high school diploma. (Data from: National Center for Education Statistics.)

85	91	91	90	90
92	89	90	84	90
86	90	88	84	93
86	87	89	92	83
82	86	92	90	89

18. **High School Graduates** The percentage of residents in the latter 25 states alphabetically with a high school diploma. (Data from: National Center for Education Statistics.)

92	85	88	91	89
90	86	90	86	90
85	86	90	82	86
93	93	85	91	91
89	89	86	92	92

19. **Bachelor's Degrees** The percentage of residents in the first 25 states alphabetically with a bachelor's degree or higher. (Data from: National Center for Education Statistics.)

23	38	32	32	41
29	38	25	23	27
27	30	33	23	34
21	27	25	30	21
32	29	28	38	28

20. **Bachelor's Degrees** The percentage of residents in the latter 25 states alphabetically with a bachelor's degree or higher. (Data from: National Center for Education Statistics.)

28	27	24	27	37
30	35	31	25	33
23	29	29	28	19
35	26	30	31	29
37	27	26	36	26

Describe the shape of each of the given histograms. (See Example 5.)

21.

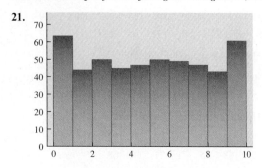

22.

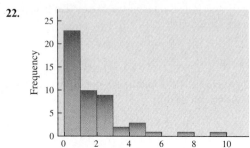

23.

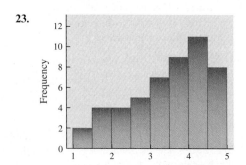

24.

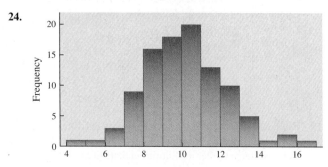

25. Student Loan Defaults The following histogram shows the student loan default rates for the 50 states and the District of Columbia reported in 2017. (Data from: www.studentloans.net.)

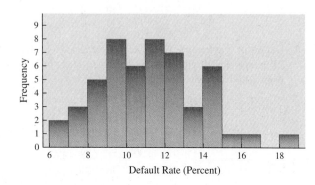

(a) Describe the shape of the histogram.

(b) How many states had a student loan default rate between 6 and 6.99%?

(c) How many states had a student loan default rate of 14% or higher?

26. Stocks The following histogram shows the stock price of 50 randomly selected stocks from the S&P 500 on March 7, 2017. (Data from: data.okfn.org.)

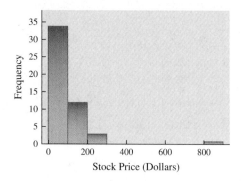

(a) Describe the shape of the histogram.

(b) How many stocks had prices between $0 and $99.99?

(c) How many stocks had prices below $200?

27. Cat Ownership The stem-and-leaf plot below summarizes the percentage of households that own cats for the 50 U.S. states (excluding Alaska). (Data from: American Veterinary Medical Association.)

Stem	Leaves
2	455
2	677777
2	8899999999
3	0001111
3	22223333333
3	4444
3	6
3	899
4	0
4	
4	
4	6
4	9

Units: 4|9 = 49%

(a) Describe the shape of the distribution.

(b) How many states have the percentage of cat ownership in the thirties?

(c) How many states have the percentage of cat ownership above 35%?

28. Personal Bankruptcies The stem-and-leaf plot below gives the number of personal bankruptcies per 100,000 residents for the 50 U.S. states and the District of Columbia. (Data from: www.worldatlas.com.)

Stem	Leaves
0	59
1	01234
1	5555666678888
2	2
2	55567899
3	00112444
3	5666779
4	0
4	567
5	12
5	
6	1

Units: 6|1 = 610

(a) Describe the shape of the distribution.

(b) How many states have 400 or more bankruptcies per 100,000 residents?

(c) How many states have 150 or less bankruptcies per 100,000 residents?

29. Test Scores The grade distribution for scores on a final exam is shown in the following stem-and-leaf plot:

Stem	Leaves
2	7
3	
3	
4	01
4	899
5	4
5	5
6	122
6	58
7	00124
7	9
8	0044
8	5679
9	00223334
9	5788

Units: 9|5 = 95%

(a) What is the shape of the grade distribution?

(b) How many students earned 90% or better?

(c) How many students earned less than 60%?

30. Chicken Production For the 20 states that produce a large number of broiler chickens, the following stem-and-leaf plot gives the production in billions of pounds for the year 2015. (Data from: U.S. Department of Agriculture.)

Stem	Leaves
0	223349
1	04447777
2	
3	7
4	5
5	
6	114
7	9

Units: 7|9 = 7.9 billion

(a) Describe the shape of the distribution.

(b) How many states had production below 1 billion pounds?

(c) How many states had production above 2 billion pounds?

✓ Checkpoint Answers

1.

Interval	Frequency
0–4	3
5–9	7
10–14	9
15–19	3
20–24	2
	Total: 24

2.

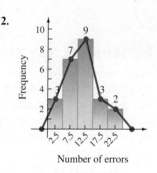

3.

Stem	Leaves
1	12444
2	478889
3	1556777889
4	224479
5	139

Units: 5|1 = $5,100

4. 59, 60, 60, 60, 61, 62, 62, 63, 63, 63, 63, 64, 64, 64, 64, 65, 65, 65, 66, 67, 67, 67, 67, 67, 69, 69, 70, 71, 71, 71, 71, 73, 74, 75, 75, 78

5. Right skewed

10.2 Measures of Center

Often, we want to summarize data numerically with a measure that represents a "typical" outcome. There are several ways to do this, and we generally call such a summary a "measure of center." In this section, we learn about the three most common measures of center: the mean, median, and mode.

Mean

The three most important measures of center are the mean, the median, and the mode. The most used of these is the mean, which is similar to the expected value of a probability distribution. The **arithmetic mean** (or just the "mean" or sometimes called the "average") of a set of numbers is the sum of the numbers, divided by the total number of numbers. We write the sum of n numbers $x_1, x_2, x_3, \ldots, x_n$ in a compact way with **summation notation,** also called **sigma notation.** With the Greek letter Σ (sigma), the sum

$$x_1 + x_2 + x_3 + \cdots + x_n$$

is written

$$x_1 + x_2 + x_3 + \cdots + x_n = \sum_{i=1}^{n} x_i.$$

In statistics, $\sum_{i=1}^{n} x_i$ is often abbreviated as just Σx_i. The symbol $\bar{x}$ (read "x-bar") is used to represent the mean of a sample.

Mean

The mean of the n numbers $x_1, x_2, x_3, \ldots, x_n$ is

$$\bar{x} = \frac{x_1 + x_2 + \cdots + x_n}{n} = \frac{\Sigma x_i}{n}.$$

TECHNOLOGY TIP Computing the mean is greatly simplified by the statistical capabilities of most scientific and graphing calculators. Calculators vary considerably in how data are entered, so read your instruction manual to learn how to enter lists of data and the corresponding frequencies. On scientific calculators with statistical capabilities, there are keys for finding most of the measures of center discussed in this section. On graphing calculators, most or all of these measures can be obtained with a single keystroke. (Look for a *one-variable statistics* option, which is often labeled 1-VAR, in the STAT menu or its CALC submenu.)

TECHNOLOGY TIP

The mean of the five numbers in Example 1 is easily found by using the $\bar{x}$ key on a scientific calculator or the one-variable statistics key on a graphing calculator. A graphing calculator will also display additional information, which will be discussed in the next section.

✓**Checkpoint 1**

Find the mean dollar amount of the following purchases of eight students selected at random at the campus bookstore during the first week of classes:

$250.56	$567.32
$45.29	$321.56
$120.22	$561.04
$321.07	$226.90

EXAMPLE 1 **Apple Revenue** The revenue, in billions of dollars, for Apple Inc. for the last five years is given in the following table. (Data from: www.morningstar.com.)

Year	2012	2013	2014	2015	2016
Revenue	157	171	183	234	216

Find the mean of revenue for the five years.

Solution Let $x_1 = 157, x_2 = 171, x_3 = 183, x_4 = 234,$ and $x_5 = 216$. Here, $n = 5$ because there are 5 numbers in the list. Thus,

$$\bar{x} = \frac{157 + 171 + 183 + 234 + 216}{5} = 192.2.$$

The mean revenue for the 5-year period is $192,200,000,000. ✓₁

The mean of data that have been arranged into a frequency distribution is found in a similar way. For example, suppose the following quiz score data are collected:

Value	Frequency
84	2
87	4
88	7
93	4
99	3
	Total: 20

The value 84 appears twice, 87 four times, and so on. To find the mean, first add 84 two times, 87 four times, and so on; or get the same result faster by multiplying 84 by 2, 87 by 4, and so on, and then adding the results. Dividing the sum by 20, the total of the frequencies, gives the mean:

$$\bar{x} = \frac{(84 \cdot 2) + (87 \cdot 4) + (88 \cdot 7) + (93 \cdot 4) + (99 \cdot 3)}{20}$$

$$= \frac{168 + 348 + 616 + 372 + 297}{20}$$

$$= \frac{1801}{20}$$

$$\bar{x} = 90.05.$$

Verify that your calculator gives the same result.

EXAMPLE 2 **Students' Ages** An instructor of a finite-mathematics class at a small liberal-arts college collects data on the age of her students. The data are recorded in the following frequency distribution:

Age	Frequency	Age × Frequency
18	12	$18 \cdot 12 = 216$
19	9	$19 \cdot 9 = 171$
20	5	$20 \cdot 5 = 100$
21	2	$21 \cdot 2 = 42$
22	2	$22 \cdot 2 = 44$
	Total: 30	Total: 573

Find the mean age.

Solution The age 18 appears 12 times, 19 nine times, and so on. To find the mean, first multiply 18 by 12, 19 by 9, and so on, to get the column "Age × Frequency," which has been added to the frequency distribution. Adding the products from this column gives a total of 573. The total from the frequency column is 30. The mean age is

$$\bar{x} = \frac{573}{30} = 19.1. \quad ✔_2$$

The mean of grouped data is found in a similar way. For grouped data, intervals are used, rather than single values. To calculate the mean, it is assumed that all of the values in a given interval are located at the midpoint of the interval. We use x_i to represent the midpoints, and f_i represents the frequencies, as shown in the next example.

✓ **Checkpoint 2**

Find $\bar{x}$ for the following frequency distribution for the variable of years of schooling for a sample of construction workers.

Years	Frequency
7	2
9	3
11	6
13	4
15	4
16	1

EXAMPLE 3   **College Tuition**   The grouped frequency distribution for annual tuition (in thousands of dollars) for the 30 community colleges described in Example 1 of Section 10.1 is as follows:

Tuition Amount (in thousands of dollars)	Frequency, f_i	Midpoint, x_i	Product, $f_i x_i$
1.0–1.49	5	1.25	6.25
1.5–1.99	0	1.75	0
2.0–2.49	1	2.25	2.25
2.5–2.99	5	2.75	13.75
3.0–3.49	1	3.25	3.25
3.5–3.99	9	3.75	33.75
4.0–4.49	4	4.25	17
4.5–4.99	2	4.75	9.5
5.0–5.49	2	5.25	10.5
5.5–5.99	1	5.75	5.75
	Total = 30		Total = 102

Find the mean from the grouped frequency distribution.

Solution   A column for the midpoint of each interval has been added. The numbers in this column are found by adding the endpoints of each interval and dividing by 2. For the interval 1.0–1.49, the midpoint is $(1.0 + 1.49)/2 = 1.245$ (which we will round to 1.25). The numbers in the product column on the right are found by multiplying each frequency by its corresponding midpoint. Finally, we divide the total of the product column by the total of the frequency column to get

$$\overline{x} = \frac{102}{30} = 3.4.$$

It is important to know that information is always lost when the data are grouped. It is more accurate to use the original data, rather than the grouped frequency, when calculating the mean, but the original data might not be available. Furthermore, the mean based upon the grouped data is typically not too different from the mean based upon the original data, and there may be situations in which the extra accuracy is not worth the effort. ✔3

NOTE   If we had used different intervals in Example 3, the mean would have come out to be a slightly different number. This is demonstrated in Checkpoint 4. ✔4

The formula for the mean of a grouped frequency distribution is as follows.

Mean of a Grouped Distribution

The mean of a distribution where x_i represents the midpoint of interval i, f_i represents the frequency of observations in interval i, and $n = \sum f_i x_i$ is

$$\overline{x} = \frac{\sum f_i x_i}{n}.$$

The mean of a random sample is a random variable, and for this reason it is sometimes called the **sample mean.** The sample mean is a random variable because it assigns a number to the experiment of taking a random sample. If a different random sample were taken, the mean would probably have a different value, with some values being more probable than

✓ **Checkpoint 3**

Find the mean of the following grouped frequency distribution for the number of classes completed thus far in the college careers of a random sample of 52 students:

Classes	Frequency
0–5	6
6–10	10
11–20	12
21–30	15
31–40	9

✓ **Checkpoint 4**

Find the mean for the college tuition data using the following intervals for the grouped frequency distribution:

Tuition Amount	Frequency
1.0–1.9	5
2.0–2.9	6
3.0–3.9	10
4.0–4.9	6
5.0–5.9	3

others. For example, if another set of 30 community colleges were selected in Example 3, the mean tuition amount might not have been 3.4.

We saw in Section 9.1 how to calculate the expected value of a random variable when we know its probability distribution. The expected value is sometimes called the **population mean,** denoted by the Greek letter μ. In other words,

$$E(x) = \mu.$$

Furthermore, it can be shown that the expected value of $\bar{x}$ is also equal to μ; that is,

$$E(\bar{x}) = \mu.$$

For instance, consider again the 30 community colleges in Example 3. We found that $\bar{x} = 3.4$, but the value of μ, the average for all tuition amounts, is unknown. If a good estimate of μ were needed, the best guess (based on these data) is 3.4.

Median

Asked by a reporter to give the average height of the players on his team, a Little League coach lined up his 15 players by increasing height. He picked out the player in the middle and pronounced this player to be of average height. This kind of average, called the **median,** is defined as the middle entry in a set of data arranged in either increasing or decreasing order. If the number of entries is even, the median is defined to be the mean of the two middle entries. The following table shows how to find the median for the two sets of data {8, 7, 4, 3, 1} and {2, 3, 4, 7, 9, 12} after each set has been arranged in increasing order.

Odd Number of Entries	Even Number of Entries
1	2
3	3
Median = 4	4 ⎫ Median = $\dfrac{4 + 7}{2}$ = 5.5
7	7 ⎭
8	9
	12

📄 **NOTE** As shown in the table, when the number of entries is even, the median is not always equal to one of the data entries.

EXAMPLE 4 Find the median number of hours worked per week

(a) for a sample of 7 male students whose work hours were

$$0, 7, 10, 20, 22, 25, 30.$$

Solution The median is the middle number, in this case 20 hours per week. (Note that the numbers are already arranged in numerical order.) In this list, three numbers are smaller than 20 and three are larger.

(b) for a sample of 11 female students whose work hours were

$$20, 0, 20, 30, 35, 30, 20, 23, 16, 38, 25.$$

Solution First, arrange the numbers in numerical order, from smallest to largest, or vice versa:

$$0, 16, 20, 20, 20, 23, 25, 30, 30, 35, 38.$$

The middle number can now be determined; the median is 23 hours per week.

(c) for a sample of 10 students of either gender whose work hours were

$$25, 18, 25, 20, 16, 12, 10, 0, 35, 32.$$

Solution Write the numbers in numerical order:

$$0, 10, 12, 16, \mathbf{18, 20}, 25, 25, 32, 35.$$

There are 10 numbers here; the median is the mean of the two middle numbers, or

$$\text{median} = \frac{18 + 20}{2} = 19.$$

The median is 19 hours per week.

 **CAUTION** Remember, the data must be arranged in numerical order before you locate the median. ✔ 5

With grouped data, we find the interval in which the median value would occur. In Example 3, there are 30 values, so the median would occur as the average of the 15th and 16th values. We have both of these values occurring in the interval between 3.5 and 3.99. Thus, the median is the midpoint of that interval—in this case 3.75. If there are an even number of observations, and the two values for which you need to take the mean fall into different intervals, then take the mean of the two midpoints.

Both the mean and the median of a sample are examples of a **statistic,** which is simply a number that gives summary information about a sample. In some situations, the median gives a truer representative or typical element of the data than the mean does. For example, suppose that in an office there are 10 salespersons, 4 secretaries, the sales manager, and Ms. Daly, who owns the business. Their annual salaries are as follows: support staff, $30,000 each; salespersons, $50,000 each; manager, $70,000; and owner, $400,000. The mean salary is

$$\bar{x} = \frac{(30,000)4 + (50,000)10 + 70,000 + 400,000}{16} = \$68,125.$$

However, since 14 people earn less than $68,125 and only 2 earn more, the mean does not seem very representative. The median salary is found by ranking the salaries by size: $30,000, $30,000, $30,000, $30,000, $50,000, $50,000, . . . , $400,000. There are 16 salaries (an even number) in the list, so the mean of the 8th and 9th entries will give the value of the median. The 8th and 9th entries are both $50,000, so the median is $50,000. In this example, the median is more representative of the distribution than the mean is.

When the data include extreme values (such as $400,000 in the preceding example), the mean may not provide an accurate picture of a typical value. So the median is often a better measure of center than the mean for data with extreme values, such as income levels and house prices. *In general, the median is a better measure of center whenever we see right-skewed or left-skewed distributions.*

Mode

Sue's scores on 10 class quizzes include one 7, two 8's, six 9's, and one 10. She claims that her average grade on quizzes is 9, because most of her scores are 9's. This kind of "average," found by selecting the most frequent entry, is called the **mode.**

EXAMPLE 5 Find the mode for the given data sets.

(a) Ages of retirement: 55, 60, 63, 63, 70, 55, 60, 65, 68, 65, 65, 71, 65, 65

Solution The number 65 occurs more often than any other, so it is the mode. It is sometimes convenient, but not necessary, to place the numbers in numerical order when looking for the mode.

(b) Total cholesterol score: 180, 200, 220, 260, 220, 205, 255, 240, 190, 300, 240

Solution Both 220 and 240 occur twice. This list has *two* modes, so it is bimodal.

✔**Checkpoint 5**

Find the median for the given heights in inches.

(a) 60, 72, 64, 75, 72, 65, 68, 70

(b) 73, 58, 77, 66, 69, 69, 66, 68, 67

▦ TECHNOLOGY TIP

Many graphing calculators display the median when doing one-variable statistics. You may have to scroll down to a second screen to find it.

✓ Checkpoint 6

Find the mode for each of the given data sets.

(a) Highway miles per gallon of an automobile: 25, 28, 32, 19, 15, 25, 30, 25

(b) Price paid for last haircut or styling: $11, $35, $35, $10, $0, $12, $0, $35, $38, $42, $0, $25

(c) Class enrollment in six sections of calculus: 30, 35, 26, 28, 29, 19

(c) Prices of new cars: $25,789, $43,231, $33,456, $19,432, $22,971, $29,876

Solution No number occurs more than once. This list has no mode. ✓6

The mode has the advantages of being easily found and not being influenced by data that are extreme values. It is often used in samples where the data to be "averaged" are not numerical. A major disadvantage of the mode is that there may be more than one, in case of ties, or there may be no mode at all when all entries occur with the same frequency.

The mean is the most commonly used measure of center. Its advantages are that it is easy to compute, it takes all the data into consideration, and it is reliable—that is, repeated samples are likely to give similar means. A disadvantage of the mean is that it is influenced by extreme values, as illustrated in the salary example.

The median can be easy to compute and is influenced very little by extremes. A disadvantage of the median is the need to rank the data in order; this can be tedious when the number of items is large.

EXAMPLE 6   **Business** A sample of 10 working adults was asked "How many hours did you work last week?" Their responses appear below. (Data from: gss.norc.org.)

$$40, 35, 43, 40, 30, 40, 45, 40, 55, 20.$$

Find the mean, median, and mode of the data.

Solution The mean number of hours worked is

$$\bar{x} = \frac{40 + 35 + 43 + 40 + 30 + 40 + 45 + 40 + 55 + 20}{10} = 38.8 \text{ hours.}$$

After the numbers are arranged in order from smallest to largest, the middle number, or median, is 40 hours.

The number 40 occurs more often than any other, so it is the mode.   ✓7

✓ Checkpoint 7

Following is a list of the number of movies seen at a theater in the last three months by nine students selected at random:

$$1, 0, 2, 5, 2, 0, 0, 1, 4$$

(a) Find the mean.

(b) Find the median.

(c) Find the mode.

Weighted Averages

Often analysts, researchers, and teachers, among others, want to count some values with more importance than other values when calculating a mean. For example, teachers in colleges and universities often assign more importance to the final examination for a course than they assign to the midterm exams. To do this, we want to calculate what is called a weighted average, or a weighted mean.

For example, if an instructor has three midterm exams that each count 20% toward the final grade and a final exam that counts 40% toward the final grade, then the final grade can be thought of as a weighted average. The weights are the amounts that each exam "counts" toward the final average expressed as a decimal. So the weights would be .2, .2, .2, and .4, respectively. The weights add up to 1.

When we calculate the weighted average, we multiply the weights by the actual exam scores and then sum them to get the final grade. We again use $x_1, x_2, x_3, \ldots, x_n$ to describe the data points and use $w_1, w_2, w_3, \ldots, w_n$ to describe the weights associated with each data point.

Weighted Average (Weighted Mean)

The weighted average of the n numbers $x_1, x_2, x_3, \ldots, x_n$, with corresponding weights $w_1, w_2, w_3, \ldots, w_n$, is $\bar{x} = w_1x_1 + w_2x_2 + w_3x_3 + \cdots + w_nx_n = \sum_{i=1}^{n} w_ix_i.$

EXAMPLE 7 **Final Grades** A teacher designs a course so that the homework counts 20% toward the final exam, quizzes count 10%, three midterm exams count 15% each, and the final exam counts 25%. Marissa earned 92% on the homework; her quiz average was 81%; her midterm exam scores were 75%, 88%, and 92%; and her final exam was 87%. Find her final average (weighted average).

Solution The weights for the weighted average are how much each assessment counts toward the final grade. Thus, the homework weight is .20, the quizzes are weighted as .10, exam 1 is weighted as .15, exam 2 is weighted as .15, exam 3 is weighted as .15, and the final exam is weighted as .25. We can make a table to correspond these weights with the scores earned on each component.

Component	Weight w_i	Score x_i	Product $w_i x_i$
Homework	.20	92	18.4
Quizzes	.10	81	8.1
Exam 1	.15	75	11.25
Exam 2	.15	88	13.2
Exam 3	.15	92	13.8
Final Exam	.25	87	21.75
			Total = 86.5

✓ **Checkpoint 8**

Manuel's professor weighs the homework 30%, quiz average 10%, four midterm exams 10% each, and the final exam 20%. Manuel earned 81% on his homework; his quiz average was 92%; his exam scores were 91, 72, 79, and 84; and his final exam was 78. Find his final average.

Thus, the final average (weighted average) for Marissa is 86.5%

10.2 Exercises

Find the mean for each data set. Round to the nearest tenth. (See Example 1.)

1. **Annual Salaries** Secretarial salaries (U.S. dollars):

   $21,900, $22,850, $24,930, $29,710, $28,340, $40,000.

2. **Annual Salaries** Starting teaching salaries (U.S. dollars):

   $38,400, $39,720, $28,458, $29,679, $33,679.

3. **Earthquakes** Earthquakes on the Richter scale:

   3.5, 4.2, 5.8, 6.3, 7.1, 2.8, 3.7, 4.2, 4.2, 5.7.

4. **Body Temperature** Body temperatures of self-classified "healthy" students (degrees Fahrenheit):

   96.8, 94.1, 99.2, 97.4, 98.4, 99.9, 98.7, 98.6.

5. **Foot Length** Lengths of foot (inches) for adult men:

   9.2, 10.4, 13.5, 8.7, 9.7.

Find the mean for each distribution. Round to the nearest tenth. (See Examples 2 and 3.)

6. **Quizzes** Scores on a quiz, on a scale from 0 to 10:

Value	7	8	9	10
Frequency	4	6	7	11

7. **Students' Ages** Age (years) of students in an introductory accounting class:

Value	19	20	21	22	23	24	28
Frequency	3	5	25	8	2	1	1

8. **Commuting Distance** Commuting distance (miles) for students at a university:

Value	0	1	2	5	10	17	20	25
Frequency	15	12	8	6	5	2	1	1

9. **Fuel Efficiency** Estimated miles per gallon of automobiles:

Value	Frequency
9	5
11	10
15	12
17	9
20	6
28	1

10–14. *Find the median of the data in Exercises 1–5. (See Example 4.)*

Find the mode or modes for each of the given lists of numbers. (See Example 5.)

15. **Children's Ages** Ages (years) of children in a day-care facility:

$$1, 2, 2, 1, 2, 2, 1, 1, 2, 2, 3, 4, 2, 3, 4, 2, 3, 2, 3.$$

16. **Health** Ages (years) in the intensive care unit at a local hospital:

$$68, 64, 23, 68, 70, 72, 72, 68.$$

17. **Students' Heights** Heights (inches) of students in a statistics class:

$$62, 65, 71, 74, 71, 76, 71, 63, 59, 65, 65, 64, 72, 71, 77, 63, 65.$$

18. **Pain Relief** Minutes of pain relief from acetaminophen after childbirth:

$$60, 240, 270, 180, 240, 210, 240, 300, 330, 360, 240, 120.$$

19. **GPA** Grade point averages for 5 students:

$$3.2, 2.7, 1.9, 3.7, 3.9.$$

20. When is the median the most appropriate measure of center?

21. When would the mode be an appropriate measure of center?

Pet Ownership *For Exercises 22 and 23, the frequency tables give the frequencies of the percentages of residents that own dogs or cats for all 50 U.S. states except Alaska. For grouped data, the modal class is the interval containing the most data values. Give the mean and modal class for each of the given collections of grouped data. (Round the midpoint to the nearest tenth.) (Data from: American Veterinary Medical Association.)*

22. The variable is the percentage of dog owners.

Interval	Frequency
10–14.99	1
15–19.99	0
20–24.99	1
25–29.99	4
30–34.99	13
35–39.99	12
40–44.99	12
45–49.99	6

23. The variable is the percentage of cat owners.

Interval	Frequency
10–14.99	1
15–19.99	0
20–24.99	1
25–29.99	18
30–34.99	22
35–39.99	4
40–44.99	1
45–49.99	2

24. To predict the outcome of the next congressional election, you take a survey of your friends. Is this a random sample of the voters in your congressional district? Explain why or why not.

Work each problem. (See Example 6.)

25. **MLB Payrolls** The following table gives the payroll (in millions of dollars) of the 10 Major League Baseball teams with the highest payrolls, as estimated by CBS Sports in 2017.

Team	Payroll
Los Angeles Dodgers	242
New York Yankees	202
Boston Red Sox	200
Detroit Tigers	200
Toronto Blue Jays	178
Texas Rangers	176
San Francisco Giants	172
Chicago Cubs	172
Washington Nationals	168
Baltimore Orioles	164

Find the following statistics for these data:

(a) mean;   (b) median.

26. **NFL Team Values** The following table gives the value (in billions of dollars) of the 10 most valued National Football League (NFL) teams, as estimated by *Forbes* in 2016.

Team	Value
Dallas Cowboys	4.20
New England Patriots	3.40
New York Giants	3.10
San Francisco 49ers	3.00
Washington Redskins	2.95
Los Angeles Rams	2.90
New York Jets	2.75
Chicago Cubs	2.70
Houston Texans	2.60
Philadelphia Eagles	2.50

Find the following statistics for these data:

(a) mean; (b) median.

27. Business The 12 movies that have earned the most revenue (in millions of dollars) from U.S. domestic box-office receipts in 2016 are given in the table. (Data from: www.boxofficemojo.com.)

Movie	Revenue
Finding Dory	486
Rogue One: A Star Wars Story	425
Captain America: Civil War	408
The Secret Life of Pets	368
The Jungle Book	364
Deadpool	363
Zootopia	341
Batman v Superman: Dawn of Justice	330
Suicide Squad	325
Doctor Strange	230
Fantastic Beasts and Where to Find Them	224
Moana	210

(a) Find the mean value in dollars for this group of movies.

(b) Find the median value in dollars for this group of movies.

28. Sirius XM Revenue The revenue (in millions of dollars) for Sirius XM Holdings Inc. from 2007 to 2016 is given in the following table. (Data from: www.morningstar.com.)

Year	Revenue
2007	922
2008	1664
2009	2473
2010	2817
2011	3015
2012	3402
2013	3799
2014	4181
2015	4570
2016	5017

(a) Find the mean and median for these data.

(b) What year's revenue is closest to the mean?

29. Dr Pepper and Snapple Revenue The revenue (in millions of dollars) for Dr Pepper Snapple Group Inc. from 2007 to 2016 is given in the following table. (Data from: www.morningstar.com.)

Year	Revenue
2007	5748
2008	5710
2009	5531
2010	5636
2011	5903
2012	5995
2013	5997
2014	6121
2015	6282
2016	6440

(a) Find the mean and median for these data.

(b) What year's revenue is closest to the mean?

Monthly Temperatures *The table gives the average monthly high and low temperatures, in degrees Fahrenheit, for Raleigh, North Carolina, over the course of a year.* (*Data from: www.weather.com.*)

Month	High	Low
January	49	30
February	53	32
March	61	40
April	71	48
May	78	57
June	84	65
July	88	69
August	86	68
September	80	62
October	70	49
November	61	42
December	52	33

Find the mean and median for each of the given subgroups.

30. The high temperatures

31. The low temperatures

For Exercises 32–33 determine the shape of the distribution from the histogram and then decide whether the mean or median is a better measure of center. If the mean is the better measure, calculate the value. If the median is the better measure, give the midpoint of the interval that contains it.

32. Market Capitalization The variable is market capitalization (in billions of dollars) of 20 randomly selected stocks from the S&P 500 on March 7, 2017. (Round to nearest tenth.) (Data from: data.okfn.org.)

Interval	Frequency
0–19.99	8
20.00–39.99	5
40.00–59.99	2
60.00–79.99	2
80.00–99.99	2
100.00–119.99	1

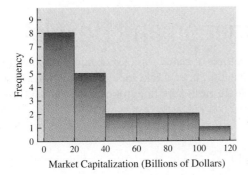

G 33. Annual Household Income The annual household income (in thousands of dollars) for 20 randomly chosen households from the 2015 American Community Survey. (Data from: www.census.gov.)

Interval	Frequency
0–24	3
25–49	4
50–74	3
75–99	2
100–124	2
125–149	3
150–174	1
175–199	1
200–224	1

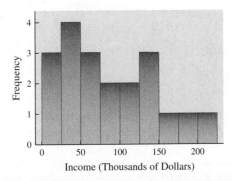

G 34. Bachelor's Degrees The stem-and-leaf plot summarizes the percentage of residents in the first 25 states alphabetically with a bachelor's degree or higher. (Data from: National Center for Education Statistics.)

Stem	Leaves
2	11333
2	557778899
3	0022234
3	888
4	1

Units: $4|1 = 41\%$

(a) Describe the shape of the distribution.

(b) Find the median.

35. Bachelor's Degrees The stem-and-leaf plot summarizes the percentage of residents in the latter 25 states alphabetically with a bachelor's degree or higher. (Data from: National Center for Education Statistics.)

Stem	Leaves
1	9
2	34
2	566677788999
3	00113
3	55677

Units: $3|9 = 39\%$

(a) Describe the shape of the distribution.

(b) Find the median.

For Exercises 36–38, calculate the weighted average. (See Example 7.) For Exercises 39 and 40, use the formula for the weighted average to solve for the final examination score.

36. Final Average Gonzalo enrolled in a class where the homework counts as 30% of the final grade, two midterm exams count as 15% each, and a final exam counts as 40% of the final grade. Gonzalo earned 88% on the homework, 85% and 81% on the two midterm exams, and 93% on the final exam. What is Gonzalo's final average?

37. Final Average Nan enrolled in a class where the homework counts as 15% of the final grade, quizzes count 15%, three midterms exams count as 12% each, and the final exam counts 34%. Nan earned 73% on her homework; 78% on her quizzes; 91%, 93%, and 83% on her three midterm exams; and 88% on the final exam. What is Nan's final average?

38. Final Average Henrietta enrolled in a class where attendance counted as 10% of the final grade, a midterm paper counted as 30% of the final grade, and a final paper counted as 60% of the final grade. She had 100% attendance. She earned a score of 85% on her midterm paper and a 91% on her final paper. What is Henrietta's final average?

39. Final Exam Score Trey enrolled in a class where attendance counted as 5% of the final grade, quizzes counted as 25%, two midterm exams counted as 20% each, and the final exam counted as 30%. Trey earned 100% on attendance, 95% on the quizzes, 81% on midterm exam 1, and 89% on midterm exam 2. If 90%

is needed to earn a grade of $A-$, what is the minimum score Trey needs to earn on the final exam in order to earn a final grade of $A-$?

40. **Final Exam Score** Mirabella enrolled in a class where attendance counted as 10% of the final grade, quizzes counted as 15%, one midterm exam counted as 35%, and the final exam counted as 40%. Mirabella earned 80% on attendance, 85% on the quizzes, and 75% on the midterm exam. If 80% is needed to earn a grade of $B-$, what is the minimum score Mirabella needs to earn on the final exam in order to earn a final grade of $B-$?

✓ **Checkpoint Answers**

1. $301.75   2. $\bar{x} = 11.75$ 3. 18.90   4. About 3.32

5. (a) 69 inches   (b) 68 inches

6. (a) 25 miles per gallon   (b) $0 and $35
   (c) No mode

7. (a) About 1.7   (b) 1   (c) 0

8. 81.7

10.3 Measures of Variation and Boxplots

The mean, median, and mode are measures of center for a list of numbers, but tell nothing about the *spread* of the numbers in the list. For example, look at the following data sets of number of times per week three people ate meals at restaurants over the course of five weeks:

Jill:	3	5	6	3	3
Miguel:	4	4	4	4	4
Sharille:	10	1	0	0	9

Each of these three data sets has a mean of 4, but the amount of dispersion or variation within the lists is different. This difference may reflect different dining patterns over time. Thus, in addition to a measure of center, another kind of measure is needed that describes how much the numbers vary.

The largest number of restaurant meals for Jill is 6, while the smallest is 3, a difference of 3. For Miguel, the difference is 0; for Sharille, it is 10. The difference between the largest and smallest number in a sample is called the **range,** one example of a measure of variation. The range is 3 for Jill, 0 for Miguel, and 10 for Sharille. The range has the advantage of being very easy to compute and gives a rough estimate of the variation among the data in the sample. However, it depends only on the two extremes and tells nothing about how the other data are distributed between the extremes.

📇 **TECHNOLOGY TIP**

Many graphing calculators show the largest and smallest numbers in a list when displaying one-variable statistics, usually on the second screen of the display.

✓ **Checkpoint 1**

Find the range for this sample of the number of miles from students' homes to college: 15, 378, 5, 210, 125.

EXAMPLE 1   **Consumer Expenses** Find the range for each given data set for a small sample of people.

(a) Price paid for last haircut (with tip): 10, 0, 15, 30, 20, 18, 50, 120, 75, 95, 0, 5

Solution The highest number here is 120; the lowest is 0. The range is the difference of these numbers, or

$$120 - 0 = 120.$$

(b) Amount spent for last vehicle servicing: 30, 19, 125, 150, 430, 50, 225

Solution Range $= 430 - 19 = 411.$ ✓₁

To find another useful measure of variation, we begin by finding the **deviations from the mean**—the differences found by subtracting the mean from each number in a distribution.

EXAMPLE 2   Find the deviations from the mean for the following sample of ages.

$$32, \quad 41, \quad 47, \quad 53, \quad 57.$$

Solution Adding these numbers and dividing by 5 gives a mean of 46 years. To find the deviations from the mean, subtract 46 from each number in the sample. For example, the first deviation from the mean is $32 - 46 = -14$; the last is $57 - 46 = 11$ years. All of the deviations are listed in the following table.

Age	Deviation from Mean
32	-14
41	-5
47	1
53	7
57	11

✓ **Checkpoint 2**

Find the deviations from the mean for the following sample of number of miles traveled by various people to a vacation location:

135, 60, 50, 425, 380.

To check your work, find the sum of the deviations. It should always equal 0. (The answer is always 0 because the positive and negative deviations cancel each other out.) ✓₂

To find a measure of variation, we might be tempted to use the mean of the deviations. However, as just mentioned, this number is always 0, no matter how widely the data are dispersed. To avoid the problem of the positive and negative deviations averaging to 0, we could take absolute values and find $\sum |x_i - \bar{x}|$ and then divide it by n to get the *mean absolute deviation*. However, statisticians generally prefer to square each deviation to get nonnegative numbers and then take the square root of the mean of the squared variations in order to preserve the units of the original data (such as inches, pounds). (Using squares instead of absolute values allows us to take advantage of some algebraic properties that make other important statistical methods much easier.) The squared deviations for the data in Example 2 are shown in the following table:

Number	Deviation from Mean	Square of Deviation
32	-14	196
41	-5	25
47	1	1
53	7	49
57	11	121

In this case, the mean of the squared deviations is

$$\frac{196 + 25 + 1 + 49 + 121}{5} = \frac{392}{5} = 78.4.$$

This number is called the **population variance,** because the sum was divided by $n = 5$, the number of items in the original list.

Since the deviations from the mean must add up to 0, if we know any 4 of the 5 deviations, the 5th can be determined. That is, only $n - 1$ of the deviations are free to vary, so we really have only $n - 1$ independent pieces of information, or *degrees of freedom*. Using $n - 1$ as the divisor in the formula for the mean gives

$$\frac{196 + 25 + 1 + 49 + 121}{5 - 1} = \frac{392}{4} = 98.$$

This number, 98, is called the **sample variance** of the distribution and is denoted s^2, because it is found by averaging a list of squares. In this case, the population and sample variances differ by quite a bit. But when n is relatively large, as is the case in real-life applications, the difference between them is rather small.

Sample Variance

The variance of a sample of n numbers $x_1, x_2, x_3, \ldots, x_n$, with mean $\overline{x}$, is

$$s^2 = \frac{\sum(x_i - \overline{x})^2}{n - 1}.$$

When computing the sample variance by hand, it is often convenient to use the following shortcut formula, which can be derived algebraically from the definition in the preceding box:

$$s^2 = \frac{\sum x_i^2 - n\overline{x}^2}{n - 1}.$$

To find the sample variance, we square the deviations from the mean, so the variance is in squared units. To return to the same units as the data, we use the *square root* of the variance, called the **sample standard deviation,** denoted s.

Sample Standard Deviation

The standard deviation of a sample of n numbers $x_1, x_2, x_3, \ldots, x_n$, with mean $\overline{x}$, is

$$\sqrt{\frac{\sum(x_i - \overline{x})^2}{n - 1}}.$$

NOTE The **population standard deviation** is

$$\sigma = \sqrt{\frac{\sum(x_i - \mu)^2}{n}},$$

where n is the population size.

TECHNOLOGY TIP When a graphing calculator computes one-variable statistics for a list of data, it usually displays the following information (not necessarily in this order, and sometimes on two screens) and possibly other information as well:

Information	Notation
Number of data entries	n or $N\Sigma$
Mean	$\overline{x}$ or mean Σ
Sum of all data entries	Σx or TOT Σ
Sum of the squares of all data entries	Σx^2
Sample standard deviation	Sx or sx or $x\sigma_{n-1}$ or SSDEV
Population standard deviation	σx or $x\sigma_n$ or PSDEV
Largest/smallest data entries	maxX/minX or MAXΣ/MINΣ
Median	Med or MEDIAN

NOTE In the rest of this section, we shall deal exclusively with the sample variance and the sample standard deviation. So whenever standard deviation is mentioned, it means "sample standard deviation," not population standard deviation.

As its name indicates, the standard deviation is the most commonly used measure of variation. The standard deviation is a measure of the variation from the mean. The size of the standard deviation indicates how spread out the data are from the mean.

Time	Square of the Time
2	4
8	64
3	9
2	4
6	36
11	121
31	961
9	81
72	1280

✓ **Checkpoint 3**

Find the standard deviation for a sample of the number of miles traveled by various people to a vacation location:

135, 60, 50, 425, 380.

EXAMPLE 3 **Mobile Phone Conversations** Find the standard deviation for the following sample of the lengths (in minutes) of eight consecutive cell phone conversations by one person:

$$2, \ 8, \ 3, \ 2, \ 6, \ 11, \ 31, \ 9.$$

Work by hand, using the shortcut variance formula on page 532.

Solution Arrange the work in columns, as shown in the table in the margin. Now use the first column to find the mean:

$$\bar{x} = \frac{\sum x_i}{8} = \frac{72}{8} = 9 \text{ minutes.}$$

The total of the second column gives $\sum x_i^2 = 1280$. The variance is

$$s^2 = \frac{\sum x_i^2 - n\bar{x}^2}{n - 1}$$

$$= \frac{1280 - 8(9)^2}{8 - 1}$$

$$\approx 90.3$$

and the standard deviation is

$$s \approx \sqrt{90.3} \approx 9.5 \text{ minutes.}$$ ✓ 3

 TECHNOLOGY TIP The screens in Figure 10.6 show two ways to find variance and standard deviation on a TI-84+ calculator: with the LIST menu and with the STAT menu. The data points are first entered in a list—here, L_5. See your instruction book for details.

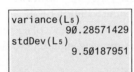

```
variance(L5)
            90.28571429
stdDev(L5)
             9.50187951
```

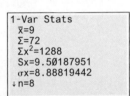

```
1-Var Stats
 x̄=9
 Σ=72
 Σx²=1288
 Sx=9.50187951
 σx=8.88819442
↓n=8
```

Figure 10.6

In a spreadsheet, enter the data in cells A1 through A8. Then, in cell A9, type "=VAR(A1:A8)" and press Enter. The standard deviation can be calculated either by taking the square root of cell A9 or by typing "=STDEV (A1:A8)" in cell A10 and pressing Enter.

⊘ **CAUTION** We must be careful to divide by $n - 1$, not n, when calculating the standard deviation of a sample. Many calculators are equipped with statistical keys that compute the variance and standard deviation. Some of these calculators use $n - 1$, and others use n for these computations; some may have keys for both. Check your calculator's instruction book before using a statistical calculator for the exercises.

One way to interpret the standard deviation uses the fact that, for many populations, most of the data are within three standard deviations of the mean. (See Section 10.4.) This implies that, in Example 3, most of the population data from which this sample is taken are between

$$\bar{x} - 3s = 9 - 3(9.5) = -19.5$$

and

$$\bar{x} + 3s = 9 + 3(9.5) = 37.5.$$

For Example 3, the preceding calculations imply that most phone conversations are less than 37.5 minutes long. This approach of determining whether sample observations are beyond 3 standard deviations of the mean is often employed in conducting quality control in many industries.

For data in a grouped frequency distribution, a slightly different formula for the standard deviation is used.

Standard Deviation for a Grouped Distribution

The standard deviation for a sample distribution with mean $\bar{x}$, where x_i is an interval midpoint with frequency f_i and $n = \Sigma f_i$, is

$$s = \sqrt{\frac{\Sigma f_i x_i^2 - n\bar{x}^2}{n-1}}.$$

The formula indicates that the product $f_i x_i^2$ is to be found for each interval. Then all the products are summed, n times the square of the mean is subtracted, and the difference is divided by one less than the total frequency—that is, by $n - 1$. The square root of this result is s, the standard deviation. The standard deviation found by this formula may (and probably will) differ somewhat from the standard deviation found from the original data.

⚠ **CAUTION**   In calculating the standard deviation for either a grouped or an ungrouped distribution, using a rounded value for the mean or variance may produce an inaccurate value.

EXAMPLE 4   **College Tuition** The following frequency distribution gives the 2017–2018 annual tuition (in thousands of dollars) for a random sample of 30 community colleges. (Data from: www.collegetuitioncompare.com.)

Find the sample standard deviation s for these data.

Solution We first need to find the mean $\bar{x}$ for these grouped data. We find the midpoint of each interval and label it x_i. We multiply the frequency by the midpoint x_i to obtain $x_i f_i$:

Tuition Amount (in thousands of dollars)	Frequency, f_i	Midpoint, x_i	Product, $x_i f_i$
1.0–1.49	5	1.25	6.25
1.5–1.99	0	1.75	0
2.0–2.49	1	2.25	2.25
2.5–2.99	5	2.75	13.75
3.0–3.49	1	3.25	3.25
3.5–3.99	9	3.75	33.75
4.0–4.49	4	4.25	17
4.5–4.99	2	4.75	9.5
5.0–5.49	2	5.25	10.5
5.5–5.99	1	5.75	5.75
	Total = 30		Total = 102

Therefore,

$$\bar{x} = \frac{102}{30} = 3.4.$$

Now that we have the mean value, we can modify our table to include columns for x_i^2 and $f_i x_i^2$. We obtain the following results:

Tuition Amount (in thousands of dollars)	Frequency, f_i	Midpoint, x_i	Product, $f_i x_i$	x_i^2	$f_i x_i^2$
1.0–1.49	5	1.25	6.25	1.5625	7.8125
1.5–1.99	0	1.75	0	3.0625	0
2.0–2.49	1	2.25	2.25	5.0625	5.0625
2.5–2.99	5	2.75	13.75	7.5625	37.8125
3.0–3.49	1	3.25	3.25	10.5625	10.5625
3.5–3.99	9	3.75	33.75	14.0625	126.5625
4.0–4.49	4	4.25	17	18.0625	72.2500
4.5–4.99	2	4.75	9.5	22.5625	45.1250
5.0–5.49	2	5.25	10.5	27.5625	55.1250
5.5–5.99	1	5.75	5.75	33.0625	33.0625
	Total = 30		Total = 102		Total = 393.3750

We now use the formula for the standard deviation with $n = 30$ to find s:

$$s = \sqrt{\frac{\sum f_i x_i^2 - n\overline{x}^2}{n-1}}$$

$$= \sqrt{\frac{393.3750 - 30(3.4)^2}{30-1}}$$

$$\approx 1.3 \qquad \boxed{4}$$

✓ Checkpoint 4

Find the standard deviation for the following grouped frequency distribution of the number of classes completed thus far in the college careers of a random sample of 52 students:

Classes	Frequency
0–5	6
6–10	10
11–20	12
21–30	15
31–40	9

Another way to interpret the standard deviation is to think of it as an *approximation* of the average deviation from the mean. In Example 4, the mean tuition was $3400 and the standard deviation was $1300. One way to think of what the value $1300 represents is as the approximate average deviation from $3400 for the 30 community colleges in the sample. Some colleges charge tuition close to $3400 and some charge tuition further from $3400, but the approximate average deviation from $3400 is $1300.

📄 **NOTE** A calculator is almost a necessity for finding a standard deviation. With a nongraphing calculator, a good procedure to follow is first to calculate $\overline{x}$. Then, for each x, square that number, and multiply the result by the appropriate frequency. If your calculator has a key that accumulates a sum, use it to accumulate the total in the last column of the table. With a graphing calculator, simply enter the midpoints and the frequencies, and then ask for the one-variable statistics.

Boxplots

Boxplots are a graphical means of presenting key characteristics of a data set. The idea is to arrange the data in increasing order and choose three numbers Q_1, Q_2, and Q_3 that divide it into four equal parts, as indicated schematically in the following diagram:

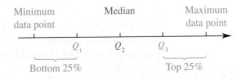

The number Q_1 is called the **first quartile,** the median Q_2 is called the **second quartile,** and Q_3 is called the **third quartile.** The minimum, Q_1, Q_2, Q_3, and the maximum are often called the **five-number summary,** and they are used to construct a boxplot, as illustrated in Examples 5 and 6.

EXAMPLE 5 The following table gives the revenue (in billions of dollars) for Home Depot and Lowe's home improvement corporations for the given years. Construct a boxplot for the Home Depot revenue data. (Data from: www.morningstar.com.)

Company	2008	2009	2010	2011	2012	2013	2014	2015	2016	2017
Home Depot	77.3	71.3	66.2	68.0	70.4	74.8	78.8	83.2	88.5	94.6
Lowe's	48.2	48.2	47.2	48.8	50.2	50.5	53.4	56.2	59.1	65.0

Solution First, we order the Home Depot revenue from low to high.

$$66.2, 68.0, 70.4, 71.3, 74.8, 77.3, 78.8, 83.2, 88.5, 94.6$$

The minimum revenue is 66.2 and the maximum is 94.6. Because there is an even number of data points, the median revenue Q_2 is 76.05 (the average of the two center entries). To find Q_1, which separates the lower 25% of the data from the rest, we first calculate

25% of n (rounded up to the nearest integer),

where n is the number of data points. Here, $n = 10$, and so .25(10) = 2.5, which rounds up to 3. Now count to the *third* revenue (in order) to get $Q_1 = 70.4$.

Similarly, since Q_3 separates the lower 75% of the data from the rest, we calculate

75% of n (rounded up to the nearest integer).

When $n = 10$, we have .75(10) = 7.5, which rounds up to 8. Count to the *eighth* observation (in order), getting $Q_3 = 83.2$.

The five key numbers for constructing the boxplot are as follows:

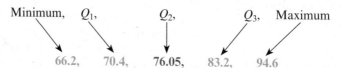

Draw a horizontal axis that will include the minimum and maximum values. Draw a box with left boundary at Q_1 and a right boundary at Q_3. Mark the location of the median Q_2 by drawing a vertical line in the box Extend a horizontal line from the vertical line at Q_1 to the minimum value and and a similar line from the vertical line at Q_3 to the maximum value as shown in Figure 10.7.

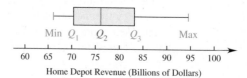

Home Depot Revenue (Billions of Dollars)

Figure 10.7

Boxplots are useful in showing the location of the middle 50% of the data. This is the region of the box between Q_1 and Q_3. Boxplots are also quite useful in comparing two distributions that are measured on the same scale, as we will see in Example 6.

EXAMPLE 6 **Business** Construct a boxplot for the Lowe's revenue data in Example 5 with the Home Depot revenue boxplot on the same graph.

Solution The Lowe's data first need to be sorted low to high:

$$47.2, 48.2, 48.2, 48.8, \mathbf{50.2}, \mathbf{50.5}, 53.4, 56.2, 59.1, 65.0.$$

The minimum is 47.2, the maximum is 65.0, and the median $Q_2 = 50.35$. Again, $n = 10$, and so

$$25\% \text{ of } n = .25(10) = 2.5, \text{ which rounds up to } 3,$$

so that $Q_1 = 48.2$ (the *third* revenue). Similarly,

$$75\% \text{ of } n = .75(10) = 7.5, \text{ which rounds up to } 8,$$

so that $Q_3 = 56.2$ (the *eighth* revenue).

We can now use the minimum, Q_1, Q_2, and Q_3, and the maximum to place the boxplot of Lowe's data above the Home Depot boxplot, as shown in Figure 10.8.

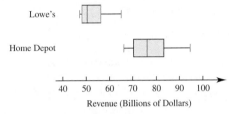

Figure 10.8

When we compare the revenue distributions in this way, we see how much more revenue Home Depot generated in the studied years than Lowe's. Because the box for Home Depot is wider, and the distance from the minimum to the maximum is also wider for Home Depot, we can also say that there is a greater degree of variability for Home Depot. ✔ 5

✓ **Checkpoint 5**

Create a boxplot for the weights (in kilograms) for eight first-year students:

88, 79, 67, 69, 58, 53, 89, 57.

TECHNOLOGY TIP

Many spreadsheets and graphing calculators can graph boxpots for single-variable data. With a calculator the procedure is similar to plotting data with the STAT PLOT menu.

10.3 Exercises

1. How are the variance and the standard deviation related?

2. Why can't we use the sum of the deviations from the mean as a measure of dispersion of a distribution?

State Expenditures *Use the following table for Exercises 3–10, which lists expenditures (in billions of dollars with the exception of Parks and Recreation, which is in tens of millions of dollars) for a random sample of seven states. Find the range and standard deviation among the seven states for each category of government expenditure. (See Examples 1–3.) (Data from: U.S. Census Bureau.)*

Type of Expenditure	State						
	AZ	CA	IL	MA	MI	NJ	PA
Education	9.7	79.8	17.7	13.5	23.5	17.1	22.9
Public Welfare	9.1	83.5	21.2	16.7	15.7	15.9	24.3
Hospitals	.9	9.8	1.1	.5	3.2	2.3	4.0
Health	2.0	8.8	2.2	1.1	1.3	1.5	3.0
Highways	1.8	12.6	5.3	2.2	2.6	3.6	6.9
Police Protection	.2	1.6	.5	.8	.4	.9	.9
Corrections	1.1	8.9	1.4	1.2	1.9	1.4	2.0
Parks and Recreation	8	55	25	21	14	21	26

3. Education
4. Public Welfare
5. Hospitals
6. Health
7. Highways
8. Police Protection
9. Corrections
10. Parks and Recreation

Education *Find the standard deviation for the grouped data in Exercises 11 and 12. (See Example 4.)*

11. Number of credits for a sample of college students:

College Credits	Frequency
0–24	4
25–49	3
50–74	6
75–99	3
100–124	5
125–149	9

12. Scores on a calculus exam:

Scores	30–39	40–49	50–59	60–69	70–79	80–89	90–99
Frequency	1	6	13	22	17	13	8

S&P 500 Stocks *The data for Exercises 13–14 come from a random sample of 10 companies that were part of the S&P 500 and information* *was current as of March 7, 2017. (Hint: Do not round the mean when calculating the standard deviation.) (Data from: data.okfn.org.)*

13. Find the mean and standard deviation of the following stock prices (rounded to the nearest dollar).

25	64	32	99	74
27	66	72	83	54

14. Find the mean and standard deviation of the following dividend yields (percent).

2.38	1.14	3.12	2.26	2.12
1.06	1.50	1.62	1.37	3.23

Find the mean and standard deviation of the market capitalization.

An application of standard deviation is given by Chebyshev's theorem. (P. L. Chebyshev was a Russian mathematician who lived from 1821 to 1894.) This theorem, which applies to any distribution of numerical data, states,

For any distribution of numerical data, at least $1 - 1/k^2$ of the numbers lie within k standard deviations of the mean.

Example *For any distribution, at least*

$$1 - \frac{1}{3^2} = 1 - \frac{1}{9} = \frac{8}{9}$$

of the numbers lie within 3 standard deviations of the mean. Find the fraction of all the numbers of a data set lying within the given numbers of standard deviations from the mean.

15. 2   **16.** 4   **17.** 1.5

In a certain distribution of numbers, the mean is 50, with a standard deviation of 6. Use Chebyshev's theorem to tell what percent of the numbers are

18. between 32 and 68;

19. between 26 and 74;

20. less than 38 or more than 62;

21. less than 32 or more than 68;

22. less than 26 or more than 74.

Movie Studios' Revenue *For Exercises 23–28, use the following table, which gives the revenue (in millions of dollars) for four major movie studios for the years 2010–2016. (Hint: do not round the mean value when calculating the standard deviation.) (Data from: www.boxofficemojo.com.)*

Year	Columbia	20th Century Fox	Warner Bros.	Paramount
2016	911	1469	1902	877
2015	966	1302	1604	675
2014	1262	1791	1563	1053
2013	1445	1064	1864	967
2012	1792	1025	1665	914
2011	1274	978	1826	1957
2010	1283	1482	1924	1714

Find the mean and standard deviation for the studio listed.

23. Columbia   **24.** 20th Century Fox

25. Warner Bros.   **26.** Paramount

27. Which studio shows greater variation: Columbia or Warner Bros.?

28. Which studio shows greater variation: 20th Century Fox or Paramount?

For Exercises 29–32, see Example 4.

29. **Household Income** A sample of 40 households from the American Housing Survey generates the following frequency table of household income (in thousands). Find the standard deviation.

Interval	0–29	30–59	60–89	90–119	120–149	150–179	180–209
Frequency	12	3	10	2	0	1	2

30. **Business** The number of hours worked in a week for 30 workers selected at random from the 2016 General Social Survey appears in the following frequency table. Find the standard deviation. (Data from: gss.norc.org.)

Interval	Frequency
0–9	2
10–19	4
20–29	0
30–39	5
40–49	8
50–59	7
60–69	3
70–79	1

Reading Scores *Shown in the following table are the reading scores of a second-grade class given individualized instruction and the reading scores of a second-grade class given traditional instruction in the same school:*

Scores	Individualized Instruction	Traditional Instruction
50–59	2	5
60–69	4	8
70–79	7	8
80–89	9	7
90–99	8	6

31. Find the mean and standard deviation for the individualized-instruction scores.

32. Find the mean and standard deviation for the traditional-instruction scores.

33. Discuss a possible interpretation of the differences in the means and the standard deviations in Exercises 35 and 36.

34. Discuss what the standard deviation tells us about a distribution.

IBM and Microsoft Revenue *The following table gives the annual revenue for IBM Corp. and Microsoft Corp. (in billions of dollars) for a 10-year period. (Data from: www.morningstar.com.) (See Examples 5 and 6.)*

Year	IBM	Microsoft
2007	99	51
2008	104	60
2009	96	58
2010	100	62
2011	107	70
2012	105	73
2013	100	78
2014	93	86
2015	82	94
2016	80	85

35. What is the five-number summary for IBM?

36. What is the five-number summary for Microsoft?

37. Construct a boxplot for IBM.

38. Construct a boxplot for Microsoft.

39. Graph the boxplots for IBM and Microsoft on the same scale. Are the distributions similar?

40. Which company has the higher median of sales over the 10-year period?

State Expenditures *Use the following table for Exercises 41–44. The table contains data from a random sample of 12 states and lists expenditures (in billions of dollars) by category spent by states and local communities. (Data from: U.S. Census Bureau.)*

State	Education	Public Welfare
AL	10.4	6.6
AZ	9.7	9.1
CA	79.8	83.5
FL	25.1	24.4
GA	17.4	11.7
IL	17.7	21.2
MA	13.5	16.7
MD	12.0	10.8
MI	23.5	15.7
NJ	17.1	15.9
PA	22.9	24.3
TX	50.4	32.8

41. What is the five-number summary for the expenditures on education?

42. What is the five-number summary for the expenditures on public welfare?

43. Construct a boxplot for the expenditures on education.

44. Construct a boxplot for the expenditures on public welfare.

State Expenditures *For Exercises 45 and 46, use the boxplots below from data similar to Exercises 41–44 for state and local community expenditures on hospitals and highways.*

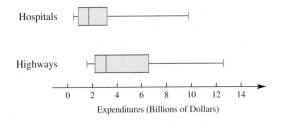

45. Estimate Q_1, Q_2, and Q_3 for spending on hospitals and highways.

46. Is there greater variability in spending on hospitals or highway? How do you know?

State Expenditures *For Exercises 47 and 48, use the boxplots below from data similar to Exercises 41–44 for state and local community expenditures on health and corrections.*

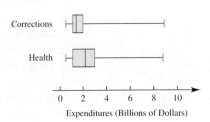

47. Is there greater variability in spending on health or corrections? How do you know?

48. Estimate Q_1, Q_2, and Q_3 for spending on health and corrections.

✓**Checkpoint Answers**

1. 373

2. Mean is 210; deviations are -75, -150, -160, 215, and 170.

3. 179.5 miles

4. 10.91 classes

5.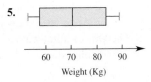

10.4 Normal Distributions

Suppose a bank is interested in improving its services to customers. The manager decides to begin by finding the amount of time tellers spend on each transaction, rounded to the nearest minute. The times for 75 different transactions are recorded, with the results shown in the following table, where the frequencies listed in the second column are divided by 75 to find the empirical probabilities:

Time	Frequency	Probability
1	3	$3/75 = .04$
2	5	$5/75 \approx .07$
3	9	$9/75 = .12$
4	12	$12/75 = .16$
5	15	$15/75 = .20$
6	11	$11/75 \approx .15$
7	10	$10/75 \approx .13$
8	6	$6/75 = .08$
9	3	$3/75 = .04$
10	1	$1/75 \approx .01$

Figure 10.9(a), on the following page, shows a histogram and frequency polygon for the data. The heights of the bars are the empirical probabilities, rather than the frequencies. The transaction times are given to the nearest minute. Theoretically, at least, they could have been timed to the nearest tenth of a minute, or hundredth of a minute, or even more precisely. In each case, a histogram and frequency polygon could be drawn. If the times are measured with smaller and smaller units, there are more bars in the histogram and the frequency polygon begins to look more and more like the curve in Figure 10.9(b) instead of a polygon. Actually, it is possible for the transaction times to take on any real-number value greater than 0. A distribution in which the outcomes can take on any real-number value within some interval is a **continuous distribution.** The graph of a continuous distribution is a curve.

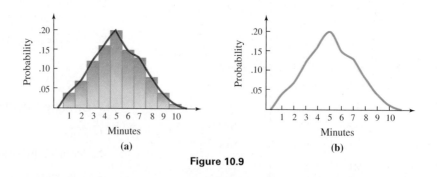

Figure 10.9

The distribution of heights (in inches) of adult women is another example of a continuous distribution, since these heights include infinitely many possible measurements, such as 53, 58.5, 66.3, 72.666, and so on. Figure 10.10 shows the continuous distribution of heights of adult women. Here, the most frequent heights occur near the center of the interval displayed.

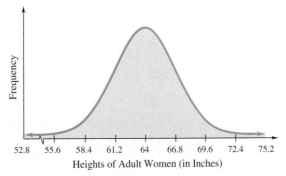

Heights of Adult Women (in Inches)

Figure 10.10

Normal Distributions

As discussed on page 515, we say that data are normal (or normally distributed) when their graph is well approximated by a bell-shaped curve. (See Figure 10.11.) We call the graphs of such distributions **normal curves.** Examples of distributions that are approximately normal are the heights of adult women and cholesterol levels in adults. We use the Greek letters μ (mu) to denote the mean and σ (sigma) to denote the standard deviation of a normal distribution.

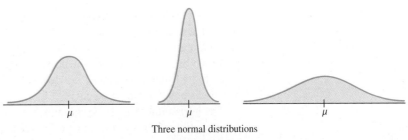

Three normal distributions

Figure 10.11

There are many normal distributions, depending on μ and σ. Some of the corresponding normal curves are tall and thin, and others short and wide, as shown in Figure 10.11. But every normal curve has the following properties:

1. Its peak occurs directly above the mean μ.

2. The curve is symmetric about the vertical line through the mean. (That is, if you fold the graph along this line, the left half of the graph will fit exactly on the right half.)

3. The curve never touches the x-axis—it extends indefinitely in both directions.

4. The area under the curve (and above the horizontal axis) is 1. (As can be shown with calculus, this is a consequence of the fact that the sum of the probabilities in any distribution is 1.)

A normal distribution is completely determined by its mean μ and standard deviation σ.[*] A small standard deviation leads to a tall, narrow curve like the one in the center of Figure 10.11, because most of the data are close to the mean. A large standard deviation means the data are very spread out, producing a flat, wide curve like the one on the right in Figure 10.11.

---

[*]As shown in more advanced courses, its graph is the graph of the function

$$f(x) = \frac{1}{\sigma\sqrt{2\pi}}e^{-(x-\mu)^2/(2\sigma^2)},$$

where $e \approx 2.71828$ is the real number introduced in Section 4.1.

Since the area under a normal curve is 1, parts of this area can be used to determine certain probabilities. For instance, Figure 10.12(a) is the probability distribution of the annual rainfall in a certain region. The probability that the annual rainfall will be between 25 and 60 inches is the area under the curve from 25 to 60. The general case, shown in Figure 10.12(b), can be stated as follows.

> The area of the shaded region under the normal curve from a to b is the probability that an observed data value will be between a and b.

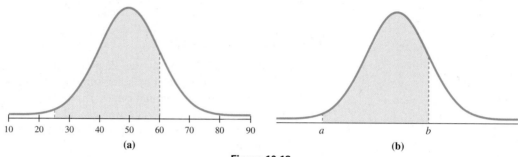

(a)   **(b)**

Figure 10.12

To use normal curves effectively, we must be able to calculate areas under portions of them. These calculations have already been done for the normal curve with mean $\mu = 0$ and standard deviation $\sigma = 1$ (which is called the **standard normal curve**) and are available in Appendix A at the back of the book. Examples 1 and 2 demonstrate how to use Appendix A to find such areas. Later, we shall see how the standard normal curve may be used to find areas under any normal curve.

The horizontal axis of the standard normal curve is usually labeled z. Since the standard deviation of the standard normal curve is 1, the numbers along the horizontal axis (the z-values) measure the number of standard deviations above or below the mean $z = 0$.

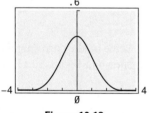

Figure 10.13

🖩 **TECHNOLOGY TIP** Some graphing calculators have the ability to graph a normal distribution, given its mean and standard deviation, and to find areas under the curve between two x-values. For an area under the curve, some calculators will give the corresponding z-value. For details, see your instruction book. (Look for "distribution" or "probability distribution.") A calculator-generated graph of the standard normal curve is shown in Figure 10.13.

EXAMPLE 1   Find the given areas under the standard normal curve.

(a)   The area between $z = 0$ and $z = 1$, the shaded region in Figure 10.14

Solution   Find the entry 1 in the z-column of Appendix A. The entry next to it in the A-column is .3413, which means that the area between $z = 0$ and $z = 1$ is .3413. Since the total area under the curve is 1, the shaded area in Figure 10.14 is 34.13% of the total area under the normal curve.

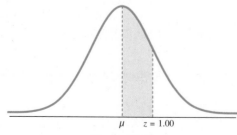

Figure 10.14

Find the percent of the area between the mean and

(a) $z = 1.51$;

(b) $z = -2.04$.

(c) Find the percent of the area in the shaded region.

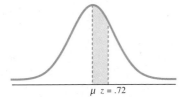

✓ **Checkpoint 2** 🖩

If your calculator can graph probability distributions and find areas, use it to find the areas requested in Example 1.

(b) The area between $z = -2.43$ and $z = 0$

Solution Appendix A lists only positive values of z. But the normal curve is symmetric around the mean $z = 0$, so the area between $z = 0$ and $z = -2.43$ is the same as the area between $z = 0$ and $z = 2.43$. Find 2.43 in the z-column of Appendix A. The entry next to it in the A-column shows that the area is .4925. Hence, the shaded area in Figure 10.15 is 49.25% of the total area under the curve. ✓₁ ✓₂

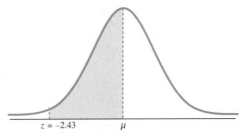

Figure 10.15

🖩 **TECHNOLOGY TIP** Because of their convenience and accuracy, graphing calculators and computers have made normal-curve tables less important. Figure 10.16 shows how part (b) of Example 1 can be done on a TI-84+ calculator using a command from the DISTR menu. The second result in the calculator screen gives the area between $-\infty$ and $z = -2.43$; the entry $-1E99$ represents $-1 \cdot 10^{99}$, which is used to approximate $-\infty$.

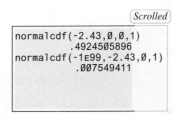

Figure 10.16

Many statistical software packages are widely used today. All of these packages are set up in a way that is similar to a spreadsheet, and they all can be used to generate normal curve values. In addition, most spreadsheets can perform a wide range of statistical calculations.

EXAMPLE 2 Use technology or Appendix A to find the percent of the total area for the given areas under the standard normal curve.

(a) The area between .88 standard deviations *below* the mean and 2.35 standard deviations *above* the mean (that is, between $z = -.88$ and $z = 2.35$)

Solution First, draw a sketch showing the desired area, as in Figure 10.17. From Appendix A, the area between the mean and .88 standard deviations below the mean is .3106. Also, the area from the mean to 2.35 standard deviations above the mean is .4906. As the figure shows, the total desired area can be found by *adding* these numbers:

$$\begin{array}{r} .3106 \\ +.4906 \\ \hline .8012. \end{array}$$

The shaded area in Figure 10.17 represents 80.12% of the total area under the normal curve.

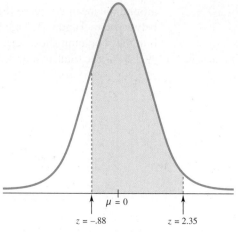

Figure 10.17

✓ Checkpoint 3

Find the given standard normal-curve areas as percentages of the total area.

(a) Between .31 standard deviations below the mean and 1.01 standard deviations above the mean

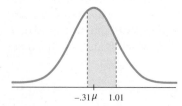

(b) Between .38 standard deviations and 1.98 standard deviations below the mean

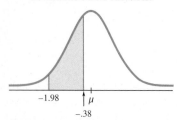

(c) To the right of 1.49 standard deviations above the mean

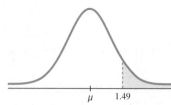

(d) What percent of the area is within 1 standard deviation of the mean? within 2 standard deviations of the mean? within 3 standard deviations of the mean? What can you conclude from the last answer?

(b) The area between .58 standard deviations above the mean and 1.94 standard deviations above the mean

Solution Figure 10.18 shows the desired area. The area between the mean and .58 standard deviations above the mean is .2190. The area between the mean and 1.94 standard deviations above the mean is .4738. As the figure shows, the desired area is found by *subtracting* one area from the other:

$$
\begin{array}{r}
.4738 \\
-.2190 \\
\hline
.2548.
\end{array}
$$

The shaded area of Figure 10.18 represents 25.48% of the total area under the normal curve.

(c) The area to the right of 2.09 standard deviations above the mean

Solution The total area under a normal curve is 1. Thus, the total area to the right of the mean is 1/2, or .5000. From Appendix A, the area from the mean to 2.09 standard deviations above the mean is .4817. The area to the right of 2.09 standard deviations is found by subtracting .4817 from .5000:

$$
\begin{array}{r}
.5000 \\
-.4817 \\
\hline
.0183.
\end{array}
$$

A total of 1.83% of the total area is to the right of 2.09 standard deviations above the mean. Figure 10.19 shows the desired area. ✓₃

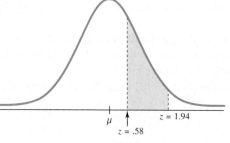

Figure 10.18

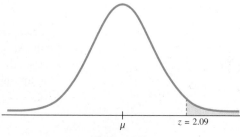

Figure 10.19

The key to finding areas under *any* normal curve is to express each number x on the horizontal axis in terms of standard deviations above or below the mean. The **z-score** for x is the number of standard deviations that x lies from the mean (positive if x is above the mean, negative if x is below the mean).

EXAMPLE 3 If a normal distribution has mean 60 and standard deviation 5, find the given z-scores.

(a) The z-score for $x = 65$

Solution Since 65 is 5 units above 60 and the standard deviation is 5, 65 is 1 standard deviation above the mean. So its z-score is 1.

(b) The z-score for $x = 52.5$

Solution The z-score is -1.5, because 52.5 is 7.5 units below the mean (since $52.5 - 60 = -7.5$) and 7.5 is 1.5 standard deviations (since $7.5/5 = 1.5$). ✓ 4

✓**Checkpoint 4**

Find each z-score, using the information in Example 3.

(a) $x = 36$

(b) $x = 55$

In Example 3(b) we found the z-score by taking the difference between 52.5 and the mean and dividing this difference by the standard deviation. The same procedure works in the general case.

Z-Score

If a normal distribution has mean μ and standard deviation σ, then the z-score for the number x is

$$z = \frac{x - \mu}{\sigma}.$$

The importance of z-scores is the following fact, whose proof is omitted.

Area under a Normal Curve

The area under a normal curve between $x = a$ and $x = b$ is the same as the area under the standard normal curve between the z-score for a and the z-score for b.

Therefore, by converting to z-scores and using a graphing calculator or Appendix A for the standard normal curve, we can find areas under any normal curve. Since these areas are probabilities (as explained earlier), we can now handle a variety of applications.

Graphing calculators, computer programs, and CAS programs (such as DERIVE) can be used to find areas under the normal curve and, hence, probabilities. The equation of the standard normal curve, with $\mu = 0$ and $\sigma = 1$, is

$$f(x) = (1/\sqrt{2\pi})e^{-x^2/2}.$$

A good approximation of the area under this curve (and above $y = 0$) can be found by using the x-interval $[-4, 4]$. However, calculus is needed to find such areas.

EXAMPLE 4 **Mileage** Dixie Office Supplies finds that its sales force drives an average of 1200 miles per month per person, with a standard deviation of 150 miles. Assume that the number of miles driven by a salesperson is closely approximated by a normal distribution.

(a) Find the probability that a salesperson drives between 1200 and 1600 miles per month.

Solution Here, $\mu = 1200$ and $\sigma = 150$, and we must find the area under the normal distribution curve between $x = 1200$ and $x = 1600$. We begin by finding the z-score for $x = 1200$:

$$z = \frac{x - \mu}{\sigma} = \frac{1200 - 1200}{150} = \frac{0}{150} = 0.$$

The z-score for $x = 1600$ is

$$z = \frac{x - \mu}{\sigma} = \frac{1600 - 1200}{150} = \frac{400}{150} = 2.67.^*$$

So the area under the curve from $x = 1200$ to $x = 1600$ is the same as the area under the standard normal curve from $z = 0$ to $z = 2.67$, as indicated in Figure 10.20. A graphing calculator or Appendix A shows that this area is .4962. Therefore, the probability that a salesperson drives between 1200 and 1600 miles per month is .4962.

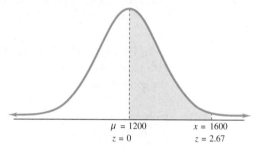

$\mu = 1200$
$z = 0$

$x = 1600$
$z = 2.67$

Figure 10.20

(b) Find the probability that a salesperson drives between 1000 and 1500 miles per month.

Solution As shown in Figure 10.21, z-scores for both $x = 1000$ and $x = 1500$ are needed.

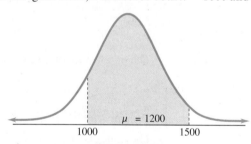

$\mu = 1200$

1000   1500

Figure 10.21

For $x = 1000$,

$$z = \frac{1000 - 1200}{150}$$

$$= \frac{-200}{150}$$

$$= -1.33.$$

For $x = 1500$,

$$z = \frac{1500 - 1200}{150}$$

$$= \frac{300}{150}$$

$$= 2.00.$$

From Appendix A , $z = 1.33$ leads to an area of .4082, while $z = 2.00$ corresponds to .4773. A total of $.4082 + .4773 = .8855$, or 88.55%, of all drivers travel between 1000 and 1500 miles per month. From this calculation, the probability that a driver travels between 1000 and 1500 miles per month is .8855.

✓**Checkpoint 5**

With the data from Example 4, find the probability that a salesperson drives between 1000 and 1450 miles per month.

📄 **NOTE** Graphing calculators and spreadsheets can calculate normal probabilities with given means and standard deviations. The answers, however, might differ slightly from our answers because the z-scores are not rounded to two decimal places.

---

*All z-scores here are rounded to two decimal places.

EXAMPLE 5 **Cholesterol** With data from a recent National Health and Nutritional Examination Survey (NHANES), we can use 186 (mg/dL) as an estimate of the mean total cholesterol level for all Americans and 41 (mg/dL) as an estimate of the standard deviation. Assuming total cholesterol levels to be normally distributed, what is the probability that an American chosen at random has a cholesterol level higher than 250? If 200 Americans are chosen at random, how many can we expect to have total cholesterol higher than 250? (Data from: www.cdc.gov/nchs/nhanes.htm.)

Solution Here, $\mu = 186$ and $\sigma = 41$. The probability that a randomly chosen American has cholesterol higher than 250 is the area under the normal curve to the right of $x = 250$. The z-score for $x = 250$ is

$$z = \frac{x - \mu}{\sigma} = \frac{250 - 186}{41} = \frac{64}{41} = 1.56.$$

From Appendix A, we see that the area to the right of 1.56 is $.5 - .4406 = .0594$, which is 5.94% of the total area under the curve. Therefore, the probability of a randomly chosen American having cholesterol higher than 250 is .0594.

With 5.94% of Americans having total cholesterol higher than 250, selecting 200 Americans at random yields

$$5.94\% \text{ of } 200 = .0594 \cdot 200 = 11.88.$$

Approximately 12 of these Americans can be expected to have a total cholesterol level higher than 250. ✓₆

✓ **Checkpoint 6**

Using the mean and standard deviation from Example 5, find the probability that an adult selected at random has a cholesterol level below 150.

📄 **NOTE** Notice in Example 5 that $P(z \geq 1.56) = P(z > 1.56)$. The area under the curve is the same whether we include the endpoint or not. Notice also that $P(z = 1.56) = 0$, because no area is included.

🛇 **CAUTION** When calculating the normal probability, it is wise to draw a normal curve with the mean and the z-scores every time. This practice will avoid confusion as to whether you should add or subtract probabilities.

As mentioned earlier, z-scores are standard deviations, so $z = 1$ corresponds to 1 standard deviation above the mean, and so on. As found in Checkpoint 3(d) of this section, 68.26% of the area under a normal curve lies within 1 standard deviation of the mean. Also, 95.46% lies within 2 standard deviations of the mean, and 99.74% lies within 3 standard deviations of the mean. These results, summarized in Figure 10.22, can be used to get quick estimates when you work with normal curves.

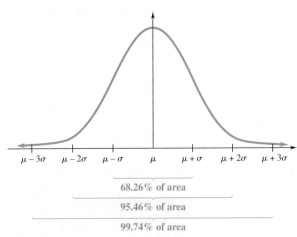

Figure 10.22

As we saw in Section 9.4, many practical experiments have only two possible outcomes. Example 3 from Section 9.4 presented a result from the U.S. Department of Education that 11.3% of student borrowers defaulted on their student loans in the first two years after college. If we use the binomial formula for 100 students chosen at random, we can create the histogram of the probabilities for the number of students who defaulted that appears in Figure 10.23.

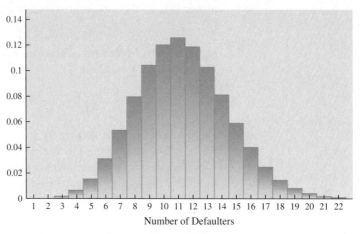

Number of Defaulters

Figure 10.23

Notice that Figure 10.23 looks very much like the normal curves we have been studying in this section. In fact, we can use the normal distribution to approximate binomial probabilities. To do this, first we need to know the mean and standard deviation of the binomial distribution. Recall from Section 9.4 that, for the binomial distribution $E(x) = np$. In Section 10.2, we referred to $E(x)$ as μ, and that notation will be used here. It is shown in more advanced courses in statistics that the standard deviation of the binomial distribution is given by $\sigma = \sqrt{np(1 - p)}$.

Mean and Standard Deviation for the Binomial Distribution

For the binomial distribution, the mean and standard deviation, respectively, are given by

$$\mu = np \quad \text{and} \quad \sigma = \sqrt{np(1 - p)},$$

where n is the number of trials and p is the probability of success on a single trial.

With a repeated binomial experiment, we count the number of successes, which will be an integer. As we see in Figure 10.23, the bars represent the probability of each integer outcome from 0 to 22. When we use a normal distribution to approximate a binomial probability, we subtract .5 from the integer that is our lower bound of interest and we add .5 to the integer of the upper bound interest. Thus, we if we wanted to approximate the probability of between 8 and 13 successes, we would use 7.5 and 13.5 as our bounds for the normal distribution.

⬡ **CAUTION** The normal-curve approximation to a binomial distribution is quite accurate *provided* that n is large and p is not close to 0 or 1. As a rule of thumb, the normal–curve approximation can be used as long as both np and $n(1 - p)$ are at least 5.

EXAMPLE 6 **Student Loan Defaults** Consider the previously discussed sample of 100 students chosen at random from a population where 11.3% of the students defaulted on their student loans within two years after college.

(a) Find the mean and standard deviation for this distribution.

Solution We have $n = 100$ and $p = .113$. Thus,

$$\mu = 100(.113) \qquad \sigma = \sqrt{100(.113)(1 - .113)}$$
$$= 11.3 \qquad\qquad = \sqrt{100(.113)(.887)}$$
$$\approx 3.17.$$

(b) Use the normal distribution to approximate the probability that between 8 and 13 (inclusive) students selected at random defaulted on their loans.

Solution We have $np = 100(.113) = 11.3$ and $n(1 - p) = 88.7$. Both numbers are greater than 5, so we can use a normal approximation. As the graph of Figure 10.24 shows, we need to find the area between 7.5 and 13.5. The z-scores corresponding to $x = 7.5$ and $x = 13.5$ are

$$z_1 = \frac{7.5 - 11.3}{3.17} = -1.20 \qquad z_2 = \frac{13.5 - 11.3}{3.17} = .69.$$

From Appendix A, a z-score of -1.20 corresponds to .3849, and a z-score of .69 corresponds to .2549. Since z_1 is on the left side of 0 and z_2 is on the right side of 0, we add these probabilities. We have

$$P(-1.20 \leq z \leq .69) = .3849 + .2549 = .6398.$$

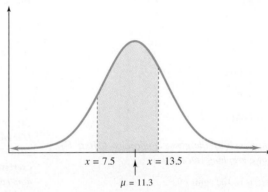

$x = 7.5$ $x = 13.5$

$\mu = 11.3$

Figure 10.24

(c) Find the probability that 6 or fewer students default on their lawns.

Solution As the graph of Figure 10.25 shows, we need to find the area below 6.5. The z-score corresponding to $x = 6.5$ is

$$z_1 = \frac{6.5 - 11.3}{3.17} = -1.51.$$

From Appendix A, we have

$$P(z \leq -1.51) = .5 - .4345 = .0655.$$

The probability of having 6 or fewer students default on their loans is approximately .0655.   ✓₇

✓ **Checkpoint 7**

On May 23, 2017, the probability that a stock advanced on the New York Stock Exchange (NYSE) was .57. If we select 50 stocks at random, find the approximate probability of the following. (Data from: www.wsj.com.)

(a) 30 or more

(b) Between 20 and 30 (inclusive)

$x = 6.5$ $\mu = 11.3$

Figure 10.25

10.4 Exercises

1. The peak in a normal curve occurs directly above _____.

2. The total area under a normal curve (above the horizontal axis) is _____.

3. How are z-scores found for normal distributions with $\mu \neq 0$ or $\sigma \neq 1$?

4. How is the standard normal curve used to find probabilities for normal distributions?

Find the percentage of the area under a normal curve between the mean and the given number of standard deviations from the mean. (See Example 1.)

5. 1.75

6. .26

7. −.43

8. −2.4

Find the percentage of the total area under the standard normal curve between the given z-scores. (See Examples 1 and 2.)

9. $z = 1.41$ and $z = 2.83$

10. $z = .64$ and $z = 2.11$

11. $z = -2.48$ and $z = -.05$

12. $z = -1.63$ and $z = -1.08$

13. $z = -3.05$ and $z = 1.36$

14. $z = -2.91$ and $z = -.51$

Find a z-score satisfying each of the given conditions. (Hint: Use Appendix A backward or a graphing calculator.)

15. 5% of the total area is to the right of z.

16. 1% of the total area is to the left of z.

17. 15% of the total area is to the left of z.

18. 25% of the total area is to the right of z.

19. For any normal distribution, what is the value of $P(x \leq \mu)$? of $P(x \geq \mu)$?

20. Using Chebyshev's theorem and the normal distribution, compare the probability that a number will lie within 2 standard deviations of the mean of a probability distribution. (See Exercises 15–22 of Section 10.3.) Explain what you observe.

21. Repeat Exercise 20, using 3 standard deviations.

Heights *Data from a recent National Health and Nutrition Examination Survey (NHANES) indicates the mean heights of adult females and males, age 18 to 60, is approximately normally distributed. The mean height for women is 63.7 inches, with a standard deviation of 2.8 inches. The mean height of men is 69.1 inches, with a standard deviation of 3.1 inches. Use this information for Exercises 22–26. (See Example 4). (Data from: www.cdc.gov/nchs/nhanes.htm.)*

Find the probability of each of the following.

22. That a female is taller than 70 inches

23. That a female is shorter than 60 inches

24. That a male is shorter than 66 inches

25. That a female is between 5 feet and 6 feet (*Hint:* Convert feet to inches.)

26. That a male is between 5 feet and 6 feet (*Hint:* Convert feet to inches.)

Assume the distributions in Exercises 27–30 are all normal. (See Examples 4 and 5.)

27. **Cholesterol** Using the data from the same study as in Example 5, we find that the average HDL cholesterol level is 52.6 mg/dL, with a standard deviation of 15.5 mg/dL. Find the probability that an individual will have an HDL cholesterol level greater than 60 mg/dL.

28. **Manufacturing** The production of cars per day at an assembly plant has mean 120.5 and standard deviation 6.2. Find the probability that fewer than 100 cars are produced on a random day.

29. **Starting Salary** Starting salaries for accounting majors have mean $53,300, with standard deviation $3,200. What is the probability an individual will start at a salary above $56,000?

30. **Driving Distance** The driving distance to work for residents of a certain community has mean 21 miles and standard deviation 3.6 miles. What is the probability that an individual drives between 10 and 20 miles to work?

Business *Scores on the Graduate Management Association Test (GMAT) are approximately normally distributed. The mean score for 2013–2015 was 552 with a standard deviation of 121. For the following exercises, find the probability that a GMAT test taker selected at random earns a score in the given range, using the normal distribution as a model. (Data from: www.gmac.com.)*

31. Between 540 and 700

32. Between 300 and 540

33. Between 300 and 700

34. Less than 400

35. Greater than 750

36. Between 600 and 700

37. Between 300 and 400

Traffic Speed *New studies by Federal Highway Administration traffic engineers suggest that speed limits on many thoroughfares are set arbitrarily and often are artificially low. According to traffic engineers, the ideal limit should be the "85th-percentile speed," the speed at or below which 85% of the traffic moves. Assuming that speeds are normally distributed, find the 85th-percentile speed for roads with the given conditions.*

38. The mean speed is 55 mph, with a standard deviation of 10 mph.

39. The mean speed is 40 mph, with a standard deviation of 5 mph.

Education *One professor uses the following system for assigning letter grades in a course:*

Grade	Total Points
A	Greater than $\mu + \frac{3}{2}\sigma$
B	$\mu + \frac{1}{2}\sigma$ to $\mu + \frac{3}{2}\sigma$
C	$\mu - \frac{1}{2}\sigma$ to $\mu + \frac{1}{2}\sigma$
D	$\mu - \frac{3}{2}\sigma$ to $\mu - \frac{1}{2}\sigma$
F	Below $\mu - \frac{3}{2}\sigma$

What percentage of the students receive the given grades?

40. A   41. B   42. C

43. Do you think the system in Exercises 40–42 would be more likely to be fair in a large first-year class in psychology or in a graduate seminar of five students? Why?

Nutrition *In nutrition, the recommended daily allowance of vitamins is a number set by the government as a guide to an individual's daily vitamin intake. Actually, vitamin needs vary drastically from person to person, but the needs are very closely approximated by a normal curve. To calculate the recommended daily allowance, the government first finds the average need for vitamins among people in the population and then the standard deviation. The recommended daily allowance is defined as the mean plus 2.5 times the standard deviation.*

44. What percentage of the population will receive adequate amounts of vitamins under this plan?

Find the recommended daily allowance for the following vitamins.

45. Mean = 550 units, standard deviation = 46 units

46. Mean = 1700 units, standard deviation = 120 units

47. Mean = 155 units, standard deviation = 14 units

48. Mean = 1080 units, standard deviation = 86 units

Education *The mean performance score of a large group of fifth-grade students on a math achievement test is 88. The scores are known to be normally distributed. What percentage of the students had scores as follows?*

49. More than 1 standard deviation above the mean

50. More than 2 standard deviations above the mean

Job Satisfaction *According to a 2016 study conducted by the Society for Human Resources Management, 38% of U.S. employees*

reported being very satisfied with their job. In the same study, 51% indicated they are satisfied but to a lesser degree. If 200 employees are selected at random, calculate the following. (See Example 6.) (Data from: www.shrm.org.)

51. The mean and standard deviation for the number of very satisfied employees (round to one decimal place)

52. The mean and standard deviation for the number of employees who are satisfied to a lesser degree (round to one decimal place)

53. The probability that between 70 and 80 employees are very satisfied

54. The probability that between 90 and 115 employees are satisfied to a lesser degree

55. The probability that at least 88 employees are very satisfied

56. The probability that 85 or less employees are satisfied to a lesser degree

Student Loan Debt *According to a report from the American Association of University Women released in 2017, 44% of female undergraduates and 39% of male undergraduates take on debt. Calculate the following if we select 75 females and 75 males at random. (Data from: www.aauw.org.)*

57. The mean and standard deviation for the number of female undergraduates who take on debt (round to the nearest tenth)

58. The mean and standard deviation for the number of male undergraduates who take on debt (round to the nearest tenth)

59. The probability that between 25 and 35 (inclusive) female undergraduates take on debt

60. The probability that between 35 and 40 (inclusive) male undergraduates take on debt

61. The probability that 27 or fewer female undergraduates take on debt

62. The probability that at least 29 male undergraduates take on debt

✓ Checkpoint Answers

1. (a) 43.45%   (b) 47.93%   (c) 26.42%

2. (a) 34.13%   (b) 49.25%

3. (a) 46.55%   (b) 32.82%   (c) 6.81%
   (d) 68.26%, 95.46%, 99.74%; almost all the data lie within 3 standard deviations of the mean.

4. (a) −4.8   (b) −1

5. .8607

6. .1894

7. (a) .3859   (b) .7106

CHAPTER 10 Summary and Review

Key Terms and Symbols

10.1 random sample
grouped frequency
distribution
histogram
frequency polygon
stem-and-leaf-plot
uniform distribution
right-skewed distribution
left-skewed distribution
normal distribution

10.2 Σ, summation (sigma)
notation
$\bar{x}$, sample mean
μ, population mean
(arithmetic) mean
median
statistic
mode

10.3 s^2, sample variance
s, sample standard deviation

σ, population standard
deviation
range
deviations from the
mean
variance
standard deviation
boxplot
quartile
five-number summary

10.4 μ, mean of a normal
distribution
σ, standard deviation of a
normal distribution
continuous distribution
normal curves
standard normal curve
z-score
normal approximation to
the binomial distribution

Chapter 10 Key Concepts

Frequency Distributions

To organize the data from a sample, we use a **grouped frequency distribution**—a set of intervals with their corresponding frequencies. The same information can be displayed with a **histogram**—a type of bar graph with a bar for each interval. The height of each bar is equal to the frequency of the corresponding interval. A **stem-and-leaf plot** presents the individual data in a similar form, so it can be viewed as a bar graph as well. Another way to display this information is with a **frequency polygon**, which is formed by connecting the midpoints of consecutive bars of the histogram with straight-line segments.

Measures of Center

The **mean** $\bar{x}$ of a frequency distribution is the expected value.

For n numbers $x_1, x_2, \ldots, x_n$,   For a grouped distribution,

$$\bar{x} = \frac{\Sigma x_i}{n} \qquad\qquad \bar{x} = \frac{\Sigma f_i x_i}{n}$$

The **median** is the middle entry (or mean of the two middle entries) in a set of data arranged in either increasing or decreasing order.

The **mode** is the most frequent entry in a set of numbers.

Measures of Variability

The **range** of a distribution is the difference between the largest and smallest numbers in the distribution.

The **sample standard deviation** s is the square root of the sample **variance**.

For n numbers,   For a grouped distribution,

$$s = \sqrt{\frac{\Sigma x_i^2 - n\bar{x}}{n-1}} \qquad s = \sqrt{\frac{\Sigma f_i x_i^2 - n\bar{x}}{n-1}}$$

Boxplots

A **boxplot** organizes a list of data using the minimum and maximum values, the median, and the first and third quartiles to give a visual overview of the distribution.

The Normal Distribution

A **normal distribution** is a continuous distribution with the following properties: The highest frequency is at the mean; the graph is symmetric about a vertical line through the mean; the total area under the curve, above the x-axis, is 1. If a normal distribution has mean μ and standard deviation σ, then the z-score for the number x is $z = \dfrac{x - \mu}{\sigma}$. The **area under a normal curve** between $x = a$ and $x = b$ is found by calculating the area under the standard normal curve from z_a to z_b.

Normal Approximation to the Binomial Distribution

The **binomial distribution** is a distribution with the following properties: For n independent repeated trials, in which the probability of success in a single trial is p, the probability of x successes is $_nC_x p^x(1 - p)^{n-x}$. The mean is $\mu = np$, and the standard deviation is

$$\sigma = \sqrt{np(1 - p)}.$$

For a large number of trials, the normal distribution can be used to approximate binomial probabilities.

Chapter 10 Review Exercises

NASDAQ 100 Stocks *For Exercises 1–10, the data come from a random sample of 20 companies of the NASDAQ 100 taken on June 8, 2017. (Data from: www.morningstar.com.)*

For Exercises 1 and 2, (a) write a frequency distribution; (b) draw a histogram.

1. The variable is the price of the stock (rounded to the nearest dollar).

123	35	76	38
45	62	50	111
180	108	65	46
66	100	64	144
57	107	48	110

2. The variable is the volume of stock (the number of shares that traded hands on a given day, in thousands).

241	7300	1700	1800
1400	1800	5700	1000
2900	411	4400	5800
2600	2900	674	2300
2500	2300	1500	4600

For Exercises 3 and 4, draw a stem-and-leaf plot for the data.

3. The variable is the forward price-to-earnings ratio rounded to the nearest integer. (The forward price-to-earnings ratio is the current stock's price over its predicted earnings per share.)

22	16	15	28
21	18	8	20
28	21	26	20
16	25	15	20
14	15	16	22

4. The variable is the price-per-sales ratio. (The price-per-sales ratio is calculated by dividing the per-share stock price by the per-share revenue.)

2.6	4.5	0.8	3.7
4.4	1.8	0.7	6.5
6.1	2.7	2.1	2.8
2.9	3.7	1.3	2.5
1.1	6.2	3.6	7.4

5. Find the mean and median for the first 10 stock prices of Exercise 1.

123, 45, 180, 66, 57, 35, 62, 108, 100, 107

6. Find the mean and median for the first 10 stock volumes of Exercise 2.

241, 1400, 2900, 2600, 2500, 7300, 1800, 411, 2900, 2300

7. Find the 5-number summary and create a boxplot for the data in Exercise 5.

8. Find the 5-number summary and create a boxplot for the data in Exercise 6.

9. The following table gives the frequency counts for the price-per-book ratio. (This ratio is calculated by dividing the current closing price of the stock by the latest quarter's book value per share.) Find the mean and median for these data. (Round midpoint to the nearest tenth.)

Interval	Frequency
2–2.99	2
3–3.99	5
4–4.99	2
5–5.99	5
6–6.99	1
7–7.99	3
8–8.99	2

10. The following table gives the frequency counts for the price/cash flow ratio. (This ratio is calculated by dividing the company's capitalization by the company's operating cash flow in the most recent fiscal year.) Find the mean and median for these data. (Round midpoint to the nearest tenth.)

Interval	Frequency
0–4.99	1
5–9.99	1
10–14.99	5
15–19.99	8
20–24.99	3
25–29.99	2

11. Discuss some reasons for organizing data into a grouped frequency distribution.

12. What do the mean, median, and mode of a distribution have in common? How do they differ? Describe each in a sentence or two.

Find the median and the mode (or modes) for each of the given data sets.

13. **Hours at Work** Hours worked for 10 adults. (Data from: gss.norc.org.)

30, 45, 16, 20, 70, 40, 48, 40, 50, 50

14. **Education Attainment** The percentage of residents with a bachelor's degree or higher for 10 states. (Data from: U.S. Census Bureau.)

22, 28, 26, 20, 30, 36, 36, 28, 26, 27

The modal class is the interval containing the most data values. Find the modal class for the distributions of the given data sets.

15. The data in Exercise 9

16. The data in Exercise 10

For the given histograms, identify the shape of the distribution.

17.

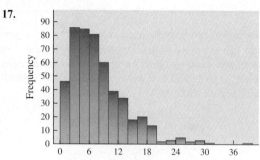

18.

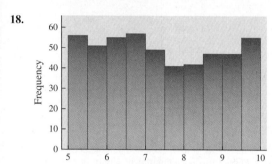

19.

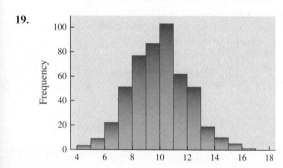

20.

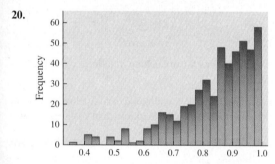

21. What is meant by the range of a distribution?

22. How are the variance and the standard deviation of a distribution related? What is measured by the standard deviation?

Find the range and the standard deviation for each of the given distributions.

23. The data in Exercise 13

24. The data in Exercise 14

Find the standard deviation for each of the given distributions.

25. The data summarized in the frequency table of Exercise 9

26. The data summarized in the frequency table of Exercise 10

27. Macy's Revenue The annual reported revenue (in billions of dollars) for Macy's Inc. is given in the following table. Find the mean and standard deviation for the 5-year period.

Year	Revenue
2013	27.7
2014	27.9
2015	28.1
2016	27.1
2017	25.8

28. Kohl's Revenue The annual reported revenue (in billions of dollars) for Kohl's Corp. is given in the following table. Find the mean and standard deviation for the 5-year period.

Year	Revenue
2013	19.3
2014	19.0
2015	19.0
2016	19.2
2017	18.7

29. Describe the characteristics of a normal distribution.

30. What is meant by a skewed distribution?

Find the given areas under the standard normal curve.

31. Between $z = 0$ and $z = 1.35$

32. To the left of $z = .38$

33. Between $z = -1.88$ and $z = 2.41$

34. Between $z = 1.53$ and $z = 2.82$

35. Find a z-score such that 8% of the area under the curve is to the right of z.

36. Why is the normal distribution not a good approximation of a binomial distribution that has a value of p close to 0 or 1?

Arm Circumference *Data from a recent National Health and Nutrition Examination Survey (NHANES) indicated that the distribution of upper arm circumference (in cm) is approximately normally distributed. The mean circumference for adult females is 38.6 cm, with a standard deviation of 2.3 cm. The mean circumference for adult males is 41.5 cm, with a standard deviation of 2.5 cm. Find the following. (Data from: www. cdc.gov/nchs/nhanes.htm.)*

37. A randomly selected female has an upper arm circumference of less than 45 cm.

38. A randomly selected female has an upper arm circumference between 35 and 45 cm.

39. A randomly selected male has an upper arm circumference of more than 38 cm.

40. A randomly selected male has an upper arm circumference between 38 and 46 cm.

Auto Sales *The table gives the number of vehicles (in thousands) sold within the United States in May 2016 and May 2017 for 12 auto manufacturers. (Data from: Wall Street Journal.)*

Auto Manufacturer	May 2016 Sales	May 2017 Sales
General Motors Corp.	240	237
Ford Motor Corp.	235	240
Chrysler LLC	192	189
Toyota Motor Sales USA	219	218
American Honda Motor Co. Inc.	147	148
Nissan North America Inc.	133	137
Hyundai Motor America	71	60
Mazda	28	26
Kia Motors America Inc.	63	59
Subaru of America Inc.	50	56
Mercedes-Benz	32	30
Volkswagen of America Inc.	29	30

41. (a) Find the mean and standard deviation of sales for the May 2016 data.

(b) Which company is closest to the mean sales for the May 2016 data?

42. (a) Find the mean and standard deviation of sales for the May 2017 data.

(b) Which company is closest to the mean sales for the May 2017 data?

43. Find the five-number summary for the May 2016 sales.

44. Find the five-number summary for the May 2017 sales.

45. Construct a boxplot for the May 2016 sales data.

46. Construct a boxplot for the May 2017 sales data.

College or University Education *A recent Gallup.com poll found that 51% of Americans who pursued or completed a college or university degree would change at least one aspect of their education experience if they could do it all over again, including their major or field of study, the institution they attended, or the type of degree they obtained. For Exercises 47–50, assume a random sample of 500 adults who pursued a college or university degree was selected. Approximate the probability of the following given events. (Round standard deviation to one decimal place.) (Data from: www.gallup.com.)*

47. More than 260 would change at least one aspect of their education experience.

48. Between 265 and 280 (inclusive) would change at least one aspect of their education experience.

49. Between 240 and 280 (inclusive) would change at least one aspect of their education experience.

50. Fewer than 240 would change at least one aspect of their education experience.

Living at Home *According to the Pew Research Center, in 2016, 15% of young adults age 25–35 were living in their parents' home. If we select 1000 young adults age 25–35 at random, find the probability of the following given events. (Round standard deviation to one decimal place.) (Data from: pewresearch.org.)*

51. More than 130 live in their parents' home.

52. Between 120 and 130 (inclusive) live in their parents' home.

53. Between 120 and 165 (inclusive) live in their parents' home.

54. 165 or fewer live in their parents' home.

Case Study 10 Standard Deviation as a Measure of Risk

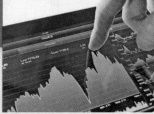

When investing, the idea of risk is difficult to define. Some individuals may deem an investment product to be of high risk if it loses all of its value in a short time, others may define a product as risky if it loses some portion of its value in a short time, and still others may define a product as risky if there is even a remote chance of it losing its value over a long period of time.

One way investors try to quantify risk is to examine the volatility of the return for an investment product. Volatility refers to how the return on the investment fluctuates from the average level of return. One simple measure of volatility is to calculate the standard deviation of the annual rate of return.

For example, if a fund has an annual return of 2% each and every year, then the mean is 2% and the standard deviation would be 0 (because there is no variation from the mean). This fund would have very low volatility. What is also interesting is that a fund that *lost* 2% each and every year would also have a standard deviation of 0 (because it too does not ever vary from its mean of –2%). Nevertheless, the standard deviation can be helpful in assessing the

risk and volatility related to an investment fund.

Let us examine two mutual funds—Fund A and Fund B. The following table lists the annual rate of return for these two funds.

Fund A	Fund B
3.2	9.1
6.6	−3.4
3.9	2.1
5.3	11.4
5.2	1.1
5.8	9.7

If we calculate the average rate of return for both funds, we obtain 5%. We can notice, however, even by casual inspection, that there is greater variability in Fund B than in Fund A. If we use the notation s_A to denote the standard deviation of fund A and s_B to denote the standard deviation of fund B, we can use the methods of Section 10.3 to obtain the following values for the standard deviation.

$$s_A = \sqrt{\frac{(3.2 - 5)^2 + (6.6 - 5)^2 + (3.9 - 5)^2 + (5.3 - 5)^2 + (5.2 - 5)^2 + (5.8 - 5)^2}{5}} \approx 1.25$$

$$s_B = \sqrt{\frac{(9.1 - 5)^2 + (-3.4 - 5)^2 + (2.1 - 5)^2 + (11.4 - 5)^2 + (1.1 - 5)^2 + (9.7 - 5)^2}{5}} \approx 5.90$$

Our calculations confirm that there is greater variability in Fund B. Thus, Fund B is said to carry more risk.

The way that standard deviation is used in the financial products industry is to examine monthly returns. For example, the website morningstar.com explains that it uses a 36-month window of monthly returns. They then calculate the average monthly return and the standard deviation from those returns. They then report these values (the mean return and the standard deviation) corrected for an annualized return rate. This should be clear in the following example.

To calculate the standard deviation for the Fidelity® Contrafund® Fund (FCNTX), we examine the monthly returns for 36 months from May 2017 to June 2014, going back in time.

Month	Return	Month	Return	Month	Return
MAY 2017	3.59	MAY 2016	1.67	MAY 2015	2.16
APR 2017	2.82	APR 2016	0.25	APR 2015	−0.84
MAR 2017	1.56	MAR 2016	5.59	MAR 2015	−0.49
FEB 2017	3.85	FEB 2016	−1.18	FEB 2015	5.98
JAN 2017	4.38	JAN 2016	−5.68	JAN 2015	−1.34
DEC 2016	0.53	DEC 2015	−1.36	DEC 2014	−0.53
NOV 2016	0.60	NOV 2015	0.64	NOV 2014	2.15
OCT 2016	−1.67	OCT 2015	7.08	OCT 2014	1.48
SEP 2016	0.43	SEP 2015	−2.06	SEP 2014	−1.13
AUG 2016	0.27	AUG 2015	−5.96	AUG 2014	4.44
JUL 2016	4.48	JUL 2015	3.47	JUL 2014	−1.44
JUN 2016	−1.51	JUN 2015	−0.30	JUN 2014	2.41

Our first step is to calculate the mean return. We find the average monthly rate of return to be .9539. We then calculate the standard deviation value and obtain 2.9361. To convert these values from monthly return rates to annual return rates, we multiply the mean by 12 and the standard deviation by $\sqrt{12}$. The annual return rate is $12(.9539) = 11.4468$, and the annual standard deviation is $\sqrt{12}(2.9361) = 10.1709$. These values are often reported rounded to two decimal places (11.45 and 10.17, respectively).

Many financial websites report these values for return rates based on monthly calculations. For example, on the morningstar.com website, clicking on a link for "Ratings and Risk" will display the mean and standard deviation values under a heading of "Volatility Measures." With this information, it is easier to make direct comparisons to other funds. For the same period, the S&P 500 had a mean return of 10.14 and a standard deviation of 10.38. Thus, we see that the Fidelity® fund had a slightly higher average return and a slightly lower level of risk.

An additional way that investors interpret the standard deviation is that they assume future returns will follow an approximate normal distribution. We know from Section 10.4 (and, in particular, Figure 10.22) that 68% of data that is normally distributed will fall within one standard deviation of the mean. With the Fidelity® fund example, we then expect that the probability of a future return being within one standard deviation of the mean of 11.45% is 68%. With the standard deviation value of 10.17, our bounds on this interval are

$$(11.45 - 10.17, 11.45 + 10.17) = (1.28, 21.62).$$

Furthermore, we expect the annual return to fall within 2 standard deviation of the mean approximately 95% of the time:

$$(11.45 - 2*10.17, 11.45 + 2*10.17) = (-8.89, 31.79).$$

The standard deviation is not a perfect measure of risk. It is possible to invest in a fund with a low standard deviation and still lose money. It is, however, a useful way to compare funds in similar categories.

Exercises

For Exercises 1–10, use the following table, which gives the monthly returns for the Vanguard PRIMECAP Fund Investor Shares (VPMCX), the S&P 500, and the sector returns for Large Growth funds. (Data from: www.morningstar.com.)

Month	Vanguard	S&P	Large Growth	Month	Vanguard	S&P	Large Growth
MAY 2017	2.95	1.41	2.53	NOV 2015	0.37	0.30	0.64
APR 2017	1.48	1.03	2.34	OCT 2015	8.84	8.44	7.87
MAR 2017	0.72	0.12	1.03	SEP 2015	−2.05	−2.47	−3.19
FEB 2017	4.25	3.97	3.58	AUG 2015	−5.81	−6.03	−6.38
JAN 2017	3.10	1.90	3.79	JUL 2015	1.51	2.10	2.95
DEC 2016	1.72	1.98	0.55	JUN 2015	−2.77	−1.94	−1.17
NOV 2016	3.88	3.70	1.65	MAY 2015	1.59	1.29	1.73
OCT 2016	−2.98	−1.82	−2.46	APR 2015	−0.29	0.96	−0.02
SEP 2016	1.76	0.02	0.64	MAR 2015	−1.27	−1.58	−0.96
AUG 2016	1.27	0.14	−0.09	FEB 2015	5.20	5.75	6.38
JUL 2016	6.54	3.69	5.00	JAN 2015	−1.56	−3.00	−1.81
JUN 2016	−2.35	0.26	−1.47	DEC 2014	−0.35	−0.25	−0.72
MAY 2016	3.10	1.80	2.24	NOV 2014	3.32	2.69	2.57
APR 2016	−0.41	0.39	−0.21	OCT 2014	2.22	2.44	2.51
MAR 2016	6.16	6.78	6.17	SEP 2014	−0.34	−1.40	−1.89
FEB 2016	−1.16	−0.13	−1.14	AUG 2014	3.89	4.00	4.21
JAN 2016	−6.57	−4.96	−7.10	JUL 2014	−0.84	−1.38	−1.53
DEC 2015	−.42	−1.58	−1.67	JUN 2014	2.44	2.07	2.46

1. Find the mean monthly return for the Vanguard fund. Convert this value to the annual return.

2. Find the mean monthly return for the S&P 500. Convert this value to the annual return.

3. Find the mean monthly return for the Large Growth funds. Convert this value to the annual return.

4. Which fund has the highest annual rate of return?

5. Find the standard deviation of the monthly returns for the Vanguard fund. Convert this value to the annual return.

6. Find the standard deviation of the monthly returns for the S&P 500. Convert this value to the annual return.

7. Find the standard deviation of the monthly returns for the Large Growth funds. Convert this value to the annual return.

8. Which fund has the highest level of risk?

9. Determine the bounds for 68% of future annual returns for the Vanguard Fund.

10. Determine the bounds for 95% of future annual returns for the Vanguard Fund.

Extended Projects

1. Investigate several mutual funds of your choice. Use a financial data website and obtain the monthly returns for the funds of your choice. Calculate the mean return and the standard deviation of the monthly data and convert the values to the annual values. Determine which funds have the highest return values and which have the lowest standard deviation values.

2. Investigate other measures of risk for investment products. Compare these measures with the standard deviation discussed here.

Areas under the Normal Curve

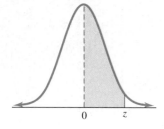

The column under A gives the proportion of the area under the entire curve that is between $z = 0$ and a positive value of z.

z	A	z	A	z	A	z	A
.00	.0000	.48	.1844	.96	.3315	1.44	.4251
.01	.0040	.49	.1879	.97	.3340	1.45	.4265
.02	.0080	.50	.1915	.98	.3365	1.46	.4279
.03	.0120	.51	.1950	.99	.3389	1.47	.4292
.04	.0160	.52	.1985	1.00	.3413	1.48	.4306
.05	.0199	.53	.2019	1.01	.3438	1.49	.4319
.06	.0239	.54	.2054	1.02	.3461	1.50	.4332
.07	.0279	.55	.2088	1.03	.3485	1.51	.4345
.08	.0319	.56	.2123	1.04	.3508	1.52	.4357
.09	.0359	.57	.2157	1.05	.3531	1.53	.4370
.10	.0398	.58	.2190	1.06	.3554	1.54	.4382
.11	.0438	.59	.2224	1.07	.3577	1.55	.4394
.12	.0478	.60	.2258	1.08	.3599	1.56	.4406
.13	.0517	.61	.2291	1.09	.3621	1.57	.4418
.14	.0557	.62	.2324	1.10	.3643	1.58	.4430
.15	.0596	.63	.2357	1.11	.3665	1.59	.4441
.16	.0636	.64	.2389	1.12	.3686	1.60	.4452
.17	.0675	.65	.2422	1.13	.3708	1.61	.4463
.18	.0714	.66	.2454	1.14	.3729	1.62	.4474
.19	.0754	.67	.2486	1.15	.3749	1.63	.4485
.20	.0793	.68	.2518	1.16	.3770	1.64	.4495
.21	.0832	.69	.2549	1.17	.3790	1.65	.4505
.22	.0871	.70	.2580	1.18	.3810	1.66	.4515
.23	.0910	.71	.2612	1.19	.3830	1.67	.4525
.24	.0948	.72	.2642	1.20	.3849	1.68	.4535
.25	.0987	.73	.2673	1.21	.3869	1.69	.4545
.26	.1026	.74	.2704	1.22	.3888	1.70	.4554
.27	.1064	.75	.2734	1.23	.3907	1.71	.4564
.28	.1103	.76	.2764	1.24	.3925	1.72	.4573
.29	.1141	.77	.2794	1.25	.3944	1.73	.4582
.30	.1179	.78	.2823	1.26	.3962	1.74	.4591
.31	.1217	.79	.2852	1.27	.3980	1.75	.4599
.32	.1255	.80	.2881	1.28	.3997	1.76	.4608
.33	.1293	.81	.2910	1.29	.4015	1.77	.4616
.34	.1331	.82	.2939	1.30	.4032	1.78	.4625
.35	.1368	.83	.2967	1.31	.4049	1.79	.4633
.36	.1406	.84	.2996	1.32	.4066	1.80	.4641
.37	.1443	.85	.3023	1.33	.4082	1.81	.4649
.38	.1480	.86	.3051	1.34	.4099	1.82	.4656
.39	.1517	.87	.3079	1.35	.4115	1.83	.4664
.40	.1554	.88	.3106	1.36	.4131	1.84	.4671
.41	.1591	.89	.3133	1.37	.4147	1.85	.4678
.42	.1628	.90	.3159	1.38	.4162	1.86	.4686
.43	.1664	.91	.3186	1.39	.4177	1.87	.4693
.44	.1700	.92	.3212	1.40	.4192	1.88	.4700
.45	.1736	.93	.3238	1.41	.4207	1.89	.4706
.46	.1772	.94	.3264	1.42	.4222	1.90	.4713
.47	.1808	.95	.3289	1.43	.4236	1.91	.4719

(*continued*)

z	A	z	A	z	A	z	A
1.92	.4726	2.42	.4922	2.92	.4983	3.42	.4997
1.93	.4732	2.43	.4925	2.93	.4983	3.43	.4997
1.94	.4738	2.44	.4927	2.94	.4984	3.44	.4997
1.95	.4744	2.45	.4929	2.95	.4984	3.45	.4997
1.96	.4750	2.46	.4931	2.96	.4985	3.46	.4997
1.97	.4756	2.47	.4932	2.97	.4985	3.47	.4997
1.98	.4762	2.48	.4934	2.98	.4986	3.48	.4998
1.99	.4767	2.49	.4936	2.99	.4986	3.49	.4998
2.00	.4773	2.50	.4938	3.00	.4987	3.50	.4998
2.01	.4778	2.51	.4940	3.01	.4987	3.51	.4998
2.02	.4783	2.52	.4941	3.02	.4987	3.52	.4998
2.03	.4788	2.53	.4943	3.03	.4988	3.53	.4998
2.04	.4793	2.54	.4945	3.04	.4988	3.54	.4998
2.05	.4798	2.55	.4946	3.05	.4989	3.55	.4998
2.06	.4803	2.56	.4948	3.06	.4989	3.56	.4998
2.07	.4808	2.57	.4949	3.07	.4989	3.57	.4998
2.08	.4812	2.58	.4951	3.08	.4990	3.58	.4998
2.09	.4817	2.59	.4952	3.09	.4990	3.59	.4998
2.10	.4821	2.60	.4953	3.10	.4990	3.60	.4998
2.11	.4826	2.61	.4955	3.11	.4991	3.61	.4999
2.12	.4830	2.62	.4956	3.12	.4991	3.62	.4999
2.13	.4834	2.63	.4957	3.13	.4991	3.63	.4999
2.14	.4838	2.64	.4959	3.14	.4992	3.64	.4999
2.15	.4842	2.65	.4960	3.15	.4992	3.65	.4999
2.16	.4846	2.66	.4961	3.16	.4992	3.66	.4999
2.17	.4850	2.67	.4962	3.17	.4992	3.67	.4999
2.18	.4854	2.68	.4963	3.18	.4993	3.68	.4999
2.19	.4857	2.69	.4964	3.19	.4993	3.69	.4999
2.20	.4861	2.70	.4965	3.20	.4993	3.70	.4999
2.21	.4865	2.71	.4966	3.21	.4993	3.71	.4999
2.22	.4868	2.72	.4967	3.22	.4994	3.72	.4999
2.23	.4871	2.73	.4968	3.23	.4994	3.73	.4999
2.24	.4875	2.74	.4969	3.24	.4994	3.74	.4999
2.25	.4878	2.75	.4970	3.25	.4994	3.75	.4999
2.26	.4881	2.76	.4971	3.26	.4994	3.76	.4999
2.27	.4884	2.77	.4972	3.27	.4995	3.77	.4999
2.28	.4887	2.78	.4973	3.28	.4995	3.78	.4999
2.29	.4890	2.79	.4974	3.29	.4995	3.79	.4999
2.30	.4893	2.80	.4974	3.30	.4995	3.80	.4999
2.31	.4896	2.81	.4975	3.31	.4995	3.81	.4999
2.32	.4898	2.82	.4976	3.32	.4996	3.82	.4999
2.33	.4901	2.83	.4977	3.33	.4996	3.83	.4999
2.34	.4904	2.84	.4977	3.34	.4996	3.84	.4999
2.35	.4906	2.85	.4978	3.35	.4996	3.85	.4999
2.36	.4909	2.86	.4979	3.36	.4996	3.86	.4999
2.37	.4911	2.87	.4980	3.37	.4996	3.87	.5000
2.38	.4913	2.88	.4980	3.38	.4996	3.88	.5000
2.39	.4916	2.89	.4981	3.39	.4997	3.89	.5000
2.40	.4918	2.90	.4981	3.40	.4997		
2.41	.4920	2.91	.4982	3.41	.4997		

Solutions to Prerequisite Skills Test and Calculus Readiness Test

Prerequisite Skills Test

1. We must find a common denominator before subtracting. Since 6 can be written as a fraction as $\frac{6}{1}$, then 2 can be the common denominator.

$$\frac{5}{2} - 6 = \frac{5}{2} - \frac{6}{1}$$

$$= \frac{5}{2} - \frac{6(2)}{2} = \frac{5}{2} - \frac{12}{2}$$

$$= \frac{5 - 12}{2} = \frac{-7}{2}.$$

2. Division of fraction implies that we multiply by the reciprocal. In this case, we multiply $\frac{1}{2}$ by the reciprocal of $\frac{2}{5}$, which is $\frac{5}{2}$. We obtain

$$\frac{1}{2} \div \frac{2}{5} = \frac{1}{2} \cdot \frac{5}{2} = \frac{1 \cdot 5}{2 \cdot 2} = \frac{5}{4}.$$

3. We rewrite 3 as a fraction as $\frac{3}{1}$ and then multiply by the reciprocal.

$$\frac{1}{3} \div \frac{3}{1} = \frac{1}{3} \cdot \frac{1}{3} = \frac{1 \cdot 1}{3 \cdot 3} = \frac{1}{9}.$$

4. We perform the operation inside the parenthesis first. We have

$$2 \div 6 = \frac{2}{6} = \frac{1}{3}.$$

Thus,

$$7 + 2 - 3(2 \div 6) = 7 + 2 - 3\left(\frac{1}{3}\right).$$

We then perform the multiplication and see that

$$3\left(\frac{1}{3}\right) = \frac{3}{1} \cdot \frac{1}{3} = \frac{3(1)}{1(3)} = \frac{3}{3} = 1.$$

So we now have

$$7 + 2 - 3\left(\frac{1}{3}\right) = 7 + 2 - 1.$$

Performing addition and subtraction from left to right, we obtain $9 - 1 = 8$.

5. We begin by performing the multiplication on the numerator. We obtain

$$\frac{2 \times 3 + 12}{1 + 5} - 1 = \frac{6 + 12}{1 + 5} - 1.$$

Next, we perform addition in the numerator and addition in the denominator.

$$\frac{6 + 12}{1 + 5} - 1 = \frac{18}{6} - 1.$$

Next, we perform the division and realize that $\frac{18}{6} = 3$, to obtain $3 - 1 = 2$.

6. False. We add the numerator values together to obtain

$$\frac{4 + 3}{3} = \frac{7}{3} \neq 5.$$

7. False. We obtain a common denominator of 35 to obtain

$$\frac{5}{7} + \frac{7}{5} = \frac{5(5)}{35} + \frac{7(7)}{35} = \frac{25}{35} + \frac{49}{35} = \frac{25 + 49}{35} = \frac{74}{35} \neq 1.$$

8. False. We write 1 as a fraction with the same denominator as $\frac{3}{5}$, namely $1 = \frac{5}{5}$. Thus, we have

$$\frac{3}{5} + 1 = \frac{3}{5} + \frac{5}{5} = \frac{8}{5} \neq \frac{6}{5}.$$

9. Since n represents the number of shoes that Alicia has, we can add 2 to n to obtain the number of shoes for Manuel. Thus we are looking for the expression $n + 2$ for the number of shoes for Manuel.

10. We can say x represents David's age. Since Selina's age is 6 more than David, Selina's age is $x + 6$. The two ages together then are

$$x + (x + 6) = 42.$$

If we combine like terms, we have

$$2x + 6 = 42.$$

We can solve for x by adding -6 to each side to obtain

$$2x + 6 - 6 = 42 - 6$$
$$2x = 36.$$

We then divide each side by 2 to obtain

$$x = 18.$$

Thus, David's age is 18 and Selina's age is $18 + 6 = 24$.

11. If the sweater is reduced by 20%, then its sale price is $100 - 20 = 80\%$ of the original price. To obtain the new price we first convert 80% to a decimal to obtain .80 and then multiply by the price of the sweater to obtain

$$.80(\$72) = \$57.60$$

12. When we plot points, the first coordinate is the value for the x-axis and the second value is for the y-axis. For points A, B, and C, we obtain

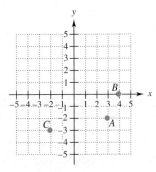

13. Similarly, we plot points D, E, and F to obtain

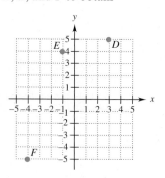

14. (a) 4.27659 has a value of 2 in the tenths place, and the value after that is 7, so we round up to obtain 4.3 to the nearest tenth.

   (b) 245.984 has a 5 in the units place and a 9 in the tenths place, so we round up the units place to obtain 246 to the nearest unit.

15. (a) 16.38572 has an 8 in the hundredths place and a 5 in the thousandths place, so we round up to obtain 16.39 to the nearest hundredth.

   (b) 1,763,304.42 has a 3 in the thousands place and the following digit is also a 3, so we do not round up and obtain 1,763,000 rounded to the nearest thousand.

16. (a) Writing 34 million dollars as a numerical value is $34 \cdot 1,000,000 = \$34,000,000$.

   (b) Writing 2.2 thousand dollars as a numerical value is $2.2 \cdot 1000 = \$2,200$.

17. (a) 17 hundred thousand as a numerical value is $17 \cdot 100,000 = 1,700,000$.

   (b) Three and a quarter billion as a numerical value is $3.25 \cdot 1,000,000,000 = 3,250,000,000$.

18. There is no solution to $\dfrac{5}{0}$ because a 0 in the denominator indicates the fraction is undefined.

19. Yes, it is. If a car is traveling 60 miles in one hour, it is traveling 60 miles in 60 minutes, which equates to 1 mile per minute.

20. -9 is greater than -900.

Calculus Readiness Test

1. $(7x^4 - 8x^3 + 2x^2 - 3) - (9x^4 - 2x^2 + 5x + 6)$
   $= (7x^4 - 9x^4) + (-8x^3 - 0) + (2x^2 - (-2x^2)) + (0 - 5x) + (-3 - 6)$
   $= -2x^4 - 8x^3 + 4x^2 - 5x - 9$
   (See Section 1.2, Example 7.)

2. $\dfrac{-6}{3x+4} + \dfrac{2}{x} = \dfrac{-6}{3x+4} \cdot \dfrac{x}{x} + \dfrac{2}{x} \cdot \dfrac{3x+4}{3x+4} = \dfrac{-6x}{x(3x+4)} + \dfrac{2(3x+4)}{x(3x+4)}$

   $= \dfrac{-6x}{3x^2+4x} + \dfrac{6x+8}{3x^2+4x} = \dfrac{8}{3x^2+4x}$
   (See Section 1.4, Example 5.)

3. $(3x + 7)(5x - 8) = (3x)(5x) + (3x)(-8) + (7)(5x) + (7)(-8)$
   $= 15x^2 - 24x + 35x - 56 = 15x^2 + 11x - 56$
   (See Section 1.2, Example 10.)

4. $(7x + 3)(6x^2 + x - 8) = 7x(6x^2 + x - 8) + 3(6x^2 + x - 8)$
   $= (42x^3 + 7x^2 - 56x) + (18x^2 + 3x - 24) = 42x^3 + 25x^2 - 53x - 24$
   (See Section 1.2, Example 8.)

5. $\dfrac{(x^5 y^{-5})^3}{(x^2 \sqrt{y})^6} = \dfrac{(x^5)^3 (y^{-5})^3}{(x^2)^6 (y^{1/2})^6} = \dfrac{x^{15} y^{-15}}{x^{12} y^3} = \dfrac{x^{15}}{x^{12} y^3 y^{15}} = \dfrac{x^{15-12}}{y^{3+15}} = \dfrac{x^3}{y^{18}}$

(See Section 1.5, Example 8.)

6. $(8x - 9)^2 = (8x - 9)(8x - 9) = 64x^2 - 72x - 72x + 81 = 64x^2 - 144x + 81$

(See Section 1.2, Example 10.)

7. $9x^2 - 49 = (3x)^2 - 7^2 = (3x + 7)(3x - 7)$

(See Section 1.3, Example 7.)

8. $9x - 4 = 8 + 7(x - 5)$

$\Rightarrow 9x - 4 = 8 + 7x - 35$

$\Rightarrow 9x - 7x = 8 - 35 - (-4)$

$\Rightarrow 2x = -23$

$\Rightarrow x = -\dfrac{23}{2}$

(See Section 1.6, Example 2.)

9. $3x^2 + 2x + 8 = 20x - 7$

$\Rightarrow 3x^2 + 2x - 20x + 8 + 7 = 0$

$\Rightarrow 3x^2 - 18x + 15 = 0$

$\Rightarrow 3(x^2 - 6x + 5) = 0$

$\Rightarrow 3(x - 5)(x - 1) = 0$

$\Rightarrow x - 5 = 0 \quad \text{or} \quad x - 1 = 0$

$\Rightarrow x = 5 \quad \text{or} \quad x = 1$

(See Section 1.7, Example 2.)

10. $x^2 - x - 6 < 0$

First solve $x^2 - x - 6 = 0 \Rightarrow (x + 2)(x - 3) = 0 \Rightarrow x = -2$ or $x = 3$. The numbers -2 and 3 divide the number line into three regions. Test one number in each region to determine that $(x + 2)(x - 3) < 0$ when $-2 < x < 3$.

(See Section 2.5, Example 3.)

11. $5e^{x-3} - 1 = 9$

$\Rightarrow 5e^{x-3} = 10$

$\Rightarrow e^{x-3} = 2$

$\Rightarrow \ln(e^{x-3}) = \ln(2)$

$\Rightarrow x - 3 = \ln(2)$

$\Rightarrow x = \ln(2) + 3$

(See Section 4.4, Example 6.)

12. $\ln(5x + 1) = 2$

$\Rightarrow e^{\ln(5x+1)} = e^2$

$\Rightarrow 5x + 1 = e^2$

$\Rightarrow 5x = e^2 - 1$

$\Rightarrow x = \dfrac{e^2 - 1}{5}$

(See Section 4.4, Example 2.)

13. $f(-2) = 5(-2)^3 - 3(-2)^2 + 9(-2) + 60$

$= 5(-8) - 3(4) + 9(-2) + 60$

$= -40 - 12 - 18 + 60$

$= -10$

(See Section 3.1, Example 8.)

14. $f(x) = 0$

$\Rightarrow 6x^2 - 7x - 20 = 0$

$\Rightarrow (3x + 4)(2x - 5) = 0$

$\Rightarrow 3x + 4 = 0 \quad \text{or} \quad 2x - 5 = 0$

$\Rightarrow x = -4/3 \quad \text{or} \quad x = 5/2$

(See Section 1.7, Example 2.)

15. $y - y_1 = m(x - x_1)$

$\Rightarrow y - (-3) = -5(x - 8)$

$\Rightarrow y + 3 = -5x + 40$

$\Rightarrow y = -5x + 37$

(See Section 2.2, Example 9.)

Answers to Selected Exercises

Chapter 1

Section 1.1 (Page 8)

1. True   **3.** Answers vary, but 2,508,429,787/798,458,000 is the best.
5. Distributive property   **7.** Identity property of addition
9. Associative property of addition   **11.** Answers vary.
13. -39 **15.** -2 **17.** 45.6   **19.** about .9167   **21.** -12
23. 0   **25.** 4   **27.** -1 **29.** $\dfrac{2040}{523}, \dfrac{189}{37}, \sqrt{27}, \dfrac{4587}{691}, 6.735, \sqrt{47}$
31. $12 < 18.5$ **33.** $x \geq 5.7$ **35.** $z \leq 7.5$ **37.** $<$ **39.** $=$
41. a lies to the right of b or is equal to b.   **43.** $c < a < b$
45.
47.
49.
51. -3 **53.** -19 **55.** $=$
57. $=$ **59.** $=$ **61.** $>$ **63.** $7 - a$ **65.** Answers vary.
67. Answers vary.   **69.** 1   **71.** 9   **73.** 4   **75.** 4.3
77. 64.1   **79.** 2.3   **81.** 2010, 2015, 2016   **83.** 2010, 2011, 2012, 2013, 2014, 2015

Section 1.2 (Page 16)

1. 1,973,822.685   **3.** 289.0991339   **5.** Answers vary.   **7.** 4^5
9. $(-6)^7$ **11.** $(5u)^{28}$ **13.** Degree 4; coefficients: $6.2, -5, 4, -3, 3.7$;
constant term 3.7   **15.** 3   **17.** $-x^3 + x^2 - 13x$
19. $-6y^2 + 3y + 6$ **21.** $-6x^2 + 4x - 4$ **23.** $-18m^3 - 54m^2 + 9m$
25. $12z^3 + 14z^2 - 7z + 5$ **27.** $12k^2 + 16k - 3$ **29.** $6y^2 + 13y + 5$
31. $18k^2 - 7kq - q^2$ **33.** $4.34m^2 + 5.68m - 4.42$ **35.** $-k + 3$
37. $R = 5000x;\ C = 200,000 + 1800x;\ P = 3200x - 200,000$
39. $R = 9750x;\ C = -3x^2 + 3480x + 259,675;$
$P = 3x^2 + 6270x - 259,675$ **41. (a)** \$673 million
(b) About \$505 million   **43. (a)** \$1384 million **(b)** About \$1022 million
45. About \$4036 million   **47.** About \$6745 million
49. False   **51.** False   **53.** About \$20,663 million
55. About \$34,896 million   **57.** 2013   **59.** $-100,375$; \$96,220;
Answers vary.   **61.** \$342,500

Section 1.3 (Page 23)

1. $12x(x - 2)$ **3.** $r(r^2 - 5r + 1)$ **5.** $6z(z^2 - 2z + 3)$
7. $(2y - 1)^2(14y - 4) = 2(2y - 1)^2(7y - 2)$
9. $(x + 5)^4(x^2 + 10x + 28)$ **11.** $(x + 1)(x + 4)$ **13.** $(x + 3)(x + 4)$
15. $(x + 3)(x - 2)$ **17.** $(x - 1)(x + 3)$ **19.** $(x - 4)(x + 1)$
21. $(z - 7)(z - 2)$ **23.** $(z + 4)(z + 6)$ **25.** $(2x - 1)(x - 4)$
27. $(3p - 4)(5p - 1)$ **29.** $(2z - 5)(2z - 3)$ **31.** $(2x + 1)(3x - 4)$
33. $(5y - 2)(2y + 5)$ **35.** $(2x - 1)(3x + 4)$ **37.** $(3a + 5)(a - 1)$
39. $(x + 9)(x - 9)$ **41.** $(3p - 2)^2$ **43.** $(r - 2t)(r + 5t)$
45. $(m - 4n)^2$ **47.** $(2u + 3)^2$ **49.** Cannot be factored.
51. $(2r + 3v)(2r - 3v)$ **53.** $(x + 2y)^2$ **55.** $(3a + 5)(a - 6)$
57. $(7m + 2n)(3m + n)$ **59.** $(y - 7z)(y + 3z)$ **61.** $(11x + 8)(11x - 8)$
63. $(a - 4)(a^2 + 4a + 16)$ **65.** $(2r - 3s)(4r^2 + 6rs + 9s^2)$
67. $(4m + 5)(16m^2 - 20m + 25)$ **69.** $(10y - z)(100y^2 + 10yz + z^2)$
71. $(x^2 + 3)(x^2 + 2)$ **73.** $b^2(b + 1)(b - 1)$
75. $(x + 2)(x - 2)(x^2 + 3)$ **77.** $(4a^2 + 9b^2)(2a + 3b)(2a - 3b)$
79. $x^2(x^2 + 2)(x^4 - 2x^2 + 4)$ **81.** Answers vary.
83. Answers vary.

Section 1.4 (Page 29)

1. $\dfrac{x}{7}$ **3.** $\dfrac{5}{7p}$ **5.** $\dfrac{5}{4}$ **7.** $\dfrac{4}{w + 6}$ **9.** $\dfrac{y - 4}{3y^2}$ **11.** $\dfrac{m - 2}{m + 3}$
13. $\dfrac{x + 3}{x + 1}$ **15.** $\dfrac{3}{16a}$ **17.** $\dfrac{3y}{x^2}$ **19.** $\dfrac{5}{4c}$ **21.** $\dfrac{3}{4}$ **23.** $\dfrac{3}{10}$
25. $\dfrac{2(a + 4)}{a - 3}$ **27.** $\dfrac{k + 2}{k + 3}$ **29.** Answers vary. **31.** $\dfrac{3}{35z}$
33. $\dfrac{4}{3}$ **35.** $\dfrac{20 + x}{5x}$ **37.** $\dfrac{3m - 2}{m(m - 1)}$ **39.** $\dfrac{37}{5(b + 2)}$
41. $\dfrac{33}{20(k - 2)}$ **43.** $\dfrac{7x - 1}{(x - 3)(x - 1)(x + 2)}$ **45.** $\dfrac{y^2}{(y + 4)(y + 3)(y + 2)}$
47. $\dfrac{x + 1}{x - 1}$ **49.** $\dfrac{-1}{x(x + h)}$ **51. (a)** $\dfrac{\pi x^2}{4x^2}$ **(b)** $\dfrac{\pi}{4}$ **53. (a)** $\dfrac{x^2}{25x^2}$ **(b)** $\dfrac{1}{25}$
55. $\dfrac{-7.2x^2 + 6995x + 230,000}{1000x}$ **57.** About \$3.8 million   **59.** No
61. \$2.55   **63.** \$6531

Section 1.5 (Page 41)

1. 49 **3.** $16c^2$ **5.** $\dfrac{32}{x^5}$ **7.** $108u^{12}$ **9.** $\dfrac{1}{7}$ **11.** $-\dfrac{1}{7776}$
13. $-\dfrac{1}{y^3}$ **15.** $\dfrac{9}{16}$ **17.** $\dfrac{b^3}{a}$ **19.** 7 **21.** About 1.55 **23.** -16
25. $\dfrac{81}{16}$ **27.** $\dfrac{4^2}{5^3}$ **29.** 4^3 **31.** 4^8 **33.** z^3 **35.** $\dfrac{p}{9}$ **37.** $\dfrac{q^5}{r^3}$
39. $\dfrac{8}{25p^7}$ **41.** $2^{5/6}p^{3/2}$ **43.** $2p + 5p^{5/3}$ **45.** $\dfrac{1}{3y^{2/3}}$ **47.** $\dfrac{a^{1/2}}{49b^{5/2}}$
49. $x^{7/6} - x^{11/6}$ **51.** $x - y$ **53.** (f) **55.** (h) **57.** (g)
59. (c) **61.** 5 **63.** 5 **65.** 21 **67.** $\sqrt{77}$ **69.** $5\sqrt{3}$
71. $-\sqrt{2}$ **73.** $15\sqrt{5}$ **75.** 3 **77.** $-3 - 3\sqrt{2}$ **79.** $4 + \sqrt{3}$
81. $\dfrac{7}{11 + 6\sqrt{2}}$ **83. (a)** 14 **(b)** 85 **(c)** 58.0   **85.** About \$11.1 billion
87. About \$11.7 billion   **89.** About \$4.2 billion   **91.** About \$5.8 billion
93. About 6.1 million   **95.** About 6.5 million

Section 1.6 (Page 50)

1. 4 **3.** 7 **5.** $-\dfrac{10}{9}$ **7.** 4 **9.** $\dfrac{40}{7}$ **11.** $\dfrac{26}{3}$ **13.** $-\dfrac{12}{5}$
15. $-\dfrac{59}{6}$ **17.** $-\dfrac{9}{4}$ **19.** $x = .72$ **21.** $r \approx -13.26$ **23.** $\dfrac{b - 5a}{2}$
25. $x = \dfrac{3b}{a + 5}$ **27.** $V = \dfrac{k}{p}$ **29.** $g = \dfrac{V - V_0}{t}$ **31.** $B = \dfrac{2A}{h} - b$
or $B = \dfrac{2A - bh}{h}$ **33.** $-2, 3$ **35.** $-8, 2$ **37.** $\dfrac{5}{2}, \dfrac{7}{2}$ **39.** $23°$
41. $71.6°$ **43.** 13.12   **45.** 2018   **47.** 2021   **49.** 2011
51. 2018   **53.** 2013   **55.** 2020   **57.** \$205.41   **59.** \$21,000
61. \$70,000 for the first plot; \$50,000 for the second
63. About 8.6 million   **65.** 301 million   **67.** $\dfrac{400}{3}L$

Section 1.7 (Page 58)

1. $-4, 14$ **3.** $0, -6$ **5.** $0, 2$ **7.** $-7, -8$ **9.** $\dfrac{1}{2}, 3$

11. $-\dfrac{1}{2}, \dfrac{1}{3}$ **13.** $\dfrac{5}{2}, 4$ **15.** $-5, -2$ **17.** $\dfrac{4}{3}, -\dfrac{4}{3}$ **19.** $0, 1$

21. $2 \pm \sqrt{7}$ **23.** $\dfrac{1 \pm 2\sqrt{5}}{4}$ **25.** $\dfrac{-7 \pm \sqrt{41}}{4}$; $-.1492, -3.3508$

27. $\dfrac{-1 \pm \sqrt{5}}{4}$; $.3090, -.8090$ **29.** $\dfrac{-5 \pm \sqrt{65}}{10}$; $.3062, -1.3062$

31. No real-number solutions   **33.** $-\dfrac{5}{2}, 1$ **35.** No real-number solutions

37. $-5, \dfrac{3}{2}$ **39.** 1   **41.** 2   **43.** $x \approx .4701$ or 1.8240

45. $x \approx -1.0376$ or $.6720$ **47.** \$43.6 thousand   **49.** 2006 ($x \approx 5.46$)
51. 2015 ($x \approx 14.42$)   **53.** \$43.19   **55.** First quarter of 2017 ($x \approx 8.14$)
57. About 1.046 ft.   **59.** (a) $x + 20$ (b) Northbound: $5x$; eastbound:
$5(x + 20)$ or $5x + 100$ (c) $(5x)^2 + (5x + 100)^2 = 300^2$
(d) About 31.23 mph and 51.23 mph   **61.** (a) $150 - x$
(b) $x(150 - x) = 5000$ (c) Length 100 m; width 50 m
63. 9 ft. by 12 ft.   **65.** 6.25 sec.   **67.** (a) About 3.54 sec.   (b) 2.5 sec.
(c) 144 ft.   **69.** (a) 2 sec.   (b) $\dfrac{3}{4}$ sec. or $\dfrac{13}{4}$ sec.   (c) It reaches the given
height twice: once on the way up and once on the way down.
71. $t = \dfrac{\sqrt{2Sg}}{g}$ **73.** $h = \dfrac{d^2\sqrt{kL}}{L}$ **75.** $R = \dfrac{-2Pr + E^2 \pm E\sqrt{E^2 - 4Pr}}{2P}$

Chapter 1 Review Exercises (Page 61)

Refer to Section	1.1	1.2	1.3	1.4	1.5	1.6	1.7
For Exercises	1–18, 81–84	19–24, 85–92	25–32	33–38, 93–96	39–60, 97–98	61–68, 99–100	69–80, 101–106

1. $0, 6$ **3.** $-12, -6, -\dfrac{9}{10}, -\sqrt{4}, 0, \dfrac{1}{8}, 6$ **5.** Commutative property
of multiplication   **7.** Distributive property   **9.** $x \geq 9$
11. $-|3 - (-2)|, -|-2|, |6 - 4|, |8 + 1|$ **13.** -1

15.   **17.** $-\dfrac{7}{9}$ **19.** $4x^4 - 4x^2 + 11x$

21. $25k^2 - 4h^2$ **23.** $9x^2 + 24xy + 16y^2$ **25.** $k(2h^2 - 4h + 5)$
27. $a^2(5a + 2)(a + 2)$ **29.** $(12p + 13q)(12p - 13q)$
31. $(3y - 1)(9y^2 + 3y + 1)$ **33.** $\dfrac{9x^2}{4}$ **35.** 4   **37.** $\dfrac{(m - 1)^2}{3(m + 1)}$
39. $\dfrac{1}{5^3}$ or $\dfrac{1}{125}$ **41.** -1 **43.** 4^3 **45.** $\dfrac{1}{8}$ **47.** $\dfrac{7}{10}$ **49.** $\dfrac{1}{5^{\frac{2}{3}}}$
51. $3^{\frac{7}{2}}a^{\frac{5}{2}}$ **53.** 3   **55.** $3pq\sqrt[3]{2q^2}$ **57.** $-21\sqrt{3}$ **59.** $\sqrt{6} - \sqrt{3}$
61. $-\dfrac{1}{3}$ **63.** No solution   **65.** $x = \dfrac{3}{8a - 2}$ **67.** $-38, 42$
69. $-7 \pm \sqrt{5}$ **71.** $\dfrac{1}{2}, -2$ **73.** $-\dfrac{3}{2}, 7$ **75.** $\pm\dfrac{\sqrt{3}}{3}$
77. $r = \dfrac{-Rp \pm E\sqrt{Rp}}{p}$ **79.** $s = \dfrac{a \pm \sqrt{a^2 + 4K}}{2}$ **81.** 82%
83. \$1536.25   **85.** 2013   **87.** 2014   **89.** 17.05 million
91. March ($x \approx 2.33$)   **93.** About 9.16 million   **95.** 2013 ($x \approx 12.67$)
97. About \$30.91 billion   **99.** 11%   **101.** About 69.9%
103. 3.2 feet   **105.** About 7.77 seconds

Case Study 1 Exercises (Page 64)

1. $218 + 508x$ **3.** Electric by \$1880   **5.** $1529.10 + 50x$
7. LG by \$29.10

Chapter 2

Section 2.1 (Page 72)

1. IV, II, I, III   **3.** Yes   **5.** No

7.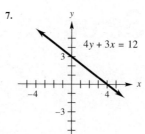

9.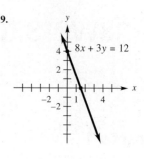

11.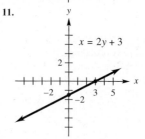

13. x-intercepts $-2.5, 3$; y-intercept 3   **15.** x-intercepts $-1, 2$;
y-intercept -2 **17.** x-intercepts 4; y-intercept 3   **19.** x-intercept 12;
y-intercept -8 **21.** x-intercepts $3, -3$; y-intercept -9
23. x-intercepts $-5, 4$; y-intercept -20 **25.** No x-intercept; y-intercept 7

27. 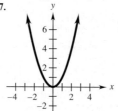   **29.**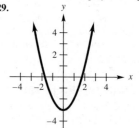

31. 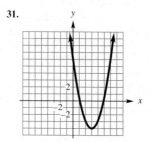   **33.**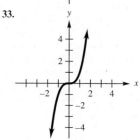

35.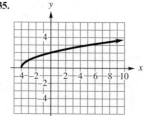

37.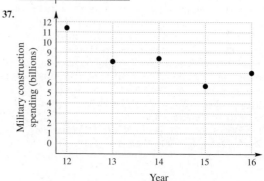

39.

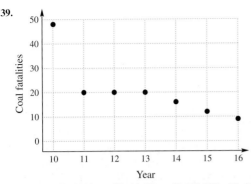

41. 1999 **43.** 2011 **45. (a)** About $1,250,000 **(b)** About $1,750,000
(c) About $4,250,000 **47. (a)** About $500,000 **(b)** About $1,000,000
(b) About $1,500,000 **49.** Beef: about 57 pounds; chicken: about
81 pounds; pork: about 45 pounds; **51.** 2009 **53.** About $115
55. Days 6, 12, 14 to 21 **57.** Day 19; about $117.50
59. Day 2; about $117.00 **61.** About $8.50

63.

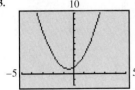

65.

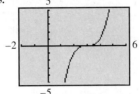

67.

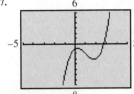

69. No;

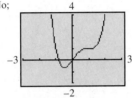

71. $x \approx -1.1038$ **73.** $x \approx 2.1017$ **75.** $x \approx -1.7521$

Section 2.2 (Page 85)

1. $-\dfrac{3}{2}$ **3.** -2 **5.** $-\dfrac{5}{2}$ **7.** Not defined **9.** $y = 4x + 5$

11. $y = -2.3x + 1.5$ **13.** $y = -\dfrac{3}{4}x + 4$ **15.** $m = 2; b = -9$

17. $m = 3; b = -2$ **19.** $m = \dfrac{2}{3}; b = -\dfrac{16}{9}$ **21.** $m = \dfrac{2}{3}; b = 0$

23. $m = 1; b = 5$ **25. (a)** C **(b)** B **(c)** B **(d)** D

27.

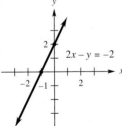

29.

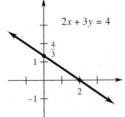

31.

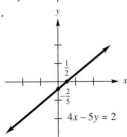

33. Perpendicular **35.** Parallel **37.** Neither **39. (a)** $\dfrac{2}{5}, \dfrac{9}{8}, -\dfrac{5}{2}$

(b) Yes **41.** $y = -\dfrac{2}{3}x$ **43.** $y = 3x - 3$ **45.** $y = 1$

47. $x = -2$ **49.** $y = 2x + 3$ **51.** $2y = 7x - 3$ **53.** $y = 5x$
55. $x = 6$ **57.** $y = 2x - 2$ **59.** $y = x - 6$ **61.** $y = -x + 2$
63. $1330.42 **65.** $6715.23 **67.** -11.7% **69.** 2016 ($x \approx 15.21$)
71. 91.2 million **73.** 2018 ($x \approx 17.5$) **75.** 13.104 minutes;
.044 minutes **77. (a)** $y = 4.5x - 44.5$ **(b)** 14% **(c)** 2017 ($x \approx 16.56$)
79. (a) $y = 300x + 500$ **(b)** $4700 **(c)** 2019 ($x \approx 18.33$)

Section 2.3 (Page 93)

1. (a) $y = \dfrac{5}{9}(x - 32)$ **(b)** $10°C$ and $23.89°C$ **3.** $463.89°C$

5. 227.92; 236.98 **7.** 13.08 million **9.** 4 ft.
11. (a) Equation 1: $-.55, -.17, 0, .35, 0$; sum $= -.37$; Equation
2: $-.12, 0, -.09, 0, -.61$; sum $= -.82$ **(b)** Equation 1: .4539;
Equation 2: .3946 **(c)** Equation 2
13. Yes. $r \approx -.9999$

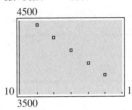

15. (a) $y = -170.8x + 6320$ **(b)** About 3246
17. (a) $y = .8x + 14.2$ **(b)** About $23 billion
19. (a) $y = -1.829x + 98.4$ **(b)** About $76.5 billion
21. (a) $y = .295x + 48$ **(b)** About $50.95 billion **(c)** Second quarter
of 2018 ($x \approx 13.56$) **(d)** $r \approx .795$

Section 2.4 (Page 100)

1. Answers vary.

3. $[-4, \infty)$

5. $(-\infty, 0)$

7. $\left(-\infty, \dfrac{10}{3}\right]$

9. $(-\infty, -8]$

11. $(-\infty, 3)$

13. $(-1, \infty)$

15. $(-\infty, 1]$

17. $\left(\dfrac{1}{5}, \infty\right)$

19. $(-5, 7)$

21. $\left[\dfrac{7}{3}, 5\right]$

23. $\left[-\dfrac{11}{2}, \dfrac{7}{2}\right]$

25. $\left[-\dfrac{17}{7}, \infty\right)$

27. $x \geq 2$ **29.** $-3 < x \leq 5$ **31.** $x \geq 400$ **33.** $x \geq 50$
35. Impossible to break even
37. $(-\infty, -7)$ or $(7, \infty)$

39. $[-5, 5]$

41. All real numbers

43. $\left(-\dfrac{3}{2}, \dfrac{5}{2}\right)$

45. $\left(-\infty, \dfrac{3}{2}\right]$ or $\left[\dfrac{1}{4}, \infty\right)$

47. $76 \leq T \leq 90$ **49.** $40 \leq T \leq 82$ **51.** $32.5 \leq P \leq 38.5$
53. $35 \leq B \leq 43$ **55.** $0 < x \leq 9325;$ $9325 < x \leq 37{,}950;$
$37{,}950 < x \leq 91{,}900;$ $91{,}900 < x \leq 191{,}650;$
$191{,}650 < x \leq 416{,}700;$ $416{,}700 < x \leq 418{,}400; x > 418{,}400$

Section 2.5 (Page 108)

1. $\left[-4, \dfrac{3}{2}\right]$

3. $(-\infty, -3)$ or $(-1, \infty)$

5. $\left[-2, \dfrac{1}{4}\right]$

7. $(-\infty, -1)$ or $\left(\dfrac{1}{4}, \infty\right)$

9. $[-6, 6]$

11. $(-\infty, 0)$ or $(16, \infty)$

13. $[-3, 0]$ or $[3, \infty)$ **15.** $[-7, -2]$ or $[2, \infty)$ **17.** $(-\infty, -5)$ or $(-1, 3)$
19. $\left(-\infty, -\dfrac{1}{2}\right)$ or $\left(0, \dfrac{4}{3}\right)$ **21.** No   **23.** $(-.1565, 2.5565)$
25. $[-2.2635, .7556]$ or $[3.5079, \infty)$ **27.** $(.5, .8393)$
29. $(-\infty, 1)$ or $[4, \infty)$ **31.** $\left(\dfrac{7}{2}, 5\right)$ **33.** $(-\infty, 2)$ or $(5, \infty)$
35. $(-\infty, -1)$ **37.** $(-\infty, -2)$ or $(0, 3)$ **39.** $[-1, .5]$
41. $(8, \infty)$ **43.** $[52, 200]$

Chapter 2 Review Exercises (Page 110)

Refer to Section	2.1	2.2	2.3	2.4	2.5
For Exercises	1–10	11–34	35–38	39–54	55–68

1. $(-2, 3), (0, -5), (3, -2), (4, 3)$

3.

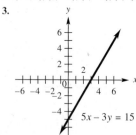

$5x - 3y = 15$

5.

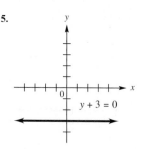

$y + 3 = 0$

7.

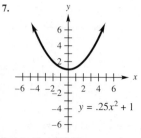

$y = .25x^2 + 1$

9. (a) About 11:30 a.m. to about 7:30 p.m.   **(b)** From midnight until about 5 a.m. and after about 10:30 p.m.   **11.** Answers vary.

13. -3 **15.** $-\dfrac{1}{4}$ **17.** 3 **19.** 0 **21.** -3

23.

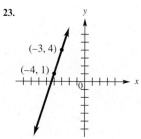

25. $3y = 2x - 13$ **27.** $4y = -5x + 17$ **29.** $x = -1$
31. $3y = 5x + 15$ **33. (a)** $y = .08x$ **(b)** Positive; answers vary.
(c) $\$1.36$ trillion   **35. (a)** $y = 590x + 43{,}376$
(b) $y = 964.6x + 40{,}833$ **(c)** $\$52{,}226; \$55{,}302$; least squares
predicts better   **(d)** $\$58{,}196$ **37. (a)** $y = 93.1x + 2034.8$
(b)

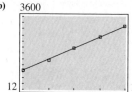

(c) Yes   **(d)** $r \approx .9993$

39. $\left(\dfrac{3}{8}, \infty\right)$ **41.** $\left(-\infty, \dfrac{1}{4}\right]$ **43.** $\left[-\dfrac{1}{2}, 2\right]$ **45.** $[-8, 8]$
47. $(-\infty, 2]$ or $[5, \infty)$ **49.** $\left[-\dfrac{9}{5}, 1\right]$ **51. (d)** $|x - 6| > 4$
53. Less than or equal to 2 gigabytes   **55.** $(-3, 2)$
57. $(-\infty, -5]$ or $\left[\dfrac{3}{2}, \infty\right)$ **59.** $(-\infty, -5]$ or $[-2, 3]$
61. $[-2, 0)$ **63.** $\left(-1, \dfrac{3}{2}\right)$ **65.** $[-19, -5)$ or $(2, \infty)$

Case Study 2 Exercises (Page 113)

1.
```
LinReg
y=ax+b
a=-.765952381
b=82.32928571
r²=.0399034604
r=-.1997585053
■
```

3. $10.24, -0.73, 4.83, -13.86, -10.49, -1.14, -1.10, 12.63$

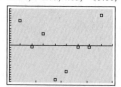

5.
```
LinReg
y=ax+b
a=.3911151515
b=-24.59715152
r²=.9918612299
r=.9959223011
■
```

7. The model predicts a negative wage.

9.
```
LinReg
y=ax+b
a=1.6242424ᴇ-5
b=.0026575758
r²=2.101177ᴇ-7
r=4.5838595ᴇ-4
```

A flat line with 0 slope is a good fit for the residuals.

Chapter 3

Section 3.1 (Page 120)

1. Function   **3.** Not a function   **5.** Function
7. Not a function   **9.** Function   **11.** $(-\infty, \infty)$ **13.** $(-\infty, \infty)$
15. $(-\infty, 5]$ **17.** All real numbers except 2
19. All real numbers except 2 and -2 **21.** All real numbers
such that $x > -4$ and $x \neq 3$ **23.** (a) 8 (b) 8 (c) 8 (d) 8
25. (a) 48 (b) 6 (c) 25.38 (d) 28.42   **27.** (a) 1 (b) $\sqrt{8}$
(c) $\sqrt{2.3}$ (d) $\sqrt{9.9}$ **29.** (a) 12 (b) 23 (c) 12.91
(d) 49.41   **31.** (a) $\dfrac{\sqrt{3}}{15}$ (b) Not defined (c) $\dfrac{\sqrt{1.7}}{6.29}$
(d) Not defined   **33.** (a) 5 (b) 10 (c) 3 (d) 13.8
35. (a) $6 - p$ (b) $6 + r$ (c) $3 - m$ **37.** (a) $\sqrt{4 - p}\,(p \leq 4)$
(b) $\sqrt{4 + r}\,(r \geq -4)$ (c) $\sqrt{1 - m}\,(m \leq 1)$ **39.** (a) $p^2 - 1$
(b) $r^2 - 1$ (c) $m^2 + 6m + 8$ **41.** (a) $\dfrac{3}{p - 1}\,(p \neq 1)$
(b) $\dfrac{3}{-r - 1}\,(r \neq -1)$ (c) $\dfrac{3}{m + 2}\,(m \neq -2)$ **43.** 2
45. $2x + h$ **47.** (a) About $16,515 billion (b) About $17,693
billion (c) About $18,282 billion   **49.** (a) At the start of 2012, the
closing price of Microsoft was $30.20. (b) At the start of 2014, the closing
price of Microsoft was $37.80. (c) At the start of 2016, the closing price
of Microsoft was $51.80.   **51.** (a) A person with a taxable income
of $100 pays $2 in state income tax. (b) A person with a taxable
income of $2500 pays $90 in state income tax. (c) A person with
a taxable income of $103,000 pays $5110 in state income tax.
53. (a) About 23.9 years (b) About 24.2 years (c) About 24.4 years
55. $f(t) = 2050 - 500t$ **57.** (a) $c(x) = 1800 + .5x$ (b) $r(x) = 1.2x$
(c) $p(x) = .7x - 1800$

59.

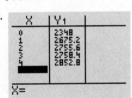

Section 3.2 (Page 130)

1.

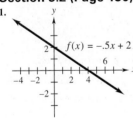

$f(x) = -.5x + 2$

3.

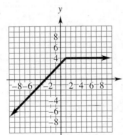

$f(x) = \begin{cases} x + 3 & \text{if } x \ 1 \\ 4 & \text{if } x > 1 \end{cases}$

5.

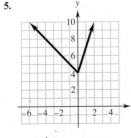

$y = \begin{cases} 4 - x & \text{if } x \leq 0 \\ 3x + 4 & \text{if } x > 0 \end{cases}$

7.

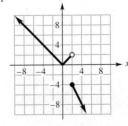

$f(x) = \begin{cases} |x| & \text{if } x < 2 \\ -2x & \text{if } x \geq 2 \end{cases}$

9.

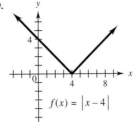

$f(x) = |x - 4|$

11.

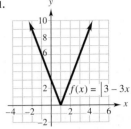

$f(x) = |3 - 3x|$

13.

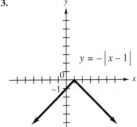

$y = -|x - 1|$

15.

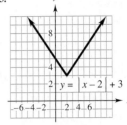

$y = |x - 2| + 3$

17.

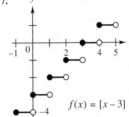

$f(x) = [x - 3]$

19.

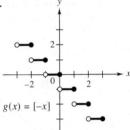

$g(x) = [-x]$

21.

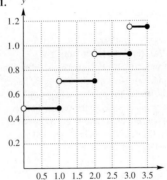

23.

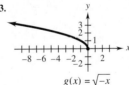

$g(x) = \sqrt{-x}$

25.

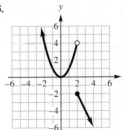

27. Function   **29.** Not a function   **31.** Function

33.

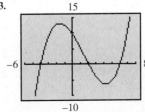

35.

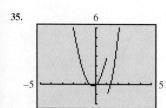

$(1, -1)$ is on the graph; $(1, 3)$ is not on the graph

37. (a) $0, $2900, $7600

(b)

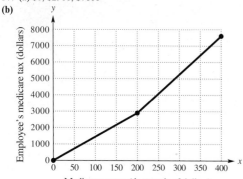

Medicare wages (thousands of dollars)

39.

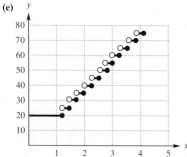

41. (a) 15% **(b)** $f(x) = -.6x + 24$ **(c)** 24%; 13.8% **(d)** Yes
43. (a) 2011 **(b)** 2012 **(c)** 2009 and 2015 **45. (a)** About $120 billion
(b) [2011, 2015] **(c)** [40, 170] **47. (a)** 44¢; 49¢ **(b)** The figure has
vertical line segments, which can't be part of the graph of a function.
(Why?) To make the figure into the graph of f, delete the vertical line
segments; then, for each horizontal segment of the graph, put a closed dot
on the left end and an open circle on the right end (as in Figure 3.8).
49. (a) $34.99 **(b)** $24.99 **(c)** $64.99 **(d)** $74.99
(e)

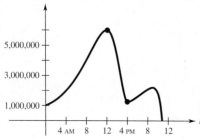

51. There are many correct answers, including the following:

Section 3.3 (Page 142)
1. Let $C(x)$ be the cost of renting a saw for x hours; $C(x) = 25 + 5x$.
3. Let $C(x)$ be the cost (in dollars) for x hours; $C(x) = 8 + 2.5x$.
5. $C(x) = 36x + 200$ **7.** $C(x) = 120x + 3800$
9. $48, $15.60, $13.80 **11.** $55.50, $11.40, $8.46
13. (a) $f(x) = -2730x + 25,035$ **(b)** $11,385 **(c)** $2730 per year
15. (a) $f(x) = -11,875x + 120,000$ **(b)** [0, 8] **(c)** $48,750
17. (a) $80,000 **(b)** $42.50 **(c)** $122,500; $1,440,000 **(d)** $122.50; $45
19. (a) $C(x) = .42x + 37$ **(b)** $37 **(c)** $457 **(d)** $457.42
(e) $.42 **(f)** $.42 **21.** $R(x) = 1.87x + 2,816,000$
23. (a) $C(x) = 10x + 750$ **(b)** $R(x) = 35x$ **(c)** $P(x) = 25x - 750$
(d) $1750 **25. (a)** $C(x) = 18x + 300$ **(b)** $R(x) = 28x$
(c) $P(x) = 10x - 300$ **(d)** $700 **27.** (a) Decreased, Increased
29. $(3, -1)$ **31.** $\left(-\dfrac{11}{4}, -\dfrac{61}{4}\right)$ **33. (a)** 200,000 policies ($x = 200$)

(b)

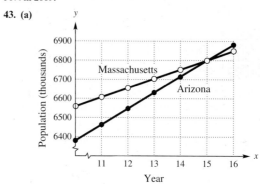

(c) Revenue: $12,500; cost: $15,000 **35. (a)** $C(x) = .126x + 1.5$
(b) $2.382 million **(c)** About 17.857 units **37.** Break-even point is
about 467 units; do not produce the item. **39.** Break-even point is about
1037 units; produce the item. **41.** The percentage is approximately
51% in 2009.

43. (a)

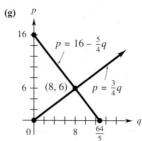

Year

(b) Yes **(c)** 2015 **45.** $140 **47.** 10 items
49. (a) $16 **(b)** $11 **(c)** $6 **(d)** 8 units **(e)** 4 units **(f)** 0 units
(g)

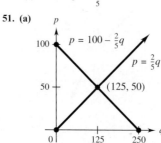

(h) 0 units **(i)** $\dfrac{40}{3}$ units **(j)** $\dfrac{80}{3}$ units
(k) See part (g). **(l)** 8 units **(m)** $6

51. (a)

(b) 125 units **(c)** 50¢
(d) Shortage of supply

53. Total cost increases when more items are made (because it includes the cost of all previously made items), so the graph cannot move downward. No; the average cost can decrease as more items are made, so its graph can move downward.

Section 3.4 (Page 154)
1. Upward   **3.** Downward   **5.** (5, 7); upward
7. (−1, −9); downward   **9.** i   **11.** k   **13.** j   **15.** f
17. $f(x) = \frac{1}{4}(x-1)^2 + 2$ **19.** $f(x) = (x+1)^2 - 2$
21. (−3, −12)   **23.** (2, −7)   **25.** x-intercepts 1, 3; y-intercept 9
27. x-intercepts $5 \pm \sqrt{5}$; y-intercept 20
29. (−2, 0), x = −2

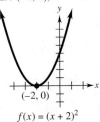

$f(x) = (x+2)^2$

31. (2, 2), x = 2

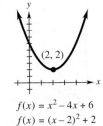

(2, 2)
$f(x) = x^2 - 4x + 6$
$f(x) = (x-2)^2 + 2$

33. (a) (13, 27.5)   **(b)** 2013   **(c)** About $28 billion
35. (a) 10 milliseconds   **(b)** 40 responses per millisecond
37. 54   **39. (a)** About 12 books   **(b)** 10 books   **(c)** About 7 books
(d) 0 books   **(e)** 5 books   **(f)** About 7 books   **(g)** 10 books
(h) About 12 books
(i)

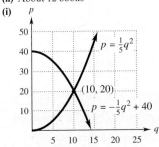

41. (a) $640   **(b)** $515   **(c)** $140
(d)

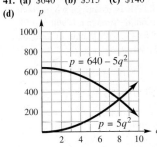

(e) 800 units   **(f)** $320   **43.** 80; $3600   **45.** 30; $1500
47. 20   **49.** 10   **51. (a)** $R(x) = 84x - 2x^2$ **(b)** 21   **(c)** $882
53. (a) $R(x) = 300x - 10x^2$ **(b)** 15 nights   **(c)** $2250
55. (a) $R(x) = (100 - x)(200 + 4x) = 20{,}000 + 200x - 4x^2$
(b)

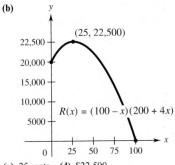

(c) 25 seats   **(d)** $22,500

57. 13 weeks; $96.10/hog   **59. (a)** $f(x) = 68x^2 + 5674$
(b) About $10,026 billion; yes, the estimate is reasonable.
61. (a) $f(x) = -37.5(x - 13)^2 + 545$ **(b)** $395; yes, the estimate is reasonable.
63. $f(x) = 29.92x^2 + 305.90x + 5580.91$; about $9943 billion

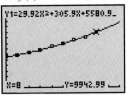

65. $f(x) = -39.71x^2 + 1032.37x - 6151.69$; about $400

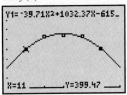

67. (a) 11.3 and 88.7   **(b)** 50   **(c)** $3000   **(d)** x < 11.3 or x > 88.7
(e) 11.3 < x < 88.7

Section 3.5 (Page 165)
1.

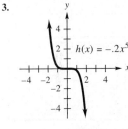

$f(x) = x^4$

3.

$h(x) = -.2x^5$

5. (a) Yes   **(b)** No   **(c)** No   **(d)** Yes   **7. (a)** Yes   **(b)** No   **(c)** Yes
(d) No   **9.** d   **11.** b   **13.** a
15.

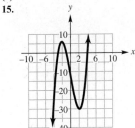

17.

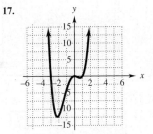

19.

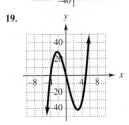

21. (a) In 2004, Home Depot revenue was about $65 billion.
(b) About $78 billion; about $68 billion; about $78 billion

(c)

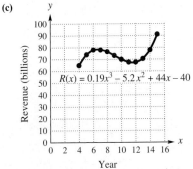

$R(x) = 0.19x^3 - 5.2x^2 + 44x - 40$

(d) Revenue fluctuates.

23. (a) In 2004, Home Depot costs were about $41 billion.
(b) About $50 billion; about $44 billion; about $51 billion
(c)   **(d)** Costs fluctuate

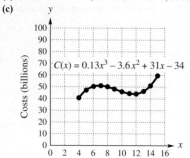

$C(x) = 0.13x^3 - 3.6x^2 + 31x - 34$

25. $P(x) = .06x^3 - 1.6x^2 + 13x - 6$; about $27 billion
27. (a) $933.33 billion   **(b)** $1200 billion   **(c)** $1145.8 billion
(d) $787.5 billion
29. (a)

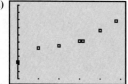

(b) $f(x) = x^3 - 33x^2 + 465x + 47274$
(c) Yes

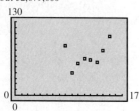

(d) About 52,679,000
31. (a) 130

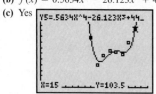

(b) $f(x) = 0.5634x^4 - 26.123x^3 + 449.82x^2 - 3403.8x + 9594$
(c) Yes

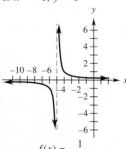

(d) 103.5; yes   **33.** $-3 \le x \le 5$ and $-20 \le y \le 5$ **35.** $-3 \le x \le 4$
and $-35 \le y \le 20$

Section 3.6 (Page 174)

1. $x = -5$, $y = 0$ **3.** $x = -\dfrac{5}{2}$, $y = 0$

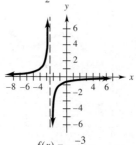

$f(x) = \dfrac{1}{x+5}$ $f(x) = \dfrac{-3}{2x+5}$

5. $x = 1$, $y = 3$

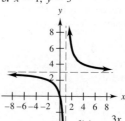

$f(x) = \dfrac{3x}{x-1}$

7. $x = -5$, $y = 1$

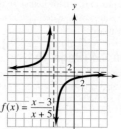

$f(x) = \dfrac{x-3}{x+5}$

9. $x = 3$; $y = -1$

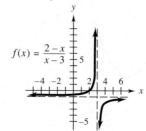

$f(x) = \dfrac{2-x}{x-3}$

11. $x = -\dfrac{1}{8}$, $y = \dfrac{1}{2}$

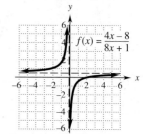

$f(x) = \dfrac{4x-8}{8x+1}$

13. $x = -4$, $x = 2$, $y = 0$

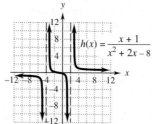

$h(x) = \dfrac{x+1}{x^2+2x-8}$

15. $x = -2$, $x = 2$, $y = 1$

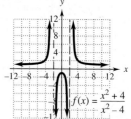

$f(x) = \dfrac{x^2+4}{x^2-4}$

17. $x = 1$, $x = -1$ **19.** $x = -1$, $x = 5$ **21. (a)** $C(x) = 100 + 5x$
(b) $\overline{C}(x) = \dfrac{100+5x}{x}$ **(c)** $25; When producing 5 pairs, the average cost
is $25 per pair. **(d)** $7; When producing 50 pairs, the average cost is
$7 per pair.   **(e)** $y = 5$; The average cost approaches $5 as more pairs
are produced.   **23. (a)** 0%; 2.5%; 4%; 4.875% In the presence of no food,
there will be no growth; with 2 units of food, the growth rate will be 2.5%;
with 8 units of food, the growth rate will be 4%; with 78 units of food, the
growth rate will be 4.875%.   **(b)** $y = 5$ **(c)** No
(d)

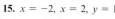

$f(x) = \dfrac{5x}{2+x}$

(e) The maximum possible growth rate
25. (a) $4300   **(b)** $10,033.33   **(c)** $17,200   **(d)** $38,700   **(e)** $81,700
(f) $210,700   **(g)** $425,700   **(h)** No
(i)

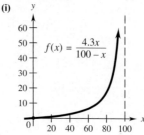

$f(x) = \dfrac{4.3x}{100-x}$

27.

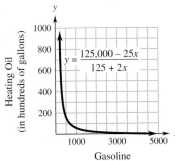

$$y = \frac{125{,}000 - 25x}{125 + 2x}$$

100,000 gal of oil;
500,000 gal of gasoline

29. (a) \$11.2 billion **(b)** \$12.5 billion **(c)** \$12 billion **(d)** \$9.9 billion
(e) 15 **(f)** 28.99%; \$12.6 billion

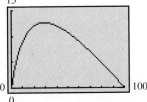

31. (a)

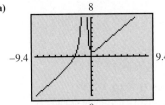

(b) No; yes

(c) Yes

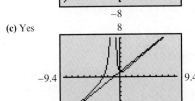

Chapter 3 Review Exercises (Page 177)

Refer to Section	3.1	3.2	3.3	3.4	3.5	3.6
For Exercises	1–12	13–24	25–36	37–60	61–70	71–80

1. Not a function   **3.** Function   **5.** Not a function   **7. (a)** 23
(b) −9 **(c)** $4p - 1$ **(d)** $4r + 3$ **9. (a)** −28 **(b)** −12
(c) $-p^2 + 2p - 4$ **(d)** $-r^2 - 3$ **11. (a)** −13 **(b)** 3 **(c)** $-k^2 - 4k$
(d) $-9m^2 + 12m$ **(e)** $-k^2 + 14k - 45$ **(f)** $12 - 5p$

13.

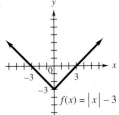

$f(x) = |x| - 3$

15.

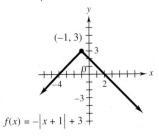

$(-1, 3)$

$f(x) = -|x + 1| + 3$

17.

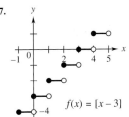

$f(x) = [x - 3]$

19.

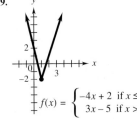

$f(x) = \begin{cases} -4x + 2 & \text{if } x \le 1 \\ 3x - 5 & \text{if } x > 1 \end{cases}$

21.

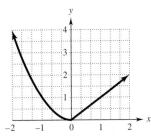

23.

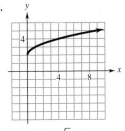

$h(x) = \sqrt{x} + 2$

25. (a)

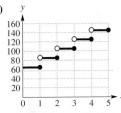

(b) Domain: $(0, \infty)$ range: {65, 85, 105, 125, ...}   **(c)** 2 days

27. (a) Decreasing **(b)** 45% **(c)** $f(x) = -x + 49$ **(d)** 32%
(e) 2016 **29. (a)** $C(x) = 30x + 60$ **(b)** \$30 **(c)** \$30.60
31. (a) $C(x) = 30x + 85$ **(b)** \$30 **(c)** \$30.85 **33. (a)** \$18,000
(b) $R(x) = 28x$ **(c)** 4500 cartridges **(d)** \$126,000 **35.** Equilibrium
quantity is 36 million subscribers at a price of \$12.95 per month.
37. Upward; (2, 6) **39.** Downward; (−1, 8)

41.

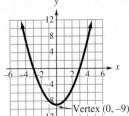

Vertex (0, −9)

43.

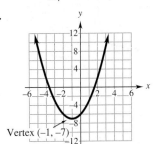

Vertex (−1, −7)

45. Vertex (−3, 14)

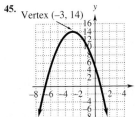

47.

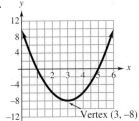

Vertex (3, −8)

49. Minimum value; -11 **51.** Maximum value; 7   **53. (a)** 71,741,000
(b) (9,74090); maximum   **(c)** 2009   **(d)** 74,090,000   **55.** 2012; 3.4%
57. (a) $f(x) = 72.875(x - 6)^2 + 997$ **(b)** \$8284.5 million
59. (a) $R(x) = 65x^2 - 724x + 3047$
(b) Yes   8000   **(c)** \$8103 million

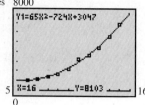

61.

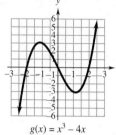

$g(x) = x^3 - 4x$

63.

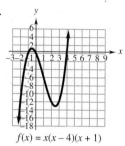

$f(x) = x(x - 4)(x + 1)$

65.

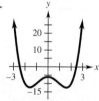

$f(x) = x^4 - 5x^2 - 6$

67. (a) $R(x) = -.194x^3 + 6.576x^2 - 69.942x + 256.288$
(b) Yes   40   **(c)** \$28.696 million; yes

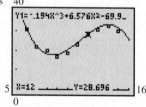

69. $P(x) = -.041x^3 + 1.439x^2 - 16.040x + 61.120$; \$4.928 million
71. $x = 3$, $y = 0$ **73.** $x = 2$, $y = 0$

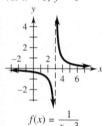

$f(x) = \dfrac{1}{x - 3}$

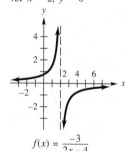

$f(x) = \dfrac{-3}{2x - 4}$

75. $x = -\dfrac{1}{2}$, $x = \dfrac{3}{2}$, $y = 0$

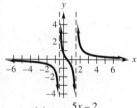

$g(x) = \dfrac{5x - 2}{4x^2 - 4x - 3}$

77. (a) About \$10.83   **(b)** About \$4.64   **(c)** About \$3.61   **(d)** About \$2.71
(e)

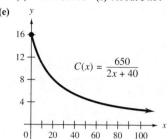

$$C(x) = \frac{650}{2x + 40}$$

79. (10, 50)

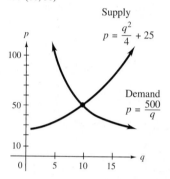

Supply $p = \dfrac{q^2}{4} + 25$

Demand $p = \dfrac{500}{q}$

Case Study 3 Exercises (Page 181)

1. 550 washing machine loads leads to a maximum profit of \$992.50.
3. \$3.25 per load

Extended Project (Page 181)

(a) 500 dryer loads   **(b)** maximum profit of \$105   **(c)** \$1.50 per dryer load

Chapter 4

Section 4.1 (Page 188)

1. Exponential   **3.** Quadratic   **5.** Exponential   **7. (a)** The graph
is entirely above the x-axis and falls from left to right, crossing the y-axis at
1 and then getting very close to the x-axis.   **(b)** (0, 1), (1, .6)
9. (a) The graph is entirely above the x-axis and rises from left to right, less
steeply than the graph of $f(x) = 2^x$.   **(b)** (0, 1), (1, $2^{.5}$) = (1, $\sqrt{2}$)
11. (a) The graph is entirely above the x-axis and falls from left to right,
crossing the y-axis at 1 and then getting very close to the x-axis.   **(b)** (0, 1),
(1, e^{-1}) ≈ (1, .367879)

13.

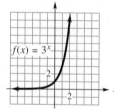

$f(x) = 3^x$

15.

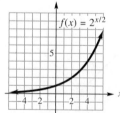

$f(x) = 2^{x/2}$

17.

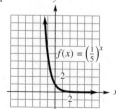

$f(x) = \left(\dfrac{1}{5}\right)^x$

19. (a)–(c)   **(d)** Answers vary.

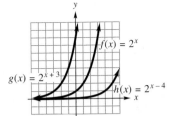

21. 2.3   **23.** .75   **25.** .31   **27. (a)** $a > 1$ **(b)** Domain: $(-\infty, \infty)$, range: $(0, \infty)$

(c)   **(d)** Domain: $(-\infty, \infty)$, range: $(-\infty, 0)$

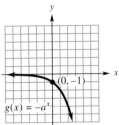

(e)   **(f)** Domain: $(-\infty, \infty)$, range: $(0, \infty)$

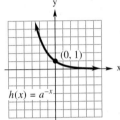

29. (a) 3   **(b)** $\dfrac{1}{3}$ **(c)** 9   **(d)** 1

31.   **33.**

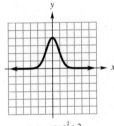 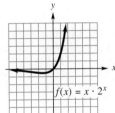

$f(x) = 2^{-x^2 + 2}$

35. (a)

t	0	1	2	3	4	5	6	7	8	9	10
y	1	1.06	1.12	1.19	1.26	1.34	1.42	1.50	1.59	1.69	1.79

(b)

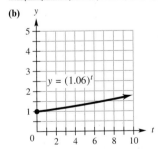

37. (a) About $141,892   **(b)** About $64.10   **39. (a)** About $616.8 billion
(b) About $729.7 billion   **(c)** About $863.1 billion   **41. (a)** About .97 kg
(b) About .75 kg   **(c)** About .65 kg   **(d)** About 24,360 years
43. $34,706.99   **45. (a)** About 6.85 billion   **(b)** About 7.20 billion
(c) About 8.78 billion   **(d)** Answers vary.   **47.** About $7.8 trillion;
about $20.6 trillion   **49.** 2027 ($x \approx 126.99$)   **51. (a)** About $2.74 trillion;
about $5.50 trillion   **(b)** 2017 $x \approx (16.96)$ **53. (a)** About $4.46 billion
(b) 2018 ($x \approx 17.59$)

Section 4.2 (Page 195)

1. (a) $752.27   **(b)** $707.39   **(c)** $432.45   **(d)** $298.98   **(e)** Answer vary.
3. (a) About 27.17 trillion cubic feet   **(b)** About 36.24 quadrillion Btus
5. (a) $f(t) = 30.48(1.074)^t$ **(b)** $71.79   **(c)** 2010   **7. (a)** $f(t) = 9.6(1.102)^t$
(b) About $98.8 billion   **(c)** About $176.9 billion   **9. (a)** Two-point:
$f(t) = 17(1.178)^t$; regression: $g(t) = 17.4(1.181)^t$ **(b)** Two-point: 324.4 trillion
yuan; 530.0 trillion yuan; regression: 347.6 trillion yuan; 572.5 trillion yuan
(c) Two-point: 2020 ($x \approx 19.998$); regression: 2020 ($x \approx 19.55$)
11. (a) Two-point: $f(t) = 257.6(.97)^t$; regression: $g(t) = 252.6(.97)^t$
(b) Two-point: about 144.4 about 127.8; regression: about 141.6; about 125.4
(c) Two-point: 2032 ($x \approx 31.07$); regression: 2031 ($x \approx 30.42$)
13. (a) About 6 items   **(b)** About 23 items   **(c)** 25 items   **15.** 2.6°C
17. (a) About 12.0%; about 47.6%   **(b)** No   **19. (a)** About $27.4 billion;
about $109.1 billion   **(b)** 2013 ($x \approx 12.15$)   **21. (a)** About 66.1%;
about 61.6%
(b)   70

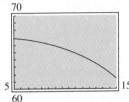

60

Section 4.3 (Page 204)

1. a^y **3.** It is missing the value that equals b^y. If that value is x, the
expression should read $y = \log_b x$.   **5.** $10^5 = 100,000$

7. $9^2 = 81$ **9.** $\log 96 = 1.9823$ **11.** $\log_3\left(\dfrac{1}{9}\right) = -2$

13. 3   **15.** 2   **17.** 3   **19.** -2 **21.** $\dfrac{1}{2}$ **23.** 8.77

25. 1.724   **27.** -4.991 **29.** Because $a^0 = 1$ for every valid base a.

31. $\log 24$ **33.** $\ln 5$ **35.** $\log\left(\dfrac{u^2 w^3}{v^6}\right)$ **37.** $\ln\left(\dfrac{(x + 2)^2}{x + 3}\right)$

39. $\dfrac{1}{2}\ln 6 + 2\ln m + \ln n$ **41.** $\dfrac{1}{2}\log x - \dfrac{5}{2}\log z$ **43.** $2u + 5v$

45. $3u - 2v$ **47.** 3.32112   **49.** 2.429777   **51.** There are many
correct answers, including $b = 1, c = 2$.
53.   **55.**

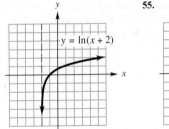 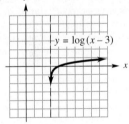

57. Answers vary.   **59.** $\ln 2.75 = 1.0116009; e^{1.0116009} = 2.75$
61. (a) 17.67 yr   **(b)** 9.01 yr   **(c)** 4.19 yr   **(d)** 2.25 yr   **(e)** The pattern is
that it takes about $72/k$ years to double at $k\%$ interest.   **63. (a)** $2089.84
(b) 3000   **(c)** Gradually increasing

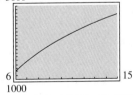

1000
65. (a) $399.83; $417.81   **(b)** 2019 ($x \approx 18.89$)
67. (a) About $2.1 billion; about $3.5 billion   **(b)** 2024 ($x \approx 23.46$)
69. (a) About 205.5 million   **(b)** 2017 ($x \approx 16.77$)

Section 4.4 (Page 212)

1. 8   **3.** 9   **5.** 11   **7.** $\dfrac{11}{6}$ **9.** $\dfrac{4}{9}$ **11.** 10   **13.** 5.2378   **15.** 10

17. $\dfrac{4 + b}{4}$ **19.** $\dfrac{10^{2-b} - 5}{6}$ **21.** Answers vary.   **23.** 4   **25.** $-\dfrac{5}{6}$

27. −4   **29.** −2   **31.** 2.3219   **33.** 2.710   **35.** −1.825

37. .597253   **39.** −.123   **41.** $\dfrac{\log d + 3}{4}$ **43.** $\dfrac{\ln b + 1}{2}$

45. 4   **47.** No solution   **49.** −4, 4   **51.** 9   **53.** 1

55. 4, −4   **57.** ±2.0789   **59.** 1.386   **61.** Answers vary.

63. (a) 2013 ($x \approx$ 12.03)   **(b)** 2016 ($x \approx$ 15.28)   **65. (a)** About $138.0 billion   **(b)** 2011 ($x \approx$ 20.06)   **67. (a)** 2021 ($x \approx$ 20.74)   **(b)** 2027 ($x \approx$ 26.87)   **69. (a)** $2.85   **(b)** 2012 ($x \approx$ 11.2)

71. (a) About $109.5 billion   **(b)** 2014 ($x \approx$ 13.8)   **73.** The seventh quarter is the third quarter of 2016 ($x \approx$ 6.44).   **75. (a)** 25 g   **(b)** About 4.95 yr   **77.** About 3689 yr old   **79. (a)** Approximately 79,432,823 i_0 **(b)** Approximately 251,189 i_0 **(c)** About 316.23 times stronger   **81. (a)** 21   **(b)** 100   **(c)** 105   **(d)** 120   **(e)** 140

Chapter 4 Review Exercises (Page 216)

Refer to Section	4.1.	4.2	4.3	4.4
For Exercises	1–10	13–14, 55–60, 63–64	11–12, 15–40, 65–68	41–54, 61–62

1. (c)   **3.** (d)   **5.** $0 < a < 1$ **7.** All positive real numbers

9.

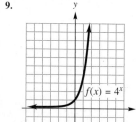

$f(x) = 4^x$

11.

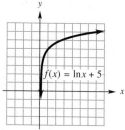

$f(x) = \ln x + 5$

13. (a) About $294 billion   **(b)** 2008 ($x \approx$ 17.95)

15. $\log 340 = 2.53148$ **17.** $\ln 45 = 3.8067$ **19.** $10^4 = 10,000$

21. $e^{4.3957} = 81.1$ **23.** 5   **25.** 8.9   **27.** $\dfrac{1}{3}$ **29.** $\log 20x^6$

31. $\log\left(\dfrac{b^3}{c^2}\right)$ **33.** 4   **35.** 97   **37.** 5   **39.** 5   **41.** −2

43. −2   **45.** 1.416   **47.** −2.807   **49.** −3.305   **51.** .747

53. 28.463   **55. (a)** C   **(b)** A   **(c)** D   **(d)** B   **57. (a)** About 400

(b) 2001 ($x \approx$ 10.11)   **59. (a)** $70.69   **(b)** The third quarter of 2015 ($x \approx$ 6.82)   **61. (a)** 63,095,734.45   **(b)** 1,000,000i_0

(c) About 63.1 times stronger   **63.** 81.25°C

65. (a) $f(x) = 25(1.057)^x$ **(b)** $g(x) = 25.08(1.052)^x$

(c) Two point: about $34.9 billion; regression: about $34.0 billion

(d) Two point: 2020 ($x \approx$ 8.48); regression: 2021 ($x \approx$ 9.21)

67. (a) $f(x) = 4.9 + .684 \ln x$ **(b)** $g(x) = 5.1 + .697 \ln x$

(c) Two point: about 6.6 million; regression: about 6.9 million

(d) Two point: March 2016 ($x \approx$ 21.5); regression: Oct 2015 ($x \approx$ 16.2)

Case Study 4 Exercises (Page 221)

1.

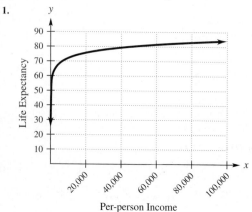

3. About 67.3 years   **5.** About $993

7.

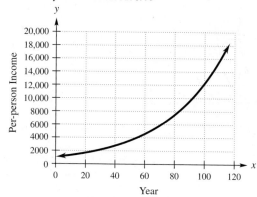

9. About $3564; about $12,251

11.

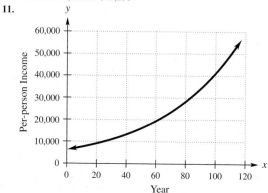

13. About $19,365; about $49,629

Chapter 5

Section 5.1 (Page 228)

1. Time and interest rate   **3.** $133   **5.** $217.48   **7.** $86.26

9. $158.82   **11.** $155.50; $9330   **13.** $256.25; $5125

15. $39; $390   **17.** $12,105   **19.** $6727.50   **21.** Answers vary.

23. $14,354.07   **25.** $15,089.46   **27.** $19,986.50; about .27018 %

29. $15,472.10; about .36065%   **31.** $19,746.50; about 5.1351%

33. $15,104.75; about 5.2335%   **35.** $652,778.08   **37.** $3090

39. About 5.0%   **41.** 4 months   **43.** $5825.24   **45.** $83,533.51

47. 3.5%   **49.** 4.25%   **51.** About 11.36%   **53.** About 102.91%

55. $13,725; yes   **57. (a)** $y_1 = 8t + 100$; $y_2 = 6t + 200$ **(b)** The graph of y_1 is a line with slope 8 and y-intercept 100. The graph of y_2 is a line with slope 6 and y-intercept 200.

(c)

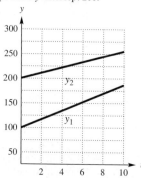

(d) The y-intercept of each graph indicates the amount invested. The slope of each graph is the annual amount of interest paid.

Section 5.2 (Page 238)

1. Answers vary.   **3.** Interest rate and total number of compounding periods

5. Answers vary.   **7.** $1223.59; $753.59   **9.** $9297.93; $2797.93

11. \$10,111.62; \$111.62   **13.** \$15,456.62; \$456.62
15. \$77,746.04; \$2746.04   **17.** About 3.75%   **19.** About 5.25%
21. \$23,824.92; \$3824.92   **23.** \$31,664.54; \$1664.54   **25.** \$10,000
27. \$20,000   **29.** 4.04%   **31.** 5.095%   **33.** \$8954.58
35. \$11,572.58   **37.** \$4203.64   **39.** \$9963.10   **41.** \$4688.71
43. \$465.85   **45.** \$1000 now   **47.** Treasury note
49. The 4.8% bond (by 11 cents).   **51.** About \$2,259,696
53. About \$21,043   **55.** 23.4 years   **57.** 14.2 years
59. About 11.9 years   **61. (a)** \$16,659.95   **(b)** \$21,472.67   **63.** \$16,500

Section 5.3 (Page 247)

1. 21.5786   **3.** \$73,167.09   **5.** \$7679.69   **7.** \$105,033.01
9. About \$205,490   **11.** About \$310,831   **13.** \$797.36
15. \$4566.33   **17.** \$152.53   **19.** 5.19%   **21.** 5.223%
23. Answers vary.   **25.** \$5069.27   **27.** \$2,133,205.87
29. \$18,128.80   **31.** \$5661.31   **33.** \$620.46   **35.** \$265.71
37. \$163,959.61   **39. (a)** \$137,895.79   **(b)** \$132,318.77   **(c)** \$5577.02
41. (a) \$69,421.34   **(b)** \$99,185.58   **43.** \$32,426.46   **45.** \$284,527.35
47. (a) \$256.08   **(b)** \$247.81   **49.** \$863.68   **51. (a)** \$1200
(b) \$3511.58   **53.** 6.5%   **55. (a)** Answers vary.   **(b)** \$24,000
(c) \$603,229   **(d)** \$84,000   **(e)** \$460,884

Section 5.4 (Page 258)

1. Answers vary.   **3.** \$8693.71   **5.** \$1,566,346.66   **7.** \$11,468.10
9. \$38,108.61   **11.** \$557.68   **13.** \$6272.14   **15.** \$446.31
17. \$11,331.18   **19.** \$589.31   **21.** \$1033.09   **23.** \$942.10
25. \$416.11; \$18,524.33   **27.** \$958.30; \$151,811.44   **29.** \$6.80
31. \$42.04   **33.** \$30,669,881   **35.** \$23,016,361   **37. (a)** \$1465.42
(b) \$214.58   **39.** \$2320.83   **41.** \$414.18; \$14,701.60   **43. (a)** \$2717.36
(b) 2   **45. (a)** \$969.75   **(b)** \$185,658.15   **(c)** \$143,510
47. About \$8143.79   **49.** \$406.53   **51.** \$663.22   **53. (a)** \$566.93
(b) \$5,818.96   **(c)** 60   **(d)** \$4794.04

55.

Payment Number	Amount of Payment	Interest for Period	Portion to Principal	Principal at End of Period
0	—	—	—	\$4000.00
1	\$1207.68	\$320.00	\$887.68	3112.32
2	1207.68	248.99	958.69	2153.63
3	1207.68	172.29	1035.39	1118.24
4	1207.70	89.46	1118.24	0

57.

Payment Number	Amount of Payment	Interest for Period	Portion to Principal	Principal at End of Period
0	—	—	—	\$7184.00
1	\$189.18	\$71.84	117.34	7066.66
2	189.18	70.67	118.51	6948.15
3	189.18	69.48	119.70	6828.45
4	189.18	68.28	120.90	6707.55

Chapter 5 Review Exercises (Page 262)

Refer to Section	5.1	5.2	5.3	5.4
For Exercises	1–14	15–28, 65–66, 71	32–41, 67–70, 75–76	29–31, 42–64, 72–74, 77–78

1. \$292.08   **3.** \$62.05   **5.** \$285; \$3420   **7.** \$7925.67
9. \$3084.05   **11.** \$78,742.54   **13.** About 4.082%
15. \$5634.15; \$2834.15   **17.** \$20,402.98; \$7499.53   **19.** \$8532.58
21. \$15,000   **23.** 5.0625%   **25.** \$18,207.65   **27.** \$1088.54
29. \$9706.74   **31.** Answers vary.   **33.** \$162,753.15
35. \$22,643.29   **37.** Answers vary.   **39.** \$2619.29   **41.** \$916.12
43. \$31,921.91   **45.** \$14,222.42   **47.** \$136,340.32   **49.** \$1194.02
51. \$38,298.04   **53.** \$3581.11   **55.** \$435.66   **57.** \$85,459.20
59. \$45.04   **61.** \$277.13   **63. (a)** \$14,320,000   **(b)** \$270,088,972.50
65. \$2298.58   **67.** \$24,818.76; \$2418.76   **69.** \$12,538.59
71. \$5596.62   **73.** \$3560.61   **75.** \$165,249.03   **77.** \$711.25

Case Study 5 Exercises (Page 266)

1. \$150,677.26   **3. (a)** 67   **(b)** \$301,354.52   **5.** 3 times; \$80,000

Chapter 6

Section 6.1 (Page 273)

1. Yes   **3.** $(-1, -4)$ **5.** $\left(\dfrac{2}{7}, -\dfrac{11}{7}\right)$ **7.** $\left(\dfrac{11}{5}, -\dfrac{7}{5}\right)$
9. $(28, 22)$ **11.** $(2, -1)$ **13.** No solution   **15.** $(4y + 1, y)$ for any real number y **17.** $(12, 6)$ **19.** $(15, 75)$; In 2015, the number of millennials and boomers were each 75 million.   **21.** About 2025
23. About 2014   **25.** No   **27.** 145 adults; 55 children
29. Plane: 550 mph; wind: 50 mph   **31. (a)** $y = x + 1$
(b) $y = 3x + 4$ **(c)** $\left(-\dfrac{3}{2}, -\dfrac{1}{2}\right)$

Section 6.2 (Page 283)

1.
$$\begin{aligned} x - 3z &= 2 \\ 2x - 4y + 5z &= 1 \\ 5x - 8y + 7z &= 6 \\ 3x - 4y + 2z &= 3 \end{aligned}$$
3.
$$\begin{aligned} x + 2y + 4z &= 3 \\ x + 2z &= 0 \\ x + 2y + \tfrac{1}{2}z &= \tfrac{3}{2} \end{aligned}$$
5.
$$\begin{aligned} x + y + 2z + 3w &= 1 \\ -y - z - 2w &= -1 \\ 3x + y + 4z + 5w &= 2 \end{aligned}$$
7.
$$\begin{aligned} x + 12y - 3z + 4w &= 10 \\ 2y + 3z + w &= 4 \\ -z &= -7 \\ 6y - 2z - 3w &= 0 \end{aligned}$$
9. $(-68, 13, -6, 3)$ **11.** $\left(20, -9, \dfrac{15}{2}, 3\right)$
13. $\begin{bmatrix} 2 & 1 & 1 & | & 3 \\ 3 & -4 & 2 & | & -5 \\ 1 & 1 & 1 & | & 2 \end{bmatrix}$
15.
$$\begin{aligned} 2x + 3y + 8z &= 20 \\ x + 4y + 6z &= 12 \\ 3y + 5z &= 10 \end{aligned}$$
17. $\begin{bmatrix} 1 & 7 & -3 & | & 5 \\ 2 & 2 & 9 & | & 3 \\ 5 & -1 & 4 & | & 8 \end{bmatrix}$
19. $\begin{bmatrix} 2 & 5 & 1 & | & -1 \\ -4 & 0 & 4 & | & 6 \\ \tfrac{3}{2} & 0 & 2 & | & -1 \end{bmatrix}$
21. $\begin{bmatrix} -4 & -3 & 1 & -1 & | & 2 \\ 0 & -4 & 7 & -2 & | & 10 \\ 0 & -2 & 9 & 4 & | & 5 \end{bmatrix}$ **23.** $\left(\dfrac{3}{2}, 17, -5, 0\right)$
25. No solution   **27.** $(12 - w, -3 - 2w, -5, w)$ for any real number w
29. Dependent   **31.** Inconsistent   **33.** Independent
35. $\left(\dfrac{3}{2}, \dfrac{3}{2}, -\dfrac{3}{2}\right)$ **37.** $(-.8, -1.8, .8)$ **39.** $(100, 50, 50)$
41. $(0, 3.5, 1.5)$ **43.** $(-3, 4, 0)$ **45.** $(1, 0, -1)$ **47.** $(-93, -31, 10)$
49. No solution   **51.** $(22 + 7z, 15 + 10z, z)$ for any real number z
53. $(-7, 5)$ **55.** No solution   **57.** $(-3, z - 17, z)$ for any real number z **59.** $(-11.5, 13.75, -2.25, -5)$ **61.** No solution
63. $(z - 3, -z + 5, z)$ for any real number z **65.** $(12, 5)$; In 2012, all three counties had a population of 5000.   **67.** 220 adults; 150 teenagers; 200 children

69.

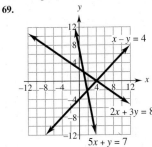

There is no point that is on all three lines.

71. $y = .75x^2 + .25x - .5$

Section 6.3 (Page 292)

1. 15 minutes; 5 miles   **3.** Purchase 200 pounds and 600 euros
5. 200 of Coca-Cola; 500 of Intel

7. (a) $x + 2y + 3z = 31$

$ y + z = 12$

$x = 4$

(b) The center made 4 free throws, 9 regular baskets, and 3 three-point baskets.
9. 100 vans, 50 small trucks, 25 large trucks   **11.** 10 units of corn;
25 units of soybeans; 40 units of cottonseed   **13.** Shirt $1.99; slacks $4.99;
sports coat $6.49   **15.** 220 adults; 150 teenagers; 200 preteens
17. $12,000 at 6%, $7000 at 7%, and $6000 at 10%   **19.** $13,750 in the
mutual fund, $27,500 in bonds, and $28,750 in the franchise   **21.** 40 lb
pretzels, 20 lb dried fruit, 80 lb nuts   **23.** Four possible solutions: (1) no
cases of A and D; 12 cases of B; 8 cases of C; (2) 1 of A; 8 of B; 9 of C; 1 of
D; (3) 2 of A; 4 of B; 10 of C; 2 of D; (4) 3 of A; none of B; 11 of C; 3 of D
25. (a) $f(x) = -.008x^2 + 1.6x + 140.8$ **(b)** 169.6 million
27. (a) $y = .01x^2 - .3x + 4.24$ **(b)** 15 platters; $1.99   **29.** 4 cup rotini,
5 cups healthy, 6 cups chunky, serving size 1.5 cups

Section 6.4 (Page 300)

1. 2×3; $\begin{bmatrix} -7 & 8 & -4 \\ 0 & -13 & -9 \end{bmatrix}$ **3.** 3×3; square matrix; $\begin{bmatrix} 3 & 0 & -11 \\ -1 & -\frac{1}{4} & 7 \\ -5 & 3 & -9 \end{bmatrix}$

5. 2×1; column matrix; $\begin{bmatrix} -7 \\ -11 \end{bmatrix}$ **7.** B is a 5×3 zero matrix

9. $\begin{bmatrix} -7 & 14 & 2 & 4 \\ 6 & -3 & 2 & -4 \end{bmatrix}$ **11.** $\begin{bmatrix} 3 & -1 & 2 \\ 3 & 1 & 5 \end{bmatrix}$ **13.** $\begin{bmatrix} -8 & -8 \\ -8 & -8 \\ -8 & -8 \end{bmatrix}$

15. Not defined   **17.** $\begin{bmatrix} -4 & 0 \\ 10 & 6 \end{bmatrix}$ **19.** $\begin{bmatrix} 0 & -8 \\ -16 & 24 \end{bmatrix}$

21. $\begin{bmatrix} 8 & 10 \\ 0 & -42 \end{bmatrix}$ **23.** $\begin{bmatrix} 4 & -\frac{7}{2} \\ 4 & \frac{21}{2} \end{bmatrix}$

25. $X + T = \begin{bmatrix} x & y \\ z & w \end{bmatrix} + \begin{bmatrix} r & s \\ t & u \end{bmatrix} = \begin{bmatrix} x+r & y+s \\ z+t & w+u \end{bmatrix}$; a 2×2 matrix

27. $X + (T + P) = \begin{bmatrix} x + (r+m) & y + (s+n) \\ z + (t+p) & w + (u+q) \end{bmatrix}$

$= \begin{bmatrix} (x+r) + m & (y+s) + n \\ (z+t) + p & (w+u) + q \end{bmatrix} = (X + T) + P$

29. $P + O = \begin{bmatrix} m+0 & n+0 \\ p+0 & q+0 \end{bmatrix} = \begin{bmatrix} m & n \\ p & q \end{bmatrix} = P$

31. Several possible correct answers, including

	Basketball	Hockey	Football	Baseball
Percent of no shows	16	16	20	18
Lost revenue per fan ($)	18.20	18.25	19	15.40
Lost annual revenue (millions $)	22.7	35.8	51.9	96.3

33.

	2010	2016
Heart	2874	4154
Lung	1759	1470
Liver	15,394	14,794
Kidney	83,919	100,467

35. (a)

	Bread	Milk	Peanut Butter	Cold Cuts
I	88	48	16	112
II	105	72	21	147
III	60	40	0	50

(b) $\begin{bmatrix} 110 & 60 & 20 & 140 \\ 140 & 96 & 28 & 196 \\ 66 & 44 & 0 & 55 \end{bmatrix}$ **(c)** $\begin{bmatrix} 198 & 108 & 36 & 252 \\ 245 & 168 & 49 & 343 \\ 126 & 84 & 0 & 105 \end{bmatrix}$

Section 6.5 (Page 310)

1. 2×2; 2×2 **3.** 3×3; 5×5 **5.** AB does not exist; 3×2

7. columns; rows   **9.** $\begin{bmatrix} 5 \\ 9 \end{bmatrix}$ **11.** $\begin{bmatrix} -2 & 4 \\ 0 & -8 \end{bmatrix}$

13. $\begin{bmatrix} -4 & 1 \\ 2 & -3 \end{bmatrix}$ **15.** $\begin{bmatrix} 3 & -5 & 7 \\ -2 & 1 & 6 \\ 0 & -3 & 4 \end{bmatrix}$ **17.** $\begin{bmatrix} 16 & 10 \\ 2 & 28 \\ 58 & 46 \end{bmatrix}$

19. $AB = \begin{bmatrix} -30 & -45 \\ 20 & 30 \end{bmatrix}$, but $BA = \begin{bmatrix} 0 & 0 \\ 0 & 0 \end{bmatrix}$.

21. $(A + B)(A - B) = \begin{bmatrix} -7 & -24 \\ -28 & -33 \end{bmatrix}$, but $A^2 - B^2 = \begin{bmatrix} -37 & -69 \\ -8 & -3 \end{bmatrix}$.

23. $(PX)T = \begin{bmatrix} (mx + nz)r + (my + nw)t & (mx + nz)s + (my + nw)u \\ (px + qz)r + (py + qw)t & (px + qz)s + (py + qw)u \end{bmatrix}$.

$P(XT)$ is the same, so $(PX)T = P(XT)$.

25. $k(X + T) = k \begin{bmatrix} x + r & y + s \\ z + t & w + u \end{bmatrix}$

$= \begin{bmatrix} k(x + r) & k(y + s) \\ k(z + t) & k(w + u) \end{bmatrix}$

$= \begin{bmatrix} kx + kr & ky + ks \\ kz + kt & kw + ku \end{bmatrix}$

$= \begin{bmatrix} kx & ky \\ kz & kw \end{bmatrix} + \begin{bmatrix} kr & ks \\ kt & ku \end{bmatrix} = kX + kT$

27. No   **29.** Yes   **31.** Yes   **33.** No inverse

35. $\begin{bmatrix} 1 & 2 \\ 1 & 1 \end{bmatrix}$ **37.** $\begin{bmatrix} 2 & 3 \\ 1 & 2 \end{bmatrix}$ **39.** $\begin{bmatrix} -4 & -2 & 3 \\ -5 & -2 & 3 \\ 2 & 1 & -1 \end{bmatrix}$

41. $\begin{bmatrix} -13 & 8 & 7 \\ -6 & 4 & 3 \\ 2 & -1 & -1 \end{bmatrix}$ **43.** No inverse   **45.** $\begin{bmatrix} 6 & -5 & 8 \\ -1 & 1 & -1 \\ -1 & 1 & -2 \end{bmatrix}$

47. $\begin{bmatrix} \frac{1}{2} & -1 & -\frac{1}{2} & \frac{1}{2} \\ \frac{1}{2} & 4 & \frac{5}{2} & -\frac{1}{2} \\ -\frac{1}{4} & -\frac{1}{2} & -\frac{1}{4} & \frac{1}{4} \\ \frac{1}{2} & -2 & -\frac{3}{2} & \frac{1}{2} \end{bmatrix}$ **49. (a)** $\begin{bmatrix} \$56 & \$52 \\ \$20 & \$22 \end{bmatrix}$

(b) Row 1 represents you, and row 2 represents your roommate.
(c) Column 1 represents Timesaver, and column 2 represents Wheels.

(d) $76; $74   **51. (a)** $AB = \begin{bmatrix} 705 & 80 \\ 121 & 14 \end{bmatrix}$ **(b)** Row 1 represents the
restaurant, and row 2 represents the food truck. Column 1 represents total
salary, and column 2 represents total benefits (in thousands) for all
employees at those locations.   **(c)** $705,000; $121,000; $80,000; $14,000

53. (a) No   **(b)** $\begin{bmatrix} 4800 \\ 3200 \\ 5600 \end{bmatrix}$ represents the revenue in each state.   **(c)** $13,600

55. (a) $A = \begin{bmatrix} 195 & 143 & 1225 & 1341 \\ 210 & 141 & 1387 & 1388 \end{bmatrix}$ **(b)** $B = \begin{bmatrix} .016 & .006 \\ .011 & .014 \\ .023 & .008 \\ .013 & .007 \end{bmatrix}$

(c) $AB = \begin{bmatrix} 50 & 22 \\ 55 & 24 \end{bmatrix}$ **(d)** The rows represent the years 2010 and 2020.
Column 1 gives the total births (in millions) in those years, and column 2
gives the total deaths (in millions).   **(e)** About 22 million; about 55 million

Section 6.6 (Page 321)

1. $\begin{bmatrix} 2 \\ -2 \end{bmatrix}$ **3.** $\begin{bmatrix} \frac{1}{2} & 1 \\ \frac{3}{2} & 1 \end{bmatrix}$ **5.** $\begin{bmatrix} -7 \\ 15 \\ -3 \end{bmatrix}$ **7.** $(-63, 50, -9)$

9. $(-31, -131, 181)$ **11.** $(-48, 8, 12)$ **13.** $(1, 0, 2, 1)$
15. $9000 and 100 transactions   **17.** 60 buffets; 600 chairs; 100 tables
19. Jeans $34.50; jacket $72; sweater $44; shirt $21.75   **21.** 2340 of the
first species, 10,128 of the second species, 224 of the third species
23. $\begin{bmatrix} 6.43 \\ 26.12 \end{bmatrix}$ **25.** About 1073 metric tons of wheat, about 1431 metric
tons of oil   **27. (a)** $\frac{7}{4}$ bushels of yams, $\frac{15}{8} \approx 2$ pigs   **(b)** 167.5 bushels
of yams, 153.75 $\approx$ 154 pigs   **29.** Gas $98 million; electric $123 million
31. (a) .40 unit of agriculture, .12 unit of manufacturing, and 3.60 units of
households   **(b)** 848 units of agriculture, 516 units of manufacturing, and
2970 units of households   **(c)** About 813 units   **33. (a)** .017 unit of
manufacturing and .216 unit of energy   **(b)** 123,725,000 pounds of
agriculture, 14,792,000 pounds of manufacturing, 1,488,000 pounds of energy

(c) 195,492,000 pounds of agriculture, 25,933,000 pounds of manufacturing, 13,580,000 pounds of energy   **35.** $532 million of natural resources, $481 million of manufacturing, $805 million of trade and services, $1185 million of personal consumption   **37.** $\begin{bmatrix} 43 \\ 100 \end{bmatrix}, \begin{bmatrix} 29 \\ 63 \end{bmatrix}, \begin{bmatrix} 54 \\ 117 \end{bmatrix}, \begin{bmatrix} 100 \\ 227 \end{bmatrix}, \begin{bmatrix} 53 \\ 121 \end{bmatrix}, \begin{bmatrix} 28 \\ 61 \end{bmatrix}$

39. Love   **41. (a)** 3 **(b)** 3 **(c)** 5 **(d)** 3

43. (a) $C =$

	Dogs	Rats	Cats	Mice
Dogs	0	1	1	1
Rats	0	0	0	1
Cats	0	1	0	1
Mice	0	0	0	0

(b) $\begin{bmatrix} 0 & 1 & 0 & 2 \\ 0 & 0 & 0 & 0 \\ 0 & 0 & 0 & 1 \\ 0 & 0 & 0 & 0 \end{bmatrix}$; C^2 gives the number of food sources once removed from the feeder.

Chapter 6 Review Exercises (Page 326)

Refer to Section	6.1	6.2	6.3	6.4	6.5	6.6
For Exercises	1–10	11–16	17–20, 79–86	21–38	39–66	67–78, 87–96

1. $(9, -14)$ **3.** $(2, -2)$ **5.** 100 shares of first stock; 300 shares of second stock   **7.** Inconsistent, no solution   **9.** $(-18, 69, 12)$
11. $(-14, 17, 11)$ **13.** No solution   **15.** $(4, 3, 2)$ **17.** 19 ones; 7 fives; 9 tens   **19.** 5 blankets; 3 rugs; 8 skirts   **21.** 2×2; square

23. 1×4; row   **25.** 2×3 **27.** $\begin{bmatrix} 162.45 & 164.44 & 163.64 \\ 31.88 & 31.94 & 31.86 \\ 51.66 & 52.19 & 51.95 \end{bmatrix}$

29. $\begin{bmatrix} -1 & -2 & 3 \\ -2 & -3 & 0 \\ 0 & -1 & -4 \end{bmatrix}$ **31.** $\begin{bmatrix} 14 & 6 \\ -4 & 4 \\ 13 & 23 \end{bmatrix}$ **33.** Not defined

35. $\begin{bmatrix} 2 & 18 \\ -4 & -12 \\ 7 & 13 \end{bmatrix}$ **37.** $\begin{bmatrix} -.08 & 25.1 \\ -1.50 & 12.6 \\ .65 & 3.1 \end{bmatrix}, \begin{bmatrix} .10 & 27.2 \\ .63 & 12.4 \\ .47 & 2.9 \end{bmatrix}, \begin{bmatrix} .02 & 52.3 \\ -.87 & 25.0 \\ 1.12 & 6.0 \end{bmatrix}$

39. $\begin{bmatrix} 14 & 56 \\ -6 & -22 \\ 19 & 79 \end{bmatrix}$ **41.** Not defined   **43.** $\begin{bmatrix} 322 & 420 \\ -126 & -166 \\ 455 & 591 \end{bmatrix}$

45. About 20 head and face injuries; about 12 concussions; about 3 neck injuries; about 55 other injuries

47. (a) $\begin{bmatrix} 60.94 & 1.00 \\ 113.22 & 3.22 \\ 119.11 & 4.24 \end{bmatrix}$ **(b)** $[500 \quad 300 \quad 200]$ **(c)** $[88,258 \quad 2314]$

49. Many correct answers, including $\begin{bmatrix} 1 & 2 \\ 3 & 4 \end{bmatrix}$ **51.** $\begin{bmatrix} -.5 & .2 \\ 0 & .2 \end{bmatrix}$

53. No inverse   **55.** $\begin{bmatrix} \frac{1}{4} & \frac{1}{2} & \frac{1}{2} \\ \frac{1}{4} & -\frac{1}{2} & \frac{1}{2} \\ -\frac{1}{12} & -\frac{1}{6} & -\frac{1}{6} \end{bmatrix}$ **57.** No inverse

59. $\begin{bmatrix} -\frac{2}{3} & -\frac{17}{3} & -\frac{14}{3} & -3 \\ \frac{1}{3} & \frac{1}{3} & \frac{1}{3} & 0 \\ -\frac{1}{3} & -\frac{10}{3} & -\frac{7}{3} & -2 \\ 0 & 2 & 1 & 1 \end{bmatrix}$ **61.** $\begin{bmatrix} -\frac{7}{19} & \frac{2}{19} \\ \frac{6}{19} & \frac{1}{19} \end{bmatrix}$ **63.** $\begin{bmatrix} -\frac{1}{12} & -\frac{1}{4} \\ \frac{5}{12} & \frac{1}{4} \end{bmatrix}$

65. $\begin{bmatrix} -1.2 & 1.1 & -.9 \\ .8 & -.4 & .6 \\ -.2 & .1 & .1 \end{bmatrix}$ **67.** $\begin{bmatrix} -5 \\ -3 \end{bmatrix}$ **69.** $\begin{bmatrix} -22 \\ -18 \\ 15 \end{bmatrix}$ **71.** $(-4, 2)$

73. $(4, 2)$ **75.** $(-1, 0, 2)$ **77.** No inverse; no solution for the system
79. 16 liters of the 9%; 24 liters of the 14%   **81.** 30 liters of 40% solution; 10 liters of 60% solution   **83.** 80 bowls; 120 plates   **85.** $12,750 at 8%; $27,250 at 8.5%; $10,000 at 11%

87. (a) $\begin{bmatrix} 1 & -\frac{1}{4} \\ -\frac{1}{2} & 1 \end{bmatrix}$ **(b)** $\begin{bmatrix} \frac{8}{7} & \frac{2}{7} \\ \frac{4}{7} & \frac{8}{7} \end{bmatrix}$ **(c)** $\begin{bmatrix} 2800 \\ 2800 \end{bmatrix}$ **89.** Agriculture $140,909; manufacturing $95,455   **91. (a)** .4 unit agriculture; .09 unit construction; .4 unit energy; .1 unit manufacturing; .9 unit transportation **(b)** 2000 units of agriculture; 600 units of construction; 1700 units of energy; 3700 units of manufacturing; 2500 units of transportation

93. (a) $\begin{bmatrix} 2 & 1 & 1 & 2 \\ 1 & 3 & 2 & 1 \\ 1 & 2 & 3 & 1 \\ 2 & 1 & 1 & 2 \end{bmatrix}$ **(b)** 2 **(c)** $\begin{bmatrix} 2 & 5 & 5 & 2 \\ 5 & 4 & 5 & 5 \\ 5 & 5 & 4 & 5 \\ 2 & 5 & 5 & 2 \end{bmatrix}$ **(d)** 2

95. $\begin{bmatrix} 54 \\ 32 \end{bmatrix}, \begin{bmatrix} 134 \\ 89 \end{bmatrix}, \begin{bmatrix} 172 \\ 113 \end{bmatrix}, \begin{bmatrix} 118 \\ 74 \end{bmatrix}, \begin{bmatrix} 208 \\ 131 \end{bmatrix}$

Case Study 6 Exercises (Page 331)

1. All Stampede Air cities except Lubbock may be reached by a two-flight sequence from San Antonio; all Stampede Air cities may be reached by a three-flight sequence.   **3.** The connection between Lubbock and Corpus Christi and the connection between Lubbock and San Antonio take three flights.

5. $B^2 = \begin{bmatrix} 2 & 1 & 1 & 1 & 1 \\ 1 & 1 & 1 & 1 & 0 \\ 1 & 1 & 1 & 1 & 0 \\ 1 & 1 & 1 & 2 & 1 \\ 1 & 0 & 0 & 1 & 4 \end{bmatrix}$; an entry is nonzero whenever there is a two-step sequence between the cities.

Chapter 7

Section 7.1 (Page 339)
1. F   **3.** A   **5.** E

7.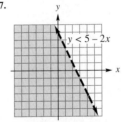

9.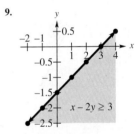

11.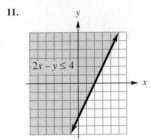

13.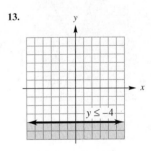

15.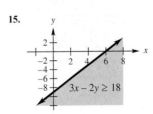

17.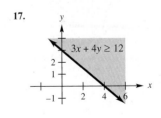

19.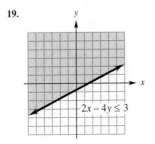

21.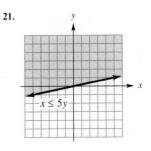

23.

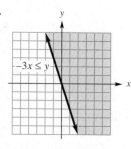

25.

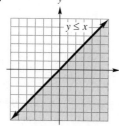

27. Answers vary.

29.

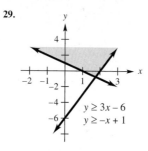

31.

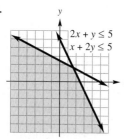

33.

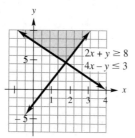

35.

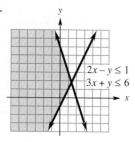

37.

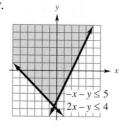

39.

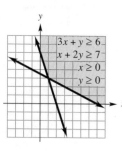

41.

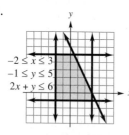

43.

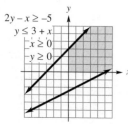

45.

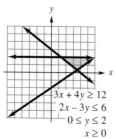

47. $x \geq 0$
$0 \leq y \leq 4$
$4x + 3y \leq 24$

49. $2 < x < 7$
$-1 < y < 3$

51.

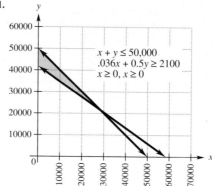

53.

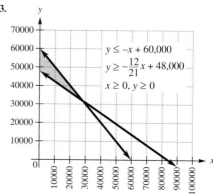

55.

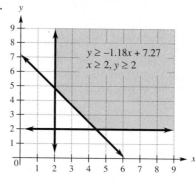

Section 7.2 (Page 347)

1. Maximum of 40 at (5, 10); minimum of 7 at (1, 1)   **3.** Maximum of 6 at (0, 12) minimum of 0 at (0, 0)   **5. (a)** No maximum; minimum of 12 at (12, 0)   **(b)** No maximum; minimum of 18 at (3, 4)   **(c)** No maximum; minimum of 21 at $\left(\frac{13}{2}, 2\right)$ **(d)** No maximum; minimum of 8 at (0, 8)
7. Maximum of 8.4 at (1.2, 1.2)   **9.** Minimum of 13 at (5, 3)
11. Maximum of 68.75 at $\left(\frac{105}{8}, \frac{25}{8}\right)$ **13.** No maximum, minimum of 9
15. Maximum of 22; no minimum   **17. (a)** (18, 2)   **(b)** $\left(\frac{12}{5}, \frac{39}{5}\right)$ **(c)** No maximum   **19.** Answers vary.

Section 7.3 (Page 352)

1. $8x + 5y \leq 110, x \geq 0,$ and $y \geq 0,$ where x is the number of canoes, and y is the number of rowboats   **3.** $250x + 750y \leq 9500, x \geq 0,$ and $y \geq 0,$ where x is the number of radio spots, and y is the number of TV ads
5. 12 chain saws; no chippers   **7.** 10 lb deluxe; 40 lb regular
9. 12 shocks; 6 brakes   **11.** 8 radio spots; 10 TV ads   **13.** 3 Brand X and 2 Brand Z, for a minimum cost $1.05   **15.** 800 type 1 and 1600 type 2,

for a maximum revenue of $272   **17.** From warehouse I, ship 60 boxes to San Jose and 250 boxes to Memphis; from warehouse II, ship 290 boxes to San Jose and none to Memphis, for a minimum cost of $136.70.   **19.** $20 million in bonds and $24 million in mutual funds, for maximum interest of $2.24 million
21. About 218 shares of General Electric and 66 shares of Procter & Gamble   **23.** 125 shares of Visa stock   **25.** About 109 shares of the Fidelity fund

Section 7.4 (Page 365)

1. (a) 3   **(b)** s_1, s_2, s_3 **(c)**
$$4x_1 + 2x_2 + s_1 \qquad\qquad s_3 = 20$$
$$5x_1 + x_2 \qquad + s_2 \qquad = 50$$
$$2x_1 + 3x_2 \qquad\qquad s_2 + s_3 = 25$$

3. (a) 3   **(b)** s_1, s_2, s_3 **(c)**
$$3x_1 - x_2 + 4x_3 + s_1 \qquad\qquad s_2 = 95$$
$$7x_1 + 6x_2 + 8x_3 \qquad + s_2 \qquad = 118$$
$$4x_1 + 5x_2 + 10x_3 \qquad\qquad s_2 + s_3 = 220$$

5.

x_1	x_2	s_1	s_2	s_3	z	
2	5	1	0	0	0	6
4	1	0	1	0	0	6
5	3	0	0	1	0	15
−5	−1	0	0	0	1	0

7.

x_1	x_2	x_3	s_1	s_2	s_3	z	
1	2	3	1	0	0	0	10
2	1	1	0	1	0	0	8
3	0	4	0	0	1	0	6
−1	−5	−10	0	0	0	1	0

9. 4 in row 2, column 2   **11.** 6 in row 3, column 1

13.

x_1	x_2	x_3	s_1	s_2	z	
−1	0	3	1	−1	0	16
1	1	$\frac{1}{2}$	0	$\frac{1}{2}$	0	20
2	0	$-\frac{1}{2}$	0	$\frac{3}{2}$	1	60

15.

x_1	x_2	x_3	s_1	s_2	s_3	z	
$-\frac{1}{2}$	$\frac{1}{2}$	0	1	$-\frac{1}{2}$	0	0	10
$\frac{3}{2}$	$\frac{1}{2}$	1	0	$\frac{1}{2}$	0	0	50
$-\frac{7}{2}$	$\frac{1}{2}$	0	0	$-\frac{3}{2}$	1	0	50
2	0	0	0	1	0	1	100

17. (a) Basic: x_3, s_2, z; nonbasic: x_1, x_2, s_1 **(b)** $x_1 = 0, x_2 = 0, x_3 = 16$, $s_1 = 0, s_2 = 29, z = 11$ **(c)** Not a maximum   **19. (a)** Basic: x_1, x_2, s_2, z; nonbasic: x_3, s_1, s_3 **(b)** $x_1 = 6, x_2 = 13, x_3 = 0, s_1 = 0, s_2 = 21, s_3 = 0$, $z = 18$ **(c)** Maximum   **21.** Maximum is 30 when $x_1 = 0, x_2 = 10$, $s_1 = 0, s_2 = 0$, and $s_3 = 16$ **23.** Maximum is 8 when $x_1 = 4, x_2 = 0$, $s_1 = 8, s_2 = 2$, and $s_3 = 0$ **25.** No maximum   **27.** No maximum
29. Maximum is 34 when $x_1 = 17, x_2 = 0, x_3 = 0, s_1 = 0$, and $s_2 = 14$ or when $x_1 = 0, x_2 = 17, x_3 = 0, s_1 = 0$, and $s_2 = 14$.   **31.** Maximum is 26,000 when $x_1 = 60, x_2 = 40, x_3 = 0, s_1 = 0, s_2 = 80$, and $s_3 = 0$.
33. Maximum is 64 when $x_1 = 28, x_2 = 16, x_3 = 0, s_1 = 0, s_2 = 28$, and, $s_3 = 0$ **35.** Maximum is 250 when $x_1 = 0, x_2 = 0, x_3 = 0, x_4 = 50$, $s_1 = 0$, and $s_2 = 50$ **37. (a)** Maximum is 24 when $x_1 = 12, x_2 = 0$, $x_3 = 0, s_1 = 0$, and $s_2 = 6$ **(b)** Maximum is 24 when $x_1 = 0, x_2 = 12$, $x_3 = 0, s_1 = 0$, and $s_2 = 18$ **(c)** The unique maximum value of z is 24, but this occurs at two different basic feasible solutions.

Section 7.5 (Page 371)

1.

x_1	x_2	s_1	s_2	s_3	
2	1	1	0	0	90
1	2	0	1	0	80
1	1	0	0	1	50
−12	−10	0	0	0	0

where x_1 is the number of units of WildSplash and x_2 is the number of units of TastyTreat.

3.

x_1	x_2	x_3	x_4	s_1	s_2	s_3	
0	0	.375	.625	1	0	0	500
0	.75	.5	.375	0	1	0	600
1	.25	.125	0	0	0	1	300
−90	−70	−60	−50	0	0	0	0

where x_1 is the number of kilograms of P, x_2 is the number of kilograms of Q, x_3 is the number of kilograms of R, and x_4 is the number of kilograms of S.

5. (a) Make no 1-speed or 3-speed bicycles; make 2295 10-speed bicycles; maximum profit is $550,800.   **(b)** 5280 units of aluminum are unused; all the steel is used.   **7. (a)** 12 minutes to the sports segment, 9 minutes to the news segment, and 6 minutes to the weather segment for a maximum of 1.32 million viewers   **(b)** $s_1 = 0$ means that all of the 27 available minutes were used; $s_2 = 0$ means that sports had exactly twice as much as the weather; $s_3 = 0$ means that the sports and the weather had a total time exactly twice the time of the news.   **9. (a)** 300 Japanese maple trees and 300 tricolor beech trees are sold, for a maximum profit of $255,000
(b) There are 200 unused hours of delivering to the client.   **11.** 4 radio ads, 6 TV ads, and no newspaper ads, for a maximum exposure of 64,800 people
13. (a) 22 fund-raising parties, no mailings, and 3 dinner parties, for a maximum of $6,200,000   **(b)** Answers vary.   **15.** 3 hours running, 4 hours biking, and 8 hours walking, for a maximum calorie expenditure of 6313 calories.
17. The juice blender should make 40 units of WildSplash and 10 units of TastyTreat, for a maximum gross income of $580.   **19.** 163.6 kilograms of food P, none of Q, 1090.9 kilograms of R, 145.5 kilograms of S; maximum is 87,454.5

Section 7.6 (Page 381)

1. $\begin{bmatrix} 3 & 1 & 0 \\ -4 & 10 & 3 \\ 5 & 7 & 6 \end{bmatrix}$ **3.** $\begin{bmatrix} 3 & 4 \\ 0 & 17 \\ 14 & 8 \\ -5 & -6 \\ 3 & 1 \end{bmatrix}$

5. Maximize $z = 4x_1 + 6x_2$
subject to $3x_1 - x_2 \le 3$
$x_1 + 2x_2 \le 5$
$x_1 \ge 0, x_2 \ge 0$.

7. Maximize $z = 18x_1 + 15x_2 + 20x_3$
subject to $x_1 + 4x_2 + 5x_3 \le 2$
$7x_1 + x_2 + 3x_3 \le 8$
$x_1 \ge 0, x_2 \ge 0, x_3 \ge 0$.

9. Maximize $z = 18x_1 + 20x_2$
subject to $7x_1 + 4x_2 \le 5$
$6x_1 + 5x_2 \le 1$
$8x_1 + 10x_2 \le 3$
$x_1 \ge 0, x_2 \ge 0$.

11. Maximize $z = 5x_1 + 4x_2 + 15x_3$
subject to $x_1 + x_2 + 2x_3 \le 8$
$x_1 + x_2 + x_3 \le 9$
$x_1 + 3x_3 \le 3$
$x_1 \ge 0, x_3 \ge 0, x_3 \ge 0$.

13. $y_1 = 0, y_2 = 4, y_3 = 0$; minimum is 4.   **15.** $y_1 = 4, y_2 = 0, y_3 = \frac{7}{3}$; minimum is 39.   **17.** $y_1 = 0, y_2 = 100, y_3 = 0$, minimum is 100.
19. $y_1 = 0, y_2 = 12, y_3 = 0$, minimum is 12.   **21.** $y_1 = 0, y_2 = 0, y_3 = 2$; minimum is 4.   **23.** $y_1 = 10, y_2 = 10, y_3 = 0$; minimum is 320.
25. $y_1 = 0, y_2 = 0, y_3 = 4$; minimum is 4.   **27.** 4 servings of A and 2 servings of B, for a minimum cost of $1.76   **29.** 28 units of regular beer and 14 units of light beer, for a minimum cost of $1,680,000
31. (a) 1 bag of feed 1, 2 bags of feed 2   **(b)** 1.4 bags of feed 1 and 1.2 bags of feed 2, for a daily cost of $6.60
33. (a) Minimize $w = 200y_1 + 600y_2 + 90y_3$
subject to $y_1 + 4y_2 \ge 1$
$2y_1 + 3y_2 + y_3 \ge 1.5$
$y_1 \ge 0, y_2 \ge 0, y_3 \ge 0$.
(b) $y_1 = .6, y_2 = .1, y_3 = 0, w = 180$ **(c)** $186

Section 7.7 (Page 391)

1. (a) Maximize $z = -5x_1 + 4x_2 - 2x_3$

subject to $\quad -2x_2 + 5x_3 - s_1 \quad\ s_2 = 8$

$\qquad 4x_1 - x_2 + 3x_3 \qquad + s_2 = 12$

$\qquad x_1 \geq 0, x_2 \geq 0, x_3 \geq 0, s_1 \geq 0, s_2 \geq 0.$

(b)
$$\begin{array}{ccccc} x_1 & x_2 & x_3 & s_1 & s_2 \\ \left[\begin{array}{ccccc|c} 0 & -2 & 5 & -1 & 0 & 8 \\ 4 & -1 & 3 & 0 & 1 & 12 \\ \hline 5 & -4 & 2 & 0 & 0 & 0 \end{array}\right] \end{array}$$

3. (a) Maximize $z = 2x_1 - 3x_2 + 4x_3$

subject to $\quad x_1 + x_2 + x_3 + s_1 \qquad\qquad = 100$

$\qquad x_1 + x_2 + x_3 \qquad - s_2 \qquad = 75$

$\qquad x_1 + x_2 \qquad\qquad\quad - s_3 = 27$

$\qquad x_1 \geq 0, x_2 \geq 0, x_3 \geq 0, s_1 \geq 0, s_2 \geq 0, s_3 \geq 0.$

(b)
$$\begin{array}{cccccc} x_1 & x_2 & x_3 & s_1 & s_2 & s_3 \\ \left[\begin{array}{cccccc|c} 1 & 1 & 1 & 1 & 0 & 0 & 100 \\ 1 & 1 & 1 & 0 & -1 & 0 & 75 \\ 1 & 1 & 0 & 0 & 0 & -1 & 27 \\ \hline -2 & 3 & -4 & 0 & 0 & 0 & 0 \end{array}\right] \end{array}$$

5. Maximize $z = -2y_1 - 5y_2 + 3y_3$

subject to $\quad y_1 + 2y_2 + 3y_3 \geq 115$

$\qquad 2y_1 + y_2 + y_3 \leq 200$

$\qquad y_1 \quad\quad + y_3 \geq 50$

$\qquad y_1 \geq 0, y_2 \geq 0, y_3 \geq 0.$

$$\begin{array}{cccccc} y_1 & y_2 & y_3 & s_1 & s_2 & s_3 \\ \left[\begin{array}{cccccc|c} 1 & 2 & 3 & -1 & 0 & 0 & 115 \\ 2 & 1 & 1 & 0 & 1 & 0 & 200 \\ 1 & 0 & 1 & 0 & 0 & -1 & 50 \\ \hline 2 & 5 & -3 & 0 & 0 & 0 & 0 \end{array}\right] \end{array}$$

7. Maximize $z = -10y_1 - 8y_2 - 15y_3$

subject to $\quad y_1 + y_2 + y_3 \geq 12$

$\qquad 5y_1 + 4y_2 + 9y_3 \geq 48$

$\qquad y_1 \geq 0, y_2 \geq 0, y_3 \geq 0.$

$$\begin{array}{ccccc} y_1 & y_2 & y_3 & s_1 & s_2 \\ \left[\begin{array}{ccccc|c} 1 & 1 & 1 & -1 & 0 & 12 \\ 5 & 4 & 9 & 0 & -1 & 48 \\ \hline 10 & 8 & 15 & 0 & 0 & 0 \end{array}\right] \end{array}$$

9. Maximum is 480 when $x_1 = 40$ and $x_2 = 0$.

11. Maximum is 114 when $x_1 = 38$, $x_2 = 0$, and $x_3 = 0$.

13. Maximum is 90 when $x_1 = 12$ and $x_2 = 3$ or when $x_1 = 0$ and $x_2 = 9$.

15. Minimum is 40 when $y_1 = 0$ and $y_2 = 20$.

17. Maximum is $133\frac{1}{3}$ when $x_1 = 33\frac{1}{3}$ and $x_2 = 16\frac{2}{3}$.

19. Minimum is 512 when $y_1 = 6$ and $y_2 = 8$.

21. Maximum is 112 when $x_1 = 0$, $x_2 = 36$, and $x_3 = 16$.

23. Maximum is 346 when $x_1 = 27$, $x_2 = 0$, and $x_3 = 73$.

25. Minimum is -600 when $y_1 = 0$, $y_2 = 0$, and $y_3 = 200$.

27. Minimum is 96 when $y_1 = 0$, $y_2 = 12$, and $y_3 = 0$.

29.
$$\begin{array}{ccccccc} y_1 & y_2 & y_3 & s_1 & s_2 & s_3 & s_4 \\ \left[\begin{array}{ccccccc|c} 1 & 1 & 1 & -1 & 0 & 0 & 0 & 10 \\ 1 & 1 & 1 & 0 & 1 & 0 & 0 & 15 \\ 1 & -\frac{1}{4} & 0 & 0 & 0 & -1 & 0 & 0 \\ -1 & 0 & 1 & 0 & 0 & 0 & -1 & 0 \\ \hline .30 & .09 & .27 & 0 & 0 & 0 & 0 & 0 \end{array}\right] \end{array}$$

31.
$$\begin{array}{cccccccccc} y_1 & y_2 & y_3 & y_4 & s_1 & s_2 & s_3 & s_4 & s_5 & s_6 \\ \left[\begin{array}{cccccccccc|c} 1 & 1 & 0 & 0 & -1 & 0 & 0 & 0 & 0 & 0 & 32 \\ 1 & 1 & 0 & 0 & 0 & 1 & 0 & 0 & 0 & 0 & 32 \\ 0 & 0 & 1 & 1 & 0 & 0 & -1 & 0 & 0 & 0 & 20 \\ 0 & 0 & 1 & 1 & 0 & 0 & 0 & 1 & 0 & 0 & 20 \\ 1 & 0 & 1 & 0 & 0 & 0 & 0 & 0 & 1 & 0 & 25 \\ 0 & 1 & 0 & 1 & 0 & 0 & 0 & 0 & 0 & 1 & 30 \\ \hline 14 & 12 & 22 & 10 & 0 & 0 & 0 & 0 & 0 & 0 & 0 \end{array}\right] \end{array}$$

33. Ship 200 barrels of oil from supplier S_1 to distributor D_1; ship 2800 barrels of oil from supplier S_2 to distributor D_1; ship 2800 barrels of oil from supplier S_1 to distributor D_2; ship 2200 barrels of oil from supplier S_2 to distributor D_2. Minimum cost is $180,400.

35. Use 1000 lb of bluegrass, 2400 lb of rye, and 1600 lb of Bermuda, for a minimum cost of $560.

37. Allot $3,000,000 in commercial loans and $22,000,000 in home loans, for a maximum return of $2,940,000.

39. Make 32 units of regular beer and 10 units of light beer, for a minimum cost of $1,632,000.

41. $1\frac{2}{3}$ ounces of ingredient I, $6\frac{2}{3}$ ounces of ingredient II, and $1\frac{2}{3}$ ounces of ingredient III produce a minimum cost of $1.55 per barrel.

43. 22 from W_1 to D_1, 10 from W_2 to D_1, none from W_1 to D_2, and 20 from W_2 to D_2, for a minimum cost of $628.

Chapter 7 Review Exercises (Page 395)

Refer to Section	7.1	7.2	7.3	7.4	7.5	7.6	7.7
For Exercises	1–10	11–16	17–18	19–32, 37	33–36, 57–59	38–46, 51–52, 61–62	47–50, 53–56, 63–64

1.

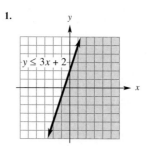

3.

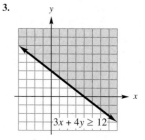

5.

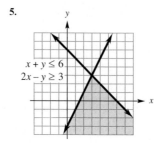

7.

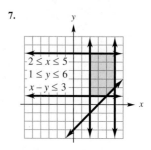

9. Let x represent the number of summary reports, and let y represent the number of inference reports. Then

$$2x + 4y \leq 15$$
$$1.5x + 2y \leq 9$$
$$x \geq 0, y \geq 0.$$

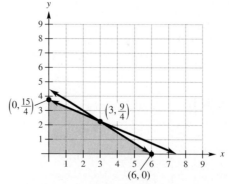

11. Maximum of 46 at $(6, 7)$; minimum of 10 at $(2, 1)$.

13. Maximum of 30 when $x = 5$ and $y = 0$.

15. Minimum of 40 when $x = 0$ and $y = 20$.

17. She should complete 3 summary projects and 2.25 inference projects for a maximum profit of $3187.50.   **19.** Donald should buy 367 (rounded) shares from the BlackRock fund for a minimum risk of about 348.

21. (a)
$$x_1 + x_2 + x_3 + s_1 \quad\quad\quad = 100$$
$$2x_1 + 3x_2 \quad\quad + s_2 \quad\quad = 500$$
$$x_1 \quad\quad + 2x_3 \quad\quad + s_3 = 350$$

(b)

$$\begin{array}{ccccccc} x_1 & x_2 & x_3 & s_1 & s_2 & s_3 & \\ \left[\begin{array}{cccccc|c} 1 & 1 & 1 & 1 & 0 & 0 & 100 \\ 2 & 3 & 0 & 0 & 1 & 0 & 500 \\ 1 & 0 & 2 & 0 & 0 & 1 & 350 \\ \hline -5 & -6 & -3 & 0 & 0 & 0 & 0 \end{array}\right] \end{array}$$

23. (a)
$$x_1 + x_2 + x_3 + s_1 \quad\quad\quad = 90$$
$$2x_1 + 5x_2 + x_3 \quad + s_2 \quad\quad = 120$$
$$x_1 + 3x_2 \quad\quad\quad + s_3 = 80$$

(b)

$$\begin{array}{ccccccc} x_1 & x_2 & x_3 & s_1 & s_2 & s_3 & \\ \left[\begin{array}{cccccc|c} 1 & 1 & 1 & 1 & 0 & 0 & 90 \\ 2 & 5 & 1 & 0 & 1 & 0 & 120 \\ 1 & 3 & 0 & 0 & 0 & 1 & 80 \\ \hline -1 & -8 & -2 & 0 & 0 & 0 & 0 \end{array}\right] \end{array}$$

25. Maximum is 80 when $x_1 = 16$, $x_2 = 0$, $x_3 = 0$, $s_1 = 12$, and $s_2 = 0$.
27. Maximum is 35 when $x_1 = 5$, $x_2 = 0$, $x_3 = 5$, $s_1 = 35$, $s_2 = 0$, and $s_3 = 0$.   **29.** Maximum of 600 when $x_1 = 0$, $x_2 = 100$, and $x_3 = 0$
31. Maximum of 225 when $x_1 = 0$, $x_2 = \frac{15}{2}$, and $x_3 = \frac{165}{2}$
33. (a) Let $x_1 =$ the number of item A, $x_2 =$ the number of item B, and $x_3 =$ the number of item C.
(b) $z = 4x_1 + 3x_2 + 3x_3$
(c) $2x_1 + 3x_2 + 6x_3 \leq 1200$
$$x_1 + 2x_2 + 2x_3 \leq 800$$
$$2x_1 + 2x_2 + 4x_3 \leq 500$$
$$x_1 \geq 0, x_2 \geq 0, x_3 \geq 0$$

35. (a) Let $x_1 =$ the number of gallons of Fruity wine and $x_2 =$ the number of gallons of Crystal wine.   **(b)** $z = 12x_1 + 15x_2$
(c) $2x_1 + x_2 \leq 110$
$$2x_1 + 3x_2 \leq 125$$
$$2x_1 + x_2 \leq 90$$
$$x_1 \geq 0, x_2 \geq 0$$

37. When there are more than 2 variables   **39.** Any standard minimization problem   **41.** Minimum of 172 at (5, 7, 3, 0, 0, 0).
43. Minimum of 640 at (7, 2, 0, 0).   **45.** Minimum of 170 when $y_1 = 0$ and $y_2 = 17$

47.
$$\left[\begin{array}{cccc|c} 5 & 10 & 1 & 0 & 120 \\ 10 & 15 & 0 & -1 & 200 \\ \hline -20 & -30 & 0 & 0 & 0 \end{array}\right]$$

49.
$$\left[\begin{array}{ccccccc|c} 1 & 1 & 2 & -1 & 0 & 0 & 0 & 48 \\ 1 & 1 & 0 & 0 & 1 & 0 & 0 & 12 \\ 0 & 0 & 1 & 0 & 0 & -1 & 0 & 10 \\ 3 & 0 & 1 & 0 & 0 & 0 & -1 & 30 \\ \hline 12 & 20 & -8 & 0 & 0 & 0 & 0 & 0 \end{array}\right]$$

51. Minimum of 427 at (8, 17, 25, 0, 0, 0)   **53.** Maximum is 480 when $x_1 = 24$ and $x_2 = 0$.   **55.** Minimum is 40 when $y_1 = 10$ and $y_2 = 0$.
57. Get 250 of A and none of B or C for a maximum profit of $1000.
59. Make 17.5 gal of Crystal and 36.25 gal of Fruity, for a maximum profit of $697.50.   **61.** Produce 660 cases of corn, no beans, and 340 cases of carrots, for a minimum cost of $15,100.   **63.** Use 1,060,000 kilograms for whole tomatoes and 80,000 kilograms for sauce, for a minimum cost of $4,500,000.

Case Study 7 Exercises (Page 399)

1. The answer in 100-gram units is 0.243037 unit of feta cheese, 2.35749 units of lettuce, 0.3125 unit of salad dressing, and 1.08698 units of tomato. Converting into kitchen units gives approximately $\frac{1}{6}$ cup feta cheese, $4\frac{1}{4}$ cups lettuce, $\frac{1}{8}$ cup salad dressing, and $\frac{7}{8}$ of a tomato.

Chapter 8

Section 8.1 (Page 407)

1. False   **3.** True   **5.** True   **7.** True   **9.** False
11. Answers vary.   **13.** $\subset$ **15.** $\not\subseteq$ **17.** $\subset$ **19.** $\subset$ **21.** 8
23. 128   **25.** $\{x | x \text{ is an integer} \leq 0 \text{ or } \geq 8\}$ **27.** Answers vary.
29. $\cap$ **31.** $\cap$ **33.** $\cup$ **35.** $\cup$ **37.** $\{b, 1, 3\}$
39. $\{d, e, f, 4, 5, 6\}$ **41.** $\{e, 4, 6\}$ **43.** $\{a, b, c, d, 1, 2, 3, 5\}$
45. All students not taking this course.   **47.** All students taking both accounting and philosophy.   **49.** C and D, A and E, C and E, D and E
51. $\{\text{Ford}\}$ **53.** $\{\text{Ford}\}$ **55.** $M \cap E$ is the set of all male employed clients.   **57.** $M' \cup S'$ is the set of all female or married clients.
59. $\{T\}$ **61.** $\{C\}$ **63.** $\{\text{Direct TV, Comcast, Charter, Dish}\}$
65. $\{\text{Charter, Dish, Verizon FIOS, Altice}\}$
67. $\{\text{Direct TV, Comcast, Charter, Dish}\}$
69. $\{\text{Dish, Verizon FIOS, Altice}\}$ **71.** $\varnothing$ **73.** $\{\text{Corn, Rice}\}$
75. $\varnothing$ **77.** $\{\text{Wheat, Corn, Rice}\}$

Section 8.2 (Page 414)

1. 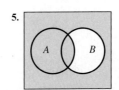   **3.**

5. 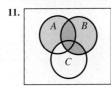   **7.** ϕ
9. 8

11. 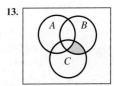   **13.**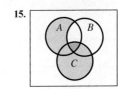

15. 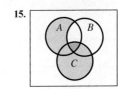   **17.** 13
19. (a) 105   **(b)** 39   **(c)** 15   **(d)** 3
21. (a) 227   **(b)** 182   **(c)** 27
23. (a) 54   **(b)** 17   **(c)** 10   **(d)** 7
(e) 15   **(f)** 3   **(g)** 12   **(h)** 1

25. (a) 290,793   **(b)** 149,167   **(c)** 81,878   **(d)** 317,794   **27. (a)** 945
(b) 55   **(c)** 1340   **(d)** 928   **29. (a)** 507   **(b)** 245   **(c)** The number of females who are completely satisfied or very satisfied with their job.
31. (a) 800   **(b)** 411   **(c)** The number of males who are completely, very, or fairly satisfied with their job.   **33.** Answers vary.   **35.** 9   **37.** 27

39. 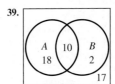   **41.**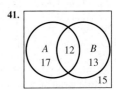

43. 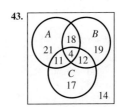   **45.**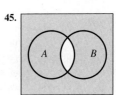

47.

49. The complement of A intersect B equals the union of the complement of A and the complement of B.
51. A union (B intersect C) equals (A union B) intersect (A union C).

Section 8.3 (Page 425)

1. Answers vary.   **3.** {January, February, March,..., December}
5. $\{0, 1, 2, \ldots, 80\}$ **7.** {Go ahead, cancel}   **9.** $\{Q_1, Q_2, Q_3, Q_4\}$
11. Answers vary.   **13.** No   **15.** Yes   **17.** No
19. $S = \{$C&K, C&L, C&J, C&T, C&N, K&L, K&J, K&T, K&N, L&J, L&T, L&N, J&T, J&N, T&N$\}$ **(a)** {C&T, K&T, L&T, J&T, T&N}
(b) {C&J, C&N, J&N}   **21.** $S = \{$forest sage and rag painting, forest sage and colorwash, evergreen and rag painting, evergreen and colorwash, opaque and rag painting, opaque and colorwash$\}$ **(a)** $\frac{1}{2}$ **(b)** $\frac{2}{3}$
23. $S = \{$10′ and beige, 10′ and forest green, 10′ and rust, 12′ and beige, 12′ and forest green, 12′ and rust$\}$ **(a)** $\frac{1}{6}$ **(b)** $\frac{1}{2}$ **(c)** $\frac{1}{3}$ **25.** $\frac{1}{2}$ **27.** $\frac{1}{3}$
29. S = {male beagle, male boxer, male collie, male Labrador, female beagle, female boxer, female collie, female Labrador}   **31.** $\frac{1}{2}$ **33.** $\frac{1}{8}$
35. $\frac{3}{4}$ **37.** $\frac{800}{4836} \approx .1654$ **39. (a)** $\frac{312}{2867} \approx .1088$ **(b)** $\frac{250}{2867} \approx .0872$
(c) $\frac{817}{2867} \approx .2850$ **41.** $\frac{35}{505} \approx .0693$ **43.** $\frac{58}{505} \approx .1149$
45. $\frac{159}{505} \approx .3149$ **47.** Possible   **49.** Not possible   **51.** Not possible

Section 8.4 (Page 433)

1. $\frac{20}{38} = \frac{10}{19}$ **3.** $\frac{28}{38} = \frac{14}{19}$ **5.** $\frac{5}{38}$ **7.** $\frac{12}{38} = \frac{6}{19}$ **9. (a)** $\frac{5}{36}$
(b) $\frac{1}{9}$ **(c)** $\frac{1}{12}$ **(d)** 0   **11. (a)** $\frac{5}{18}$ **(b)** $\frac{5}{12}$ **(c)** $\frac{1}{3}$ **13.** $\frac{1}{3}$
15. (a) $\frac{1}{2}$ **(b)** $\frac{2}{5}$ **(c)** $\frac{3}{10}$ **17. (a)** .62   **(b)** .32   **(c)** .11   **(d)** .81
19. (a) .3962   **(b)** .7952   **(c)** .2048   **(d)** .9464   **21. (a)** .2600
(b) .4580   **(c)** .8373   **23.** 1:5   **25.** 2:1   **27. (a)** 1:4   **(b)** 8:7
(c) 4:11   **29.** 2:7   **31.** 19:81   **33.** 1:19   **35.** $\frac{31}{100}$ **37.** $\frac{23}{100}$
39. About .8950   **41.** About .0594   **43.** About .9307
45. About .6416   **47.** About .8053   **49.** About .2535
51. About .0411   **53.** About .8464   **55. (a)** Answers vary but should be close to .2778.   **(b)** Answers vary, but should be close to .4167.
57. (a) Answers vary but should be close to .0463.   **(b)** Answers vary but should be close to .2963.   **59.** $\frac{3}{4}$

Section 8.5 (Page 446)

1. $\frac{1}{3}$ **3.** 1   **5.** $\frac{1}{6}$ **7.** $\frac{4}{17}$ **9.** $\frac{32}{2652} \approx .0121$ **11.** Answers vary.
13. Answers vary.   **15.** No   **17.** Yes   **19.** No, yes
21. About .0787   **23.** About .1266   **25.** About .9401
27. About .0204   **29.** About .5065   **31.** About .0737
33. About .1648   **35.** .527   **37.** .042   **39.** .857   **41.** Dependent
43. $P(C) = .0800, P(D) = .0050, P(C \cap D) = .0004, P(C|D) = .08$; yes; independent   **45.** .05   **47.** About .8019   **49.** About .7961
51. About .1497   **53.** About .8232   **55.** Delta   **57.** .4955
59. .999985   **61. (a)** 3 backups **(b)** Answers vary.   **63.** Dependent or "No"

Section 8.6 (Page 452)

1. $\frac{1}{3}$ **3.** .0488   **5.** .5122   **7.** .4706   **9.** About .0931
11. About .1826   **13.** About .3880   **15.** About .1604
17. About .7367   **19.** .519   **21.** About .4856   **23.** About .3328
25. About .5543   **27.** About .3696   **29.** About .4457
31. About .5009   **33.** .5335   **35.** .4947

Chapter 8 Review Exercises (Page 454)

Refer to Section	8.1	8.2	8.3	8.4	8.5	8.6
For Exercises	1–23	24–30	31–48	49–54, 65–74, 79–82	55–64, 75–78, 89–96	83–88

1. False   **3.** False   **5.** True   **7.** True   **9.** {New Year's Day, Martin Luther King Jr.'s Birthday, Presidents' Day, Memorial Day, Independence Day, Labor Day, Columbus Day, Veterans' Day, Thanksgiving, Christmas}   **11.** $\{1, 2, 3, 4\}$ **13.** $\{B_1, B_2, B_3, B_6, B_{12}\}$
15. $\{A, C, E\}$ **17.** $\{A, B_3, B_6, B_{12}, C, D, E\}$ **19.** Female students older than 22   **21.** Females or students with a GPA > 3.5 **23.** Non–finance majors who are 22 or younger
25.

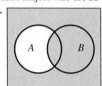

27. 4   **29.** 43   **31.** $\{1, 2, 3, 4, 5, 6\}$ **33.** {(Rock, Rock), (Rock, Pop), (Rock, Alternative), (Pop, Rock), (Pop, Pop), (Pop, Alternative), (Alternative, Rock), (Alternative, Pop), (Alternative, Alternative)}
35. {(Dell, Epson), (Dell, HP), (Apple, Epson), (Apple, HP), (HP, Epson), (HP, HP)}   **37.** $E' \cap F'$ or $(E \cup F)'$ **39.** A probability cannot be > 1.   **41.** Answers vary.   **43.** About .2157   **45.** About .8384
47. About .7843   **49.** About .4255   **51.** About .1030
53. About .8066   **55.** About .1694   **57.** About .1894
59. 12 years; 13 – 16 years
61.

	N_2	T_2
N_1	N_1N_2	N_1T_2
T_1	T_1N_2	T_1T_2

63. $\frac{1}{2}$ **65.** $\frac{5}{36} \approx .1389$ **67.** $\frac{5}{18} \approx .2778$ **69.** $\frac{1}{3} \approx .3333$
71. .79   **73.** .66   **75.** .7   **77.** Answers vary.   **79.** About .5427
81. 1363:1137   **83.** $\frac{19}{22} \approx .8636$ **85.** $\frac{1}{3} \approx .3333$ **87.** .0646
89. .271   **91.** .625   **93.** .690   **95.** No. Answers vary.

Case Study 8 Exercises (Page 458)

1. .001   **3.** About .331   **5.** About .2094

Chapter 9

Section 9.1 (Page 465)

1.

Number of Boys	0	1	2	3	4
$P(x)$	.063	.25	.375	.25	.063

3.

Number of Queens	0	1	2	3
$P(x)$	.7826	.2042	.0130	.0002

5.

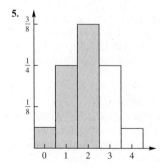

7.

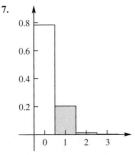

9. 4   **11.** 5.4   **13.** 2.7   **15.** 2.5   **17.** $-\dfrac{1}{19} \approx -.0526$

19. $-\$.50$ **21.** $-\$.97$ **23.** $-\$1.67$ **25.** 1.1319

27. Not valid, cannot have a probability < 0 **29.** Valid   **31.** .15

33. .25   **35.** Many correct answers, including .15, .15   **37.** \$4550

39. .513   **41. (a)** Drug A \$217; Drug B \$229   **(b)** Drug A

43. (a) £1,500,000   **(b)** £1,640,000

Section 9.2 (Page 477)

1. 12   **3.** 56   **5.** 8   **7.** 24   **9.** 84   **11.** 1716

13. 6,375,600   **15.** 240,240   **17.** 8568   **19.** 40,116,600

21. $\dfrac{24}{0}$ is undefined.   **23.** 1,669,536   **25.** 576   **27.** 1 billion in

theory; however, some numbers will never be used (such as those beginning with 000); yes   **29.** 1 billion   **31.** 120   **33. (a)** 160; 8,000,000

(b) Some, such as 800, 900, and so on, are reserved.   **35.** 1600

37. Answers vary.   **39.** 95,040   **41.** 12,144   **43.** 863,040

45. 272   **47. (a)** 210   **(b)** 210   **49.** 330   **51.** Answers vary.

53. (a) 9   **(b)** 6   **(c)** 3; yes   **55. (a)** 20,349   **(b)** 462   **(c)** 6600

(d) 19,887   **57. (a)** 300   **(b)** 13,800   **59. (a)** 1,120,529,256

(b) 806,781,064,320   **61.** 81   **63.** Not possible; 4 initials

65. (a) 10   **(b)** 0   **(c)** 1   **(d)** 10   **(e)** 30   **(f)** 15   **(g)** 0

67. 3,247,943,160   **69. (a)** 2520   **(b)** 840   **(c)** 5040

71. (a) 479,001,600   **(b)** 6   **(c)** 27,720

Section 9.3 (Page 484)

1. About .4224   **3.** $\dfrac{5}{8}$ **5.** $\dfrac{5}{28}$ **7.** About .00005   **9.** About .0163

11. About .5630   **13.** 1326   **15.** About .8507   **17.** About .7647

19. About .9412   **21.** Answers vary.   **23.** About .1396

25. (a) 8.9×10^{-10} **(b)** 1.2×10^{-12} **27.** 1.17×10^{-17}

29. (a) About .4072   **(b)** About .0771   **(c)** About .7511

31. (a) About .0015   **(b)** About .0033   **(c)** About .3576

33. $1 - \dfrac{{}_{365}P_{100}}{(365)^{100}}$ **35.** About .0031   **37.** About .3083

Section 9.4 (Page 490)

1. About .1499   **3.** About .0030   **5.** About .9970   **7.** About .2194

9. About .0192   **11.** About .9808   **13.** $\dfrac{1}{32}$ **15.** $\dfrac{13}{16}$

17. Answers vary.   **19.** About .2029   **21.** About .1818

23. About .0355   **25.** About .1115   **27.** 155   **29.** .2468

31. .1927   **33.** .3504   **35.** 3.85   **37.** About .8237   **39.** 42

41. About .0424   **43.** 44

Section 9.5 (Page 498)

1. Yes   **3.** Yes   **5.** No   **7.** No   **9.** No   **11.** No

13. Not a transition diagram   **15.** $\begin{bmatrix} .6 & .20 & .20 \\ .9 & .02 & .08 \\ .4 & 0 & .6 \end{bmatrix}$

17. Yes   **19.** Yes   **21.** $\left[\dfrac{4}{11}, \dfrac{7}{11}\right]$ **23.** [.675, .325]

25. [.302, .350, .348]   **27.** [.230, .278, .492]

29. $A^2 = \begin{bmatrix} .23 & .21 & .24 & .17 & .15 \\ .26 & .18 & .26 & .16 & .14 \\ .23 & .18 & .24 & .19 & .16 \\ .19 & .19 & .27 & .18 & .17 \\ .17 & .2 & .26 & .19 & .18 \end{bmatrix}$

$A^3 = \begin{bmatrix} .226 & .192 & .249 & .177 & .156 \\ .222 & .196 & .252 & .174 & .156 \\ .219 & .189 & .256 & .177 & .159 \\ .213 & .192 & .252 & .181 & .162 \\ .213 & .189 & .252 & .183 & .163 \end{bmatrix}$

$A^4 = \begin{bmatrix} .2205 & .1916 & .2523 & .1774 & .1582 \\ .2206 & .1922 & .2512 & .1778 & .1582 \\ .2182 & .1920 & .2525 & .1781 & .1592 \\ .2183 & .1909 & .2526 & .1787 & .1595 \\ .2176 & .1906 & .2533 & .1787 & .1598 \end{bmatrix}$

$A^5 = \begin{bmatrix} .21932 & .19167 & .25227 & .17795 & .15879 \\ .21956 & .19152 & .25226 & .17794 & .15872 \\ .21905 & .19152 & .25227 & .17818 & .15898 \\ .21880 & .19144 & .25251 & .17817 & .15908 \\ .21857 & .19148 & .25253 & .17824 & .15918 \end{bmatrix}$; .17794

31. (a) $\begin{bmatrix} .85 & .15 \\ .40 & .60 \end{bmatrix}$ **(b)** [.517, .483]   **(c)** $\left[\dfrac{8}{11}, \dfrac{3}{11}\right]$ **33.** [.883, .117]

35. (a) $\begin{matrix} & R & W \\ R & \begin{bmatrix} .75 & .25 \\ .50 & .50 \end{bmatrix} \\ W & \end{matrix}$ **(b)** [.75, .25]   **(c)** [.667, .333]   **(d)** $\left[\dfrac{2}{3}, \dfrac{1}{3}\right]$

37. (a) [.5607, .43361, .00569]   **(b)** [.473, .526, .002]

39. (a) $\begin{bmatrix} .85 & .10 & .05 \\ 0 & .8 & .2 \\ 0 & 0 & 1 \end{bmatrix}$ **(b)** 42,500; 5000; 2500   **(c)** 36,125; 8250; 5625

(d) 30,706; 10,213; 9081   **(e)** 26,100; 11,241; 12,659   **(f)** [0 0 1]

41. (a) 37.5% Verizon; 31.9% AT&T; 30.6% other   **(b)** 39.3% Verizon; 31.6% AT&T; 29.0% other   **(c)** 40.8% Verizon; 31.4% AT&T; 27.8% other   **(d)** 45.75% Verizon, 31.6% AT&T, 22.64% other

43. (a) About .2073 **(b)** About .2865   **45. (a)** About 85%

(b) 1.28%, 5.13%, 9.62%, 83.97%

Section 9.6 (Page 503)

1. (a) Coast   **(b)** Highway   **(c)** Highway; \$38,000   **(d)** Coast

3. (a) Do not upgrade.   **(b)** Upgrade.   **(c)** Do not upgrade; \$10,600.

5. (a) $\begin{matrix} & \text{Fails} & \text{Doesn't Fail} \\ \text{Overhaul} & \begin{bmatrix} -\$8600 & -\$2600 \\ -\$6000 & \$0 \end{bmatrix} \\ \text{Don't Overhaul} & \end{matrix}$ **(b)** Don't overhaul

the machine.   **7. (a)** $\begin{matrix} & \text{No Rain} & \text{Rain} \\ \text{Rent tent} & \begin{bmatrix} \$2500 & \$1500 \\ \$3000 & \$0 \end{bmatrix} \\ \text{Do not rent tent} & \end{matrix}$

(b) Rent tent because the expected value is \$2100.   **9.** Environment, 15.3

Chapter 9 Review Exercises (Page 506)

Refer to Section	9.1	9.2	9.3	9.4	9.5	9.6
For Exercises	1–12	13–21, 31–33	22–30	33–38	39–44	45–46

1. (a)

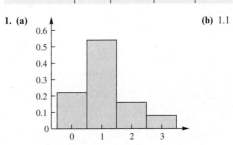

(b) 1.1

3. (a)   **(b)** 0

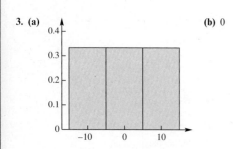

3.

Interval	Frequency
0–14.99	6
15.00–29.99	11
30.00–44.99	2
45.00–59.99	0
60.00–74.99	0
75.00–89.99	1

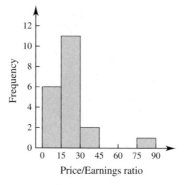

5. 1.833   **7. (a)**

x	0	1	2
$P(x)$	$\frac{7}{15}$	$\frac{7}{15}$	$\frac{1}{15}$

(b) .6   **9.** \$1.29

11. .213   **13.** 40,320   **15.** 220   **17.** 427,518,000
19. (a) 14,905,800   **(b)** 14,905,800   **21.** Answers vary.
23. .0588   **25.** .9955   **27.** .0182   **29.** .2182
31. (a) $1, n, \dfrac{n(n-1)}{2}, 1$ **(b)** $_nC_0 + {_nC_1} + {_nC_2} + \cdots + {_nC_n}$

33.

x	0	1	2	3	4	5	6	7
$P(x)$	0.0071	0.0510	0.1572	0.2695	0.2771	0.1710	0.0586	0.0086

$E(x) = 3.549$

35. (a) About .0002   **(b)** About .9944   **(c)** About .6134

37. (a)

x	0	1	2	3	4	5
$P(x)$	0.0051	0.0478	0.1792	0.3357	0.3144	0.1178

(b) About .9471   **(c)** $E(x) = 3.26$ **(d)** Answers vary.   **39.** Yes
41. Yes   **43. (a)** [.25, .29, .46]   **(b)** [.2345, .277, .4795]
(c) [.240, .278, .482]   **45. (a)** Oppose   **(b)** Oppose   **(c)** Oppose; 2700
(d) Oppose; 1400

5. Answers vary, but one way to create the intervals appears here.

Interval	Frequency
0–24	3
25–49	4
50–74	3
75–99	2
100–124	2
125–149	3
150–174	1
175–199	1
200–224	1

Case Study 9 Exercises (Page 510)

1. +.2031   **3.**

x	0	1	2	3	4
$P(x)$	.30832	.43273	.21264	.04325	.00306

5. \$.19438

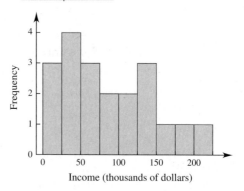

7. Answers vary, but one way to create the intervals appears here.

Interval	Frequency
0–49	1
50–99	3
100–149	6
150–199	4
200–249	2
250–299	1
300–349	0
350–399	2
400–449	0
450–500	1

Chapter 10

Section 10.1 (Page 516)

1.

Interval	Frequency
0–39.99	5
40.00–79.99	9
80.00–119.99	1
120.00–159.99	2
160.00–199.99	2
200.00–239.99	1

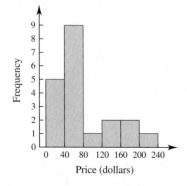

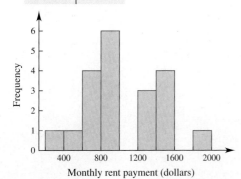

Monthly electric bill (dollars)

9. Answers vary, but one way to create the intervals appears here.

Interval	Frequency
200–399	1
400–599	1
600–799	4
800–999	6
1000–1199	0
1200–1399	3
1400–1599	4
1600–1799	0
1800–1999	1

Monthly rent payment (dollars)

11.

Stem	Leaves
2	57
3	239
4	3
5	2449
6	4667
7	2245
8	3
9	9

Units: 9|9 = $99

13.

Stem	Leaves
5	4
5	
5	
6	0111
6	22223
6	444455
6	66
6	89

Units: 6|8 = 68 inches

15.

Stem	Leaves
0	566
1	134577
2	00234679
3	14
4	
5	
6	
7	9

Units: 7|9 = 79

17.

Stem	Leaves
8	23
8	445
8	6667
8	8999
9	00000011
9	2223

Units: 9|2 = 92%

19.

Stem	Leaves
2	11333
2	557778899
3	0022234
3	888
4	1

Units: 4|1 = 41%

21. Uniform   **23.** Left skewed   **25.** (a) Normal (b) 2 (c) 9
27. (a) Right skewed (b) 26 (c) 7   **29.** (a) Left skewed (b) 12 (c) 8

Section 10.2 (Page 526)

1. $27,955   **3.** 4.8   **5.** 10.3   **7.** 21.2   **9.** 14.8   **11.** $33,679
13. 98.5   **15.** 2   **17.** 65, 71   **19.** No mode   **21.** Answers vary.
23. $\bar{x} \approx 31.28$; 30 − 34.99   **25.** (a) $187.4 million   (b) $177 million
27. (a) $339.5 million   (b) $352 million   **29.** (a) Mean $5936.3 million;
median $5949 million   (b) 2011   **31.** $\bar{x} = 49.6$; median = 48.5
33. Right skewed; median = $74.5 thousand   **35.** (a) Normal
(b) Median = 29%   **37.** 84.61   **39.** 90.84

Section 10.3 (Page 537)

1. Answers vary.   **3.** 70.1; about 24.1   **5.** 9.3; about 3.2
7. 10.8; about 3.8   **9.** 7.8; about 2.8   **11.** 45.2

13. $\bar{x} = 59.6$; $s \approx 24.9$ **15.** $\frac{3}{4}$ **17.** $\frac{5}{9}$ **19.** 93.75%

21. 11.1%   **23.** $\bar{x} \approx 1276.14286$; $s \approx 295.6$
25. $\bar{x} = 1764$; $s \approx 149.6$ **27.** Columbia   **29.** $\bar{x} = 60.5$; $s = 52.1$
31. $\bar{x} = 80.17$; $s = 12.2$ **33.** Answers vary.
35. Min = 80; $Q_1 = 93$; $Q_2 = 99.5$; $Q_3 = 104$; Max = 107
37.

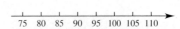

IBM

Revenue (billions of dollars)

39.

Microsoft

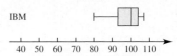

IBM

Revenue (billions of dollars)

The distributions are not similar. IBM generally has higher revenue and less variability.
41. Min = 9.7; $Q_1 = 12$; $Q_2 = 17.55$; $Q_3 = 23.5$; Max = 79.8
43.

Education

Expenditures (billions of dollars)

45. Answers vary, but they should be close to the following: Hospitals $Q_1 = .9$; $Q_2 = 1.7$; $Q_3 = 3.2$; Highways $Q_1 = 2.2$; $Q_2 = 3.1$; $Q_3 = 6.6$.
47. Health, because the box is wider.

Section 10.4 (Page 550)

1. The mean   **3.** Answers vary.   **5.** 45.99%   **7.** 16.64%   **9.** 7.7%
11. 47.35%   **13.** 91.20%   **15.** 1.64 or 1.65   **17.** −1.04   **19.** .5; .5
21. .889; .997   **23.** .0934   **25.** .9051   **27.** .3156   **29.** .2004
31. .4286   **33.** .8700   **35.** .0505   **37.** .0850   **39.** 45.2 mph
41. 24.17%   **43.** Answers vary.   **45.** 665 units   **47.** 190 units
49. 15.87%   **51.** $\mu = 76$; $\sigma \approx 6.9$ **53.** .5686   **55.** .0475
57. $\mu = 33$; $\sigma \approx 4.3$ **59.** .6952   **61.** .1003

Chapter 10 Review Exercises (Page 553)

Refer to Section	10.1	10.2	10.3	10.4
For Exercises	1–4, 11, 17–20, 30	5–6, 9–10, 12–16	7–8, 21–28, 41–46	29, 31–40, 47–54

1. Answers vary, but one way to create the intervals appears here.

Interval	Frequency
20–39.99	2
40–59.99	5
60–79.99	5
80–99.99	0
100–119.99	5
120–139.99	1
140–159.99	1
160–179.99	0
180–199.99	1

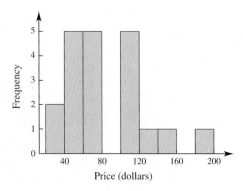

3.

Stem	Leaves
0	8
1	
1	
1	4555
1	666
1	8
2	00011
2	22
2	5
2	6
2	88

Units: 2|8 = 28

5. Mean $88.3; median $83
7. Min = 35, $Q_1 = 57$, $Q_2 = 83$, $Q_3 = 108$, Max = 180

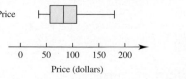

9. Mean 5.25; median 5.5   **11.** Answers vary.   **13.** 42.5 hours; 40, 50 hours   **15.** $3 - 3.99$ and $5 - 5.99$ **17.** Right skewed
19. Normal   **21.** Answers vary.   **23.** Range = 54; $s = 15.8$
25. $s \approx 1.9$ **27.** $\bar{x} = 27.32, s \approx .93$ **29.** Answers vary.
31. .4115   **33.** .9620   **35.** 1.41   **37.** .9973   **39.** .9192
41. (a) Mean is about 119.9167, $s \approx 84.5$ (b) Nissan
43. Min = 28, $Q_1 = 32$, $Q_2 = 102$, $Q_3 = 192$, max = 240

45.

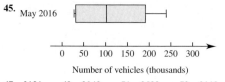

47. .3121   **49.** .9049   **51.** .9582   **53.** .9112

Case Study 10 Exercises (Page 556)
1. 1.032; 12.38   **3.** .8064; 9.68   **5.** 3.2555; 11.28   **7.** 3.2307; 11.19
9. (1.10, 23.66)

Index of Companies, Products, and Agencies

Index of Applications

Subject Index

Photo Credits

Finite Math Review

Compound Amount

If PV dollars are invested at interest rate i per period, the **compound amount** (future value, FV) after n compounding periods is $FV = PV(1 + i)^n$.

Present and Future Value

An **ordinary annuity** of n payments of PMT dollars each at the end of consecutive interest periods with interest compounded at a rate of i per period has **future value**
$$FV = PMT\left[\frac{(1 + i)^n - 1}{i}\right] \text{ and \textbf{present value} } PV = PMT\left[\frac{1 - (1 + i)^{-n}}{i}\right].$$

Matrix Operations

The **matrix** version of the elimination method uses the following **matrix row operations** to obtain the augmented matrix of an equivalent system. They correspond to using elementary row operations on a system of equations.

1. Interchange any two rows.

2. Multiply each element of a row by a nonzero constant.

3. Replace a row by the sum of itself and a constant multiple of another row in the matrix.

Union Rule

For any events E and F from a sample space S: $P(E \cup F) = P(E) + P(F) - P(E \cap F)$.

Properties of Probability

Let S be a sample space consisting of n distinct outcomes $s_1, s_2, \ldots, s_n$. An acceptable probability assignment consists of assigning to each outcome s_i a number p_i (the probability of s_i) according to the following rules.

1. The probability of each outcome is a number between 0 and 1:
$$0 \leq p_1 \leq 1, 0 \leq p_2 \leq 1, \ldots, 0 \leq p_n \leq 1$$

2. The sum of the probabilities of all possible outcomes is 1:
$$p_1 + p_2 + p_3 + \cdots + p_n = 1$$

Bayes' Formula

For any events E and $F_1, F_2, \ldots F_n$, from a sample space S, where $F_1 \cup F_2 \cup \cdots \cup F_n = S$,
$$P(F_i|E) = \frac{P(F_i) \cdot P(E|F_i)}{P(F_1) \cdot P(E|F_1) + \cdots + P(F_n) \cdot P(E|F_n)}.$$